T0271460

Quantum Mechanics

An Experimentalist's Approach

Eugene D. Commins takes an experimentalist's approach to quantum mechanics, preferring to use concrete physical explanations over formal, abstract descriptions to address the needs and interests of a diverse group of students. Keeping physics at the foreground and explaining difficult concepts in straightforward language, Commins examines the many modern developments in quantum physics, including Bell's inequalities, locality, photon polarization correlations, the stability of matter, Casimir forces, geometric phases, Aharonov-Bohm and Aharonov-Casher effects, magnetic monopoles, neutrino oscillations, neutron interferometry, the Higgs mechanism, and the electroweak standard model. The text is self-contained, covering the necessary background on atomic and molecular structure in addition to the traditional topics. Developed from the author's well-regarded course notes for his popular first-year graduate course at the University of California, Berkeley, instruction is supported by over 160 challenging problems to illustrate concepts and provide students with ample opportunity to test their knowledge and understanding, with solutions available online for instructors at www.cambridge.org/commins.

EUGENE D. COMMINS is Professor Emeritus at UC Berkeley's Department of Physics, where he has been a faculty member since 1960. His main area of research is experimental atomic physics. He is a member of the National Academy of Sciences, a Fellow of the American Association for the Advancement of Science, a Fellow of the American Physical Society, and he has been awarded several prizes for his teaching, including the American Association of Physics Teachers Ørsted Medal in 2005, its most prestigious award for notable contributions to physics teaching. He is the author (with Philip H. Buckshaum) of the monograph *Weak Interactions of Leptons and Quarks* (Cambridge University Press, 1983).

Quantum Mechanics

An Experimentalist's Approach

EUGENE D. COMMINS
Professor of Physics, Emeritus
University of California, Berkeley

CAMBRIDGE
UNIVERSITY PRESS

University Printing House, Cambridge CB2 8BS, United Kingdom

One Liberty Plaza, 20th Floor, New York, NY 10006, USA

477 Williamstown Road, Port Melbourne, VIC 3207, Australia

314-321, 3rd Floor, Plot 3, Splendor Forum, Jasola District Centre, New Delhi - 110025, India

103 Penang Road, #05-06/07, Visioncrest Commercial, Singapore 238467

Cambridge University Press is part of the University of Cambridge.

It furthers the University's mission by disseminating knowledge in the pursuit of education, learning and research at the highest international levels of excellence.

www.cambridge.org
Information on this title: www.cambridge.org/9781107063990

First published 2014
Reprinted 2015

A catalogue record for this publication is available from the British Library

Library of Congress Cataloging in Publication data
Commins, Eugene D., author.
Quantum mechanics : an experimentalist's approach / Eugene D. Commins, Professor of Physics, Emeritus, University of California, Berkeley.
pages cm
Includes bibliographical references and index.
ISBN 978-1-107-06399-0 (hardback)
1. Quantum theory. I. Title.
QC174.12.C646 2014
530.12–dc23 2014002491

ISBN 978-1-107-06399-0 Hardback

Additional resources for this publication at www.cambridge.org/commins

Contents

Preface

This book developed from lecture notes that I wrote and rewrote while teaching the graduate course in quantum mechanics at Berkeley many times and to many hundreds of students between 1965 and 2010. It joins a crowded field of well-established quantum mechanics texts. I hope that by virtue of its contents and approach, this book may add something distinctive and be of use to physics students and to working physicists.

I am grateful for the encouragement I have received from scores of Berkeley students and from Berkeley colleagues D. Budker, E. L. Hahn, J. D. Jackson, H. Steiner, M. Suzuki, E. Wichmann, and the late S. J. Freedman, who was once one of my Ph.D. students, then a highly respected colleague, and always a devoted and loyal friend. I thank P. Bucksbaum, A. Cleland, T. Sleator, and H. Stroke for trying out at least part of my lecture notes on students at other institutions and B. C. Regan, D. DeMille, L. Hunter, P. Drell, J. Welch, and I. Ratowsky for their support and friendship. I am sincerely grateful to Vince Higgs, editor at Cambridge University Press, for his crucial encouragement and support. I also thank Sara Werden at Cambridge University Press in New York, and Jayashree, project manager, and her co-workers at Newgen in Chennai, India, for their unfailing courtesy and expert professionalism.

Finally, I am profoundly grateful to my older son, David, for his unswerving support during dark times for both of us. This book is dedicated to the memories of three who are no longer with us: my wife, Ulla, younger son, Lars, and nephew, Bill.

1 Introduction

1.1 What this book is about

Quantum mechanics is an extraordinarily successful theory. The quantum mechanical description of the structures and spectra of atoms and molecules is virtually complete, and in principle, this provides the basis for understanding all of chemistry. Quantum mechanics gives detailed insight into many thermal, electrical, magnetic, optical, and elastic properties of condensed materials, including superconductivity, superfluidity, and Bose-Einstein condensation. Quantum mechanics underlies the theory of nuclear structure, nuclear reactions, and radioactive decay. Quantum electrodynamics (QED), an outgrowth of quantum mechanics and special relativity, is a very successful and detailed description of the interaction of charged leptons (i.e., electrons, muons, and tau leptons) with the electromagnetic radiation field. More generally, relativistic quantum field theory, the extension of quantum mechanics to relativistic fields, is the basis for all successful theoretical attempts so far to describe the phenomena of elementary particle physics.

We assume that you, the reader, have some elementary knowledge of quantum mechanics and that you know something about the historical development of the subject and its main principles and methods. We take advantage of this background, after a brief mathematical review in Chapter 2, by stating the rules of quantum mechanics in Chapter 3. An advantage of this approach is that all the rules are set forth in one place so that we can focus on them. In Chapter 3 we also describe application of the rules to several real physical situations, most significantly experiments with photon polarizations. Following some development of wave mechanics (Chapter 4), we illustrate the rules with additional examples (Chapter 5). We then develop the theory further in subsequent chapters, giving as many examples as we can from the physical world.

Our choice of topics is determined to a large extent by diverse student needs. Some students plan a career in theoretical physics, but most will work in experimental physics or will use quantum mechanics in some other branch of science or technology. Many will never take a subsequent course in elementary particle physics or quantum field theory. Yet most students want to know, and should know, something about the most interesting and important modern developments in quantum physics, even if time or preparation does not permit going into full detail about many topics. Thus, in addition to standard material, which can be found in a large number of existing textbooks, we include discussions of Bell's inequality and photon polarization correlations, neutron interferometry, the Aharonov-Bohm effect, neutrino oscillations, the path integral method, second quantization for fermions, the stability of matter, quantization of the electromagnetic radiation field, the Casimir-Polder effect, the Lamb shift, the adiabatic theorem and geometric phases, relativistic wave equations and especially the Dirac equation, the Dirac field, elementary QED, and a lengthy chapter on quantum mechanics of weak interactions, including an introduction to the electroweak standard model. The choice of topics is also influenced by my background and experience: I was trained as an experimentalist and have spent my entire research career in experimental physics.

The rules of quantum mechanics are remarkably successful in accounting for all experimental results to which they have been applied. However, because of the unique way in which probabilistic concepts appear, particularly in one rule (the so-called collapse postulate), controversy about the foundations of quantum mechanics has existed from the very beginning, and it continues today [see, e.g., Laloe (2012)]. Indeed, if we insist that quantum mechanics should apply not only to a microscopic system such as an electron or an atom but also to the macroscopic apparatus employed to measure that system's properties and the environment that is coupled to the apparatus, the collapse postulate is in conflict with another essential rule that describes how an isolated quantum mechanical system evolves continuously in time. This thorny issue is called the *quantum measurement problem*, and it has troubled many thoughtful persons, including two of the great founders of quantum theory, Albert Einstein and Erwin Schroedinger, and in more recent times the distinguished physicists John S. Bell and Stephen L. Adler, among many others. A summary of the quantum measurement problem and of several attempts to resolve it is given in Chapter 25.

Before we start, let us remark briefly on notation and units. Throughout this book, when the symbol e refers to electric charge it means the *magnitude* of the electronic charge, a positive quantity. The actual charge of the electron is $-e$. If we refer to a nonspecific electric charge that might or might not be e or $-e$, we use the symbol q.

It is not practical for us to work with a single system of units. Instead, we try to employ units that are most appropriate for the topic at hand. Initially, this may seem confusing and discouraging to the student, but it is a fact of life that a practicing physicist must learn to be conversant with several different unit systems. For the most part, we use Heaviside-Lorentz units (hlu system) for general discussions of nonrelativistic quantum mechanics. The hlu and cgs systems are the same, except that if a given electric charge has numerical value q_{cgs} in the cgs system, it has the value $q_{\text{hlu}} = \sqrt{4\pi}q_{\text{cgs}}$ in the hlu system, and similarly for currents, magnetic moments, electric dipole moments, and other electromagnetic sources. On the other hand, if a given electric field has numerical value $\mathcal{E}_{\text{cgs}}$ in the cgs system, it has numerical value $\mathcal{E}_{\text{hlu}} = \mathcal{E}_{\text{cgs}}/\sqrt{4\pi}$ in the hlu system, and similarly for magnetic fields and scalar and vector potentials. We employ atomic units (defined in Section 8.5) for atomic and molecular physics and natural units for relativistic quantum mechanics and field theory, with hlu conventions for electric and magnetic sources and fields. (This natural unit system is defined in Section 15.7 and is used extensively in Chapters 19–24). Although Système International (SI) units are familiar to many students and are convenient for practical engineering and technology, they are awkward and inconvenient for quantum mechanics, especially for relativistic quantum mechanics, so we avoid them.

1.2 A very brief summary of the antecedents of quantum mechanics

Although the invention of quantum mechanics occurred in the remarkably short time interval from 1925 through 1928, this burst of creativity was the culmination of a twenty-five-year gestation period (1900–1925). During that era, the failures of classical physics to account for a wide range of important physical phenomena were revealed, and the need for radical new explanations of these phenomena became increasingly evident. In the following paragraphs we briefly summarize some of the most important achievements of the period from 1900 through 1925. [For a detailed history, see Jammer (1966)]. Here and in the rest of this book we encourage the reader to pay attention to the interplay between experiment and theory that has been so essential for the invention and development of quantum mechanics.

The question of how to account theoretically for the frequency spectrum of black-body radiation had been discussed in the last decades of the nineteenth century, but it gained urgency by 1900 because of accurate measurements of the spectrum by a number of experimentalists, notably H. Rubens and F. Kurlbaum. In that era, the energy per unit volume per hertz of black-body radiation at frequency ν in a cavity at absolute temperature T was predicted by the classical Rayleigh-Jeans formula to be

$$u_\nu = \frac{8\pi\nu^2 k_B T}{c^3} \quad \text{(Rayleigh – Jeans formula)} \tag{1.1}$$

where k_B is Boltzmann's constant, and c is the velocity of light. This formula not only disagreed with the observations of Rubens and Kurlbaum, but when integrated over all frequencies, it led to the nonsensical conclusion that the total energy of radiation in a cavity of any finite volume at any finite temperature is infinite. Max Planck (1900) introduced the quantum of action h in late 1900 to obtain a new formula[1] for u_ν:

$$u_\nu = \frac{8\pi h\nu^3}{c^3}\frac{1}{\exp(h\nu/k_B T)-1} \quad \text{(Planck's law)} \tag{1.2}$$

Planck's law agrees with experiment, and in the limit where $k_B T / \nu \gg h$, it reduces to the Rayleigh-Jeans formula. Planck later called his great achievement an act of desperation, and for some years after 1900, he struggled without success to find an explanation for the existence of h within the laws of classical physics.

Albert Einstein recognized the significance of Planck's law more deeply than Planck himself. Einstein was thus motivated to suggest a corpuscular description of electromagnetic radiation (Einstein 1905). He proposed that the corpuscles (later called *photons*) have energy $E = h\nu$, where ν is the radiation frequency, and he employed this idea in his theory of the photoelectric effect. Convincing experimental evidence was obtained in support of this theory by a number of investigators, most notably Robert Millikan, in the decade following 1905 (Millikan 1916). Nevertheless, many physicists found it difficult to reconcile the idea of discrete photons with the highly successful and universally accepted wave theory of classical electromagnetism. Thus Einstein's corpuscular description gained adherents only very slowly. However, in 1923, Arthur H. Compton made careful observations of x-ray–electron scattering, and he gave a successful kinematic description of this scattering (now called the *Compton effect*) by considering the relativistic collision of a photon with an electron, where both are regarded as particles

[1] The presently accepted value of h, now called *Planck's constant*, is $h = 6.62606957(29) \times 10^{-27}$ erg·s.

(Compton 1923). Compton showed that a photon not only carries energy $E = h\nu$ but also linear momentum; that is,

$$p = \frac{h\nu}{c} \tag{1.3}$$

His results finally convinced the community of physicists to accept wave-particle duality for electromagnetic radiation. What we mean by this duality is that electromagnetic radiation has wavelike properties or particlelike properties depending on what sort of observation is made.

The specific heats of solids presented a problem somewhat related to that of the black-body spectrum. In 1819, these specific heats were predicted classically by DuLong and Petit to be a constant independent of temperature. However, by the end of the nineteenth century, it became clear that while measured specific heats agree with the DuLong-Petit law at relatively high temperatures, they tend toward zero as $T \to 0$. This behavior was explained by Einstein (1911) and in more detail by Peter Debye (1912) as well as by Max Born and Theodore von Karman (Born and von Karman 1912, 1913). Their theory, which invoked quantization of lattice vibrations of solids, was a natural outgrowth of the early quantum theory of black-body radiation.

Ernest Rutherford used the results of alpha-particle scattering experiments to propose the nuclear atom model (Rutherford 1911). Needless to say, an atom in this model consists of a massive and very compact nucleus about which atomic electrons circulate in orbital motion. According to classical physics, such electrons should radiate electromagnetic waves because of their centripetal acceleration, and a simple classical estimate shows that they should lose energy and spiral into the nucleus in times of order 10^{-15} s. However, atoms are stable, so it is obvious that the classical description is very wrong. In 1913, Niels Bohr recognized this, as well as the fact that no combination of fundamental constants in classical physics can yield a natural length scale for an atom, whereas $4\pi\hbar^2/m_e e^2 \approx 10^{-8}$ cm does provide such a scale. (Here m_e is the electron mass, and e is the magnitude of electron charge in the hlu system.) Employing the concepts of quantized stationary (nonradiating) orbits and radiative transitions between them, where h plays a crucial role, Bohr constructed his model of atomic hydrogen (Bohr 1913). He thereby successfully accounted for the frequencies of optical transitions in atomic hydrogen and in singly ionized helium. His model quickly gained wide acceptance in part because of convincing supportive evidence from the experiments of James Franck and Gustav Herz (Franck and Herz 1914). Here electrons from a thermionic source were accelerated in an evacuated tube containing a low density of atomic vapor (e.g., sodium, potassium, thallium, mercury, etc.). If the electron kinetic energy was sufficiently low, only elastic collisions between electrons and atoms occurred. However, if the electron kinetic energy was high enough to excite a transition from the ground state of an atom to an excited state, the electron suffered an inelastic collision with corresponding energy loss, and fluorescence was observed as the excited atom decayed back to the ground state.

Bohr's model was elaborated by Arnold Sommerfeld (1916), who derived a formula for the fine-structure splittings in hydrogen and singly ionized helium by applying quantization conditions to classical Keplerian orbits of the electron and by including an important relativistic correction. Sommerfeld's formula agreed (albeit fortuitously) with spectroscopic observations of the fine structure and thus the Bohr-Sommerfeld model was taken seriously for about a decade as a plausible way to understand atomic structure.

The fund of experimental data concerning atomic spectra grew very rapidly in the first decades of the twentieth century, thanks to the efforts of many optical and x-ray spectroscopists. Attention naturally was drawn to the problem of assigning Bohr-Sommerfeld quantum numbers to hundreds of newly observed energy levels in scores of atoms. Of special interest were the quantum numbers of atoms in their ground states because this was obviously related to the role of atomic structure in building up the periodic table. Here Edmund C. Stoner made a valuable contribution in October 1924 by publishing an authoritative classification of such quantum numbers (Stoner 1924). Stoner's conclusions came to the attention of Wolfgang Pauli, who used them to formulate the extremely important exclusion principle at the end of 1924 (Pauli 1925).

Observations and analyses of the Zeeman effect played an especially significant role in the elucidation of atomic energy level quantum numbers. Following Peter Zeeman's pioneering measurements of the splitting of sodium spectral lines in a magnetic field (Zeeman 1897), Henrik A. Lorentz gave what appeared to be a correct theoretical explanation based on classical electrodynamics in the same year (Lorentz 1897). This was called the *normal* Zeeman effect. However, as more observations with higher resolution were carried out on many spectral lines in various atoms, it became apparent that the normal Zeeman effect is the exception rather than the rule. Instead, the *anomalous* Zeeman effect is typical, in which more complicated patterns of level splittings occur. For years, the anomalous Zeeman effect remained a mystery because all efforts to explain it failed. Finally, the puzzle was resolved with invention of the concept of electron spin by George Uhlenbeck and Samuel Goudsmit in November 1925 (Uhlenbeck and Goudsmit 1925, 1926). Earlier in 1925, Ralph Kronig had conceived of the same idea, but he was discouraged by adverse criticism and withdrew his proposal. [For a brief history of electron spin, see Commins (2012).]

Next we turn to the phenomenon of wave-particle duality for *material particles* (i.e., electrons, protons, atoms, etc.). First, let us recall relation (1.3) between momentum and frequency established by Compton for the photon. Employing the familiar expression $\lambda = c/\nu$ relating wavelength and frequency, we see that (1.3) implies

$$\lambda = \frac{h}{p} \tag{1.4}$$

In 1923, Louis de Broglie (1923, 1924) made the extremely important suggestion that each material particle is associated with a wave such that if the momentum of the particle is p, the wavelength of the corresponding "matter" wave is also given by (1.4). Fragmentary experimental evidence supporting de Broglie's hypothesis was already available in 1921 from results obtained by C. Davisson and C. H. Kunsman on the scattering of electrons from a nickel surface (Davisson 1921). By 1927, Davisson and L. Germer (1927) and, independently, G. Thomson and A. Reid (Thomson and Reid 1927; Thomson 1928) provided convincing evidence for relation (1.4) from electron diffraction experiments. Since then, the validity of (1.4) for material particles has been demonstrated precisely in many different experiments using a variety of material particles (e.g., neutrons and neutral atoms) as well as electrons.

Even in the absence of any formal quantitative theory, it is natural to assume that a particle is most likely to be located where the amplitude of its corresponding de Broglie wave packet is large. However, if the momentum and hence the wavelength are reasonably well defined, the wave packet must extend over many wavelengths, in which case the position of the particle is quite uncertain. Conversely, if the position is well defined, the wave packet must be confined to a small region of space, and therefore, it must be a superposition of components with many different wavelengths. Hence the momentum is very uncertain. The de Broglie relation (1.4) thus implies that it is impossible to determine simultaneously and precisely the position and the conjugate momentum of a particle.

This qualitative statement is made more precise by the uncertainty principle, which was formulated by Werner Heisenberg (1927) from consideration of a variety of thought experiments in which one tries to measure the position and momentum of a particle but where relation (1.4) applies not only to the particle in question but also to a photon that might be used in the measurement process. According to the uncertainty principle, the uncertainties Δx and Δp_x associated with a simultaneous measurement of coordinate x and conjugate momentum p_x, respectively, satisfy the inequality

$$\Delta x \Delta p_x \geq \frac{\hbar}{2} \tag{1.5}$$

where $\hbar = h/2\pi$. Although in classical mechanics the state at any given time of an isolated system of N particles, each with f degrees of freedom, is determined by specifying Nf generalized coordinates and Nf corresponding generalized momenta, the uncertainty principle tells us that this specification cannot be done precisely. A coordinate and the corresponding momentum are incompatible observables.

Intuitively, it is clear that because not only material particles but also photons obey the de Broglie relation (1.4), there should be an uncertainty principle for the electromagnetic field. Indeed, this is so (Jordan and Pauli 1928), although the uncertainty relation for electromagnetic field components is more complicated than for nonrelativistic material particles. We need not be concerned with such complications here. The main point for our present discussion is that the classical prescription for specifying a state of the electromagnetic field at any given time, by giving each component of the electric and magnetic fields at every point in space, cannot always be achieved.

It is easy to see from the de Broglie relation and the uncertainty principle that the Bohr-Sommerfeld model has a fatal defect, for in that model one starts in any given situation by finding the possible classical orbits of an electron or electrons and then selects from those orbits the ones that satisfy the Bohr-Sommerfeld quantization conditions. However, given the incompatibility of coordinate and conjugate momentum, and specifically the uncertainty relation (1.5), such orbits are in general not observable and indeed have no meaning, especially for states such as a ground state, that have small quantum numbers. In fact, looking back on the Bohr-Sommerfeld model from the viewpoint of quantum mechanics, and using the Wentzel-Kramers-Brillouin (WKB) approximation, one can show that the Bohr-Sommerfeld quantization conditions are valid only in the limiting case in which the potential energy varies very slowly over distances comparable with the linear dimensions of an electron wave packet [see, e.g., equation (9.18)].

Although the Bohr-Sommerfeld model was recognized for this and other reasons to be defective and was eventually replaced by quantum mechanics, it turned out that the Bohr-Sommerfeld quantum numbers did not have to be discarded wholesale; rather, some of these numbers could be retained if given new interpretations and new names. Consequently, after quantum mechanics was invented, the results of many analyses of atomic and molecular spectra carried out before 1925 could be salvaged, including most interpretations of Zeeman-effect data, Stoner's very useful contribution, and the exclusion principle itself.

We have seen in this section that deep and broad flaws appeared in the classical picture of the atomic world in the first quarter of the twentieth century. These flaws were so fundamental and serious that it would be necessary to replace the entire classical edifice with a radically different theory – quantum mechanics. It should be no surprise that these radically new concepts required a new mathematical language that was quite different from the mathematical language of classical physics. It turned out that the natural mathematical language of quantum mechanics is the theory of linear vector spaces and, in particular, Hilbert spaces. Therefore, before we discuss the rules of quantum mechanics in Chapter 3, we review and summarize some of the most important features of Hilbert spaces in Chapter 2.

2 Mathematical Review

In this chapter we summarize the most important definitions and theorems concerning Hilbert spaces that are relevant for quantum mechanics. Much of the material that follows is quite elementary and is probably well known to most readers. We discuss it mainly to establish a common language and notation. The reader will notice as we proceed that our standards of rigor are low and would be scorned by a proper mathematician. For example, we omit any discussion of convergence when considering infinite-dimensional spaces.

2.1 Linear vector spaces

A linear vector space S consists of certain elements $|u\rangle, |v\rangle, \ldots$ called *vectors* together with a field of ordinary numbers (sometimes called *c-numbers*) $a, b, c, \ldots$. In quantum mechanics, the latter are the complex numbers, and we deal with complex vector spaces. The vectors $|u\rangle, |v\rangle, \ldots$ and the numbers $a, b, c, \ldots$ satisfy the following rules:

1. *Vector addition is defined.* If $|u\rangle$ and $|v\rangle$ are members of S, there exists another vector $|w\rangle$, also a member of S, such that

$$|w\rangle = |u\rangle + |v\rangle \tag{2.1}$$

2. *Vector addition is commutative.*

$$|u\rangle + |v\rangle = |v\rangle + |u\rangle \qquad \text{for all } |u\rangle, |v\rangle \tag{2.2}$$

3. *There exists a null vector* $|0\rangle$ *or simply* 0 such that

$$|u\rangle + |0\rangle = |0\rangle + |u\rangle = |u\rangle \qquad \text{for any } |u\rangle \tag{2.3}$$

4. *Multiplication of a vector* $|u\rangle$ *by any c-number* a *is defined.*

$$|u'\rangle = a|u\rangle \tag{2.4}$$

is in the same "direction" (along the same ray) as $|u\rangle$.

5. *The following distributive law holds.*

$$a(|u\rangle + |v\rangle) = a|u\rangle + a|v\rangle \tag{2.5}$$

2.2 Subspaces

A vector space may contain subspaces. A *subspace* is a subclass of the space, itself having the properties of a vector space. For example, ordinary Euclidean 3-space contains as subspaces all the straight lines passing through the origin and all the two-dimensional planes that pass through the origin. All subspaces possess the null vector in common. They may or may not possess other vectors in common. If they do not, they are said to be *orthogonal subspaces*.

2.3 Linear independence and dimensionality

Vectors $|u_1\rangle, |u_2\rangle, \ldots, |u_n\rangle$ are by definition linearly independent if and only if the equation

$$a_1|u_1\rangle + a_2|u_2\rangle + \cdots + a_n|u_n\rangle = 0 \tag{2.6}$$

has no solution except for the trivial solution

$$a_1 = a_2 = \cdots = a_n = 0 \tag{2.7}$$

Suppose that in a certain space S there are n linearly independent vectors $|u_1\rangle, |u_2\rangle, \ldots, |u_n\rangle$, but any $n + 1$ vectors are linearly dependent. Then, by definition, the space is n-dimensional. The number n may be finite, denumerably infinite, or even continuously infinite. In most of the following discussion, we pretend that the space in question has finite n, but the results we obtain can be extended in a natural way to the other two cases. Particular problems associated with infinite dimensionality will be dealt with as we come to them (see, e.g., Section 2.14).

In an n-dimensional space where n is finite, n linearly independent vectors $|u_1\rangle, |u_2\rangle, \ldots, |u_n\rangle$ are said to span the space or form a basis for the space. This means that any vector $|w\rangle$ can be expressed as a linear combination of the $|u_i\rangle$; that is,

$$|w\rangle = \sum_{i=1}^{n} a_i |u_i\rangle \tag{2.8}$$

where the a_i are complex numbers.

2.4 Unitary spaces: The scalar product

A *unitary space* is one in which for any two vectors $|u\rangle, |v\rangle$ the scalar product $\langle u \mid v\rangle$ is defined as a complex number with the following properties:

1. $\langle u \mid v\rangle = \overline{\langle v \mid u\rangle}$, where the bar means complex conjugate (thus $\langle u \mid u\rangle$ is real).

2. $\langle u | av \rangle = a \langle u | v \rangle$. Thus,

$\langle au | v \rangle = \overline{\langle v | au \rangle} = a^* \langle u | v \rangle$

where the asterisk means complex conjugate.

3. $\langle u | v + w \rangle = \langle u | v \rangle + \langle u | w \rangle$.
4. $\langle u | u \rangle > 0$ unless $|u\rangle = 0$.

We also use the following terminology:

$$\langle u | u \rangle = \text{ norm of } |u\rangle \qquad \sqrt{\langle u | u \rangle} = \text{length of } |u\rangle$$

Also, if $|w\rangle \neq 0, |u\rangle \neq 0$, but $\langle w | u \rangle = 0$, $|u\rangle$ and $|w\rangle$ are said to be *orthogonal.*

2.5 Formation of an orthonormal basis: Completeness – definition of Hilbert spaces

Given a set of linearly independent vectors $|u_1\rangle, |u_2\rangle, \ldots, |u_n\rangle$ in a unitary space, we can construct an orthonormal basis as follows: form the unit vector

$$|\phi_1\rangle = \frac{|u_1\rangle}{\sqrt{\langle u_1 | u_1 \rangle}} \tag{2.9}$$

Next, form

$$|\phi_2\rangle = a|u_2\rangle + b|\phi_1\rangle \tag{2.10}$$

with a, b chosen so that $\langle \phi_1 | \phi_2 \rangle = 0$ and $\langle \phi_2 | \phi_2 \rangle = 1$. That is,

$$\langle \phi_1 | \phi_2 \rangle = a \langle \phi_1 | u_2 \rangle + b = 0 \tag{2.11}$$

which yields $b = -a \langle \phi_1 | u_2 \rangle$ and thus

$$|\phi_2\rangle = a\left[|u_2\rangle - |\phi_1\rangle \langle \phi_1 | u_2 \rangle\right] \tag{2.12}$$

with a chosen so that $\langle \phi_2 | \phi_2 \rangle = 1$. Next, define $|\phi_3\rangle$ by

$$|\phi_3\rangle = a'\left[|u_3\rangle - |\phi_1\rangle \langle \phi_1 | u_3 \rangle - |\phi_2\rangle \langle \phi_2 | u_3 \rangle\right] \tag{2.13}$$

with a' chosen to normalize $|\phi_3\rangle$, and so on. The result of this *Schmidt process* is a basis of orthogonal unit vectors (orthonormal vectors) $|\phi_1\rangle, \ldots, |\phi_n\rangle$.

A set of basis vectors in a unitary space is said to be *complete* if any vector in the space can be expressed as a linear combination of the basis members. For finite-dimensional vector spaces, any set that spans the space is complete. If the space has infinite dimensionality, questions of limits and convergence arise. We ignore such questions and content ourselves with the following simplified definition of a Hilbert space: it is a unitary space where even if the dimensionality is infinite, it is still possible to form a complete basis.

2.6 Expansion of an arbitrary vector in terms of an orthonormal basis

Suppose that the $|\phi_i\rangle$, $i = 1, \ldots, n$, form a complete orthonormal basis in an n-dimensional Hilbert space. (Here we assume that n is finite or denumerably infinite. If the dimension is continuously infinite, we must replace sums by integrals, indices $i, j, \ldots$ by continuous parameters, and each Kronecker delta by a Dirac delta function. This is explained in Section 2.14). An arbitrary vector $|w\rangle$ can be expressed as a linear combination of the $|\phi_i\rangle$, that is,

$$|w\rangle = \sum_{i=1}^{n} a_i |\phi_i\rangle \tag{2.14}$$

Then we have

$$\langle \phi_j | w\rangle = \sum_i a_i \langle \phi_j | \phi_i\rangle = \sum_i a_i \delta_{ij} = a_j$$

where δ_{ij} is a Kronecker delta. Hence we can write

$$|w\rangle = \sum_i |\phi_i\rangle\langle \phi_i | w\rangle \tag{2.15}$$

The scalar product of two arbitrary vectors $|v\rangle, |w\rangle$ can be expressed in terms of their expansion coefficients as follows: let

$$|w\rangle = \sum_{i=1}^{n} a_i |\phi_i\rangle$$

and

$$|v\rangle = \sum_{j=1}^{n} b_j |\phi_j\rangle$$

Then

$$\langle v | w\rangle = \sum_{i,j} b_j^* a_i \langle \phi_j | \phi_i\rangle = \sum_{i,j} b_j^* a_i \delta_{ij} = \sum_i b_i^* a_i \tag{2.16}$$

2.7 The Cauchy-Schwarz inequality

Consider any two vectors $|u\rangle, |v\rangle$. We now prove the following important inequality attributed to Schwarz but actually discussed first by Cauchy:

$$\langle u \mid u\rangle \cdot \langle v \mid v\rangle \geq |\langle u \mid v\rangle|^2 \tag{2.17}$$

where equality is achieved only if $|u\rangle = a|v\rangle$ for some constant a. To prove (2.17), let

$$|u\rangle = \sum a_i |\phi_i\rangle \quad \text{and} \quad |v\rangle = \sum b_i |\phi_i\rangle$$

where the $|\phi_i\rangle$ form a complete orthonormal basis. Then

$$|\langle u|v\rangle|^2 = \sum_{i,j} a_i^* b_i a_j b_j^*$$

and

$$\langle u|u\rangle\langle v|v\rangle = \sum_{i,j} a_i^* a_i b_j^* b_j$$

Thus

$$\langle u|u\rangle\langle v|v\rangle - |\langle u|v\rangle|^2 = \sum_{i,j} a_i^* b_j^* \left(a_i b_j - a_j b_i\right) \equiv P \tag{2.18}$$

Because i, j are dummy indices in the sum on the right-hand side of (2.18), we can interchange i and j, which yields

$$\langle u|u\rangle\langle v|v\rangle - |\langle u|v\rangle|^2 = \sum_{i,j} a_j^* b_i^* \left(a_j b_i - a_i b_j\right) \equiv Q \tag{2.19}$$

Thus

$$\begin{aligned}\langle u|u\rangle\langle v|v\rangle - |\langle u|v\rangle|^2 &= \frac{1}{2}(P+Q) = \frac{1}{2}\sum_{i,j} \overline{\left(a_i b_j - a_j b_i\right)} \times \left(a_i b_j - a_j b_i\right) \\ &= \frac{1}{2}\sum_{i,j} |a_i b_j - a_j b_i|^2 \geq 0\end{aligned} \tag{2.20}$$

and equality is achieved only if $a_i b_j = a_j b_i$ for all i, j, which would imply $|u\rangle = \text{const} \cdot |v\rangle$.

The reader will readily see that the Cauchy-Schwarz inequality is a generalization of the following simple inequality for two vectors in Euclidean 3-space: the absolute value of the scalar product of these two vectors is less than or equal to the products of their lengths, and equality is achieved if and only if the angle between these two vectors is zero. Several other inequalities of importance in quantum mechanics are derived in Appendix A. These are of particular value for analyzing the stability of matter, which is discussed in Section 14.1.

2.8 Linear operators

An operator A transforms a vector $|u\rangle$ into another vector $|v\rangle$. It achieves a mapping of the vectors in the space S. We write $|v\rangle = A|u\rangle$. An operator is linear if and only if it satisfies the following conditions:

$$Aa|u\rangle = aA|u\rangle \qquad \text{and} \qquad A(|u\rangle + |v\rangle) = A|u\rangle + A|v\rangle$$

for any vectors $|u\rangle, |v\rangle$. The expression $AB|u\rangle$ means operate first on $|u\rangle$ with B: $|w\rangle = B|u\rangle$. Then operate on $|w\rangle$ with A: $|v\rangle = A|w\rangle$. If for all $|u\rangle$, $AB|u\rangle = BA|u\rangle$, we write $[A, B] \equiv AB - BA = 0$. Then A and B are said to *commute*. It is easy to think of operators that do not commute. For example, in three-dimensional Euclidean space, B could be a rotation about the x-axis and A a rotation about the y-axis.

2.9 Inverse of a linear operator

If $|w\rangle = A|u\rangle$, consider the operator A^{-1} such that $|u\rangle = A^{-1}|w\rangle$ for all $|u\rangle$. A^{-1} is the inverse of A, and

$$AA^{-1} = A^{-1}A = I$$

where I is the identity operator, which has the property $I|u\rangle = |u\rangle$ for all $|u\rangle$. Not all operators have inverses: those that do are called *nonsingular*, and those that don't are *singular*. For example, consider a projection operator P that maps all vectors $\mathbf{r}$ in three-dimensional Euclidean space onto their projections in the xy-plane. P is obviously singular – it has no inverse.

If A, B, and C are nonsingular operators and $A = BC$, how can we express A^{-1} in terms of B^{-1} and C^{-1}? Clearly,

$$C^{-1}B^{-1}BC = I$$

Hence we must have $A^{-1} = C^{-1}B^{-1}$.

2.10 The adjoint operator

While discussing the scalar product of two vectors, we could have introduced the idea of a dual vector space with vectors $\langle u|$ in one-to-one correspondence with vectors $|u\rangle$ in the original space. Let A be a linear operator in the original space. We also may consider the application of operator A to a dual vector $\langle u|$. The latter operation is defined by requiring that for any two vectors $|u\rangle,\ |v\rangle$,

$$(\langle u|A)|v\rangle = \langle u|(A|v\rangle) \tag{2.21}$$

where the left-hand side is the scalar product of $\langle u|A$ with $|v\rangle$, and the right-hand side is the scalar product of $\langle u|$ with $A|v\rangle$. Given this equality, it is not necessary to write the parentheses, so we drop them in what follows. Now suppose that $A|u\rangle = |w\rangle$. The operator that transforms $\langle u|$ into $\langle w|$ is called the *adjoint of A* and is denoted by the symbol $A^\dagger$: $\langle u|A^\dagger = \langle w|$. For our purposes, it is unnecessary to distinguish between the concepts of adjoint and Hermitian conjugate, so we shall henceforth call $A^\dagger$ the *Hermitian conjugate of A*.

Because $\langle v \mid w\rangle = \overline{\langle w \mid v\rangle}$, we have

$$\langle v \mid A \mid u\rangle = \overline{\langle u \mid A^\dagger \mid v\rangle} \tag{2.22}$$

Also, we can easily show that $(AB)^\dagger = B^\dagger A^\dagger$ as follows: let $|w\rangle = A^\dagger|u\rangle$ and $|x\rangle = B|v\rangle$. Then

$$\langle w \mid x\rangle = \langle u \mid AB \mid v\rangle = \overline{\langle x \mid w\rangle} = \overline{\langle v \mid B^\dagger A^\dagger \mid u\rangle}$$

It follows from (2.22) that

$$(AB)^\dagger = B^\dagger A^\dagger \tag{2.23}$$

A is said to be *self-adjoint* or *Hermitian* if $A = A^\dagger$. For a Hermitian operator A,

$$\langle v \mid A \mid u\rangle = \overline{\langle u \mid A \mid v\rangle} \tag{2.24}$$

For example, the expression $|u\rangle\langle u|$ is a Hermitian operator. For any $|w\rangle, |v\rangle$, we have

$$\langle w \mid u\rangle\langle u \mid v\rangle = \overline{\langle u \mid w\rangle}\,\overline{\langle v \mid u\rangle} = \overline{\langle v \mid u\rangle\langle u \mid w\rangle}$$

2.11 Eigenvectors and eigenvalues

In general, if $|v\rangle = A|u\rangle$, then $|v\rangle$ is in a different *direction* than $|u\rangle$. However, there might exist certain vectors $|u_1\rangle, |u_2\rangle, \ldots$, called the *characteristic vectors* or *eigenvectors of A*, that have the property

$$A|u_i\rangle = \lambda_i |u_i\rangle \tag{2.25}$$

where the λ_i are c-numbers called the *eigenvalues of A*. The collection of all the λ_i for given A is called the *eigenvalue spectrum of A*. We now prove a theorem that is simple but very important for quantum mechanics.

The eigenvalues of a Hermitian operator are real, and the eigenvectors are mutually orthogonal.

Proof: Let $A|u_n\rangle = \lambda_n |u_n\rangle$. Then

$$\langle u_n | A | u_n \rangle = \lambda_n \langle u_n | u_n \rangle \tag{2.26}$$

Now take the complex conjugate of both sides of (2.26). Noting that $\langle u_n | A | u_n \rangle = \overline{\langle u_n | A | u_n \rangle}$ and that $\langle u_n | u_n \rangle$ is always real, we see that λ_n also must be real. Furthermore,

$$\langle u_m | A | u_n \rangle = \lambda_n \langle u_m | u_n \rangle \tag{2.27}$$

and

$$\overline{\langle u_n | A | u_m \rangle} = \lambda_m^* \overline{\langle u_n | u_m \rangle} = \lambda_m \langle u_m | u_n \rangle \tag{2.28}$$

However, the left-hand sides of (2.27) and (2.28) are equal; therefore,

$$(\lambda_m - \lambda_n)\langle u_m | u_n \rangle = 0 \tag{2.29}$$

Thus, if $\lambda_m \neq \lambda_n$, $\langle u_m | u_n \rangle = 0$. So far we have proved that the eigenvalues of a Hermitian operator are all real and that the eigenvectors corresponding to distinct eigenvalues are orthogonal. However, we can also consider all the eigenvectors that belong to the same eigenvalue λ. Suppose that there exist k such linearly independent "degenerate" eigenvectors. These form a k-dimensional subspace S_k of the original vector space because any linear combination of such eigenvectors is also in S_k. The basis of S_k may now be rendered orthonormal by the Schmidt process. Thus all eigenvectors of a Hermitian operator are orthogonal or can be made so.

2.12 Projection operators and completeness

Consider a vector space S with a complete orthonormal basis consisting of vectors $|u_i\rangle$, $i = 1, \ldots, n$. As stated previously, completeness means that an arbitrary vector $|w\rangle$ can be expressed as

$$|w\rangle = \sum_{i=1}^{n} |u_i\rangle\langle u_i | w\rangle \tag{2.30}$$

The Hermitian operator $P_i = |u_i\rangle\langle u_i|$ is a *projection operator* or *projector* that acts on $|w\rangle$ to give the component of $|w\rangle$ along the direction of $|u_i\rangle$. From (2.30) the sum of all these projectors is the identity operator

$$\sum_{i=1}^{n} |u_i\rangle\langle u_i| = \sum_{i=1}^{n} P_i = I \tag{2.31}$$

Equation (2.31) is known as the *completeness relation*. For a vector space with continuously infinite dimensionality, the basis vectors are labeled by one or more continuous parameters λ, and the completeness relation is written as

$$\int |\lambda\rangle\langle\lambda| \, d\lambda = I \tag{2.32}$$

2.13 Representations

In (2.30), the complex numbers $\langle u_i \mid w\rangle$ represent vector $|w\rangle$ with respect to the basis $|u_i\rangle$. If the vector space dimension is finite or denumerably infinite, these numbers can be arranged in a column matrix as follows:

$$\begin{pmatrix} \langle u_1 \mid w\rangle \\ \langle u_2 \mid w\rangle \\ \cdots \end{pmatrix} \tag{2.33}$$

Similarly, we can expand a dual vector in terms of the basis of dual vectors $\langle u_i|$; that is,

$$\langle v| = \sum_i \langle v \mid u_i\rangle\langle u_i| \tag{2.34}$$

The row matrix

$$\left(\langle v \mid u_1\rangle \quad \langle v \mid u_2\rangle \quad \ldots\right) \tag{2.35}$$

represents $\langle v|$ in this basis. The scalar product $\langle v \mid w\rangle$ is obtained by multiplying the column matrix of (2.33) on the left by the row matrix of (2.35) according to the usual rule.

Next, consider an operator A. We write

$$A = IAI = \sum_{i,j} |u_i\rangle\langle u_i \mid A \mid u_j\rangle\langle u_j| \tag{2.36}$$

The complex numbers $\langle u_i \mid A \mid u_j\rangle$ form a square matrix that represents A with respect to the basis of vectors $|u_i\rangle$. Suppose that each $|u_i\rangle$ is an eigenvector of a Hermitian operator Q, with corresponding eigenvalue q_i; that is,

$$Q|u_i\rangle = q_i|u_i\rangle$$

The matrix that represents Q with respect to the basis $|u_i\rangle$ is obviously diagonal, and the diagonal elements are the eigenvalues q_1, q_2, ... A convenient way to write Q is as follows:

$$\begin{aligned} Q = IQI &= \sum_{i,j} |u_i\rangle\langle u_i | Q | u_j\rangle\langle u_j| \\ &= \sum_{i,j} |u_i\rangle q_j \langle u_i | u_j\rangle\langle u_j| \\ &= \sum_i q_i |u_i\rangle\langle u_i| \end{aligned} \tag{2.37}$$

2.14 Continuously infinite dimension: The Dirac delta function

As we have seen in (2.32), the completeness relation for a space with continuously infinite dimension is

$$I = \int |\lambda\rangle\langle\lambda|\, d\lambda$$

where λ is a continuous parameter that labels the basis states. (For simplicity, we assume here that λ is a single parameter.) Consider an arbitrary vector $|u\rangle$. We write

$$|u\rangle = I|u\rangle = \int |\lambda\rangle\langle\lambda|u\rangle\, d\lambda$$

and thus

$$\langle\lambda'|u\rangle = \int \langle\lambda'|\lambda\rangle\langle\lambda|u\rangle\, d\lambda \tag{2.38}$$

The quantity $f(\lambda) = \langle\lambda|u\rangle$ is a continuous function of λ of functional form determined by $|u\rangle$. Thus (2.38) can be written as

$$f(\lambda') = \int \langle\lambda'|\lambda\rangle f(\lambda)\, d\lambda \tag{2.39}$$

To determine $\langle\lambda'|\lambda\rangle$, we introduce the (one-dimensional) Dirac delta function. Strictly speaking, the delta function $\delta(x - y)$ is not a proper function at all but rather a distribution: it is defined by its effect when multiplied by another function $f(x)$ in an integral; that is,

$$f(y) = \int_{-\infty}^{\infty} f(x)\delta(x-y)\, dx \tag{2.40}$$

However, we can think heuristically of the delta function as a Gaussian or some other sharply peaked function taken in the limit as we reduce its width to zero while at the same time keeping the area under its curve equal to unity. The delta function has the following properties:

$$\delta(x-y) = 0 \qquad \text{if } x \neq y \tag{2.41}$$

$$\left.\frac{d^n f(x)}{dx^n}\right|_{x=y} = (-1)^n \int_{-\infty}^{\infty} f(x) \frac{d^n}{dx^n} [\delta(x-y)]\, dx \tag{2.42}$$

$$\int f(x)\delta[g(x)]\, dx = \sum_i \frac{f(x_i)}{|\partial g/\partial x|_{x_i}} \tag{2.43}$$

where in (2.43) the x_i are the real roots of $g(x)$. A useful integral representation of the delta function is obtained from the Fourier integral theorem. Here $\delta(\omega)$ is the Fourier transform of the function $F(t) = 1$; that is,

$$\delta(\omega) = \frac{1}{2\pi} \lim_{T\to\infty} \int_{-T}^{T} e^{i\omega t}\, dt \tag{2.44}$$

Equation (2.44) also can be written as

$$\delta(\omega) = \frac{1}{\pi} \lim_{T\to\infty} \frac{\sin \omega T}{\omega} \tag{2.45}$$

Another useful representation of the delta function is

$$\delta(\omega) = \frac{1}{\pi} \lim_{T\to\infty} \frac{\sin^2 \omega T}{\omega^2 T} \tag{2.46}$$

Comparing (2.45) with (2.46), we obtain the formula

$$\delta^2(\omega) = \delta(\omega)\delta(0) \tag{2.47}$$

Now, returning to (2.39) and comparing it with (2.40), we see that

$$\langle \lambda' | \lambda \rangle = \delta(\lambda - \lambda') \tag{2.48}$$

This is to be compared with the orthonormality relation for a discrete set of basis vectors:

$$\langle \phi_i | \phi_j \rangle = \delta_{ij}$$

2.15 Unitary transformations

The choice of a complete basis set for a vector space is not unique. For example, in three-dimensional Euclidean space we can choose as a basis the unit vectors $\hat{i}, \hat{j}, \hat{k}$ that lie along the x, y, and z directions, respectively. However, we can equally well choose three other unit vectors obtained from $\hat{i}, \hat{j}, \hat{k}$ by, for example, making an arbitrary rotation about the origin. Consider a space S spanned by the complete orthonormal set of vectors $|u_i\rangle$ and also spanned by a different complete orthonormal set $|u_i'\rangle$. Let the one-to-one mapping from the set $|u_i\rangle$ to the set $|u_i'\rangle$ be achieved by an operator U; that is,

$$U|u_i\rangle = |u_i'\rangle \qquad i = 1,\ldots,n \tag{2.49}$$

Then we also have

$$\langle u_i|U^\dagger = \langle u_i'| \tag{2.50}$$

Now

$$U = UI = U\sum_i |u_i\rangle\langle u_i| = \sum |u_i'\rangle\langle u_i| \tag{2.51}$$

and

$$U^\dagger = IU^\dagger = \sum_j |u_j\rangle\langle u_j|U^\dagger = \sum |u_j\rangle\langle u_j'| \tag{2.52}$$

Hence

$$\begin{aligned} UU^\dagger &= \sum_{i,j} |u_i'\rangle\langle u_i \,|\, u_j\rangle\langle u_j'| = \sum_{i,j} |u_i'\rangle\delta_{ij}\langle u_j'| \\ &= \sum_i |u_i'\rangle\langle u_i'| = I \end{aligned}$$

where the last step follows because the set $|u_i'\rangle$ is complete. Therefore,

$$UU^\dagger = U^\dagger U = I \tag{2.53}$$

or, equivalently,

$$U^\dagger = U^{-1} \tag{2.54}$$

An operator with this property is called *unitary*. Unitary transformations are important in quantum mechanics because they are norm-preserving: if $|u'\rangle = U|u\rangle$, then $\langle u' \,|\, u'\rangle = \langle u \,|\, U^\dagger U \,|\, u\rangle = \langle u \,|\, u\rangle$.

Next, consider some operator A such that $A|u\rangle = |v\rangle$, whereas $U|u\rangle = |u'\rangle$ and $U|v\rangle = |v'\rangle$. There is some operator A' that transforms $|u'\rangle$ into $|v'\rangle$. What is the relationship between A' and A? We have

$$A'|u'\rangle = |v'\rangle = U|v\rangle = UA|u\rangle = UAU^{-1}|u'\rangle$$

Thus

$$A' = UAU^{-1} = UAU^\dagger \tag{2.55}$$

This formula gives the transformation rule for operator A under unitary transformation U. For Hermitian operators, this has the following consequence: consider a Hermitian operator

H that has a complete set of orthonormal eigenvectors $|h_i\rangle$ with corresponding eigenvalues η_i; that is,

$$H|h_i\rangle = \eta_i |h_i\rangle$$

As noted previously, H can be written as

$$H = \sum_i \eta_i |h_i\rangle\langle h_i| \tag{2.56}$$

Now consider another orthonormal basis in the same space consisting of the vectors $|u_i\rangle$. In general, the matrix of H with respect to this new basis is not diagonal. However, there is a related operator H' the matrix of which is diagonal in the new basis. It is easily found as follows: we write

$$H' = UHU^\dagger = \sum_{i,j} |u_i\rangle\langle h_i | H | h_j\rangle\langle u_j| = \sum_i \eta_i |u_i\rangle\langle u_i| \tag{2.57}$$

Comparing (2.56) with (2.57), we see that H is a weighted sum of projectors in the $|h_i\rangle$ basis, whereas H' is a weighted sum of projectors in the $|u_i\rangle$ basis. The weights are the same in both cases: they are the eigenvalues of H, which are also the eigenvalues of H'.

Suppose that we do not know the eigenvalues η_k of operator H that appear in $H|h_k\rangle = \eta_k |h_k\rangle$, but we do know the elements of the nondiagonal matrix; that is,

$$H_{ij} = \langle u_i|H|u_j\rangle \tag{2.58}$$

How can we find the eigenvalues η? We write

$$H|h_k\rangle = \sum_j H|u_j\rangle\langle u_j|h_k\rangle = \eta_k |h_k\rangle$$

Thus

$$\sum_j \langle u_i|H|u_j\rangle\langle u_j|h_k\rangle = \eta_k \langle u_i|h_k\rangle \tag{2.59}$$

This can be written as

$$(H - \eta I)X = 0 \tag{2.60}$$

where X is a column matrix of elements $\langle u_i|h_k\rangle$, and $H - \eta I$ is a square matrix in the $|u_i\rangle$ basis. Expression (2.60) is a system of n homogeneous linear equations in n unknowns that has a solution if and only if the determinant of the coefficients (i.e., the determinant of $H - \eta I$) vanishes.[1] For finite n, the left-hand side of the equation

[1] Det $(H - \eta I)$ is called the *secular determinant* because equations such as (2.60) have been used since Euler's time in celestial mechanics to calculate secular (long-period) perturbations of planetary orbits.

$$\det(H - \eta I) = 0 \tag{2.61}$$

is a polynomial in η; thus, when we find the roots of that polynomial, we have found the eigenvalues.

2.16 Invariants

Under a unitary transformation U, we have seen that $A \to A' = UAU^{-1}$. However, certain quantities associated with A remain invariant as follows:

- The *trace* of a matrix is defined as the sum of its diagonal elements. As is well known, the trace of a product of matrices is invariant under cyclic permutation of those matrices. Thus

$$\mathrm{tr}(A') = \mathrm{tr}(UAU^{-1}) = \mathrm{tr}(U^{-1}UA) = \mathrm{tr}(A)$$

  Therefore, the trace of any Hermitian matrix, diagonal or not, is the sum of its diagonal elements, and this sum remains invariant when the matrix is diagonalized by a unitary transformation.

- The determinant is also an invariant because:

$$\begin{aligned}\det(A') &= \det(U)\cdot\det(A)\cdot\det(U^{-1}) \\ &= \det(U)\cdot\det(U^{-1})\cdot\det(A) = \det(A)\end{aligned}$$

2.17 Simultaneous diagonalization of Hermitian matrices

We now encounter another elementary theorem of linear algebra that is very important in quantum mechanics:

Two Hermitian matrices A, B are diagonalized by the same unitary transformation U if and only if A and B commute.

Proof: Consider $A' = UAU^\dagger$ and $B' = UBU^\dagger$. First, suppose that A' and B' are both diagonal. Then they obviously commute. Hence

$$\begin{aligned}AB &= U^{-1}A'UU^{-1}B'U = U^{-1}A'B'U = U^{-1}B'A'U \\ &= U^{-1}B'UU^{-1}A'U \\ &= BA\end{aligned}$$

Therefore, if A and B are diagonalized by the same unitary transformation, they commute. Conversely, suppose that $AB = BA$. Then $A'B' = B'A'$ because

$$\begin{aligned}UABU^{-1} &= UAU^{-1}UBU^{-1} = A'B' \\ &= UBAU^{-1} = B'A'\end{aligned}$$

Then, by the rule for matrix multiplication, we have for any i, k

$$\sum_j A'_{ij} B'_{jk} = \sum_n B'_{in} A'_{nk} \tag{2.62}$$

We can always find a unitary transformation that diagonalizes any given Hermitian matrix. Thus we choose U so that A' is diagonal. Then (2.62) becomes

$$\left(A'_{ii} - A'_{kk}\right) B'_{ik} = 0 \tag{2.63}$$

If all the eigenvalues of A are distinct, then the factor in parentheses on the left-hand side of (2.63) is nonzero for $i \neq k$, in which case $B'_{ik} = 0$, so B' is also diagonal. It remains to analyze the case where some of the eigenvalues of A are the same. Thus consider a particular eigenvalue λ of A that is m-fold degenerate. Associated with λ there is an $m \times m$ diagonal submatrix λI_m of A' (where I_m is the $m \times m$ identity matrix), and there is also an $m \times m$ Hermitian submatrix of B' that is in general nondiagonal. However, we can always find a further unitary transformation V on the submatrix of B' that brings it to diagonal form, and because $V \lambda I_m V^\dagger = \lambda I_m$, we do not disturb the diagonal form of A'. Thus, finally, we have proved that if and only if A and B commute, they are simultaneously diagonalized by the same unitary transformation.

2.18 Functions of an operator

How can we define a function f of an operator A? In almost all situations of interest for quantum mechanics, A is a Hermitian operator with a complete set of eigenstates $|u_n\rangle$ and

$$A|u_n\rangle = \lambda_n |u_n\rangle$$

where the λ_n are the eigenvalues of A. As shown previously, A can be written as

$$A = \sum_n \lambda_n |u_n\rangle\langle u_n|$$

We define $f(A)$ by the formula

$$f(A) = \sum_n f(\lambda_n)|u_n\rangle\langle u_n|$$

Alternatively, $f(A)$ can be defined by the power-series expansion of f; that is,

$$f(A) = \sum a_n A^n \tag{2.64}$$

where the a_n are numerical coefficients. For example,

$$\exp(A) = \sum_{n=0}^{\infty} \frac{1}{n!} A^n \tag{2.65}$$

We must be careful about the order of noncommuting operators when constructing functions of two or more operators. For example, consider the expression

$$e^{A}Be^{-A} \tag{2.66}$$

where A and B are two operators that do not necessarily commute. We now show that

$$e^{A}Be^{-A} = B+[A,B]+\frac{1}{2!}\left[A,[A,B]\right]+\frac{1}{3!}\left[A,\left[A,[A,B]\right]\right]+\cdots \tag{2.67}$$

This identity is very useful for many applications in quantum mechanics. To demonstrate it, consider the function

$$f(\lambda) = e^{\lambda A}Be^{-\lambda A} \tag{2.68}$$

where λ is a continuous parameter. Differentiating both sides of (2.68), we have

$$\frac{\partial f}{\partial \lambda} = e^{\lambda A}ABe^{-\lambda A} - e^{\lambda A}BAe^{-\lambda A} = e^{\lambda A}\left[A,B\right]e^{-\lambda A} \tag{2.69}$$

Differentiating once again, we obtain

$$\frac{\partial^2 f}{\partial \lambda^2} = e^{\lambda A}A\left[A,B\right]e^{-\lambda A} - e^{\lambda A}\left[A,B\right]Ae^{-\lambda A} = e^{\lambda A}\left[A,\left[A,B\right]\right]e^{-\lambda A} \tag{2.70}$$

and so on. Inserting these derivatives into the MacLaurin expansion

$$f(\lambda) = f(0)+\lambda f'(0)+\frac{1}{2!}\lambda^2 f''(0)+\cdots \tag{2.71}$$

and setting $\lambda = 1$, we obtain the result (2.67). Note that if $\left[A,B\right]=0$, $f(\lambda)=B$, whereas if $\left[A,B\right]=C$, where C is a c-number, $f(\lambda) = B + \lambda C$.

Next, consider a unitary operator $U(\varepsilon)$ that differs infinitesimally from the identity I; that is,

$$U(\varepsilon) = I - i\varepsilon K \tag{2.72}$$

Here ε is a real parameter that is small enough that we can neglect quantities of order ε^2 in what follows, and K is another operator. K must be Hermitian for the following reason:

$$\begin{aligned} I = U^{\dagger}U &= \left(I+i\varepsilon K^{\dagger}\right)\left(I-i\varepsilon K\right) \\ &= I+i\varepsilon\left(K^{\dagger}-K\right) \end{aligned} \tag{2.73}$$

Let $x = n\varepsilon$ be a finite real quantity obtained by multiplying n factors of ε together. We can take the limits $n \to \infty$, $\varepsilon \to 0$ in such a way that x is fixed. Also let $U(x)$ be the following unitary transformation:

$$U(x) = \prod U(\varepsilon) = \lim_{n\to\infty}\left(I - i\frac{x}{n}K\right)^n = \exp(-ixK) \tag{2.74}$$

Here we see that a unitary transformation of this type can be expressed as the exponential of a Hermitian operator.

Problems for Chapter 2

2.1. (a) A skew-Hermitian operator A is defined by the property $A^\dagger = -A$. Prove that A can have at most one real eigenvalue (which may be degenerate).

(b) Prove that the commutator $[A,B] \equiv AB - BA$ of two Hermitian operators A, B is skew-Hermitian or zero.

(c) Show that the equation $[A,B] = iqI$, where q is a real constant and I is the identity matrix, cannot be satisfied by any finite-dimensional Hermitian matrices A, B.

2.2. Consider the functions $u_n(x) = x^n$ with $n = 0, 1, 2, \ldots$. These form a basis for a vector space consisting of all real analytic functions of x on the real line in the interval $-1 \le x \le +1$. From the functions $u_n(x) = x^n$, we can form an orthonormal basis of functions $\phi_n(x)$ by means of the Schmidt process. Here we define the scalar product of two "vectors" $f(x), g(x)$ by

$$\langle f|g\rangle = \frac{1}{2}\int_{-1}^{1} f(x)g(x)\,dx$$

Find the orthonormal basis functions ϕ_n, $n = 0, 1, 2, 3$. What well-known functions are these?

2.3. Let $f(x), g(x)$ be two real functions. Show that

$$\int_{x_1}^{x_2} f^2(x)\,dx \cdot \int_{x_1}^{x_2} g^2(x)\,dx \ge \left[\int_{x_1}^{x_2} f(x)g(x)\,dx\right]^2$$

This is the Cauchy-Schwarz inequality for real integrals.

2.4. (a) Let G and H be two Hermitian operators, and assume that the eigenvectors of each form a complete orthonormal set. Show that if G, H each have positive eigenvalues, then $\mathrm{tr}(GH) > 0$.

(b) Let H be a Hermitian operator, the eigenvectors of which form a complete orthonormal set and the eigenvalues of which are all positive. Prove that for any two vectors $|u\rangle, |\mathrm{v}\rangle$,

$$|\langle u|H|\mathrm{v}\rangle|^2 \le \langle u|H|u\rangle\langle \mathrm{v}|H|\mathrm{v}\rangle$$

Also prove that $\mathrm{tr}(H) > 0$.

2.5. Prove that if $|u\rangle$ is a vector in a Hilbert space and A is a Hermitian operator on that space, then

$$A|u\rangle = \frac{\langle u|A|u\rangle}{\langle u|u\rangle}|u\rangle + \Delta A|w\rangle$$

where $|w\rangle$ is some vector orthogonal to $|u\rangle$, and $\Delta A = \sqrt{\langle A^2\rangle - \langle A\rangle^2}$.

2.6. Let A be a Hermitian operator with a complete set of eigenvectors $|n\rangle$ and associated eigenvalues λ_n. Suppose that there are two Hermitian operators B, C such that $[A,B]=[A,C]=0$, but $[B,C]\neq 0$. Prove that

(a) At least one of the eigenvalues of A must be degenerate.

(b) If $[B,C]=iq$, where q is a real constant, the entire spectrum of A is degenerate, and the degeneracy is of infinite degree.

2.7. Show that if a linear operator A satisfies any two of the following conditions, it also satisfies the remaining condition:

A is Hermitian
A is unitary
$A^2 = I$

2.8. Show that if a matrix M is *idempotent*, that is, $M^2 = M$, and it is not the identity matrix, then its determinant is zero.

2.9. Show that $\det(e^A) = \exp[\mathrm{tr}(A)]$, and state the condition or conditions that must hold for matrix A so that your proof is valid.

2.10. Show that if A is an arbitrary 2×2 matrix, then

$$A^2 = \mathrm{tr}(A)\cdot A - \det A \cdot I$$

where I is the 2×2 identity matrix. This is a special case of the well-known *Cayley-Hamilton theorem* for $N \times N$ matrices.

2.11. Two matrices A, B satisfy the following equations:

$$A^2 = 0 \qquad AA^\dagger + A^\dagger A = I \qquad B = A^\dagger A$$

(a) Show that $B^2 = B$.

(b) Obtain explicit expressions for A and B in a representation in which B is diagonal, assuming that B is nondegenerate. Can A be diagonalized in any representation?

2.12. Consider four linearly independent nonsingular Hermitian $N \times N$ matrices A_i, $i = 1, 2, 3, 4$, that satisfy the relations

$$A_i A_j + A_j A_i = 2\delta_{ij} I$$

where I is the $N \times N$ identity matrix. Show that N must be an even number greater than 2.

2.13. The three 2×2 Pauli spin matrices

$$\sigma_x = \begin{pmatrix} 0 & 1 \\ 1 & 0 \end{pmatrix} \qquad \sigma_y = \begin{pmatrix} 0 & -i \\ i & 0 \end{pmatrix} \qquad \sigma_z = \begin{pmatrix} 1 & 0 \\ 0 & -1 \end{pmatrix}$$

play an important role in many areas of quantum mechanics. The $\sigma_{x,y,z}$ and the 2×2 identity matrix

$$I = \begin{pmatrix} 1 & 0 \\ 0 & 1 \end{pmatrix}$$

together span the four-dimensional vector space of all 2 × 2 matrices. It is easy to verify from Pauli spin matrices that

$$\sigma_x^2 = \sigma_y^2 = \sigma_z^2 = I$$

and also that

$$\sigma_x \sigma_y = i\sigma_z$$

and cyclic permutations. These two relations can be combined into the following compact form:

$$\sigma_i \sigma_j = \delta_{ij} I + i\varepsilon_{ijk}\sigma_k$$

where δ_{ij} is a Kronecker delta, and ε_{ijk} is the completely antisymmetric unit 3-tensor. Here $\varepsilon_{ijk} = +1$ for $ijk = 1,2,3$ and even permutations thereof; $\varepsilon_{ijk} = -1$ for odd permutations of 1, 2, 3; and $\varepsilon_{ijk} = 0$ whenever two or more of the indices ijk are equal.

The Pauli spin matrices shown earlier constitute the "standard" representation of the Pauli matrices. Another representation is obtained by writing

$$\sigma'_{x,y,z} = U\sigma_{a,y,z}U^{-1}$$

where U is a unitary 2 × 2 matrix. Because there are an infinite number of possible unitary 2 × 2 matrices, there are an infinite number of representations of the Pauli spin matrices.

(a) Show that it is impossible to construct a nonvanishing 2 × 2 matrix that anticommutes with each of the three Pauli matrices.

(b) Show that it is impossible to find a representation of the Pauli matrices where all three are real or where two are pure imaginary and one is real.

(c) Let us define $\boldsymbol{\sigma} = \sigma_x\hat{i} + \sigma_y\hat{j} + \sigma_z\hat{k}$, where, as usual, $\hat{i}, \hat{j}, \hat{k}$ are unit vectors along the x-, y-, and z-axes, respectively, in Euclidean 3-space. Let $\boldsymbol{A}$, $\boldsymbol{B}$ be any two vector operators that commute with $\boldsymbol{\sigma}$. Show that the following identity holds:

$$\boldsymbol{\sigma}\cdot\boldsymbol{A}\boldsymbol{\sigma}\cdot\boldsymbol{B} = \boldsymbol{A}\cdot\boldsymbol{B} + i\boldsymbol{\sigma}\cdot\boldsymbol{A}\times\boldsymbol{B}$$

(d) Let $\hat{\boldsymbol{n}}$ be a unit vector in an arbitrary direction in Euclidean 3-space and θ an arbitrary angle. Show that

$$\exp(i\boldsymbol{\sigma}\cdot\hat{\boldsymbol{n}}\,\theta) = I\cos\theta + i\boldsymbol{\sigma}\cdot\hat{\boldsymbol{n}}\sin\theta$$

The preceding two identities are important in many quantum mechanical applications.

2.14. Let $A(x)$ be an operator that depends on the continuous parameter x, and let dA/dx be the derivative of $A(x)$ with respect to x. Note that A and dA/dx do not necessarily commute. Derive the following identity:

$$e^{-iA}\frac{d}{dx}\left(e^{iA}\right) = i\sum_{n=0}^{\infty}\frac{(-i)^n}{(n+1)!}A^n\left\{\frac{dA}{dx}\right\}$$

where $A^0\{B\} = B$, $A^1\{B\} = [A,B]$, $A^2\{B\} = [A,[A,B]]$, and so on.

2.15. Sum rules are frequently very useful in atomic, nuclear, and particle physics. We give some examples of sum rules and their application to physical situations in a problem in Chapter 16. For now, we are concerned with a mathematical identity that is useful for calculating sum rules. Consider two Hermitian operators H and A. H is the Hamiltonian of a physical system (e.g., an atom), and $|n\rangle$, E_n are a typical eigenvector and corresponding eigenvalue of H: $H|n\rangle = E_n|n\rangle$. A is any Hermitian operator other than H. We are interested in evaluating the following sum in closed form:

$$S_p = \sum_k (E_k - E_n)^p \langle n|A^\dagger|k\rangle\langle k|A|n\rangle$$

where p is a nonnegative integer.

(a) By employing the identity

$$e^{iHt}Ae^{-iHt} = A + it[H,A] + \frac{1}{2!}(it)^2[H,[H,A]] + \ldots$$

where t is a real variable, show that for any given eigenvector $|n\rangle$ of H,

$$S_p = \langle n|A^\dagger A_p|n\rangle$$

and where $A_p \equiv [H, A_{p-1}]$ with $A_0 = A$.

(b) Show that

$$\langle n|A^\dagger A_p|n\rangle = \langle n|A_m^\dagger A_{p-m}|n\rangle$$

where $m \le p$ is a nonnegative integer, $A_m^\dagger = [A_{m-1}^\dagger, H]$, and $A_0^\dagger = A^\dagger$.

2.16. Starting from equation (2.40), obtain the results (2.44), (2.46), and (2.47).

3 The Rules of Quantum Mechanics

3.1 Statement of the rules

When the rules of quantum mechanics were formulated, they expressed a revolutionary break with the past. Because they were something totally new, they could not be derived from the old principles of classical theoretical physics, which, as we saw in Chapter 1, gave no predictions or erroneous predictions for many experimental results. Instead, the rules of quantum mechanics had to be obtained by intuition and inspiration from experimental results.

The first fundamental concept we encounter in formulating the rules is that of a *dynamical system*. As in classical physics, this usually means an isolated system such as a free electron, free neutron, free molecule, and so on. However, if we assume that quantum mechanics has general validity, it also must apply to macroscopic systems containing vast numbers of particles, and as a practical matter, it is impossible to isolate a macroscopic system from environmental influences, except in very special circumstances.

A *dynamical variable* or *observable* is a quantity associated with the system that can be measured. Some examples are position of a particle, linear momentum, orbital angular momentum, energy, and electric charge, all of which have classical analogues. However, in quantum physics, there are also many observables with no classical analogues, such as intrinsic spin, parity, isospin, lepton number, baryon number, and so on.

Next we come to the concept *state of a system*. Because in quantum mechanics we deal with probabilities of experimental outcomes, this phrase refers to an ensemble of which the system in question is a member. What sort of ensemble? Not a real collection of electrons that are somewhat similar to an electron of interest, as in a beam of electrons emitted from a hot filament, but rather an ideal ensemble with an infinite number of members consisting of the system in question and mental copies of it, all prepared in exactly the same way by some suitable experimental procedure. The mathematical characterization of the state of the system is equivalent to a complete list of instructions for carrying out this experimental preparation.

Bearing these remarks in mind, we come to the first rule:

> *Rule 1: A dynamical system corresponds to a Hilbert space in such a way that a definite state of the system corresponds to a definite ray in the space.*

The state is said to be $|u\rangle$ if it corresponds to the ray in the direction of vector $|u\rangle$. Because multiplication of $|u\rangle$ by an arbitrary complex number results in another vector $|u'\rangle$ associated with the same ray, it is assumed that $|u'\rangle$ also describes the same state. Even if we choose a definite norm for $|u\rangle$, it is still uncertain by an arbitrary phase factor; thus two vectors in the Hilbert space differing only by a phase factor correspond to the same physical state. Given any two vectors $|u\rangle, |v\rangle$ associated with distinct states, we may form

$$|w\rangle = |u\rangle + |v\rangle \tag{3.1}$$

which is also in the Hilbert space and therefore corresponds to another possible state of the system. For this reason, Rule 1 is frequently called the *linear superposition principle*. Note that although the absolute phases of two state vectors have no physical meaning, once those phases are chosen, the relative phase of the two vectors added together as in (3.1) does have meaning and can be of crucial importance.

An isolated quantum mechanical system can evolve with time, and this variation with time, which is sometimes called *unitary evolution*, is assumed to be continuous and causal. (For example, the spin magnetic moment of an electron precesses continuously and causally in an external magnetic field.) Let the state at some initial time t_0 be $|u(t_0)\rangle$, and at some later time let it be $|u(t)\rangle$. We assume that there exists a continuous and linear time-development operator $U(t,t_0)$ that transforms $|u(t_0)\rangle$ into $|u(t)\rangle$ for $t > t_0$; that is,

$$|u(t)\rangle = U(t,t_0)|u(t_0)\rangle \tag{3.2}$$

We require that the norm of $|u\rangle$ be independent of time so that probability is conserved, a point that will be clarified shortly. Hence U must be unitary. Furthermore, for very small $t - t_0$, U must differ only infinitesimally from the identity if it is to be continuous. Thus we write

$$U(t_0 + \delta t, t_0) = I - iK\delta t$$

where K is Hermitian. Because U is dimensionless, K must have the dimension of inverse time = frequency. It is also intuitively clear that K should contain the fundamental constant $\hbar$, which has dimension action = energy × time. Thus we assume that

$$U(t_0 + \delta t, t_0) = I - \frac{i}{\hbar}H\delta t \tag{3.3}$$

where the Hermitian operator H, which has the dimension of energy, is the Hamiltonian of the system. Inserting (3.3) in (3.2), we obtain

$$|u(t_0 + \delta t)\rangle = \left(I - \frac{i}{\hbar}H\delta t\right)|u(t_0)\rangle$$

Now we drop the subscript and divide both sides of this equation by δt to arrive at

$$H|u(t)\rangle = i\hbar\frac{\partial|u(t)\rangle}{\partial t} \tag{3.4}$$

which is the well-known time-dependent Schroedinger equation. Equation (3.2) may be employed once again in (3.4) to obtain

$$i\hbar\partial_t\left[U(t,t_0)|u(t_0)\rangle\right] = HU(t,t_0)|u(t_0)\rangle$$

which implies

$$\frac{\partial}{\partial t}U(t,t_0) = -\frac{i}{\hbar}HU(t,t_0) \tag{3.5}$$

If, as is often the case, the Hamiltonians at two different times commute (i.e., if $[H(t_1), H(t_2)] = 0$ for any t_1, t_2), then it is easy to integrate both sides of (3.5) to obtain

$$U(t,t_0) = \exp\left[-\frac{i}{\hbar}\int_{t_0}^{t} H(t')\,dt'\right] \tag{3.6a}$$

and, if H is independent of the time,

$$U(t,t_0) = \exp\left[-\frac{i}{\hbar}H(t-t_0)\right] \tag{3.6b}$$

However, if the Hamiltonians at different times do not commute, (3.6a) must be replaced by

$$U(t,t_0) = \exp\left[-\frac{i}{\hbar}\int_{t_0}^{t} H(t')\,dt'\right]_+ \tag{3.7}$$

where the subscript + means that operators are time-ordered; that is, an operator at an earlier time stands to the right of one at a later time. This is visualized by expanding the exponential in a power series:

$$\exp\left[-\frac{i}{\hbar}\int_{t_0}^{t} H(t')\,dt'\right]_+ = I - \frac{i}{\hbar}\int_{t_0}^{t} H(t')\,dt' + \left(\frac{i}{\hbar}\right)^2 \int_{t_0}^{t} H(t_1)\,dt_1 \int_{t_0}^{t_1} H(t')\,dt'$$
$$-\left(\frac{i}{\hbar}\right)^3 \int_{t_0}^{t} H(t_2)\,dt_2 \int_{t_0}^{t_2} H(t_1)\,dt_1 \int_{t_0}^{t_1} H(t')\,dt' + \cdots)$$

The derivation of this power series and details concerning time ordering are given in Section 23.1. The foregoing discussion of unitary evolution is summarized as follows:

Rule 2: An isolated quantum state evolves continuously and causally in time, as described by the equation

$$|u(t)\rangle = U(t,t_0)|u(t_0)\rangle \tag{3.8}$$

where

$$U(t,t_0) = \exp\left[-\frac{i}{\hbar}\int_{t_0}^{t} H(t')\,dt'\right]_+ \tag{3.9}$$

and H is the Hamiltonian operator, which is Hermitian. The content of (3.9) is also expressed by the time-dependent Schroedinger equation

$$H|u(t)\rangle = i\hbar\frac{\partial|u(t)\rangle}{\partial t} \tag{3.10}$$

We next come to the concept of *measurement*. In quantum mechanics, this means interaction of the system in question with a second system, called the *apparatus*. Often the latter is a complicated macroscopic piece of laboratory equipment with many degrees of freedom, and most of these degrees of freedom are coupled in an uncontrollable way to the environment. Usually the apparatus contains a recording device, such as a pointer on a dial, or a counter that can be read and noted by a human, and usually this recording device and much of the rest of the apparatus is described to an adequate approximation by classical laws. (However, we must always keep in mind that quantum mechanics should apply to the apparatus and to the environment.) In a measurement, we assume that the apparatus is initially prepared in a "ready" state. Then, as a result of interaction with the system, something happens to the apparatus so that a record is left of the event of interest. Because many degrees of freedom of the apparatus are coupled uncontrollably to the environment, this change in the apparatus is irreversible. Thus, although the equations describing the evolution of an isolated quantum system are time-reversal-symmetric in almost all cases, the irreversibility inherent in the measurement process defines a unique sense of time in quantum mechanics.

Because a measurement generally disturbs the system, it leads to a change of state; thus the state vector $|u\rangle$ is transformed by the measurement into some other vector $|v\rangle$. We know that such a transformation is produced by applying some operator to a vector $|u\rangle$. Therefore, it is natural to associate each observable with an operator on the Hilbert space. For simplicity, we confine ourselves to linear operators. Furthermore, the result of any physical measurement is always a real number. Because the eigenvalues of Hermitian operators are real, it is reasonable to make the following assumption:

Rule 3: Each observable of a system is associated with a Hermitian operator on the Hilbert space of the system. We assume that the eigenstates of each observable form a complete set.

We also assume the following:

Rule 4: The result of any given measurement of an observable is an eigenvalue of the corresponding Hermitian operator.

Next, let us consider the results obtained when we make repeated measurements of the same observable A on a system that has always been prepared in the same way, in state $|u\rangle$, prior to the measurement. The mean value $\bar{A} \equiv \langle A\rangle$ of those measurements of A is a real number, which clearly must depend linearly on operator A and on the ray defined by $|u\rangle$ but not on the norm of $|u\rangle$. Thus it is reasonable to assume that

Rule 5: The average $\bar{A} \equiv \langle A\rangle$ of a number of measurements of A is

$$\bar{A} \equiv \langle A\rangle = \frac{\langle u|A|u\rangle}{\langle u|u\rangle} \tag{3.11}$$

This is frequently called the *expectation value* of A. Now either $|u\rangle$ is an eigenstate of A or it is not. If it is, $|u\rangle = |\phi_i\rangle$, where $A|\phi_i\rangle = \lambda_i |\phi_i\rangle$. Then

$$\bar{A} = \frac{\langle \phi_i | A | \phi_i \rangle}{\langle \phi_i | \phi_i \rangle} = \lambda_i \tag{3.12}$$

The mean-square deviation of measurements of A in this case is

$$\begin{aligned} \Delta^2 &= \langle A^2 \rangle - \langle A \rangle^2 = \frac{\langle \phi_i | AA | \phi_i \rangle}{\langle \phi_i | \phi_i \rangle} - \lambda_i^2 \\ &= \lambda_i^2 - \lambda_i^2 = 0 \end{aligned} \tag{3.13}$$

Therefore, in this case, all measurements give the same result: λ_i. The other possibility is that $|u\rangle$ is not an eigenstate of A. Because the eigenstates of A form a complete set, we can always express $|u\rangle$ as a superposition of the $|\phi_i\rangle$; that is,

$$|u\rangle = \sum |\phi_i\rangle \langle \phi_i | u \rangle$$

Thus

$$\bar{A} = \sum_{i,j} \frac{\langle u | \phi_i \rangle \langle \phi_i | A | \phi_j \rangle \langle \phi_j | u \rangle}{\langle u | u \rangle} = \frac{\sum_i \lambda_i |\langle u | \phi_i \rangle|^2}{\sum_i |\langle u | \phi_i \rangle|^2} \tag{3.14}$$

This is interpreted as follows:

> *Rule 6: The quantities $\langle u | \phi_i \rangle$ are regarded as probability amplitudes, and $p_i = |\langle u | \phi_i \rangle|^2$ is assumed to be the probability for achieving result λ_i on a given measurement*

$$\bar{A} = \frac{\sum_i p_i \lambda_i}{\sum_i p_i}$$

Rule 6 is where probability first appears in quantum mechanics. In all other applications of probability (e.g., in medical statistics, gambling, and classical statistical mechanics), actual events are assumed to be deterministic, and probability enters only because we do not have complete information about these events. However, we assume that such information could be obtained, at least in principle. For example, consider classical kinetic theory applied to a gas of N hard spheres in a container. According to Newtonian mechanics, if we could specify the position and momentum of each hard sphere at time $t = 0$, and if we knew all the forces, then it would be possible in principle to calculate the motion of each hard sphere for all future times, including collisions of the spheres with one another and with the walls. Of course, if $N >> 1$, such a calculation would be prohibitively lengthy and difficult, and as a practical matter, we could not know all the initial conditions. Thus, for all practical purposes, we must be content with a statistical description.

In quantum mechanics, probability appears in a different way. Certain observables, for example, a coordinate and its conjugate momentum, are incompatible, which means that it is impossible in principle to determine them simultaneously and precisely.

Because in general a measurement disturbs the system, it is natural for us to ask: What becomes of the state vector as the result of measurement of observable A? This question is addressed by the next rule.

> *Rule 7: A measurement of observable A resulting in eigenvalue λ_i projects the state vector from $|u\rangle$ to that subspace of the Hilbert space associated with λ_i. If the eigenvalue is nondegenerate, the subspace is just a single eigenstate $|\phi_i\rangle$ associated with λ_i. If $|u\rangle$ is already an eigenstate of A, it is not changed by the measurement. If $|u\rangle$ is not an eigenstate of A, then in the nondegenerate case it "collapses" to $|\phi_i\rangle$ with probability $|\langle u\,|\,\phi_i\rangle|^2$.*

Rule 7 is called the *collapse* or *state-vector reduction postulate*. The change in state that occurs here is stochastic and nonlinear, as opposed to the deterministic and linear evolution that appears in Rule 2. In fact, Rules 2 and 7 are incompatible if we insist that quantum mechanics should describe not only an isolated microsystem (e.g., an electron or an atom) but also the experimental apparatus used to observe that system and the environment surrounding that apparatus. This is the as-yet-unresolved quantum measurement problem (also called the *macroscopic objectification problem*) discussed in Chapter 25.

Rule 7 implies the following concerning the measurement of two observables A, B. Suppose that we measure observable A and obtain the result λ_i. The measurement throws the system into the eigenstate $|\phi_i\rangle$ of A that corresponds to λ_i. Suppose that this is immediately followed by a measurement of B. If $[A,B]=0$, then A and B are simultaneously diagonalized by the same unitary transformation. Thus $|\phi_i\rangle$ is also an eigenstate of B, which means that measurement of B does not disturb the system further; it remains in state $|\phi_i\rangle$. Thus an immediately subsequent measurement of A is certain to give the result λ_i. However, if $[A,B]\neq 0$, the $|\phi_i\rangle$ are not in general eigenstates of B. In this case, a measurement of B projects the system from $|\phi_i\rangle$ to some eigenstate of B. Here it is not certain that a subsequent measurement of A will yield the same result as the first one.

If $[A, B] = 0$, A and B are said to be compatible. Otherwise, they are incompatible. If A, B, C, ... form a complete set of commuting (compatible) observables, then we know everything there is to know about the state of the system when we specify the eigenvalue of each of A, B, C, ... for that state. This is called a *complete set* of quantum numbers for the state.

Assuming that $\langle u(t)\,|\,u(t)\rangle = 1$, the expectation value of an observable A is $\bar{A} = \langle A\rangle = \langle u\,|\,A\,|\,u\rangle$, and

$$\frac{d\langle A\rangle}{dt} = \left(\frac{d\langle u|}{dt}\right)A|u\rangle + \langle u|A\left(\frac{d|u\rangle}{dt}\right) + \left\langle u\,\middle|\,\frac{\partial A}{\partial t}\,\middle|\,u\right\rangle \tag{3.15}$$

where the third term on the right-hand side arises from the explicit dependence on time, if any, of A. Employing $H|u\rangle = i\hbar\partial_t|u\rangle$ and $\langle u|H = -i\hbar\partial_t\langle u|$, we obtain from (3.15)

$$\begin{aligned}\frac{d\langle A\rangle}{dt} &= \frac{i}{\hbar}\left[\langle u\,|\,HA\,|\,u\rangle - \langle u\,|\,AH\,|\,u\rangle\right] + \left\langle u\,\middle|\,\frac{\partial A}{\partial t}\,\middle|\,u\right\rangle \\ &= \frac{1}{i\hbar}\langle u\,|\,[A,H]\,|\,u\rangle + \left\langle u\,\middle|\,\frac{\partial A}{\partial t}\,\middle|\,u\right\rangle\end{aligned} \tag{3.16}$$

We have assumed in (3.16) that $|u(t)\rangle$ is time dependent (the Schroedinger picture), but it is possible to carry out a unitary transformation S (itself time dependent) to a new basis in which $|u\rangle$ is constant, and all the time dependence is carried by the operators. (This is called the *Heisenberg picture*). Because $|u(t)\rangle = U(t,t_0)|u(t_0)\rangle$, it is clear that we should choose

$$S(t,t_0) = U^\dagger(t,t_0) = \exp\left\{\left[-\frac{i}{\hbar}\int_{t_0}^{t} H(t')\,dt'\right]\right\}^\dagger \tag{3.17}$$

In the Heisenberg picture we have $A' = SAS^\dagger = U^\dagger AU$. Thus

$$\frac{dA'}{dt} = \frac{1}{i\hbar}[A',H] + \frac{\partial A'}{\partial t} \tag{3.18}$$

which is called the *Heisenberg equation.*

3.2 Photon polarizations

We now illustrate the foregoing rules by considering some experiments that can be performed with visible light. The apparatus consists of a light source, a detector consisting of an ordinary photomultiplier connected to an amplifier and an ammeter or a counter, some optical attenuators, and linear and circular polarizers and analyzers. We start with an intense beam of coherent light directed along the z-axis. The situation is then well described by classical electromagnetic theory. First, we insert a linear polarizer, assumed ideal, in the path of the beam (Figure 3.1).

After the polarizer, the light beam has definite polarization, with electric vector parallel to the polarizer axis and given by

$$\mathcal{E} = \mathcal{E}_0 \hat{i} \exp[i(kz - \omega t)] \tag{3.19}$$

where $\mathcal{E}_0$ is the electric field amplitude, $\hat{i}$ is a unit vector in the x-direction, $k = \omega / c$ is the wave number, and ω is the angular frequency of the light beam, which is assumed to be monochromatic. The intensity of the beam is proportional to $\mathcal{E}_0^2$.

Next, we insert a second linear polarizer (called an *analyzer*). Its plane is also normal to the z-axis, but the analyzer polarization axis x' makes an angle θ with respect to the x-axis (Figure 3.2).

Light emerging from the analyzer is polarized along the x' axis, and the electric field amplitude is $\mathcal{E}_0' = \mathcal{E}_0 \cos\theta$. Thus the intensity measured at the detector is proportional to $\mathcal{E}_0^2 \cos^2\theta$.

We can easily perform other well-known polarization experiments of classical optics. For example, we might insert a quarter-wave ($\lambda/4$) plate after the first polarizer (Figure 3.3).

The "fast" (f) and "slow" (s) axes are indicated in Figure 3.3. One has

$$\hat{i} = \frac{1}{\sqrt{2}}\left(\hat{f} + \hat{s}\right) \qquad \hat{j} = \frac{1}{\sqrt{2}}\left(\hat{s} - \hat{f}\right)$$

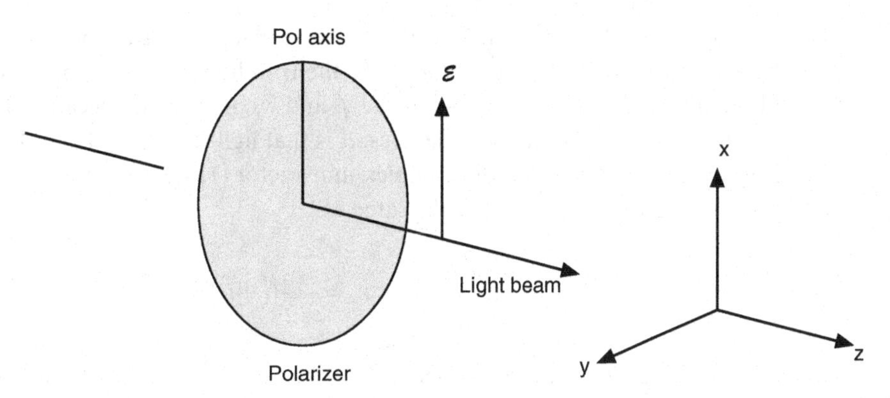

Figure 3.1 Light beam directed along the *z* axis through a linear polarizer.

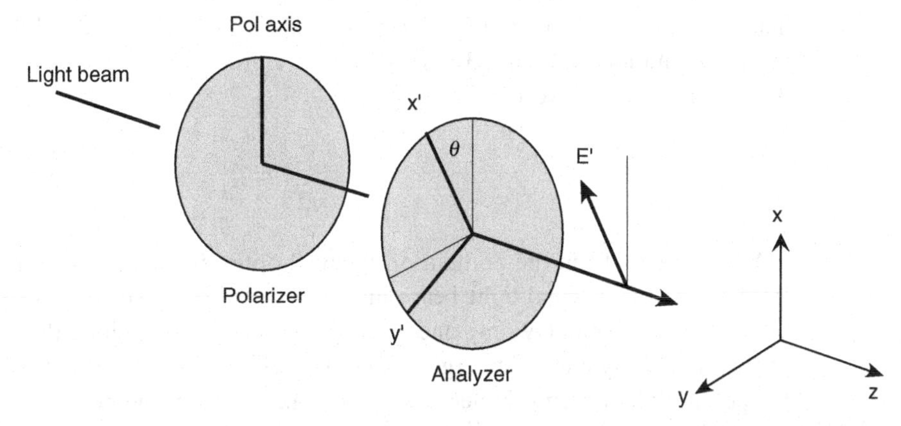

Figure 3.2 Light beam directed along the *z* axis through a linear polarizer and an analyzer.

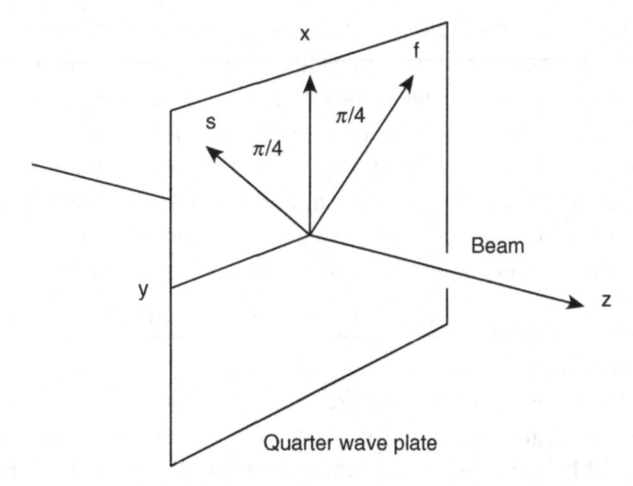

Figure 3.3 Light beam passing through a quarter-wave plate.

Suppose that light entering the quarter-wave plate is in a state of linear polarization described by the unit vector $\hat{i} = \left(1/\sqrt{2}\right)\left(\hat{f} + \hat{s}\right)$. As the traveling light wave proceeds through the plate, a phase difference develops between the $\hat{f}$ and $\hat{s}$ components because the latter correspond to different indices of refraction. The result is that light emerging from the $\lambda/4$ plate is in a polarization state described by the complex unit vector $\left(1/\sqrt{2}\right)\left(\hat{f} + i\hat{s}\right)$, which apart from an unimportant overall phase factor is the same as

$$\hat{\varepsilon}_+ = \frac{1}{\sqrt{2}}\left(\hat{i} + i\hat{j}\right) \tag{3.20}$$

Here the electric vector rotates in the sense of a right-handed screw as the light propagates in the $+z$-direction. Thus the light beam has positive helicity. (In the confusing language of classical optics, it is left circularly polarized. The latter is defined by the sense of rotation of the electric vector according to an observer looking *into* the oncoming beam.) By rotating the original linear polarizer so that its polarization axis is along the y-axis while keeping the $\lambda/4$ plate in the orientation of Figure 3.3, we could have produced a light beam with negative helicity, described by the unit vector

$$\hat{\varepsilon}_- = \frac{1}{\sqrt{2}}\left(\hat{i} - i\hat{j}\right) \tag{3.21}$$

Now we pass from the domain of classical optics to the quantum regime by reducing the intensity of the original light beam until the detector registers individual photons, one at a time. At first, we employ only one linear polarizer, oriented along the x-axis as in Figure 3.1. In this case, every photon emerging from the polarizer is in the same polarization state: $|x\rangle$. If the polarizer had been oriented along the y-axis instead, the polarization state of the emerging photons would have been $|y\rangle$. Ignoring all variables that characterize the light beam except for polarization, we have a two-dimensional Hilbert space spanned by only two orthonormal state vectors that we can choose as $|x\rangle$ and $|\mathrm{y}\rangle$. The Hermitian operators corresponding to the polarizer oriented along the directions x, y are $\hat{P}_x = |x\rangle\langle x|$ and $\hat{P}_y = |y\rangle\langle y|$, respectively. We suppose that with only a single linear polarizer present, as in Figure 3.1, we obtain an average of N counts per second at the detector.

Next, we insert the analyzer (Figure 3.2), but initially we choose the angle $\theta = 0$ so that the primed and unprimed axes coincide. Because the polarizer and analyzer are assumed to be ideal, there is no absorption, and the average counting rate is still N per second, and the polarization state of photons emerging from the analyzer is still $|x\rangle$. To state this in other words, the Hermitian operator for the analyzer oriented along x is $\hat{A}_x = \hat{P}_x = |x\rangle\langle x|$. Because $\langle x|\hat{A}_x|x\rangle = \langle x|x\rangle\langle x|x\rangle = 1$, the photon beam suffers no loss of intensity in passing through the analyzer.

Next, we rotate the analyzer about the z-axis, making $\theta \neq 0$. We observe that the average counting rate is reduced from N to $N' = n\cos^2\theta$. Evidently some photons are not being transmitted by the analyzer, and those that are transmitted must be in a new polarization state $|x'\rangle$ that is a superposition of the states $|x\rangle$ and $|y\rangle$. In addition to $|x'\rangle$, there must exist an orthogonal state $|y'\rangle$.

Because the relative phase of the states $|x\rangle, |y\rangle$ is arbitrary, we can choose the phase so that

$$\begin{aligned} |x'\rangle &= |x\rangle\langle x|x'\rangle + |y\rangle\langle y|x'\rangle = \cos\theta|x\rangle + \sin\theta|y\rangle \\ |y'\rangle &= |x\rangle\langle x|y'\rangle + |y\rangle\langle y|y'\rangle = -\sin\theta|x\rangle + \cos\theta|y\rangle \end{aligned} \tag{3.22}$$

The Hermitian operator corresponding to the analyzer oriented along the x'-axis is $\hat{A}_{x'} = |x'\rangle\langle x'|$, and $\langle x|\hat{A}_{x'}|x\rangle = \langle x|x'\rangle\langle x'|x\rangle = \cos^2 6$.

If we consider a given photon with polarization state $|x\rangle$ between the polarizer and the analyzer, can we predict in advance whether it will be transmitted by the analyzer? According to the rules of quantum mechanics, we cannot. We can only calculate the probability of transmission, and it is $\cos^2\theta$. Rejection by the analyzer is equivalent to projection of $|x\rangle$ onto $|y'\rangle$, which occurs with probability $\sin^2\theta$.

The choice of basis states for our two-dimensional Hilbert space is arbitrary and a matter of convenience. We have chosen the linear polarization states $|x\rangle, |y\rangle$, but we could equally well have chosen the helicity eigenstates, namely,

$$\begin{aligned} |+\rangle &= \frac{1}{\sqrt{2}}(|x\rangle + i|y\rangle) \\ |-\rangle &= \frac{1}{\sqrt{2}}(|x\rangle - i|y\rangle) \end{aligned} \tag{3.23}$$

If we set up an experiment with a linear polarizer aligned along x followed by a $\lambda/4$ plate as in Figure 3.3, photons emerging from the linear polarizer would be in the state

$$|x\rangle = \frac{1}{\sqrt{2}}(|+\rangle + |-\rangle) \tag{3.24}$$

After the $\lambda/4$ plate, the normalized photon polarization state would be $|+\rangle$. We show later that the states $|+\rangle, |-\rangle$ correspond to photon angular momentum $\pm\hbar$, respectively, with respect to the direction of propagation (see Appendix C).

3.3 Polarization correlations, locality, and Bell's inequalities

We now discuss two-photon polarization correlation experiments that have actually been performed and that bear on significant questions concerning the rules of quantum mechanics. Such experiments and their interpretations have stimulated controversial debate since the publication of an important paper by Einstein, Podolsky, and Rosen (1935). Our presentation here must anticipate later discussions of atomic physics and the quantum theory of angular momentum, but only in such an elementary way that there should be no serious difficulties for the reader. We consider an optical experiment involving the calcium atom, which has 20 electrons arranged in the following shells in the ground state:

$$\left[1s^2 2s^2 2p^6 3s^2 3p^6\right] 4s^2 \tag{3.25}$$

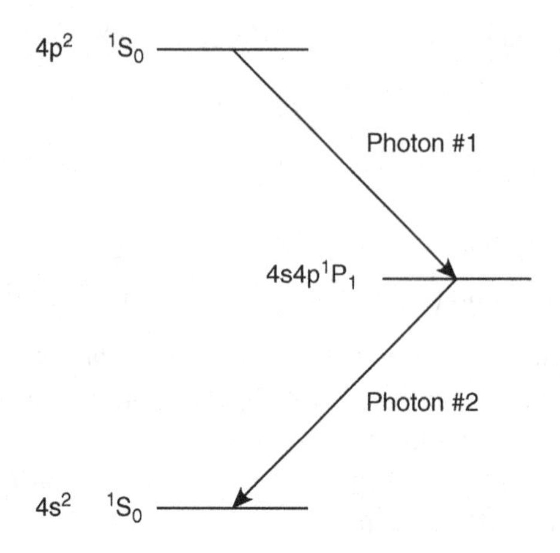

Figure 3.4 Relevant calcium energy levels.

In (3.25) we place the symbols for the innermost 18 electrons in square brackets to emphasize that these electrons form an inert spherically symmetric inner core with zero angular momentum. For present purposes, all the relevant properties of calcium are determined by the two outermost valence electrons. In the ground state, they are both in the $4s$ shell (orbital angular momentum equal to zero). In addition, the spins of these two electrons are opposed, so the net electron spin S in the ground state is also zero. We designate the ground state by the symbol

$$(4s^2)\ {}^1S_0$$

In general, the superscript on the S (1 in this expression) refers to the total spin multiplicity $2S + 1$, whereas the capital letter S (not to be confused with spin S) signifies that the total orbital angular momentum is $L = 0$. If this capital letter were P, D, …, it would signify that the total orbital angular momentum is $L = 1, 2, \ldots$, respectively. The subscript refers to the total angular momentum J, where $\mathbf{J} = \mathbf{L} + \mathbf{S}$. In addition to the ground state, we consider several excited states of calcium, shown in Figure 3.4.

In the experiment of interest, calcium atoms are prepared in the $(4p^2)\ {}^1S_0$ excited state, and they decay via the intermediate $(4s4p)\ {}^1P_1$ state to the ground state by spontaneous emission of two successive visible photons γ_1, γ_2. The initial and intermediate state mean lifetimes are only a few nanoseconds. The goal is to observe the polarizations of γ_1, γ_2 in coincidence (i.e., where the relative timing of the two detectors is arranged to ensure that both photons are radiated by the same atom) and also where the two photons are emitted in opposite directions along the z-axis (Figure 3.5).

For simplicity, we assume in what follows that each polarizer is ideal in the sense that it transmits 100 percent of the "right" polarization and 0 percent of the "wrong" polarization. We also assume that each detector is 100 percent efficient. Actual polarizers and detectors fall short of these ideals, and appropriate corrections must be made, but for present purposes, we can ignore such complications.

Because the initial and final atomic angular momenta are $J_i = J_f = 0$, the two photons must carry off zero total angular momentum. Thus, if γ_1 is in a helicity eigenstate $|+\rangle$, the helicity

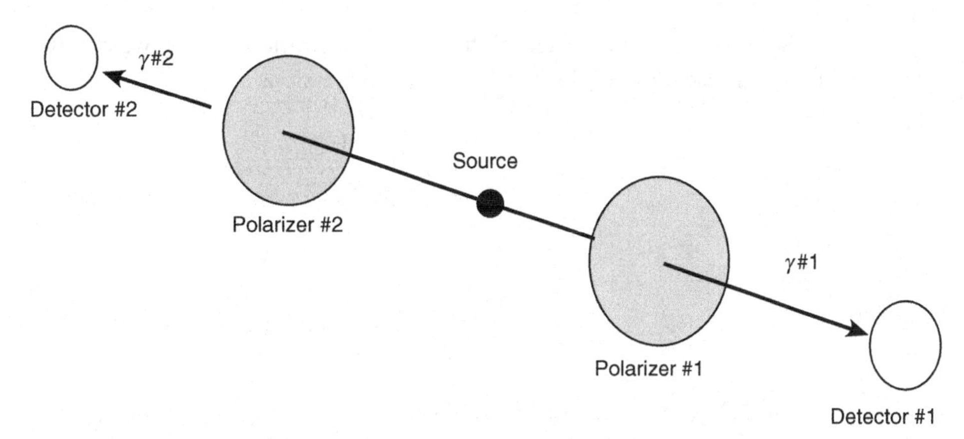

Figure 3.5 Photons γ_1, γ_2 propagating in opposite directions.

state of γ_2 also must be $|+\rangle$ because the linear momenta of the photons are in opposite directions. Similarly, if γ_1 is in helicity eigenstate $|-\rangle$, the state of γ_2 also must be $|-\rangle$. Thus the following two-photon state vectors satisfy the requirements imposed by conservation of angular momentum and are possible a priori:

$$|+_1,+_2\rangle \qquad |-_1,-_2\rangle \tag{3.26}$$

Which of these is the actual state vector, or is it some combination of the two? To answer this question, we need additional information pertaining to the spatial inversion symmetry (called the *parity*) of the initial and final atomic states. If under inversion of spatial coordinates $(x \to -x,\ y \to -y,\ z \to -z)$ a state remains invariant, it is said to have even or positive parity, whereas if the state changes sign, it has odd or negative parity. It can be shown that a nondegenerate eigenstate of an inversion-symmetric Hamiltonian must have definite parity, even or odd (see Section 6.4). One can also show that the parity of a multielectron atomic state is $(-1)^{\sum \ell_i}$, where ℓ_i is the orbital angular momentum of the ith electron, and the sum is taken over all electrons. From this one can see that the parities of the initial and final atomic states in Figure 3.4 are both even. Because parity is known to be conserved in electromagnetic interactions, the two-photon polarization state also must have even parity. However, under spatial inversion, the linear momentum of a photon reverses, but its angular momentum does not; hence its helicity reverses. Therefore, under a parity transformation, each of the two-photon polarization states in (3.26) transforms to the other state. Thus the positive-parity two-photon polarization state must be

$$|\psi\rangle = \frac{1}{\sqrt{2}}\left[|+_1,+_2\rangle + |-_1,-_2\rangle\right] \tag{3.27}$$

where we have included a factor of $2^{-1/2}$ for normalization. A state such as $|\psi\rangle$ in (3.27), which consists of a sum of two or more product states and which cannot be expressed as a single product state, is called *entangled*.

Because we are interested in the linear polarizations of $\gamma_{1,2}$, we rewrite (3.27) in terms of linear polarization states by employing the expressions

$$|\pm_1\rangle = \frac{1}{\sqrt{2}}|x_1 \pm iy_1\rangle \tag{3.28}$$

and

$$|\pm_2\rangle = \frac{1}{\sqrt{2}}|x_2 \mp iy_2\rangle \tag{3.29}$$

[In (3.29), the $\mp$ sign on the right-hand side takes into account the fact that the momentum of γ_2 is directed along the $-z$-axis]. Substituting (3.29) and (3.28) into (3.27), we obtain

$$\begin{aligned} |\psi\rangle &= \frac{1}{2\sqrt{2}}|(x_1 + iy_1)(x_2 - iy_2) + (x_1 - iy_1)(x_2 + iy_2)\rangle \\ &= \frac{1}{\sqrt{2}}|x_1x_2 + y_1y_2\rangle \end{aligned} \tag{3.30}$$

Now we are ready to consider several important questions.

1. *Is photon γ_1 in a state of definite linear polarization?* If this were so, then the counting rate of detector 1 by itself (without concern for detector 2) would depend on the angular orientation of polarizer 1. However, experiment shows that the counting rate for photons γ_1 passing through polarizer 1 is independent of the angular orientation of polarizer 1 and is half the rate detected when polarizer 1 is removed from the apparatus. Quantum mechanics accounts for this as follows: let us orient polarizer 1 along the x'-axis. Then passage of γ_1 through polarizer 1 is associated with application of operator $\hat{P}_{x1'} = |x_1'\rangle\langle x_1'|$ to $|\psi\rangle$; that is,

$$|x_1'\rangle\langle x_1'|\psi\rangle = \frac{1}{\sqrt{2}}\left[\cos\theta|x_1'x_2\rangle + \sin\theta|x_1'y_2\rangle\right] \tag{3.31}$$

Thus the probabilities for obtaining the two distinct results $x_1'x_2$, $x_1'y_2$ are predicted by quantum mechanics to be ½ cos² θ, ½ sin² θ, respectively. Because at the moment we are not interested in the polarization of γ_2, we must add these probabilities to obtain the total probability that γ_1 is transmitted through polarizer 1. Obviously, this total probability is ½, independent of θ; hence photon 1 has no definite linear polarization. By a similar argument, photon 2 has no definite polarization.

2. *What is the coincidence counting rate when photon γ_1 passes through polarizer 1 oriented along axis x', whereas photon γ_2 passes through polarizer 2 oriented along x?* Experiment shows that this rate is ½ cos² θ times the coincidence rate when both polarizers are removed and where θ is the angle between axes x and x'. Quantum mechanics accounts for this result by application of the projection operators $\hat{P}_{x2}$ and $\hat{P}_{x1'}$ to $|\psi\rangle$; that is,

$$\begin{aligned} \hat{P}_{x2}\hat{P}_{x1'}|\psi\rangle &= \frac{1}{\sqrt{2}}|x2\rangle\langle x2|\left[\cos\theta|x1'x2\rangle + \sin\theta|x1'y2\rangle\right] \\ &= \frac{1}{\sqrt{2}}\cos\theta|x1'x2\rangle \end{aligned}$$

Hence, according to quantum mechanics, the probability of coincidence is

$$p_C = \frac{1}{2}\cos^2\theta \tag{3.32}$$

in accord with the experimental result.

We have just seen that no matter what the angular orientation of polarizer 1, the probability that photon γ_1 is transmitted through this polarizer is ½. Similarly, the probability that γ_2 is transmitted through polarizer 2 is ½, independent of the orientation of polarizer 2. However, the experimental results [and (3.32)] also tell us that if polarizers 1 and 2 are both oriented along the same axis (say, the x-axis) and a specific γ_1 is observed to pass through polarizer 1, we can predict with certainty that the corresponding γ_2 will also pass through polarizer 2. This is true even if polarizers 1 and 2 are separated by a very large distance.

Now it is plausible to assume that if the separation between polarizers is very large, whatever happens at polarizer 1 (be it transmission or rejection of γ_1), can have no physical effect at polarizer 2, and vice versa. This is called the *locality assumption*, and it seems reasonable to assume it provisionally and to consider the consequences. This same assumption was made by Einstein, Podolsky, and Rosen in their discussion of a thought experiment different in detail from the present experiment but equivalent as far as the main points are concerned.

If polarizers 1 and 2 are both aligned along x, and γ_1 is observed to pass through polarizer 1 – *and locality is valid* – then because the passage of γ_1 through polarizer 1 cannot affect the conditions for transmission of γ_2 through polarizer 2, γ_2 must have possessed x linear polarization *before* it arrived at polarizer 2. The same conclusion also should apply when the roles of γ_1 and γ_2 are reversed: if a particular γ_2 passes through polarizer 2 and locality is valid, we can be certain that the corresponding γ_1 possessed x linear polarization before it reached polarizer 1. Therefore, because the probability of transmission of photons γ_1 through polarizer 1 is ½, no matter what the orientation of polarizer 1 (and similarly for photons γ_2 and polarizer 2), locality implies that some γ_1, γ_2 pairs are endowed with x-polarization from birth (although this assignment is evidently random from pair to pair), whereas an equal number are endowed with y-polarization. In other words, acceptance of locality implies that there is a random but deterministic "hidden variable" in the two-photon system that fixes what the polarization of each photon in a given pair must be. According to quantum mechanics, no such statement is possible because no assumption about locality is made, and all that we can know about the polarizations of γ_1, γ_2 is contained in $|\psi\rangle$ of equation (3.30).

J. S. Bell (1966) made a very important contribution to this discussion by showing that the assumption of locality, which implies the existence of a deterministic (if random) hidden variable, can be put to experimental test. Bell proved that if a local hidden variable description were in fact operative, it would necessarily imply results for various correlation experiments that are in conflict with the predictions of quantum mechanics. Specifically, for the two-photon polarization correlation experiment in calcium that we have just analyzed, any local-hidden-variable theory would yield the following inequality (derived in Appendix B):

$$\frac{R(\pi/8)}{R_0} - \frac{R(3\pi/8)}{R_0} \le \frac{1}{4} \quad \text{(Bell's inequality)} \tag{3.33}$$

where R_0 is the coincidence rate with both polarizers removed, $R(\theta)$ is that with both polarizers in place, and as usual, θ is the angle between the polarizer axes. However, from (3.32) we see that according to quantum mechanics,

$$\frac{R(\pi/8)}{R_0} = \frac{1}{4}\left(1+\cos\frac{\pi}{4}\right) = \frac{1}{4}\left(1+\frac{1}{\sqrt{2}}\right)$$

and

$$\frac{R(3\pi/8)}{R_0} = \frac{1}{4}\left(1+\cos\frac{3\pi}{4}\right) = \frac{1}{4}\left(1-\frac{1}{\sqrt{2}}\right)$$

Hence

$$\frac{R(\pi/8)}{R_0} - \frac{R(3\pi/8)}{R_0} = \frac{\sqrt{2}}{4} \quad \text{(quantum mechanics)} \tag{3.34}$$

It is obvious from (3.34) that the quantum mechanical prediction conflicts with Bell's inequality (3.33). As already stated, the experimental results agree with quantum mechanics. The first definitive results on calcium were obtained by S. J. Freedman and J. F. Clauser (1972). A similar experiment using mercury instead of calcium was performed several years later by E.S. Fry and R. C. Thompson (1976) with results in agreement with those of Freedman and Clauser. Very elegant and precise observations were subsequently made by A. Aspect and coworkers (Aspect 1982) and, since then, by many other experimenters using systems other than calcium and a variety of experimental techniques. Although there still exist some minor caveats, the results of these experiments enable us to conclude that for a hidden-variable description to account for the experimental results we have described, it would have to be nonlocal in character. Of course, quantum mechanics itself is nonlocal because a measurement of the polarization of either photon of the γ_1, γ_2 pair implies instantaneous collapse of the state $|\psi\rangle$ in (3.30).

3.4 Larmor precession of a spin ½ particle in a magnetic field

Although we have not yet discussed the quantum theory of angular momentum, most readers know that for a particle of spin ½ (e.g., an electron, proton, or neutron) the spin operator is

$$\boldsymbol{S} = \frac{\hbar}{2}\boldsymbol{\sigma} \tag{3.35}$$

where $\boldsymbol{\sigma} = \sigma_x\hat{i} + \sigma_y\hat{j} + \sigma_z\hat{k}$, and $\sigma_{x,y,z}$ are the Pauli spin matrices

$$\sigma_x = \begin{pmatrix} 0 & 1 \\ 1 & 0 \end{pmatrix} \quad \sigma_y = \begin{pmatrix} 0 & -i \\ i & 0 \end{pmatrix} \quad \sigma_z = \begin{pmatrix} 1 & 0 \\ 0 & -1 \end{pmatrix} \tag{3.36}$$

They have the following properties:

1. Each σ_i $(i = x, y, z)$ is Hermitian.
2. $$\mathrm{tr}(\sigma_i) = 0 \tag{3.37}$$
3. $$\sigma_i \sigma_j = \delta_{ij} I + i\varepsilon_{ijk}\sigma_k \tag{3.38}$$

In the second term on the right-hand side of (3.38), ε_{ijk} is the completely antisymmetric unit 3 tensor, with values +1 for ijk = 1, 2, 3 or cyclic permutations thereof; –1 for ijk = odd permutations of 1, 2, 3; and 0 when any two of the indices ijk are the same. Thus $\sigma_x\sigma_y = -\sigma_y\sigma_x = i\sigma_z$, with similar expressions obtained by cyclic permutation of x, y, and z.

The electron, proton, and neutron each have a magnetic dipole moment associated with their spin. The electron-spin magnetic dipole operator is

$$\boldsymbol{\mu}_e = -g_s \frac{\mu_B}{\hbar}\boldsymbol{S} = -\frac{g_s}{2}\mu_B \boldsymbol{\sigma} \tag{3.39a}$$

where $\mu_B = e\hbar/2m_ec = 9.27408 \times 10^{-21}$ erg/G is the Bohr magneton in cgs units, and $g_s = 2(1 + a_e)$. The factor 2 arises naturally from Dirac's equation in relativistic quantum mechanics. The quantity a_e is a small (but very important) quantum electrodynamic correction that we need not be concerned with at present. In Heaviside-Lorentz units, the formula for the Bohr magneton is the same, $\mu_B = e\hbar/2m_ec$, but the numerical value is larger by a factor $\sqrt{4\pi}$ simply because in hlu system the Bohr magneton is measured in erg/$\left(\mathrm{G}\sqrt{4\pi}\right)$. The proton-spin magnetic dipole operator is

$$\boldsymbol{\mu}_p = g_p \frac{\mu_N}{\hbar}\boldsymbol{S} = \frac{g_p}{2}\mu_N \boldsymbol{\sigma} \tag{3.39b}$$

Here $\mu_N = (m_e/m_p)\mu_B$ is the nuclear Bohr magneton, and $g_p/2 = 2.79$ is a numerical factor that accounts for the experimentally observed proton magnetic moment. The Hamiltonian for the interaction of the proton's spin magnetic moment with an external magnetic field $\boldsymbol{B}$ is

$$H = -\boldsymbol{\mu}_p \cdot \boldsymbol{B} = -\frac{g_p}{2}\mu_N \boldsymbol{\sigma}\cdot\boldsymbol{B} \tag{3.40}$$

To investigate the behavior of the spin in a constant field $\boldsymbol{B}$, we start with equation (3.16), here applied to the ith component of $\boldsymbol{S}$:

$$\begin{aligned}\frac{d\langle S_i\rangle}{dt} &= \frac{1}{i\hbar}\langle[S_i, H]\rangle = -\frac{1}{i\hbar}\left\langle\left[S_i, \frac{g_p}{2}\mu_N\sigma_j B_j\right]\right\rangle \\ &= i\frac{g_p}{4}\mu_N B_j\left\langle[\sigma_i,\sigma_j]\right\rangle \\ &= -\frac{g_p}{2}\mu_N B_j\varepsilon_{ijk}\langle\sigma_k\rangle\end{aligned} \tag{3.41}$$

where we employ the repeated index summation convention. Written in vector notation, (3.41) is

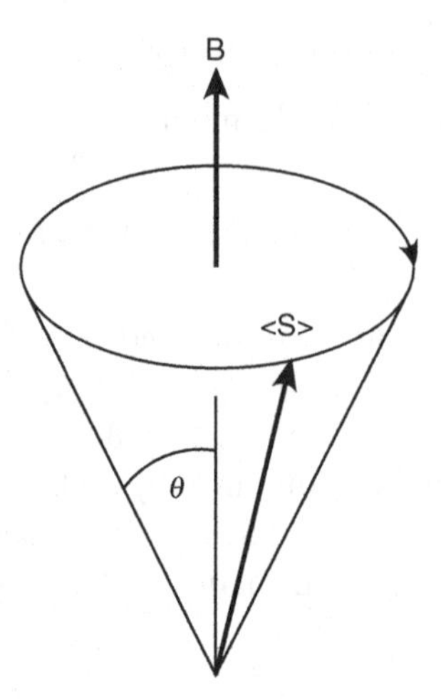

Figure 3.6 Schematic diagram showing precession of the expectation value of a spin vector about a magnetic field.

$$\frac{d\langle \boldsymbol{S}\rangle}{dt} = \langle \boldsymbol{\mu}_p \rangle \times \boldsymbol{B} \tag{3.42}$$

The quantity on the right-hand side of (3.42) is the torque on $\langle \boldsymbol{S}\rangle$ in field $\boldsymbol{B}$, and (3.42) simply means that the expectation value $\langle \boldsymbol{S}\rangle$ follows the classical Newtonian equation torque equals rate of change of angular momentum.

Taking into account that $\boldsymbol{\mu}_p = (g_p/\hbar)\mu_N \boldsymbol{S}$ and defining $\boldsymbol{\omega} = (g_p/\hbar)\mu_N \boldsymbol{B} \equiv \gamma \boldsymbol{B}$, we rewrite (3.42) in the useful alternative form

$$\frac{d\langle \boldsymbol{S}\rangle}{dt} = -\gamma \boldsymbol{B} \times \langle \boldsymbol{S}\rangle = -\boldsymbol{\omega} \times \langle \boldsymbol{S}\rangle \tag{3.43}$$

This implies that $\langle \boldsymbol{S}\rangle$ precesses around $\boldsymbol{B}$ with angular frequency ω, as shown in Figure 3.6.

If $\boldsymbol{B} = B_0\hat{k}$ is in the z-direction and at time $t = 0$, $\langle \boldsymbol{S}\rangle$ is in the x-direction, the precession cone half-angle is $\Theta = \pi/2$, and the precession cone becomes a circle in the xy-plane. We consider how the state vector evolves in time as $\langle \boldsymbol{S}\rangle$ precesses. The operator σ_z has two eigenvectors

$$\chi_+ = \begin{pmatrix} 1 \\ 0 \end{pmatrix} \quad \text{and} \quad \chi_- = \begin{pmatrix} 0 \\ 1 \end{pmatrix}$$

corresponding to eigenvalues ± 1, respectively (i.e., spin "up" and spin "down" respectively), with respect to the z-axis. It is easy to show that the eigenvectors of σ_x are $(\chi_+ + \chi_-)/\sqrt{2}$ and $(\chi_+ - \chi_-)/\sqrt{2}$ with eigenvalues ± 1, respectively. Thus, at $t = 0$,

$$\psi(0) = \frac{\chi_+ + \chi_-}{\sqrt{2}} = \frac{1}{\sqrt{2}}\begin{pmatrix} 1 \\ 1 \end{pmatrix}$$

To find $\psi(t)$ for $t > 0$, we use the time-development operator

$$\begin{aligned}\psi(t) &= U(t,0)\psi(0)\\ &= \exp\left(-\frac{i}{\hbar}Ht\right)\psi(0) = \exp\left(\frac{i\omega t}{2}\sigma_z\right)\psi(0)\\ &= \left(I\cos\frac{\omega t}{2} + i\sigma_z\sin\frac{\omega t}{2}\right)\psi(0)\\ &= \frac{1}{\sqrt{2}}\begin{pmatrix} e^{i\omega t/2}\\ e^{-i\omega t/2}\end{pmatrix}\end{aligned} \tag{3.44}$$

When $t = \pi/\omega$, $\langle \boldsymbol{S}\rangle$ lies along $-x$, and

$$\psi = \frac{i}{\sqrt{2}}\begin{pmatrix}1\\-1\end{pmatrix}$$

After a full period of precession when $t = 2\pi/\omega$, $\langle \boldsymbol{S}\rangle$ has returned to its initial orientation along x, but

$$\psi(t) = \frac{-1}{\sqrt{2}}\begin{pmatrix}1\\1\end{pmatrix} = -\psi(0)$$

In general, one can show that a rotation of 2π about any axis causes a sign change in a half-integral spin function (see Chapter 7). For example, if the initial state is χ_+, a rotation of the spin about the y-axis by angle β yields the state $\chi_+\cos(\beta/2) + \chi_-\sin(\beta/2)$. For spin ½, this phenomenon has been demonstrated experimentally (see Section 5.5).

3.5 The density operator

In many circumstances, we do not have sufficient information to say that a physical system is in a single quantum state. Nevertheless, we may be able to deduce useful conclusions by employing the *density operator*, which is somewhat analogous to the density in phase space in classical statistical mechanics. In this section we introduce the main properties of the density operator and show how it is applied to the description of photon polarizations.

Suppose that there exists a complete orthonormal set of states $|u_i\rangle$, and for a given physical system, we know only enough to assign a probability $g_i < 1$ that any given $|u_i\rangle$ is occupied by the system of interest. We are then dealing with what is called a *statistical ensemble*. The density operator is defined as[1]

$$\rho = \sum_i g_i |u_i\rangle\langle u_i| \tag{3.45}$$

where $\sum_i g_i = 1$. The density operator has the following properties:

[1] In the definition of ρ, it is not always necessary that the various $|u_i\rangle$ be mutually orthogonal.

1. It is Hermitian: $\rho^\dagger = \rho$.
2. If we define a *generalized* expectation value of any observable A as

$$\langle A \rangle = \sum_j g_j \langle u_j | A | u_j \rangle$$

then

$$\langle A \rangle = tr(A\rho) = tr(\rho A) \tag{3.46}$$

To prove (3.46), we write

$$\begin{aligned}\langle A \rangle &= \sum_j g_j \langle u_j | A | u_j \rangle \\ &= \sum_{i,j} g_j \langle u_j | \phi_i \rangle \langle \phi_i | A | u_j \rangle = \sum_{i,j} g_j \langle \phi_i | A | u_j \rangle \langle u_j | \phi_i \rangle = tr(A\rho)\end{aligned}$$

In the special case $A = I$, (3.46) reduces to

$$\mathrm{tr}\rho = 1 \tag{3.47}$$

3.
$$\mathrm{tr}(\rho^2) = \sum_i g_i^2 \le 1 \tag{3.48}$$

Proof:

$$\rho^2 = \sum_{i,j} g_i g_j |u_i\rangle\langle u_i | u_j \rangle\langle u_j| = \sum_i g_i^2 |u_i\rangle\langle u_i|$$

The inequality in (3.48) becomes an equality only in the special case where a single probability g corresponding to a single state $|u\rangle$ is equal to unity and all other g_i vanish. In this special case, $\rho = |u\rangle\langle u|$, $\rho^2 = \rho$, and ρ is said to be *idempotent*, and the density-operator description of quantum mechanics becomes equivalent to the conventional state-vector description.

4. If each state $|u_i\rangle$ is an energy eigenstate with energy eigenvalue E_i and the system in question is in thermal equilibrium with a heat bath at absolute temperature T, then from elementary statistical mechanics,

$$g_i = \frac{\exp(-E_i / kT)}{Z} \tag{3.49}$$

where $Z = \sum_i \exp(-E_i / kT)$ is the partition function. Here the free energy F is

$$F = U - TS = -kT \ell n Z \tag{3.50}$$

where $U = \sum_i g_i E_i$ is the average energy, and S is the entropy. From (3.50), we have

$$\begin{aligned} S &= k\ell n Z + U/T \\ &= -k\sum_i g_i \ell n\left(\frac{e^{-E_i/kT}}{Z}\right) \\ &= -k\sum_i g_i \ell n g_i \end{aligned} \tag{3.51}$$

We can generalize result (3.51) by defining the dimensionless von Neumann entropy of an arbitrary statistical ensemble as

$$S_{vN} = -\mathrm{tr}(\rho \ell n \rho) \tag{3.52}$$

where $\ell n\rho \equiv \sum_i \ell n(g_i)|u_i\rangle\langle u_i|$.

5. If the system Hamiltonian is H, then

$$H|u_i\rangle = i\hbar\partial|u_i\rangle/\partial t \quad \text{and} \quad \langle u_i|H = -i\hbar\partial\langle u_i|/\partial t$$

Thus from (3.45) we have

$$\frac{d\rho}{dt} = \frac{1}{i\hbar}[H,\rho] \tag{3.53}$$

which is the equation of motion of the density operator. It is similar to the Heisenberg equation for an observable A that has no explicit time dependence, except for the fact that the sign of the commutator is opposite in the two equations.

6. The representative of ρ with respect to a given basis of states $|\phi_i\rangle$ in an n-dimensional Hilbert space is an $n \times n$ matrix called the *density matrix M*. In the pure-state case where $\rho = |u\rangle\langle u|$ with $|u\rangle = \sum_i a_i|\phi_i\rangle$ and where the a_i are expansion coefficients, the elements of M are

$$M_{ij} = a_i a_j^* \tag{3.54}$$

As is well known, the set of all $n \times n$ matrices forms an n^2-dimensional vector space. Thus, if the Hilbert space is two-dimensional, as is the case for spin ½ or for photon polarizations, one can express any density matrix as a linear combination of four linearly independent 2×2 matrices. It is convenient to choose the latter in a standard way:

$$I = \begin{pmatrix} 1 & 0 \\ 0 & 1 \end{pmatrix} \quad \sigma_x = \begin{pmatrix} 0 & 1 \\ 1 & 0 \end{pmatrix} \quad \sigma_y = \begin{pmatrix} 0 & -i \\ i & 0 \end{pmatrix} \quad \sigma_z = \begin{pmatrix} 1 & 0 \\ 0 & -1 \end{pmatrix} \tag{3.55}$$

Hence an arbitrary 2×2 density matrix M can be written as

$$M = \frac{1}{2}(aI + b\sigma_x + c\sigma_y + d\sigma_z) \tag{3.56}$$

where a, b, c, and d are numerical constants. Assuming that $\mathrm{tr}(M)=1$, and making use of (3.37) and $\mathrm{tr}(I)=2$, we have $a=1$. Next, because M is a scalar under rotation in coordinate space, the coefficients b, c, and d must constitute the x, y, z components, respectively, of a vector $\boldsymbol{P}$. Hence we can write

$$M=\frac{1}{2}(I+\boldsymbol{P}\cdot\boldsymbol{\sigma}) \tag{3.57}$$

and therefore

$$M^2=\frac{1}{4}(I+\boldsymbol{P}\cdot\boldsymbol{\sigma})^2=\frac{1}{4}(I+2\boldsymbol{P}\cdot\boldsymbol{\sigma}+\boldsymbol{\sigma}\cdot\boldsymbol{P}\boldsymbol{\sigma}\cdot\boldsymbol{P}) \tag{3.58}$$

Now, employing the repeated index summation convention, we have

$$\begin{aligned}\boldsymbol{\sigma}\cdot\boldsymbol{P}\boldsymbol{\sigma}\cdot\boldsymbol{P}&=\sigma_i P_i \sigma_j P_j=\sigma_i\sigma_j P_i P_j\\&=\left(\delta_{ij}I+i\varepsilon_{ijk}\sigma_k\right)P_iP_j=|\boldsymbol{P}|^2 I\end{aligned} \tag{3.59}$$

from which it follows that

$$M^2=\frac{1}{4}\left(1+|\boldsymbol{P}|^2\right)I+\frac{1}{2}\boldsymbol{P}\cdot\boldsymbol{\sigma} \tag{3.60}$$

Recalling that $M^2 = M$ for a pure state, and comparing (3.57) and (3.60), we see that $|\boldsymbol{P}| = 1$. In other words, $\boldsymbol{P} = \hat{\boldsymbol{p}}$ is a unit vector for a pure state.

The density matrix M for a statistical ensemble of polarization states is the weighted sum of pure polarization state density matrices M_i:

$$M=\frac{1}{2}(I+\boldsymbol{P}\cdot\boldsymbol{\sigma})=\sum_i M_i=\sum_i g_i\frac{1}{2}(I+\hat{\boldsymbol{p}}_i\cdot\boldsymbol{\sigma})$$

It follows that $\boldsymbol{P}=\sum_i g_i\hat{\boldsymbol{p}}_i$; hence, if the various $\hat{\boldsymbol{p}}_i$ are oriented in different directions, we must have $|\boldsymbol{P}| < 1$ for a statistical ensemble. Because the only experimentally accessible quantities related to polarization of a beam of light are the components of $\boldsymbol{P}$, and because for a statistical ensemble there are in general many ways to choose the weights g_i and/or the directions of the $\hat{\boldsymbol{p}}_i$ to form a given $\boldsymbol{P}$, the individual g_i, $\hat{\boldsymbol{p}}_i$ cannot in general be determined by experiment, even in principle, for a statistical ensemble.

To describe photon polarizations explicitly, let $|+\rangle=1/\sqrt{2}\,|x+iy\rangle$ be represented by the 2-spinor $\begin{pmatrix}1\\0\end{pmatrix}$ and $|-\rangle=1/\sqrt{2}\,|x-iy\rangle$ be represented by $\begin{pmatrix}0\\1\end{pmatrix}$. Then $|x\rangle$ and $|y\rangle$ are represented by

$$\frac{1}{\sqrt{2}}\begin{pmatrix}1\\1\end{pmatrix} \quad\text{and}\quad \frac{1}{\sqrt{2}}\begin{pmatrix}-i\\i\end{pmatrix}$$

respectively. Using this representation, we give a few examples of density matrices as follows:

- Pure state $|+\rangle$. $M=\begin{pmatrix}1\\0\end{pmatrix}(1\quad 0)=\begin{pmatrix}1&0\\0&0\end{pmatrix}=\frac{1}{2}(I+\sigma_z)$; $\hat{p}=\hat{k}$
- Pure state $|-\rangle$. $M=\begin{pmatrix}0\\1\end{pmatrix}(0\quad 1)=\begin{pmatrix}0&0\\0&1\end{pmatrix}=\frac{1}{2}(I-\sigma_z)$; $\hat{p}=-\hat{k}$
- Pure state: Linear polarization along the x'-axis, where $x'=x\cos\theta+y\sin\theta$:

$$\rho=|x'\rangle\langle x'|=|x\cos\theta+y\sin\theta\rangle\langle x\cos\theta+y\sin\theta|$$
$$=\cos^2\theta|x\rangle\langle x|+\sin^2\theta|y\rangle\langle y|+\cos\theta\sin\theta(|x\rangle\langle y|+|y\rangle\langle x|)$$
$$M=\frac{1}{2}\cos^2\theta\begin{pmatrix}1&1\\1&1\end{pmatrix}+\frac{1}{2}\sin^2\theta\begin{pmatrix}1&-1\\-1&1\end{pmatrix}+\frac{i}{2}\cos\theta\sin\theta\left[\begin{pmatrix}-1&-1\\1&1\end{pmatrix}-\begin{pmatrix}-1&1\\-1&1\end{pmatrix}\right]$$
$$=\frac{1}{2}\left(I+\cos 2\theta\sigma_x+\sin 2\theta\sigma_y\right)\qquad \hat{p}=\cos 2\theta\hat{i}+\sin 2\theta\hat{j}$$

  Here note that whereas the polarization axis and $\hat{p}$ are both in the xy-plane, they are inclined at angles θ and 2θ, respectively, with respect to the x-axis. This is not surprising when we realize that there is no physical distinction between linear polarization along any direction and linear polarization in the opposite direction. Thus, for example, the density matrices should be, and are, identical for linear polarization along x and along $-x$.
- Statistical ensemble: Incoherent mixture of right and left circular polarization (unpolarized light):

$$\rho=\frac{1}{2}(|+\rangle\langle+|)+\frac{1}{2}(|-\rangle\langle-|)$$
$$M=\frac{1}{2}\begin{pmatrix}1&0\\0&0\end{pmatrix}+\frac{1}{2}\begin{pmatrix}0&0\\0&1\end{pmatrix}=\frac{1}{2}I\qquad P=0$$

Problems for Chapter 3

3.1. Consider a spin-½ particle S that interacts with a "measuring apparatus" that is another spin-½ particle A. At time $t=0$, A is in the state

$$\begin{pmatrix}1\\0\end{pmatrix}_A$$

and S is in the state

$$\begin{pmatrix}a\\b\end{pmatrix}_S$$

where a, b are two complex numbers, with $|a|^2+|b|^2=1$. During the time interval $0\le t\le T$, S and A are coupled by means of the interaction Hamiltonian

$$H = \frac{1}{4}\sigma_z^S \sigma_y^A g(t)$$

where $g(t)$ is a continuous real function that is nonzero only for $0 \le t \le T$ and satisfies

$$\int_0^T g(t)\,dt = \pi\hbar$$

Also, σ_z^S, σ_y^A are Pauli spin matrices that refer to S and A, respectively, and thus commute with one another. Find the state vector for the combined system S, A at time T.

3.2. Consider the density matrix for a particle of spin ½ and magnetic moment μ in a magnetic field, where the Hamiltonian is

$$H = -\frac{1}{2} g\mu \boldsymbol{\sigma} \boldsymbol{\cdot} \boldsymbol{B}$$

and where g is a constant. Using the equation of motion of the density operator, find the motion of the polarization vector $\hat{\boldsymbol{p}} = \langle \boldsymbol{\sigma} \rangle$, and compare it with the classical equation of motion of a spinning magnetic dipole in a magnetic field.

3.3. An electron and a positron can form a bound system called a *positronium* (Ps). In the ground $^1S_0$ state, a positronium atom decays by annihilation of e^+ and e^- to two photons with equal and opposite linear momenta in the positronium rest frame. Each photon has energy $E \approx 0.511$ MeV = electron rest energy because the binding energy of the positronium atom is negligible by comparison. Because in the ground state, the positronium atom has zero total angular momentum, the two photons carry off zero total angular momentum. Furthermore, as can be shown from equations 20.65 and 22.11–22.14, the parity of the positronium ground state is odd, which means that under spatial inversion the state vector changes sign.Because parity is conserved in the annihilation process, the parity of the two-photon final state is also odd.

(a) Consider a photon emitted along $+z$ and another emitted along $-z$. Show that the two-photon final state is of the form

$$|\psi\rangle = \frac{1}{\sqrt{2}}\left[|+\rangle_1 |+\rangle_2 - |-\rangle_1 |-\rangle_2\right] \tag{1}$$

where $|+\rangle_{1,2}$ signify a positive-helicity photon emitted along $\pm z$, and $|-\rangle_{1,2}$ denotes a negative-helicity photon emitted along $\pm z$. How are the linear polarizations of the two photons correlated?

(b) Because the polarization correlation of the two annihilation photons is somewhat analogous to that of the two photons in the calcium experiment described earlier in this chapter, it seems at first that observation of the polarization of these annihilation photons could yield another significant test of Bell's inequality. However, the only practical way to detect polarization of 0.5-MeV gamma rays is by Compton scattering. This was actually done in an early experiment by Wu and Shaknov [*Phys. Rev.* **77**, 136 (1950)], where the correlation in the planes of polarization of the annihilation photons $(\boldsymbol{k}_1, \boldsymbol{k}_2)$ in $^1S_0$ positronium decay was detected by Compton scattering at a scattering angle $\theta = 90°$ (Figure 3.7). It can be shown that at $\theta = \pi/2$, each scattered photon has energy $E/2$, and the probability that an annihilation photon becomes a Compton-scattered photon is proportional to

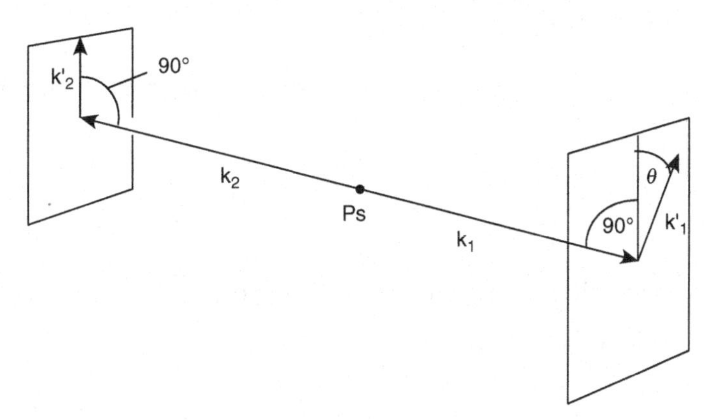

Figure 3.7 Orientation of the vectors $k_{1,2}$, $k'_{1,2}$ in the Wu-Shaknov experiment.

$$1+8\left(\hat{\varepsilon}\cdot\hat{\varepsilon}'\right)^2 \tag{2}$$

where $\hat{\varepsilon}$, $\hat{\varepsilon}'$ are unit vectors in the direction of annihilation and scattered photon linear polarization, respectively.

The plane defined by the $\boldsymbol{k}$ vectors of a given incident and scattered photon pair (e.g., the plane defined by $\boldsymbol{k}_1, \boldsymbol{k}_1'$) may be rotated through an angle θ with respect to the plane defined by $\boldsymbol{k}_2, \boldsymbol{k}_2'$. Using expression (2), and the symmetry of the two-photon final state, show that the coincidence rate for detection of the two scattered photons is proportional to

$$R(\theta) = 1 - \frac{8}{13}\cos^2\theta \tag{3}$$

In deriving this result, it is essential to recognize that the scattering processes $(1 \to 1')$ and $(2 \to 2')$ are mutually incoherent. Why is this so?

(c) Unfortunately, the analyzing power in this positronium experiment is insufficient to test Bell's inequality: the factor 8/13 in equation (3) is too small. To see this, consider Bell's inequality given in expression (3.33) of this chapter for the calcium experiment and for the angles $\theta = \pi/8$ and $\theta = 3\pi/8$. For the Wu-Shaknov experiment, the appropriate modification of (3.33) is

$$R\left(\frac{3\pi}{8}\right) - R\left(\frac{\pi}{8}\right) - \frac{1}{2} \le 0 \tag{4}$$

Check the numbers to show that the quantum mechanical result (3) does not violate the local hidden-variable inequality (4).

3.4. Consider three particles described by the wave function

$$\psi = f(\boldsymbol{r}_1, \boldsymbol{r}_2, \boldsymbol{r}_3)(u_1 u_2 u_3 - \mathrm{v}_1 \mathrm{v}_2 \mathrm{v}_3)$$

where the spatial wave function $f(\boldsymbol{r}_1, \boldsymbol{r}_2, \boldsymbol{r}_3)$ is such that the particles are widely separated, and the spin states u and v are eigenstates of σ_z satisfying

$$\begin{array}{lll} \sigma_x u = \mathrm{v} & \sigma_y u = i\mathrm{v} & \sigma_z u = u \\ \sigma_x \mathrm{v} = u & \sigma_y \mathrm{v} = -iu & \sigma_z \mathrm{v} = -\mathrm{v} \end{array}$$

(a) Verify that

$$\sigma_{1x}\sigma_{2y}\sigma_{3y}\psi = \sigma_{1y}\sigma_{2x}\sigma_{3y}\psi = \sigma_{1y}\sigma_{2y}\sigma_{3x}\psi = \psi \quad (1)$$

that

$$\sigma_{1x}\sigma_{2x}\sigma_{3x}\psi = -\psi \quad (2)$$

and that the four operators $(\sigma_{1x}\sigma_{2y}\sigma_{3y}), (\sigma_{1y}\sigma_{2x}\sigma_{3y}), (\sigma_{1y}\sigma_{2y}\sigma_{3x})$, and $(\sigma_{1x}\sigma_{2x}\sigma_{3x})$ commute.

(b) Show that

$$(\sigma_{1x}\sigma_{2y}\sigma_{3y})(\sigma_{1y}\sigma_{2x}\sigma_{3y})(\sigma_{1y}\sigma_{2y}\sigma_{3x}) = -(\sigma_{1x}\sigma_{2x}\sigma_{3x}) \quad (3)$$

(c) The following argument seems very convincing. We may measure either σ_x or σ_y for each particle without disturbing the other particles. Let the results of these measurements be called m_x and m_y, respectively. From (2), we can predict with certainty that if σ_x is measured for all three particles, the result must be

$$m_{1x}m_{2x}m_{3x} = -1 \quad (4)$$

Therefore, the value of σ_x can be predicted with certainty for any one of the particles by measuring σ_x for the other two particles. From (1), we can also predict with certainty the value of σ_x on any one particle by measuring σ_y on the other two particles:

$$m_{1x}m_{2y}m_{3y} = +1 \quad (5)$$

and by cyclic permutation,

$$m_{1y}m_{2x}m_{3y} = +1 \quad (6)$$

and

$$m_{1y}m_{2y}m_{3x} = +1 \quad (7)$$

However, the product of equations (4) through (7) gives a contradiction:

$$+1 = -1$$

What error was made in our assumptions and/or logic?

3.5. In this problem we consider a thought experiment proposed in 1993 by L. Hardy. It illustrates the peculiar and counterintuitive nature of entangled states in a different way than the usual Bell's inequality experiments. Hardy's thought experiment makes use of a source S and two detectors D_L and D_R (L, R for left, right, respectively; Figure 3.8).

Each detector has two modes 1, 2 determined by the position of a switch $K_{L,R}$. Each detector is equipped with a light that can flash either green or red. An experimental trial begins when the observer presses a button that launches a pair of correlated particles from source S; one particle goes to the left, and the other to the right. After they have been emitted from

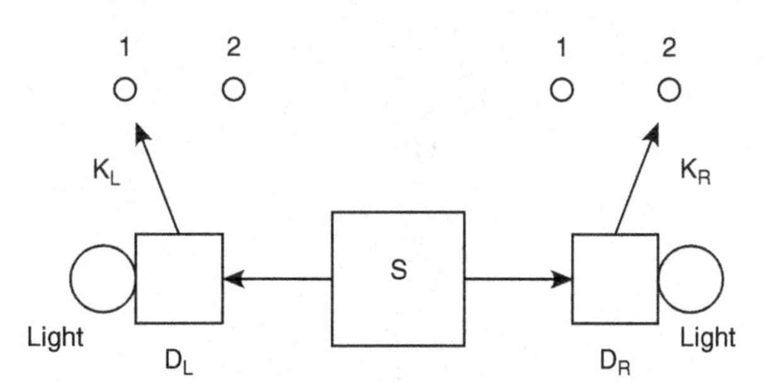

Figure 3.8 Schematic diagram of Hardy's thought experiment.

the source but before they arrive at their respective detectors, the observer flips one coin to determine the position of K_L and another to determine the position of K_R. The arrival of a particle at D_L is indicated by the flashing of a green or red light there, similarly for the arrival of the other particle at D_R. The outcome of a given trial is specified by giving the positions of the two switches and the colors of the lights that flashed; for example, (1G2R) signifies that K_L was in position 1 and D_L flashed green, whereas K_R was in position 2 and D_R flashed red.

The observer repeats the experiment, recording the outcome for each trial, and finds the following results after many trials:

i. When both switches are in position 1, both lights never flash red: 1R1R never occurs.
ii. When the switches are in different positions, both lights never flash green: 1G2G and 2G1G never occur.
iii. In a nonzero fraction of the trials, 2G2G does occur.

It is tempting to make the following classical analysis: something in the common origin of the particles must be responsible for the observed correlations. Because the switches $K_{L,R}$ are not set until after the particles leave the source, whatever features the particles possess cannot depend on how the switches are set. Also, we can safely assume that D_L can only respond to the particle on the left, whereas D_R can only respond to the particle on the right. Then, because any trial could be a 12 or a 21 trial, whenever one of the particles is such as to allow a type 2 detector to flash green, the other particle must be such as to make a type 1 detector flash red. This follows from (ii). Then, in any of the occasional 22 trials where both detectors flash green, both particles must be of the variety to make a type 1 detector flash red. In other words, had both switches been set to position 1 in these trials, the outcome 1R1R would have been observed. However, according to (i), 1R1R is *never* observed. Thus there is no way to explain the results by a classical argument.

However, it is indeed possible in principle to set up such an experiment and to get the results (i), (ii), and (iii), but we must use quantum mechanics to describe the two-particle system. Because we never obtain the outcome 1R1R, we can assume that the two-particle quantum state is of the form

$$|\psi\rangle = \alpha|1\text{R}1\text{G}\rangle + \beta|1\text{G}1\text{R}\rangle + \gamma|1\text{G}1\text{G}\rangle \tag{1}$$

with $\alpha, \beta,$ and γ constants where $|\alpha|^2 + |\beta|^2 + |\gamma|^2 = 1$.

Table 3.1 Probabilities for various outcomes in Hardy's thought experiment.

1G1G	z^3
1G1R	z^2
1R1G	z^2
1R1R	0
1G2G	0
1G2R	z
1R2G	z^3
1R2R	z^4
2G1G	0
2G1R	z^3
2R1G	z
2R1R	z^4
2G2G	$z^5 = p$
2G2R	z^4
2R2G	z^4
2R2R	z

(a) Show that because 1G2G and 2G1G never occur, we must have

$$\begin{aligned} \alpha\langle 2\text{G}|1\text{R}\rangle + \gamma\langle 2\text{G}|1\text{G}\rangle = 0 \\ \beta\langle 2\text{G}|1\text{R}\rangle + \gamma\langle 2\text{G}|1\text{G}\rangle = 0 \end{aligned} \tag{2}$$

(b) It must be possible to express $|2\text{G}\rangle$ as a linear combination of the states $|1\text{G}\rangle$ and $|1\text{R}\rangle$. Also, $|2\text{R}\rangle$ must be an orthogonal linear combination of the same states:

$$\begin{aligned} |2\text{G}\rangle &= q^{1/2}|1\text{G}\rangle + \sqrt{1-q}\,|1\text{R}\rangle \\ |2\text{R}\rangle &= -\sqrt{1-q}\,|1\text{G}\rangle + q^{1/2}|1\text{R}\rangle \end{aligned} \tag{3}$$

where $0 < q < 1$. Show that because outcome 2G2G sometimes occurs, that is, $|\langle 2\text{G}2\text{G}|\psi\rangle|^2 = p \neq 0$, it follows that

$$p = \frac{q^2(1-q)^2}{1-q^2} \tag{4}$$

(c) Show that when p is maximized in (4), the probabilities of the various outcomes are given in Table 3.1, where $z = (\sqrt{5}-1)/2$.

4 The Connection between the Fundamental Rules and Wave Mechanics

4.1 The de Broglie relation

In Chapter 1 we mentioned the de Broglie relation

$$ƛ = \frac{\lambda}{2\pi} = \frac{\hbar}{p} \tag{4.1}$$

which implies that the position of a particle and its conjugate momentum are incompatible observables. Let us define the operators for position q and momentum p in one spatial dimension as $\hat{q}, \hat{p}$, respectively. Because $\hat{q}$ and $\hat{p}$ are incompatible, they do not commute. Now the commutator of two Hermitian operators is always anti-Hermitian; thus

$$[\hat{q}, \hat{p}] = iB$$

where B is a Hermitian operator. The simplest choice for B is a real constant with dimension of action, and as we shall see, the choice $B = \hbar$ yields agreement with the de Broglie relation. Indeed, a convenient starting point for the entire development of wave mechanics is the fundamental commutation relation

$$[\hat{q}, \hat{p}] = i\hbar \tag{4.2a}$$

in one spatial dimension or, more generally,

$$[\hat{x}_i, \hat{p}_j] = i\hbar\delta_{ij} \tag{4.2b}$$

where $i, j = 1, 2, 3$ for three spatial dimensions. We now proceed with the development of wave mechanics from (4.2a), starting with the uncertainty principle.

4.2 The uncertainty principle

Consider any two observables A and B that satisfy $[A, B] = i\hbar$. If we measure both observables simultaneously, what is the minimum combined uncertainty $\Delta A \Delta B$? First form two new Hermitian operators

$$A_0 = A - \langle A \rangle$$
$$B_0 = B - \langle B \rangle$$

Then

$$(\Delta A)^2 = \langle A^2 \rangle - \langle A \rangle^2 = \langle A_0^2 \rangle$$

and similarly, $(\Delta B)^2 = \langle B_0^2 \rangle$. We are interested in a simultaneous measurement of A_0^2 and B_0^2, which means that we must calculate the expectation values of both quantities with respect to the same state $|u\rangle$, which we assume is normalized to unity. Hence

$$(\Delta A)^2 (\Delta B)^2 = \langle u|A_0 A_0|u\rangle \langle u|B_0 B_0|u\rangle$$

From the Cauchy-Schwarz inequality, we have

$$(\Delta A)^2 (\Delta B)^2 = \langle u|A_0 A_0|u\rangle \langle u|B_0 B_0|u\rangle \geq |\langle u|A_0 B_0|u\rangle|^2$$

Now

$$A_0 B_0 = \frac{1}{2}(A_0 B_0 + B_0 A_0) + \frac{1}{2}(A_0 B_0 - B_0 A_0)$$

The first term in parentheses on the right-hand side is one-half the anticommutator $\{A_0, B_0\}$ of A_0 and B_0, and $\{A_0, B_0\}$ is a Hermitian operator. Its expectation value is therefore a real number α. Thus

$$(\Delta A)^2 (\Delta B)^2 \geq \frac{1}{4}|\alpha + i\hbar|^2 \geq \frac{\hbar^2}{4}$$

Hence

$$(\Delta A)(\Delta B) \geq \frac{\hbar}{2} \tag{4.3}$$

which implies that

$$\Delta q \Delta p \geq \frac{\hbar}{2} \tag{4.4}$$

We are often concerned with the time-energy uncertainty relation as well; that is,

$$\Delta E \Delta t \geq \frac{\hbar}{2} \tag{4.5}$$

This means that measurement of the energy of a system to precision ΔE requires a time greater than or equal to $\hbar/(2\Delta E)$. Relation (4.5) also applies to an excited state of an atom, molecule, nucleus, or elementary particle that decays spontaneously with mean lifetime τ. Because τ is finite, the excited-state energy cannot be sharp; instead, it must have a natural width $\Delta E \approx \hbar/(2\tau)$. This is discussed in more detail in Chapter 16.

Intuitively it seems that (4.5) should be closely related to (4.3), yet it obviously has a somewhat different status because A and B are operators, whereas in quantum mechanics t is not an operator but a parameter. Nevertheless, we can derive (4.5) by essentially the same method that led to (4.3) as follows: let A be any observable that is not explicitly time dependent, but in place of B, choose the Hamiltonian H. In this case, B_0 is replaced by $H_0 = H - \langle H \rangle = H - E$. Using the Cauchy-Schwarz inequality just as we did in deriving (4.3), we arrive at the inequality

$$(\Delta A)^2 (\Delta H)^2 \geq \frac{1}{4} \left| \langle u | [A, H] | u \rangle \right|^2 \tag{4.6}$$

Now we recall equation (3.16), which expresses the time derivative of the expectation value of an observable A in terms of the commutator of A with the Hamiltonian. If A has no explicit time dependence, (3.16) yields

$$\langle u | [A, H] | u \rangle = i\hbar \frac{d\langle A \rangle}{dt} \tag{4.7}$$

Substituting (4.7) in (4.6) and taking the square root of both sides of the resulting equation, we obtain

$$\Delta H \cdot \Delta T \geq \frac{\hbar}{2}$$

where ΔT is defined as

$$\Delta T = \frac{\Delta A}{\left| \dfrac{d\langle A \rangle}{dt} \right|}$$

Qualitatively, ΔT is the time required for the expectation value of A to change by an amount comparable with the root-mean-square (rms) dispersion in A. Thus ΔT is the shortest time required for a significant change in $\langle A \rangle$. Of course, there may be many observables A associated with the system of interest, and ΔT is not necessarily the same for all of them. However, it is reasonable to assume that the smallest of the times ΔT should be the characteristic time for the system to evolve substantially from its initial state.

4.3 Eigenvalues and eigenvectors of $\hat{q}$, $\hat{p}$

To analyze the eigenvalues and eigenvectors of $\hat{q}$, $\hat{p}$, we first consider some arbitrary operator $A(\hat{q}, \hat{p})$ that is a function of $\hat{q}$, $\hat{p}$ defined by a power-series expansion; that is,

$$A = \sum_{mn} a_{mn} \hat{q}^m \hat{p}^n$$

where the a_{mn} are constants. Now

$$[\hat{q}, \hat{p}^n] = ni\hbar\hat{p}^{n-1} \tag{4.8a}$$

which is easily shown by induction, starting with the basic commutation relation (4.2a). Similarly,

$$[\hat{p}, \hat{q}^m] = -mi\hbar\hat{q}^{m-1} \tag{4.8b}$$

Employing (4.8a), we have

$$[\hat{q}, A] = i\hbar\sum_{mn} na_{mn}\hat{q}^m\hat{p}^{n-1} = i\hbar\frac{\partial A}{\partial \hat{p}} \tag{4.9a}$$

and, similarly,

$$[\hat{p}, A] = -i\hbar\frac{\partial A}{\partial \hat{q}} \tag{4.9b}$$

Now consider the eigenvectors $|q\rangle$ and associated eigenvalues q that appear in the equation

$$\hat{q}|q\rangle = q|q\rangle \tag{4.10a}$$

Similarly, consider

$$\hat{p}|p\rangle = p|p\rangle \tag{4.10b}$$

It is useful to define the unitary translation operator $S(x)$, where x is an arbitrary real number with dimension of length

$$S(x) = \exp\left(-\frac{i}{\hbar}x\hat{p}\right)$$

(Note that because x is a real number, it commutes with $\hat{p}$.) From (4.9a), we have

$$[\hat{q}, S(x)] = i\hbar\frac{\partial S(x)}{\partial \hat{p}} = xS(x) \tag{4.11}$$

Applying both sides of (4.11) on the left to an eigenvector $|q\rangle$, we obtain

$$\hat{q}S(x)|q\rangle - S(x)\hat{q}|q\rangle = xS(x)|q\rangle$$

which yields

$$\hat{q}[S(x)|q\rangle] = (q+x)[S(x)|q\rangle] \tag{4.12}$$

Comparing (4.12) to (4.10a), we see that $S(x)|q\rangle$ is also an eigenvector of $\hat{q}$, with eigenvalue $q+x$. Because x is an arbitrary real number, this implies that the spectrum of eigenvalues of $\hat{q}$ is continuous and covers the entire real line. A similar remark applies to the eigenvalues of $\hat{p}$.

4.4 Wave functions in coordinate and momentum space

Because $\hat{q}$ and $\hat{p}$ are observables, we assume that their eigenvectors each form a complete set, which means that

$$\int |q\rangle\langle q|\, dq = I$$

and

$$\int |p\rangle\langle p|\, dp = I$$

Thus, for any state vector $|\Psi\rangle$, we can write

$$|\Psi\rangle = \int |q\rangle\langle q|\Psi\rangle\, dq = \int |q\rangle\psi(q)\, dq \tag{4.13}$$

Here $\psi(q) = \langle q|\Psi\rangle$ is the coordinate wave function; it represents $|\Psi\rangle$ in coordinate space. Multiplying both sides of (4.13) on the left by $\langle q'|$, we obtain

$$\langle q'|\Psi\rangle = \psi(q') = \int \langle q'|q\rangle\psi(q)\, dq$$

Therefore,

$$\langle q'|q\rangle = \delta(q-q')$$

where $\delta(q-q')$ is the one-dimensional Dirac delta function. In addition to the coordinate representative of $|\Psi\rangle$, we also consider the momentum representative

$$|\Psi\rangle = \int |p\rangle\langle p|\Psi\rangle\, dp = \int |p\rangle\phi(p)\, dp$$

where $\phi(p)$ is the momentum space wave function. What is the relationship between ψ and ϕ? We have

$$\psi(q) = \langle q|\Psi\rangle = \int \langle q|p\rangle\phi(p)\, dp \tag{4.14}$$

and

$$\phi(p) = \langle p|\Psi\rangle = \int \langle p|q\rangle\psi(q)\, dq \tag{4.15}$$

We now use the commutation relation $[\hat{q}, \hat{p}] = i\hbar$ to show that

$$\langle q|p\rangle = \frac{1}{\sqrt{2\pi\hbar}} e^{ipq/\hbar}$$

which implies that (4.14) is a Fourier integral transform and (4.15) is its inverse. We start with

$$\langle q|p\rangle = \left\langle q\left|\frac{[\hat{q},\hat{p}]}{i\hbar}\right|p\right\rangle = \frac{1}{i\hbar}\left[qp\langle q|p\rangle - \langle q|\hat{p}\hat{q}|p\rangle\right]$$

This can be written as

$$\begin{aligned}\langle q|p\rangle\left(1-\frac{qp}{i\hbar}\right) &= -\frac{1}{i\hbar}\langle q|\hat{p}\hat{q}|p\rangle \\ &= -\frac{1}{i\hbar}\int \langle q|\hat{p}\hat{q}|q'\rangle\langle q'|p\rangle\, dq' \\ &= -\frac{1}{i\hbar}\int \langle q|\hat{p}|q'\rangle q'\langle q'|p\rangle\, dq'\end{aligned} \tag{4.16}$$

To analyze the latter expression, note that for small $\Delta q'$,

$$S(\Delta q') = \exp\left(\frac{\hat{p}\Delta q'}{i\hbar}\right) \approx I + \frac{\hat{p}\Delta q'}{i\hbar}$$

and

$$\langle q|S(\Delta q')|q'\rangle = \langle q|q' + \Delta q'\rangle = \langle q|q'\rangle + \frac{\Delta q'}{i\hbar}\langle q|\hat{p}|q'\rangle$$

This implies that

$$\langle q|\hat{p}|q'\rangle = i\hbar\frac{\langle q|q'+\Delta q'\rangle - \langle q|q'\rangle}{\Delta q'} = i\hbar\frac{\partial}{\partial q'}\delta(q-q') \tag{4.17}$$

Substituting (4.17) into the last integrand in (4.16), we obtain

$$\begin{aligned}\langle q|p\rangle\left(1-\frac{qp}{i\hbar}\right) &= -\int q'\langle q'|p\rangle\frac{\partial}{\partial q'}\delta(q-q')\, dq' \\ &= \frac{\partial}{\partial q}\left(q\langle q|p\rangle\right) = \langle q|p\rangle + q\frac{\partial}{\partial q}\langle q|p\rangle\end{aligned}$$

which reduces to

$$\frac{\partial}{\partial q}\langle q|p\rangle = \frac{i}{\hbar}p\langle q|p\rangle$$

Integrating this equation, we arrive at

$$\langle q|p\rangle = C\cdot\exp\left(\frac{i}{\hbar}qp\right)$$

The constant of integration C is determined by normalization; that is,

$$\begin{aligned}\langle p|p'\rangle &= \delta(p-p') = \int \langle p|q\rangle\langle q|p'\rangle\, dq \\ &= |C|^2 \int \exp\left[\frac{i}{\hbar}q(p-p')\right] dq \\ &= 2\pi\hbar\delta(p-p')|C|^2\end{aligned}$$

Thus $|C|^2 = (2\pi\hbar)^{-1}$. Because we can choose otherwise arbitrary phases so that C is real and positive, we arrive at the result

$$\langle q|p\rangle = \frac{1}{\sqrt{2\pi\hbar}}e^{ipq/\hbar}$$

Thus

$$\psi(q) = \frac{1}{\sqrt{2\pi\hbar}}\int_{-\infty}^{\infty}\phi(p)e^{iqp/\hbar}\, dp \tag{4.18}$$

$$\phi(p) = \frac{1}{\sqrt{2\pi\hbar}}\int_{-\infty}^{\infty}\psi(q)e^{-ipq/\hbar}\, dq \tag{4.19}$$

These formulas are easily generalized to three spatial dimensions; that is,

$$\psi(\mathbf{r}) = \frac{1}{(2\pi\hbar)^{3/2}}\int \phi(\mathbf{p})e^{i\mathbf{p}\cdot\mathbf{r}/\hbar}\, d^3\mathbf{p} \tag{4.20}$$

$$\phi(\mathbf{p}) = \frac{1}{(2\pi\hbar)^{3/2}}\int \psi(\mathbf{r})e^{-i\mathbf{p}\cdot\mathbf{r}/\hbar}\, d^3\mathbf{r}. \tag{4.21}$$

With the aid of the foregoing Fourier transforms, we can superpose plane-wave states of definite momentum to construct an arbitrary wave packet in coordinate space, and conversely, we can start with such a wave packet and resolve it into plane-wave states. Equations (4.18) and (4.19) or (4.20) and (4.21) imply the de Broglie relation.

4.5 Expectation values of operators in coordinate and momentum representation

Consider a state $|\Psi\rangle$ with $\langle\Psi|\Psi\rangle = 1$. For any operator A, we have

$$\langle A\rangle = \iint \langle \Psi|q'\rangle\langle q'|A|q\rangle\langle q|\Psi\rangle\, dqdq' = \iint \psi^*(q')\langle q'|A|q\rangle \psi(q)\, dqdq'$$

Suppose that $A = A(\hat{q}) = \sum_n a_n \hat{q}^n$. Then

$$\langle q'|A|q\rangle = \sum_n a_n \langle q'|\hat{q}^n|q\rangle = \delta(q'-q)\sum_n a_n q^n$$

Hence

$$\langle A(\hat{q})\rangle = \int \psi^*(q) A(q)\psi(q)\, dq$$

where $A(q) = \sum a_n q^n$ is an ordinary function of the real variable q.

Suppose, instead, that A is a function of $\hat{p}$, for example, $A = \hat{p}$. Recalling (4.17), we have

$$\langle q' \mid \hat{p} \mid q\rangle = i\hbar \frac{\partial}{\partial q}\delta(q-q')$$

Thus

$$\begin{aligned}\langle \hat{p}\rangle &= i\hbar \iint \psi^*(q')\left[\frac{\partial}{\partial q}\delta(q-q')\right]\psi(q)\, dqdq' \\ &= -i\hbar \int \psi^*(q)\frac{\partial \psi(q)}{\partial q}\, dq\end{aligned}$$

Similarly, in three dimensions, one has

$$\langle \hat{\boldsymbol{p}}\rangle = -i\hbar \int \psi^*(\boldsymbol{r})\nabla \psi(\boldsymbol{r})\, d^3\boldsymbol{r}$$

In other words, in Cartesian coordinates, the $\hat{p}_{x,y,z}$ are represented by the differential operators $-i\hbar(\partial/\partial x, \partial/\partial y, \partial/\partial z)$. By similar reasoning, in momentum space, the position operator in Cartesian coordinates takes the following form:

$$\hat{\boldsymbol{r}} \rightarrow i\hbar\nabla_p = i\hbar\left(\frac{\partial}{\partial p_x}, \frac{\partial}{\partial p_y}, \frac{\partial}{\partial p_z}\right)$$

In a curvilinear coordinate system, the differential operators representing the various components of $\boldsymbol{p}$ are more complicated [see, e.g., Leaf (1979)]. Let the curvilinear coordinates be (q^1, q^2, q^3) with base vectors $(\boldsymbol{e}_1, \boldsymbol{e}_2, \boldsymbol{e}_3)$. A small displacement is given by $d\boldsymbol{x} = \boldsymbol{e}_n dq^n$, where we use the repeated index summation convention. Let the metric tensor be g_{ij}, and $J = \sqrt{g}$, where $g = \det(g_{ij})$, and J is the Jacobian for the volume element in the curvilinear coordinate system. It can be shown that

$$p_n = -i\hbar\left(\frac{\partial}{\partial q^n} + \frac{1}{\sqrt{J}}\frac{\partial\sqrt{J}}{\partial q^n}\right) \tag{4.22}$$

and that this operator is Hermitian. In almost all applications, one employs an orthogonal set of curvilinear coordinates, in which case the components of the metric tensor are

$$g_{11} = h_1^2 \qquad g_{22} = h_2^2 \qquad g_{33} = h_3^2$$

and $g_{ij} = 0$ for $i \neq j$. Here $J = h_1 h_2 h_3$. For example, for spherical polar coordinates,

$$\begin{array}{lll} q^1 = r & q^2 = \theta & q^3 = \phi \\ h_1 = 1 & h_2 = r & h_3 = r \sin\theta \end{array}$$

and

$$J = r^2 \sin\theta$$

Thus (4.22) yields

$$\begin{aligned} p_r &= -i\hbar\left(\frac{\partial}{\partial r} + \frac{1}{r}\right) \\ p_\theta &= -i\hbar\left(\frac{\partial}{\partial\theta} + \frac{1}{2}\cot\theta\right) \\ p_\phi &= -i\hbar\frac{\partial}{\partial\phi} \end{aligned}$$

4.6 Choosing the Hamiltonian. The Schroedinger wave equation

How do we formulate the quantum-mechanical Hamiltonian for a given physical system, and once that is accomplished, how do we recover the well-known Schroedinger wave equation from the formalism we have developed so far? We begin by considering the simple classical Hamiltonian for motion of a nonrelativistic particle in one spatial dimension; that is,

$$H_c = \frac{p_c^2}{2m} + V(q_c)$$

We replace the classical variables q_c, p_c by the quantum-mechanical operators $\hat{q}, \hat{p}$, respectively, to arrive at the quantum-mechanical Hamiltonian

$$H_{QM} = \frac{\hat{p}^2}{2m} + V(\hat{q})$$

We now employ the latter in the time-dependent Schroedinger equation; that is,

$$H_{QM}|\Psi\rangle = \left[\frac{\hat{p}^2}{2m} + V(\hat{q})\right]|\Psi\rangle = i\hbar|\dot{\Psi}\rangle$$

Multiplying this on the left by $\langle q|$, we obtain

$$\frac{1}{2m}\langle q|\hat{p}^2|\Psi\rangle + \langle q|V(\hat{q})|\Psi\rangle = i\hbar\langle q|\dot{\Psi}\rangle = i\hbar\dot{\psi}$$

Now, employing the completeness relation, we have

$$\begin{aligned}\langle q|V(\hat{q})|\Psi\rangle &= \int \langle q|V(\hat{q})|q'\rangle\langle q'|\Psi\rangle\, dq' \\ &= \int V(q')\langle q|q'\rangle\psi(q')\, dq' \\ &= \int V(q')\delta(q-q')\psi(q')\, dq' \\ &= V(q)\psi(q)\end{aligned}$$

Also from the completeness relation and (4.17), we obtain

$$\begin{aligned}\frac{1}{2m}\langle q|\hat{p}^2|\Psi\rangle &= \frac{1}{2m}\iint \langle q|\hat{p}|q'\rangle\langle q'|\hat{p}|q''\rangle\, \psi(q'')\, dq dq' \\ &= -\frac{\hbar^2}{2m}\iint\left[\frac{\partial}{\partial q'}\delta(q-q')\right]\left[\frac{\partial}{\partial q''}\delta(q'-q'')\right]\psi(q'')\, dq' dq'' \\ &= \frac{\hbar^2}{2m}\int\left[\frac{\partial}{\partial q'}\delta(q-q')\right]\frac{\partial\psi(q')}{\partial q'}\, dq' \\ &= -\frac{\hbar^2}{2m}\frac{\partial^2\psi(q)}{\partial q^2}\end{aligned}$$

Thus we obtain the familiar Schroedinger wave equation

$$-\frac{\hbar^2}{2m}\psi'' + V(q) = i\hbar\dot{\psi}$$

In more complicated situations, but when the system still has a classical analogue, we start with the following basic considerations in classical mechanics. There a physical system is described by generalized coordinates q_i and generalized velocities $\partial q_i / \partial t$. (From now on, we drop the subscript c for classical variables unless it causes confusion to do so.) The kinetic energy T is a well-defined function of these coordinates and velocities, and it obeys Lagrange's equations; that is,

$$\frac{d}{dt}\frac{\partial T}{\partial \dot{q}_i} - \frac{\partial T}{\partial q_i} = Q_i$$

where the Q_i are generalized forces. Frequently, the latter may be derivable from an ordinary potential V; that is,

$$Q_i = -\frac{\partial V}{\partial q_i}$$

where V depends on the q_i but not on the $\dot{q}_i$. In this case, we obtain

$$\frac{d}{dt}\frac{\partial T}{\partial \dot{q}_i}-\frac{\partial T}{\partial q_i}=-\frac{\partial V}{\partial q_i}$$

These equations can be rewritten in terms of the Lagrangian $L = T - V$ as

$$\frac{d}{dt}\frac{\partial L}{\partial \dot{q}_i}-\frac{\partial L}{\partial q_i}=0 \tag{4.23}$$

However, it may happen that the potential (now called U) has a velocity-dependent part. If we can write

$$Q_i=\frac{d}{dt}\frac{\partial U}{\partial \dot{q}_i}-\frac{\partial U}{\partial q_i} \tag{4.24}$$

then by redefining the Lagrangian as $L = T - U$, we once again obtain equation (4.23). We also define the generalized (canonical) momenta by

$$p_i=\frac{\partial L}{\partial \dot{q}_i}$$

and the classical Hamiltonian by

$$H_c=\sum_i \dot{q}_i p_i - L$$

Consider a classical observable A_c that can be expressed as a function of the coordinates q_i, the momenta p_i, and the time t. Then

$$\frac{dA_c}{dt}=\sum_i\left[\frac{\partial A_c}{\partial q_i}\frac{\partial q_i}{\partial t}+\frac{\partial A_c}{\partial p_i}\frac{\partial p_i}{\partial t}\right]+\frac{\partial A_c}{\partial t} \tag{4.25}$$

However, Hamilton's equations are

$$\frac{\partial H_c}{\partial p_i}=\dot{q}_i \qquad \frac{\partial H_c}{\partial q_i}=-\dot{p}_i$$

Substituting these in (4.25), we obtain

$$\begin{aligned}\frac{dA_c}{dt}&=\sum_i\left[\frac{\partial A_c}{\partial q_i}\frac{\partial H_c}{\partial p_i}-\frac{\partial H_c}{\partial q_i}\frac{\partial A_c}{\partial p_i}\right]+\frac{\partial A_c}{\partial t}\\&=\{A_c,H_c\}+\frac{\partial A_c}{\partial t}\end{aligned} \tag{4.26}$$

where $\{A_c, H_c\}$ is a *Poisson bracket*. Equation (4.26) resembles the Heisenberg equation (3.18). More generally, there is almost always a strong similarity between the Poisson bracket of any pair of classical canonical variables and the commutator of the corresponding quantum-mechanical operators. For example, consider the Poisson bracket of the canonical coordinate q_j and the canonical momentum p_j. It is

$$\{q_j, p_k\} = \sum_i \frac{\partial q_j}{\partial q_i}\frac{\partial p_k}{\partial p_i} = \delta_{jk}$$

whereas the corresponding commutator is

$$[\hat{q}_j, \hat{p}_k] = i\hbar\delta_{jk}$$

The similarity is so general that it is usually understood as a rule: to obtain the commutator from the Poisson bracket, multiply the latter by $i\hbar$, and change the {} to [].[1] The general rule for obtaining the quantum-mechanical Hamiltonian is as before: write the classical Hamiltonian in terms of the canonically conjugate coordinates and momenta, and then replace these by the corresponding quantum operators.

We now consider an important example: a particle of mass m and electric charge q in an external electromagnetic field. In classical electrodynamics, the *Lorentz force* in cgs or hlu units is

$$\boldsymbol{F} = q\left(\boldsymbol{\mathcal{E}} + \frac{1}{c}\boldsymbol{v}\times\boldsymbol{B}\right) \tag{4.27}$$

The electric field $\boldsymbol{\mathcal{E}}$ and magnetic field $\boldsymbol{B}$ can be expressed in terms of the scalar potential Φ and vector potential $\boldsymbol{A}$; that is,

$$\boldsymbol{\mathcal{E}} = -\nabla\Phi - \frac{1}{c}\frac{\partial \boldsymbol{A}}{\partial t}$$

$$\boldsymbol{B} = \nabla\times\boldsymbol{A}$$

Substituting these expressions in (4.27) and employing the identity

$$\mathbf{v}\times(\nabla\times\boldsymbol{A}) = \nabla(\mathbf{v}\cdot\boldsymbol{A}) - (\mathbf{v}\cdot\nabla)\boldsymbol{A}$$

we obtain the ith Cartesian component of the Lorentz force; that is,

$$F_i = q\left\{-\frac{\partial\Phi}{\partial x_i} - \frac{1}{c}\left[(\mathbf{v}\cdot\nabla)A_i + \frac{\partial A_i}{\partial t}\right] + \frac{1}{c}\mathbf{v}\cdot\frac{\partial \boldsymbol{A}}{\partial x_i}\right\} \tag{4.28}$$

However,

$$\frac{dA_i}{dt} = (\mathbf{v}\cdot\nabla)A_i + \frac{\partial A_i}{\partial t}$$

[1] However, the rule is not universally valid. See D. Giulini, "That Strange Procedure Called Quantization," in *Aspects of Quantum Gravity: From Theory to Experimental Search*, D. Giulini, C. Kiefer, and C. Lammerzahl, eds. (Berlin: Springer Verlag, 2003).

Thus (4.28) becomes

$$F_i = q\left(-\frac{\partial \Phi}{\partial x_i} - \frac{1}{c}\frac{dA_i}{dt} + \frac{1}{c}\mathbf{v}\bullet\frac{\partial A}{\partial x_i}\right) \tag{4.29}$$

Comparing (4.29) with (4.24), we see that $\boldsymbol{F}$ is derivable from the velocity-dependent potential

$$U = q\Phi - \frac{q}{c}\mathbf{v}\bullet \boldsymbol{A}$$

Hence the Lagrangian is

$$L = T - U = T - q\Phi + \frac{q}{c}\mathbf{v}\bullet \boldsymbol{A}$$

and

$$p_i = \frac{\partial L}{\partial \dot{x}_i} = \frac{\partial T}{\partial \dot{x}_i} + \frac{q}{c}A_i \tag{4.30}$$

The first term on the far right-hand side of (4.30) is called the *mechanical momentum* or *kinetic momentum*, whereas p_i itself is the *canonical momentum.*

The Hamiltonian is

$$\begin{aligned} H_c &= \sum \dot{x}_i p_i - L \\ &= \sum \dot{x}_i \frac{\partial T}{\partial \dot{x}_i} + \frac{q}{c}\mathbf{v}\bullet \boldsymbol{A} - T + q\Phi - \frac{q}{c}\mathbf{v}\bullet \boldsymbol{A} \\ &= \sum \dot{x}_i \frac{\partial T}{\partial \dot{x}_i} - T + q\Phi \end{aligned} \tag{4.31}$$

Often T is a homogeneous quadratic function of the velocities, in which case (4.31) becomes

$$H_c = 2T - T + q\Phi = \frac{m}{2}\left(\dot{x}^2 + \dot{y}^2 + \dot{z}^2\right) + q\Phi \tag{4.32}$$

From (4.30), each component of canonical momentum is

$$p_i = m\dot{x}_i + \frac{q}{c}A_i \qquad i = 1, 2, 3 \text{ for } x, y, z$$

We write (4.32) in terms of the canonical momentum $\boldsymbol{p}$ as follows:

$$H_c = \frac{1}{2m}\left(\boldsymbol{p} - \frac{q}{c}\boldsymbol{A}(\boldsymbol{r},t)\right)^2 + q\Phi(\boldsymbol{r},t) \tag{4.33}$$

Thus, in coordinate representation, the quantum-mechanical Hamiltonian is

$$H = \frac{1}{2m}\left(-i\hbar\nabla - \frac{q}{c}A\right)^2 + q\Phi \tag{4.34}$$

and the time-dependent Schroedinger wave equation is

$$H\psi = \frac{1}{2m}\left(-i\hbar\nabla - \frac{q}{c}A\right)^2 \psi + q\Phi\psi = i\hbar\frac{\partial\psi}{\partial t} \tag{4.35}$$

In the following sections we develop some consequences of this very important equation. Before we do, however, let's note something quite obvious: no argument by analogy from classical mechanics can tell us how to construct terms in the quantum Hamiltonian that have no classical analogue. For example, if a particle with charge q and mass m also has intrinsic spin (a nonclassical observable) and as a consequence a spin magnetic moment, then, at least in non-relativistic quantum mechanics, we must add a term to the Hamiltonian in (4.34) to account for the interaction of that spin magnetic moment with an external electromagnetic field. Such nonclassical terms can only be found by intuition (guessing), from appeals to symmetry, and from experimental clues.

4.7 General properties of Schroedinger's equation: The equation of continuity

Because $\langle \hat{r}^n \rangle = \int \psi^*(\boldsymbol{r})\boldsymbol{r}^n\psi(\boldsymbol{r})\, d^3\boldsymbol{r}$ for any value of n, $\rho = \psi^*\psi$ is the probability density for finding the particle of interest, and $\rho d\tau$ is the probability of finding the particle in volume $d\tau$. Because the particle should be somewhere, we should always require $\int \rho\, d\tau = 1$. It's not always possible to normalize the wave function in this manner, for example, when the wave function is a plane wave extending over all space. However, this difficulty is only an apparent one. In reality, the wave is always bounded in space (although it may extend over a very large region). When we replace it by a plane wave, we are making a convenient approximation and saying that the momentum spread is negligible for the problem in question.

How does the probability density vary with time? We write the Schroedinger equation (4.35) as follows:

$$\frac{-\hbar^2}{2m}\nabla^2\psi + \frac{i\hbar q}{2mc}\boldsymbol{A}\cdot\nabla\psi + \frac{i\hbar q}{2mc}\nabla\cdot(\boldsymbol{A}\psi) + \frac{q^2A^2}{2mc^2}\psi + q\Phi\psi = i\hbar\dot{\psi} \tag{4.36a}$$

The complex conjugate equation is

$$\frac{-\hbar^2}{2m}\nabla^2\psi^* - \frac{i\hbar q}{2mc}\boldsymbol{A}\cdot\nabla\psi^* - \frac{i\hbar q}{2mc}\nabla\cdot(\boldsymbol{A}\psi^*) + \frac{q^2A^2}{2mc^2}\psi^* + q\Phi\psi^* = -i\hbar\dot{\psi}^* \tag{4.36b}$$

Multiplying (4.36a) by ψ^* and (4.36b) by ψ, and subtracting the latter equation from the former, we obtain

$$-\frac{\hbar^2}{2m}(\psi^*\nabla^2\psi-\psi\nabla^2\psi^*)+\frac{i\hbar q}{2mc}(\psi^* \boldsymbol{A}\cdot\nabla\psi+\psi\boldsymbol{A}\cdot\nabla\psi^*) \\ +\frac{i\hbar q}{2mc}\left[\psi^*\nabla\cdot(\boldsymbol{A}\psi)+\psi\nabla\cdot(\boldsymbol{A}\psi^*)\right]=i\hbar\frac{\partial}{\partial t}(\psi^*\psi)=i\hbar\frac{\partial\rho}{\partial t} \tag{4.37}$$

Now $(\psi^*\nabla^2\psi-\psi\nabla^2\psi^*)=\nabla\cdot(\psi^*\nabla\psi-\psi\nabla\psi^*)$, and

$$\psi^*\nabla\cdot(\boldsymbol{A}\psi)+\psi\nabla\cdot(\boldsymbol{A}\psi^*)=2\nabla\cdot(\psi^*\boldsymbol{A}\psi)-(\psi^*\boldsymbol{A}\cdot\nabla\psi+\psi\boldsymbol{A}\cdot\nabla\psi^*)$$

Thus, defining the probability current density as

$$\boldsymbol{j}=\frac{\hbar}{2mi}(\psi^*\nabla\psi-\psi\nabla\psi^*)-\frac{q}{mc}\boldsymbol{A}\psi^*\psi \tag{4.38}$$

we see that (4.37) reduces to

$$\nabla\cdot\boldsymbol{j}+\frac{\partial\rho}{\partial t}=0 \tag{4.39}$$

which is the equation of continuity. It is analogous to the equation of continuity for electric charge and current densities in classical electrodynamics. Whereas that equation implies conservation of charge, (4.39) implies conservation of probability, no matter how ψ changes with time. If $\boldsymbol{A}=0$, (4.38) reduces to

$$\boldsymbol{j}=\frac{\hbar}{2mi}(\psi^*\nabla\psi-\psi\nabla\psi^*) \tag{4.40}$$

In the classical Hamiltonian (4.34), the canonical momentum is not uniquely defined because the electromagnetic potentials are themselves not unique: with an arbitrary real scalar function χ, we can always make a gauge transformation; that is,

$$\Phi\to\Phi'=\Phi+\frac{1}{c}\frac{\partial\chi}{\partial t} \\ \boldsymbol{A}\to\boldsymbol{A}'=\boldsymbol{A}-\nabla\chi \tag{4.41}$$

This transformation leaves the electric and magnetic fields invariant. It is easy to verify that in quantum mechanics, Schroedinger's equation (4.35) is covariant with respect to gauge transformations in the following sense: if we make the replacements given in (4.41) and also replace ψ by $\psi'=\psi\exp\left[-i(q\chi/\hbar c)\right]$, then ψ' satisfies

$$\frac{1}{2m}\left(-i\hbar\nabla-\frac{q}{c}\boldsymbol{A}'\right)^2\psi'+q\Phi'\psi'=i\hbar\frac{\partial\psi'}{\partial t}$$

The probability current density $\boldsymbol{j}$ in (4.38) remains invariant in this transformation. Finally, we consider the Schroedinger equation (4.35) in the important case where $\boldsymbol{A}$ and $V = q\Phi$ are independent of time. We try a solution of the form $\psi(\boldsymbol{r},t) = u(\boldsymbol{r}) f(t)$ and obtain

$$\frac{f}{2m}\left(-i\hbar\nabla - \frac{q}{c}\boldsymbol{A}\right)^2 u + ufV = i\hbar u\frac{\partial f}{\partial t}$$

Dividing both sides of this equation by uf, we have

$$\frac{1}{2m}\frac{1}{u}\left(-i\hbar\nabla - \frac{q}{c}\boldsymbol{A}\right)^2 u + V = i\hbar\frac{\dot{f}}{f} \tag{4.42}$$

The left-hand side of (4.42) is a function of $\boldsymbol{r}$ only, and the right-hand side is a function of t only; hence both must be equal to a constant, which is the energy E. In fact, we have

$$\frac{\dot{f}}{f} = -\frac{i}{\hbar}E$$

which yields $f(t) = f(0)e^{-iEt/\hbar}$. Thus

$$\frac{1}{2m}\left(-i\hbar\nabla - \frac{q}{c}\boldsymbol{A}\right)^2 u + Vu = Eu \tag{4.43}$$

which is the time-independent Schroedinger wave equation. Because the constant $f(0)$ can be absorbed in u, $\rho = \psi^*\psi = u^*u$, which is independent of the time. Therefore, $\nabla \cdot \boldsymbol{j} = 0$, so there is no net flow of probability into or out of any closed volume. In other words, a state of well-defined energy is a stationary state.

Equation (4.43) is an eigenvalue equation for the Hermitian operator $(1/2m)\left[-i\hbar\nabla - (q/c)\boldsymbol{A}\right]^2$ on the Hilbert space of spatial wave functions $u(\boldsymbol{r})$. Hence solutions u corresponding to distinct values of E are orthogonal, and degenerate solutions corresponding to a common eigenvalue E can be rendered orthogonal by the Schmidt process. Of course, here orthogonality means that if u and u' are two distinct solutions, then $\int u^* u' \, d\tau = 0$. Furthermore, the solutions $u_n \cdot \exp(-iE_n t/\hbar)$ form a complete set. This means that any solution ψ of the time-dependent Schroedinger equation (4.33) for time-independent potentials $\boldsymbol{A}$ and V can be expressed as a superposition of the $u_n \cdot \exp(-iE_n t/\hbar)$; that is,

$$\psi(\boldsymbol{r},t) = \sum_n c_n u_n(\boldsymbol{r}) \exp(-iE_n t/\hbar)$$

It is easy to show that for time-independent potentials, the coefficients c_n appearing in the sum are independent of $\boldsymbol{r}$ and of t.

4.8 Galilean invariance of the Schroedinger wave equation

To describe any natural phenomenon in classical or quantum physics, one needs a reference frame. Inertial reference frames possess a special property in classical physics – in such a frame, a body subject to no external forces describes uniform rectilinear motion. Given one such frame F, we can form another F' by moving with constant velocity v in the x-direction with respect to F. Assuming that the origins of F and F' coincide at $t = 0$ and that all motions are nonrelativistic, the coordinates of a space-time point in F' are related to those of the same space-time point in F by the Galilean transformation

$$\begin{aligned} x' &= x - vt \\ y' &= y \\ z' &= z \\ t' &= t \end{aligned} \tag{4.44}$$

If the Schroedinger equation is to have general validity in nonrelativistic quantum mechanics, its form must be the same in both frames; in other words, the wave equation must be covariant with respect to Galilean transformations. We now work out the consequences of this covariance using one spatial dimension for simplicity.

Let the Schroedinger equations (including a scalar velocity-independent potential V) be

$$-\frac{\hbar^2}{2m}\frac{\partial^2\psi(x,t)}{\partial x^2}+V(x,t)\psi(x,t)=i\hbar\frac{\partial\psi(x,t)}{\partial t} \tag{4.45}$$

and

$$-\frac{\hbar^2}{2m}\frac{\partial^2\chi(x',t')}{\partial x'^2}+V(x',t')\chi(x',t')=i\hbar\frac{\partial\chi(x',t')}{\partial t'} \tag{4.46}$$

in frames F and F', respectively. It is natural to require that probabilities for physical processes be the same in both frames. Then, because the Jacobian of the transformation (4.44) is unity, probability densities also must be the same. In other words, we must have

$$|\psi(x,t)|^2 = |\chi(x',t')|^2$$

which implies

$$\psi(x,t) = e^{ig(x',t')}\chi(x',t') \tag{4.47}$$

where g is a real function. To find g, we substitute (4.47) into (4.45) and rewrite the derivatives in (4.45) in terms of x', t' using

$$\frac{\partial}{\partial x}=\frac{\partial}{\partial x'} \qquad \frac{\partial}{\partial t}=\frac{\partial}{\partial t'}-v\frac{\partial}{\partial x'}$$

After some simple algebra, this yields

$$-\frac{\hbar^2}{2m}\left\{\frac{\partial^2\chi}{\partial x'^2}+2i\frac{\partial g}{\partial x'}\frac{\partial \chi}{\partial x'}+\left[i\frac{\partial^2 g}{\partial x'^2}-\left(\frac{\partial g}{\partial x'}\right)^2\right]\chi\right\}+V(x'+\mathrm{v}t',t')\chi$$
$$=i\hbar\frac{\partial\chi}{\partial t'}-i\hbar\mathrm{v}\frac{\partial\chi}{\partial x'}-\hbar\left(\frac{\partial g}{\partial t'}-\mathrm{v}\frac{\partial g}{\partial x'}\right)\chi \tag{4.48}$$

Comparison of (4.48) with (4.46) yields

$$\frac{\partial g}{\partial x'}=\frac{m\mathrm{v}}{\hbar} \tag{4.49}$$

and

$$-i\frac{\hbar}{2m}\frac{\partial^2 g}{\partial x'^2}+\frac{\hbar}{2m}\left(\frac{\partial g}{\partial x'}\right)^2+\frac{\partial g}{\partial t'}-\mathrm{v}\frac{\partial g}{\partial x'}=0 \tag{4.50}$$

However, when (4.49) is taken into account, (4.50) simplifies to

$$\frac{\partial g}{\partial t'}=\frac{m\mathrm{v}^2}{2\hbar} \tag{4.51}$$

Thus, from (4.49) and (4.51), we obtain

$$g=\frac{m\mathrm{v}x'}{\hbar}+\frac{m\mathrm{v}^2}{2\hbar}t'=\frac{m\mathrm{v}x}{\hbar}-\frac{m\mathrm{v}^2}{2\hbar}t$$

For example, consider a free particle with definite momentum p in the x-direction as observed in F. The wave function ψ is

$$\psi=\frac{1}{\sqrt{2\pi\hbar}}\exp\left(\frac{ipx}{\hbar}\right)\exp\left(-\frac{i}{\hbar}\frac{p^2}{2m}t\right) \tag{4.52}$$

Therefore,

$$\begin{aligned}\chi&=e^{-ig}\psi\\&=\frac{1}{\sqrt{2\pi\hbar}}\exp\left[-i\left(\frac{mvx'}{\hbar}+\frac{mv^2}{2\hbar}t'\right)\right]\exp\left[\frac{i(x'+vt')p}{\hbar}\right]\exp\left(-\frac{i}{\hbar}\frac{p^2}{2m}t'\right)\\&=\frac{1}{\sqrt{2\pi\hbar}}\exp\left[\frac{i(p-mv)x'}{\hbar}\right]\exp\left[-\frac{i}{\hbar}\frac{(p-mv)^2}{2m}t'\right]\end{aligned} \tag{4.53}$$

Comparing (4.52) with (4.53), we see that the general forms of the wave functions ψ and χ are the same. However, to go from (4.52) to (4.53), it is necessary not only to make the replacements $x\to x'$, $t\to t'$ but also to make the replacement $p\to p-mv$.

This reveals a profound difference between de Broglie waves and waves in classical mechanics. As a classical example, consider water waves propagating in the $+x$-direction with straight parallel wave fronts on the ocean surface. Let the wavelength be λ according to an observer O in frame F who employs coordinates x and t. Now consider another observer O' in frame F' who flies over the ocean surface with nonrelativistic velocity v in the $+x$-direction with respect to F. The latter employs coordinates $x' = x - vt$, $t' = t$ but still observes the same wavelength λ. However, if two such observers describe a de Broglie wave propagating in the $+x$-direction, and if the momentum in frame F is p, the momentum in frame F' is $p' = p - mv$; hence the wavelength is

$$\lambda' = \frac{h}{p - mv} = \frac{\lambda}{1 - \frac{mv}{p}}$$

4.9 Ehrenfest's equations and the classical limit

We have shown previously that if A is an observable that does not depend explicitly on the time, and $|u\rangle$ is a state of the system of interest that is normalized to unity, then

$$\frac{d\langle A\rangle}{dt} = \frac{1}{i\hbar}\langle u|[A, H]|u\rangle$$

Let $H = \left(\hat{p}^2/2m\right) + V\left(\hat{q}\right)$ and $A = \hat{p}$. Then

$$\frac{d\langle\hat{p}\rangle}{dt} = \frac{1}{i\hbar}\left\langle u\left|\left[\hat{p}, V\right]\right|u\right\rangle = -\left\langle\frac{\partial V}{\partial\hat{q}}\right\rangle$$

Similarly,

$$\frac{d\langle\hat{q}\rangle}{dt} = \frac{1}{i\hbar}\left\langle u\left|\left[\hat{q}, \frac{\hat{p}^2}{2m}\right]\right|u\right\rangle = \frac{1}{m}\langle\hat{p}\rangle$$

These are known as *Ehrenfest's equations*. They are obviously analogous to the corresponding equations of classical mechanics; that is,

$$\dot{\boldsymbol{p}}_c = -\nabla V_c = \boldsymbol{F}_c$$

where $\boldsymbol{F}_c$ is a force, and

$$\dot{\boldsymbol{q}}_c = \frac{1}{m^c} = \boldsymbol{v}$$

When are Ehrenfest's equations identical to the classical equations of motion, and when do they represent something different? If the two sets of equations are identical, then $\langle\hat{q}\rangle, \langle\hat{p}\rangle$ play the same roles as $\boldsymbol{q}_c, \boldsymbol{p}_c$, respectively, which means that

$$\left\langle \frac{\partial V}{\partial \boldsymbol{q}_c} \right\rangle = \frac{\partial V\left(\langle \hat{\boldsymbol{q}} \rangle\right)}{\partial \langle \hat{\boldsymbol{q}} \rangle} \tag{4.54}$$

Under what circumstances is (4.54) valid? Suppose that V is a continuous and differentiable function. Then we can write the expansion

$$V(\boldsymbol{q}_c) = V(\boldsymbol{q}_0) + a_1(\boldsymbol{q}_c - \boldsymbol{q}_0) + a_2(\boldsymbol{q}_c - \boldsymbol{q}_0)^2 + a_3(\boldsymbol{q}_c - \boldsymbol{q}_0)^3 + \cdots$$

where the a_n are numerical coefficients, and $\boldsymbol{q}_0$ is a constant vector. Meanwhile,

$$V(\hat{\boldsymbol{q}}) = V(\boldsymbol{q}_0) + a_1\left(\hat{\boldsymbol{q}} - \boldsymbol{q}_0\right) + a_2\left(\hat{\boldsymbol{q}} - \boldsymbol{q}_0\right)^2 + a_3\left(\hat{\boldsymbol{q}} - \boldsymbol{q}_0\right)^3 + \cdots$$

Thus

$$\left\langle \frac{\partial V}{\partial \boldsymbol{q}_c} \right\rangle = a_1 + 2a_2(\langle \boldsymbol{q}_c \rangle - \boldsymbol{q}_0) + 3a_3\langle (\boldsymbol{q}_c - \boldsymbol{q}_0)^2 \rangle + \cdots \tag{4.55}$$

and

$$\frac{\partial V\left(\langle \hat{\boldsymbol{q}} \rangle\right)}{\partial \langle \hat{\boldsymbol{q}} \rangle} = a_1 + 2a_2(\langle \boldsymbol{q} \rangle - \boldsymbol{q}_0) + 3a_3(\langle \boldsymbol{q} \rangle - \boldsymbol{q}_0)^2 + \cdots \tag{4.56}$$

Comparing (4.55) with (4.56), we see that for $\langle \hat{\boldsymbol{q}} \rangle, \langle \hat{\boldsymbol{p}} \rangle$ to describe a classical orbit, V must vary sufficiently slowly over the entire wave packet that terms of third and higher order can be ignored in its Taylor expansion. For example, consider the electron in a hydrogen atom. For each stationary state with low principal quantum number n, the spatial wave function extends over a region of space in which the Coulomb potential varies rapidly, so terms of third and higher order in the Taylor expansion of V cannot be ignored. Hence a wave packet constructed from a superposition of stationary states with low n values cannot resemble a particle that describes a classical Bohr orbit. For very large principal quantum numbers, the electron wave packet can be confined to a region of space over which the potential varies slowly, and the picture of a classical Bohr orbit becomes more realistic.

Problems for Chapter 4

4.1. Starting from the basic commutation relation $[\hat{q}, \hat{p}] = i\hbar$ and using mathematical induction, show that $[\hat{q}, \hat{p}^n] = ni\hbar\hat{p}^{n-1}$ and $[\hat{p}, \hat{q}^m] = -mi\hbar\hat{q}^{m-1}$, where $n \geq 2, m \geq 2$ are positive integers.

4.2. Let $\hat{q}$ and $\hat{p}$ be position and momentum operators, respectively, in one dimension, and let $f(\hat{q}, \hat{p})$ be an operator that can be expressed as a power series in $\hat{q}$ and $\hat{p}$. What uncertainty relation holds for $\hat{q}$ and f? That is, what is $\Delta q \cdot \Delta f$?

4.3. The Heisenberg uncertainty relation for a coordinate and its conjugate momentum

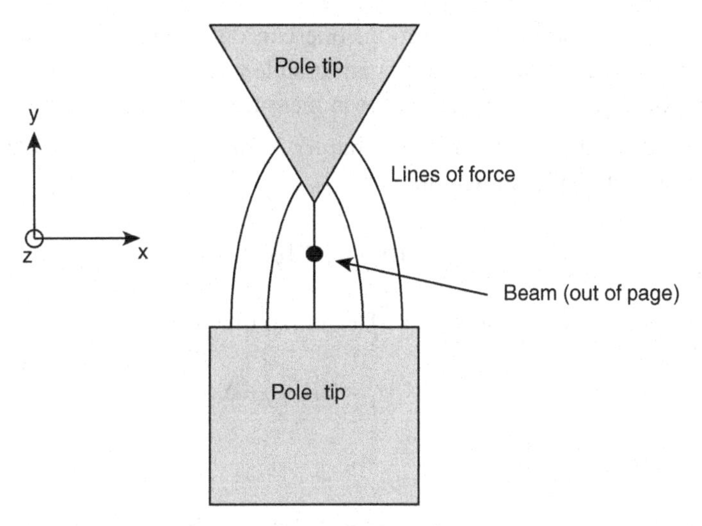

Figure 4.1 Schematic diagram of Stern-Gerlach experiment. The magnet extends for a considerable length in the z-direction.

$$\Delta x \Delta p \geq \frac{\hbar}{2}$$

is derived from the commutation relation $[x, p] = i\hbar$ with the aid of the Cauchy-Schwarz inequality. From consideration of the conditions that must be satisfied for that inequality to become an equality, show that the minimum uncertainty wave packet must be a Gaussian.

4.4. The magnetic field $\boldsymbol{B} = \boldsymbol{B}_1 + \boldsymbol{B}_2$ produced by a free electron is partly due to its motion ($\boldsymbol{B}_1$) and partly due to its intrinsic spin magnetic moment ($\boldsymbol{B}_2$). To determine the magnetic moment of a free electron from a measurement of the field strength produced by it, two conditions must be satisfied:

$$(a)\ |\boldsymbol{B}_2| >> |\boldsymbol{B}_1|$$

(b) The electron must be localized in a region Δr that is much smaller than the distance r to the point of observation.

Show that the simultaneous satisfaction of these two conditions is incompatible with the uncertainty principle.

4.5. In a Stern-Gerlach experiment, one prepares a beam of particles with rather well-defined momentum (e.g., in the z-direction) and imposes an inhomogeneous magnetic field (in the y-direction). The beam initially consists of equal numbers of atoms with magnetic moments along $\pm y$. The inhomogeneous magnetic field causes the force

$$F_y = \pm\mu \frac{\partial B_y}{\partial y}$$

which deflects the partial beams with magnetic moment $\mu > 0$ ($\mu < 0$) up (down), respectively (Figure 4.1).

By measuring the beam deflection, one can determine the magnetic moment. Using the uncertainty principle, show that it is impossible to measure the spin magnetic moment of the free electron by this method. (It has been measured very precisely by other methods.)

4.6. As mentioned in Problem 2.5 in Chapter 2, for any Hermitian operator A and any state $|\psi\rangle$ of unit norm, it can be shown that

$$A|\psi\rangle = \langle A\rangle|\psi\rangle + \Delta A|\psi_1\rangle \tag{1}$$

where $\langle\psi|\psi_1\rangle = 0$. Thus, for any two Hermitian operators A, B,

$$A|\psi\rangle = \langle A\rangle|\psi\rangle + \Delta A|\psi_{1A}\rangle \tag{2}$$

and

$$B|\psi\rangle = \langle B\rangle|\psi\rangle + \Delta B|\psi_{1B}\rangle \tag{3}$$

where $\langle\psi|\psi_{1A}\rangle = \langle\psi|\psi_{1B}\rangle = 0$.

(a) Show that

$$\langle[A,B]\rangle = 2i\Delta A\Delta B\,\mathrm{Im}\langle\psi_{1A}|\psi_{1B}\rangle \tag{4}$$

and

$$\langle\{A,B\}\rangle = 2\langle A\rangle\langle B\rangle + 2\Delta A\Delta B\,\mathrm{Re}\langle\psi_{1A}|\psi_{1B}\rangle \tag{5}$$

(b) Show that (4) yields the uncertainty principle for A, B and that (5) yields the inequality

$$\Delta A\Delta B \geq \left|\frac{1}{2}\langle\{A,B\}\rangle - \langle A\rangle\langle B\rangle\right| \tag{6}$$

4.7. (a) Consider a physical system in state $|\psi(0)\rangle$ at time $t = 0$ and for which the energy E is not exact but has some constant indeterminacy ΔE. How much time T is required for the system to evolve from $|\psi(0)\rangle$ to a state $|\psi(T)\rangle$ such that $|\psi(T)\rangle$ and $|\psi(0)\rangle$ are orthogonal? From the time-energy uncertainty relation, it is easy to guess the approximate answer: $T \sim (\hbar/\Delta E)$. In this problem, use the relation that has already been mentioned in Problem 4.6 and in Problem 2.5, namely,

$$A|\psi\rangle = \langle A\rangle|\psi\rangle + \Delta A|\psi_1\rangle \quad \text{with } \langle\psi|\psi_1\rangle = 0$$

to show that

$$T \geq \frac{\pi\hbar}{2\Delta E} \tag{1}$$

You may find it useful to start with consideration of $d/dt\left|\langle\psi(0)|\psi(t)\rangle\right|^2$.

(b) Show that for a spin magnetic moment precessing in the xy-plane about a uniform magnetic field in the z-direction,

$$T = \frac{\pi\hbar}{2\Delta E}$$

4.8. In quantum mechanics, it is sometimes useful and convenient to employ the Wigner distribution function, also known as the *phase-space distribution* (PSD) *function*. Let $\psi(x,t)$ be a coordinate-space wave function that satisfies the time-dependent Schroedinger equation; that is,

$$-\frac{1}{2m}\frac{\partial^2\psi}{\partial x^2} + V(x)\psi = i\frac{\partial\psi}{\partial t} \tag{1}$$

Here we use one space dimension for simplicity and choose units where $\hbar = 1$. The corresponding PSD function is defined as

$$W(x,p,t) = \frac{1}{\pi}\int \psi^*(x+y,t)\psi(x-y,t)e^{2ipy}\,dy \tag{2}$$

Here x and p are both c-numbers, but they can still be interpreted as coordinate and conjugate momentum, respectively, without violating the uncertainty principle.
(a) Show that the probability density in coordinate space is

$$\rho(x,t) = |\psi(x,t)|^2 = \int W(x,p,t)\,dp \tag{3}$$

(b) Show that the probability density in momentum space is

$$\sigma(p,t) = |\phi(p,t)|^2 = \int W(x,p,t)\,dx \tag{4}$$

Here $\phi(p,t)$ is the momentum wave function corresponding to $\psi(x,t)$.
(c) Let $W_1(x,p,t)$, $W_2(x,p,t)$ be two PSD functions corresponding to two independent wave functions $\psi_1(x,t)$, $\psi_2(x,t)$, respectively. Show that

$$|\langle\psi_1|\psi_2\rangle|^2 = 2\pi\iint W_1^*(x,p,t)W_2(x,p,t)\,dxdp \tag{5}$$

4.9. This problem concerns the density operator

$$\rho = \sum_i g_i |u_i\rangle\langle u_i|$$

The density matrix

$$M_{jk} = \sum_i g_i \langle\phi_j|u_i\rangle\langle u_i|\phi_k\rangle$$

represents ρ with respect to a particular orthonormal basis $\{|\phi\rangle\}$. Often it is useful to choose for that basis the eigenvectors of the coordinate operator, in which case the density matrix for a single spatial dimension becomes

$$M(x,y) = \sum_i g_i \langle x|u_i\rangle\langle u_i|y\rangle = \sum_i g_i \psi_i(x)\psi_i^*(y) \tag{1}$$

where ψ is a coordinate wave function. Sometimes we encounter a continuous probability distribution $g(a)$ instead of discrete probabilities g_i. Then the sum in (1) is replaced by an integral over the continuous parameter a; that is,

$$M(x,y) = \int g(a)\psi(x,a)\psi^*(y,a)\,da \tag{2}$$

In this problem, we consider a particular example of (2) that can be worked out exactly. Let us assume that the wave function $\psi(x,a)$ is a Gaussian of fixed width displaced from the origin by amount a; that is,

$$\psi(x,a) = \frac{1}{(2\pi s^2)^{1/4}} \exp\left[-\frac{(x-a)^2}{4s^2}\right] \tag{3}$$

and that the displacement a is described by a Gaussian probability distribution

$$g(a) = \frac{1}{\sqrt{2\pi\sigma^2}} \exp\left[-\frac{a^2}{2\sigma^2}\right] \tag{4}$$

where s and σ are positive real constants. Show that

$$M(x,y) = \frac{1}{\sqrt{2\pi s^2}} \frac{1-z}{1+z} \exp\left\{-\left[\frac{(1+z^2)(X^2+Y^2)-4zXY}{2(1-z^2)}\right]\right\} \tag{5}$$

where

$$x = 2^{1/2} s\sqrt{\frac{1+z}{1-z}}X \qquad y = 2^{1/2} s\sqrt{\frac{1+z}{1-z}}Y$$

and

$$\frac{\sigma^2}{2s^2} = \frac{2z}{(1-z)^2}$$

Here z with $0 \le z < 1$ is a convenient parameter that characterizes the magnitude of σ/s.

5 Further Illustrations of the Rules of Quantum Mechanics

5.1 The neutron interferometer

For our next illustration of the rules, we consider experiments that have been performed with an interferometer and collimated beams of free neutrons generated in a reactor. A typical neutron kinetic energy in such a beam is of order 0.03 eV, which corresponds to a velocity $v \approx 3.7 \times 10^5$ cm/s and a de Broglie wavelength $\lambda \approx 10^{-8}$ cm. The interferometer, shown in Figure 5.1, is fabricated from a single crystal of pure silicon, originally in the form of a right circular cylinder about 10 cm long and 8 cm in diameter. Portions of the cylinder are cut away to leave three "wings" A, B, and C. The silicon atoms in one wing, though separated from those in the adjacent wing by several centimeters, form part of an essentially perfect lattice that includes all three wings. The interatomic lattice spacing is $a \approx 10^{-8}$ cm, comparable with the neutron de Broglie wavelength, and the lattice is approximately 10^9 atoms in length.

The neutron beam, with collimated cross-sectional area ≈ 1 cm^2, strikes the interferometer as shown in Figure 5.2. One finds experimentally that for a given incident neutron momentum, two beams (I and II) emerge from wing A if and only if the angle of incidence $\theta \approx 0.35$ rad is chosen correctly within the narrow range $\delta\theta \approx 5 \times 10^{-6}$ rad. The existence of beams I and II could be demonstrated by placing detectors at points F and G. Moreover, the waves associated with beams I and II are coherent, with definite relative phase, a fact that can be demonstrated by bringing the two beams I'' and II' together to interfere in wing C so that interference effects are observed at detector D. We emphasize that the coherence mentioned here pertains to interference between different portions of an individual neutron wave packet. The wave packets of distinct neutrons are, of course, mutually incoherent.

To understand how the interferometer works, we first consider scattering of a neutron by a single silicon nucleus (Figure 5.3).

Although the incident neutron de Broglie wave is really a packet, it is described well enough for present purposes by a plane wave. The silicon nucleus is a localized short-range force center, and the scattered neutron wave is spherically symmetric. However, there are many silicon nuclei in the perfect rigid lattice, and the scattered spherical wave from each of these has a definite relative phase. These waves therefore interfere very much as electromagnetic waves do when scattered from individual rulings of a diffraction grating. In fact, the silicon crystal is a three-dimensional diffraction grating for the neutron wave. It is not difficult to show that the scattered waves interfere constructively only when the *Bragg condition*, familiar from x-ray diffraction, is satisfied; that is,

$$\sin\theta = \frac{\lambda}{2a} \tag{5.1}$$

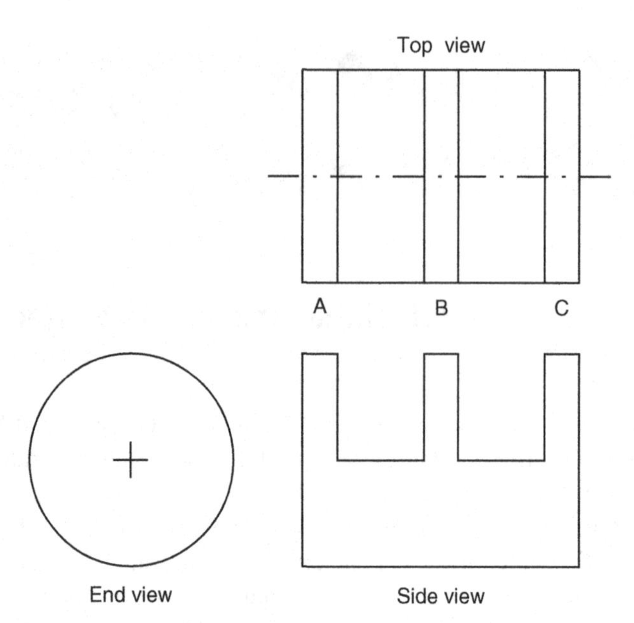

Figure 5.1 Sketch of the single-crystal silicon neutron interferometer.

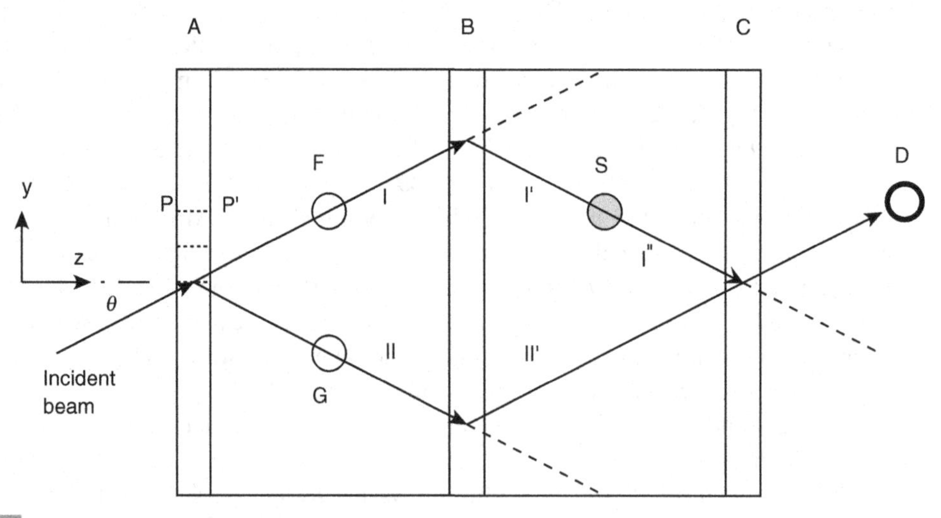

Figure 5.2 Top view of neutron beams in the neutron interferometer.

(In the present case, the relevant scattering planes *PP'* are as indicated in Figure 5.2, and the scattering is actually of the Laue type.)

To analyze the situation simply but in a way that retains the essential features, we describe the incident neutron wave by the expression

$$\psi_i = \chi \exp\left(ik_y y + ik_z z\right) \tag{5.2}$$

Here χ is a two-component spinor describing the neutron spin, we have ignored a time-dependent factor and an overall multiplicative constant $k_y / k_z = \tan\theta$, and the y origin of

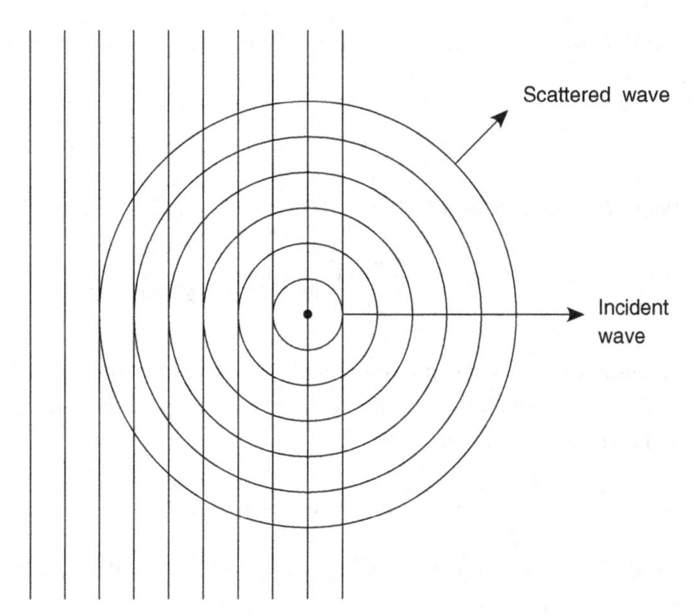

Figure 5.3 Scattering of a plane neutron wave by a silicon nucleus.

coordinates is chosen to coincide with a crystal plane. We follow the consequences of (5.2) by writing

$$\exp(ik_y y) = \cos\left(k_y y\right) + i\sin\left(k_y y\right)$$

In this expression, the cosine standing wave has a large amplitude and the sine standing wave vanishes at each crystal plane. The effect of the Laue scattering is thus to shift the relative phase of these two partial waves as the net wave propagates through the crystal. Therefore, at the exit of wing A we have

$$\begin{aligned}\psi &= \chi \exp(ik_z z)\left[cos(k_y y+\beta)+i\sin k_y y\right]\\ &= \chi\frac{\exp(ik_z z)}{2}\left[\left(e^{i\beta}+1\right)e^{ik_y y}+\left(e^{-i\beta}-1\right)e^{-ik_y y}\right]\\ &= \psi_I + \psi_{II}\end{aligned} \tag{5.3}$$

where

$$\psi_I = \frac{\chi}{2}\left(e^{i\beta}+1\right)e^{ik_z z}e^{ik_y y} = \chi u e^{ik_z z}e^{ik_y y} \tag{5.4}$$

$$\psi_{II} = \frac{\chi}{2}\left(e^{-i\beta}-1\right)e^{ik_z z}e^{-ik_y y} = \chi v e^{ik_z z}e^{-ik_y y} \tag{5.5}$$

and where β is the phase shift – a real number. Thus we obtain two coherent waves $\psi_{I,II}$ that propagate in the directions indicated in Figure 5.2. A similar analysis occurs for each of the beams I and II at wing B. Beam I splits into two coherent parts, of which one is beam I':

$$\psi_{I'} = \chi vu e^{ik_z z} e^{-ik_y y} \tag{5.6}$$

whereas beam II also divides into two parts, one of which is beam II':

$$\psi_{II'} = \chi \frac{\left(e^{i\beta} - 1\right)}{2} v e^{ik_z z} e^{ik_y y} = \chi w v e^{ik_z z} e^{ik_y y} \tag{5.7}$$

Between wings B and C we may introduce a phase shifter S in the path of beam I' (see Figure 5.2). The effect of S is represented quite generally by a multiplicative factor $a \cdot e^{i\delta}$, where a and δ are real numbers with $0 \le a \le 1$; that is,

$$\psi_{I'} \to \psi_{I''} = a \cdot e^{i\delta} \psi_{I'} \tag{5.8}$$

Beams I'' and II' now undergo Laue scattering in wing C, and each produces a component that enters detector D. The contribution from I'' at D is

$$\psi_{I''}(D) = \chi w v u a \cdot e^{i\delta} e^{ik_y y} e^{ik_z z}$$

whereas the contribution from beam II' at D is

$$\psi_{II'}(D) = \chi u w v \cdot e^{ik_y y} e^{ik_z z}$$

Thus the net wave function at D is

$$\psi(D) = \chi w v u \cdot e^{ik_y y} e^{ik_z z} \left(1 + a \cdot e^{i\delta}\right) \tag{5.9}$$

The counting rate of detector D is proportional to the probability density $|\psi(D)|^2$ for the neutron at D and is thus proportional to

$$\mathcal{P} = \left|1 + a \cdot e^{i\delta}\right|^2 = 1 + a^2 + 2a \cdot \cos\delta \tag{5.10}$$

$\mathcal{P}$ is analogous to the relative intensity as a function of relative optical path length in a Mach-Zehnder interferometer in classical optics. For example, if $a = 1$ (no absorption at S but only a phase shift), $\mathcal{P}$ varies from 4 to 0 as δ varies from 0 to π.

When S is a magnet, the situation is particularly interesting. The magnetic field strength can be chosen so that a neutron spin (with spin magnetic moment equal to –1.91 nuclear Bohr magnetons) precesses through an angle of 2π in S. As shown in Section 3.4, this implies that χ is replaced by $-\chi$ in beam I''; that is, $a = 1$ and $\delta = \pi$. Hence we have destructive interference between beams I'' and II' resulting in $\mathcal{P} = 0$. Observation of this effect constitutes a good experimental demonstration of the change of sign in the spin function associated with a 2π rotation of spin-½ spinors (Werner et al. 1975).

Another interesting way to shift the relative phase of beams I' and II' is illustrated in Figure 5.4; it also has been observed. Here the beams I, I', II, and II' are arranged in a vertical plane with I' and II horizontal, and the difference in height between them equal to y.

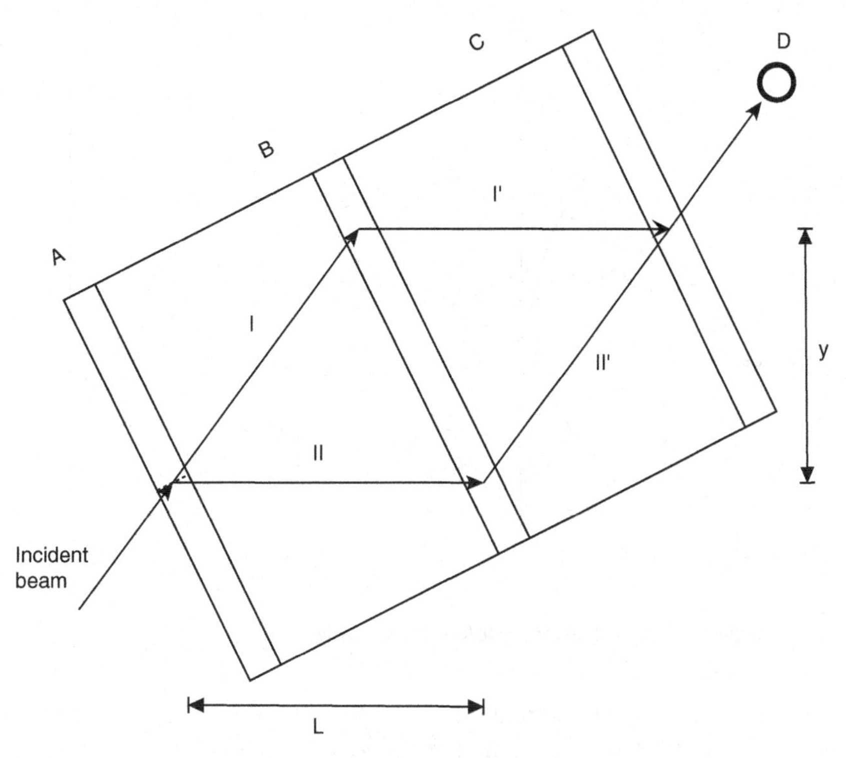

Figure 5.4 Neutron interferometer oriented for detection of gravitationally induced phase shift.

Schroedinger's wave equation for a particle in an external scalar potential that depends only on the time is

$$-\frac{\hbar^2}{2m}\nabla^2\psi + V(t)\psi = i\hbar\dot{\psi} \tag{5.11}$$

Let $\psi(\boldsymbol{r},t) = f(\boldsymbol{r})F(t)$. Then $\dot{F}/F = (1/i\hbar)V(t) + \text{const}$, so

$$F(t) = F(0)\exp\left[-\frac{i}{\hbar}\int_0^t V(t)\,dt\right] \tag{5.12}$$

In the present example, $V = m_g gy$, where m_g is the neutron gravitational mass, and g is the acceleration of gravity at Earth's surface. Note that in (5.11), m is the neutron inertial mass. Of course, according to the principle of equivalence, $m = m_g$. The phase difference between beams I' and II' observed at detector D is $\Delta\phi = -(m_g gyL/\hbar v)$, where v is the neutron velocity.

5.2 Aharonov-Bohm effect

This quantum-mechanical effect, discussed by Ehrenberg and Siday (1949), Aharonov and Bohm (1959), and Furry and Ramsey (1960) and first observed by Chambers (1960), is an

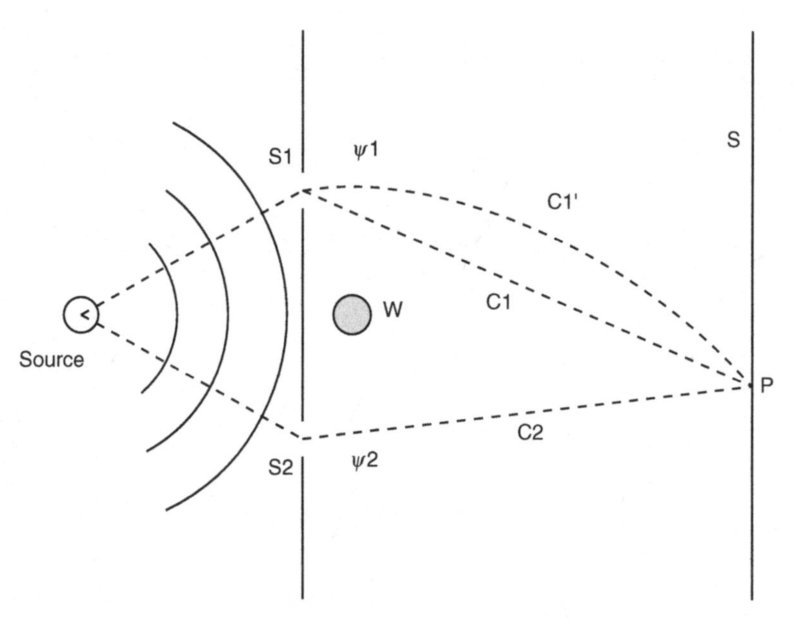

Figure 5.5 Schematic diagram of apparatus for observing the Aharonov-Bohm effect.

interesting counterintuitive manifestation of the electromagnetic vector potential $\boldsymbol{A}$. Figure 5.5 illustrates a setup for observation of the effect.

It includes a source of electrons and a pair of slits S_1, S_2 that split each electron wave packet into two coherent portions ψ_1, ψ_2 that meet and interfere at screen S. In the region between, there is a long thin magnetized whisker (labeled W) that acts like an ideal solenoid, providing a confined magnetic induction field $\boldsymbol{B}$ perpendicular to the page, with total flux Φ. Inside the whisker, ψ_1 and ψ_2 are negligible, and wherever ψ_1 and ψ_2 are significantly different from zero, $\boldsymbol{B}$ is negligible. Nevertheless, the interference fringes shift whenever $\boldsymbol{B}$ is changed, revealing a phase difference between ψ_1 and ψ_2 equal to $-e\Phi/\hbar c$. This happens even though the electron does not encounter $\boldsymbol{B}$ directly and there is no Lorentz force. How can we explain the effect?

We begin by recalling the time-dependent Schroedinger wave equation for an electron with charge $-e$ in an external vector potential $\boldsymbol{A}$ and where the scalar potential vanishes; that is,

$$\frac{1}{2m}\left(-i\hbar\nabla+\frac{e}{c}\boldsymbol{A}\right)^2\psi=i\hbar\dot{\psi} \tag{5.13}$$

Let us assume that when $\boldsymbol{A}=0$, the solution ψ_0 to (5.13) satisfying fixed boundary conditions is known. We now show that if $\boldsymbol{A}$ is independent of the time, the solution to (5.13) satisfying the same boundary conditions but including $\boldsymbol{A}$ is

$$\psi(\boldsymbol{r},t)=\psi_0(\boldsymbol{r},t)\exp\left(\frac{-ie}{\hbar c}\int_{r_0}^{r}\boldsymbol{A}\cdot d\vec{\ell}\right) \tag{5.14}$$

where $\int_{r_0}^{r}\boldsymbol{A}\cdot d\vec{\ell}$ is a line integral from some fixed point $\boldsymbol{r}_0$ to the point $\boldsymbol{r}$ of interest.

To demonstrate (5.14), we first consider the expression

$$\left(-i\hbar\nabla+\frac{e}{c}A(r)\right)\psi_0\exp\left[\frac{-ie}{\hbar c}\int_{r_0}^{r}A\bullet d\vec{\ell}\right]=\left(-i\hbar\nabla\psi_0+i\hbar\frac{ie}{\hbar c}\psi_0 A+\frac{e}{c}\psi_0 A\right)\exp\left[\frac{-ie}{\hbar c}\int_{r_0}^{r}A\bullet d\vec{\ell}\right]$$
$$=(-i\hbar\nabla\psi_0)\exp\left[\frac{-ie}{\hbar c}\int_{r_0}^{r}A\bullet d\vec{\ell}\right] \tag{5.15}$$

Note that on the right-hand side of the first line there are two terms proportional to A that cancel one another. We now apply the operator $(1/2m)\left[-i\hbar\nabla+(e/c)A(r)\right]$ to the last line of (5.15). In the same way as in (5.15), this yields

$$\frac{1}{2m}\left[-i\hbar\nabla+\frac{e}{c}A(r)\right]^2\psi=-\frac{\hbar^2}{2m}(\nabla^2\psi_0)\exp\left(\frac{-ie}{\hbar c}\int_{r_0}^{r}A\bullet d\vec{\ell}\right)$$
$$=i\hbar\dot{\psi}_0\exp\left(\frac{-ie}{\hbar c}\int_{r_0}^{r}A\bullet d\vec{\ell}\right)$$

which completes the proof.

We now apply (5.14) to the wave packets $\psi_{1,2}$ in Figure 5.5. When there is no $\boldsymbol{B}$ field, let these wave packets be ψ_{10},ψ_{20}, respectively. Then, when $\boldsymbol{B}$ is present, we have at point P on the screen

$$\psi_1=\psi_{10}\exp\left(\frac{-ie}{\hbar c}\int_{C1}A\bullet d\vec{\ell}\right) \tag{5.16}$$

$$\psi_2=\psi_{20}\exp\left(\frac{-ie}{\hbar c}\int_{C2}A\bullet d\vec{\ell}\right) \tag{5.17}$$

where $C_{1,2}$ are the paths of integration shown in Figure 5.5. Thus

$$\psi_1+\psi_2=\exp\left(\frac{-ie}{\hbar c}\int_{C1}A\bullet d\vec{\ell}\right)\left[\psi_{10}+\psi_{20}\exp\left(\frac{-ie}{\hbar c}\oint A\bullet d\vec{\ell}\right)\right] \tag{5.18}$$

where the loop integral indicates integration around the closed path ($C2$–$C1$). Now, from Stokes' theorem,

$$\oint A\bullet d\vec{\ell}=\int\nabla\times A\bullet\vec{n}\,dS=\int \boldsymbol{B}\bullet\hat{n}\,dS=\Phi$$

is the magnetic flux enclosed by the loop ($C2$–$C1$). Thus the intensity at the screen is proportional to

$$|\psi_1+\psi_2|^2=\left|\psi_{10}+\psi_{20}\exp\left(\frac{-ie}{\hbar c}\Phi\right)\right|^2 \tag{5.19}$$

Obviously, the interference term in this expression depends on Φ, and the phase shift is

$$\delta = \frac{-e\Phi}{\hbar c} \tag{5.20}$$

We note that Φ does not depend on the precise choice of path. For example, in Figure 5.5 we could have chosen any path $C1'$ instead of $C1$ so long as the loop integral around $C1$–$C1'$ contains no flux. Also, the quantity $\oint \boldsymbol{A} \cdot d\vec{\ell} = \Phi$ is independent of the choice of gauge for $\boldsymbol{A}$ because under a gauge transformation, $\boldsymbol{A} \to \boldsymbol{A}' = \boldsymbol{A} - \nabla\chi$, where χ is some scalar function. However, the line integral of the gradient of any scalar function around a closed path is zero.

5.3 A digression on magnetic monopoles

In ordinary electrodynamics, there are free electric charges (electric monopoles); however, although there is no known prohibition against free magnetic charges (magnetic monopoles), there is no experimental evidence for their existence. Despite this lack of evidence, interest in magnetic monopoles persists. In this section we want to discuss some of the properties of magnetic monopoles that follow from very simple classical considerations and then show how their possible existence is connected to the Aharonov-Bohm effect.

To begin, consider a magnetic monopole g and an electric charge e separated by a distance a (Figure 5.6). We now show, as was first done by J. J. Thomson in 1900, that the angular momentum in the field generated by g and e is directed along the line from e to g, and the magnitude of this angular momentum is independent of the magnitude of a.

In Heaviside-Lorentz units, the electric field at point of observation Q is $\boldsymbol{\mathcal{E}} = e\boldsymbol{r}/4\pi r^3$, and the magnetic field at Q due to g is $\boldsymbol{B} = g\boldsymbol{R}/4\pi R^3$. The momentum density in the electromagnetic field at Q is

$$\boldsymbol{P} = \frac{1}{c}\boldsymbol{\mathcal{E}} \times \boldsymbol{B} = \frac{1}{(4\pi)^2} \frac{eg}{c\, r^3 R^3} \boldsymbol{r} \times \boldsymbol{R}$$

However, $\boldsymbol{R} = \boldsymbol{r} - \boldsymbol{a}$; thus

$$\boldsymbol{P} = -\frac{1}{(4\pi)^2} \frac{eg}{c\, r^3 R^3} \boldsymbol{r} \times \boldsymbol{a} \tag{5.21}$$

The angular momentum density in the field is

$$\begin{aligned} \boldsymbol{j} = \boldsymbol{r} \times \boldsymbol{P} &= -\frac{1}{(4\pi)^2} \frac{eg}{c\, r^3 R^3} \boldsymbol{r} \times (\boldsymbol{r} \times \boldsymbol{a}) \\ &= \frac{1}{(4\pi)^2} \frac{eg}{c\, r^3 R^3} \left[r^2 \boldsymbol{a} - (\boldsymbol{r} \cdot \boldsymbol{a}) \boldsymbol{r} \right] \end{aligned} \tag{5.22}$$

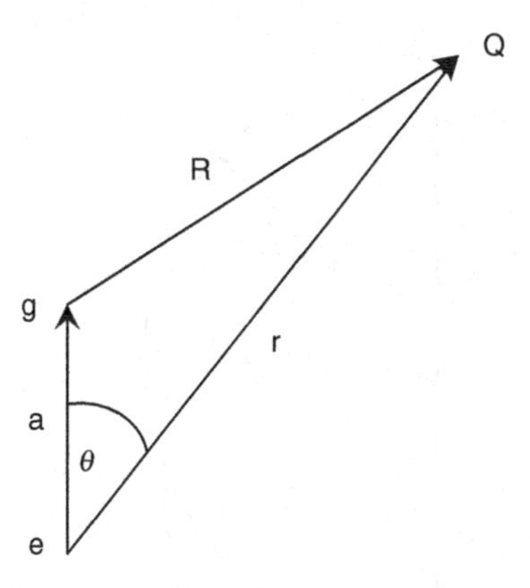

Figure 5.6 Vectors for describing the field of the monopole-charge pair

Let us denote the total angular momentum of the field by $\boldsymbol{S}$; it is usually called the *spin* of the *eg* pair. Because by symmetry only the component of $\boldsymbol{j}$ parallel to $\boldsymbol{a}$ makes a contribution to $\boldsymbol{S}$, we have

$$\begin{aligned} |\boldsymbol{S}| &= \int \boldsymbol{j}\cdot\hat{\boldsymbol{a}}\,d\tau = \frac{eg}{8\pi c}\int \frac{a}{R^3} r\,dr \sin^3\theta\, d\theta \\ &= \int_0^\infty r\,dr \int_0^\pi \frac{a\sin^3\theta\, d\theta}{\left(r^2+a^2-2ar\cos\theta\right)^{3/2}} \end{aligned} \tag{5.23}$$

Thus $\boldsymbol{S}$ is not only parallel to $\boldsymbol{a}$ but, as can easily be verified by making the substitution $x = r/a$ in the integrand, $|\boldsymbol{S}|$ is independent of a. Evaluation of the integral involves some effort, but we can avoid this by making a very elementary calculation to determine the magnitude of $\boldsymbol{S}$ (Figure 5.7).

In Figure 5.7, consider two electric charges $e/2$ on opposite sides of g. By symmetry, the angular momentum in the field (the spin of the $e/2{:}g{:}e/2$ triplet) is zero. Now, move the upper electric charge $e/2$ on the indicated semicircular arc to the lower charge in time T at constant speed $v = \pi R/T$, where R is the arc radius. The Lorentz force on the moving charge in the magnetic field due to g is out of the page, and it has the magnitude

$$F = \frac{e}{2}\frac{v}{c}B = \frac{e}{2c}\frac{\pi R}{T}\frac{g}{4\pi R^2}$$

To keep the moving charge in the plane of the page, we must exert an equal and opposite force $F' = -F$. The torque about the vertical axis due to F' is directed upward, and it has magnitude

$$\tau = R\sin\phi\,|F'| = \frac{e}{8c}\frac{g}{T}\sin\phi$$

We integrate this torque to find the total change in the angular momentum; that is,

$$\Delta S = \int_0^T \tau\, dt = \frac{eg}{8\pi c}\int_0^\pi \frac{\pi}{T}\frac{R}{v}\sin\phi\, d\phi = \frac{eg}{4\pi c}$$

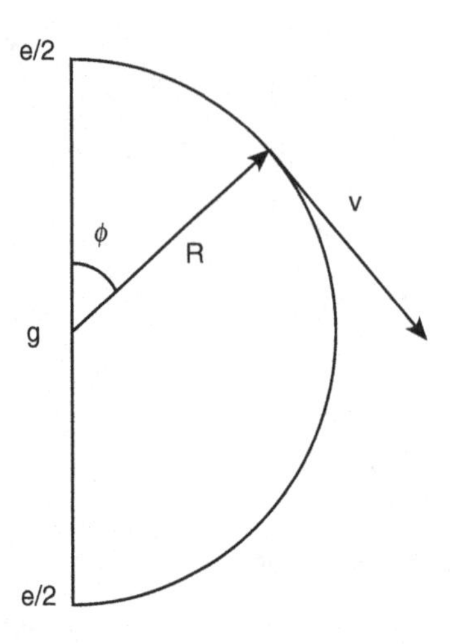

Figure 5.7 Semi-circular path of charge *e*/2

Thus the final angular momentum is directed from the two superposed charges *e*/2 to *g*, and it has magnitude

$$S = \frac{eg}{4\pi c} \tag{5.24}$$

Because spin is quantized and the smallest nonzero value of spin is $\hbar/2$, we might guess that

$$\frac{eg}{c} = 2\pi\hbar$$

and therefore that

$$g = \frac{2\pi\hbar c}{e} = \frac{e}{2\alpha} \tag{5.25}$$

where

$$\alpha = \frac{e^2}{4\pi\hbar c} = \frac{1}{137.036}$$

is the dimensionless fine-structure constant. This is an interesting conclusion, but so far it is just based on a guess, and we should find a more convincing argument. For this purpose, consider a hypothetical Aharonov-Bohm experiment illustrated in Figure 5.8. Two electron packets originate at P_0 and meet again to interfere at P, which is diametrically opposed to P_0. The paths $C1$ and $C2$ are semicircular arcs that together define the equator of a sphere. At the center of the sphere we place a magnetic monopole g. From the symmetry of this arrangement, it is obvious that there cannot be any observable phase shift $\boldsymbol{\delta}$ due to the presence of g. Nevertheless, it is instructive to calculate (in spherical polar coordinates) the vector potential associated with the magnetic field $\boldsymbol{B} = g\boldsymbol{R}/4\pi R^3$ of the monopole.

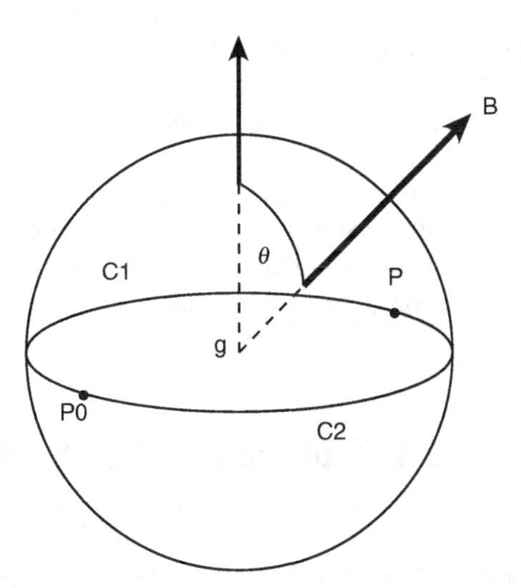

Figure 5.8 Split electron wave packet travels from P_0 to P along semi-circular paths C_1, C_2 in field of magnetic monopole g located at center of circle.

We have

$$B = \frac{g}{4\pi R^2} = (\nabla \times A)_R = \frac{1}{R\sin\theta}\left[\frac{\partial(\sin\theta A_\phi)}{\partial\theta} - \frac{\partial A_\theta}{\partial\phi}\right] \tag{5.26}$$

Symmetry about the polar axis allows us to set the second term on the right-hand side of (5.26) to zero. We then integrate (5.26) to obtain

$$A_\phi = \frac{g}{4\pi R}\frac{k - \cos\theta}{\sin\theta} \tag{5.27}$$

where k is a constant of integration. Thus, in the equatorial plane, $A_\phi = gk/4\pi R$, and the line integral $\oint A \cdot d\vec{\ell}$ taken once around the equator (starting and ending, for example, at P_0 in Figure 5.8) must be the total flux through the loop. Because the total flux emanating from g is g, half of which is in the northern and half in the southern hemisphere, k must be ± 1. If we choose $k = -1$, A_φ has a singularity at $\theta = 0$, whereas if we choose $k = +1$, A_φ has a singularity at $\theta = \pi$. The only way to avoid a singularity is to "patch" two different vector potentials together, for example, as follows:

$$\theta \le \frac{\pi}{2}: \quad A_1 = \frac{g}{4\pi R}\frac{1 - \cos\theta}{\sin\theta}\hat{\phi}$$

$$\theta \ge \frac{\pi}{2}: \quad A_2 = -\frac{g}{4\pi R}\frac{1 + \cos\theta}{\sin\theta}\hat{\phi}$$

At $\theta = \pi/2$, either A_1 or A_2 can be chosen, with the result that at the equatorial plane $\Phi = \pm g/2$. The resulting Aharonov-Bohm phase shift is $\delta = \pm ge/2\hbar c$. This is obviously an absurd and ambiguous result unless

$$\frac{ge}{2\hbar c} = -\frac{ge}{2\hbar c} + 2n\pi \tag{5.28}$$

where n is an integer. Equation (5.28) implies that

$$g = \frac{2\pi n\hbar c}{e} = \frac{ne}{2\alpha} \tag{5.29}$$

For $n = 1$, (5.29) is the same result as (5.25). The foregoing considerations show how possible values of g and e are not independent. In fact, the existence of just one magnetic monopole in the universe would imply that electric charge is quantized (as all electric charges appear to be).

5.4 Neutrino mixing and oscillations

According to present understanding, all matter (with the possible exception of dark matter) consists of two fundamental types of spin-½ objects – leptons and quarks. The known leptons fall into three families, also known as *generations* or *flavors*:

$$\begin{pmatrix} \nu_e \\ e^- \end{pmatrix} \quad \begin{pmatrix} \nu_\mu \\ \mu^- \end{pmatrix} \quad \begin{pmatrix} \nu_\tau \\ \tau^- \end{pmatrix} \tag{5.30}$$

Each generation consists of a neutrino (ν_e, ν_μ, ν_τ) and a charged lepton (e^-, μ^-, τ^-). Although the charged leptons experience gravitational, weak, and electromagnetic interactions, the neutrinos evidently participate only in gravitational and weak interactions. For each lepton, there is a corresponding antilepton; for example, corresponding to the electron, there is the positron, and corresponding to the neutrino ν_e, there is an antineutrino $\bar{\nu}_e$.

What experimental facts distinguish one neutrino flavor from another? It is found experimentally that when a charged pion decays by weak interaction to a muon and a neutrino, that is,

$$\pi^+ \to \mu^+ \nu_\mu \tag{5.31}$$

the neutrino so produced (named ν_μ) can induce the reaction

$$\nu + n \to p + \mu^- \tag{5.32}$$

but not the analogous reactions

$$\nu + n \to p + e^- \tag{5.33}$$

and

$$\nu + n \to p + \tau^- \tag{5.34}$$

Meanwhile, neutrinos from the decay $\pi^+ \to e^+ \nu_e$ are called ν_e because they can induce reaction (5.33) but not (5.32) or (5.34). These and similar findings are summarized by the empirical law of lepton number conservation as follows: we define a lepton number L_e to be $L_e = +1$ for

e^- or ν_e, $L_e = -1$ for e^+ or $\bar{\nu}_e$, and $L_e = 0$ for all other particles. We also define lepton numbers L_μ and L_τ in an analogous way for the second and third generations, respectively. Before 1998, it appeared to be an experimental fact with no known exceptions that each lepton number is conserved in all reactions. For example, the positive muon decays as follows:

$$\mu^+ \to e^+ \nu_e \bar{\nu}_\mu$$

Here $L_\mu = -1$, $L_e = 0$ for both the initial and final states, so both numbers are conserved in the decay. By contrast, although the decay

$$\mu^+ \to e^+ \gamma$$

would be consistent with conservation of energy, linear momentum, and angular momentum, this decay does not occur (the branching ratio is less than 4×10^{-11}). Here neither L_e nor L_μ is conserved.

The following question arises concerning neutrinos: Do they have mass, or are they, like photons, strictly massless particles that travel with the speed of light? And there is a related question that arises as follows: the neutrino states that participate in weak interactions by which neutrinos are created or absorbed are called *weak-interaction eigenstates* and are denoted by the symbols

$$|\nu_e\rangle \quad |\nu_\mu\rangle \quad |\nu_\tau\rangle \quad |\bar{\nu}_e\rangle \quad |\bar{\nu}_\mu\rangle \quad |\bar{\nu}_\tau\rangle$$

and each such state is associated with definite values of L_e, L_μ, and L_τ. However, if neutrinos have nonzero mass, *are the weak-interaction eigenstates associated with definite mass, or is it possible that a state associated with definite mass is some linear combination of weak-interaction eigenstates?* (Roughly speaking, this is analogous to the following: we can speak about photon linear polarization eigenstates $|x\rangle, |y\rangle$, but neither is an eigenstate of angular momentum in the z-direction. The latter are the helicity eigenstates $|+\rangle, |-\rangle$, each of which is a linear combination of $|x\rangle$ and $|y\rangle$.)

What is the motivation for asking the italicized question in the last paragraph? We do not go into details here, but we assure the reader that there are very good reasons for asking the question, not the least of which is that an analogous "mixing" effect is known to occur in the quark sector. Here, in order to illustrate the fundamental rules of quantum mechanics, we only discuss the consequence of assuming that such neutrino mixing does occur. (In fact, the experimental evidence since 1998 for neutrino mixing is very convincing.) For simplicity, we confine ourselves to just two lepton generations and assume that there are two distinct neutrino states, denoted by $|\nu_1\rangle, |\nu_2\rangle$, associated with definite and distinct masses m_1 and m_2, respectively; that each such state is a linear combination of the weak-interaction eigenstates $|\nu_e\rangle$ and $|\nu_\mu\rangle$; and that these two linear combinations are mutually orthogonal. Clearly, we are once again dealing with a two-dimensional Hilbert space, the basis for which can be $|\nu_1\rangle, |\nu_2\rangle$ or, alternatively, $|\nu_e\rangle$, $|\nu_\mu\rangle$. Assuming that each of these states has unit norm, there is some unitary transformation that takes us from one basis to the other, and this is described quite generally by the equations

$$\begin{aligned}|\nu_1\rangle &= \cos\theta|\nu_e\rangle + \sin\theta|\nu_\mu\rangle \\ |\nu_2\rangle &= -\sin\theta|\nu_e\rangle + \cos\theta|\nu_\mu\rangle\end{aligned} \tag{5.35}$$

where θ is the mixing angle. It is convenient to represent the states $|\nu_1\rangle, |\nu_2\rangle$ by the spinors $\begin{pmatrix}1\\0\end{pmatrix}, \begin{pmatrix}0\\1\end{pmatrix}$, respectively. Then an arbitrary linear combination of these states

$a|\nu_1\rangle + b|\nu_2\rangle$ is represented by the spinor $\psi = \begin{pmatrix}a\\b\end{pmatrix}$. Transformation from the basis $|\nu_1\rangle, |\nu_2\rangle$ to the basis $|\nu_e\rangle, |\nu_\mu\rangle$ is accomplished by the unitary matrix

$$U = \begin{pmatrix}\cos\theta & -\sin\theta \\ \sin\theta & \cos\theta\end{pmatrix} \tag{5.36}$$

Thus $\psi' = U\psi$, where ψ' is expressed in terms of the $|\nu_e\rangle, |\nu_\mu\rangle$ basis.

Now suppose that a neutrino has well-defined linear momentum $\boldsymbol{p}$. Then, according to the well-known formula of special relativity, a neutrino state of definite mass m must have definite energy

$$E = \sqrt{m^2c^4 + \boldsymbol{p}^2c^2}$$

Actually, it is convenient in what follows to choose units where $\hbar = c = 1$, in which case the preceding formula becomes

$$E = \sqrt{\boldsymbol{p}^2 + m^2}$$

Furthermore, in all practical experiments with neutrinos, the first term inside the square root is always much larger than the second term (i.e., the neutrino kinetic energy always greatly exceeds its rest energy). Hence we can make the following expansions for the energies E_1, E_2 associated with states $|\nu_1\rangle, |\nu_2\rangle$, respectively:

$$\begin{aligned}E_1 &\approx |\boldsymbol{p}| + \frac{m_1^2}{2|\boldsymbol{p}|} \\ E_2 &\approx |\boldsymbol{p}| + \frac{m_2^2}{2|\boldsymbol{p}|}\end{aligned} \tag{5.37}$$

In the $|\nu_1\rangle, |\nu_2\rangle$ basis, the Hamiltonian is

$$\begin{aligned}H &= \begin{pmatrix}E_1 & 0 \\ 0 & E_2\end{pmatrix} \\ &= \frac{E_1 + E_2}{2}\begin{pmatrix}1 & 0 \\ 0 & 1\end{pmatrix} + \frac{E_1 - E_2}{2}\begin{pmatrix}1 & 0 \\ 0 & -1\end{pmatrix} \\ &= \frac{E_1 + E_2}{2} I + \omega\sigma_3\end{aligned} \tag{5.38}$$

where $\omega = (E_1 - E_2)/2 = (m_1^2 - m_2^2)/4|\boldsymbol{p}| \cong (m_1^2 - m_2^2)/4E$. From now on we ignore the first term on the right-hand side of the last line of (5.38) because it produces no observable effects. Thus $H = \omega\sigma_3$.

In the $|\nu_e\rangle, |\nu_\mu\rangle$ basis, the Hamiltonian is

$$\begin{aligned} H' = UHU^{-1} &= \omega\begin{pmatrix} \cos 2\theta & \sin 2\theta \\ \sin 2\theta & -\cos 2\theta \end{pmatrix} \\ &= \omega[\cos 2\theta\, \sigma_3 + \sin 2\theta\, \sigma_1] \\ &= \omega\hat{\boldsymbol{n}}\boldsymbol{\cdot}\boldsymbol{\sigma} \end{aligned} \tag{5.39}$$

where $\hat{\boldsymbol{n}} = \sin 2\theta\hat{i} + \cos 2\theta\hat{k}$. The time evolution of ψ' is then given by

$$\begin{aligned} \psi'(t) &= \exp[-iH't]\,\psi'(0) \\ &= e^{-i\omega\hat{\boldsymbol{n}}\boldsymbol{\cdot}\boldsymbol{\sigma}t}\,\psi'(0) \\ &= \left[I\cos\omega t - i\hat{\boldsymbol{n}}\boldsymbol{\cdot}\boldsymbol{\sigma}\sin\omega t\right]\psi'(0) \end{aligned} \tag{5.40}$$

Suppose that at time $t = 0$, the neutrino is created as ν_e. Then $\psi'(0) = \begin{pmatrix} 1 \\ 0 \end{pmatrix}$, and (5.40) becomes

$$\psi'(t) = \begin{pmatrix} \cos\omega t - i\cos 2\theta\sin\omega t \\ -i\sin 2\theta\sin\omega t \end{pmatrix} \tag{5.41}$$

Thus the transition probability from ν_e to ν_μ is

$$P(\nu_e \to \nu_\mu) = \sin^2 2\theta \sin^2\left(\frac{\Delta m^2}{4E}t\right) \tag{5.42}$$

where $\Delta m^2 = m_1^2 - m_2^2$. Because the neutrinos in all these experiments are highly relativistic, they travel a distance $L \cong ct$ in time t. However, recalling that $c = 1$ in the present units, we replace t by L to obtain

$$P(\nu_e \to \nu_\mu) = \sin^2 2\theta \sin^2\left(\frac{\Delta m^2}{4E}L\right) \tag{5.43}$$

This formula describes *neutrino oscillations*. For such oscillations to occur,

1. The mixing angle must satisfy the condition $0 < \theta < \pi/2$.
2. The masses m_1 and m_2 must be different.

As we have mentioned, the experimental evidence is now very convincing that neutrino mixing and oscillations occur. Clearly, the transformation $\nu_e \to \nu_\mu$ described by (5.43) and the general phenomenon of neutrino oscillations involving all three generations imply that the lepton numbers $L_{e,\mu,\tau}$ are not separately conserved.

Problems for Chapter 5

5.1. We show in Section 5.3 that the angular momentum $\boldsymbol{S}$ in the electromagnetic field associated with an electric charge e and a magnetic monopole g has magnitude $eg/(4\pi c)$ (independent of the distance between e and g) and is directed from the charge to the monopole. In this classical problem, which may give you some insight into the behavior of a charge-monopole pair, we consider the nonrelativistic motion of a point charge e of mass m in the magnetic field of a fixed (i.e., extremely massive) monopole g. It is convenient to place g at the origin and define the position of e as $\boldsymbol{r}(t)$.

(a) Let $\boldsymbol{L}$ be the orbital angular momentum of e with respect to the origin. Show that the speed v of e is constant, that $\boldsymbol{L} \cdot \boldsymbol{S} = 0$, and that $\boldsymbol{J} = \boldsymbol{L} + \boldsymbol{S}$ is a constant of the motion. Also show that if r_0 is the distance of the closest approach of e to the origin and we choose $t = 0$ when this distance is achieved, then

$$r^2 = r_0^2 + v^2 t^2$$

(b) Choose spherical polar coordinates with $\boldsymbol{J}$ along the polar axis and $\boldsymbol{r}$ the position vector of e. Show that

$$\cos\theta = \frac{b}{\sqrt{b^2 + v^2 r_0^2}}$$

where $b = -eg/(4\pi mc)$. Because θ is constant, the orbit of e lies on a cone with vertex at the origin, symmetry axis $\boldsymbol{J}$, and half-angle θ.

(c) At remote times in the past, when e is approaching g but still far away, choose the initial value of the azimuthal angle $\phi = 0$. Show that

$$\phi(t) = \frac{\frac{\pi}{2} + Arc\tan\frac{vt}{r_0}}{\sin\theta}$$

5.2. Figure 5.9 shows a hypothetical experiment somewhat analogous to that employed in the Aharonov-Bohm effect. A source S emits a succession of electrons. Each electron wave packet is split into two coherent parts ψ_1, ψ_2 by slits S_1, S_2. Wave packet ψ_1 (ψ_2) passes through a long conducting tube T_1 (T_2), respectively. While each packet is well inside its respective tube, a voltage difference ΔV is applied to the tubes by means of a battery and a switch for a time t. The battery is disconnected well before either packet emerges from its tube. The packets meet at detector D. How does the intensity of the resulting wave at D depend on ΔV and t?

5.3. Whereas the Aharonov-Bohm (AB) effect concerns the wave function of a charged particle in the presence of a vector potential A, the Aharonov-Casher (AC) effect (Aharonov and Casher 1984; Cimmino et al. 1989) is related to the behavior of a neutral particle with a spin magnetic moment (e.g., a neutron) that moves in an electric field. The AC effect has actually been demonstrated experimentally by means of a neutron interferometer in a setup shown schematically in Figure 5.10.

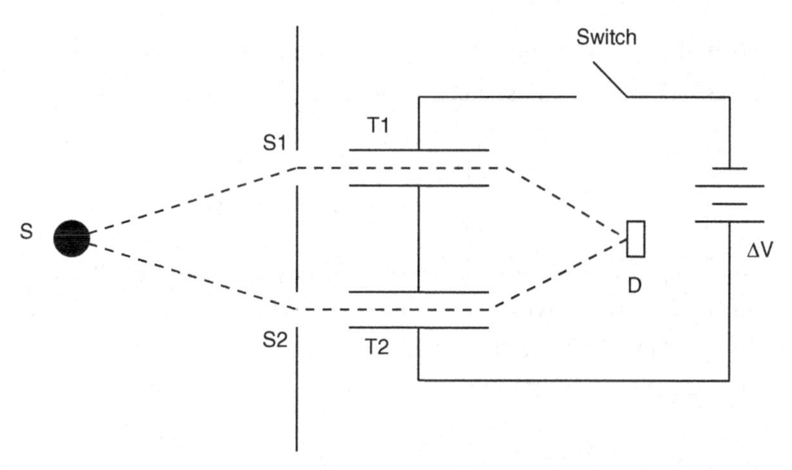

Figure 5.9 Schematic diagram of Aharonov-Bohm-like experiment described in Problem 5.2.

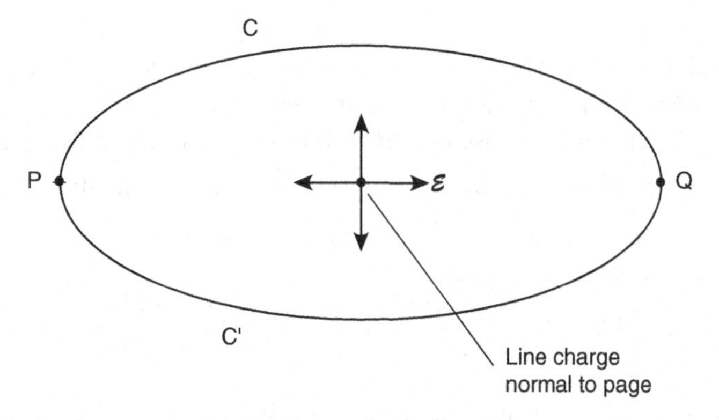

Figure 5.10 Schematic diagram of apparatus for observing Aharonov-Casher effect

A neutron wave is split into two coherent parts at P, one of which takes path C and the other C'. These parts are reunited at Q. Normal to the page is an effective line charge with charge per unit length λ that generates a cylindrically symmetric electric field $\mathcal{E}$. Find the effective non-relativistic Lagrangian that describes the motion of the neutron in this field. This Lagrangian should be linear in $\mathcal{E}$ and should contain the neutron mass, velocity, and spin magnetic moment μ_n. Show that the phase shift between the two neutron waves at Q arising from λ is

$$\delta = \pm \frac{\lambda \mu_n}{\hbar c}$$

with ± for neutron spin up (down).

5.4. This problem concerns neutrino oscillations. Using the two-component model of neutrino oscillations, we demonstrate the following result in Section 5.4: if a neutrino with linear momentum p is created at time $t = 0$ in the state $|\nu_e\rangle$, the probability $P_{\nu_e \to \nu_\mu}$ that it will be found in the state $|\nu_\mu\rangle$ at a time $t > 0$ is given by the formula

$$P_{\nu_e \to \nu_\mu} = \sin^2 2\theta \sin^2 \frac{\Delta m^2}{4p} t \qquad (1)$$

where θ is the mixing angle defining the unitary transformation between weak-interaction eigenstates $|\nu_e\rangle, |\nu_\mu\rangle$ and mass eigenstates $|\nu_1\rangle, |\nu_2\rangle$; that is,

$$|\nu_e\rangle = \cos\theta|\nu_1\rangle - \sin\theta|\nu_2\rangle$$
$$|\nu_\mu\rangle = \sin\theta|\nu_1\rangle + \cos\theta|\nu_2\rangle$$

Also, $\Delta m^2 = m_1^2 - m_2^2$, and in (1) we employ units where $\hbar = c = 1$. Because in all practical cases the neutrino is ultrarelativistic, we can write $p = E$ and $t = L/c = L$, where L is the linear distance from the point of origin.

(a) Show that in units where Δm^2 is measured in eV^2/c^4, L is in kilometers and E is in GeV, (1) becomes

$$P_{\nu_e \to \nu_\mu} = \sin^2 2\theta \sin^2\left(\frac{1.27\Delta m^2}{E}L\right) \tag{2}$$

(b) Formulas (1) and (2) and the entire discussion of Section 5.4 refer to neutrino oscillations in vacuum. When neutrinos pass through matter, there is an additional feature that must be considered. This happens because matter contains electrons, and these electrons interact with the state $|\nu_e\rangle$ but not with the state $|\nu_\mu\rangle$ to produce virtual intermediate vector bosons W^-. It can be shown that this results in the following change in the Hamiltonian H' of equation (5.35):

$$H' = \omega\begin{pmatrix}\cos 2\theta & \sin 2\theta \\ \sin 2\theta & -\cos 2\theta\end{pmatrix} \Rightarrow \omega\begin{pmatrix}\cos 2\theta & \sin 2\theta \\ \sin 2\theta & -\cos 2\theta\end{pmatrix} + \frac{NG_F}{\sqrt{2}}\sigma_3 \tag{3}$$

where N is the electron number density, G_F is Fermi's weak interaction coupling constant, $\omega = \Delta m^2/(4E)$, and we have reverted to units where $\hbar = c = 1$. If you do not know about weak interactions, don't worry – just accept (3) as given and consider what follows as an exercise in quantum mechanics.

Show that when the new term proportional to N is included in (3), the Hamiltonian matrix is no longer diagonal in the basis $|\nu_1\rangle, |\nu_2\rangle$. Instead, show that the new basis in which the Hamiltonian is diagonal is $|\nu_{1m}\rangle, |\nu_{2m}\rangle$, where

$$\begin{aligned}|\nu_e\rangle &= \cos\theta_m|\nu_{1m}\rangle - \sin\theta_m|\nu_{2m}\rangle \\ |\nu_\mu\rangle &= \sin\theta_m|\nu_{1m}\rangle + \cos\theta_m|\nu_{2m}\rangle\end{aligned} \tag{4}$$

and

$$\tan 2\theta_m = \frac{\omega\sin 2\theta}{\frac{NG_F}{\sqrt{2}} + \omega\cos 2\theta}$$

Note that even if θ were small, θ_m could be large: $\theta_m = \pi/4$ if $N = -\sqrt{2}\omega\cos 2\theta / G_F$. A neutrino born near the center of the Sun as ν_e passes through a region of very large but rapidly decreasing electron density in its flight outward. When it encounters a density that satisfies the last condition, the probability for transformation into a ν_μ is enhanced. This is called the *Mikheyev-Smirnov-Wolfenstein* (MSW) *effect*; (Wolfenstein 1978).

6 Further Developments in One-Dimensional Wave Mechanics

6.1 Free-particle green function. Spreading of free-particle wave packets

An arbitrary free-particle wave packet can be synthesized by superposing plane waves. Using one spatial dimension for simplicity, we consider how such a wave packet evolves with time. Let the state vector at time $t = 0$ be $|\Psi(0)\rangle$. Then the initial wave function in coordinate space is

$$\psi(x_0,0) = \langle x_0 \mid \Psi(0)\rangle$$

At a time $t \geq 0$, we have

$$\psi(x,t) = \langle x \mid U(t,0) \mid \Psi(0)\rangle = \left\langle x \mid \exp\left(-\frac{i}{\hbar}\frac{\hat{p}^2}{2m}t\right) \mid \Psi(0)\right\rangle \tag{6.1}$$

where, as usual, $\hat{p}$ denotes the momentum operator in one dimension. We expand $|\Psi(0)\rangle$ in plane-wave states of definite momentum; that is,

$$|\Psi(0)\rangle = \int |p\rangle\langle p \mid \Psi(0)\rangle\, dp = \int |p\rangle \phi(p,0)\, dp$$

Then (6.1) becomes

$$\begin{aligned}
\psi(x,t) &= \int \left\langle x \mid \exp\left(-\frac{i}{\hbar}\frac{\hat{p}^2}{2m}t\right) \mid p \right\rangle \phi(p,0)\, dp \\
&= \int \exp\left(-\frac{i}{\hbar}\frac{p^2}{2m}t\right)\langle x \mid p\rangle \phi(p,0)\, dp \\
&= \frac{1}{\sqrt{2\pi\hbar}} \int \exp\left(\frac{i}{\hbar}px\right)\exp\left(-\frac{i}{\hbar}\frac{p^2}{2m}t\right)\phi(p,0)\, dp
\end{aligned} \tag{6.2}$$

Now

$$\phi(p,0) = \frac{1}{\sqrt{2\pi\hbar}} \int \exp\left(-\frac{i}{\hbar}px_0\right)\psi(x_0)\, dx_0$$

Substitution into (6.2) yields

$$\psi(x,t) = \frac{1}{2\pi\hbar}\int_{-\infty}^{\infty} dp \int_{-\infty}^{\infty} \exp\left\{\frac{i}{\hbar}\left[p(x-x_0) - \frac{p^2 t}{2m}\right]\right\}\psi(x_0,0)\, dx_0 \tag{6.3}$$

Frequently, it is useful to write (6.3) as follows:

$$\psi(x,t)=\int_{-\infty}^{\infty} G_0(x,t;x_0,0)\,\psi(x_0,0)\,dx_0 \tag{6.4}$$

where

$$G_0(x,t;x_0,0)=\frac{1}{2\pi\hbar}\int_{-\infty}^{\infty}\exp\left\{\frac{i}{\hbar}\left[p(x-x_0)-\frac{p^2t}{2m}\right]\right\}dp \tag{6.5}$$

is the free-particle Green function for the Schroedinger equation. It is clear from (6.4) and/or (6.5) that in the limit as $t\to 0$, $G_0(x,t;x_0,0)\to\delta(x-x_0)$. For $t>0$, the integral on the right-hand side of (6.5) is easily evaluated by completing the square in the exponent of the integrand; that is,

$$\frac{p^2t}{2m}-p(x-x_0)=\frac{1}{2m}\left[p-\frac{m(x-x_0)}{t}\right]^2-\frac{m(x-x_0)^2}{2t}$$

Hence

$$G_0(x,t;x_0,0)=\frac{1}{2\pi\hbar}\exp\left[\frac{im(x-x_0)^2}{2\hbar t}\right]\int_{-\infty}^{\infty}\exp\left\{-\frac{it}{2m\hbar}\left[p-\frac{m(x-x_0)^2}{t}\right]\right\}dp$$

With use of

$$\int_{-\infty}^{\infty}\exp(-iap^2)\,dp=\sqrt{\frac{\pi}{ia}}$$

this becomes

$$G_0(x,t;x_0,0)=\sqrt{\frac{m}{2\pi i\hbar t}}\exp\left[\frac{im(x-x_0)^2}{2\hbar t}\right]$$

The initial time need not be $t=0$. If instead it is t_0, then

$$G_0(x,t;x_0,t_0)=\sqrt{\frac{m}{2\pi i\hbar T}}\exp\left[\frac{im(x-x_0)^2}{2\hbar T}\right] \tag{6.6}$$

where $T=t-t_0\geq 0$. This is readily generalized to three spatial dimensions; that is,

$$G_0(\boldsymbol{r},t;\boldsymbol{r}_0,t_0)=\left(\frac{m}{2\pi i\hbar T}\right)^{3/2}\exp\left[\frac{im(\boldsymbol{r}-\boldsymbol{r}_0)^2}{2\hbar T}\right] \tag{6.7}$$

Note that Schroedinger's equation for a free particle

$$\nabla^2 \psi = \frac{2m}{i\hbar} \frac{\partial \psi}{\partial t} \tag{6.8}$$

is similar to the diffusion equation of classical transport theory; that is,

$$\nabla^2 u = \frac{1}{D} \frac{\partial u}{\partial t} \tag{6.9}$$

Where, for example, u is the concentration of dissolved matter in some solvent, and D is the diffusion coefficient. We need only replace D by $i\hbar / 2m$ in (6.9) to obtain (6.8). Thus it should be no surprise that the Green function for diffusion is related to G_0 in (6.6) or (6.7) by the same substitution.

As an example of the application of (6.6), we consider the Gaussian wave packet

$$\psi(x_0, 0) = \frac{1}{(2\pi a^2)^{1/4}} \exp\left(-\frac{x_0^2}{4a^2}\right) \tag{6.10}$$

The mean-square dispersion in x_0 is

$$\begin{aligned} [\Delta x_0(0)]^2 &= \langle x_0^2 \rangle - \langle x_0 \rangle^2 \\ &= \langle x_0^2 \rangle = \int_{-\infty}^{\infty} x_0^2 |\psi(x_0, 0)|^2 \, dx_0 = a^2 \end{aligned}$$

Hence, at $t = 0$, the root-mean-square (rms) dispersion in x_0 is

$$\Delta x_0(0) = a$$

The momentum-space wave function at $t = 0$ is easily found to be

$$\phi(p, 0) = \left(\frac{2a^2}{\pi\hbar^2}\right)^{1/4} \exp\left(-\frac{a^2 p^2}{\hbar^2}\right)$$

which yields

$$\Delta p(0) = \frac{\hbar}{2a}$$

Therefore,

$$\Delta x_0(0) \cdot \Delta p(0) = \frac{\hbar}{2} \tag{6.11}$$

Thus (6.10) is a minimum-uncertainty packet. In fact, it can be shown that if (6.11) holds, the corresponding wave packet is necessarily Gaussian (see Problem 4.3). Now, employing (6.6) with (6.10), we calculate the time evolution of ψ. The result is

$$\psi(x,t)=\frac{1}{(2\pi)^{1/4}}\frac{1}{\left[a+(i\hbar t/2ma)\right]^{1/2}}\exp\left[-\frac{x^2}{4a^2+(2i\hbar t/m)}\right]$$

which yields

$$|\psi(x,t)|^2=\frac{1}{\sqrt{2\pi\left[a^2+(\hbar t/2ma)^2\right]}}\exp\left\{-\frac{x^2}{2\left[a^2+(\hbar^2t^2/4m^2a^2)\right]}\right\} \tag{6.12}$$

From (6.12) it is evident that

$$\Delta x(t)=\sqrt{a^2+\frac{\hbar^2t^2}{4m^2a^2}} \tag{6.13}$$

Clearly, the narrower the packet is at $t = 0$, the faster it spreads for finite t of either sign. This can be understood intuitively: the original packet of width a contains momentum components up to $|p| \sim \hbar/2a$, and a particle with such momentum and mass m moves a distance $\hbar t/2ma$ in time t. Of course, although Δx is time dependent, $\Delta p = \Delta p(0)$ is independent of t. Equation (6.13) is only a special example of the following general result: $(\Delta x)^2$ is a quadratic function of time for any arbitrary square-integrable free-particle wave packet.

6.2 Two-particle wave functions: Relative motion and center-of-mass motion

We now consider two particles with masses $m_{1,2}$ and coordinates $\mathbf{r}_{1,2}$, respectively, and assume that their interaction is described by a central potential $V(|\mathbf{r}_1-\mathbf{r}_2|)$ that depends only on the distance between the particles. Working in one spatial dimension for simplicity, we write the Schroedinger equation

$$-\frac{\hbar^2}{2m_1}\frac{\partial^2}{\partial x_1^2}\psi(x_1,x_2,t)-\frac{\hbar^2}{2m_2}\frac{\partial^2}{\partial x_2^2}\psi(x_1,x_2,t)+V(|x_1-x_2|)\psi(x_1,x_2,t)=i\hbar\frac{\partial\psi}{\partial t} \tag{6.14}$$

It is useful to define the relative coordinate x and the center-of-mass coordinate X by

$$x = x_1 - x_2$$

$$X=\frac{m_1x_1+m_2x_2}{m_1+m_2}$$

Then

$$\frac{\partial}{\partial x_1}=\frac{\partial x}{\partial x_1}\frac{\partial}{\partial x}+\frac{\partial X}{\partial x_1}\frac{\partial}{\partial X}=\frac{\partial}{\partial x}+\frac{m_1}{m_1+m_2}\frac{\partial}{\partial X}$$

and, similarly,

$$\frac{\partial}{\partial x_2} = -\frac{\partial}{\partial x} + \frac{m_2}{m_1 + m_2}\frac{\partial}{\partial X}$$

Rewriting (6.14) in terms of x and X and doing simple algebra, we obtain

$$-\frac{\hbar^2}{2\mu}\frac{\partial^2}{\partial x^2}\psi(x,X,t) - \frac{\hbar^2}{2M}\frac{\partial^2}{\partial X^2}\psi(x,X,t) + V(|x|)\psi(x,X,t) = i\hbar\frac{\partial\psi}{\partial t} \tag{6.15}$$

where $\mu = m_1 m_2/(m_1 + m_2)$ is the reduced mass, and $M = m_1 + m_2$. Equation (6.15) possesses solutions of the form $\psi(x,X,t) = \phi(x,t)\chi(X,t)$, where

$$-\frac{\hbar^2}{2\mu}\frac{\partial^2\phi}{\partial x^2} + V\phi = i\hbar\dot{\phi} \tag{6.16}$$

and

$$-\frac{\hbar^2}{2M}\frac{\partial^2\chi}{\partial X^2} = i\hbar\dot{\chi} \tag{6.17}$$

The equation of relative motion (6.16) contains the potential, which in some cases is associated with bound states ϕ. Equation (6.17) describes free-particle motion of the center of mass. These conclusions are easily generalized from one to three spatial dimensions.

6.3 A theorem concerning degeneracy

In one dimension, the time-independent Schroedinger equation is

$$u'' + \varepsilon u - U(x)u = 0$$

where $\varepsilon = 2mE/\hbar^2$ and $U(x) = 2mV(x)/\hbar^2$. Consider two solutions u and w corresponding to the same energy; that is,

$$u'' - U(x)u = -\varepsilon u \tag{6.18}$$

$$w'' - U(x)w = -\varepsilon w \tag{6.19}$$

and assume that $U(x)$ is real and finite for all x. Multiplying (6.18) by w, (6.19) by u, and subtracting one of the resulting equations from the other, we obtain

$$wu'' - uw'' = 0$$

Integrating this equation once, we have

$$wu' - uw' = \text{const} \tag{6.20}$$

If $u = w = 0$ at $x = \infty$ and/or at $x = -\infty$, the constant on the right-hand side of (6.20) is zero, and $w = \text{const}\cdot u$. When one has two linearly independent solutions corresponding to the same eigenvalue, these solutions are degenerate. In the present case, however, w is proportional to u; thus we have only a single nondegenerate solution u. Because both u and u^* are solutions and u is nondegenerate, we must have

$$u = cu^* \tag{6.21}$$

where c is a constant. Taking the complex conjugate of both sides of (6.21), we obtain

$$u^* = c^* u = |c|^2 u^*$$

hence $c = \exp(i\theta)$, where θ is an arbitrary real constant. Thus u must be real (apart from the arbitrary constant phase factor). A bound-state wave function is one that vanishes at $x = +\infty$ and at $x = -\infty$. The foregoing theorem implies that in a one-dimensional potential that is finite everywhere, every bound-state wave function is nondegenerate and real.

6.4 Space-inversion symmetry and parity

A potential is space-inversion symmetric if $V(\boldsymbol{r}) = V(-\boldsymbol{r})$. A Hamiltonian with such a potential is itself symmetric: $H(\boldsymbol{r}) = H(-\boldsymbol{r})$. We now investigate the eigenvalue equation for such a Hamiltonian. We have $H(\boldsymbol{r})u(\boldsymbol{r}) = Eu(\boldsymbol{r})$ and $H(-\boldsymbol{r})u(\boldsymbol{r}) = Eu(\boldsymbol{r})$. Replacing $\boldsymbol{r}$ by $-\boldsymbol{r}$ in this last equation, we obtain

$$H(\boldsymbol{r})u(-\boldsymbol{r}) = Eu(-\boldsymbol{r})$$

Therefore, $u(\boldsymbol{r})$ and $u(-\boldsymbol{r})$ are both solutions for the same eigenvalue E. If u is nondegenerate, we must have

$$u(\boldsymbol{r}) = cu(-\boldsymbol{r})$$

where c is a constant. Replacing $\boldsymbol{r}$ by $-\boldsymbol{r}$, we see that $c^2 = 1$, and therefore, $c = \pm 1$. Thus each solution must have definite parity; that is,

$$\begin{aligned} c = +1:\quad & u(\boldsymbol{r}) = +u(-\boldsymbol{r}) \qquad \text{Positive (even) parity} \\ c = -1:\quad & u(\boldsymbol{r}) = -u(-\boldsymbol{r}) \qquad \text{Negative (odd) parity} \end{aligned}$$

Every bound state in a symmetric one-dimensional potential that is finite everywhere is nondegenerate and therefore has definite parity, either even or odd. In the next few sections we discuss several elementary examples in one-dimensional wave mechanics that illustrate the foregoing theorems and several other principles of interest.

6.5 Potential step

The potential step is shown in Figure 6.1. Here $U = -U_0$ for $x < 0$ and $U = 0$ for $x \geq 0$.
Suppose that $\varepsilon > 0$. Then, writing $k_1^2 = \varepsilon + U_0$, $k_2^2 = \varepsilon$, we have

$$\begin{aligned} x<0: &\quad u_-'' + k_1^2 u_- = 0 \\ x>0: &\quad u_+'' + k_2^2 u_+ = 0 \end{aligned}$$

where u_-, u_+ are the solutions for $x < 0$, $x > 0$, respectively. The general forms of these solutions are

$$u_- = Ae^{ik_1x} + Be^{-ik_1x}$$

and

$$u_+ = Ce^{ik_2x} + De^{-ik_2x}$$

If a wave is incident from the left, we may set $A = 1$ and $D = 0$. In this case, there is a wave reflected back to the left at $x = 0$ ($B \neq 0$) and a wave transmitted to the right at $x = 0$ ($C \neq 0$). The coefficients B, C are determined by the boundary conditions at $x = 0$, which are that $u(x)$ and $u'(x)$ must be continuous. Thus we obtain

$$\begin{aligned} u_-(0) &= 1 + B = C = u_+(0) \\ u_-'(0) &= ik_1(1-B) = ik_2C = u_+'(0) \end{aligned}$$

Hence

$$u_- = e^{ik_1x} + \frac{k_1 - k_2}{k_1 + k_2} e^{-ik_1x}$$

and

$$u_+ = \frac{2k_1}{k_1 + k_2} e^{ik_2x}$$

The probability current density for the incident wave is

$$\begin{aligned} j_{inc} &= \frac{\hbar}{2mi}\left[e^{-ik_1x} ik_1 e^{ik_1x} - (-ik_1) e^{-ik_1x} e^{ik_1x}\right] \\ &= \frac{\hbar k_1}{m} = \frac{p_1}{m} = v_1 \end{aligned}$$

The reflected probability current density is

$$j_{\text{refl}} = -\left(\frac{k_1 - k_2}{k_1 + k_2}\right)^2 j_{\text{inc}}$$

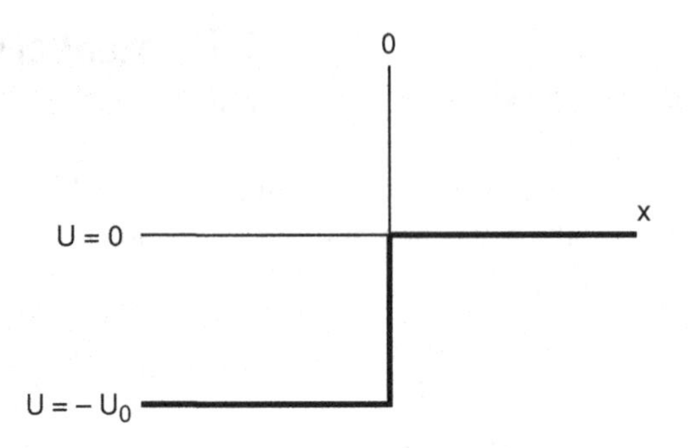

Figure 6.1 Potential step

We define the reflection coefficient as

$$R \equiv \left| \frac{j_{refl}}{j_{inc}} \right| = \left(\frac{k_1 - k_2}{k_1 + k_2} \right)^2 \tag{6.22}$$

The transmitted probability current density is

$$j_{trans} = \frac{4k_1^2}{(k_1 + k_2)^2} \frac{\hbar k_2}{m}$$

and the transmission coefficient is defined as

$$T \equiv \left| \frac{j_{trans}}{j_{inc}} \right| = \frac{4k_1 k_2}{(k_1 + k_2)^2} \tag{6.23}$$

From (6.22) and (6.23) we see that $R + T = 1$, which obviously must hold for conservation of probability. In the limit as $\varepsilon = k_2^2 \to 0$, $R \to 1$ and $T \to 0$. However, when $\varepsilon \to +\infty$, $k_1 = k_2$ and $R = 0$, $T = 1$.

There is another linearly independent solution where the incident wave comes from the right. Here $A = 0$, $D = 1$, and

$$B = \frac{2k_2}{k_1 + k_2} \qquad C = \frac{k_2 - k_1}{k_2 + k_1}$$

It is easy to show that in this case $T = 4k_1k_2/(k_1 + k_2)^2$, just as in (6.23). This is analogous to reflection of a normally incident electromagnetic wave at a plane vacuum-dielectric interface. There also the transmission coefficient is the same no matter what the direction of the incident wave. The theorem of Section 6.3 does not apply to the example just discussed because neither of the two solutions vanishes at $\pm\infty$.

Next, consider the potential step of Figure 6.1, with $-U_0 < \varepsilon < 0$. In this case, we write $k_2^2 = -\varepsilon > 0$, and the Schroedinger equations become

$$x < 0: \quad u_-'' + k_1^2 u_- = 0$$
$$x > 0: \quad u_+'' - k_2^2 u_+ = 0$$

For an incident wave $\exp(ik_1 x)$ from the left, we have

$$u_- = e^{ik_1 x} + Be^{-ik_1 x}$$
$$u_+ = Ce^{-k_2 x}$$

Application of the boundary conditions yields

$$C = \frac{2k_1}{k_1 + ik_2} \qquad B = \frac{k_1 - ik_2}{k_1 + ik_2}$$

which result in

$$u_- = \frac{2k_1}{k_1 + ik_2} \cos k_1 x - \frac{2k_2}{k_1 + ik_2} \sin k_1 x$$

and

$$u_+ = \frac{2k_1}{k_1 + ik_2} e^{-k_2 x}$$

Here we have a standing wave in the classically allowed region ($x < 0$) and an exponentially decaying amplitude in the classically forbidden region ($x > 0$). In the present example, where $-U_0 < \varepsilon < 0$, the theorem of Section 6.3 is applicable.

6.6 One-dimensional rectangular barrier

For the one-dimensional barrier shown in Figure 6.2, let $k_1^2 = \varepsilon > 0$ and $k_2^2 = U_0 - \varepsilon > 0$. The Schroedinger equations for regions I, II, and III are

$$\text{I,III:} \quad u'' + k_1^2 u = 0$$
$$\text{II:} \quad u'' - k_2^2 u = 0$$

Assuming a wave incident from the left, we have the solutions

$$\text{I:} \quad u = e^{ik_1 x} + Be^{-ik_1 x}$$
$$\text{II:} \quad u = Ce^{-k_2 x} + De^{k_2 x}$$
$$\text{III:} \quad u = Se^{ik_1 x}$$

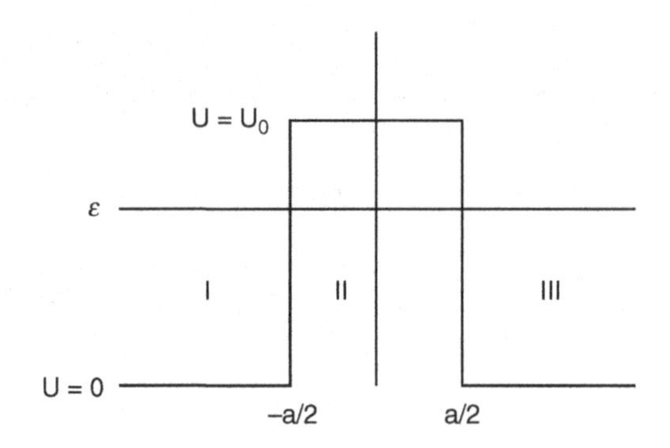

Figure 6.2 Rectangular potential barrier

where *B, C, D,* and *S* are constants. By employing the boundary conditions and straightforward algebra, we find

$$S = \frac{4ik_1k_2e^{-ik_2a}}{e^{-k_2a}\left(k_2+ik_1\right)^2 - e^{k_2a}\left(k_2-ik_1\right)^2}$$

Obviously, when $a \to 0,\ S \to 1$. For $k_2a >> 1$, S becomes

$$S \approx -\frac{4ik_1k_2}{\left(k_2-ik_1\right)^2}e^{-ik_1a}e^{-k_2a}$$

The transmission coefficient in this case is

$$T = \left|S\right|^2 = \frac{16k_1^2k_2^2}{\left(k_1^2+k_2^2\right)^2}e^{-2k_2a}$$

This example is an elementary illustration of barrier penetration.

6.7 One-dimensional rectangular well

As an elementary example of the parity theorem of Section 6.4, we consider the potential and energy illustrated in Figure 6.3. Writing $k_2^2 = -\varepsilon,\ k_1^2 = \varepsilon + U_0$, we have the Schroedinger equations:

$$\begin{aligned} &\text{I,III:} \quad u'' - k_2^2u = 0 \\ &\text{II:} \quad u'' + k_1^2u = 0 \end{aligned}$$

Because the bound-state solutions must have even or odd parity, we need only consider the solutions in regions II and III and the boundary conditions at $x = a/2$. In region II, the even solutions are of the form

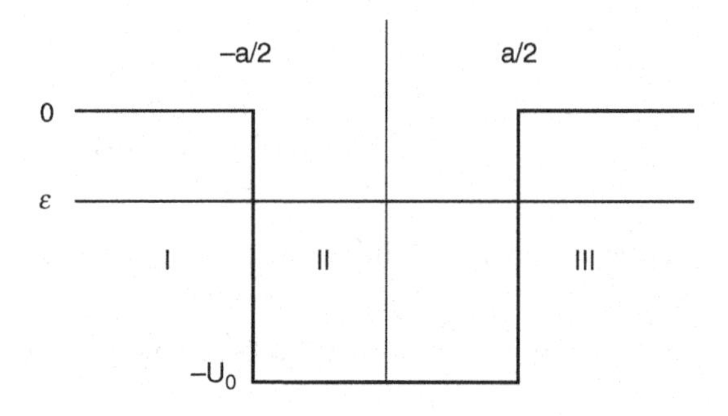

Figure 6.3 Rectangular potential well: $\varepsilon < 0$

$$u_{\text{II,even}} = A\cos k_1 x$$

and the odd solutions are of the form

$$u_{\text{II,odd}} = B\sin k_1 x$$

In the limit where $U_0 \to \infty$, $u = 0$ for $|x| \geq a/2$. In this case, the even-solution eigenvalues are determined by the condition

$$\cos\left(\frac{k_1 a}{2}\right) = 0$$

Hence the possible energies are given by

$$\varepsilon' = \varepsilon + U_0 = \frac{\pi^2}{a^2}, \quad \frac{9\pi^2}{a^2}, \quad \frac{25\pi^2}{a^2}, \ldots, \quad \left[\frac{(2n+1)\pi}{a}\right]^2, \ldots$$

Meanwhile, the odd-solution eigenvalues, determined by the condition

$$\sin\left(\frac{k_1 a}{2}\right) = 0$$

are given by

$$\varepsilon' = \varepsilon + U_0 = \frac{4\pi^2}{a^2}, \quad \frac{16\pi^2}{a^2}, \quad \frac{36\pi^2}{a^2}, \ldots$$

In order of increasing energy, the solutions alternate between even and odd parity. The qualitative character of these solutions remains the same even if we distort the shape of the well considerably.

Now, returning to the finite rectangular well of Figure 6.3, we have for the lowest-energy solution

$$|x| < \frac{a}{2}: \quad u = A\cos k_1 x$$
$$x > \frac{a}{2}: \quad u = Be^{-k_2 x}$$

Imposing the boundary conditions at $x = a/2$, we obtain

$$A\cos\frac{k_1 a}{2} = B\exp\left(-\frac{k_2 a}{2}\right) \tag{6.24}$$

and

$$-k_1 A\sin\frac{k_1 a}{2} = -k_2 B\exp\left(-\frac{k_2 a}{2}\right) \tag{6.25}$$

Division of (6.25) by (6.24) yields the condition

$$\tan\frac{k_1 a}{2} = \frac{k_2}{k_1}$$

or

$$\tan\sqrt{\frac{\varepsilon a^2}{4}} = \sqrt{\frac{U_0}{\varepsilon} - 1}$$

With the substitution $\omega^2 = \varepsilon a^2/4$, this becomes

$$\tan\omega = \sqrt{\frac{U_0 a^2}{4\omega^2} - 1} \tag{6.26}$$

The solutions to this transcendental equation are displayed in Figure 6.4 as the intersections of $y = \tan\omega$ and $y = f(\omega) = \sqrt{(U_0 a^2/4\omega^2) - 1}$. Note that a solution exists for any value of $U_0 a^2$; thus, no matter how shallow the well, there is always at least one bound state. The energy of the first odd-parity solution is determined by the transcendental equation:

$$\cot\omega = -f(\omega)$$

and no solution to this equation exists unless $U_0 \geq (\pi/a)^2$. Thus, if the well is sufficiently shallow, there is no odd-parity bound state.

Next, we consider the finite rectangular well for positive energy, as in Figure 6.5. Here it is convenient to define $k_1^2 = \varepsilon + U_0$, $k_2^2 = \varepsilon$. If we choose a wave $\exp(ik_2 x)$ incident from the left, the solutions are

$$x < -\frac{a}{2}: \quad u = \exp(ik_2 x) + B\exp(-ik_2 x)$$
$$|x| \leq \frac{a}{2}: \quad u = C\exp(ik_1 x) + D\exp(-ik_1 x)$$
$$x > \frac{a}{2}: \quad u = S\exp(ik_2 x)$$

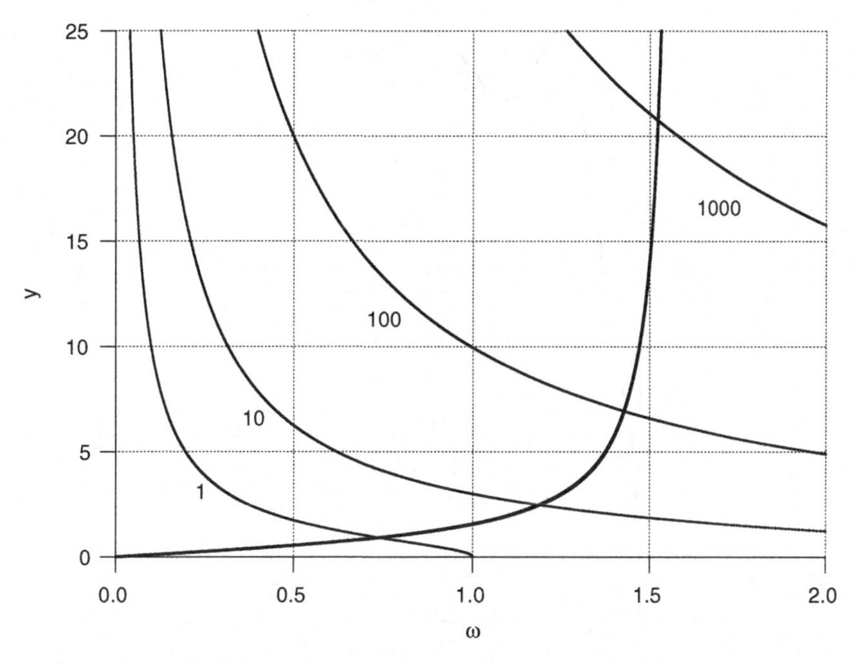

Figure 6.4 $y = \tan\omega$ (heavy curve) and $y = f(\omega)$ for $U_0a^2/4 = 1, 10, 100, 1{,}000$.

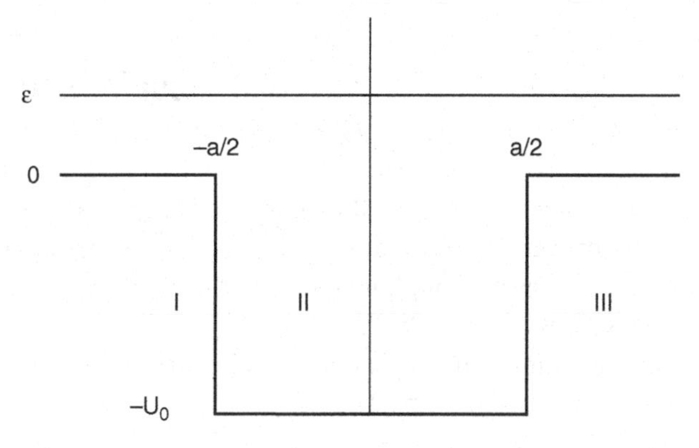

Figure 6.5 Rectangular potential well: $\varepsilon > 0$

Making use of the boundary conditions, we find, after some algebra, that the transmission coefficient is

$$T = |S|^2 = \frac{1}{\cos^2(k_1 a) + \dfrac{(k_1^2 + k_2^2)^2}{4k_1^2 k_2^2}\sin^2(k_1 a)} \tag{6.27}$$

In Figure 6.6 we plot T versus εa^2 for $U_0 a^2 = 100$. The sequence of maxima and local minima in T provides an elementary example of resonance in scattering.

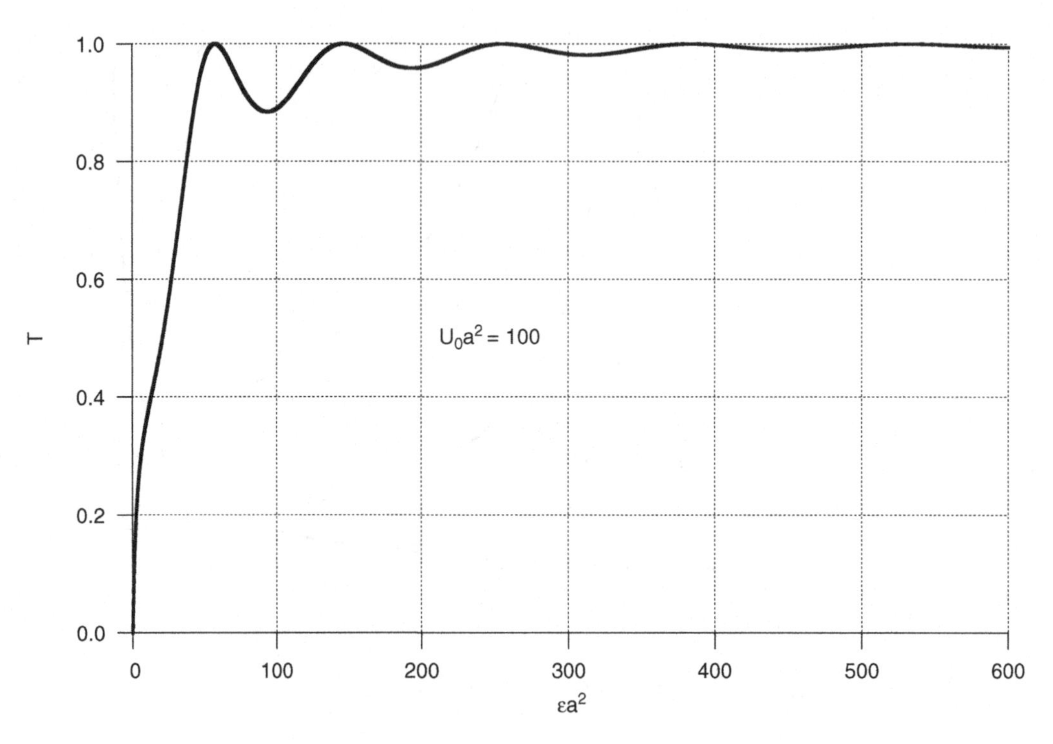

Figure 6.6 Transmission coefficient for the potential well of Figure 6.5 with $U_0a^2 = 100$, plotted as a function of εa^2

6.8 Double wells

Figure 6.7 shows a double well with infinitely high walls. Two obviously degenerate solutions u_1 and u_2, with even and odd parity, respectively, are also shown. The theorem of Section 6.3 does not apply here even though u_1 and u_2 both vanish at $\pm\infty$ because the potential barrier between the two wells is infinitely high.

Because u_1 and u_2 are degenerate, any linear combination of these solutions is also a solution with the same energy. In particular,

$$u_+ = \frac{1}{\sqrt{2}}(u_1 + u_2)$$

and

$$u_- = \frac{1}{\sqrt{2}}(u_1 - u_2)$$

correspond to a particle localized in the left or right well, respectively. If we make the barrier between the left and right wells finite, as in Figure 6.8, the solutions u_1, u_2 are no longer degenerate, and it is easy to see by going to the limit where the central barrier disappears that $E_2 > E_1$.

We can still form

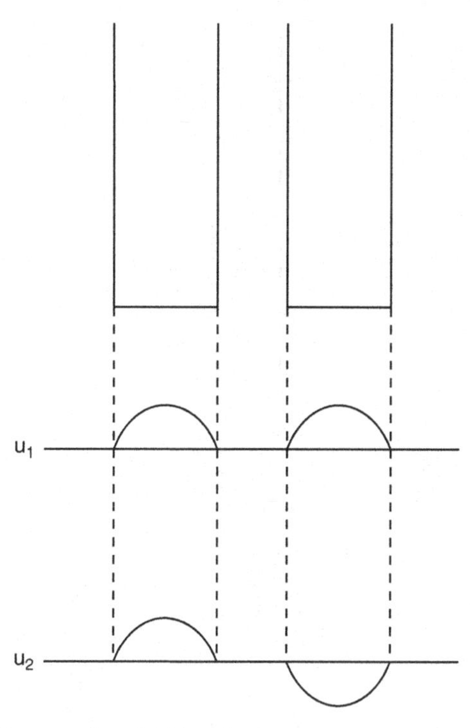

Figure 6.7 Ground state wave functions in a double rectangular well of infinite height

$$u_{\pm}(x,t)=\frac{1}{\sqrt{2}}\left[u_1(x)e^{-iE_1t/\hbar}\pm u_2(x)e^{-iE_2t/\hbar}\right]$$

but because $E_1 \neq E_2$, neither u_+ nor u_- is a stationary state. Instead, whereas $u_+(x,0), u_-(x,0)$ describe a particle that is more or less localized in the left (right) well, respectively, the particle tunnels to the opposite well and back again with an oscillation period $T = 2\pi\hbar/(E_2 - E_1)$ in each case. Such oscillations, frequently called *quantum beats*, occur whenever one has a coherent superposition of two or more states with different energies. Such beats have widespread and diverse physical manifestations. We have already seen two important examples (Larmor precession in Section 3.4 and neutrino oscillations in Section 5.4). In what follows, we give two additional examples.

6.9 Ammonia molecule

In the NH_3 molecule, three hydrogen atoms are located at the vertices of an equilateral triangle, and the nitrogen atom oscillates back and forth on a line perpendicular to the plane of the triangle (Figure 6.9a). The effective potential seen by the nitrogen atom is sketched in Figure 6.9b. The stationary states of ammonia do not correspond to N on one side (u_+) or the other side (u_-) of the plane defined by the equilateral triangle but rather to the symmetric and

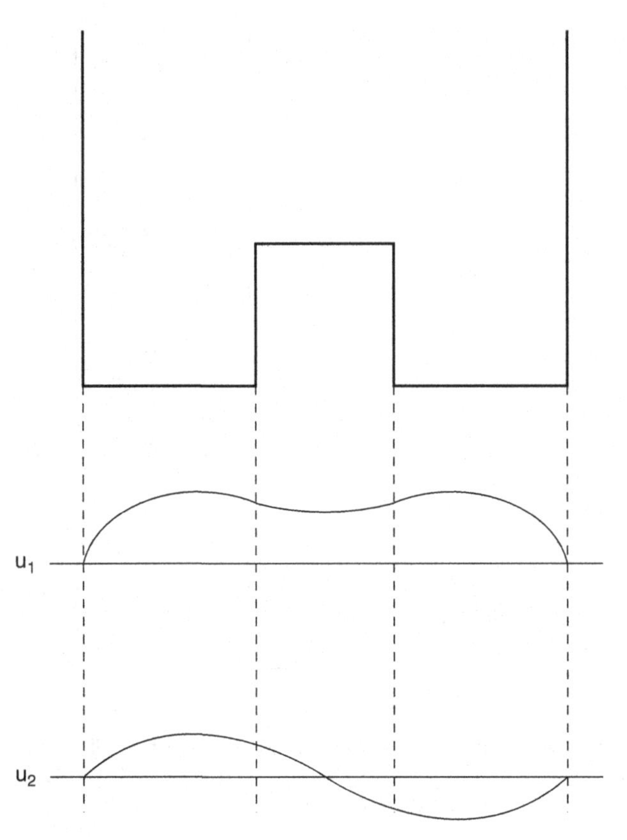

Figure 6.8 Sketch of lowest symmetric wave function (u_1) and antisymmetric wave function (u_2) for a rectangular well of infinite height with symmetric finite rectangular barrier.

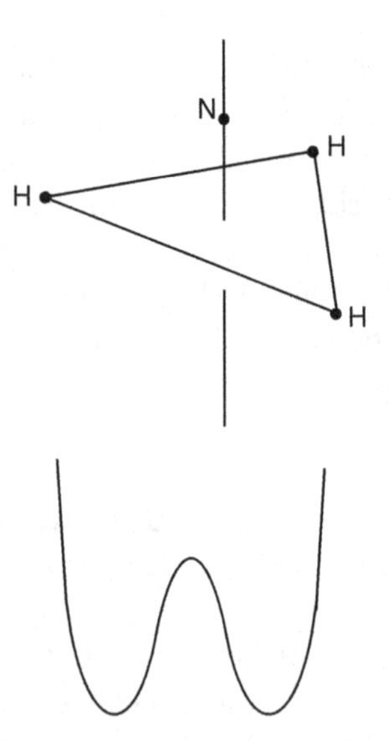

Figure 6.9 a) Sketch of ammonia molecule (NH_3) b) Sketch of potential energy for nitrogen motion along axis perpendicular to plane defined by hydrogen atoms

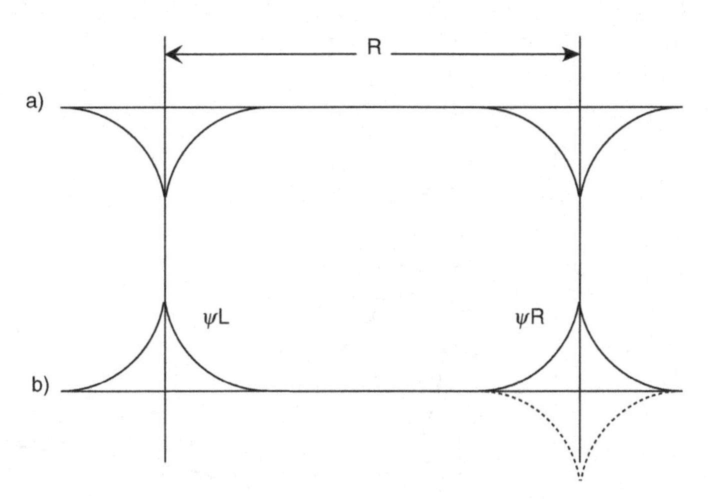

Figure 6.10 a) Sketch of potential energy for electron motion in Coulomb field of two widely separated protons b) Sketch of ground state electronic wave function along internuclear axis

antisymmetric combinations u_1, u_2, respectively, of u_+, u_-. The first MASER (a forerunner of all LASERs), invented in 1954 by C. H. Townes and coworkers, was a molecular beam apparatus that used a transition between a pair of states u_1, u_2 of NH_3 at a frequency of approximately 24 GHz.

6.10 Hydrogen molecular ion

An analogous situation occurs in the H_2^+ molecule. We first consider a hydrogen atom in the ground state and also an isolated proton separated from the atom by a very large distance R. The electronic wave function is $\psi_L = N\exp(-r_1/\ a_0)$, where r_1 is the distance between the electron and the proton in the initial atom; $a_0 = 4\pi\hbar^2 / m_e e^2$ is the Bohr radius; and N is a normalization constant. The potential seen by the electron is plotted in Figure 6.10a, and the wave function ψ_L is plotted on the left side of Figure 6.10b. However, another solution ψ_R also exists: it corresponds to the electron centered about the second proton.

When the distance R is very large, we can form the essentially degenerate symmetric and antisymmetric solutions

$$\psi_S = \frac{1}{\sqrt{2}}(\psi_L + \psi_R)$$
$$\psi_A = \frac{1}{\sqrt{2}}(\psi_L - \psi_R)$$

If R is decreased sufficiently so that there is overlap of ψ_L and ψ_R, ψ_S and ψ_A are no longer degenerate; instead, the electronic energy of ψ_A is greater than that of ψ_S. The total energy of the molecule, which includes the Coulomb repulsion energy of the two protons in addition to the electronic energy, is sketched in Figure 6.11.

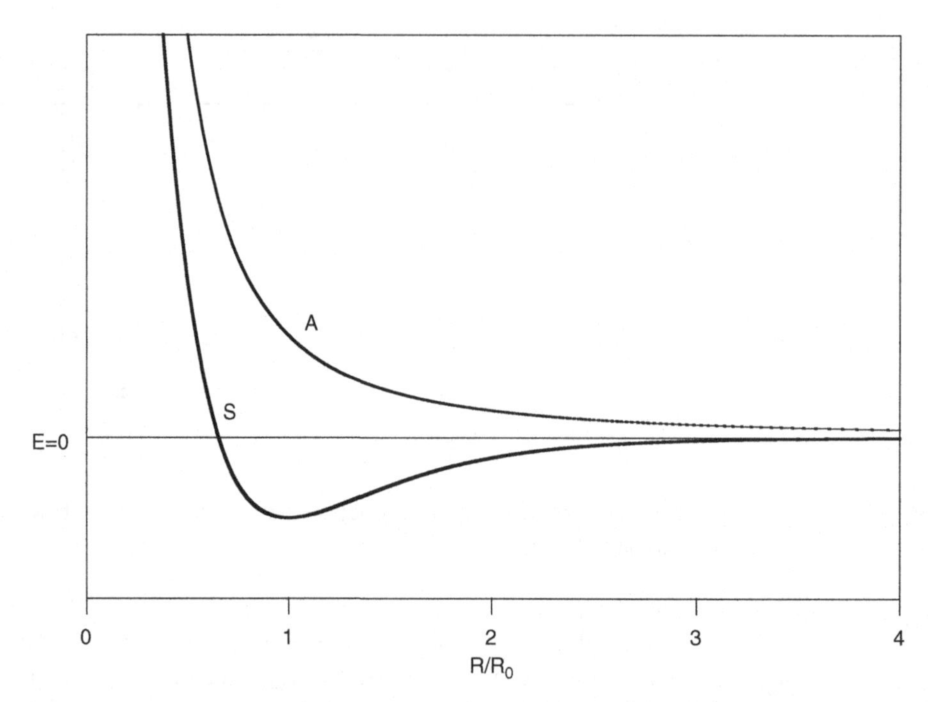

Figure 6.11 Sketch of lowest symmetric (S) and antisymmetric (A) energies for H_2^+ plotted as a function of internuclear separation R in units of equilibrium internuclear separation R_0. Coulomb repulsion energy of the protons is included.

6.11 Periodic potentials: Bloch's theorem

Periodic potentials of the form $V(x) = V(x+a)$, where a is a real constant, are obviously of importance in condensed-matter physics, where one deals with the properties of crystalline solids. Let us assume that we have such a potential, in which case the Hamiltonian is also periodic: $H(x) = H(x+a)$. Let $\psi(x)$ be an energy eigenfunction satisfying

$$H(x)\psi(x) = E\psi(x)$$

and let T be the translation operator such that

$$T\psi(x) = \psi(x+a)$$

T is clearly unitary, and also

$$\begin{aligned} TH(x)\psi(x) &= TE\psi(x) = ET\psi(x) \\ &= E\psi(x+a) = H(x+a)\psi(x+a) \\ &= H(x)T\psi(x) \end{aligned}$$

Thus $[T, H] = 0$, which implies that the eigenstates of H are also eigenstates of T. Let the eigenvalues of T be λ; that is,

$$T\psi(x) = \lambda\psi(x)$$

and define k by $\lambda = \exp(ika)$. Then, because T is unitary, k must be real. Defining

$$u_k(x) = \exp(-ikx)\psi(x)$$

we have

$$\begin{aligned} u_k(x+a) &= e^{-ikx}e^{-ika}\psi(x+a) \\ &= e^{-ikx}e^{-ika}T\psi(x) \\ &= e^{-ikx}e^{-ika}e^{ika}\psi(x) \\ &= u_k(x) \end{aligned}$$

Hence $u_k(x)$ is periodic, and $\psi(x)$ can always be expressed as

$$\psi(x) = e^{ikx}u_k(x) \tag{6.28}$$

This result is known as *Bloch's theorem*, and the expression on the right-hand side of (6.28) is frequently called a *Bloch wave*. Note that $-\pi/a \le k \le \pi/a$ is a sufficient range for k because k and $k + 2\pi/a$ yield the same eigenvalue $\exp(ika)$.

A simple and useful example of the foregoing is the Kronig-Penney model, where

$$V(x) = \sum_{n=-\infty}^{\infty} V_0\delta(x - na)$$

and V_0 is a positive constant. We seek a solution of the form (6.28) with energy E. In the region $0 < x < a,\ V = 0$ and

$$\psi = Ae^{iqx} + Be^{-iqx}$$

where $E = (\hbar q)^2/2m$. Thus

$$u = A\exp[i(q-k)x] + B\exp[-i(k+q)x]$$

The usual boundary conditions determine A, B, and q. After straightforward algebra, we find that

$$\cos ka = \cos qa + \frac{maV_0}{\hbar^2}\frac{\sin qa}{qa}$$

Let us define

$$f(y) = \cos y + \frac{maV_0}{\hbar^2}\frac{\sin y}{y} \tag{6.29}$$

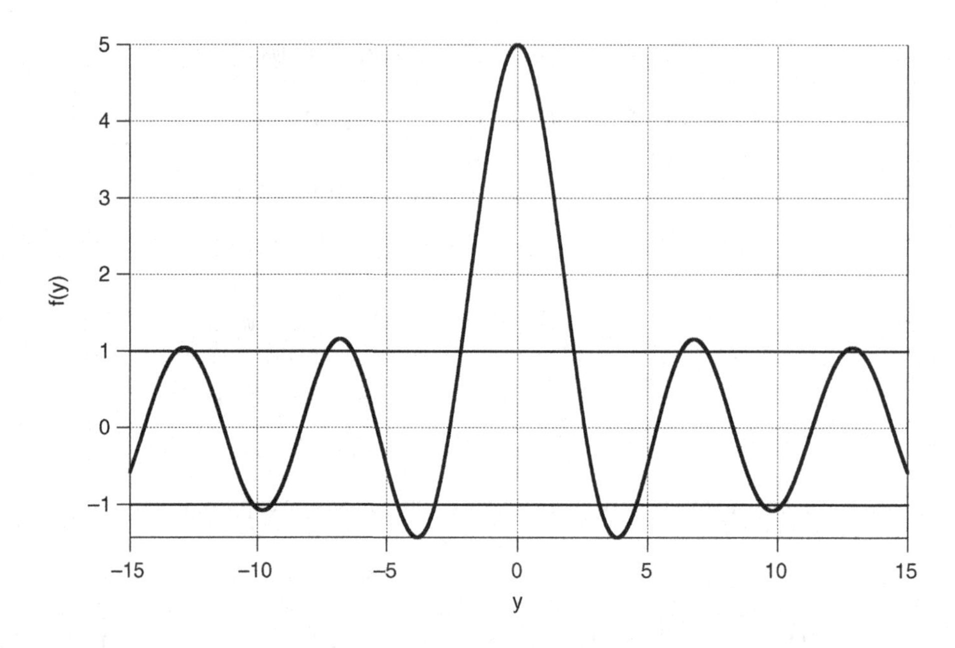

Figure 6.12 $f(y)$ from equation (6.29) with $(maV_0)/\hbar^2 = 4$ is plotted versus y. The allowed values of $q = y/a$ are those for which $|f(y)| \le 1$. The permissible energies (energy bands) are values $E = (\hbar^2 q^2)/2m$ for which q is allowed.

Then values of $q = y/a$ are allowed for which $|f(y)| \le 1$ (see Figure 6.12). The permissible energies are values $E = (\hbar q)^2/2m$, for which q is allowed. This is a simple example of energy bands.

6.12 Particle in a uniform field

We now consider the motion of a particle of mass m in a uniform force field. This could be a charge q of mass m in a uniform electric field or a particle of mass m in a uniform gravitational field with gravitational acceleration g. Choosing the latter as an example, we have the time-independent Schroedinger equation

$$\frac{-\hbar^2}{2m}\frac{\partial^2 \psi}{\partial y^2} + mgy\psi = E\psi$$

Making the substitutions $\alpha = 2mE/\hbar^2$, $\beta = 2gm^2/\hbar^2$, we obtain

$$\frac{\partial^2 \psi}{\partial y^2} + (\alpha - \beta y)\psi = 0 \qquad (6.30)$$

With the change of variable $z = \beta^{-2/3}(\beta y - \alpha)$, (6.30) is transformed to the Airy differential equation; that is,

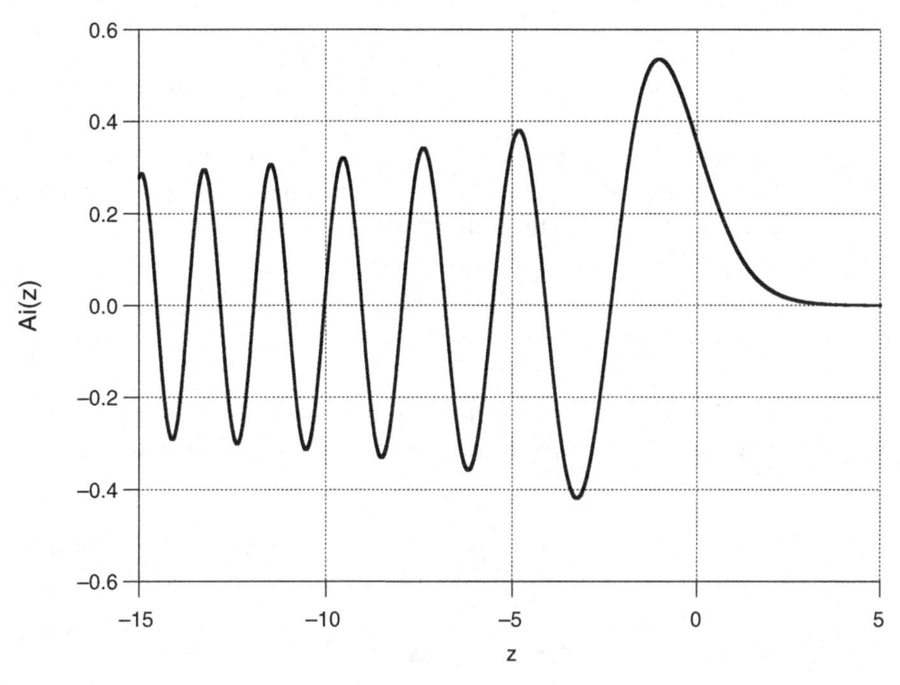

Figure 6.13 The Airy function.

$$\frac{\partial^2 \psi}{\partial z^2} - z\psi = 0 \tag{6.31}$$

a regular solution of which is the Airy function defined for $z \geq 0$; that is,

$$\begin{aligned} \psi(z) = Ai(z) &= \frac{\sqrt{z}}{3}\left[I_{-1/3}\left(\frac{2}{3}z^{3/2}\right) - I_{1/3}\left(\frac{2}{3}z^{3/2}\right)\right] \\ \psi(-z) = Ai(-z) &= \frac{\sqrt{z}}{3}\left[J_{-1/3}\left(\frac{2}{3}z^{3/2}\right) - J_{1/3}\left(\frac{2}{3}z^{3/2}\right)\right] \end{aligned} \tag{6.32}$$

where $J_\nu(z)$ is an ordinary Bessel function of order ν, and $I_\nu(z) = \exp(-i\pi\nu/2)J_\nu(iz)$. The Airy function is plotted in Figure 6.13.

For large z, the following asymptotic formulas are useful:

$$Ai(-z) \rightarrow \frac{1}{\pi^{1/2}}\frac{1}{z^{1/4}}\sin\left(\frac{2}{3}z^{3/2} + \frac{\pi}{4}\right) \tag{6.33}$$

$$Ai(z) \rightarrow \frac{1}{2\pi^{1/2}}\frac{\exp(-2/3\,z^{3/2})}{z^{1/4}} \tag{6.34}$$

Thus the solution to the time-dependent Schroedinger equation for energy E is

$$\psi(y,t) = Ai\left[\left(\frac{2m^2g}{\hbar^2}\right)^{1/3}\left(y - \frac{E}{mg}\right)\right]e^{-iEt/\hbar} \tag{6.35}$$

It is instructive to consider the same problem from the viewpoint of the principle of equivalence, which states that a uniform gravitational field is equivalent to an accelerated frame of reference. By making the transformation

$$y' = y - y_0 + \frac{1}{2}gt^2$$
$$t' = t$$

from frame F with coordinates (y, t) to frame F' with coordinates $(y', t' = t)$, we eliminate the gravitational force. Writing the time-dependent Schroedinger equation in terms of the new variable y', we obtain

$$-\frac{\hbar^2}{2m}\frac{\partial^2\psi}{\partial y'^2} + mg\left(y' + y_0 - \frac{gt^2}{2}\right)\psi = i\hbar\left(gt\frac{\partial}{\partial y'} + \frac{\partial}{\partial t}\right)\psi \tag{6.36}$$

Because there is no force in the accelerated frame, we try a solution to (6.36) of the form

$$\psi(y',t) = u(y',t)e^{i\lambda(y',t)} \tag{6.37}$$

where $u(y',t)$ is a solution to the free-particle equation

$$-\frac{\hbar^2}{2m}\frac{\partial^2 u}{\partial y'^2} = i\hbar\frac{\partial u}{\partial t} \tag{6.38}$$

Substituting (6.37) in (6.36), carrying out some algebra, and setting $y_0 = E / mg$, we arrive at

$$u(y',t) = Ai\left[\left(\frac{2m^2g}{\hbar^2}\right)^{1/3}\left(y' - \frac{gt^2}{2}\right)\right]\exp\left[i\left(\frac{mgt}{\hbar}y' - \frac{mg^2t^3}{3\hbar}\right)\right] \tag{6.39}$$

The wave packet (6.35) in frame F and the free-particle packet (6.39) in frame F' appear at first sight to pose a paradox. In frame F, a classical particle accelerates downward with acceleration g, but solution (6.35) to the Schroedinger equation for definite energy E is stationary: the corresponding probability density is time independent, and the probability current density is zero. On the other hand, in the accelerated frame, a classical particle is stationary, but the free-particle wave packet $u(y',t)$ of (6.39) accelerates. [Furthermore, $u(y',t)$ does not spread as time elapses, but because u is not square integrable, this does not contradict the principle, stated in Section 6.1, that the mean square dispersion in the coordinate of any square-integrable free-particle packet must increase quadratically with time.]

The resolution of the apparent paradox becomes clear when we realize that to obtain a wave packet in frame F that behaves like a classical accelerating particle, we must create a superposition of solutions (6.35) over a range of E values. If this were done and we were then to transform to the accelerated frame F', we would obtain a square-integrable free-particle packet that did not accelerate, and the mean square dispersion in y' would grow quadratically with time. For example, see Problem 6.5.

6.13 One-dimensional simple harmonic oscillator

6.13.1 Hamiltonian, eigenvectors, and eigenvalues

The simple harmonic oscillator plays a central role in the development of quantum mechanics and quantum field theory and is the single most important topic in this chapter. We begin with the classical Hamiltonian

$$H_c = \frac{p^2}{2m} + \frac{m\omega^2}{2} q^2$$

where $m\omega^2$ is the *spring constant*, and q and p are the classical coordinate and momentum, respectively. Replacing the latter by the corresponding quantum-mechanical operators $\hat{q}, \hat{p}$, which satisfy

$$[\hat{q}, \hat{p}] = i\hbar$$

we obtain the quantum Hamiltonian

$$H = \frac{\hat{p}^2}{2m} + \frac{m\omega^2}{2} \hat{q}^2 \tag{6.40}$$

It is convenient to make a scale change by the substitutions

$$\hat{Q} = \left(\frac{m\omega}{\hbar}\right)^{1/2} \hat{q} \quad \text{and} \quad \hat{P} = \left(\frac{1}{m\omega\hbar}\right)^{1/2} \hat{p} \tag{6.41}$$

which result in the new commutation relation

$$\left[\hat{Q}, \hat{P}\right] = i \tag{6.42}$$

Expressing the Hamiltonian (6.40) in terms of these new operators, we have

$$H = \frac{\hbar\omega}{2}\left(\hat{P}^2 + \hat{Q}^2\right) = \hbar\omega\mathcal{H} \tag{6.43}$$

where

$$\mathcal{H} = \frac{1}{2}\left(\hat{P}^2 + \hat{Q}^2\right)$$

At this point, one could solve Schroedinger's wave equation in coordinate space by the conventional methods of ordinary differential equations to find the bound-state wave functions and corresponding energy eigenvalues, and this is done in many older texts. However, it is more instructive to treat the problem algebraically, as we do in what follows, by defining the *destruction operator* a and the *creation operator* $a^\dagger$; that is,

$$a = \frac{1}{\sqrt{2}}\left(\hat{Q} + i\hat{P}\right) \tag{6.44}$$

and

$$a^\dagger = \frac{1}{\sqrt{2}}\left[\hat{Q} - i\hat{P}\right] \tag{6.45}$$

These defining formulas, together with (6.42), easily lead to

$$\left[a, a^\dagger\right] = 1 \tag{6.46}$$

and

$$\mathcal{H} = a^\dagger a + \frac{1}{2} = N + \frac{1}{2} \tag{6.47}$$

where $N = a^\dagger a$. From (6.46), we also obtain

$$\left[N, a\right] = a^\dagger a a - a a^\dagger a = -a \tag{6.48}$$

and

$$\left[N, a^\dagger\right] = a^\dagger \tag{6.49}$$

Let $|n\rangle$ be a nonnull eigenvector of N with corresponding eigenvalue n; that is,

$$N|n\rangle = n|n\rangle$$

At this initial stage, we can only say that n must be a real number. However, defining $|u\rangle = a|n\rangle$, we have

$$\langle n|N|n\rangle = \langle u \mid u\rangle = n\langle n \mid n\rangle$$

Therefore, $n = 0$ if $|u\rangle$ is a null vector; otherwise, n must be positive. On the other hand, we can apply (6.48) to $|n\rangle$; that is,

$$Na|n\rangle = aN|n\rangle - a|n\rangle = (n-1)a|n\rangle$$

and this implies that $a|n\rangle$ is also an eigenvector of N, with eigenvalue $n - 1$. Repeating this argument, we find that $a^2|n\rangle, a^3|n\rangle, \ldots$ are eigenvectors of N with eigenvalues $n - 2$, $n - 3$, …, respectively. This, in turn, implies that we eventually arrive at a negative eigenvalue of N, which we have shown is impossible, or there is some $n' = n - k$ where k is a nonnegative integer and $a|n'\rangle = 0$. In the latter case, the eigenvalue corresponding to $|n'\rangle$ is $n' = 0$, as has been demonstrated. Here negative eigenvalues are avoided because further applications of a continue to yield the null vector. We conclude that the eigenvalues of N must be zero and the positive integers. That this list of integers extends to infinity is seen by applying (6.49) to $|n\rangle$. The latter implies that $a^\dagger|n\rangle$ is also an eigenvector of N with eigenvalue $n + 1$, and this process also can be repeated, yielding arbitrarily large positive-integer eigenvalues. Finally, from (6.43) and (6.47), we see that the energy eigenvalues are

$$E_n = \left(n + \frac{1}{2}\right)\hbar\omega \qquad n = 0, 1, 2, 3, \ldots \tag{6.50}$$

6.13.2 Normalization of eigenstates

We next consider normalization of the eigenvectors $|n\rangle$. Let us begin by choosing $|0\rangle$ such that $\langle 0 | 0\rangle = 1$. Defining $|1\rangle = a^\dagger|0\rangle$, we have

$$\langle 1|1\rangle = \langle 0|aa^\dagger|0\rangle = \langle 0|a^\dagger a|0\rangle + \langle 0|0\rangle = \langle 0|0\rangle = 1$$

Next, we define $|2\rangle = \left(1/\sqrt{2}\right)a^\dagger|1\rangle$. Then

$$\langle 2|2\rangle = \frac{1}{2}\langle 1|aa^\dagger|1\rangle = \frac{1}{2}\langle 1|N|1\rangle + \frac{1}{2}\langle 1|1\rangle = 1$$

Next, we define $|3\rangle = \left(1/\sqrt{3}\right)a^\dagger|2\rangle$. Then, by a similar argument, we find that $\langle 3|3\rangle = 1$. In general, we have

$$\begin{aligned} |n\rangle &= \frac{1}{\sqrt{n}}a^\dagger|n-1\rangle \\ &= \frac{1}{\sqrt{n(n-1)}}\left(a^\dagger\right)^2|n-2\rangle \\ &= \ldots \\ &= \frac{1}{\sqrt{n!}}a^{\dagger n}|0\rangle \end{aligned} \tag{6.51}$$

Similarly,

$$\frac{1}{\sqrt{n!}}a^n|n\rangle = |0\rangle \tag{6.52}$$

From (6.51) and (6.52) it is easy to show that

$$\langle n'|a|n\rangle = n^{1/2}\delta_{n',n-1} \tag{6.53}$$

and

$$\langle n'|a^\dagger|n\rangle = (n+1)^{1/2}\,\delta_{n',n+1} \tag{6.54}$$

6.13.3 Wave functions in coordinate representation

The coordinate wave functions corresponding to states $|n\rangle$ are $u_n(Q) = \langle Q|n\rangle$. To find the explicit forms of these functions, we recall that $a|0\rangle = \left(1/\sqrt{2}\right)\left(\hat{Q}+i\hat{P}\right)|0\rangle = 0$. Also recalling that in coordinate representation, $\hat{p} \to -i\hbar\nabla$, and taking into account the scale change of (6.41), we have

$$\left(Q + \frac{\partial}{\partial Q}\right)u_0(Q) = 0$$

The normalized solution of this differential equation is

$$u_0(Q) = \frac{1}{\pi^{1/4}}e^{-Q^2/2} \tag{6.55}$$

From (6.45) and (6.51), we also obtain

$$u_n(Q) = \frac{1}{\sqrt{2^n n!}}\left(Q - \frac{\partial}{\partial Q}\right)^n u_0(Q) \tag{6.56}$$

In Figure 6.14a–e we plot $u_n(Q)$ versus Q for $n = 0, 1, 2, 3, 4$. Obviously, the even-parity functions ($n = 0, 2, 4, \ldots$) alternate in energy with the odd-parity functions ($n = 1, 3, 5, \ldots$).

6.13.4 Harmonic oscillator and Heisenberg equation

We have seen that if A is an operator in the Schroedinger representation, the corresponding operator in the Heisenberg representation is $A' = U^\dagger A U$, where $U = \exp(-iHt/\hbar)$. Also, if A has no explicit time dependence, A' satisfies the Heisenberg equation

$$\frac{dA'}{dt} = \frac{1}{i\hbar}[A', H]$$

We apply the latter to $a' = U^\dagger a U$ to get

$$\frac{da'}{dt} = \frac{1}{i\hbar}[a', H] = \frac{\hbar\omega}{i\hbar}\left[a', \left(N + \frac{1}{2}\right)\right] = -i\omega[a', N]$$

Because N commutes with the Hamiltonian, we have

$$\begin{aligned}[a', N] &= U^\dagger a U N - N U^\dagger a U \\ &= U^\dagger [a, N] U = U^\dagger a U = a'\end{aligned}$$

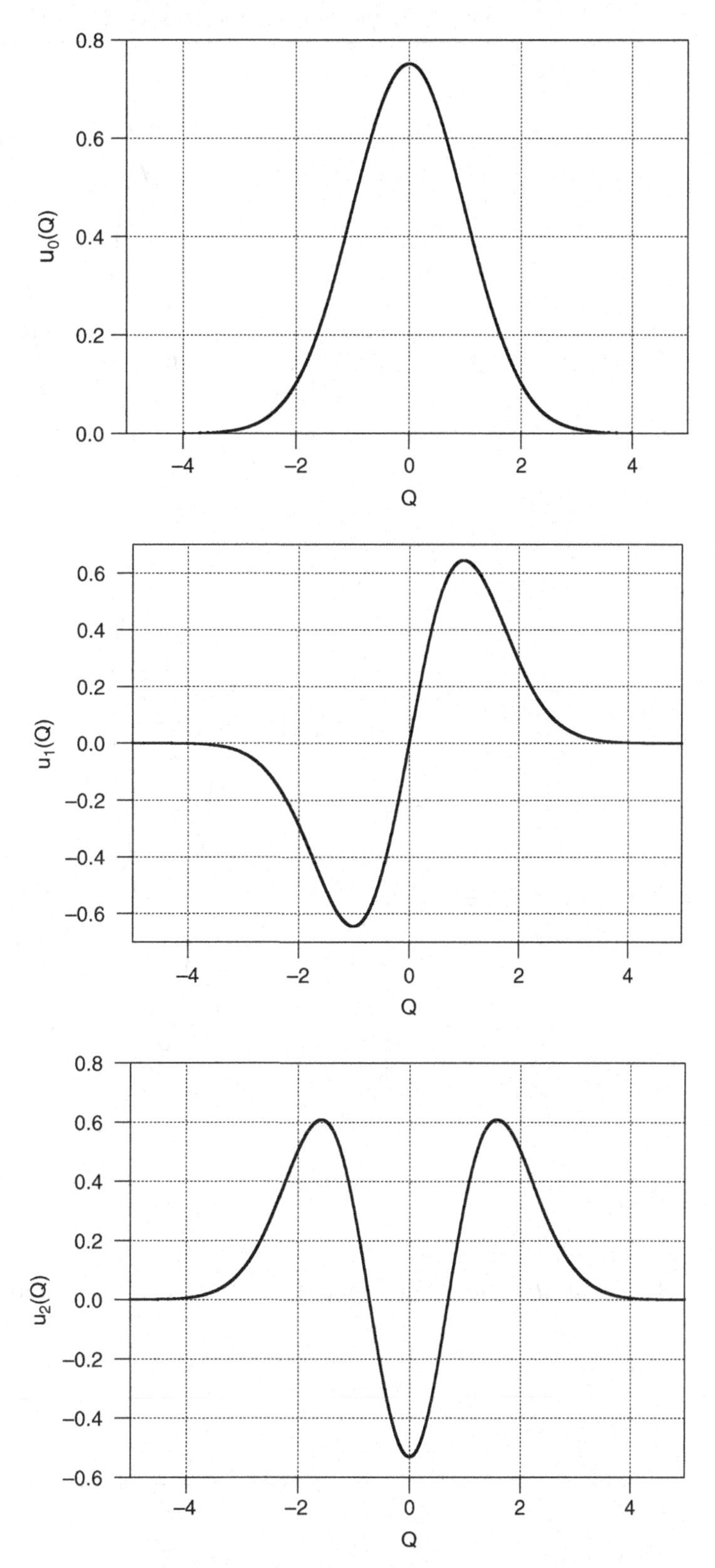

Figure 6.14 (a) $u_0(Q)=(1/\pi^{1/4})e^{-Q^2/2}$. (b) $u_1(Q)=\sqrt{2}(Q/\pi^{1/4})e^{-Q^2/2}$. (c) $u_2(Q)=\left[(2Q^2-1)/\sqrt{2}\pi^{1/4}\right]e^{-Q^2/2}$. (d) $u_3(Q)=\left[(4Q^3-6Q)/\sqrt{12}\pi^{1/4}\right]e^{-Q^2/2}$. (e) $u_4(Q)=\left[(4Q^4-12Q^2+3)/\sqrt{24}\pi^{1/4}\right]e^{-Q^2/2}$.

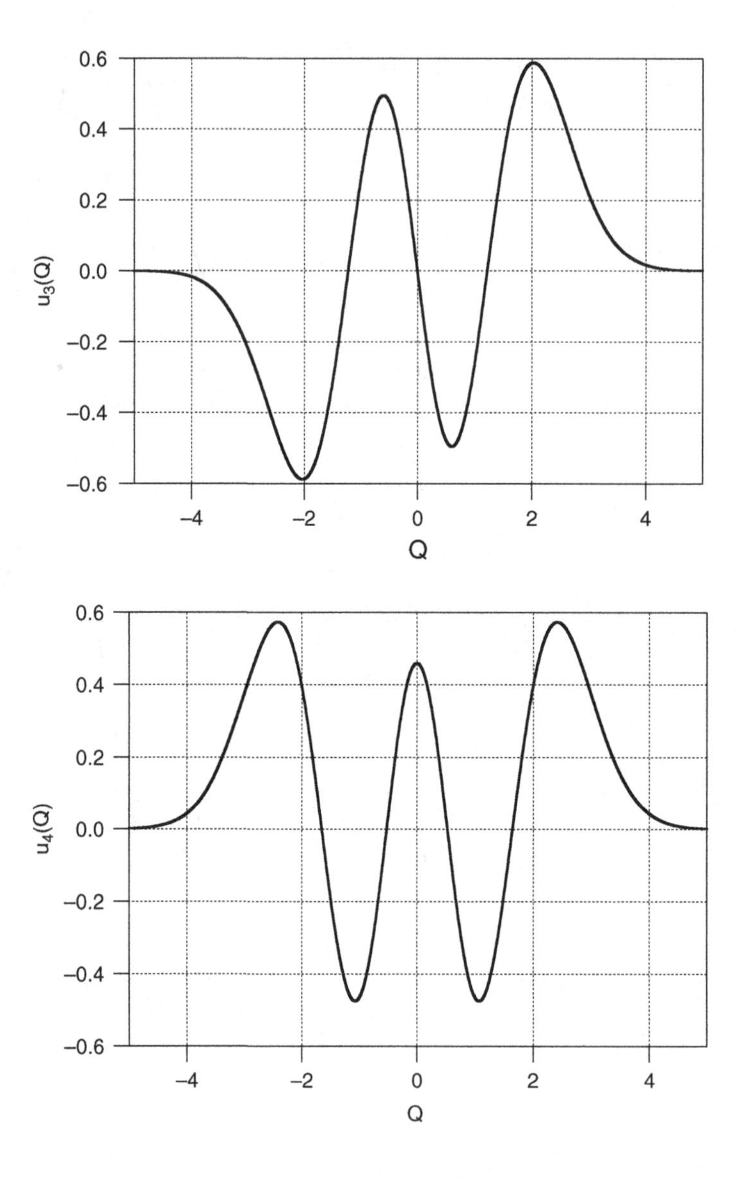

Figure 6.14 (*cont.*)

Hence $da'/dt = -i\omega a'$, and similarly, $da'^{\dagger}/dt = i\omega a'^{\dagger}$. Integrating these equations, we obtain

$$a'(t) = e^{-i\omega t}a'(0) = e^{-i\omega t}a \tag{6.57}$$

and

$$a'^{\dagger}(t) = e^{i\omega t}a'^{\dagger}(0) = e^{i\omega t}a^{\dagger} \tag{6.58}$$

Now

$$\hat{Q}' = \frac{1}{\sqrt{2}}\left(a' + a'^{\dagger}\right) \qquad \hat{P}' = \frac{i}{\sqrt{2}}\left(a'^{\dagger} - a'\right)$$

Thus

$$\hat{Q}'(t) = \cos\omega t\hat{Q}'(0) + \sin\omega t\hat{P}'(0)$$

and

$$\hat{P}'(t) = \cos\omega t\hat{P}'(0) - \sin\omega t\hat{Q}'(0)$$

Taking into account the scaling relations (6.41), we finally obtain

$$\hat{q}'(t) = \hat{q}'(0)\cos\omega t + \frac{1}{m\omega}\hat{p}'(0)\sin\omega t \quad (6.59)$$

and

$$\hat{p}'(t) = \hat{p}'(0)\cos\omega t - m\omega\hat{q}'(0)\sin\omega t \quad (6.60)$$

It can be seen that these relations are exactly the same as the analogous relations for the classical harmonic oscillator variables q_c and p_c.

6.13.5 Oscillating wave packets: Classical limit for the harmonic oscillator

We usually think of a classical harmonic oscillator as a localized mass that oscillates back and forth about the origin. On the other hand, each $|n\rangle$ state is stationary, with a probability density $\rho_n = |u_n(Q)|^2$ that is constant in time, exhibiting no motion. Clearly, then, a state $|\Psi(t)\rangle$ that describes classical simple harmonic motion cannot be a single $|n\rangle$ state but must be an appropriate superposition of $|n\rangle$ states formed in such a way that the associated probability density $\rho = |\langle Q|\Psi\rangle|^2$ executes simple harmonic oscillation about the origin.

In fact, we can construct such a *coherent* state $|\Psi(t)\rangle$ by displacing the ground state $|0\rangle$ from the origin in one-dimensional coordinate space and then following its time evolution. The displacement s (along the Q axis and of arbitrary magnitude) is achieved by means of the translation operator

$$S(s) = \exp\left(-i\hat{P}s\right) = \exp\left[\frac{s}{\sqrt{2}}(a^\dagger - a)\right] \quad (6.61)$$

Formula (6.61) is inconvenient because the noncommuting operators a and $a^\dagger$ both appear in the exponent on the right-hand side. Thus, as a first step toward our goal of constructing $|\Psi(t)\rangle$, we write (6.61) in a more convenient form by employing two theorems, the first of which was already discussed in Section 2.17 as follows: let A and B be two operators that do not necessarily commute. Then

$$e^A B e^{-A} = B + [A, B] + \frac{1}{2!}[A,[A, B]] + \ldots \quad (6.62)$$

If $[A, B] = C$, where C is a c-number, (6.62) reduces to

$$e^{A}Be^{-A} = B + C$$

In this case, the second theorem states that

$$e^{A+B}e^{-B}e^{-A} = \exp\left\{-\frac{[A,B]}{2}\right\} = e^{-C/2} \tag{6.63}$$

To prove (6.63), write $F(\lambda) = e^{\lambda(A+B)}e^{-\lambda B}e^{-\lambda A}$, where λ is a continuous parameter. Then

$$\frac{\partial F(\lambda)}{\partial \lambda} = (A+B)F(\lambda) - e^{\lambda(A+B)}Be^{-\lambda(A+B)}F(\lambda) - e^{\lambda(A+B)}e^{-\lambda B}Ae^{\lambda B}e^{-\lambda B}e^{-\lambda A}$$

However,

$$e^{\lambda(A+B)}Be^{-\lambda(A+B)} = B + \lambda C$$

and

$$\begin{aligned} e^{\lambda(A+B)}e^{-\lambda B}Ae^{\lambda B}e^{-\lambda B}e^{-\lambda A} &= e^{\lambda(A+B)}(A+\lambda C)e^{-\lambda B}e^{-\lambda A} \\ &= \lambda CF + (A - \lambda C)F = AF \end{aligned}$$

Therefore,

$$\frac{\partial F(\lambda)}{\partial \lambda} = -\lambda CF$$

Because $F(0) = 1$, $F(\lambda)$ is a c-number for any λ, and in fact,

$$F(\lambda) = \exp\left(-\lambda^2 \frac{C}{2}\right)$$

Thus, setting $\lambda = 1$, we obtain (6.63). Application to the right-hand side of (6.61) yields the desired convenient formula for S; that is,

$$S(s) = \exp\left(-\frac{s^2}{4}\right)\exp\left(\frac{s}{\sqrt{2}}a^\dagger\right)\exp\left(-\frac{s}{\sqrt{2}}a\right) \tag{6.64}$$

which we now use to displace $|0\rangle$. Bearing in mind that $a|0\rangle = 0$, we have

$$\begin{aligned} |\Psi(0)\rangle = S(s)|0\rangle &= e^{-s^2/4}\sum_0^\infty \frac{1}{n!}\left(\frac{s}{\sqrt{2}}\right)^n (a^\dagger)^n|0\rangle \\ &= e^{-s^2/4}\sum_0^\infty \frac{1}{\sqrt{n!}}\left(\frac{s}{\sqrt{2}}\right)^n |n\rangle \end{aligned} \tag{6.65}$$

The probability of finding $|\Psi(0)\rangle$ in a particular state $|n\rangle$ is

$$p_n = \left|\langle n|\Psi(0)\rangle\right|^2 = \frac{1}{n!}\left(\frac{s^2}{2}\right)^n \exp\left(-\frac{s^2}{2}\right) \tag{6.66}$$

This is a Poisson distribution with mean value

$$\bar{n} = \frac{s^2}{2} \tag{6.67}$$

and rms dispersion

$$\Delta n = \sqrt{\overline{n^2} - \bar{n}^2} = \sqrt{\bar{n}} = \frac{s}{\sqrt{2}} \tag{6.68}$$

To determine the time evolution of $|\Psi\rangle$, we write

$$\begin{aligned} |\Psi(t)\rangle &= U(t,0)|\Psi(0)\rangle = US(s)|0\rangle \\ &= \exp\left(-\frac{s^2}{4}\right)\left[U\exp\left(\frac{s}{\sqrt{2}}a^\dagger\right)U^\dagger\right]\left[U\exp\left(-\frac{s}{\sqrt{2}}a\right)U^\dagger\right]U|0\rangle \end{aligned} \tag{6.69}$$

Now

$$U(t,0)|0\rangle = \exp\left(-i\frac{E_0 t}{\hbar}\right)|0\rangle = \exp\left(-i\frac{\omega t}{2}\right)|0\rangle$$

Also,

$$U\exp\left(\frac{s}{\sqrt{2}}a^\dagger\right)U^\dagger = \sum_{k=0}^{\infty}\frac{s^k}{2^{k/2}k!}U\left(a^\dagger\right)^k U^\dagger$$

However,

$$U\left(a^\dagger\right)^k U^\dagger = (Ua^\dagger U^\dagger)(Ua^\dagger U^\dagger)\cdots(Ua^\dagger U^\dagger)$$

where the product on the right-hand side contains k factors. Thus

$$U\exp\left(\frac{s}{\sqrt{2}}a^\dagger\right)U^\dagger = \exp\left[U\left(\frac{s}{\sqrt{2}}a^\dagger\right)U^\dagger\right]$$

and similarly,

$$U\exp\left(-\frac{s}{\sqrt{2}}a\right)U^\dagger = \exp\left[-U\left(\frac{s}{\sqrt{2}}a\right)U^\dagger\right]$$

Furthermore, multiplying (6.57) and, similarly, (6.58) on the left by U and on the right by $U^\dagger$, we obtain

$$Ua^\dagger U^\dagger = a^\dagger e^{-i\omega t}$$

and

$$UaU^\dagger = ae^{i\omega t}$$

Therefore, (6.69) can be written

$$\begin{aligned}|\Psi(t)\rangle &= e^{-i\omega t/2}\exp\left(-\frac{s^2}{4}\right)\left[\exp\left(\frac{s}{\sqrt{2}}a^\dagger e^{-i\omega t}\right)\right]\left[\exp\left(-\frac{s}{\sqrt{2}}ae^{i\omega t}\right)\right]|0\rangle \\ &= e^{-i\omega t/2}\exp\left[\frac{s}{\sqrt{2}}\left(a^\dagger e^{-i\omega t}-ae^{i\omega t}\right)\right]|0\rangle\end{aligned} \tag{6.70}$$

Now,

$$\begin{aligned}\frac{1}{\sqrt{2}}\left(a^\dagger e^{-i\omega t}-ae^{i\omega t}\right) &= \frac{1}{\sqrt{2}}\left[\left(a^\dagger - a\right)\cos\omega t - i\left(a^\dagger + a\right)\sin\omega t\right] \\ &= -i\left(\hat{P}\cos\omega t + \hat{Q}\sin\omega t\right)\end{aligned}$$

Therefore, (6.70) becomes

$$\begin{aligned}|\Psi(t)\rangle &= e^{-i\omega t/2}\exp\left[-is\left(\hat{P}\cos\omega t + \hat{Q}\sin\omega t\right)\right]|0\rangle \\ &= e^{-i\omega t/2}e^{i\left(s^2\sin 2\omega t\right)/4}e^{-is\hat{Q}\sin\omega t}e^{-is\hat{P}\cos\omega t}|0\rangle\end{aligned}$$

This yields the coordinate-space wave function

$$\psi(Q,t) = \langle Q|\Psi(t)\rangle = e^{-i\omega t/2}e^{i\left(s^2\sin 2\omega t\right)/4}e^{-isQ\sin\omega t}\left\langle Q\left|e^{-is\hat{P}\cos\omega t}\right|0\right\rangle$$

with corresponding probability density

$$\rho(Q,t) = |\psi(Q,t)|^2 = \left|\left\langle Q\left|e^{-is\hat{P}\cos\omega t}\right|0\right\rangle\right|^2 \tag{6.71}$$

Thus, as is obvious from the right-hand side of (6.71), the coherent wave packet oscillates back and forth about the origin with amplitude s and angular frequency ω, retaining the shape of the ground-state wave packet as time elapses. Such behavior is unique to the simple harmonic oscillator.

To appreciate the significance of the foregoing for description of the simple harmonic oscillator in the classical limit, it is instructive to work out a numerical example. From (6.67), the energy of the state $n=\bar{n}$ that contributes most to the coherent wave packet is

$$E_{\bar{n}} = \hbar\omega\left(\bar{n}+\frac{1}{2}\right) = \hbar\omega\left(\frac{s^2}{2}+\frac{1}{2}\right) \approx \hbar\omega\frac{s^2}{2}$$

where in the last step we assume that $s \gg 1$. Because $Q=(m\omega/\hbar)^{1/2}q$, it is convenient to express s as

$$s = \left(\frac{m\omega}{\hbar}\right)^{1/2} d$$

where d is the amplitude of simple harmonic motion in ordinary (cgs) units. Thus we have

$$E_{\bar{n}} \approx \hbar\omega \frac{m\omega}{2\hbar} d^2 = \frac{m\omega^2 d^2}{2}$$

This is the classical energy of a mass point m executing simple harmonic motion with angular frequency ω and amplitude d. For example, consider a classical oscillator with $m = 1$ g, $d = 1$ cm, and $\omega = 1$ rad/s. Then

$$E_{\bar{n}} = \frac{1}{2} = \frac{\hbar s^2}{2}$$

which implies that

$$s = \frac{1}{\sqrt{\hbar}} \approx 3 \times 10^{13}$$

as well as

$$\bar{n} = \frac{s^2}{2} \approx 5 \times 10^{26}$$

and

$$\Delta n = \sqrt{\bar{n}} \approx 2 \times 10^{13}$$

Although $\Delta n >> 1$ in this example, the fractional spread in n is extremely small; that is,

$$\frac{\Delta n}{\bar{n}} \approx 4 \times 10^{-14}$$

Thus the classical energy, while not exactly sharp, is defined to a precision better than one part in 10^{13}. Hence, in this example, the classical description of simple harmonic motion is accurate for all practical purposes.

6.13.6 Shape invariance and potentials with exact solutions

There are not many one-dimensional attractive potentials for which exact analytic eigenfunctions are known, and only a few of these are relevant to real physical situations. Nevertheless, it is worth our while to summarize a generalization of the harmonic oscillator creation and destruction operators that allows us to generate one or more new potentials and their associated exact solutions from a known potential and its exact solutions. For this purpose, consider a Hamiltonian with potential $V_-(x)$ and known bound-state eigenfunctions $u_0, u_1, \ldots$ with corresponding energies $E_0, E_1, \ldots$, the latter arranged in increasing order. For convenience, we choose the zero of energy, so $E_0 = 0$. Then, in units where $\hbar = m = 1$, we have

$$\left[-\frac{1}{2}\frac{\partial^2}{\partial x^2}+V_-(x)\right]u_0=0 \tag{6.72}$$

Hence $V_- = u_0''/2u_0$, so we can write

$$H_- = \frac{1}{2}\left(-\frac{\partial^2}{\partial x^2}+\frac{u_0''}{u_0}\right)$$

We now define two differential operators

$$A=\frac{1}{\sqrt{2}}\left(\frac{\partial}{\partial x}-\frac{u_0'}{u_0}\right) \quad \text{and} \quad A^\dagger = \frac{1}{\sqrt{2}}\left(-\frac{\partial}{\partial x}-\frac{u_0'}{u_0}\right) \tag{6.73}$$

that are the analogues of the harmonic oscillator destruction and creation operators, respectively. It is easy to verify that $H_- = A^\dagger A$. Also, defining $H_+ = AA^\dagger$, it is easy to show that

$$H_+ = H_- - \frac{\partial}{\partial x}\left(\frac{u_0'}{u_0}\right)$$

We now define a superpotential $W(x)$ by

$$W(x) = -\frac{1}{\sqrt{2}}\frac{u_0'}{u_0}$$

From the foregoing relationships, the next five equations follow immediately:

$$u_0(x) = \text{const} \bullet \exp\left[-\sqrt{2}\int^x W(x)\,dx\right]$$

$$A = \frac{1}{\sqrt{2}}\frac{\partial}{\partial x}+W(x)$$

$$A^\dagger = -\frac{1}{\sqrt{2}}\frac{\partial}{\partial x}+W(x)$$

$$V_-(x) = W^2 - \frac{1}{\sqrt{2}}\frac{\partial W}{\partial x}$$

and

$$V_+(x) = W^2 + \frac{1}{\sqrt{2}}\frac{\partial W}{\partial x}$$

Now we show that except for $E_0 = 0$, the eigenvalues of H_- and H_+ are identical. Recall that $H_- u_n = E_n u_n$, and let the analogous eigenvalue equation for H_+ be $H_+ \mathrm{v}_n = F_n \mathrm{v}_n$. Then

$$\begin{aligned} H_+(Au_n) &= AA^\dagger Au_n \\ &= A(A^\dagger Au_n) \\ &= AE_n u_n \\ &= E_n(Au_n) \end{aligned}$$

which implies that Au_n is an eigenstate of H_+ with eigenvalue E_n. Similarly,

$$H_-(A^\dagger \mathrm{v}_n) = A^\dagger AA^\dagger \mathrm{v}_n = A^\dagger F_n \mathrm{v}_n = F_n A^\dagger \mathrm{v}_n$$

hence $A^\dagger \mathrm{v}_n$ is an eigenstate of H_- with eigenvalue F_n. Also $Au_0 = 0$. Thus, although the shapes of the potentials V_- and V_+ are generally quite different, their eigenvalue spectra are the same (with the exception of $E_0 = 0$), and these spectra are said to be *shape invariant*. The method is illustrated by the following simple example of the square-well potential with infinite walls:

$$\begin{aligned} V(x) &= 0 \qquad && |x| \le \frac{\pi}{2} \\ V(x) &= +\infty \qquad && |x| > \frac{\pi}{2} \end{aligned}$$

The eigenfunctions and corresponding energy eigenvalues are, of course, known as follows:

n	Eigenfunction[a]	Energy	Shifted Energy
0	$\cos x$	1/2	0
1	$\sin 2x$	4/2	3/2
2	$\cos 3x$	9/2	8/2
.			
.			
.			
2k	$\cos(2k+1)x$	$(2k+1)/2$	$2k(k+1)$
	$\sin 2nx$	$2n^2$	$2n^2 - 1/2$

[a]Not normalized.

In the last column of the table we list the shifted energy that occurs when we make the replacement $V(x) \to V_-(x) = V(x) - ½$. With the understanding that all remarks apply to the region $|x| \le \pi/2$, the superpotential is

$$W = -\frac{1}{\sqrt{2}}\frac{u_0'}{u_0} = \frac{1}{\sqrt{2}}\tan x$$

Thus the new potential is

$$V_+ = W^2 + \frac{1}{\sqrt{2}}W' = \frac{1}{2}(\tan^2 x + \sec^2 x) = \frac{1}{2} + \tan^2 x$$

The eigenstates of H_+ are $\mathrm{v}_0 = Au_1$, $\mathrm{v}_1 = Au_2$, $\mathrm{v}_2 = Au_3$, and so forth. Because

$$A = \frac{1}{\sqrt{2}}\left(\frac{\partial}{\partial x} - \frac{u_0'}{u_0}\right) = \frac{1}{\sqrt{2}}\left(\frac{\partial}{\partial x} + \tan x\right)$$

we have

$$v_0 = \text{const} \bullet \cos^2 x$$

$$v_1 = \text{const} \bullet \sin x \cos^2 x$$

and so on. Sometimes it is useful to iterate the process by treating $V_+ + \delta$ as a new V_-, where δ is a suitable energy shift.

6.14 Path integral method

Because the Schroedinger equation

$$-\frac{\hbar^2}{2m}\nabla^2\psi + V(\boldsymbol{r},t)\psi = i\hbar\frac{\partial\psi}{\partial t} \tag{6.74}$$

is first order in time, it is possible to find $\psi(\boldsymbol{r},t)$ for all $t > t_0$ if we are given $\psi(\boldsymbol{r}_0,t_0)$ for all $\boldsymbol{r}_0$. The relationship between $\psi(\boldsymbol{r},t)$ and $\psi(\boldsymbol{r}_0,t_0)$ can be expressed in terms of a Green function G and an integral equation. In one spatial dimension, we have

$$\psi(x,t) = \int G(x,t;x_0,t_0)\psi(x_0,t_0)\,dx_0 \tag{6.75}$$

In the special case of a free particle, we have already found $G = G_0$ [see Section 6.1, equation (6.5)]. To obtain G when a potential V is present, we introduce the path integral method, which was initiated by P. A. M. Dirac in the 1930s, independently described by E. Stueckelberg and developed by R. P. Feynman. For most problems in quantum mechanics, the actual calculation of a wave function from (6.75) or its generalization to three spatial dimensions using the path integral method is more complicated than straightforward integration of the Schroedinger equation by analytical and/or numerical methods. The path integral method is especially unsuited to bound-state problems or to problems involving particles with spin. However, it is useful in many scattering problems for developing general insights into the connection between quantum mechanics and classical mechanics and in quantum statistical mechanics. Finally, when suitably generalized, the path integral method plays an important role in the formulation of modern field theories.

To begin our discussion of the path integral method, we divide the time interval t–t_0 in (6.75) into $n + 1$ small intervals ε as shown in Figure 6.15 and define

$$\begin{aligned}
t_0 &= t_0 \\
t_1 &= t_0 + \varepsilon \\
t_2 &= t_0 + 2\varepsilon \\
&\vdots \\
t_n &= t_0 + n\varepsilon \\
t_{n+1} &= t_0 + (n+1)\varepsilon = t
\end{aligned}$$

Using the completeness relation repeatedly, we write

$$\begin{aligned}
\psi(x,t) &= \langle x|\Psi(t)\rangle = \int \langle x|e^{-iH(t_{n+1}-t_n)/\hbar}|x_n\rangle\langle x_n|\Psi(t_n)\rangle\, dx_n \\
&= \int dx_n \int dx_{n-1} \langle x|e^{-iH\varepsilon/\hbar}|x_n\rangle\langle x_n|e^{-iH\varepsilon/\hbar}|x_{n-1}\rangle\langle x_{n-1}|\Psi(t_{n-1})\rangle \\
&\ \ \vdots \\
&= \int dx_n \int dx_{n-1} \ldots \int dx_1 \int dx_0 \\
&\qquad \langle x|e^{-iH\varepsilon/\hbar}|x_n\rangle\langle x_n|e^{-iH\varepsilon/\hbar}|x_{n-1}\rangle \ldots \langle x_1|e^{-iH\varepsilon/\hbar}|x_0\rangle\langle x_0|\Psi(t_0)\rangle
\end{aligned} \tag{6.76}$$

Noting that $\psi(x_0,t_0) = \langle x_0|\Psi(t_0)\rangle$, we see that (6.75) and (6.76) yield

$$G(x,t;x_0,t_0) = \int dx_n \int dx_{n-1} \cdots \int dx_1 \langle x_{n+1}|e^{-iH\varepsilon/\hbar}|x_n\rangle \cdots \langle x_1|e^{-iH\varepsilon/\hbar}|x_0\rangle \tag{6.77}$$

where $x_{n+1} = x$. Because ε is infinitesimal, we write: $U(\varepsilon) = e^{-iH\varepsilon/\hbar} \approx 1 - iH\varepsilon/\hbar$, and we assume that H can be expressed as a power series in operators $\hat{x}$ and $\hat{p}$. Thus we have

$$\begin{aligned}
\langle x_{n+1}|U(\varepsilon)|x_n\rangle &= \int \left\langle x_{n+1}\left|\exp\left[-\frac{i}{\hbar}H(\hat{x},\hat{p})\varepsilon\right]\right|p_{n+1}\right\rangle\langle p_{n+1}|x_n\rangle\, dp_{n+1} \\
&= \int \exp\left[-\frac{i}{\hbar}H(x_{n+1},p_{n+1})\varepsilon\right]\langle x_{n+1}|p_{n+1}\rangle\langle p_{n+1}|x_n\rangle\, dp_{n+1} \\
&= \frac{1}{2\pi\hbar}\int \exp\left[-\frac{i}{\hbar}H(x_{n+1},p_{n+1})\varepsilon\right]\exp\left[\frac{i}{\hbar}p_{n+1}(x_{n+1}-x_n)\right] dp_{n+1}
\end{aligned}$$

with similar expressions for all the other factors in the integrand of (6.77). Thus we obtain

$$G(x,t;x_0,t_0) = \frac{1}{(2\pi\hbar)^{n+1}}\prod_{i=1}^{n}\int dx_i \prod_{j=1}^{n+1}\int dp_j \exp\left\{\frac{i}{\hbar}\varepsilon\sum_{k=1}^{n+1}\left[p_k\left(\frac{x_k - x_{k-1}}{\varepsilon}\right) - H(x_k,p_k)\right]\right\} \tag{6.78}$$

In many cases of interest, the Hamiltonian can be written as

$$H(x_k,p_k) = \frac{p_k^2}{2m} + V(x_k)$$

An important exception occurs when there is a velocity-dependent potential, as in electrodynamics, but we ignore that case for present purposes. Then (6.78) becomes

$$\begin{aligned}
G(x,t;x_0,t_0) &= \frac{1}{(2\pi\hbar)^{n+1}}\prod_{i=1}^{n}\int \exp\left[-\frac{i\varepsilon}{\hbar}V(x_i)\right] dx_i \\
&\quad \prod_{j=1}^{n+1}\int dp_j \exp\left\{\frac{i}{\hbar}\varepsilon\sum_{k=1}^{n+1}\left[p_k\left(\frac{x_k - x_{k-1}}{\varepsilon}\right) - \frac{p_k^2}{2m}\right]\right\}
\end{aligned} \tag{6.79}$$

The $n + 1$ integrals over the momenta are readily evaluated by completing the square in each exponent, just as in Section 6.1. We thus have

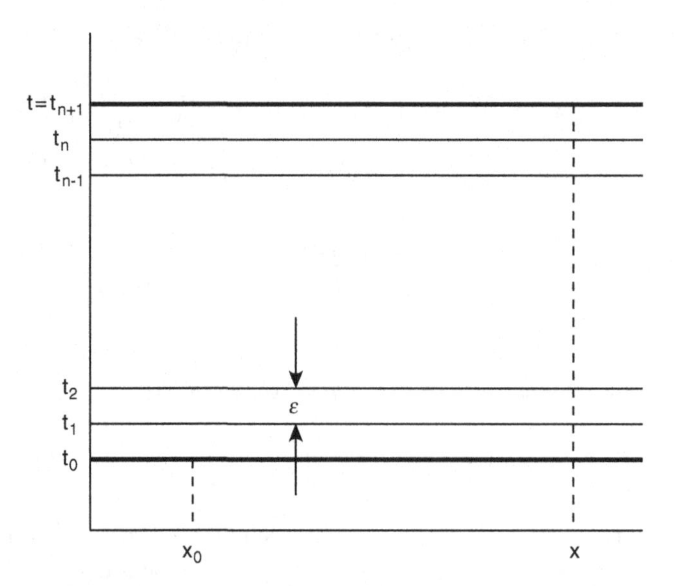

Figure 6.15 The time interval $t-t_0$ divided into $n+1$ equal sub-intervals of magnitude ε

$$\frac{1}{(2\pi\hbar)}\int_{-\infty}^{\infty}dp_j\exp\left\{\frac{i}{\hbar}\varepsilon\left[p_j\left(\frac{x_j-x_{j-1}}{\varepsilon}\right)-\frac{p_j^2}{2m}\right]\right\}=\sqrt{\frac{m}{2\pi i\hbar\varepsilon}}\exp\left[\frac{im}{2\hbar}\frac{\left(x_j-x_{j-1}\right)^2}{\varepsilon}\right]$$

Hence (6.79) becomes

$$G(x,t;x_0,t_0)=\left(\frac{m}{2\pi i\hbar\varepsilon}\right)^{(n+1)/2}\prod_{i=1}^{n}\int\exp\left\{\frac{i}{\hbar}\sum_{j=1}^{n+1}\varepsilon\left[\frac{m\left(x_j-x_{j-1}\right)^2}{\varepsilon^2}-V\left(x_j\right)\right]\right\}dx_i \tag{6.80}$$

In the limit as $\varepsilon\to 0$ and $n\to\infty$, $\left(x_j-x_{j-1}\right)/\varepsilon$ becomes dx_j/dt, and the sum in the exponent of (6.80) becomes the integral

$$\int\left[m\frac{\dot{x}^2}{2}-V(x)\right]dt=\int_{x_0}^{x'}Ldt=S$$

where L is the Lagrangian and S is the action. Thus, finally, we have

$$G(x,t;x_0,t_0)=\int[dx]\exp\left(\frac{i}{\hbar}S\right) \tag{6.81}$$

where

$$\int[dx]\equiv\lim_{\substack{n\to\infty\\ \varepsilon\to 0}}\left(\frac{m}{2\pi i\hbar\varepsilon}\right)^{(n+1)/2}\prod_{i=1}^{n}\int dx_i \tag{6.82}$$

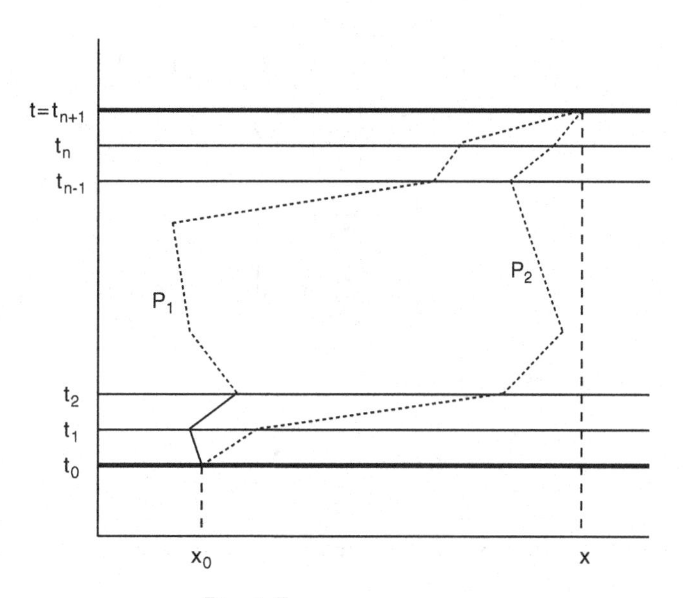

Figure 6.16 The Green function is the sum over all paths of $\exp\left[(i/\hbar)S\right]$. Two paths $P_{1,2}$ are shown.

It is easy to show that for the special case of the free particle, (6.81) reduces to (6.6) of Section 6.1.

What is the meaning of (6.81)? It states that we should construct the action, multiply it by $i/\hbar$, exponentiate, and then integrate over all values of $x_1, x_2, \ldots, x_n$. A given choice of $x_1, x_2, \ldots, x_n$ corresponds to a given path between fixed end points x_0 and x. Another choice of $x_1, x_2, \ldots, x_n$ corresponds to another path with the same end points. Thus integration over all values of $x_1, x_2, \ldots, x_n$ means summation of $\exp\left[(i/\hbar)S\right]$ over all paths between x_0 and x, which is illustrated in Figure 6.16.

In classical mechanics, the action S is always very large compared with $\hbar$. Thus $\exp\left[(i/\hbar)S\right]$ generally oscillates very rapidly when we go from one path to a nearby path. Contributions from neighboring paths thus tend to cancel, except in the immediate vicinity of that path, where S takes an extreme value (most often a minimum). Hence the classical path is the path of least action. In quantum mechanics, S is comparable with $\hbar$, and $\exp\left[(i/\hbar)S\right]$ varies slowly from one path to another. Therefore, many paths contribute to G. The reader will see that we have formulated a way to calculate the propagation of matter waves in nonrelativistic wave mechanics according to Huyghens' principle. Similar considerations are relevant in the transition from geometric optics to wave optics, where the principle of least action is replaced by Fermat's principle.

Before proceeding further, we note that the Green function can be expressed in a useful alternative form. Suppose that the eigenstates of the Hamiltonian are $|\Phi_n\rangle$ with corresponding eigenvalues E_n and spatial wave functions $\phi_n(x) = \langle x|\Phi_n\rangle$. Using the completeness relation, we have

$$\begin{aligned}\psi(x,t) &= \langle x|\Psi(t)\rangle = \langle x|U(t,t_0)|\Psi(t_0)\rangle \\ &= \sum_n \left\langle x\left|\exp\left[-\frac{i}{\hbar}H(t-t_0)\right]\right|\Phi_n\right\rangle\langle\Phi_n|\Psi(t_0)\rangle \\ &= \int dx_0 \sum_n \left\langle x\left|\exp\left[-\frac{i}{\hbar}E_n(t-t_0)\right]\right|\Phi_n\right\rangle\langle\Phi_n|x_0\rangle\langle x_0|\Psi(t_0)\rangle \\ &= \int dx_0 \sum_n \left\langle x\left|\exp\left[-\frac{i}{\hbar}E_n(t-t_0)\right]\right|\Phi_n\right\rangle\langle\Phi_n|x_0\rangle\psi(x_0,t_0)\end{aligned}$$

Thus the Green function is

$$G(x,t;x_0,t_0) = \sum_n \exp\left[-\frac{i}{\hbar}E_n(t-t_0)\right]\phi_n(x)\phi_n^*(x_0) \tag{6.83}$$

Of course, in (6.83), Σ may mean integration over a continuous parameter and/or summation over discrete values.

Now let us look more carefully at the action; that is,

$$S = \int_{t_0}^{t} L(x,\dot{x})\,d\tau$$

The change in S that occurs when we change from a given path to a closely neighboring path is

$$\begin{aligned}\delta S &= \int_{t_0}^{t}\left(\frac{\partial L}{\partial x}\delta x + \frac{\partial L}{\partial \dot{x}}\delta \dot{x}\right)d\tau \\ &= \int_{t_0}^{t}\left[\frac{\partial L}{\partial x}\delta x + \frac{\partial L}{\partial \dot{x}}\frac{\partial}{\partial t}(\delta x)\right]d\tau\end{aligned} \tag{6.84}$$

Here δx means variation of the path by an independent infinitesimal amount for each value of τ subject to the constraint that the end points of the path at t_0 and t are fixed. Equation (6.84) may be rewritten as follows:

$$\delta S = \int_{t_0}^{t'}\left[\frac{\partial L}{\partial x}\delta x + \frac{\partial}{\partial \tau}\left(\frac{\partial L}{\partial \dot{x}}\delta x\right) - \left(\frac{\partial}{\partial \tau}\frac{\partial L}{\partial \dot{x}}\right)\delta x\right]d\tau \tag{6.85}$$

The second term on the right hand side of (6.85) is the integral of a derivative, and it vanishes because $\delta x(t) = \delta x(t_0) = 0$. Thus,

$$\delta S = \int_{t_0}^{t}\left[\frac{\partial L}{\partial x} - \left(\frac{\partial}{\partial \tau}\frac{\partial L}{\partial \dot{x}}\right)\right]\delta x\,d\tau \tag{6.86}$$

Also, for the classical path, $\delta S_c = 0$. Because this holds for arbitrary variations δx, we must have

$$\frac{\partial L}{\partial x} - \left(\frac{\partial}{\partial \tau} \frac{\partial L}{\partial \dot{x}} \right) = 0 \tag{6.87}$$

for the classical path. Of course, equation (6.87) is Lagrange's equation (the classical equation of motion), and it is obeyed on the classical path but not in general on other paths. Denoting the classical path by $x_c(\tau)$, we specify another path by $x(\tau) = x_c(\tau) + y(\tau)$, where $y(t_0) = y(t) = 0$. The action for this path is

$$S = \int_{t_0}^{t} \left[\frac{m}{2} (\dot{x}_c + \dot{y})^2 - V(x_c + y) \right] d\tau$$

$$= \int_{t_0}^{t} \left[\frac{m}{2} \dot{x}_c^2 - V(x_c) \right] d\tau + \int_{t_0}^{t} \left(m\dot{x}_c \dot{y} - \left.\frac{\partial V}{\partial x}\right|_{x_c} y \right) d\tau + \int_{t_0}^{t} \left(\frac{m}{2} \dot{y}^2 - \frac{1}{2!} \left.\frac{\partial^2 V}{\partial x^2}\right|_{x_c} y^2 + \cdots \right) d\tau$$

Here the first integral on the right-hand side is the classical action S_c, and the second integral vanishes because its integrand, after partial integration of the first term, is

$$-y \bullet \left(\frac{d}{d\tau} \frac{\partial L}{\partial \dot{x}} - \frac{\partial L}{\partial x} \right)_{x_c}$$

Thus

$$\exp\left(\frac{iS}{\hbar} \right) = \exp\left(\frac{iS_c}{\hbar} \right) \exp\left[\frac{i}{\hbar} \int_{t_0}^{t} \left(\frac{m}{2} \dot{y}^2 - \frac{1}{2!} \left.\frac{\partial^2 V}{\partial x^2}\right|_{x_c} y^2 + \ldots \right) d\tau \right] \tag{6.88}$$

We now integrate (6.88) over all paths with fixed end points $x_0(t_0)$, $x(t)$ to obtain the Green function. In this sum over paths, the classical path is fixed, and only y varies. Hence

$$G = \exp\left(\frac{iS_c}{\hbar} \right) \int_{0}^{0} [dy] \exp\left(\frac{iS'}{\hbar} \right) \tag{6.89}$$

where the inferior and superior limits of integration are written to indicate that $y(t_0) = y(t) = 0$ for all paths, and

$$S' = \int_{t_0}^{t} \left(\frac{m}{2} \dot{y}^2 - \frac{1}{2!} \left.\frac{\partial^2 V}{\partial x^2}\right|_{x_c} y^2 + \ldots \right) d\tau \tag{6.90}$$

If $\partial^2 V / \partial x^2$ is a constant, all terms of order y^3 and higher vanish in the integrand of (6.90). Then, because $y(t_0) = y(t) = 0$, the sum over all paths in (6.89) depends only on t_0 and t and can be expressed as $F(t, t_0)$, where the form of the function F depends on V. Thus

$$G = F(t, t_0) \exp\left(\frac{iS_c}{\hbar} \right) \tag{6.91}$$

We now give three examples where (6.91) applies.

The free particle

We already know the Green function for a free particle: it is given by (6.6). The Lagrangian is $L = m\dot{x}^2/2$, and on the classical path, $m\ddot{x} = 0$. It is easy to show from this that

$$S_c = \frac{m}{2}\frac{(x-x_0)^2}{(t-t_0)}$$

Thus (6.91) yields

$$G_0 = F_0(t,t_0)\exp\left[\frac{im}{2\hbar}\frac{(x-x_0)^2}{(t-t_0)}\right] \tag{6.92}$$

Comparison with (6.6) yields

$$F_0(t,t_0) = \sqrt{\frac{m}{2\pi i\hbar(t-t_0)}} \tag{6.93}$$

Particle in a Uniform Field

Here the Lagrangian is $L = (m/2)\dot{x}^2 - mgx$, and the classical equation of motion is $\ddot{x} = -g$. Assuming that $t_0 = 0$, we easily obtain

$$S_c = \frac{mx^2}{2t} - \frac{mgxt}{2} - \frac{mg^2t^3}{24}$$

As for $F(t,0)$, it is the same as for a free particle because $\partial^2 V/\partial x^2 = 0$. Therefore,

$$G = \sqrt{\frac{m}{2\pi i\hbar t}}\exp\left[\frac{im}{2\hbar}\left(\frac{x^2}{t} - gxt - \frac{g^2t^3}{12}\right)\right] \tag{6.94}$$

Harmonic oscillator

Once again, we assume that $t_0 = 0$. The Lagrangian is $L = (m/2)(\dot{x}^2 - \omega^2 x^2)$, and the classical equation of motion is $\ddot{x} + \omega^2 x = 0$. Thus, on the classical path,

$$x = A\cos\omega\tau + B\sin\omega\tau$$

where $A = x_0$ and $B = (x - x_0\cos\omega t)/\sin\omega t$. From this it is straightforward to show that the classical action is

$$S_c = \frac{m\omega}{2}\left[(x^2 + x_0^2)\frac{\cos\omega t}{\sin\omega t} - \frac{2x_0 x}{\sin\omega t}\right] \tag{6.95}$$

Now G satisfies the Schroedinger equation

$$-\frac{\hbar^2}{2m}\frac{\partial^2 G}{\partial x^2}+V(x)G=i\hbar\frac{\partial G}{\partial t}.$$

Because $G=F(t,0)\exp(iS_c/\hbar)$, we have

$$\frac{\partial G}{\partial t}=\frac{\dot{F}}{F}G+\frac{i}{\hbar}\dot{S}_c G$$

$$\frac{\partial G}{\partial x}=\frac{i}{\hbar}\frac{\partial S_c}{\partial x}G$$

$$\frac{\partial^2 G}{\partial x^2}=\left[\frac{i}{\hbar}\frac{\partial^2 S_c}{\partial x^2}-\frac{1}{\hbar^2}\left(\frac{\partial S_c}{\partial x}\right)^2\right]G$$

It follows that

$$\frac{1}{2m}\left(\frac{\partial S_c}{\partial x}\right)^2+\frac{\partial S_c}{\partial t}+V(x)=i\hbar\left(\frac{1}{2m}\frac{\partial^2 S_c}{\partial x^2}+\frac{\dot{F}}{F}\right) \tag{6.96}$$

However, it is easy to verify that the left-hand side of (6.96) is zero. Thus

$$\frac{\dot{F}}{F}=-\frac{1}{2m}\frac{\partial^2 S_c}{\partial x^2}=-\frac{\omega}{2}\cot\omega t$$

This immediately yields

$$F=\sqrt{\frac{\text{const}\bullet\omega}{\sin\omega t}}$$

In the limit where $\omega\to 0$, F must reduce to F_0. Therefore, we finally obtain

$$G=\sqrt{\frac{m\omega}{2\pi i\hbar\sin\omega t}}\exp\left\{\frac{im\omega}{2\hbar}\left[\left(x^2+x_0^2\right)\frac{\cos\omega t}{\sin\omega t}-\frac{2x_0 x}{\sin\omega t}\right]\right\} \tag{6.97}$$

We return to the path integral method and apply it to the theory of scattering in Chapter 18.

Problems for Chapter 6

6.1. Prove that $(\Delta x)^2$ is a quadratic function of time for any arbitrary square-integrable free-particle wave packet.

6.2. In this one-dimensional scattering problem for the Schroedinger equation, a particle of mass m has energy $E > 0$. There is a potential barrier of arbitrary shape $V(x)$ such that $V(x) \neq 0$ only in the interval $a < x < b$. Prove that the transmission coefficient is the same whether the particle is incident on the barrier from the left or the right.

6.3. A particle of mass m and energy E is confined in the potential well

$$\begin{aligned} V &= 0 \qquad -\frac{a}{2} \leq x \leq \frac{a}{2} \\ V &= +\infty \qquad |x| > \frac{a}{2} \end{aligned}$$

Find the average force exerted by the particle on the wall at $x = a/2$, and compare your result with that expected in classical mechanics. *Hint:* First work out the problem for a potential well of finite depth; then take the limit as the depth becomes infinite.

6.4. A particle of mass m is confined in the potential

$$V(x) = k|x|^{3/2} \qquad -\infty < x < \infty \tag{1}$$

where $k = \sqrt{2m/\hbar^2}$. The bound-state energies $E_0 < E_1 < E_2 < E_3 < \ldots$ can be calculated accurately by solving Schroedinger's equation numerically. Now consider a particle of mass m bound in the potential

$$\begin{aligned} V(x) &= \infty \qquad x < 0 \\ V(x) &= kx^{3/2} \qquad x > 0 \end{aligned} \tag{2}$$

with k as before. What are the energy eigenvalues for this potential, expressed in terms of the energy eigenvalues for the potential in (1)?

6.5. The Green function for a particle in a uniform gravitational field is

$$G(x',T;x_0,0) = \sqrt{\frac{m}{2\pi i\hbar T}} \exp\left\{\frac{im}{2\hbar}\left[\frac{(x'-x_0)^2}{T} - g(x'+x_0)T - \frac{g^2T^3}{12}\right]\right\} \tag{1}$$

where g is the gravitational acceleration. Suppose that at $t = 0$ the initial wave function is

$$\psi(x_0,0) = \frac{1}{(2\pi a^2)^{1/4}} \exp\left(-\frac{x_0^2}{4a^2}\right) \tag{2}$$

where a is a constant. Find the wave function $\psi(x',T)$ that evolves from $\psi(x_0,0)$, and show that it describes a wave packet that undergoes uniform acceleration g. Find the corresponding wave function in the accelerated reference frame by making the appropriate Galilean transformation. Discuss the spreading of these wave functions.

6.6. Here we study a simple one-dimensional quantum-mechanical system that resembles the hydrogen molecular ion. A particle of mass m is in a potential

$$V(x) = V_0\left[\delta(x-a) + \delta(x+a)\right]$$

where $V_0 < 0$.

(a) Find the expression that determines the bound-state energy for even-parity states, and determine graphically how many even-parity bound states exist.

(b) What is the wave function for the lowest even-parity bound state? Sketch this function for large, intermediate, and small a.

(c) Repeat parts (a) and (b) for odd parity. For what values of V_0 is there at least one such bound state?

(d) Find the even- and odd-parity binding energies for $m|V_0|a >> \hbar^2$. Explain physically why these energies move closer together as $a \to \infty$.

6.7. Here we consider a Schroedinger Hamiltonian that depends on a continuous real parameter λ so that the time-independent Schroedinger equation reads

$$H(\lambda)|\psi(\lambda)\rangle = E(\lambda)|\psi(\lambda)\rangle$$

(a) Show that

$$\frac{\partial E}{\partial \lambda} = \left\langle \psi \left| \frac{\partial H}{\partial \lambda} \right| \psi \right\rangle \tag{1}$$

where we assume that $|\psi\rangle$ is normalized to unity. The simple and useful result (1) is known as the *Feynman-Hellmann theorem.*

(b) Find a one-dimensional potential $V(x)$ that gives bound states with energy eigenvalue separations (*splittings*) that are independent of particle mass. *Hint:* Use the Feynman-Hellmann theorem, and consider the quantity

$$\frac{d\langle \hat{x}\hat{p}\rangle}{dt}$$

where $\hat{x}$ and $\hat{p}$ are the coordinate and momentum operators, respectively.

6.8. A one-dimensional harmonic oscillator is in the nth quantum state. Prove that

$$\Delta x \Delta p = \left(n + \frac{1}{2}\right)\hbar$$

6.9. (a) A one-dimensional harmonic oscillator is in the normalized state

$$|\psi(t=0)\rangle = A_0|0\rangle + B_0|1\rangle$$

at time $t = 0$, where A_0 and B_0 are constants. Find the values of A_0 and B_0 that maximize $\langle \psi(0)|Q|\psi(0)\rangle$, and find the latter quantity for these values of A_0 and B_0.

(b) Find $|\psi(t > 0)\rangle$ and $\langle \psi(t)|Q|\psi(t)\rangle$ in the Schroedinger picture.

(c) Find $\Delta Q^2 = \overline{Q^2} - \overline{Q}^2$ as a function of t.

6.10. In Section 6.13.5 we show that displacement of the one-dimensional oscillator ground state $|0\rangle$ from the origin leads to a coherently oscillating wave packet. Use the time-development operator and the displacement operator to show that the same is true for any oscillator stationary state $|n\rangle$.

6.11. Using the creation-destruction operator method, find the energy levels and wave functions of a two-dimensional harmonic oscillator with potential

$$V = \frac{1}{2}m\omega_0^2\left(x^2 + y^2\right)$$

Find the degeneracy of each level. Write out the wave functions of the ground state and each of the first excited states. Identify the latter with eigenstates of angular momentum in the z-direction.

6.12. A particle of charge e and mass m is in a uniform magnetic field $\boldsymbol{B}$ in the z-direction. Choose the vector potential to be axially symmetric, and find the energy eigenvalues and eigenstates by the method of creation and destruction operators. Classically, a charged particle in a uniform magnetic field in the z-direction executes uniform motion along the z-axis and circular motion in the xy-plane. The center of the circle is not determined from the equation of motion but from initial conditions. How do you make the correspondence between the classical situation just described and the quantum-mechanical solution? Calculate the degeneracy per unit area in the xy-plane of the lowest-energy solution for motion in the xy-plane.

6.13. Employing completeness of the one-dimensional simple harmonic oscillator eigenstates $|n\rangle$ and the parity of their corresponding spatial wave functions $u_n(x)$, show that the parity operator P applicable to all one-dimensional spatial wave functions $\psi(x)$ can be written in the explicit form

$$P = \exp(i\pi N)$$

where N is the one-dimensional simple harmonic oscillator number operator.

6.14. In discussions of the quantized simple harmonic oscillator, the concept of a *squeezed state* is interesting and useful, especially with regard to nonlinear optics, quantum limits on noise, and so on. In this problem we try to develop an elementary understanding of squeezed states. Consider the operator

$$S(r) = \exp\left[\frac{r}{2}\left(a^2 - a^{\dagger 2}\right)\right]$$

where r is a real parameter, and $a = \left(1/\sqrt{2}\right)(Q + iP)$ and $a^\dagger = \left(1/\sqrt{2}\right)(Q - iP)$ are the usual simple harmonic oscillator destruction and creation operators.

(a) Show that $S(r)$ is unitary and that it can be written as $S(r) = S_1 = S_2$, where

$$S_1 = \exp(r/2)\exp(irQP) \qquad S_2 = \exp(-r/2)\exp(irPQ)$$

(b) In the coordinate representation $Q \to Q,\ P \to -i\partial/\partial Q$, show that S_1 can be written as

$$S_1 = e^{r/2}\exp\left(r\frac{\partial}{\partial z}\right)$$

where $z = \ell n Q$.

(c) Apply this operator to the ground-state wave function $u_0(Q) = \pi^{-1/4}\exp(-Q^2/2)$ to obtain a new wave function $\chi_0(Q)$. Compare ΔQ and ΔP for the two functions u_0 and χ_0.

(d) Find how the quantity $\Delta Q \Delta P$ varies with time. Do this by using the Green function for the harmonic oscillator [see equation (6.97)] to calculate the time development of the squeezed coordinate wave function. Also follow an analogous procedure for the corresponding squeezed momentum wave function.

(e) Consider two independent simple harmonic oscillators described by the operators $a,a^\dagger$ and $b,b^\dagger$, where $[a,a^\dagger]=[b,b^\dagger]=1$ and $[a,b]=[a,b^\dagger]=0$. Let

$$a = \frac{1}{\sqrt{2}}(X + iP_X) \qquad b = \frac{1}{\sqrt{2}}(Y + iP_Y)$$

Again working in coordinate representation, show that the operator $T(r) = \exp[r(ab - a^\dagger b^\dagger)]$ squeezes one mode at the expense of the other. It may be helpful to use the new coordinates

$$x_1 = 2^{-1/2}(x - y) \qquad x_2 = 2^{-1/2}(x + y)$$

What are the coordinate and momentum uncertainties at time $t = 0$?

6.15. A particle of unit mass moves in a certain one-dimensional potential $V_-(x)$. The bound states $u_0, u_1, \ldots, u_n$ have energies $E_0, E_1, \ldots, E_n$, respectively, in order of increasing energy. It is known that $u_0 = N_0 \operatorname{sec}h^p(\beta x)$, where N_0 is a normalization constant, and β and $p > 0$ are real parameters.

(a) Find the superpotential $W(x)$.

(b) Find V_- and its supersymmetric partner V_+.

(c) Show that V_- is shape invariant, and obtain a general formula for the E_n. What restriction applies to p so that there exist n bound states?

(d) Find u_1 up to a normalization constant.

(e) Use the result obtained in (c) to discuss the potential $V = -V_0 \operatorname{sec}h^2(\beta x)$, where V_0 is a positive constant, and show that its bound-state eigenvalues for $m = \hbar = 1$ are given by

$$E_n = -\frac{\beta^2}{8}\left[-(1+2n) + \sqrt{1 + \frac{8V_0^2}{\beta^2}}\right]^2$$

7 The Theory of Angular Momentum

7.1 Transformations and invariance

Consider a physical system in state $|u\rangle$ as described by observer O in reference frame F. We perform a transformation $\boldsymbol{T}$ that might be a rotation, a translation, a Lorentz boost, or a spatial inversion. We may think of transforming the system itself, for example, by rotating it (this is called the *active case*), or we may leave the system intact and transform to a new reference frame F' with observer O' (this is the *passive case*). In either case, the state vector after the transformation is no longer $|u\rangle$; it becomes some other state $|u'\rangle$. We write

$$\boldsymbol{T}\left[|u\rangle\right]=|u'\rangle \tag{7.1}$$

The square brackets on the left-hand side of (7.1) are inserted to emphasize that $\boldsymbol{T}$ is not an operator on the Hilbert space but an actual transformation.

We are particularly interested in transformations that leave the laws of nature invariant. It is important to be clear exactly what we mean by such invariance. Suppose that in addition to the initial state $|u\rangle$ we also consider an observable A with eigenstates $|\phi_i\rangle$ and corresponding eigenvalues λ_i; that is,

$$A|\phi_i\rangle = \lambda_i|\phi_i\rangle$$

We know that when the system is initially prepared in state $|u\rangle$ (assumed to be normalized to unity), the probability of achieving result λ_i on any given measurement of A is $p_i = |\langle\phi_i|u\rangle|^2$. Under $\boldsymbol{T}$, not only $|u\rangle$ but also the $|\phi_i\rangle$ are modified; that is,

$$\boldsymbol{T}\left[|\phi_i\rangle\right]=|\phi_i'\rangle$$

Invariance of the laws of nature under transformation $\boldsymbol{T}$ means that the probabilities must be the same before and after the transformation. Thus we require that

$$|\langle\phi_i|u\rangle|^2 = |\langle\phi_i'|u'\rangle|^2 \tag{7.2}$$

Now, corresponding to each transformation $\boldsymbol{T}$, we assume that there exists some operator T on the Hilbert space

$$T|u\rangle=|u'\rangle \qquad T|\phi_i\rangle=|\phi_i'\rangle$$

Hence we can write (7.2) as

$$\left|\langle\phi_i|u\rangle\right|^2 = \left|\langle\phi_i|T^\dagger T|u\rangle\right|^2 \tag{7.3}$$

which implies that $T^\dagger T = \pm I$. If $T^\dagger T = +I$, then T is unitary, which is the case that occurs most frequently. It applies to all transformations that can be built up continuously from those that differ only infinitesimally from the identity, such as rotations, spatial displacements, time displacements, and proper Lorentz transformations. It even applies to some discrete transformations such as spatial inversion and charge conjugation. If $T^\dagger T = -I$, the operator T is said to be *antiunitary*. We will see in Chapter 22 that the time-reversal transformation corresponds to an antiunitary operator. For the present, however, we confine ourselves to the unitary case, and in fact, for almost all this chapter we specialize to rotations. Thus, instead of a generalized transformation $\boldsymbol{T}$ that corresponds to a generalized unitary operator T, we are concerned almost exclusively with a rotation $\boldsymbol{R}$ about some axis in three-dimensional Euclidean space corresponding to a unitary rotation operator R. Our rotations will always be active (rotation of the system rather than the reference frame) unless otherwise noted.

7.2 Rotation group: Angular-momentum operators

The set of all rotations $\boldsymbol{R}$ in three-dimensional Euclidean space forms a group that is in one-to-one correspondence (is isomorphic to) the group $SO(3)$ of all real orthogonal 3×3 matrices with determinant equal to plus unity. Thus we shall frequently refer to a given $\boldsymbol{R}$ by its corresponding matrix in $SO(3)$. Corresponding to each rotation $\boldsymbol{R}$, there is also an operator R on the Hilbert space of physical states. Hence the operators R also form a group [but it is not necessarily isomorphic to $SO(3)$, as we shall see].

A given rotation is specified by the angle of rotation and the axis about which it occurs (in particular, by a unit vector $\hat{n}$ along this axis). The expression

$$\cdots \boldsymbol{R}_{\hat{n}_3}(\theta_3)\boldsymbol{R}_{\hat{n}_2}(\theta_2)\boldsymbol{R}_{\hat{n}_1}(\theta_1)$$

means "first rotate about $\hat{n}_1$ by angle θ_1, then rotate about $\hat{n}_2$ by angle θ_2, then rotate about $\hat{n}_3$ by angle θ_3, and so on." The corresponding operator is

$$\cdots R_{\hat{n}_3}(\theta_3)R_{\hat{n}_2}(\theta_2)R_{\hat{n}_1}(\theta_1)$$

which means "first operate with $R_{\hat{n}_1}(\theta_1)$, then with $R_{\hat{n}_2}(\theta_2)$, then with $R_{\hat{n}_3}(\theta_3)$, and so on."

We now consider a rotation about some axis $\hat{n}$ by an infinitesimally small angle ε. Then $\boldsymbol{R}$ differs only infinitesimally from the 3×3 identity matrix, and the corresponding R differs only infinitesimally from the identity operator I. Thus we can write

$$R_{\hat{n}}(\varepsilon) = I - \frac{i}{\hbar}\varepsilon J_{\hat{n}} \tag{7.4}$$

$J_{\hat{n}}$ is defined by (7.4) and is often called the *generator* of $R_{\hat{n}}(\varepsilon)$. $J_{\hat{n}}$ has the same dimensions as $\hbar$ (energy $\times$ time = action = angular momentum), and $J_{\hat{n}}$ is frequently called the *angular momentum* operator for axis $\hat{n}$. Because R is unitary, $J_{\hat{n}}$ is Hermitian. A finite rotation by angle θ about axis $\hat{n}$ can be built up from a succession of infinitesimal rotations (7.4) as follows: let $\theta = n\varepsilon$. Then

$$\begin{aligned} R_{\hat{n}}(\theta) &= R_{\hat{n}}(\varepsilon) R_{\hat{n}}(\varepsilon) \cdots R_{\hat{n}}(\varepsilon) \\ &= \left(I - i\frac{\varepsilon}{\hbar} J_{\hat{n}} \right)^n \end{aligned}$$

Taking the limit as $n \to \infty$ and $\varepsilon \to 0$ and employing the identity

$$\lim_{n\to\infty}\left(1 - \frac{x}{n}\right)^n = e^{-x}$$

we obtain

$$R_{\hat{n}}(\theta) = \exp\left(-\frac{i}{\hbar}\theta J_{\hat{n}} \right) \tag{7.5}$$

In most of the discussions to follow, it is convenient to choose units where $\hbar = 1$, in which case (7.5) becomes

$$R_{\hat{n}}(\theta) = \exp(-i\theta J_{\hat{n}}) \tag{7.6}$$

7.3 Commutation relations for angular-momentum operators

Once the operators $J_{\hat{n}}$ are defined as in (7.4), the entire quantum theory of angular momentum follows not from any other quantum-mechanical assumptions or assertions but merely from the geometric properties of three-dimensional Euclidean space, in particular, from the fact that rotations about different axes do not commute. To illustrate this, we consider four successive infinitesimal rotations as follows:

1. Rotation about x by angle ε:

$$\boldsymbol{R}_x(\varepsilon) = \begin{pmatrix} 1 & 0 & 0 \\ 0 & 1-\varepsilon^2/2 & -\varepsilon \\ 0 & \varepsilon & 1-\varepsilon^2/2 \end{pmatrix}$$

2. Rotation about y by angle ε:

$$\boldsymbol{R}_y(\varepsilon) = \begin{pmatrix} 1-\varepsilon^2/2 & 0 & \varepsilon \\ 0 & 1 & 0 \\ -\varepsilon & 0 & 1-\varepsilon^2/2 \end{pmatrix}$$

3. Rotation about x by angle $-\varepsilon$:

$$\boldsymbol{R}_x(-\varepsilon)=\begin{pmatrix}1 & 0 & 0\\ 0 & 1-\varepsilon^2/2 & \varepsilon\\ 0 & -\varepsilon & 1-\varepsilon^2/2\end{pmatrix}$$

4. Rotation about y by angle $-\varepsilon$:

$$\boldsymbol{R}_y(-\varepsilon)=\begin{pmatrix}1-\varepsilon^2/2 & 0 & -\varepsilon\\ 0 & 1 & 0\\ \varepsilon & 0 & 1-\varepsilon^2/2\end{pmatrix}$$

In these matrices and in what follows, we retain terms of order ε^2 but discard terms of order ε^3 and higher. Now, carrying out simple algebra, we calculate the product of all four matrices to obtain

$$\boldsymbol{R}_y(-\varepsilon)\boldsymbol{R}_x(-\varepsilon)\boldsymbol{R}_y(\varepsilon)\boldsymbol{R}_x(\varepsilon)=\begin{pmatrix}1 & \varepsilon^2 & 0\\ -\varepsilon^2 & 1 & 0\\ 0 & 0 & 1\end{pmatrix}=\boldsymbol{R}_z(-\varepsilon^2) \tag{7.7}$$

There is a corresponding relationship between rotation operators; that is,

$$R_y(-\varepsilon)R_x(-\varepsilon)R_y(\varepsilon)R_x(\varepsilon)=R_z(-\varepsilon^2) \tag{7.8}$$

Now, using (7.6), we express each of these operators in terms of the $J_{\hat{n}}$ to order ε^2; that is,

$$\left(I+i\varepsilon J_y-\frac{1}{2}\varepsilon^2 J_y^2\right)\left(I+i\varepsilon J_x-\frac{1}{2}\varepsilon^2 J_x^2\right)\left(I-i\varepsilon J_y-\frac{1}{2}\varepsilon^2 J_y^2\right)\left(I-i\varepsilon J_x-\frac{1}{2}\varepsilon^2 J_x^2\right)$$
$$=\left(I+i\varepsilon^2 J_z\right)$$

We multiply out the four factors on the left-hand side, paying attention to the order of the operators and discarding all terms of order ε^3 and higher. Thus we find

$$J_xJ_y-J_yJ_x=\left[J_x,J_y\right]=iJ_z \tag{7.9}$$

Similarly, by cyclic permutation, we have

$$\left[J_y,J_z\right]=iJ_x \tag{7.10}$$

and

$$\left[J_z,J_x\right]=iJ_y \tag{7.11}$$

The last three relations are expressed by the single formula

$$\left[J_i,J_j\right]=i\varepsilon_{ijk}J_k \tag{7.12}$$

where ε_{ijk} is the completely antisymmetric unit 3-tensor, and we use the repeated index summation convention. If, instead, we start with (7.5), where $\hbar$ is written explicitly, (7.12) becomes

$$\left[J_i, J_j\right] = i\varepsilon_{ijk}\hbar J_k \tag{7.13}$$

The important commutation relations (7.12) and (7.13) form the basis for the entire quantum theory of angular momentum.

7.4 Properties of the angular-momentum operators

It is convenient to define the operators $J_\pm = J_x \pm iJ_y$ and to rewrite the commutation relations (7.12) in terms of them. We have

$$\left[J_\pm, J_z\right] = \left[J_x \pm iJ_y, J_z\right] = -iJ_y \mp J_x = \mp J_\pm \tag{7.14}$$

and

$$\left[J_+, J_-\right] = \left[J_x + iJ_y, J_x - iJ_y\right] = i\left[J_y, J_x\right] - i\left[J_x, J_y\right] = 2J_z \tag{7.15}$$

We also define the operator

$$\boldsymbol{J}^2 = J_x^2 + J_y^2 + J_z^2 \tag{7.16}$$

It is easy to verify that

$$J_x^2 + J_y^2 = \frac{1}{2}\left(J_+J_- + J_-J_+\right) \tag{7.17}$$

Thus, from (7.15), we obtain

$$\boldsymbol{J}^2 = J_-J_+ + J_z\left(J_z + 1\right) \tag{7.18}$$

and

$$\boldsymbol{J}^2 = J_+J_- + J_z\left(J_z - 1\right) \tag{7.19}$$

Let the eigenstates of $\boldsymbol{J}^2$ be called $|j\rangle$. Then, using the repeated index summation convention for $i = 1, 2, 3 = x, y, z$, we have

$$\langle j|\boldsymbol{J}^2|j\rangle = \langle j|J_iJ_i|j\rangle = \langle w_i|w_i\rangle \geq 0$$

where $|w_i\rangle = J_i|j\rangle$, and we have used the fact that J_i is Hermitian. Because the eigenvalues of $\boldsymbol{J}^2$ are nonnegative, we can call them $j(j+1)$ with $j \geq 0$; that is,

$$J^2|j\rangle = j(j+1)|j\rangle \qquad j \geq 0 \tag{7.20}$$

Now $\boldsymbol{J}^2$ commutes separately with J_x, J_y, and J_z. For example, from (7.14) and (7.18) we have

$$\begin{aligned}[J_z, \boldsymbol{J}^2] &= J_z J_- J_+ - J_- J_+ J_z \\ &= (J_z J_- J_+ - J_- J_z J_+) + (J_- J_z J_+ - J_- J_+ J_z) \\ &= J_- J_+ - J_- J_+ = 0\end{aligned}$$

and similarly, $[J_x, \boldsymbol{J}^2] = [J_y, \boldsymbol{J}^2] = 0$. Thus we can form simultaneous eigenstates of $\boldsymbol{J}^2$ and one of the three operators J_x, J_y, and J_z. Choosing J_z for this role and naming the eigenvalue of J_z as m, we have

$$\boldsymbol{J}^2|jm\rangle = j(j+1)|jm\rangle \tag{7.21a}$$

and

$$J_z|jm\rangle = m|jm\rangle \tag{7.21b}$$

Of course, if $\hbar$ is exhibited explicitly, these equations become

$$\boldsymbol{J}^2|jm\rangle = \hbar^2 j(j+1)|jm\rangle \tag{7.22a}$$

and

$$J_z|jm\rangle = \hbar m|jm\rangle \tag{7.22b}$$

Note that because $[J_x, J_z] \neq 0$ and $[J_y, J_z] \neq 0$, the $|j,m\rangle$ are not in general eigenstates of J_x or J_y. We now determine the possible values of j and m. From (7.14), we have

$$\begin{aligned}J_z J_+|jm\rangle &= J_+|jm\rangle + J_+ J_z|jm\rangle \\ &= (m+1)J_+|jm\rangle\end{aligned} \tag{7.23}$$

and

$$\begin{aligned}J_z J_-|jm\rangle &= -J_-|jm\rangle + J_- J_z|jm\rangle \\ &= (m-1)J_-|jm\rangle\end{aligned} \tag{7.24}$$

Thus $J_+|jm\rangle$ and $J_-|jm\rangle$ are each eigenstates of J_z with eigenvalues $m \pm 1$, respectively. Also, because J_+, J_- each commute with $\boldsymbol{J}^2$, $J_+|jm\rangle$ and $J_-|jm\rangle$ are each eigenstates of $\boldsymbol{J}^2$ with eigenvalue $j(j+1)$. Hence we can write

$$J_\pm|jm\rangle = C_\pm^{jm}|j, m \pm 1\rangle \tag{7.25}$$

where it remains to determine the coefficients $C_{\pm}^{jm}$. From (7.18) and (7.19), we have

$$J_{-}J_{+}|jm\rangle = [j(j+1)-m(m+1)]|jm\rangle \tag{7.26}$$

and

$$J_{+}J_{-}|jm\rangle = [j(j+1)-m(m-1)]|jm\rangle \tag{7.27}$$

Multiplying each of these last two equations on the left by $\langle jm|$ and normalizing each of the states $|jm\rangle$ to unity, we have

$$\langle jm|J_{-}J_{+}|jm\rangle = [j(j+1)-m(m+1)] \tag{7.28}$$

and

$$\langle jm|J_{+}J_{-}|jm\rangle = [j(j+1)-m(m-1)] \tag{7.29}$$

Now, the left-hand side of (7.28) is zero only if $J_{+}|jm\rangle = 0$; otherwise, it is positive. Similarly, the left-hand side of (7.29) is positive unless $J_{-}|jm\rangle = 0$. Therefore, we have

$$m(m+1) \le j(j+1) \tag{7.30}$$

and

$$m(m-1) \le j(j+1) \tag{7.31}$$

On the other hand, for given j and starting with a particular value of m, we can apply J_{+} repeatedly to generate states $|j,m+1\rangle, |j,m+2\rangle, \ldots$. Thus we eventually violate inequality (7.30) unless the maximum value of m is equal to j, in which case further application of J_{+} results in the null vector. Furthermore, starting with $m_{\max} = j$, we may apply J_{-} repeatedly to generate the states $|j,j-1\rangle, |j,j-2\rangle, \ldots$, thus eventually violating inequality (7.31) unless $m_{\min} = -j$, in which case this process also terminates. To summarize, inequalities (7.30) and (7.31) require that for given j, m is restricted to the $2j+1$ values

$$j, \quad j-1, \quad j-2, \quad \ldots, \quad -j+1, \quad -j$$

This condition is satisfied if and only if j is integral or half-integral

$$j = 0, \frac{1}{2}, 1, \frac{3}{2}, \ldots$$

As for the coefficients $C_{\pm}^{jm}$, it is clear from (7.28) and (7.29) that

$$\left|C_{\pm}^{jm}\right|^2 = j(j+1) - m(m \pm 1)$$

Furthermore, we are free to choose the phases of the orthogonal states $|j,m\rangle$ so that the coefficients $C_{\pm}^{jm}$ are all real and positive. Thus, finally, we obtain

$$J_{+}|jm\rangle = \sqrt{j(j+1) - m(m+1)}\,|j,m+1\rangle \tag{7.32}$$

and

$$J_{-}|jm\rangle = \sqrt{j(j+1) - m(m-1)}\,|j,m-1\rangle \tag{7.33}$$

7.5 Rotation matrices

We now consider an arbitrary rotation about the origin in three-dimensional Euclidean space. It is convenient to decompose this rotation in a standard way into three successive rotations through the Euler angles α, β, and γ, as shown in Figure 7.1.

As Figure 7.1 shows, the net rotation can be expressed as

$$\boldsymbol{R}_{\hat{n}}(\theta) = \boldsymbol{R}_{Z}(\gamma)\,\boldsymbol{R}_{u}(\beta)\,\boldsymbol{R}_{z}(\alpha) \tag{7.34}$$

However, because for any matrix M the application of an orthogonal transformation with matrix Q results in

$$M' = QMQ^{-1}$$

we have

$$\boldsymbol{R}_{Z}(\gamma) = \boldsymbol{R}_{u}(\beta)\,\boldsymbol{R}_{z}(\gamma)\,\boldsymbol{R}_{u}^{-1}(\beta) \tag{7.35}$$

and

$$\boldsymbol{R}_{u}(\beta) = \boldsymbol{R}_{z}(\alpha)\,\boldsymbol{R}_{y}(\beta)\,\boldsymbol{R}_{z}^{-1}(\alpha) \tag{7.36}$$

Substituting (7.35) and (7.36) into (7.34) and taking advantage of the fact that rotations about the same axis commute, we obtain

$$\begin{aligned}\boldsymbol{R}_{\hat{n}}(\theta) &= \boldsymbol{R}_{u}(\beta)\,\boldsymbol{R}_{z}(\gamma)\,\boldsymbol{R}_{u}^{-1}(\beta)\,\boldsymbol{R}_{u}(\beta)\,\boldsymbol{R}_{z}(\alpha)\\ &= \left[\boldsymbol{R}_{z}(\alpha)\,\boldsymbol{R}_{y}(\beta)\,\boldsymbol{R}_{z}^{-1}(\alpha)\right]\left[\boldsymbol{R}_{z}(\gamma)\,\boldsymbol{R}_{u}^{-1}(\beta)\,\boldsymbol{R}_{u}(\beta)\right]\boldsymbol{R}_{z}(\alpha)\\ &= \boldsymbol{R}_{z}(\alpha)\,\boldsymbol{R}_{y}(\beta)\,\boldsymbol{R}_{z}(\gamma)\end{aligned} \tag{7.37}$$

In (7.37), the net rotation is expressed in terms of the space axes, which are far more convenient to employ than the axes z, u, and Z of (7.34). Also note that the order of Euler angle rotations in (7.37) – γ, then β, and finally α – is the reverse of that in (7.34).

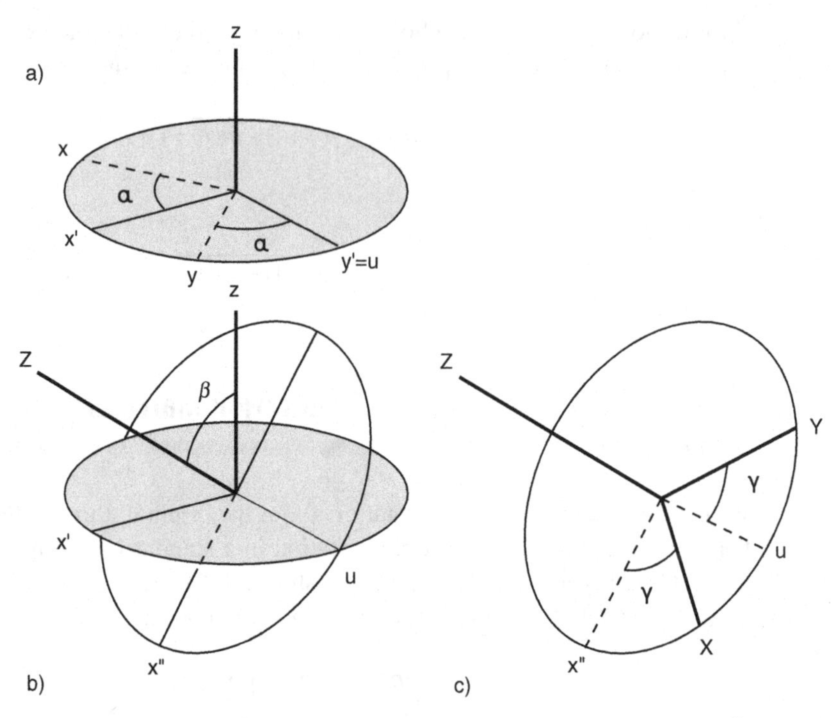

Figure 7.1 An arbitrary rotation about the origin can be decomposed into successive rotations through the Euler angles α, β, and γ. In parts (a)–(c), x, y, and z are axes fixed in space, whereas X, Y, and Z are axes fixed in the rotated object (*body axes*). (a) Rotation by angle α about z changes x to x' and y to $y' = u$. (b) Rotation by angle β about u changes z to Z and x' to x''. (c) Rotation by angle γ about Z changes x'' to X and u to Y.

Because there exists a correspondence between actual rotations $\boldsymbol{R}$ in three-dimensional Euclidean space and rotation operators R in Hilbert space, (7.37) yields the following result:

$$\begin{aligned} R_{\hat{n}}(\theta) &= R_z(\alpha)R_y(\beta)R_z(\gamma) \\ &= \exp(-iJ_z\alpha)\exp(-iJ_y\beta)\exp(-iJ_z\gamma) \end{aligned} \tag{7.38}$$

Now consider a physical system initially in state $|jm\rangle$. Rotation of the system about axis $\hat{n}$ through angle θ results in the new state $|\psi'\rangle = R_{\hat{n}}(\theta)|jm\rangle$. We expand $|\psi'\rangle$ in a complete set of angular-momentum eigenstates $|j',m'\rangle$ as follows:

$$R|jm\rangle = \sum_{j',m'} |j'm'\rangle\langle j'm'|R|jm\rangle \tag{7.39}$$

However, because R commutes with $\boldsymbol{J}^2$, the only terms in the sum that contribute are those for which $j' = j$. Thus the sum in (7.39) runs only over m'; that is,

$$R|jm\rangle = \sum_{m'=-j}^{j} |jm'\rangle\langle jm'|R|jm\rangle \tag{7.40}$$

The coefficients in this expansion

$$D^j_{m',m}(\alpha,\beta,\gamma) = \langle jm' | e^{-iJ_z\alpha} e^{-iJ_y\beta} e^{-iJ_z\gamma} | jm \rangle \tag{7.41}$$

constitute the elements of the unitary $(2j+1)\times(2j+1)$ rotation matrix. Because $|jm\rangle$ and $|jm'\rangle$ are both eigenstates of J_z, we can write (7.41) as follows:

$$D^j_{m',m}(\alpha,\beta,\gamma) = e^{-i(m'\alpha+m\gamma)} d^j_{m',m}(\beta) \tag{7.42}$$

where

$$d^j_{m',m}(\beta) = \langle jm' | e^{-iJ_y\beta} | jm \rangle \tag{7.43}$$

We now consider as examples the rotation matrices for $j = 1/2$ and $j = 1$.

$j = 1/2$

For $j = 1/2$, there are two $|jm\rangle$ states: $m = +1/2$ and $m = -1/2$. It is convenient to represent these by the column spinors

$$\chi_{+1/2} = \begin{pmatrix} 1 \\ 0 \end{pmatrix} \tag{7.44}$$

and

$$\chi_{-1/2} = \begin{pmatrix} 0 \\ 1 \end{pmatrix} \tag{7.45}$$

In this representation, it is obvious that

$$J_z = \frac{1}{2}\sigma_z = \frac{1}{2}\begin{pmatrix} 1 & 0 \\ 0 & -1 \end{pmatrix} \tag{7.46}$$

From (7.32) and (7.33), we also have

$$J_+ \begin{pmatrix} 0 \\ 1 \end{pmatrix} = \begin{pmatrix} 1 \\ 0 \end{pmatrix} \qquad J_+ \begin{pmatrix} 1 \\ 0 \end{pmatrix} = 0$$

and

$$J_- \begin{pmatrix} 1 \\ 0 \end{pmatrix} = \begin{pmatrix} 0 \\ 1 \end{pmatrix} \qquad J_- \begin{pmatrix} 0 \\ 1 \end{pmatrix} = 0$$

Clearly, the matrix representations for $J_\pm$ are

$$J_+ = \begin{pmatrix} 0 & 1 \\ 0 & 0 \end{pmatrix} \tag{7.47}$$

and

$$J_- = \begin{pmatrix} 0 & 0 \\ 1 & 0 \end{pmatrix} \tag{7.48}$$

from which it follows that

$$J_x = \frac{1}{2}(J_+ + J_-) = \frac{1}{2}\sigma_x = \frac{1}{2}\begin{pmatrix} 0 & 1 \\ 1 & 0 \end{pmatrix} \tag{7.49}$$

and

$$J_y = \frac{i}{2}(J_- - J_+) = \frac{1}{2}\sigma_y = \frac{1}{2}\begin{pmatrix} 0 & -i \\ i & 0 \end{pmatrix} \tag{7.50}$$

Hence the matrix

$$d^{1/2}_{m',m}(\beta) = \left\langle \frac{1}{2} m' \middle| e^{-iJ_y\beta} \middle| \frac{1}{2} m \right\rangle$$

is

$$\exp\left(-i\frac{\beta}{2}\sigma_y\right) = I \cdot \cos\frac{\beta}{2} - i\sigma_y \cdot \sin\frac{\beta}{2} = \begin{pmatrix} \cos\frac{\beta}{2} & -\sin\frac{\beta}{2} \\ \sin\frac{\beta}{2} & \cos\frac{\beta}{2} \end{pmatrix} \tag{7.51}$$

Thus the full rotation matrix for $j = 1/2$ is

$$D^{1/2}_{m',m}(\alpha,\beta,\gamma) = \begin{pmatrix} e^{-i(\alpha+\gamma)/2}\cos\frac{\beta}{2} & -e^{-i(\alpha-\gamma)/2}\sin\frac{\beta}{2} \\ e^{i(\alpha-\gamma)/2}\sin\frac{\beta}{2} & e^{i(\alpha+\gamma)/2}\cos\frac{\beta}{2} \end{pmatrix} \tag{7.52}$$

The set of all such matrices for the full range of parameters α, β, and γ is the group $SU(2)$ of 2×2 unitary matrices with determinant equal to plus unity. This is an example of a Lie group, named after the nineteenth-century Norwegian mathematician Sophus Lie, who made systematic investigations of the properties of continuous groups.

Obviously, there is a close relationship between the groups $SU(2)$ and $SO(3)$, but it is not a one-to-one correspondence (isomorphism) because of the appearance in (7.52) of the half-angles $\alpha/2$, $\beta/2$, and $\gamma/2$. A given rotation in Euclidean 3-space (specified by α, β, and γ) and the corresponding $SO(3)$ matrix are unchanged if any of the angles α, β, and γ are incremented by 2π. However, any single such increment results in the replacement of the matrix in (7.52) by its negative. Thus there are two distinct matrices in $SU(2)$ for each matrix in $SO(3)$.

Consider a particle with $j = 1/2$ (e.g., an electron, proton, or neutron), and suppose that initially it has spin "up" along the z-axis; that is,

$$\chi = \begin{pmatrix} 1 \\ 0 \end{pmatrix}$$

We now rotate the spin about the y-axis through angle β by applying the matrix in (7.51) to χ. The result is

$$\chi' = \begin{pmatrix} \cos\frac{\beta}{2} \\ \sin\frac{\beta}{2} \end{pmatrix} \tag{7.53}$$

Thus, when $\beta = 2\pi$, $\chi' = -\chi$. Note that this double-valuedness has already been discussed in Sections 3.4 and 5.1, and in the latter section, a direct experimental demonstration of the phenomenon by means of neutron interferometry was described.

$j = 1$

Here we have three states (m =1, 0, –1) that are conveniently represented by the 3-spinors

$$\chi_1 = \begin{pmatrix} 1 \\ 0 \\ 0 \end{pmatrix} \quad \chi_0 = \begin{pmatrix} 0 \\ 1 \\ 0 \end{pmatrix} \quad \chi_{-1} = \begin{pmatrix} 0 \\ 0 \\ 1 \end{pmatrix}$$

In this representation, we clearly have

$$J_z = \begin{pmatrix} 1 & 0 & 0 \\ 0 & 0 & 0 \\ 0 & 0 & -1 \end{pmatrix} \tag{7.54}$$

Also, for this representation, (7.32) and (7.33) yield

$$J_+ = \begin{pmatrix} 0 & \sqrt{2} & 0 \\ 0 & 0 & \sqrt{2} \\ 0 & 0 & 0 \end{pmatrix} \tag{7.55}$$

and

$$J_- = \begin{pmatrix} 0 & 0 & 0 \\ \sqrt{2} & 0 & 0 \\ 0 & \sqrt{2} & 0 \end{pmatrix} \tag{7.56}$$

from which it follows that

$$J_x = \frac{1}{\sqrt{2}}\begin{pmatrix} 0 & 1 & 0 \\ 1 & 0 & 1 \\ 0 & 1 & 0 \end{pmatrix} \tag{7.57}$$

and

$$J_y = \frac{1}{\sqrt{2}}\begin{pmatrix} 0 & -i & 0 \\ i & 0 & -i \\ 0 & i & 0 \end{pmatrix} \tag{7.58}$$

From (7.58) we can calculate the rotation matrix $d^1_{m',m}(\beta)$ by writing the power-series expansion

$$\exp(-i\beta J_y) = I + (-i)\beta J_y + (-i)^2 \frac{\beta^2}{2!} J_y^2 + (-i)^3 \frac{\beta^3}{3!} J_y^3 + \cdots \tag{7.59}$$

It is easy to verify that in the present representation,

$$J_y^2 = \frac{1}{2}\begin{pmatrix} 1 & 0 & -1 \\ 0 & 2 & 0 \\ -1 & 0 & 1 \end{pmatrix} \tag{7.60}$$

and $J_y^3 = J_y$. Hence (7.59) can be written in closed form as

$$\exp(-i\beta J_y) = I - iJ_y \sin\beta + J_y^2(\cos\beta - 1) \tag{7.61}$$

Substituting (7.58) and (7.60) into (7.61), we obtain

$$d^1_{m',m}(\beta) = \begin{pmatrix} \frac{1}{2}(1+\cos\beta) & -\frac{1}{\sqrt{2}}\sin\beta & \frac{1}{2}(1-\cos\beta) \\ \frac{1}{\sqrt{2}}\sin\beta & \cos\beta & -\frac{1}{\sqrt{2}}\sin\beta \\ \frac{1}{2}(1-\cos\beta) & \frac{1}{\sqrt{2}}\sin\beta & \frac{1}{2}(1+\cos\beta) \end{pmatrix} \tag{7.62}$$

Wigner (1959, chap. 15) has derived the following general formula for $D^j_{m',m}(\alpha,\beta,\gamma)$, valid for any j:

$$\begin{aligned} D^j_{m',m}(\alpha,\beta,\gamma) = \sum_k (-1)^k & \frac{\sqrt{(j+m)!(j-m)!(j+m')!(j-m')!}}{(j-m-k)!(j+m'-k)!k!(k+m-m')!} \\ & \times e^{-im'\alpha} e^{-im\gamma} \left(\cos\frac{\beta}{2}\right)^{2j+m'-m-2k} \left(\sin\frac{\beta}{2}\right)^{2k+m-m'} \end{aligned} \tag{7.63}$$

In (7.63), k ranges between the larger of the numbers 0 and $m' - m$ and the smaller of the numbers $j - m$ and $j + m'$ so that all the factorials are nonnegative. It can be seen from (7.63) that for all half-integral j, the correspondence between the unitary matrices D and the 3×3 matrices of $SO(3)$ is 2:1, whereas for all integral j, it is 1:1.

Therefore, if we denote any state with integral angular momentum by $|u\rangle$, any state with half-integral angular momentum by $|w\rangle$, and the rotation operator for a rotation about any axis by 2π with the symbol $R_{2\pi}$, then $R_{2\pi}|u\rangle = |u\rangle$ and $R_{2\pi}|w\rangle = -|w\rangle$. At the same time, it is physically reasonable to assume that for any observable A whatsoever, including the Hamiltonian, $R_{2\pi}AR_{2\pi}^{-1} = A$. (This does not contradict the experimental result discussed in Section 5.1.) It follows that

$$\langle u|A|w\rangle = \langle u|R_{2\pi}^{\dagger}R_{2\pi}AR_{2\pi}^{-1}R_{2\pi}|w\rangle = -\langle u|A|w\rangle = 0$$

Hence the relative phase η between the states $|u\rangle$ and $|w\rangle$ in the superposition $|\psi\rangle = |u\rangle + e^{i\eta}|w\rangle$ is unobservable. This is an example of what is called a *superselection rule.*

7.6 Magnetic resonance: The rotating frame – Rabi's formula

Magnetic resonance, a very important experimental method that is widely employed in physics, chemistry, biology, and medicine, provides an instructive example of some of the ideas developed so far in this chapter. The magnetic resonance technique was invented by I. I. Rabi at Columbia University in 1937 (with help from C. Gorter of the Netherlands and clarification by the teenager J. Schwinger). Rabi and coworkers at Columbia used the method in atomic and molecular beam experiments to measure nuclear spins and moments, Zeeman and Stark effects, hyperfine structure splittings, and other properties of atoms, molecules, and nuclei. E. Purcell at Harvard and F. Bloch at Stanford independently succeeded in applying nuclear magnetic resonance (NMR) to bulk samples shortly after the end of World War II. Important advances were made in the early 1950s by many investigators, notably by N. F. Ramsey in atomic/molecular beam magnetic resonance and by E. L. Hahn in NMR. Since then, the method continues to develop in many diverse directions. Here we confine ourselves to a discussion of magnetic resonance for spin-½ particles and specifically the proton.

Magnetic resonance experiments typically make use of the magnetic field

$$\boldsymbol{B} = B_1\left[(\cos\omega' t)\hat{i} - (\sin\omega' t)\hat{j}\right] + B_0\hat{k} \tag{7.64}$$

which consists of a constant component B_0 in the z-direction and a component of magnitude B_1 that rotates in the xy-plane with angular frequency ω'. (For particles where the spin and magnetic moment are antiparallel, $\boldsymbol{B}_1$ is chosen to rotate in the opposite sense.) As we discussed in Section 3.4, the equation of motion in the laboratory frame for the expectation value of the proton spin in a magnetic field is

$$\begin{aligned}\frac{d\langle \boldsymbol{S}\rangle}{dt} &= \langle \boldsymbol{\mu}\rangle \times \boldsymbol{B} \\ &= -\gamma \boldsymbol{B} \times \langle \boldsymbol{S}\rangle = -\boldsymbol{\omega} \times \langle \boldsymbol{S}\rangle\end{aligned} \tag{7.65}$$

where $\boldsymbol{\omega} = (g_p/\hbar)\mu_N \boldsymbol{B} \equiv \gamma \boldsymbol{B}$, and $\boldsymbol{B}$ is given by (7.64) in the present discussion. In a frame F' that rotates with $\boldsymbol{B}_1$ about the z-axis (i.e., rotates with angular velocity ω' with respect to the laboratory frame), the time derivative, denoted by $\partial\langle \boldsymbol{S}\rangle/\partial t$, is related to $d\langle \boldsymbol{S}\rangle/dt$ by a well-known kinematic formula originally obtained by Euler; that is,

$$\begin{aligned}\frac{\partial\langle \boldsymbol{S}\rangle}{\partial t} &= \frac{d\langle \boldsymbol{S}\rangle}{dt} - \boldsymbol{\omega}' \times \langle \boldsymbol{S}\rangle \\ &= -(\gamma \boldsymbol{B} + \boldsymbol{\omega}') \times \langle \boldsymbol{S}\rangle\end{aligned} \tag{7.66}$$

In the rotating frame F', $\boldsymbol{B}_1$ is a constant vector. Also, from (7.66), because $\boldsymbol{\omega}'$ is oriented in the –z-direction, the effective z-component of the magnetic field is

$$B_{0,\text{eff}} = \frac{1}{\gamma}(\gamma B_0 - \omega') = \frac{1}{\gamma}(\omega_0 - \omega') \tag{7.67}$$

where $\omega_0 = \gamma B_0$. Thus the resultant effective magnetic field in F' is as shown in Figure 7.2. In the rotating frame, $\langle \boldsymbol{S}\rangle$ precesses about $\boldsymbol{B}_{\text{eff}}$ with angular frequency

$$\Omega = \gamma B_{\text{eff}} = \sqrt{\gamma^2 B_1^2 + (\omega_0 - \omega')^2} \tag{7.68}$$

Also, if $\langle \boldsymbol{S}\rangle$ is oriented along the $+z$-axis at $t = 0$, the precession cone half-angle Θ is given by

$$\sin\Theta = \frac{B_1}{B_{\text{eff}}} = \frac{\gamma B_1}{\sqrt{(\gamma B_1)^2 + (\omega_0 - \omega')^2}} \tag{7.69}$$

The precession cone is illustrated in Figure 7.3.

Because $\langle \boldsymbol{S}\rangle$ lies along the z-axis at $t = 0$, $\psi(0) = \begin{pmatrix}1\\0\end{pmatrix}$; $(m = +1/2)$. However, at time $t > 0$, $\langle \boldsymbol{S}\rangle$ lies along OC in Figure 7.3; hence $\langle \boldsymbol{S}\rangle$ tips downward by angle β with respect to the z-axis, and therefore, $\psi(t)$ contains a component proportional to $\begin{pmatrix}0\\1\end{pmatrix}$; $(m = -1/2)$. From (7.53), it is clear that the probability of finding the proton with $m = -1/2$ at time t (which we call the *transition probability* $P_{1/2\to-1/2}$) is

$$P_{1/2\to-1/2} = \sin^2\frac{\beta}{2}$$

However, from Figure 7.3,

$$\begin{aligned}\sin\frac{\beta}{2} &= \frac{1}{2}\frac{\overline{AC}}{\overline{OA}} = \frac{\overline{AB}}{\overline{OA}} = \frac{\overline{AP}}{\overline{OA}}\sin\frac{\phi}{2} \\ &= \sin\Theta\sin\frac{\Omega t}{2}\end{aligned} \tag{7.70}$$

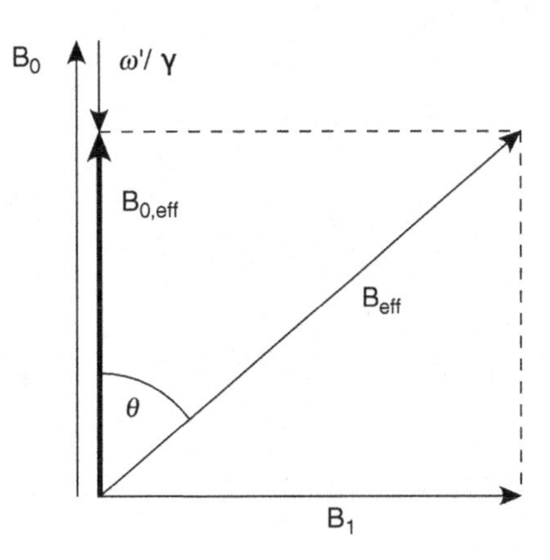

Figure 7.2 Magnetic field vectors in the rotating frame.

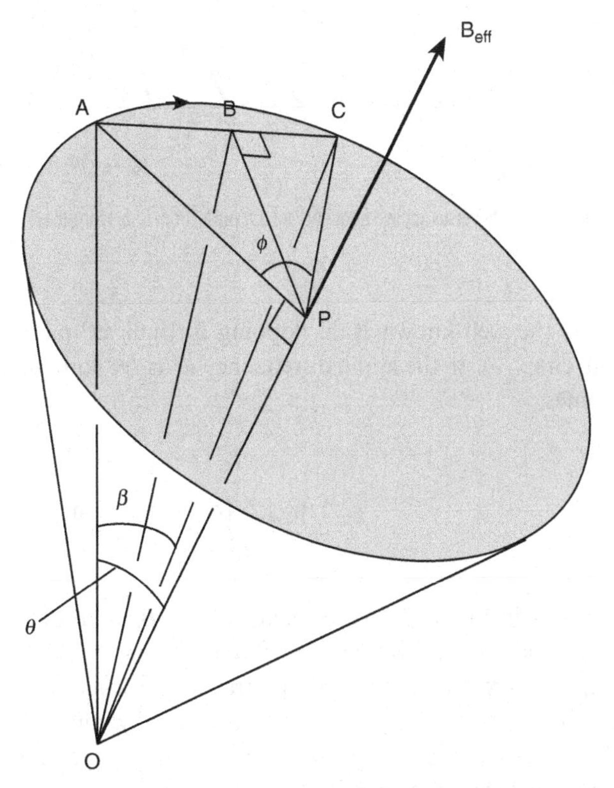

Figure 7.3 Sketch of the precession cone. *OP* is the direction of $\boldsymbol{B}_{\text{eff}}$ in rotating frame *F*′. $\langle \boldsymbol{S} \rangle$ lies along *OA* (vertically upward) at time $t = 0$, but $\langle \boldsymbol{S} \rangle$ precesses about *OP* and is aligned along *OC* at time $t > 0$. Thus *OC* is tipped downward by angle β with respect to *OA*. The precession angle is $\phi = \Omega t$.

Thus we obtain

$$\begin{aligned} P_{1/2\rightarrow -1/2} &= \sin^2\Theta \sin^2\frac{\Omega t}{2} \\ &= \frac{(\gamma B_1)^2}{(\gamma B_1)^2 + (\omega_0 - \omega')^2}\sin^2\left[\sqrt{(\gamma B_1)^2 + (\omega_0 - \omega')^2}\,\frac{t}{2}\right] \end{aligned} \tag{7.71}$$

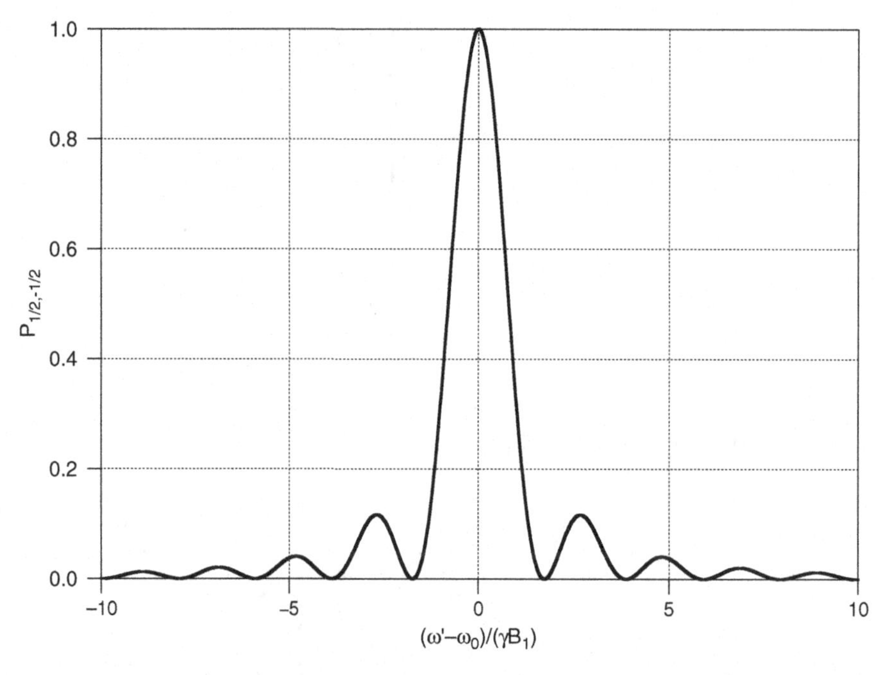

Figure 7.4 Transition probability versus applied frequency for a π pulse, from Rabi's formula (7.71).

This is the well-known Rabi flopping formula of magnetic resonance (Rabi 1937; Ramsey 1956, chap. V). If the applied frequency ω' is set equal to ω_0 (the resonance condition), (7.71) becomes

$$P_{1/2\to-1/2}(\omega'=\omega_0)=\sin^2\left(\frac{\gamma B_1 t}{2}\right) \tag{7.72}$$

Here $B_{0,\text{eff}}$ in Figure 7.2 is reduced to zero, $\boldsymbol{B}_{\text{eff}}=\boldsymbol{B}_1$, and $\Theta=\pi/2$. Thus the precession cone becomes a circle on resonance. If the rotating field is turned on at $t=0$ and off again at $t=\pi/\gamma B_1$ (a *π pulse*), the resulting transition probability is unity at resonance. In Figure 7.4 we plot $P_{1/2\to-1/2}$ versus applied frequency for a π pulse.

Rabi's formula (7.71) can be generalized to apply to any value of J, integral or half-integral, by using a formula originally derived for another purpose by E. Majorana (1932). Equation (7.71) also can be derived directly from Schroedinger's time-dependent equation $H\psi=i\hbar\dot{\psi}$. Here we obtain two first-order linear differential equations that couple the upper and lower components of ψ. These two equations can be reduced to a single second-order differential equation, for example, for the lower component. This can be solved subject to the initial conditions to yield (7.71). Although this is a useful elementary exercise in solving differential equations, it does not provide the same insight as the geometric method we have employed in the foregoing discussion.

7.7 Orbital angular momentum

In classical mechanics, a particle with linear momentum $\boldsymbol{p}$ and position $\boldsymbol{r}$ relative to some origin O has orbital angular momentum $\boldsymbol{L} = \boldsymbol{r} \times \boldsymbol{p}$ with respect to O. In quantum mechanics, we adopt the same definition for $\boldsymbol{L}$, but now $\boldsymbol{r}$ and $\boldsymbol{p}$ are noncommuting operators. In units where $\hbar = 1$, we have $[x, L_y] = [x, zp_x - xp_z] = iz$ and cyclic permutations thereof. Hence

$$[x_i, L_j] = i\varepsilon_{ijk} x_k \tag{7.73}$$

Also, $[p_x, L_y] = [p_x, zp_x - xp_z] = ip_z$. Thus

$$[p_i, L_j] = i\varepsilon_{ijk} p_k \tag{7.74}$$

Finally, $[L_x, L_y] = [(yp_z - zp_y), (zp_x - xp_z)] = -iyp_x + ixp_y = iL_z$. Hence

$$[L_i, L_j] = i\varepsilon_{ijk} L_k \tag{7.75}$$

The latter implies that $\boldsymbol{L}$ is a quantum-mechanical angular-momentum operator with all the usual properties; that is,

$$\boldsymbol{L}^2 |\ell m_\ell\rangle = \ell(\ell+1)|\ell m_\ell\rangle \tag{7.76}$$

$$L_z |\ell m_\ell\rangle = m_\ell |\ell m_\ell\rangle \tag{7.77}$$

$$L_\pm |\ell m_\ell\rangle = \sqrt{\ell(\ell+1) - m_\ell(m_\ell \pm 1)}\,|\ell, m_\ell \pm 1\rangle \tag{7.78}$$

and so on. However, for orbital angular momentum, there is an addition restriction: ℓ must be integral. To explain this, we define the operators

$$a = \frac{1}{\sqrt{2}}(x + ip_x) \qquad b = \frac{1}{\sqrt{2}}(y + ip_y)$$
$$c = \frac{1}{\sqrt{2}}(a + ib) \qquad d = \frac{1}{\sqrt{2}}(a - ib)$$

Then $[a, a^\dagger] = [b, b^\dagger] = [c, c^\dagger] = [d, d^\dagger] = 1$, and $[c, d] = [c, d^\dagger] = 0$. Also, it is easy to verify that

$$L_z = i(b^\dagger a - a^\dagger b) = d^\dagger d - c^\dagger c \tag{7.79}$$

From the theory of the simple harmonic oscillator (see Section 6.13.1), we know that the eigenvalues of $d^\dagger d$ and of $c^\dagger c$ are the nonnegative integers. Hence the eigenvalues m_ℓ must be integral, which implies that ℓ is also integral.

It is frequently useful to write $\boldsymbol{L}$ in coordinate representation and to employ spherical polar coordinates; that is,

$$x = r\sin\theta\cos\phi$$
$$y = r\sin\theta\sin\phi$$
$$z = r\cos\theta$$

Then

$$L_z = -i\left(x\frac{\partial}{\partial y} - y\frac{\partial}{\partial x}\right) = -i\frac{\partial}{\partial \phi} \tag{7.80}$$

Defining $\psi_{\ell m_\ell}$ to be the coordinate representative of $|\ell, m_\ell\rangle$, we see that (7.77) and (7.80) imply

$$\psi_{\ell,m_\ell} = f_{\ell,m_\ell}(\theta)e^{im_\ell\phi} \tag{7.81}$$

where $f_{\ell,m_\ell}(\theta)$ is a function to be determined. Now

$$\boldsymbol{L}^2 = L_-L_+ + L_z(L_z + 1)$$

and

$$L_\pm = e^{\pm i\phi}\left(\pm\frac{\partial}{\partial\theta} + i\cot\theta\frac{\partial}{\partial\phi}\right) \tag{7.82}$$

Therefore,

$$\boldsymbol{L}^2\psi_{\ell,m_l} = -\left[\frac{1}{\sin\theta}\frac{\partial}{\partial\theta}\left(\sin\theta\frac{\partial}{\partial\theta}\right) + \frac{1}{\sin^2\theta}\frac{\partial^2}{\partial\phi^2}\right]\psi_{l,m_\ell} = \ell(\ell+1)\psi_{l,m_\ell} \tag{7.83}$$

Substitution of (7.81) into (7.83) yields the associated Legendre differential equation; that is,

$$\frac{\partial}{\partial(\cos\theta)}\left[\sin^2\theta\frac{\partial f}{\partial(\cos\theta)}\right] + \left[\ell(\ell+1) - \frac{m_\ell^2}{\sin^2\theta}\right]f = 0 \tag{7.84}$$

This equation possesses a solution that is analytic for all values of θ: the associated Legendre polynomial $P_\ell^{m_\ell}(\theta)$. Thus we obtain

$$\psi_{\ell,m_\ell} = N_{\ell,m_\ell}P_\ell^{m_\ell}(\theta)e^{im\phi} \tag{7.85}$$

where N_{ℓ,m_ℓ} is a normalizing factor. A standard choice for the latter [see equation (7.88)] yields the spherical harmonics

$$\psi_{\ell,m_\ell} = Y_\ell^{m_\ell}(\theta,\phi) = \sqrt{\frac{2\ell+1}{4\pi}\frac{(\ell-m_\ell)!}{(\ell+m_\ell)!}}P_\ell^{m_\ell}(\theta)e^{im_\ell\phi} \tag{7.86}$$

here defined for $m_\ell \geq 0$, and

$$Y_\ell^{-m_\ell} = (-1)^{m_\ell} \left(Y_\ell^{m_\ell}\right)^* \tag{7.87}$$

The spherical harmonic normalization and orthogonality condition is

$$\int \left(Y_{\ell'}^{m'}\right)^* Y_\ell^m d\Omega = \delta_{\ell\ell'} \delta_{mm'} \tag{7.88}$$

where $d\Omega = \sin\theta d\theta d\phi$. The first few spherical harmonics are

$$Y_0^0 = \frac{1}{\sqrt{4\pi}} \tag{7.89}$$

$$\begin{aligned} Y_1^1 &= -\sqrt{\frac{3}{8\pi}} \sin\theta \, e^{i\phi} \\ Y_1^0 &= \sqrt{\frac{3}{4\pi}} \cos\theta \\ Y_1^{-1} &= \sqrt{\frac{3}{8\pi}} \sin\theta \, e^{-i\phi} \end{aligned} \tag{7.90}$$

and

$$\begin{aligned} Y_2^2 &= \sqrt{\frac{15}{32\pi}} \sin^2\theta \, e^{2i\phi} & Y_2^{-2} &= \sqrt{\frac{15}{32\pi}} \sin^2\theta \, e^{-2i\phi} \\ Y_2^1 &= -\sqrt{\frac{15}{8\pi}} \sin\theta\cos\theta \, e^{i\phi} & Y_2^{-1} &= \sqrt{\frac{15}{8\pi}} \sin\theta\cos\theta \, e^{-i\phi} \\ Y_2^0 &= \sqrt{\frac{5}{4\pi}} \left(\frac{3}{2}\cos^2\theta - \frac{1}{2}\right) \end{aligned} \tag{7.91}$$

7.8 Addition of angular momenta: Vector coupling coefficients

A physical system with angular momentum operator $\boldsymbol{J}$ may consist of two parts with angular momentum operators $\boldsymbol{J}_1$ and $\boldsymbol{J}_2$; that is,

$$\boldsymbol{J} = \boldsymbol{J}_1 + \boldsymbol{J}_2 \tag{7.92}$$

The hydrogen atom in the ground state provides an example because the electron and proton both have spin-½. In what follows, we assume that in addition to the usual commutation rules satisfied separately by $\boldsymbol{J}_1$ and $\boldsymbol{J}_2$, we also have

$$\left[J_{1i}, J_{2j}\right] = 0 \tag{7.93}$$

which means that each component of the angular momentum of one subsystem can be measured without disturbing any component of the other subsystem. Given (7.93), we can form simultaneous eigenstates of $\boldsymbol{J}_1^2, J_{1z}$ and $\boldsymbol{J}_2^2, J_{2z}$, which are denoted by $|j_1m_1\rangle|j_2m_2\rangle$. For given j_1, j_2, the eigenstates $|jm\rangle$ of $\boldsymbol{J}^2$ and J_z can be expressed as linear combinations of the $|j_1m_1\rangle|j_2m_2\rangle$; that is,

$$|jm\rangle = \sum_{m_1,m_2} |j_1m_1\rangle|j_2m_2\rangle\langle j_1m_1j_2m_2|jm\rangle \tag{7.94}$$

The coefficients $\langle j_1m_1j_2m_2|jm\rangle$ are elements of a unitary matrix and are called *vector coupling coefficients* or *Clebsch-Gordan coefficients*. Knowledge of the properties of these coefficients and an understanding of how to use them are essential for the analysis of many important problems in quantum mechanics. We now summarize the most important of these properties.

1. Application of the operator $J_z = J_{1z} + J_{2z}$ to both sides of (7.94) yields

$$\sum_{m_1,m_2} (m - m_1 - m_2)|j_1m_1\rangle|j_2m_2\rangle\langle j_1m_1j_2m_2|jm\rangle = 0 \tag{7.95}$$

Because the product states $|j_1m_1\rangle|j_2m_2\rangle$ are linearly independent for different values of m_1, m_2, (7.95) implies that

$$(m - m_1 - m_2)\langle j_1m_1j_2m_2|jm\rangle = 0$$

Hence, if $\langle j_1m_1j_2m_2|jm\rangle \neq 0$, we must have $m = m_1 + m_2$; consequently, the sum on the right-hand side of (7.94) is actually over one variable only, for example m_1, in which case, $m_2 = m - m_1$.

2. Consider the scalar product

$$\begin{aligned}\langle j'm'|jm\rangle &= \delta_{j',j}\delta_{m',m} \\ &= \sum_{m_1,m_1'} \langle j'm'|j_1m_1'j_2(m'-m_1')\rangle\langle j_1m_1'|j_1m_1\rangle\langle j_2(m'-m_1')|j_2(m-m_1)\rangle\langle j_1m_1j_2(m-m_1)|jm\rangle\end{aligned} \tag{7.96}$$

Because $\langle j_1m_1'|j_1m_1\rangle = \delta_{m_1',m_1}$ and $\langle j_1(m'-m_1')|j_2(m-m_1)\rangle = \delta_{m'-m_1',m-m_1}$, (7.96) reduces to the orthogonality relation

$$\sum_{m_1} \langle j'm|j_1m_1j_2m_2\rangle\langle j_1m_1j_2m_2|jm\rangle = \delta_{j',j} \tag{7.97}$$

Actually, the vector coupling coefficients are real up to a common arbitrary phase factor. Thus we can write

$$\langle j'm|j_1m_1j_2m_2\rangle = \langle j_1m_1j_2m_2|j'm\rangle \tag{7.98}$$

3. The following inverse relation holds:

$$|j_1m_1\rangle|j_2m_2\rangle = \sum_{j} \langle j_1m_1j_2m_2|jm\rangle|jm\rangle \tag{7.99}$$

This is easily shown by multiplying both sides of (7.99) by $\langle j_1 m_1 j_2 m_2 | j' m \rangle$, summing over m_1, and using (7.97).

4. A second orthogonality relation

$$\sum_j \langle j_1 m_1 j_2 m_2 | jm \rangle \langle j_1 m_1' j_2 m_2' | jm' \rangle = \delta_{m_1, m_1'} \delta_{m, m'} \tag{7.100}$$

is easily shown from (7.99).

5. Assuming that $j_1 \geq j_2$, the possible values of j are

$$j = j_1 + j_2, \quad j_1 + j_2 - 1, \quad \ldots, \quad j_1 - j_2$$

Proof: Given j_1 and j_2, the maximum values of m_1 and m_2 are j_1 and j_2, respectively. Thus the maximum value of m and therefore of j is $j_1 + j_2$. The state $|j = j_1 + j_2, m = j_1 + j_2\rangle$ can be expressed in only one way in terms of $|j_1, m_1 = j_1\rangle |j_2, m_2 = j_2\rangle$; that is,

$$|j_1 + j_2, j_1 + j_2\rangle = |j_1 j_1\rangle |j_2 j_2\rangle \tag{7.101}$$

In other words, the single-vector coupling coefficient in this case is +1. There are $2(j_1 + j_2) + 1$ separate m_J states belonging to the multiplet $j = j_1 + j_2$. In particular, we consider $m = j_1 + j_2 - 1$. The state $|j_1 + j_2, j_1 + j_2 - 1\rangle$ is a linear combination of $|j_1 j_1\rangle |j_2, j_2 - 1\rangle$ and $|j_1, j_1 - 1\rangle |j_2 j_2\rangle$. Because these last two product states are orthogonal, there is a second linearly independent combination of them that must correspond to $|j = j_1 + j_2 - 1, m = j_1 + j_2 - 1\rangle$. The entire multiplet of states with $j = j_1 + j_2 - 1$ has $2(j_1 + j_2 - 1) + 1$ separate m components. Continuing in this manner, we construct the multiplets $j = j_1 + j_2, j_1 + j_2 - 1$, and so on until we have exhausted all independent linear combinations of $|j_1 m_1\rangle |j_2 m_2\rangle$, the total number of which is $(2j_1 + 1)(2j_2 + 1)$. Because

$$[2(j_1 + j_2) + 1][2(j_1 + j_2 - 1) + 1] \cdots [2(j_1 - j_2) + 1] = (2j_1 + 1)(2j_2 + 1)$$

we conclude that $j_{\min} = j_1 - j_2$.

6. The vector coupling coefficients are real and can be generated recursively. Instead of giving a general proof, we illustrate this and some other properties with three examples.

We first consider the case $j_1 = j_2 = 1/2$ and introduce the notation $\alpha = \begin{pmatrix} 1 \\ 0 \end{pmatrix}, \beta = \begin{pmatrix} 0 \\ 1 \end{pmatrix}$. (Here we must not confuse the spinors α, β with the Euler angles α, β!) There are two possible values of j: $j = 1$ with $m = 1, 0, -1$ and $j = 0$ with $m = 0$. Obviously,

$$|j = 1, m = 1\rangle = \alpha_1 \alpha_2 \tag{7.102}$$

Recalling (7.33), we apply the lowering operator $J_- = J_{1-} + J_{2-}$ to both sides of (7.102) and divide both sides of the resulting equation by $\sqrt{2}$ to obtain

$$|10\rangle = \frac{1}{\sqrt{2}}(\alpha_1 \beta_2 + \alpha_2 \beta_1) \tag{7.103}$$

To find $|1,-1\rangle$, we may apply the lowering operator $J_- = J_{1-} + J_{2-}$ once again to both sides of (7.103) or simply rotate the states on both sides of (7.102) about the y-axis through π radians. This yields

$$|1-1\rangle = \beta_1\beta_2 \tag{7.104}$$

The state $|0,0\rangle$ must be a linear combination of $\alpha_1\beta_2$ and $\beta_1\alpha_2$ that is orthogonal to (7.103). Thus, apart from an arbitrary overall phase, we have

$$|00\rangle = \frac{1}{\sqrt{2}}(\alpha_1\beta_2 - \beta_1\alpha_2) \tag{7.105}$$

This result also can be obtained by starting with the linear combination

$$|00\rangle = a\alpha_1\beta_2 + b\beta_1\alpha_2$$

where a and b are coefficients to be determined, and applying the raising operator [see equation (7.32)] or the lowering operator to both sides. Because $J_\pm|0,0\rangle = 0$, we must have $a = -b$; furthermore, normalization requires $a^2 + b^2 = 1$. Hence $a = -b = 2^{-1/2}$ with a specific choice of overall phase.

Note that the three $j = 1$ states [(7.102) through (7.104)] are symmetric with respect to exchange of spinors 1 and 2, whereas the $j = 0$ state is antisymmetric. This is an additional general property of vector coupling coefficients for $j_1 = j_2$: the $j = 2j_1,\ 2j_1 - 2, \ldots$ multiplets are symmetric under exchange of j_1 and j_2, whereas the $2j_1 - 1,\ 2j_1 - 3, \ldots$ multiplets are exchange antisymmetric.

Our next example concerns $j_1 = 1, j_2 = 1/2$, for which $j = 3/2$ or $1/2$. We start with

$$\left|\frac{3}{2}\frac{3}{2}\right\rangle = |11\rangle_1\,\alpha_2 \tag{7.106}$$

and apply the lowering operator to both sides to obtain

$$\sqrt{\frac{3}{2}\frac{5}{2} - \frac{3}{2}\frac{1}{2}}\left|\frac{3}{2}\frac{1}{2}\right\rangle = \sqrt{2}|10\rangle_1\,\alpha_2 + |11\rangle_1\,\beta_2$$

which gives

$$\left|\frac{3}{2}\frac{1}{2}\right\rangle = \sqrt{\frac{2}{3}}|10\rangle_1\,\alpha_2 + \sqrt{\frac{1}{3}}|11\rangle_1\,\beta_2 \tag{7.107}$$

Further applications of the lowering operator yield

$$\left|\frac{3}{2} -\frac{1}{2}\right\rangle = \sqrt{\frac{1}{3}}|1-1\rangle_1\,\alpha_2 + \sqrt{\frac{2}{3}}|10\rangle_1\,\beta_2 \tag{7.108}$$

and

$$\left|\frac{3}{2}-\frac{3}{2}\right\rangle = |1-1\rangle_1\,\beta_2 \tag{7.109}$$

Next, consider the $j = 1/2$ doublet. The state $|j = 1/2, m = 1/2\rangle$ must be orthogonal to (7.107). Hence, with a conventional but arbitrary choice of overall phase, we have

$$\left|\frac{1}{2}\frac{1}{2}\right\rangle = \sqrt{\frac{1}{3}}|10\rangle_1\,\alpha_2 - \sqrt{\frac{2}{3}}|11\rangle_1\,\beta_2 \tag{7.110}$$

To find the vector coupling coefficients for $|j = 1/2, m = \text{-}1/2\rangle$, we can apply the lowering operator to both sides of (7.110), or we can rotate the states on both sides of (7.110) about the y-axis by π radians. Recalling (7.51) and (7.62), we note that under this rotation, $|11\rangle \to |1-1\rangle$, $|10\rangle \to -|10\rangle$, $\alpha \to \beta$, and $\beta \to -\alpha$. Hence we obtain

$$\left|\frac{1}{2}-\frac{1}{2}\right\rangle = -\sqrt{\frac{1}{3}}|10\rangle_1\,\beta_2 + \sqrt{\frac{2}{3}}|1-1\rangle_1\,\alpha_2 \tag{7.111}$$

As a final example, consider $j_1 = j_2 = 1$. In this case, the possible multiplets are

$$\begin{array}{ll} j=2 & \text{(a quintuplet)} \\ j=1 & \text{(a triplet)} \\ j=0 & \text{(a singlet)} \end{array}$$

The vector coupling coefficients for $j = 2$ are generated in a straightforward manner by repeated application of the lowering operator, starting with the state $|j,m\rangle = |2,2\rangle$; that is,

$$\begin{aligned}
|22\rangle &= |11\rangle_1|11\rangle_2 \\
|21\rangle &= \frac{1}{\sqrt{2}}\left[|11\rangle_1|10\rangle_2 + |10\rangle_1|11\rangle_2\right] \\
|20\rangle &= \frac{1}{\sqrt{6}}\left[|11\rangle_1|1-1\rangle_2 + |1-1\rangle_1|11\rangle_2 + 2|10\rangle_1|10\rangle_2\right] \\
|2-1\rangle &= \frac{1}{\sqrt{2}}\left[|1-1\rangle_1|10\rangle_2 + |10\rangle_1|1-1\rangle_2\right] \\
|2-2\rangle &= |1-1\rangle_1|1-1\rangle_2
\end{aligned}$$

The state $|j = 1, m = 1\rangle$ must be orthogonal to $|21\rangle$; hence, with an arbitrary choice of overall phase, we have

$$|11\rangle = \frac{1}{\sqrt{2}}\left[|11\rangle_1|10\rangle_2 - |10\rangle_1|11\rangle_2\right] \tag{7.112}$$

Application of the lowering operator to both sides of (7.112) yields

$$|10\rangle = \frac{1}{\sqrt{2}}\left[|11\rangle_1|1-1\rangle_2 - |1-1\rangle_1|11\rangle_2\right] \tag{7.113}$$

Note that no term proportional to $|10\rangle_1|10\rangle_2$ appears on the right-hand side of (7.113); this is consistent with the fact that the $j = 1$ states are antisymmetric with respect to exchange of j_1 and j_2. The third member of the triplet is

$$|1-1\rangle = \frac{1}{\sqrt{2}}\left[|10\rangle_1|1-1\rangle_2 - |1-1\rangle_1|10\rangle_2\right] \tag{7.114}$$

Finally, the state $|j = 0, m = 0\rangle$ must be orthogonal to $|j = 2, m = 0\rangle$ and $|j = 1, m = 0\rangle$. It is easy to show that

$$|00\rangle = \frac{1}{\sqrt{3}}\left[|11\rangle_1|1-1\rangle_2 + |1-1\rangle_1|11\rangle_2 - |10\rangle_1|10\rangle_2\right] \tag{7.115}$$

which is obviously symmetric with respect to the exchange of j_1 and j_2.

7. 3-*j* *Symbols*. Certain symmetries of the vector coupling coefficients are conveniently expressed in terms of the Wigner 3-*j* symbol

$$\begin{pmatrix} j_1 & j_2 & j_3 \\ m_1 & m_2 & m_3 \end{pmatrix} \tag{7.116}$$

which is defined as follows:

$$\begin{pmatrix} j_1 & j_2 & j_3 \\ m_1 & m_2 & m_3 \end{pmatrix} = (-1)^{j_1 - j_2 - m_3}(2j_3 + 1)^{-1/2}\langle j_1 m_1 j_2 m_2 | j_3 - m_3\rangle \tag{7.117}$$

It can be shown that an even permutation of the columns of a 3-*j* symbol leaves its value unchanged; that is,

$$\begin{pmatrix} j_1 & j_2 & j_3 \\ m_1 & m_2 & m_3 \end{pmatrix} = \begin{pmatrix} j_3 & j_1 & j_2 \\ m_3 & m_1 & m_2 \end{pmatrix} = \begin{pmatrix} j_2 & j_3 & j_1 \\ m_2 & m_3 & m_1 \end{pmatrix} \tag{7.118}$$

but an odd permutation of the columns of (7.116) or replacement of the lower row of (7.116) by $-m_1, -m_2, -m_3$ is equivalent to multiplication of (7.116) by $(-1)^{j_1+j_2+j_3}$.

8. *General formula for vector coupling coefficients.* A general formula for the vector coupling coefficients has been derived in a variety of ways by Wigner (1959), by Schwinger, and by Racah (1942); that is,

$$\begin{aligned}\langle j_1 m_1 j_2 m_2 | jm\rangle &= \delta_{m_1+m_2,m}\left[\frac{(2j+1)(j_1+j_2-j)!(j_1-j_2+j)!(-j_1+j_2+j)!}{(j_1+j_2+j+1)!}\right]^{1/2} \\ &\times \sum_n (-1)^n \frac{[(j_1+m_1)!(j_1-m_1)!(j_2+m_2)!(j_2-m_2)!(j+m)!(j-m)!]^{1/2}}{n!(j_1+j_2-j-n)!(j_1-m_1-n)!(j_2+m_2-n)!(j-j_2+m_1+n)!(j-j_1-m_2+n)!}\end{aligned} \tag{7.119}$$

Here the sum is over all values of the integer n for which none of the factorials is negative. Unfortunately, (7.119) is quite complicated for use in most practical applications. However, tables of vector coupling coefficients are found in many standard textbooks and in the “Particle Physics” booklet published every few years by the American Institute of Physics.

9. *The Clebsch-Gordan series and its inverse.* Rotation matrices and vector coupling coefficients are related by the following useful formula, called the *Clebsch-Gordan series*:

$$D^{j_1}_{m_1'm_1}(\alpha,\beta,\gamma)D^{j_2}_{m_2'm_2}(\alpha,\beta,\gamma)=\sum_j\langle j_1m_1j_2m_2|jm\rangle\langle j_1m_1'j_2m_2'|jm'\rangle D^j_{m'm}(\alpha,\beta,\gamma) \quad (7.120)$$

In the sum on the right-hand side of (7.120), j runs from j_1+j_2 to $|j_1-j_2|$. The inverse formula is

$$D^j_{m'm}(\alpha,\beta,\gamma)=\sum_{\substack{m_1',m_1\\m_2',m_2}}\langle j_1m_1j_2m_2|jm\rangle\langle j_1m_1'j_2m_2'|jm'\rangle D^{j_1}_{m_1'm_1}(\alpha,\beta,\gamma)D^{j_2}_{m_2'm_2}(\alpha,\beta,\gamma) \quad (7.121)$$

Equation (7.121) can be used to generate rotation matrix elements for higher j from known values of rotation matrix elements for lower j.

7.9 Definition of irreducible spherical tensor operators

In many physical problems, it is advantageous to express the relevant quantities in such a way that their rotational symmetry is emphasized. For example, whereas the position vector in Cartesian coordinates is

$$\boldsymbol{r}=x\hat{i}+y\hat{j}+z\hat{k}$$

it is often more convenient to write it as follows:

$$\begin{aligned}\boldsymbol{r}&=-\frac{-(x+iy)}{\sqrt{2}}\bullet\frac{(\hat{i}-i\hat{j})}{\sqrt{2}}-\frac{(x-iy)}{\sqrt{2}}\bullet\frac{-(\hat{i}+i\hat{j})}{\sqrt{2}}+z\hat{k}\\&=-\frac{-(x+iy)}{\sqrt{2}}\hat{\varepsilon}^{(-1)}-\frac{(x-iy)}{\sqrt{2}}\hat{\varepsilon}^{(1)}+z\hat{\varepsilon}^{(0)}\\&=-r^{(1)}\hat{\varepsilon}^{(-1)}-r^{(-1)}\hat{\varepsilon}^{(1)}+r^{(0)}\hat{\varepsilon}^{(0)}\end{aligned} \quad (7.122)$$

where $r^{(1)}=-(x+iy)/\sqrt{2}$, $r^{(0)}=z$, and $r^{(-1)}=(x-iy)/\sqrt{2}$ are proportional to the spherical harmonics Y_1^1,Y_1^0,Y_1^{-1}, respectively, and are said to be the components of the rank-one irreducible spherical coordinate tensor. More generally, consider a collection of $2L+1$ operators T_L^M, where L is a fixed nonnegative integer and $M=L,L-1,\ldots,-L$. By definition, the T_L^M are components of an irreducible spherical tensor of rank L if under a rotation with operator R,

$$RT_L^MR^{-1}=\sum_{M'}D^L_{M'M}T_L^{M'} \quad (7.123)$$

This transformation rule is obviously analogous to that for the spherical harmonics; that is,

$$RY_L^M=\sum_{M'}D^L_{M'M}Y_L^{M'} \quad (7.124)$$

Irreducible spherical tensors of rank 0 and 1 are called *scalars* and *vectors*, respectively. Irreducible spherical tensor operators of half-integral rank do not exist – and for a very good reason. See, for example, Problem 11.3.

Consider two irreducible spherical tensors $P_{L_1}^{M_1}(X)$ and $Q_{L_2}^{M_2}(Y)$, where X and Y are the variables on which P and Q depend. Then it can be shown that

$$T_L^M(X,Y) = \sum_{M_1,M_2} \langle L_1M_1L_2M_2|LM\rangle P_{L_1}^{M_1} Q_{L_2}^{M_2} \tag{7.125}$$

is an irreducible tensor of rank L, where $L = L_1 + L_2,\ L_1 + L_2 - 1,\ \ldots, |L_1 - L_2|$. This combination rule is obviously analogous to the rule for combining angular-momentum states $|L_1M_1\rangle$ and $|L_2, M_2\rangle$ to form states $|LM\rangle$. A special case of (7.125) occurs when $L_1 = L_2$ and $L = 0$; that is,

$$T_0^0(X,Y) = \sum_{M_1} \langle L_1M_1L_1 - M_1|00\rangle P_{L_1}^{M_1} Q_{L_1}^{-M_1} \tag{7.126}$$

Substitution of $\langle L_1M_1L_1 - M_1|00\rangle = (-1)^{L_1-M_1}/\sqrt{2L_1+1}$ into (7.126) yields

$$\begin{aligned} T_0^0(X,Y) &= \frac{(-1)^{L_1}}{\sqrt{2L+1_1}} \sum_{M_1} (-1)^{M_1} P_{L_1}^{M_1} Q_{L_1}^{-M_1} \\ &= \frac{(-1)^{L_1}}{\sqrt{2L+1_1}} I \end{aligned}$$

where

$$I = \sum_{M_1} (-1)^{M_1} P_{L_1}^{M_1} Q_{L_1}^{-M_1} \equiv \mathbf{P}\cdot\mathbf{Q} \tag{7.127}$$

is the scalar product of the tensors P and Q. For example, suppose that $L_1 = L_2 = 1$.Then

$$\begin{aligned} P_1^1 &= -\frac{P_x + iP_y}{\sqrt{2}} \qquad & Q_1^1 &= -\frac{Q_x + iQ_y}{\sqrt{2}} \\ P_1^{-1} &= \frac{P_x - iP_y}{\sqrt{2}} \qquad & Q_1^{-1} &= \frac{Q_x - iQ_y}{\sqrt{2}} \\ P_1^0 &= P_z \qquad & Q_1^0 &= Q_z \end{aligned}$$

and (7.127) becomes

$$\begin{aligned} I &= -\frac{1}{2}(-P_x - iP_y)(Q_x - iQ_y) - \frac{1}{2}(P_x - iP_y)(-Q_x - iQ_y) + P_zQ_z \\ &= P_xQ_x + P_yQ_y + P_zQ_z \end{aligned}$$

7.10 Commutation rules for irreducible spherical tensors

We now show that

$$[J_z, T_L^M] = MT_L^M \tag{7.128}$$

$$[J_+, T_L^M] = \sqrt{L(L+1) - M(M+1)}\; T_L^{M+1} \tag{7.129}$$

$$[J_-, T_L^M] = \sqrt{L(L+1) - M(M-1)}\; T_L^{M-1} \tag{7.130}$$

First, consider an infinitesimal rotation about the z-axis; that is,

$$R_z(\varepsilon) T_L^M R_z^{-1}(\varepsilon) = \sum_{M'} D_{M'M}^L(\varepsilon,0,0) T_L^{M'} \tag{7.131}$$

Because $D_{M'M}^L(\varepsilon,0,0) = e^{-iM'\varepsilon}\delta_{M'M} \approx (1-iM'\varepsilon)\delta_{M'M}$, (7.131) gives

$$(I - i\varepsilon J_z) T_L^M (I + i\varepsilon J_z) \simeq (1 - i\varepsilon M) T_L^M$$

Comparing the left- and right-hand sides of this expression to order ε, we obtain (7.128).
 Let us make an infinitesimal rotation about y instead; that is,

$$(I - i\varepsilon J_y) T_L^M (1 + i\varepsilon J_y) = \sum_{M'} d_{M'M}^L(\varepsilon) T_L^{M'} \tag{7.132}$$

Now

$$\begin{aligned} d_{M'M}^L(\varepsilon) &= \langle LM'|I - i\varepsilon J_y|LM\rangle \\ &= \delta_{M'M} - \frac{i\varepsilon}{2i}\left(\langle LM'|J_+|LM\rangle - \langle LM'|J_-|LM\rangle\right) \\ &= \delta_{M'M} - \frac{\varepsilon}{2}\left(C_+^{LM}\delta_{M',M+1} - C_-^{LM}\delta_{M',M-1}\right) \end{aligned}$$

Comparing the left- and right-hand sides of (7.132), we obtain

$$i[J_y, T_L^M] = \frac{1}{2}\left(C_+^{LM} T_L^{M+1} - C_-^{LM} T_L^{M-1}\right) \tag{7.133}$$

Similarly, an infinitesimal rotation about x yields

$$[J_x, T_L^M] = \frac{1}{2}\left(C_+^{LM} T_L^{M+1} + C_-^{LM} T_L^{M-1}\right) \tag{7.134}$$

Combining (7.133) and (7.134), we arrive at (7.129) and (7.130).

The angular-momentum operators themselves form an irreducible spherical tensor of rank one; that is,

$$\begin{aligned} J_1 &= -\frac{J_x + iJ_y}{\sqrt{2}} \\ J_0 &= J_z \\ J_{-1} &= \frac{J_x - iJ_y}{\sqrt{2}} \end{aligned} \tag{7.135}$$

and the commutation rules [(7.128)–(7.130)] can be expressed as

$$[J_0, T_L^M] = MT_L^M \tag{7.136}$$

$$[J_1, T_L^M] = -\frac{1}{\sqrt{2}}\sqrt{L(L+1) - M(M+1)}\ T_L^{M+1} \tag{7.137}$$

$$[J_{-1}, T_L^M] = \frac{1}{\sqrt{2}}\sqrt{L(L+1) - M(M-1)}\ T_L^{M-1} \tag{7.138}$$

7.11 Wigner-Eckart theorem

Let us multiply both sides of (7.128) on the left by $\langle j'm'|$ and on the right by $|jm\rangle$. This yields

$$(m' - m)\langle j'm'|T_L^M|jm\rangle = M\langle j'm'|T_L^M|jm\rangle$$

Hence $\langle j'm'|T_L^M|jm\rangle \neq 0$ only if $m' = m + M$. Next, we do the same thing with both sides of (7.129) and (7.130). This yields the recursion relations

$$C_{\mp}^{jm'}\langle j', m' \mp 1|T_L^M|jm\rangle - C_{\pm}^{jm}\langle j'm'|T_L^M|j, m \pm 1\rangle = C_{\pm}^{LM}\langle j'm'|T_L^{M\pm 1}|jm\rangle \tag{7.139}$$

Equations (7.139) are analogous to the following recursion relations for the vector coupling coefficients:

$$C_{\mp}^{jm'}\langle jmLM|j', m' \mp 1\rangle - C_{\pm}^{jm}\langle j, m \pm 1LM|j'm'\rangle = C_{\pm}^{LM}\langle jmLM \pm 1|j'm'\rangle \tag{7.140}$$

This implies that the matrix element $\langle j'm'|T_L^M|jm\rangle$ is proportional to the following corresponding vector coupling coefficient:

$$\langle j'm'|T_L^M|jm\rangle = \langle j'\|T_L\|j\rangle\langle jmLM|j'm'\rangle \tag{7.141}$$

where the proportionality factor $\langle j' \| T_L \| j \rangle$, called the *reduced matrix element*, depends on $j, L,$ and j' but is independent of m, M, and m'. Equation (7.141) is the Wigner-Eckart theorem, one of the most important results in the quantum theory of angular momentum. Its significance stems from the fact that it separates the matrix element into a factor (the reduced matrix element) containing the specific dynamical features of the operator T and a factor (the vector coupling coefficient) that is geometric in nature and is solely related to rotational symmetry.

7.12 Consequences of the Wigner-Eckart theorem

7.12.1 Selection rules

It follows immediately from (7.141) that the condition $\langle j'm'|T_L^M|jm\rangle \neq 0$ requires not only

$$m + M = m' \tag{7.142}$$

but also the *triangle rule*; that is,

$$j' = j + L, \quad j' = j + L - 1, \quad \ldots, \quad j' = |j - L| \tag{7.143}$$

Thus, for example,

- A tensor of zero rank can only connect states of the same j.
- A tensor of the first rank can only connect states with $j' = j$ or $j' = j \pm 1$, but it cannot connect $j = 0$ to $j' = 0$.
- A tensor of the second rank can only connect states such that $j' = j$, $j \pm 1$, and $j \pm 2$, but it cannot connect states with $j = j' = 0$, $j = j' = 1/2$, or $j = 1, j' = 0$, or vice versa.

7.12.2 Static moments

The diagonal matrix element of an irreducible tensor is

$$\langle jm|T_L^0|jm\rangle = \langle jmL0|jm\rangle\langle j \| T_L \| j\rangle \tag{7.144}$$

For example, T_L might be the magnetic moment operator μ (a first-rank tensor). In this case, because $\langle jm10|jm\rangle = m/\sqrt{j(j+1)}$, (7.144) becomes

$$\langle jm|\mu_z|jm\rangle = \frac{m}{\sqrt{j(j+1)}}\langle j \| \mu \| j\rangle \tag{7.145}$$

In particular, when $m = j$, the matrix element on the left-hand side of (7.145) is by definition the *magnetic moment* μ (i.e., the quantity found in tables). Thus

$$\langle j \| \mu \| j \rangle = \frac{\sqrt{j(j+1)}}{j}\mu$$

and therefore

$$\langle jm|\mu_z|jm\rangle = \frac{m}{j}\mu \tag{7.146}$$

Next, consider (7.144) in the case $L = 2$. Here, for example, we might be dealing with the electric quadrupole moment operator Q. Because

$$\langle jm20|jm\rangle = \frac{3m^2 - j(j+1)}{\sqrt{(2j-1)(j+1)(2j+3)}} \tag{7.147}$$

(7.144) yields

$$\langle jm|Q_2^0|jm\rangle = \frac{3m^2 - j(j+1)}{\sqrt{(2j-1)(j+1)(2j+3)}}\langle j \| Q \| j\rangle$$

The quadrupole moment Q listed in tables is

$$Q \equiv \langle jj|Q_2^0|jj\rangle = \frac{3j^2 - j(j+1)}{\sqrt{(2j-1)(j+1)(2j+3)}}\langle j \| Q \| j\rangle \tag{7.148}$$

Thus

$$\langle jm|Q_2^0|jm\rangle = \frac{3m^2 - j(j+1)}{3j^2 - j(j+1)}Q \tag{7.149}$$

Equation (7.149) implies that a particle cannot possess a quadrupole moment unless $j \geq 1$.

7.12.3 Projection theorem for static moments: The Lande *g*-factor

An electron has spin $\boldsymbol{S}$ and orbital angular momentum $\boldsymbol{L}$, and these add to form the total angular momentum $\boldsymbol{J} = \boldsymbol{L} + \boldsymbol{S}$. The total magnetic moment is

$$\boldsymbol{\mu} = \boldsymbol{\mu}_S + \boldsymbol{\mu}_L = -\mu_B(g_s\boldsymbol{S} + g_\ell\boldsymbol{L}) \tag{7.150}$$

In the semiclassical vector model of atomic physics that preceded quantum mechanics, $\boldsymbol{J}$, $\boldsymbol{L}$, and $\boldsymbol{S}$ were regarded as ordinary vectors, not operators. Because $g_s \neq g_\ell$, $\boldsymbol{\mu}$ and $\boldsymbol{J}$ were not collinear, and $\boldsymbol{\mu}$ was resolved into two components, one parallel and the other perpendicular to $\boldsymbol{J}$; that is,

$$\boldsymbol{\mu} = \boldsymbol{\mu}_\parallel + \boldsymbol{\mu}_\perp$$

According to the vector model, only $\boldsymbol{\mu}_{\parallel}$ contributed to the expectation value of $\boldsymbol{\mu}$ because the contribution of $\boldsymbol{\mu}_{\perp}$ averaged to zero; hence, through a clever but mysterious argument, the effective magnetic moment turned out to be

$$\boldsymbol{\mu} = -g_J \mu_B \boldsymbol{J}$$

where g_J is the Lande g-factor; that is,

$$g_J = 1 + \frac{1}{2j(j+1)}\left[j(j+1) + s(s+1) - \ell(\ell+1)\right] \tag{7.151}$$

With the aid of a corollary to the Wigner-Eckhart theorem known as the *projection theorem*, we can sweep away the debris of the old vector model and give a clear derivation of the Lande g-factor. The projection theorem states that if T_1^M is any first-rank tensor, then

$$\langle jm|T_1^0|jm\rangle = \frac{m}{j(j+1)}\langle j \,\|\, \boldsymbol{J}\boldsymbol{\cdot}\boldsymbol{T} \,\|\, j\rangle \tag{7.152}$$

where $\boldsymbol{J}\boldsymbol{\cdot}\boldsymbol{T}$ is the scalar product of the first-rank tensors $\boldsymbol{J}$ and $\boldsymbol{T}$. First, we shall assume (7.152) and use it to derive the Lande g-factor; then we shall derive (7.152) from the Wigner-Eckart theorem. We start with the magnetic moment operator defined in (7.150). From (7.152), we have

$$\begin{aligned}
\langle jm|\mu_z|jm\rangle &= \frac{m}{j(j+1)}\langle j \,\|\, \boldsymbol{J}\boldsymbol{\cdot}\boldsymbol{\mu} \,\|\, j\rangle \\
&= -\frac{m}{j(j+1)}\mu_B\langle j \,\|\, (\boldsymbol{L}+\boldsymbol{S})\boldsymbol{\cdot}(g_\ell \boldsymbol{L} + g_s \boldsymbol{S}) \,\|\, j\rangle \\
&= -\frac{m}{j(j+1)}\mu_B\left[g_\ell\langle j \,\|\, (\boldsymbol{L}^2 + \boldsymbol{L}\boldsymbol{\cdot}\boldsymbol{S}) \,\|\, j\rangle + g_s\langle j \,\|\, (\boldsymbol{S}^2 + \boldsymbol{L}\boldsymbol{\cdot}\boldsymbol{S}) \,\|\, j\rangle\right].
\end{aligned}$$

Now

$$\begin{aligned}
\langle j \,\|\, \boldsymbol{L}^2 \,\|\, j\rangle &= \ell(\ell+1) \\
\langle j \,\|\, \boldsymbol{S}^2 \,\|\, j\rangle &= s(s+1) \\
\langle j \,\|\, \boldsymbol{L}\boldsymbol{\cdot}\boldsymbol{S} \,\|\, j\rangle &= \frac{1}{2}\langle j \,\|\, \boldsymbol{J}^2 - \boldsymbol{L}^2 - \boldsymbol{S}^2 \,\|\, j\rangle \\
&= \frac{1}{2}[j(j+1) - \ell(\ell+1) - s(s+1)]
\end{aligned}$$

Therefore,

$$\langle jm|\mu_z|jm\rangle = -\frac{m\mu_B}{j(j+1)}\boldsymbol{\cdot}
\left\{g_\ell\left[\frac{j(j+1)+\ell(\ell+1)-s(s+1)}{2}\right] + g_s\left[\frac{j(j+1)+s(s+1)-\ell(\ell+1)}{2}\right]\right\}$$

The Lande g-factor g_J is defined by

$$\langle jm|\mu_z|jm\rangle = -g_J\mu_B\langle jm|J_z|jm\rangle = -g_J\mu_B m$$

Thus we obtain

$$g_J = \frac{1}{j(j+1)}\left\{g_\ell\left[\frac{j(j+1)+\ell(\ell+1)-s(s+1)}{2}\right]+g_s\left[\frac{j(j+1)+s(s+1)-\ell(\ell+1)}{2}\right]\right\} \tag{7.153}$$

Now $g_\ell = 1$ and $g_s = 2(1+a_e)$, where a_e is a small quantum electrodynamic correction. If we ignore a_e, (7.153) yields the desired result (7.151).

We now prove the projection theorem (7.152). Consider the first-rank tensor consisting of the angular-momentum operators themselves; that is,

$$J_1, \quad J_0, \quad J_{-1}$$

We know that $\langle j'm'|J_0|jm\rangle = m\delta_{j'j}\delta_{m'm}$. On the other hand, the Wigner-Eckart theorem gives

$$\langle jm|J_0|jm\rangle = \frac{m}{\sqrt{j(j+1)}}\langle j\|J\|j\rangle$$

Hence

$$\langle j\|J\|j\rangle = \sqrt{j(j+1)} \tag{7.154}$$

For any first-rank tensor T_1^M, we have

$$\boldsymbol{J}\cdot\boldsymbol{T} = \sum_\mu(-1)^\mu J_\mu T_1^{-\mu} = \sum_\mu(-1)^\mu\left[J_\mu, T_1^{-\mu}\right]+\sum_\mu(-1)^\mu T_1^{-\mu}J_\mu \tag{7.155}$$

However, the commutation relations [(7.136)–(7.138)] imply that the first sum on the right-hand side of (7.155) vanishes. Thus

$$\boldsymbol{J}\cdot\boldsymbol{T} = \sum_\mu(-1)^\mu J_\mu T_1^{-\mu} = \sum_\mu(-1)^\mu T_1^{-\mu}J_\mu \tag{7.156}$$

Therefore,

$$\begin{aligned}\langle jm'|J_M\boldsymbol{J}\cdot\boldsymbol{T}|jm\rangle &= \sum_\mu(-1)^\mu\langle jm'|J_M T_1^{-\mu}J_\mu|jm\rangle \\ &= \sum_{\substack{\mu\\ j_1m_1\\ j_2m_2}}(-1)^\mu\langle jm'|J_M|j_1m_1\rangle\langle j_1m_1|T_1^{-\mu}|j_2m_2\rangle\langle j_2m_2|J_\mu|jm\rangle\end{aligned}$$

Now

$$\langle jm'|J_M|j_1m_1\rangle = \delta_{j,j_1}\langle j_1m_1 1M|jm'\rangle\sqrt{j(j+1)}$$

$$\left\langle j_2 m_2 \left| J_\mu \right| jm \right\rangle = \delta_{j,j_2} \left\langle jm1\mu \middle| j_2 m_2 \right\rangle \sqrt{j(j+1)}$$

whereas

$$\left\langle jm_1 \left| T_1^{-\mu} \right| jm_2 \right\rangle = \left\langle jm_2 1 - \mu \middle| jm_1 \right\rangle \langle j \parallel T_1 \parallel j \rangle$$

Thus

$$\begin{aligned} &\left\langle jm' \left| J_M \boldsymbol{J} \cdot \boldsymbol{T} \right| jm \right\rangle \\ &= j(j+1) \sum_{\mu, m_1, m_2} (-1)^\mu \left\langle jm_1 1M \middle| jm' \right\rangle \left\langle jm_2 1 - \mu \middle| jm_1 \right\rangle \left\langle jm1\mu \middle| jm_2 \right\rangle \langle j \parallel T_1 \parallel j \rangle \end{aligned} \tag{7.157}$$

For the vector coupling coefficients to be nonzero, we require

$$\begin{aligned} m + \mu &= m_2 \\ m_2 - \mu &= m_1 \\ m_1 + M &= m' \end{aligned}$$

The first two of these relations imply that for given μ, m_1 and m_2 are fixed; hence

$$\begin{aligned} &\left\langle jm' \left| J_M \boldsymbol{J} \cdot \boldsymbol{T} \right| jm \right\rangle \\ &= j(j+1) \left\langle jm1M \middle| jm' \right\rangle \langle j \parallel T_1 \parallel j \rangle \sum_\mu (-1)^\mu \left\langle j, m+\mu, 1, -\mu \middle| jm \right\rangle \left\langle jm1\mu \middle| j, m+\mu \right\rangle \end{aligned}$$

Because $\left\langle jm1\mu \middle| j, m+\mu \right\rangle = (-1)^\mu \left\langle j, m+\mu, 1, -\mu \middle| jm \right\rangle$, the sum over μ reduces to unity by orthonormality of the vector coupling coefficients. Hence we finally obtain

$$\begin{aligned} \left\langle jm' \left| J_M \boldsymbol{J} \cdot \boldsymbol{T} \right| jm \right\rangle &= j(j+1) \left\langle jm1M \middle| jm' \right\rangle \langle j \parallel T_1 \parallel j \rangle \\ &= j(j+1) \left\langle jm' \left| T_1^M \right| jm \right\rangle \end{aligned} \tag{7.158}$$

In particular, if $M = 0$, then $m' = m$, and we recover (7.152).

7.12.4 Emission and absorption of radiation and the Wigner-Eckart theorem

This is the most important domain for application of the Wigner-Eckart theorem. We illustrate by means of a simple example: a hydrogen atom (with a spinless electron for simplicity), where we suppose that initially the atom is in the $2p$ state with $m_\ell = 1, 0,$ or -1. The atom decays spontaneously to the $1s$ state (with $m_\ell = 0$). We want to calculate the probabilities for emission of the photon at various angles with respect to the z-axis for various photon polarizations. Each transition probability W is proportional to $|M|^2$, where M is a matrix element (sometimes called an *amplitude*); that is,

$$M = \left\langle n\ell m = 100; \gamma \left| T \right| n\ell m = 21m_\ell \right\rangle \tag{7.159}$$

Here $|n\ell m = 21m_\ell\rangle$ is the initial atomic state, $|n\ell m = 100, \gamma\rangle$ is the final state consisting of the H atom in the ground state and the emitted photon, and T is the transition operator. The latter is related to the Hamiltonian describing the interaction between the atomic electron and the radiation field, and we study it in detail in Chapter 16. For the present, we merely state what can be shown: T is a scalar (rotationally invariant) operator that conserves parity, and it can be expressed as a sum of products of irreducible tensors S_L and K_L; that is,

$$T = \sum_{L=1}^{\infty} S_L \bullet K_L = \sum_{L=1}^{\infty} \sum_{M=-L}^{L} (-1)^M S_L^M K_L^{-M} \tag{7.160}$$

where S_L refers only to the atom, and K_L refers only to the radiation. The right-hand side of (7.160) is frequently called the *multipole expansion*. Inserting (7.160) into (7.159), we obtain

$$M = \sum_{L=1}^{\infty} \sum_{M=-L}^{L} (-1)^M \langle 100|S_L^M|21m\rangle_a \langle \gamma|K_L^{-M}|0\rangle_r \tag{7.161}$$

Here the first matrix element in each product refers only to the atom (subscript a), whereas the second refers only to the radiation (subscript r). In the latter matrix element, $|0\rangle$ refers to the initial vacuum state of the radiation field (no photons present). This state is spherically symmetric (rotationally invariant).

According to the Wigner-Eckart theorem,

$$\langle 100|S_L^M|21m\rangle = \langle 1, m, L, M|0,0\rangle\langle 0\|S_L\|1\rangle$$

In this expression, the vector coupling coefficient vanishes unless $L = 1, M = -m$. Hence the double sum in (7.161) reduces to just one term.

$$M = (-1)^{-m} \langle 100|S_1^{-m}|21m\rangle_a \langle \gamma|K_1^m|0\rangle_r \tag{7.162}$$

This implies, again, from the Wigner-Eckart theorem, that the photon must carry off one unit of angular momentum, with z projection equal to m. Thus we state formally by the Wigner-Eckart theorem what we know intuitively: angular momentum is conserved in the decay.

Returning to (7.159), we consider M in the special case where the photon momentum is along the $+z$-axis and the photon helicity is positive. Let M be called A in this special case; that is,

$$A = \langle (100)_a ; \gamma_+(+z)|T|(211)_a ; 0\rangle \tag{7.163}$$

Here conservation of angular momentum requires that $m = +1$. Some other cases of special interest are as follows:

$$B = \langle (100)_a ; \gamma_-(-z)|T|(211)_a ; 0\rangle \tag{7.164}$$

$$C = \langle (100)_a ; \gamma_+(-z)|T|(21-1)_a ; 0\rangle \tag{7.165}$$

$$D = \langle (100)_a ; \gamma_- (+z) | T | (21-1)_a ; 0 \rangle \tag{7.166}$$

The quantities A, B, C, and D are related to one another by simple symmetry considerations as follows: let $R_y(\pi)$ be a rotation by π about the y-axis, and let P be the parity transformation operator. Because T is invariant under rotations and spatial inversions, we have

$$\begin{aligned} A &= \langle (100)_a ; \gamma_+ (+z) | R_y^{-1}(\pi) [R_y(\pi) T R_y^{-1}(\pi)] R_y(\pi) | (211)_a ; 0 \rangle \\ &= \langle (100)_a ; \gamma_+ (-z) | T | (21-1)_a ; 0 \rangle \\ &= C \end{aligned} \tag{7.167}$$

Also, because $P|(211)_a\rangle = -|(211)_a\rangle$, where P is the parity operator,

$$\begin{aligned} A &= \langle (100)_a ; \gamma_+ (+z) | P^{-1} (P T P^{-1}) P | (211)_a ; 0 \rangle \\ &= -\langle (100)_a ; \gamma_- (-z) | T | (211)_a ; 0 \rangle \\ &= -B \end{aligned} \tag{7.168}$$

and

$$\begin{aligned} D &= \langle (100)_a ; \gamma_- (+z) | P^{-1} (P T P^{-1}) P | (21-1)_a ; 0 \rangle \\ &= -\langle (100)_a ; \gamma_+ (-z) | T | (21-1)_a ; 0 \rangle \\ &= -C \end{aligned} \tag{7.169}$$

Hence

$$A = -B = C = -D \tag{7.170}$$

Now consider a new axis z' inclined with respect to z by angle θ. What is the amplitude for emission of a photon with positive (or negative) helicity and momentum along z' if the initial atomic state has $m = 1$, 0, or –1 with respect to the original z-axis? We write

$$\begin{aligned} M' &= \langle (100)_a ; \gamma_\pm (z') | T | (21m)_a ; 0 \rangle \\ &= \langle (100)_a ; \gamma_\pm (z') | R_y^{-1}(-\theta) [R_y(-\theta) T R_y^{-1}(-\theta)] R_y(-\theta) | (21m)_a ; 0 \rangle \\ &= \langle (100)_a ; \gamma_\pm (z) | T R_y(-\theta) | (21m)_a ; 0 \rangle \end{aligned} \tag{7.171}$$

Now, in the standard representation where the orbital angular-momentum states $|\ell = 1; m = 1, 0, -1\rangle$ are represented by

$$|1,1\rangle = \begin{pmatrix} 1 \\ 0 \\ 0 \end{pmatrix} \quad |1,0\rangle = \begin{pmatrix} 0 \\ 1 \\ 0 \end{pmatrix} \quad |1,-1\rangle = \begin{pmatrix} 0 \\ 0 \\ -1 \end{pmatrix}$$

Table 7.1 Relative probabilities for photon emission along z'

Initial m (w.r.t. z)	Relative probability for emission along z'	
	Helicity +	Helicity −
$+1$	$\frac{\lvert A\rvert^2}{4}(1+\cos\theta)^2$	$\frac{\lvert A\rvert^2}{4}(1-\cos\theta)^2$
0	$\frac{\lvert A\rvert^2}{2}\sin^2\theta$	$\frac{\lvert A\rvert^2}{2}\sin^2\theta$
-1	$\frac{\lvert A\rvert^2}{4}(1-\cos\theta)^2$	$\frac{\lvert A\rvert^2}{4}(1+\cos\theta)^2$

we have

$$R_y(-\theta)|21m\rangle = \begin{pmatrix} \frac{1+\cos\theta}{2} & \frac{1}{\sqrt{2}}\sin\theta & \frac{1-\cos\theta}{2} \\ -\frac{1}{\sqrt{2}}\sin\theta & \cos\theta & \frac{1}{\sqrt{2}}\sin\theta \\ \frac{1-\cos\theta}{2} & -\frac{1}{\sqrt{2}}\sin\theta & \frac{1+\cos\theta}{2} \end{pmatrix} |21m\rangle$$

Inserting this last expression into the right-hand side of (7.171) and carrying out simple algebra, we obtain the entries in Table 7.1.

The total emission rate for given initial m, regardless of photon helicity, is obtained by summing over both helicities and integrating over the solid angle. We thus find that for each m, the total emission rate is proportional to $(8\pi/3)|A|^2$. Of course, each initial m state must have the same decay rate. Otherwise, we could start out with an "unpolarized" sample of atoms in the $2p$ state, that is, with an equal population in each m level. Then, after some time, these populations would be unequal. In this way, a particular direction in space would be singled out, contrary to our general notion that space is isotropic.

Next, suppose that we have an ensemble of $2p$ atoms with initial populations in the $m = 1, 0, -1$ states of a, b, c, respectively. Then the rates of emission in direction z' with positive (negative) photon helicity are proportional to

$$+\text{ Helicity:}\quad a\left(\frac{1+\cos\theta}{2}\right)^2 + \frac{b}{2}\sin^2\theta + c\left(\frac{1-\cos\theta}{2}\right)^2$$

$$-\text{ Helicity:}\quad a\left(\frac{1-\cos\theta}{2}\right)^2 + \frac{b}{2}\sin^2\theta + c\left(\frac{1+\cos\theta}{2}\right)^2$$

In particular, if $a = b = c$ (unpolarized initial sample), the rates for each helicity are the same, and they are independent of θ. In this case, the radiation is unpolarized and isotropic.

In the foregoing we have been able to derive everything of interest (except for the constant A itself) from symmetry considerations alone, which include use of the Wigner-Eckart theorem, properties of rotation matrices, and rotational as well as parity invariance of T. To obtain the constant A, we need more than symmetry; detailed dynamical considerations are necessary.

7.13 *SU*(*n*)

$SU(n)$ is the group of all $n\times n$ unitary matrices U for which $\det U = +1$. We have seen in this chapter that the quantum theory of angular momentum is based on the group $SU(2)$. The latter is also important for the concept of isospin, discussed in Section 11.10. $SU(3)$ plays a very significant role in particle physics. Here we outline some of the main properties of $SU(n)$ that are relevant for these physical applications.

We start with the fact that any unitary matrix U can be written $U = e^{iH}$, where H is a Hermitian matrix. It is convenient to express H as a linear combination of standard linearly independent matrices $F_j = \frac{1}{2}\lambda_j$; that is,

$$H = \alpha^j F_j = \frac{1}{2}\alpha^j \lambda_j \tag{7.172}$$

Here the α^j are numerical parameters, and we use the repeated index summation convention. Because $\det U = +1$,

$$\operatorname{tr}(\lambda_j) = 0 \tag{7.173}$$

In general, there are n^2 linearly independent $n \times n$ matrices; however, the number of such traceless matrices λ_j is $n^2 - 1$. One simple (nonunique) prescription for finding the λ_j is to start with the $n - 1$ diagonal matrices; that is,

$$\begin{pmatrix} 1 & 0 & 0 & \\ 0 & -1 & 0 & \\ 0 & 0 & 0 & \\ & & & \ddots \end{pmatrix} \quad \frac{1}{\sqrt{3}}\begin{pmatrix} 1 & 0 & 0 & 0 & \\ 0 & 1 & 0 & 0 & \\ 0 & 0 & -2 & 0 & \\ 0 & 0 & 0 & 0 & \\ & & & & \ddots \end{pmatrix} \quad \cdots$$

Next, we form the $(n^2 - n)/2$ off-diagonal matrices with 1 in a given off-diagonal position, 1 in the transposed position, and zeros elsewhere, and we also form analogous matrices with $-i$ in the given off-diagonal position, i in the transposed position, and zeros elsewhere. Altogether these matrices satisfy the conditions

$$\begin{aligned} \operatorname{tr}\lambda_j &= 0 \\ \operatorname{tr}\lambda_j\lambda_k &= 2\delta_{jk} \end{aligned}$$

For $n = 2$, there are three λ matrices.

$$\lambda_1 = \begin{pmatrix} 0 & 1 \\ 1 & 0 \end{pmatrix} \quad \lambda_2 = \begin{pmatrix} 0 & -i \\ i & 0 \end{pmatrix} \quad \lambda_3 = \begin{pmatrix} 1 & 0 \\ 0 & -1 \end{pmatrix}$$

Of course, these are just the Pauli spin matrices, and the $F_j = \frac{1}{2}\lambda_j$ satisfy the commutation relations

$$[F_i, F_j] = i\varepsilon_{ijk} F_k \tag{7.174}$$

familiar for the spin operators from the theory of angular momentum. Relations (7.174) are satisfied not only by 2×2 matrices (the fundamental representation) but also by matrices of dimension $(2J+1)\times(2J+1)$, with $J = 1, 3/2, 2, \ldots$. In the fundamental representation we introduce two types of spinors:

1. Covariant or column spinors:

$$\psi = \begin{pmatrix} \psi_1 \\ \psi_2 \end{pmatrix} \tag{7.175}$$

which transform according to the rule

$$\psi_i' = U_i^k \psi_k \tag{7.176}$$

2. Contravariant or row spinors:

$$\bar{\phi} = (\bar{\phi}_1 \quad \bar{\phi}_2) \tag{7.177}$$

which transform as follows:

$$\bar{\phi}^{k\prime} = \bar{\phi}^i U_i^{k\dagger} \tag{7.178}$$

A covariant 2-spinor is transformed to a contravariant 2-spinor and vice versa with the aid of the completely antisymmetric unit tensors ε_{ij} and ε^{ij}. For example, $\phi_i = \varepsilon_{ij}\bar{\phi}^j$ is a covariant spinor of the form

$$\phi = \begin{pmatrix} \bar{\phi}_2 \\ -\bar{\phi}_1 \end{pmatrix} \tag{7.179}$$

Higher representations of $SU(2)$ are built up systematically from the fundamental representation by forming tensor products of the basic 2-spinors and then by symmetrizing and antisymmetrizing. Here we make use of an important theorem, not proved here, that states that once a multispinor has been broken into parts with different permutation symmetries, it has been decomposed uniquely into irreducible representations of $SU(n)$. For example, in $SU(2)$, consider the tensor product of two covariant spinors ψ and η; that is,

$$\psi_i \eta_j = \frac{1}{2}(\psi_i \eta_j + \psi_j \eta_i) + \frac{1}{2}(\psi_i \eta_j - \psi_j \eta_i) \tag{7.180}$$

On the left-hand side of (7.180) there are four distinct quantities because i, j independently take the values 1, 2. On the right-hand side, the first (symmetric) term has three independent elements and corresponds to $J = 1$ in the theory of angular momentum. The second term is nonzero only if $i \neq j$; it consists of one component and corresponds to $J = 0$. Symbolically, we write

$$2 \otimes 2 = 3 + 1 \tag{7.181}$$

In the theory of angular momentum, it is unnecessary to employ covariant spinors of the form (7.179). However, for isospin, we frequently need to construct multiplets with antiparticles as well as particles, and here (7.179) is necessary.

In $SU(3)$, there are eight distinct matrices λ_i, two of which are diagonal; that is,

$$\begin{aligned}
&\lambda_1 = \begin{pmatrix} 0 & 1 & 0 \\ 1 & 0 & 0 \\ 0 & 0 & 0 \end{pmatrix} \quad \lambda_2 = \begin{pmatrix} 0 & -i & 0 \\ i & 0 & 0 \\ 0 & 0 & 0 \end{pmatrix} \quad \lambda_3 = \begin{pmatrix} 1 & 0 & 0 \\ 0 & -1 & 0 \\ 0 & 0 & 0 \end{pmatrix} \\
&\lambda_4 = \begin{pmatrix} 0 & 0 & 1 \\ 0 & 0 & 0 \\ 1 & 0 & 0 \end{pmatrix} \quad \lambda_5 = \begin{pmatrix} 0 & 0 & -i \\ 0 & 0 & 0 \\ i & 0 & 0 \end{pmatrix} \\
&\lambda_6 = \begin{pmatrix} 0 & 0 & 0 \\ 0 & 0 & 1 \\ 0 & 1 & 0 \end{pmatrix} \quad \lambda_7 = \begin{pmatrix} 0 & 0 & 0 \\ 0 & 0 & i \\ 0 & -i & 0 \end{pmatrix} \\
&\lambda_8 = \sqrt{\frac{1}{3}} \begin{pmatrix} 1 & 0 & 0 \\ 0 & 1 & 0 \\ 0 & 0 & -2 \end{pmatrix}
\end{aligned} \tag{7.182}$$

Writing $F_j = \frac{1}{2}\lambda_j$ as before, we obtain the commutation relations for the F_j matrices by explicit computation from (7.182); that is,

$$\left[F_i, F_j\right] = i f_{ij}^k F_k \tag{7.183}$$

where the $SU(3)$ structure constants f_{ij}^k are given in Table 7.2.

It is easy to verify the following relations:

$$f_{ij}^k = -f_{ji}^k \tag{7.184}$$

and

$$f_{ij}^\ell f_{\ell k}^m + f_{ki}^\ell f_{\ell j}^m + f_{jk}^\ell f_{\ell i}^m = 0 \tag{7.185}$$

These relations and the structure constants are representation independent. For a number of applications, it is also useful to define the following operators:

Table 7.2 Nonzero $SU(3)$ structure constants

i	j	k	f_{ij}^k
1	2	3	1
1	4	7	1/2
1	5	6	−1/2
2	4	6	1/2
2	5	7	1/2
3	4	5	1/2
3	6	7	−1/2
4	5	8	$\sqrt{\frac{3}{2}}$
6	7	8	$\sqrt{\frac{3}{2}}$

$$\begin{aligned} T_\pm &= F_1 \pm iF_2 & T_3 &= F_3 \\ U_\pm &= F_6 \pm iF_7 & U_3 &= -\frac{1}{2}F_3 + \frac{\sqrt{3}}{2}F_8 \\ V_\pm &= F_4 \mp iF_5 & V_3 &= -\frac{1}{2}F_3 - \frac{\sqrt{3}}{2}F_8 \end{aligned} \tag{7.186}$$

Nine new operators are defined here, but only eight are independent because $T_3 + U_3 + V_3 = 0$.

Problems for Chapter 7

7.1. (a) A simultaneous eigenstate of J^2 and J_z is denoted by $|jm\rangle$. Show that the expectation values of J_x and J_y for this state are zero.

(b) Show that if any operator commutes with two components of an angular-momentum operator, it must commute with the third component.

(c) Can states of a system be found for which the root-mean-square (rms) deviations $\Delta J_x, \Delta J_y$, and ΔJ_z are simultaneously zero?

(d) In units where $\hbar = 1$, show that

$$\Delta J_x \cdot \Delta J_y \geq \frac{1}{2}|\langle J_z \rangle| \tag{1}$$

and cyclic permutations.

(e) Show that

$$(\Delta J_x)^2 + (\Delta J_y)^2 \geq |\langle J_z \rangle| \tag{2}$$

(f) For a state $|jm\rangle$, show that the inequalities (1) and (2) become equalities if and only if $m=-j$ or $m=j$.

7.2. (a) In Section 7.6, we discussed magnetic resonance. There we showed by means of the rotating coordinate method that if a particle with spin-½ and gyromagnetic ratio $\gamma = g\mu_N/\hbar$ is initially in the state $m=+1/2$ and is exposed for a time t to the magnetic field

$$\boldsymbol{B} = B_1\left[\hat{x}\cos\omega t - \hat{y}\sin\omega t\right] + B_0\hat{z} \tag{1}$$

then the probability for a transition to the state $m=-1/2$ is given by Rabi's formula; that is,

$$P_{1/2\to-1/2} = \frac{(\gamma B_1)^2}{(\gamma B_1)^2 + (\omega-\omega_0)^2}\sin^2\left[\sqrt{(\gamma B_1)^2+(\omega-\omega_0)^2}\,\frac{t}{2}\right] \tag{2}$$

where $\omega_0 = \gamma B_0$. Derive (2) by solving Schroedinger's equation $H\psi = i\hbar\dot{\psi}$, where $H = -(\hbar\gamma/2)\boldsymbol{\sigma}\boldsymbol{\cdot}\boldsymbol{B}$ and $\psi = \begin{bmatrix} a(t) \\ b(t) \end{bmatrix}$ with $a(0)=1$, $b(0)=0$. (Thus Schroedinger's equation is really two coupled first-order differential equations.)

(b) A particle initially in the state $J=1, m=0$ and with gyromagnetic ratio γ is exposed to the magnetic field described in (1). Using the rotating coordinate method and the rotation matrix for $J=1$, find the transition probabilities to arrive in the states $m=\pm1$.

7.3. We know that in coordinate representation, the eigenfunctions of the orbital angular momentum operators L^2, L_z are the spherical harmonics. What are the eigenfunctions of these operators in momentum representation?

7.4. Find the magnetic moment in nuclear magnetons of the following nuclei, assuming that the spin-magnetic moments of the proton and neutron are $\mu_p = 2.79$, $\mu_n = -1.91$, respectively.

(a) ^{15}N: Here one proton in a $p_{1/2}$ state is missing from an otherwise complete shell.

(b) ^{17}O: Here one neutron in a $d_{5/2}$ state exists outside closed shells.

(c) What would the magnetic moment of the deuteron be for each of the states

$$^3S_1 \qquad ^3P_1 \qquad ^3D_1$$

7.5. The nuclear reaction $^2H + ^3H \to ^4He + n$ is frequently used as a practical source of fast neutrons. Most commonly, one has a beam of deuterons (spin 1) at an energy of about 100 keV impinging on a solid target containing tritium nuclei (spin ½). The latter are at rest in the laboratory frame. At this energy, the relative momentum of the deuteron and triton is so low and the range of nuclear forces so short that the orbital angular momentum of relative motion is $\ell = 0$. Thus, a priori, the total angular momentum of the compound nuclear state could be $J=1/2$ or $J=3/2$. In fact, however, the reaction of interest occurs in the $J=3/2$ channel. Assume this and also that the neutron has spin ½, the alpha particle has spin 0, and parity is conserved in this strong interaction. Show that if the incoming deuterons are polarized in the spin state $m=+1$, then the neutron relative intensity angular distributions are given by the formulas

$$S_+(\theta) = 1 + 3\cos^2\theta$$
$$S_-(\theta) = 9\sin^2\theta$$

where θ is the polar angle, and $S_\pm$ refer to the relative intensities of the $m_s = \pm\frac{1}{2}$ neutron waves with respect to the z-axis. Because these angular distributions are distinct, the outgoing neutrons are in general polarized.

7.6. In this chapter we mention the Clebsch-Gordan series and its inverse. One of these two useful formulas is

$$D^{j_1}_{m_1'm_1}(\alpha,\beta,\gamma)\,D^{j_2}_{m_2'm_2}(\alpha,\beta,\gamma) = \sum_j \langle j_1m_1j_2m_2|jm\rangle\langle j_1m_1'j_2m_2'|jm'\rangle D^{j}_{m'm}(\alpha,\beta,\gamma) \qquad (1)$$

In the sum on the right-hand side of (1), j runs from $j_1 + j_2$ to $|j_1 - j_2|$. The other formula is

$$D^{j}_{m'm}(\alpha,\beta,\gamma) = \sum_{\substack{m_1',m_1\\ m_2',m_2}} \langle j_1m_1j_2m_2|jm\rangle\langle j_1m_1'j_2m_2'|jm'\rangle D^{j_1}_{m_1'm_1}(\alpha,\beta,\gamma)D^{j_2}_{m_2'm_2}(\alpha,\beta,\gamma) \qquad (2)$$

(a) Starting from the definition of the rotation matrices and the vector coupling coefficients, derive these formulas.
(b) Use (1) or (2) to derive the $D^{J=1}_{m'm}$ from the $D^{J=1/2}_{m'm}$.
(c) From (1) and/or (2) derive the addition theorem for spherical harmonics; that is,

$$Y^0_\ell(\theta,0) = \sqrt{\frac{4\pi}{2\ell+1}} \sum_{m=-\ell}^{\ell} Y^{m*}_\ell(\theta_1,\phi_1)Y^m_\ell(\theta_2,\phi_2)$$

where θ is the angle between the rays defined by θ_1,ϕ_1 and θ_2,ϕ_2.

8 Wave Mechanics in Three Dimensions: Hydrogenic Atoms

8.1 General properties of solutions to Schroedinger's wave equation

We now consider some general properties of solutions to

$$-\frac{\hbar^2}{2m}\nabla^2\psi + V\psi = E\psi \tag{8.1}$$

assuming throughout this chapter, unless otherwise noted, that $V(r) \to 0$ as $r \to \infty$. The spectrum of eigenvalues E in (8.1) may be discrete and/or continuous. A solution ψ corresponding to a discrete eigenvalue E is normalizable; that is,

$$\int |\psi|^2 \, d\tau = 1 \tag{8.2}$$

Thus $\psi \to 0$ more rapidly than $r^{-3/2}$ as $r \to \infty$, and ψ is therefore a bound state. A solution corresponding to E in the continuous spectrum is not normalizable, and in this case ψ extends to infinity.

A state with $E < 0$ necessarily belongs to the discrete spectrum because $E = \langle T \rangle + \langle V \rangle$, where T is the kinetic energy operator, and $\langle T \rangle > 0$. Thus, although $V(r) \to 0$ as $r \to \infty$, $\langle V \rangle < E < 0$, and therefore, ψ must be a bound state. Conversely, the existence of a single bound state implies that $V < 0$ in some finite region of space.

The character of important physical problems is quite different for the two cases $E < 0$ and $E > 0$. For bound states, our usual goal is to solve (8.1) analytically or numerically, subject to the boundary condition $\psi \to 0$ as $r \to \infty$, to find the discrete eigenvalues $E < 0$ and the corresponding eigenfunctions. These theoretical results are most often compared with experimental data obtained by spectroscopic techniques. When $E > 0$, we are usually interested in collisions between two or more particles or scattering of an incident particle (projectile) by a force center (target). Here the initial wave function and the initial kinetic energy $E > 0$ of the projectile are specified in advance. Interaction between the projectile and the force center generates an outgoing scattered wave, the radial probability current density of which can be measured at a detector located a macroscopic distance from the force center. At that large distance, the potential has fallen to zero; thus we are interested in calculating the asymptotic properties of the scattered wave as a function of E and the scattering angle or angles. These calculations yield a theoretical scattering cross section that one tries to compare with the results of scattering experiments. In many cases, this offers the only practical means to gain useful knowledge about the interaction between the projectile and the target.

8.2 Power-law potentials

Consider a potential with the property that as $r \to 0$, $V \to -kr^{-s}$, where k and s are positive constants. Suppose that there is a bound state with wave function centered at $r = 0$ and with spatial extent of approximately r_0. Then $\Delta p \approx \hbar/r_0$ and $\langle T\rangle \approx \hbar^2/(2mr_0^2)$, whereas $\langle V\rangle \approx -k/r_0^s$. If $s > 2$, $E = \langle T\rangle + \langle V\rangle$ becomes arbitrarily large and negative for sufficiently small r_0. In other words, the energy E has no finite lower bound. If $s < 2$, the discrete spectrum does possess a finite lower bound. The case $s = 2$ requires special treatment (see Problem 8.3).

Next, consider a potential with the property that as $r \to \infty$, $V \to -kr^{-s}$. Suppose that we have a wave function that is localized in a shell-like region between r_0 and $r_0 + \Delta$, where $\Delta \ll r_0$. (For example, this might correspond to a particle with large orbital angular momentum in an attractive central potential.) Here $\Delta p \approx \hbar/\Delta$ and $\langle T\rangle \approx \hbar^2/(2m\Delta^2)$, whereas $\langle V\rangle \approx -k/r_0^s$. Imagine that we increase r_0, keeping Δ/r_0 fixed. Then, if $s < 2$, E must be negative for sufficiently large r_0. Thus there exist bound states with ψ having appreciable size in some region arbitrarily far from the origin. In other words, there are bound states with arbitrarily small negative E and hence an infinite number of bound states reaching a limit point at $E = 0$. On the contrary, for $s > 2$, there is a last bound state with finite negative E.

8.3 Radial Schroedinger equation for a central potential

We have already seen in Section 6.2 that if two particles at positions $\boldsymbol{r}_1, \boldsymbol{r}_2$ with masses m_1, m_2 interact through a central potential $V(|\boldsymbol{r}_1 - \boldsymbol{r}_2|)$, the Schroedinger equation reduces to two separate wave equations, one describing free-particle motion of the center of mass, that is,

$$\boldsymbol{R} = \frac{m_1\boldsymbol{r}_1 + m_2\boldsymbol{r}_2}{M}$$

where $M = m_1 + m_2$, and another, that is,

$$-\frac{\hbar^2}{2\mu}\nabla^2 u(\boldsymbol{r},t) + V(r)u(\boldsymbol{r},t) = i\hbar\frac{\partial u}{\partial t} \tag{8.3}$$

describing the relative motion of the two particles. Here $\mu = m_1 m_2/M$ is the reduced mass, and $\boldsymbol{r} = \boldsymbol{r}_1 - \boldsymbol{r}_2$. In (8.3), it is obviously convenient to employ spherical polar coordinates

$$\begin{aligned} x &= r\sin\theta\cos\phi \\ y &= r\sin\theta\sin\phi \\ z &= r\cos\theta \end{aligned} \tag{8.4}$$

for which the Laplacian is

$$\nabla^2 = \frac{1}{r^2}\frac{\partial}{\partial r}\left(r^2\frac{\partial}{\partial r}\right) + \frac{1}{r^2 \sin\theta}\left[\frac{\partial}{\partial\theta}\left(\sin\theta\frac{\partial}{\partial\theta}\right)\right] + \frac{1}{r^2\sin^2\theta}\frac{\partial^2}{\partial\phi^2} \tag{8.5}$$

From (7.83), we have

$$L^2\psi = -\left[\frac{1}{\sin\theta}\frac{\partial}{\partial\theta}\left(\sin\theta\frac{\partial}{\partial\theta}\right) + \frac{1}{\sin^2\theta}\frac{\partial^2}{\partial\phi^2}\right]\psi \tag{8.6}$$

Thus the time-independent Schroedinger equation for relative motion in the central potential $V(r)$ is

$$\frac{\partial^2\psi(r,\theta,\phi)}{\partial r^2} + \frac{2}{r}\frac{\partial\psi}{\partial r} - \frac{1}{r^2}L^2\psi + \frac{2\mu}{\hbar^2}\left[E - V(r)\right]\psi = 0 \tag{8.7}$$

This equation is simplified by separation of variables. We write $\psi = R(r)Y_\ell^m(\theta,\phi)$ and use $L^2Y_\ell^m = \ell(\ell+1)Y_\ell^m$ to obtain the radial Schroedinger equation

$$R'' + \frac{2}{r}R' + \frac{2\mu}{\hbar^2}\left\{E - \left[V(r) + \frac{\hbar^2\ell(\ell+1)}{2\mu r^2}\right]\right\}R = 0 \tag{8.8}$$

where the prime signifies differentiation with respect to r. It is often convenient to make the substitution $\chi(r) = rR(r)$ in (8.8), which yields

$$\chi'' + \frac{2\mu}{\hbar^2}(E - V_{\text{eff}})\chi = 0 \tag{8.9}$$

where

$$V_{\text{eff}} = V(r) + \frac{\hbar^2\ell(\ell+1)}{2\mu r^2} \tag{8.10}$$

The second term on the right-hand side of (8.10) is called the *centrifugal potential.* If $|V(r)| \to \infty$ more slowly than r^{-2} as $r \to 0$, $R(0)$ is finite, and $\chi(0) = 0$. Also, for such a potential, the second term on the left-hand side of (8.9) is dominated at small r by the centrifugal potential when $\ell > 0$; that is,

$$\chi'' \approx \frac{\ell(\ell+1)}{r^2}\chi$$

This implies that for small r, $\chi \to a_\ell r^{\ell+1}, R \to a_\ell r^\ell$, where a_ℓ is a coefficient.

8.4 Virial theorem

The *virial theorem* relates $\langle T \rangle$ and $\langle V \rangle$ for a bound state $|\psi\rangle$ in a central potential $V(r)$. Let A be any observable. Then, because $H|\psi\rangle = E|\psi\rangle$,

$$\langle\psi|AH - HA|\psi\rangle = (E-E)\langle\psi|A|\psi\rangle = 0 \tag{8.11}$$

We choose $A = \boldsymbol{r}\boldsymbol{\cdot}\boldsymbol{p}$ and write $H = (\boldsymbol{p}^2/2\mu) + V(r)$. Then, using the repeated index summation convention, we have

$$\begin{aligned}\left[\boldsymbol{r}\boldsymbol{\cdot}\boldsymbol{p}, \frac{\boldsymbol{p}^2}{2\mu}\right] &= \frac{1}{2\mu}(x_i p_i p_k p_k - p_k p_k x_i p_i) \\ &= \frac{1}{2\mu}(x_i p_k p_k p_i - p_k p_k x_i p_i) \\ &= \frac{i\hbar}{\mu}\boldsymbol{p}^2 = 2i\hbar T\end{aligned} \tag{8.12}$$

Also,

$$[\boldsymbol{r}\boldsymbol{\cdot}\boldsymbol{p}, V(r)] = -i\hbar r\frac{\partial V}{\partial r} \tag{8.13}$$

Employing (8.12) and (8.13) in (8.11), we obtain

$$2\langle T \rangle = \left\langle r\frac{\partial V}{\partial r}\right\rangle \tag{8.14}$$

If $V = -kr^{-s}$, we have $r(\partial V/\partial r) = -sV$, and (8.14) yields

$$\langle T \rangle + \frac{s}{2}\langle V \rangle = 0 \tag{8.15}$$

This together with $E = \langle T \rangle + \langle V \rangle$ gives

$$\langle T \rangle = -\frac{s}{2-s}E \tag{8.16}$$

and

$$\langle V \rangle = \frac{2}{2-s}E \tag{8.17}$$

For the important special case of the Coulomb potential ($s = 1$), we thus obtain

$$\langle T \rangle = -E \qquad \langle V \rangle = 2E$$

8.5 Atomic units: Bound states of a hydrogenic atom in spherical coordinates

We next consider the possible states of an electron bound in the Coulomb field of a nucleus with atomic number Z and infinite mass. Obviously, this is one of the most important problems in wave mechanics. Here and in all of atomic and molecular physics, atomic units are far more convenient than cgs, Heaviside-Lorentz, or SI units. In atomic units,

- One unit of mass = electron mass = $m_e = 9.108 \times 10^{-28}$ g.
- One unit of length = Bohr radius = $a_0 = \hbar^2/m_e e^2 = 5.2917\times 10^{-9}$ cm, where e is in cgs units.
- One unit of velocity = $\alpha c = e^2/\hbar = 2.1877\times 10^8$ cm/s, where c = velocity of light in vacuum = 2.9979×10^{10} cm/s.

Thus

- One unit of time = $a_0/\alpha c = 2.4189\times 10^{-17}$ s.

Because $m_e = a_0 = 1$ and $\alpha c = 1$, we have

- One unit of electric charge = $e = 4.8029\times 10^{-10}$ esu.
- One unit of action = $\hbar = 1.0544 \times 10^{-27}$ erg · s.
- One unit of energy = $e^2/a_0 = 27.2$ eV $= 2$ R$_{\text{H}}$.
- One unit of electric field = $e/a_0^2 = 5.142\times 10^9$ V/cm.

We note in passing that in relativistic quantum mechanics it is convenient to choose "natural" units for which $m_e = \hbar = c = 1$. The unit of energy in this system is $m_e c^2 = 0.511$ MeV. Furthermore, within the natural unit system it is convenient to employ Heaviside-Lorentz units (hlu) for electromagnetic quantities and, in particular, for electric charge. Here $e_{\text{hlu}} = \sqrt{4\pi}e_{\text{cgs}}$, and because $\alpha = e_{\text{hlu}}^2/4\pi\hbar c = 137.036^{-1}$ is dimensionless, we have

- One unit of electric charge in natural units = $e = \sqrt{4\pi\alpha}$.

This unit system is discussed in detail in Section 15.7.

In spherical polar coordinates and atomic units, the radial Schroedinger equation for an electron bound in the potential $V = -Z/r$ is

$$R'' + \frac{2}{r}R' + \left[2E + \frac{2Z}{r} - \frac{\ell(\ell+1)}{r^2}\right]R = 0 \tag{8.18}$$

with $E < 0$. In the limit of very large r, (8.18) reduces to

$$R'' + 2ER = 0$$

which has the solutions

$$R = \exp(\pm\varepsilon r) \tag{8.19}$$

where $\varepsilon = (-2E)^{1/2} > 0$. For bound states, we must choose the minus sign in (8.19); thus we try solutions to (8.18) of the form

$$R(r) = f(r)\exp(-\varepsilon r) \tag{8.20}$$

where we require $f(r)$ to vary slowly for large r. Substitution of (8.20) in (8.18) yields

$$f'' + 2\left(\frac{1}{r} - \varepsilon\right)f' + \left[2\left(\frac{Z-\varepsilon}{r}\right) - \frac{\ell(\ell+1)}{r^2}\right]f = 0 \tag{8.21}$$

At very small r, we know that f is proportional to r^{ℓ}; thus we substitute the power series

$$f = r^{\ell}\sum_{\nu=0}^{\infty} a_\nu r^\nu \tag{8.22}$$

in (8.21). This yields

$$\sum_{\nu=0}^{\infty} a_\nu \left\{\left[(\ell+\nu)(\ell+\nu+1) - \ell(\ell+1)\right]r^{\ell+\nu-2} - 2\left[\varepsilon(\ell+\nu+1) - Z\right]r^{\ell+\nu-1}\right\} = 0 \tag{8.23}$$

We set the coefficient of each power ν of r equal to zero for $\nu > 0$. This yields the recursion relation

$$\frac{a_\nu}{a_{\nu-1}} = 2\frac{\varepsilon(\ell+\nu) - Z}{(\ell+\nu)(\ell+\nu+1) - \ell(\ell+1)} \tag{8.24}$$

Either the series on the right-hand side of (8.22) terminates at a largest value of ν, or it does not. If it does not, (8.24) yields

$$\lim_{\nu\to\infty} a_\nu = \frac{2\varepsilon}{\ell+\nu+1}a_{\nu-1}$$

and thus

$$a_\nu = \text{const}\bullet\frac{(2\varepsilon)^{\ell+\nu+1}}{(\ell+\nu+1)!}$$

This would result in $f = \text{const}\bullet\exp(2\varepsilon r)$; hence $R \to \text{const}\bullet\exp(\varepsilon r)$ for large r, which is impermissible. Hence the series on the right-hand side of (8.22) must terminate with $\nu_{\max} = n - \ell - 1$, where n, the principal quantum number, is a positive integer. This requires that for given n, $\ell_{\max} = n - 1$. It also requires that $\varepsilon = Z/n$, which implies

$$E_{n\ell} = -\frac{Z^2}{2n^2} \tag{8.25}$$

This is the well-known *Balmer formula* (in atomic units) for the bound-state energies of an electron in the Coulomb potential $-Z/r$. Because for given n the quantum number ℓ can take the values 0, 1, 2, …, n – 1, (8.25) implies that for given n the states with these values of ℓ are all degenerate. This phenomenon, sometimes called *accidental degeneracy*, is not at all accidental but is a manifestation of an interesting symmetry that is special for the Coulomb potential and the three-dimensional harmonic oscillator (see Section 8.7). It is also related to the fact that Schroedinger's time-independent wave equation for the Coulomb potential can be separated not only in spherical polar coordinates but also in parabolic coordinates (see Section 8.6).

To find the bound-state radial eigenfunctions, we return to (8.24) and make the substitution $Z = n\varepsilon$ to obtain

$$\frac{a_\nu}{a_{\nu-1}} = -2\varepsilon \frac{n-\ell-\nu}{\nu(2\ell+\nu+1)} \tag{8.26}$$

Defining $\rho = 2\varepsilon r = 2Zr/n$, we employ (8.26) and (8.22) in (8.20) to arrive after straightforward manipulations at

$$R(\rho) = c\times(2\varepsilon)^{3/2} e^{-\rho/2}\rho^{\ell} F\left[-(n-\ell-1), 2\ell+2, \rho\right] \tag{8.27}$$

where c is a normalization constant, and

$$F(a,b,x) = 1 + \frac{a}{1!b}x + \frac{a(a+1)}{2!b(b+1)}x^2 + \cdots \tag{8.28}$$

is the confluent hypergeometric function. It can be shown that the following formula is equivalent to (8.27):

$$R = -c\frac{(2\varepsilon)^{3/2}}{(n+\ell)!^2}(2\ell+1)!(n-\ell-1)!e^{-\rho/2}\rho^{\ell} L_{n+\ell}^{2\ell+1}(\rho) \tag{8.29}$$

where $L_{n+\ell}^{2\ell+1}(\rho)$ is an associated Laguerre polynomial defined by

$$L_\lambda^\mu = \frac{d^\mu}{d\rho^\mu} L_\lambda \tag{8.30}$$

where

$$L_\lambda = e^{\rho}\frac{d^\lambda}{d\rho^\lambda}\left(\rho^\lambda e^{-\rho}\right) \tag{8.31}$$

is a Laguerre polynomial. It also can be shown by straightforward integration that the normalization coefficient in (8.27) and (8.29) is

$$c = \frac{1}{(2\ell+1)!}\sqrt{\frac{(n+\ell)!}{(n-\ell-1)!2n}} \tag{8.32}$$

In atomic units, the first few bound-state radial functions are as follows:

$$
\begin{aligned}
R_{10} &= 2Z^{3/2}e^{-Zr} \\
R_{20} &= \frac{1}{\sqrt{2}}Z^{3/2}\left(1-\frac{Zr}{2}\right)e^{-Zr/2} \\
R_{30} &= \frac{2}{3\sqrt{3}}Z^{3/2}\left(1-\frac{2Zr}{3}+\frac{2Z^2r^2}{27}\right)e^{-Zr/3} \\
R_{21} &= \frac{Z^{5/2}}{2\sqrt{6}}re^{-Zr/2} \\
R_{31} &= \frac{8Z^{5/2}}{27\sqrt{6}}r\left(1-\frac{Zr}{6}\right)e^{-Zr/3} \\
R_{32} &= \frac{4}{81\sqrt{30}}Z^{7/2}r^2e^{-Zr/3}
\end{aligned}
\tag{8.33}
$$

The mean values of various powers of r for hydrogenic wave functions

$$
\langle r^k \rangle = \frac{\int_0^\infty r^k R_{n\ell}^2 r^2\, dr}{\int_0^\infty R_{n\ell}^2 r^2\, dr} \tag{8.34}
$$

are useful quantities in the analysis of many physical problems. These can be evaluated by explicit use of (8.27) or (8.29) in straightforward calculations. One finds

$$
\langle r \rangle = \frac{1}{2Z}\left[3n^2 - \ell(\ell+1)\right] \tag{8.35}
$$

$$
\langle r^2 \rangle = \frac{n^2}{2Z^2}\left[5n^2 + 1 - 3\ell(\ell+1)\right] \tag{8.36}
$$

$$
\langle r^{-1} \rangle = \frac{Z}{n^2} \tag{8.37}
$$

$$
\langle r^{-2} \rangle = \frac{Z^2}{n^3(\ell+1/2)} \tag{8.38}
$$

$$
\langle r^{-3} \rangle = \frac{Z^3}{n^3\ell(\ell+1/2)(\ell+1)} \tag{8.39}
$$

The probability density at the origin for $\ell = 0$ (s) states is also a useful quantity; that is,

$$
|\psi(0)|_{n0}^2 = \frac{Z^3}{\pi n^3} \tag{8.40}
$$

The analysis we have outlined in this section was first carried out by Erwin Schroedinger in 1926. Since then, several algebraic methods have been developed for solving the bound-state

Coulomb problem, but they do not appear to be simpler, more transparent, or more useful than Schroedinger's original method.

8.6 Hydrogenic bound states in parabolic coordinates

We have already noted that Schroedinger's equation with $V = -Z/r$ is separable in parabolic as well as spherical polar coordinates and that this fact is related to the accidental degeneracy. Parabolic coordinates are defined as

$$\begin{aligned} \xi &= r + z = r(1+\cos\theta) \\ \eta &= r - z = r(1+\cos\theta) \\ \phi &= \phi \end{aligned} \tag{8.41}$$

The surfaces of constant $\xi(\eta)$ are confocal paraboloids of revolution with the origin as focus and opening in the direction of negative (positive) z, respectively. In parabolic coordinates, Schroedinger's equation in atomic units with $V = -Z/r$ is

$$\frac{1}{\xi+\eta}\left[\frac{\partial}{\partial\xi}\left(\xi\frac{\partial\psi}{\partial\xi}\right)+\frac{\partial}{\partial\eta}\left(\eta\frac{\partial\psi}{\partial\eta}\right)\right]+\frac{1}{4\xi\eta}\frac{\partial^2\psi}{\partial\phi^2}+\left[\frac{E}{2}+\frac{Z}{\xi+\eta}\right]\psi = 0 \tag{8.42}$$

We try a solution to (8.42) of the form

$$\psi = u_1(\xi)u_2(\eta)e^{\pm im\phi} \qquad m \geq 0 \tag{8.43}$$

Multiplying both sides of (8.42) by $(\xi+\eta)$ and making the separation of variables, we obtain the following ordinary differential equations:

$$\frac{d}{d\xi}\left(\xi\frac{du_1}{d\xi}\right)+\left(\frac{1}{2}E\xi+Z_1-\frac{m^2}{4\xi}\right)u_1 = 0 \tag{8.44}$$

and

$$\frac{d}{d\eta}\left(\eta\frac{du_2}{d\eta}\right)+\left(\frac{1}{2}E\eta+Z_2-\frac{m^2}{4\eta}\right)u_2 = 0 \tag{8.45}$$

where $Z_1 + Z_2 = Z$. By means of the substitutions $x = \varepsilon\xi$ and

$$u_1 = e^{-\varepsilon\xi/2}\xi^{m/2}f_1(\xi) \tag{8.46}$$

with $\varepsilon = \sqrt{-2E}$ as before, we arrive at

$$x\frac{d^2 f_1}{dx^2} + (m+1-x)\frac{df_1}{dx} + \left(\frac{Z_1}{\varepsilon} - \frac{m+1}{2}\right) f_1 = 0 \tag{8.47}$$

This equation has the solution

$$f_1 = L^m_{n_1+m}(x) \tag{8.48}$$

where

$$n_1 = \frac{Z_1}{\varepsilon} - \frac{1}{2}(m+1) \tag{8.49}$$

must be a nonnegative integer so that f_1 remains finite at large x. Similar results are found for u_2. Then, defining

$$n = n_1 + n_2 + m + 1 \tag{8.50}$$

and solving (8.49) for ε, we once again obtain the Balmer formula

$$E = -\frac{Z^2}{2n^2} \tag{8.51}$$

The normalized eigenfunctions in parabolic coordinates for $m \geq 0$ are

$$\psi_{n_1,n_2,m} = \frac{e^{\pm im\phi}}{\sqrt{\pi n}} \frac{(n_1!)^{1/2}(n_2!)^{1/2}}{[(n_1+m)!]^{3/2}[(n_2+m)!]^{3/2}} \varepsilon^{m+3/2} e^{-\varepsilon(\xi+\eta)/2} (\xi\eta)^{m/2} L^m_{n_1+m}(\varepsilon\xi) L^m_{n_2+m}(\varepsilon\eta) \tag{8.52}$$

In general, these functions are asymmetric with respect to the plane $z = 0$; for $n_1 > n_2$ ($n_1 < n_2$), most of the electron charge distribution is in the half-space $z > 0$ ($z < 0$), respectively.

Any single eigenstate in spherical polar coordinates with given n, ℓ, m can be expressed as a linear combination of eigenstates in parabolic coordinates, where the latter satisfy the condition $n_1 + n_2 = n - m - 1$. Conversely, a single eigenfunction in parabolic coordinates with given n_1, n_2, $m \geq 0$ and sign of the exponent $\pm im\phi$ can be expressed as a linear combination of eigenfunctions in spherical coordinates. One can verify that just as in the case of spherical polar coordinates, there are n^2 distinct degenerate eigenfunctions in parabolic coordinates for a given n.

8.7 Bound states of hydrogenic atoms and *O*(4) symmetry

We have mentioned that the accidental degeneracy associated with the Coulomb Hamiltonian is a manifestation of an underlying symmetry. It is useful to start our discussion of that symmetry by considering the motion of a unit point mass in the potential $V = -Z/r$ according to nonrelativistic classical mechanics. We know that every bound orbit is an ellipse with the force

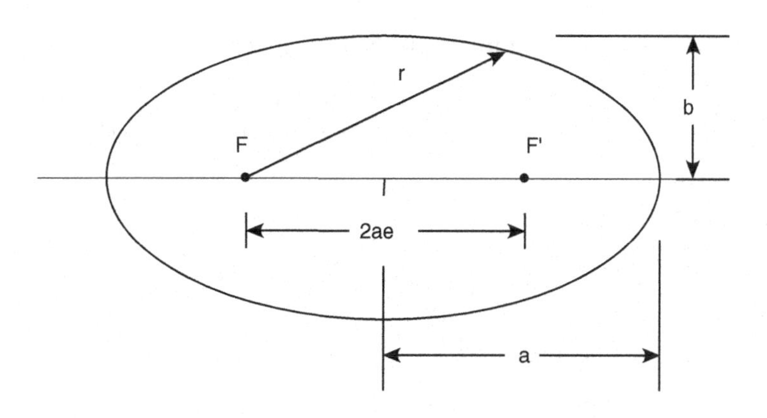

Figure 8.1 Bound orbit of a unit point mass in the potential $V=-Z/r$ according to nonrelativistic classical mechanics. The orbit is an ellipse with the force center at focus F. If the semimajor and semiminor axes are labeled a and b, respectively, the eccentricity is $e=\sqrt{a^2-b^2}/a$, the total energy is $E=-Z/2a$, and the orbital angular momentum is a constant vector perpendicular to the plane of the orbit with magnitude $|\boldsymbol{L}|=(Za)^{1/2}(1-e^2)^{1/2}$.

center at one focus. Let the semimajor axis and the eccentricity of the ellipse be a and e, respectively (Figure 8.1).

It is easy to show that the total energy for such an orbit is

$$E=-\frac{Z}{2a} \tag{8.53}$$

independent of the eccentricity and that the orbital angular momentum with respect to the force center is a constant vector $\boldsymbol{L}$ perpendicular to the orbit plane of magnitude

$$|\boldsymbol{L}|=(Za)^{1/2}(1-e^2)^{1/2} \tag{8.54}$$

There is an additional conserved vector $\boldsymbol{M}_0$, named the *Runge-Lenz vector* after two twentieth-century physicists, although it has been known since the eighteenth century and was discussed by Laplace and Hamilton, among others. It is defined as

$$\boldsymbol{M}_0=\boldsymbol{p}\times\boldsymbol{L}-Z\frac{\boldsymbol{r}}{r} \tag{8.55}$$

By making use of $\boldsymbol{L}=\boldsymbol{r}\times\boldsymbol{p}$ and the equation of motion

$$\frac{d\boldsymbol{p}}{dt}=\ddot{\boldsymbol{r}}=-\frac{Z\mathrm{r}}{r^3}$$

it is easy to show that $d\boldsymbol{M}_0/dt=0$. It is also easy to see that $\boldsymbol{M}_0\cdot\boldsymbol{L}=0$; hence $\boldsymbol{M}_0$ lies in the orbit plane (in fact, along the major axis). Its magnitude is found most easily by considering the case where r is maximum (apogee); thus, where $r=a(1+e)$ and $\boldsymbol{r}\cdot\boldsymbol{p}=0$. We find

$$|\boldsymbol{M}_0|=Ze \tag{8.56}$$

and it is convenient to rescale the Runge-Lenz vector by defining

$$\boldsymbol{M} = \sqrt{\frac{a}{Z}}\boldsymbol{M}_0 = \sqrt{-\frac{1}{2E}}\boldsymbol{M}_0 \tag{8.57}$$

Then

$$(\boldsymbol{L}+\boldsymbol{M})^2 = \boldsymbol{L}^2 + \boldsymbol{M}^2 = Za(1-e^2) + Zae^2 = Za$$

which implies that

$$E = -\frac{Z}{2a} = -\frac{Z^2}{2(\boldsymbol{L}+\boldsymbol{M})^2} = -\frac{Z^2}{2(\boldsymbol{L}^2+\boldsymbol{M}^2)} \tag{8.58}$$

This classical expression resembles the Balmer formula [(8.25) and (8.51)] for the bound-state energies of a hydrogenic atom and suggests that we try to identify n^2 in the Balmer formula with an eigenvalue of a quantum-mechanical operator defined in place of $\boldsymbol{L}^2 + \boldsymbol{M}^2$.

To this end, we return to (8.55) and try to construct a Hermitian quantum-mechanical operator analogous to $\boldsymbol{M}_0$. Noting that $(\boldsymbol{p}\times\boldsymbol{L})^\dagger = -\boldsymbol{L}\times\boldsymbol{p} \neq \boldsymbol{p}\times\boldsymbol{L}$, we define the Hermitian operator as

$$\boldsymbol{M}_0 = \frac{1}{2}(\boldsymbol{p}\times\boldsymbol{L} - \boldsymbol{L}\times\boldsymbol{p}) - Z\frac{\boldsymbol{r}}{r} \tag{8.59}$$

Then, making use of the commutation relations $[x_i, L_j] = i\varepsilon_{ijk}x_k$ and $[p_i, L_j] = i\varepsilon_{ijk}p_k$ [see (7.73) and (7.74)] and $H = \boldsymbol{p}^2/2 - Z/r$, we obtain the following results after straightforward algebra:

$$[\boldsymbol{M}_0, H] = 0 \tag{8.60}$$

$$\boldsymbol{M}_0 \cdot \boldsymbol{L} = 0 \tag{8.61}$$

and

$$\boldsymbol{M}_0 \cdot \boldsymbol{M}_0 = 2H(\boldsymbol{L}^2 + 1) + Z^2 \tag{8.62}$$

Once again, we rescale, this time by means of the definition

$$\boldsymbol{M} = \sqrt{-\frac{1}{2E}}\boldsymbol{M}_0 \tag{8.63}$$

The following commutation rules may then be verified by direct calculation:

$$[M_i, M_j] = i\varepsilon_{ijk}L_k \tag{8.64}$$

and

$$[M_i, L_j] = i\varepsilon_{ijk} M_k \tag{8.65}$$

which hold in addition to the usual relations

$$[L_i, L_j] = i\varepsilon_{ijk} L_k \tag{8.66}$$

These relations suggest that we introduce a fourth fictitious Euclidean coordinate x_4 as well as the corresponding fictitious momentum p_4 such that

$$[x_i, p_j] = i\delta_{ij} \qquad i, j = 1, \ldots, 4 \tag{8.67}$$

and

$$\begin{aligned} L_1 &= x_2 p_3 - x_3 p_2 \\ L_2 &= x_3 p_1 - x_1 p_3 \\ L_3 &= x_1 p_2 - x_2 p_1 \\ M_1 &= x_1 p_4 - x_4 p \\ M_2 &= x_2 p_4 - x_4 p_2 \\ M_3 &= x_3 p_4 - x_4 p_3 \end{aligned} \tag{8.68}$$

It is also convenient to define the operators $\boldsymbol{I} = (\boldsymbol{L} + \boldsymbol{M})/2$ and $\boldsymbol{K} = (\boldsymbol{L} - \boldsymbol{M})/2$. These satisfy the relations

$$[I_i, I_j] = \frac{1}{4}[L_i + M_i, L_j + M_j] = \frac{1}{2} i\varepsilon_{ijk}(L_k + M_k) = i\varepsilon_{ijk} I_k \tag{8.69}$$

and, similarly,

$$[K_i, K_j] = i\varepsilon_{ijk} K_k \tag{8.70}$$

Because $\boldsymbol{L}$ and $\boldsymbol{M}$ each commute with H, $\boldsymbol{I}$ and $\boldsymbol{K}$ do also; furthermore, $[I_i, K_j] = 0$ for all i and j. Thus the eigenstates of H are simultaneously eigenstates of $\boldsymbol{I}^2$ and of $\boldsymbol{K}^2$ with eigenvalues $I(I+1)$ and $K(K+1)$, respectively, where $I, K = 0, 1/2, 1, 3/2, \ldots$. In fact, because $\boldsymbol{M} \cdot \boldsymbol{L} = 0$, we have $\boldsymbol{I}^2 = \boldsymbol{K}^2$; hence, recalling (8.62) and (8.63),

$$\begin{aligned} \boldsymbol{I}^2 + \boldsymbol{K}^2 = 2\boldsymbol{I}^2 &= \frac{1}{2}(\boldsymbol{L}^2 + \boldsymbol{M}^2) \\ &= \frac{1}{2}\left[\boldsymbol{L}^2\left(1 - \frac{H}{E}\right) - \frac{H}{E} - \frac{Z^2}{2E}\right] \end{aligned} \tag{8.71}$$

Therefore,

$$E = -\frac{Z^2}{2(2I+1)^2} \tag{8.72}$$

Because the possible values of I are 0, ½, 1, 3/2, …, we may define the principal quantum number as $n = 2I + 1$, which is consistent with the usual definition $n = \ell_{\max} + 1$. Thus we see that (8.72) is the Balmer formula.

The foregoing analysis was first carried out by W. Pauli in 1926 at about the same time that Schroedinger solved the bound-state problem by the conventional analytical method. Pauli's analysis reveals the $O(4)$ symmetry of the Coulomb Hamiltonian, which is so-named because $\boldsymbol{L}^2 + \boldsymbol{M}^2$ is a generalization to four-dimensional Euclidean space from $\boldsymbol{L}^2$ in three dimensions.

Coulomb wave functions for $E > 0$ are discussed in Chapter 18 in connection with scattering.

Problems for Chapter 8

8.1. The radial equation for the H atom can be transformed into the radial equation of a two-dimensional isotropic harmonic oscillator. To do this, we replace r by $\lambda\rho^2/2$, where λ is a constant to be determined. Also we write $R_{n,\ell}(r) = F(\rho)/\rho$.

(a) Show that $F(\rho)$ obeys the radial equation of a two-dimensional harmonic oscillator of frequency

$$\omega = \sqrt{-2\lambda^2 \frac{E}{m}}$$

with angular momentum $2\ell + 1$ and energy $2e^2\lambda$, where E is the energy of the hydrogen level.

(b) Deduce the Balmer formula for hydrogen energies, and deduce their degeneracies.

(c) Construct the normalized ground-state wave function of hydrogen by this procedure.

8.2. The deuteron is a bound state of a proton and a neutron and is the simplest compound nucleus. We construct a useful elementary theoretical model of the deuteron by assuming that the short-range attractive force between proton and neutron is derived from the central potential

$$V(r) = -V_0 \exp(-r/a) \tag{1}$$

where V_0 and a are positive constants.

(a) Consider the radial Schroedinger equation for $\ell = 0$ and this potential. Change variables from r to $z = \exp(-r/2a)$, and show that Bessel's equation results. From an analysis of the boundary conditions on ψ as a function of z, derive the following requirement for the existence of a bound state:

$$V_0 a^2 \geq \frac{1.45\hbar^2}{m_p} \tag{2}$$

where m_p is the proton mass.

(b) Because at low energies the attractive force between two nucleons arises from pion exchange, it is reasonable to choose for a the Compton wavelength of the pion $a = 1.4 \times 10^{-13}$ cm. In this case, what does (2) yield for V_0 in mega-electron-volts (MeV)?

(c) In the reasonable first approximation that the low-energy neutron-proton attractive force is central, experiment shows that the magnitude of this force is strongest when the spins of the nucleons are parallel. Indeed, the only bound state of the deuteron is ${}^3S_1$, where the neutron and proton spins couple together to give a total spin $S = 1$, whereas the orbital angular momentum of relative motion is $L = 0$. When the spins are antiparallel, the force is weaker, and it turns out that the ${}^1S_0$ state is not bound.

However the central force approximation is not exactly valid. There is a small contribution to the potential arising from a noncentral "tensor" component. The primary experimental evidence for this is the deuteron's nonzero electric quadrupole moment, which reveals that there must be a small admixture of the state ${}^3D_1$ in the nuclear wave function. The value of the quadrupole moment implies that the probabilities of finding the deuteron in the ${}^3D_1$ (${}^3S_1$) state are $w = 0.04,\ 1 - w = 0.96$, respectively.

Given these facts, employ an argument involving a generalization of the Lande g-factor to show that the magnetic moment of the deuteron is

$$\mu_D = \mu_p + \mu_n - \frac{3}{2}w\left(\mu_p + \mu_n - \frac{1}{2}\mu_N\right)$$

where μ_N is the nuclear Bohr magneton, and

$$\mu_p = 2.79274\mu_N$$
$$\mu_n = -1.91314\mu_N$$

are the proton, neutron spin magnetic moments, respectively. Check your answer against the experimental value of the deuteron magnetic moment

$$\mu_D = 0.857438\mu_N$$

Note that in first approximation, a nuclear magnetic moment is the vector sum of the spin magnetic moments of the individual protons and neutrons and the magnetic moments arising from the orbital motion of the protons. There is no contribution from orbital motion of the neutrons because a neutron has no net charge.

8.3. A particle of mass m moves in the attractive potential

$$V(r) = -\frac{q}{r^2}$$

where q is a real positive constant, and we employ units where $m = \hbar = 1$. Analyze the solutions to the Schroedinger equation and show that when $q \le \frac{1}{8} + \frac{1}{2}\ell(\ell+1)$, there are no bound states and when $q > \frac{1}{8} + \frac{1}{2}\ell(\ell+1)$, there is an infinite number of bound states with no finite lower bound to the energy.

8.4. A particle of mass m is bound in the potential $V(r) = \frac{1}{2}m\omega^2 r^2$ (isotropic three-dimensional harmonic oscillator). Show that the Schroedinger equation can be separated not only in Cartesian coordinates but also in cylindrical coordinates and in spherical polar coordinates. Find the degeneracy of each energy level. Find the eigenfunctions in spherical coordinates corresponding to the energy level that has a degeneracy of 10.

8.5. Consider bound states in a spherically symmetric potential $V(r)$ such that $|V(r)| \to \infty$ more slowly than $1/r^2$ as $r \to 0$. We know that the one-dimensional radial Schroedinger equation is

$$\chi'' + f(r)\chi = 0 \qquad (1)$$

where

$$f(r) = \frac{2\mu}{\hbar^2}\left[E - V(r) - \frac{\hbar^2 \ell(\ell+1)}{2\mu r^2}\right]$$

and as was shown in Section 8.3,

$$\lim_{r \to 0} \chi = a_\ell \, r^{\ell+1}$$

where a_ℓ is a constant.

(a) Multiply (1) by $r^q \chi'(r)$, where $q \geq -2\ell$, and integrate both sides from ε to ∞, where ε is a positive infinitesimal that we shall momentarily set equal to zero. Making use of the fact that χ and χ' go to zero sufficiently rapidly at ∞ for bound states, and employing integration by parts several times, show that when $\varepsilon \to 0$, the following result is obtained:

$$(2\ell+1)^2 a_\ell^2 \delta_{-2\ell,q} = -\left\langle 2qr^{q-1} f + r^q f' + \frac{1}{2} q(q-1)(q-2) r^{q-3} \right\rangle \qquad (2)$$

(b) Use (2) with $q = \ell = 0$ to obtain a relation between $|\psi(0)|^2$ and $\langle \partial V / \partial r \rangle$.

(c) Use (2) with $q = 1$ to derive the virial theorem.

(d) For the case of the Coulomb potential, show that (2) can be used to obtain useful formulas for $\langle r^p \rangle$, where p is any positive integer, and to obtain useful recursive formulas for $\langle r^p \rangle$ from $\langle r^{-2} \rangle$ when $p = -3, -4, \ldots$ and $\ell > 0$. [Note that equation (2) is not sufficient by itself to yield $\langle r^{-2} \rangle$.]

8.6. (a) Consider a bound-state wave function $\psi(\boldsymbol{r})$ with corresponding probability density $\rho = \psi^* \psi$ and probability current density

$$\boldsymbol{j} = \frac{\hbar}{2mi}(\psi^* \nabla \psi - \psi \nabla \psi^*)$$

Assume that ρ satisfies the condition

$$\int \rho \boldsymbol{r} \times \hat{\boldsymbol{n}} \, ds = 0 \qquad (1)$$

where the quantity on the left-hand side of (1) is a surface integral taken over a surface at ∞, and $\hat{\boldsymbol{n}}$ is a unit normal vector to the surface. Show that the expectation value of the orbital angular-momentum operator $\boldsymbol{L}$ is

$$\langle \boldsymbol{L} \rangle = m \int \boldsymbol{r} \times \boldsymbol{j} \, d\tau \qquad (2)$$

where the integral on the right-hand side of (2) is a volume integral taken over all space.
(b) Now, considering the description of a particle with spin-½ in nonrelativistic quantum mechnics, let us assume that $\psi = \phi(\boldsymbol{r})\chi$, where ϕ is a bounded spatial wave function and χ is a two-component spinor. We write the expectation value of the spin operator as

$$\langle \boldsymbol{S} \rangle = \frac{\hbar}{2}\int \psi^* \boldsymbol{\sigma} \psi \; d\tau \tag{3}$$

Show that

$$\psi^* \boldsymbol{\sigma} \psi = \frac{1}{2}\boldsymbol{r}\times\left[\nabla\times\left(\psi^*\boldsymbol{\sigma}\psi\right)\right] - \frac{1}{2}\nabla\left[\boldsymbol{r}\cdot\psi^*\boldsymbol{\sigma}\psi\right] + \frac{1}{2}\sum_{i=1}^{3}\frac{\partial}{\partial x_i}\left[x_i\left(\psi^*\boldsymbol{\sigma}\psi\right)\right] \tag{4}$$

and show that the last two terms on the right-hand side of (4) make no contribution to the integral in (3). Thus, in analogy to (2), (3) can be written as

$$\langle \boldsymbol{S} \rangle = m\int \boldsymbol{r}\times \boldsymbol{j}_S \; d\tau \tag{5}$$

Here $\boldsymbol{j}_S$, the *spin current density*, is $\boldsymbol{j}_S = \nabla\times \boldsymbol{V}_S$, where

$$\boldsymbol{V}_S = \frac{\hbar}{4m}\psi^* \boldsymbol{\sigma}\psi \tag{6}$$

(c) Calculate the spin current density in atomic units for the $2^2s_{1/2}$, $m_s = 1/2$ state of atomic hydrogen.

9 Time-Independent Approximations for Bound-State Problems

9.1 Variational method

Very few Hamiltonians in quantum mechanics are exactly solvable, and therefore, approximations must be used to treat most problems of physical interest. We start with the *variational method.* Consider a Hamiltonian H with a spectrum of discrete energy eigenvalues E_n arranged in ascending order $E_0 < E_1 < E_2 < \cdots$ and with corresponding eigenfunctions u_n, where the latter form a complete orthonormal set; that is,

$$\langle u_n | u_m \rangle = \delta_{nm}$$

Although the energies may be known from experiment, the eigenfunctions are usually not known because the Hamiltonian is too complicated to solve exactly. Let us approximate the ground-state wave function by a trial function $\psi(\lambda)$ that depends on one or more continuous real parameters λ. Because the $|u_n\rangle$ form a complete set, we have

$$\begin{aligned} \langle \psi | H | \psi \rangle &= \sum_{n,m} \langle \psi | u_n \rangle \langle u_n | H | u_m \rangle \langle u_m | \psi \rangle \\ &= \sum_n E_n |\langle u_n | \psi \rangle|^2 \\ &\geq E_0 \sum_n |\langle u_n | \psi \rangle|^2 = E_0 \langle \psi | \psi \rangle \end{aligned}$$

Thus, for any trial function ψ whatsoever, we have

$$\frac{\langle \psi | H | \psi \rangle}{\langle \psi | \psi \rangle} \geq E_0 \tag{9.1}$$

and equality is achieved only when $\psi = u_0$. Suppose that we choose a definite functional form for the wave function ψ in terms of the parameter(s) λ. Then we vary λ and seek a minimum on the left-hand side of (9.1). When that minimum is found, we have obtained the best trial function of that particular functional form.

To give a trivial example, we consider the ground state of a hydrogenic ion of atomic number Z. Of course, in this particular example, we know the energy $E_{1s} = -Z^2/2$ (in atomic units), and we also know the wave function

$$u_{1s} = \left(\frac{Z^3}{\pi} \right)^{1/2} e^{-Zr}$$

However, suppose that we did not know the wave function but were only able to guess that it is exponential in form; that is,

$$\psi = \left(\frac{\lambda^3}{\pi}\right)^{1/2} e^{-\lambda r}$$

where we have included the first factor so that ψ is normalized to unity. We now calculate $\langle \psi | H | \psi \rangle$.

$$\langle \psi | H | \psi \rangle = -\frac{1}{2}\int \psi^* \nabla^2 \psi \, d\tau - Z \int \psi^* \frac{1}{r} \psi \, d\tau$$

The integrals are easily evaluated, with the following result:

$$\langle \psi | H | \psi \rangle = \frac{\lambda^2}{2} - \lambda Z \tag{9.2}$$

Taking the derivative of this expression and setting it equal to zero, we find that $\langle \psi | H | \psi \rangle$ reaches the minimum $-Z^2/2$ when $\lambda = Z$.

The variational method has many important applications, some of which are discussed in subsequent chapters and their problems. However, the method has a fundamental limitation: even a rather poor trial function can yield a close upper bound on the experimentally determined ground-state energy. Thus achievement of the latter does not necessarily indicate that we have a good trial-wave function. The reason is as follows: suppose that the discrepancy between ψ and u_0 is of first order in a small quantity δ; that is,

$$\psi = u_0 + \delta \cdot u'$$

where $\langle u_0 | u_0 \rangle = \langle u' | u' \rangle = 1$ and $\langle u' | u_0 \rangle = 0$. Then

$$\begin{aligned}\langle \psi | H | \psi \rangle &= \langle u_0 + \delta \cdot u' | H | u_0 + \delta \cdot u' \rangle \\ &= \langle u_0 | H | u_0 \rangle + \delta^* \langle u' | H | u_0 \rangle + \delta \langle u_0 | H | u' \rangle + |\delta|^2 \langle u' | H | u' \rangle\end{aligned}$$

In this expression, the terms in δ^* and δ vanish because of the orthogonality of u' and u_0; hence, although the error in ψ is of first order in δ, the error in $\langle \psi | H | \psi \rangle$ is only of second order in δ.

9.2 Semiclassical (WKB) approximation

We next consider the Wentzel-Kramers-Brillouin (WKB) approximation, which was actually discussed in a classical context, long before quantum mechanics, by Liouville and Rayleigh. The method is useful for finding approximate solutions to the time-independent Schroedinger equation in situations where the potential varies slowly over distances comparable with the wavelength. To introduce the WKB method and show how it is applied, we need only one spatial dimension. Thus we start with the Schroedinger equation

$$\frac{d^2\psi}{dx^2} + k^2(x)\psi = 0 \tag{9.3}$$

where $k^2(x) = (2m/\hbar^2)[E - V(x)]$, and we try a solution of the form

$$\psi(x) = e^{if(x)} \tag{9.4}$$

Substitution of (9.4) in (9.3) yields

$$if'' - f'^2 + k^2(x) = 0 \tag{9.5}$$

If k^2 is a constant, (9.5) has the solution

$$f' = \pm k \tag{9.6}$$

with $f'' = 0$; hence $\psi = e^{\pm ikx}$. If $k(x)$ varies very slowly, we expect that the first term on the left-hand side of (9.5) should be much smaller in magnitude than the other two terms, even if it is no longer zero. Hence (9.6) still should be approximately correct; therefore, $f'' \approx \pm k'$. Thus, from (9.5), a better approximation for f'^2 is

$$\begin{aligned} f'^2 &= k^2 \pm ik' \\ &= k^2\left(1 \pm i\frac{k'}{k^2}\right) \end{aligned} \tag{9.7}$$

This yields

$$f' = \pm k\sqrt{1 \pm i\frac{k'}{k^2}} \approx \pm k + i\frac{k'}{2k} \tag{9.8}$$

Integrating both sides of (9.8), we obtain

$$if \approx \pm i\int k\,dx - \frac{1}{2}\ln(k) \tag{9.9}$$

Hence

$$\psi \approx \frac{A}{\left|2m[E - V(x)]\right|^{1/4}} \exp\left[\pm\frac{i}{\hbar}\int\sqrt{2m[E - V(x)]}\,dx\right] \tag{9.10}$$

where A is a constant. The analysis that yielded (9.9) can be iterated further, resulting in correction terms of higher order. However, for most applications of physical interest, such terms are of negligible importance.

Result (9.10) is justified only if the second term on the right-hand side of (9.9) is much smaller in magnitude than the first. However, this is not the case in the immediate neighborhood of

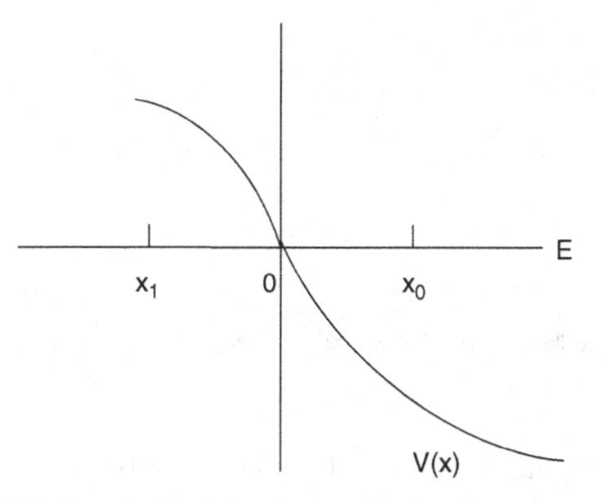

Figure 9.1 Sketch of a one dimensional potential $V(x)$ with energy E such that the classically allowed region is $x \geq 0$.

the classical turning point, where $k(x) = 0$, and thus the WKB approximation fails at that location.

Equation (9.10) really describes two independent solutions. An actual wave function in the WKB approximation is an appropriate linear combination of the two solutions with coefficients determined by boundary conditions. For example, consider how the WKB approximation (9.10) is applied to a situation where the classically allowed region $E > V(x)$ lies to the right of $x = 0$ in Figure 9.1.

If we assume that $E < V(x)$ for all $x < 0$, $\psi \to 0$ as $|x_1|$ increases; hence, recalling Section 6.3, ψ must be nondegenerate and real apart from an arbitrary phase factor. Thus, for $x_0 > 0$, (9.10) implies

$$\psi = \frac{A}{p^{1/2}} \sin\left(\frac{1}{\hbar}\int_0^{x_0} p\,dx + \alpha\right) \tag{9.11}$$

where $p(x) = \sqrt{2m[E - V(x)]}$, and α is a real constant. To determine α, we compare (9.11) with the asymptotic form of the exact solution for a linearized potential in the neighborhood of $x = 0$. By this we mean that a sufficiently small value of x_0 is chosen so that $V(x_0)$ may be replaced by a linear function with the slope and intercept of $V(x)$ at $x = 0$, but x_0 is large enough so that the solution for the linear potential reaches its asymptotic form [recall equation (6.33)]. Clearly, these two conditions can be satisfied simultaneously only if $V(x)$ varies sufficiently slowly.

Thus we assume that $V(x)$ is replaced by $V_0(x) = E - F_0 x$, where $F_0 = -\partial V/\partial x|_{x=0}$. Then

$$\psi'' + \frac{2m}{\hbar^2}\left[E - V_0(x)\right]\psi = \psi'' + \frac{2m}{\hbar^2}F_0 x\psi = 0$$

The substitution $z = \left(2mF_0/\hbar^2\right)^{1/3} x \equiv \beta x$ yields the Airy differential equation

$$\frac{d^2\psi}{dz^2} + z\psi = 0 \tag{9.12}$$

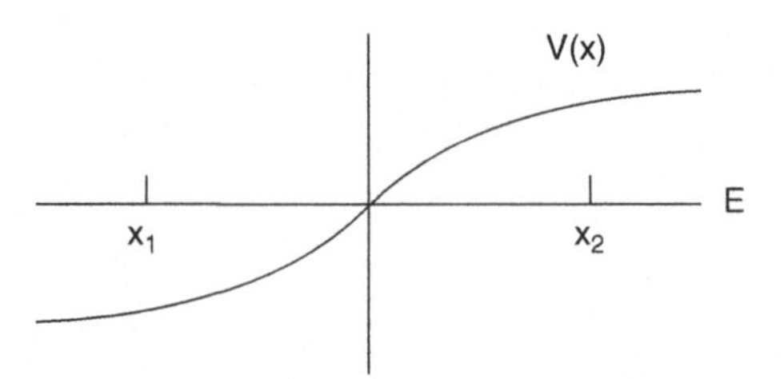

Figure 9.2 Sketch of a one dimensional potential $V(x)$ with energy E such that the classically allowed region is $x \leq 0$.

The asymptotic solution to (9.12) is given by (6.33); that is,

$$\psi = \frac{\text{const}}{z^{1/4}} \sin\left(\frac{2}{3} z^{3/2} + \frac{\pi}{4}\right) \tag{9.13}$$

and it is easy to verify that $\frac{2}{3} z^{3/2} = (1/\hbar)\int_0^x \sqrt{2m[E - V_0(x)]}\,dx$. Comparing (9.13) with (9.11), we thus obtain

$$\psi(x_0) = \frac{\text{const}}{\sqrt{p(x_0)}} \sin\left(\frac{1}{\hbar}\int_0^{x_0} p\,dx + \frac{\pi}{4}\right) \tag{9.14}$$

If the classically allowed region lies to the left of $x = 0$ as in Figure 9.2, similar considerations yield

$$\psi(x_1) = \frac{A}{\sqrt{p(x_1)}} \sin\left(\frac{1}{\hbar}\int_{x_1}^{0} p\,dx + \frac{\pi}{4}\right) \tag{9.15}$$

Furthermore, from the asymptotic properties (6.34) of the Airy function in the classically forbidden region, we obtain

$$\psi(x_2) = \frac{A}{2\sqrt{|p(x_2)|}} \exp\left(-\frac{1}{\hbar}\int_0^{x_2} |p|\,dx\right) \tag{9.16}$$

where the constant A is the same in (9.15) and (9.16).

We now show that the WKB approximation yields the Bohr-Sommerfeld quantization condition for a potential well with classical turning points x_1, x_2 and a classically allowed region $x_1 < x < x_2$, as shown in Figure 9.3.

From (9.14) and (9.15) we have

$$\psi(x) = \frac{A}{\sqrt{p(x)}} \sin\left(\frac{1}{\hbar}\int_{x_1}^{x} p\,dx + \frac{\pi}{4}\right) = \frac{A'}{\sqrt{p(x)}} \sin\left(\frac{1}{\hbar}\int_{x}^{x_2} p\,dx + \frac{\pi}{4}\right) \tag{9.17}$$

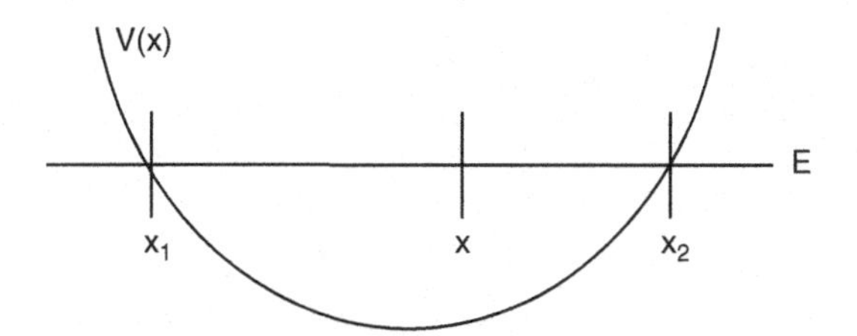

Figure 9.3 One dimensional potential well $V(x)$ with energy E such that classically allowed region is $x_1 \le x \le x_2$.

Because $\int_x^{x_2} p\,dx = \int_{x_1}^{x_2} p\,dx - \int_{x_1}^{x} p\,dx$, (9.17) yields the conditions $A = (-1)^m A'$ and

$$\int_{x_1}^{x_2} p\,dx = \left(m + \frac{1}{2}\right)\pi\hbar \tag{9.18}$$

where m is a nonnegative integer. Result (9.18), the quantization condition from the old Bohr-Sommerfeld model, is reasonably accurate if V varies slowly compared with the wave function of interest.

The WKB method provides a useful (if rather complicated) approximate formula for hydrogenic radial wave functions when the principal quantum number is large: $n \gg 1$ and also $n \gg \ell$. Here we start with the radial equation in atomic units; that is,

$$\frac{d^2\chi}{dr^2} + \left[-\frac{Z^2}{n^2} + \frac{2Z}{r} - \frac{\ell(\ell+1)}{r^2}\right]\chi = 0 \tag{9.19}$$

The quantity in brackets, $k^2(r)$, is positive between the classical turning points $r_{1,2}$; that is,

$$r_{1,2} = \frac{n^2}{Z} \pm \frac{n}{Z}\sqrt{n^2 - \ell(\ell+1)} \tag{9.20}$$

In the region $r_1 < r < r_2$, the normalized WKB approximation for χ is

$$\begin{aligned}\chi &\approx Z\sqrt{\frac{2}{\pi n^3}}\left[\frac{2Z}{r} - \frac{Z^2}{n^2} - \frac{(l+1/2)^2}{r^2}\right]^{-1/4} \sin\left[\int_{r_1}^{r}\sqrt{\frac{2Z}{x} - \frac{Z^2}{n^2} - \frac{(l+1/2)^2}{x^2}}\,dx + \frac{\pi}{4}\right]\\ &= Z\sqrt{\frac{2}{\pi n^3}}\left[\frac{2Z}{r} - \frac{Z^2}{n^2} - \frac{(l+1/2)^2}{r^2}\right]^{-1/4} \cos\left\{\sqrt{2Zr - \frac{Z^2r^2}{n^2} - (l+1/2)^2} + n\arcsin\left[\frac{Zr - n^2}{n\sqrt{n^2 - (l+1/2)^2}}\right]\right.\\ &\quad \left. - (l+1/2)\arcsin\left[\frac{n}{Zr}\frac{Zr - (l+1/2)^2}{\sqrt{n^2 - (l+1/2)^2}}\right] + (n - \ell - 1)\frac{\pi}{2}\right\}\end{aligned} \tag{9.21}$$

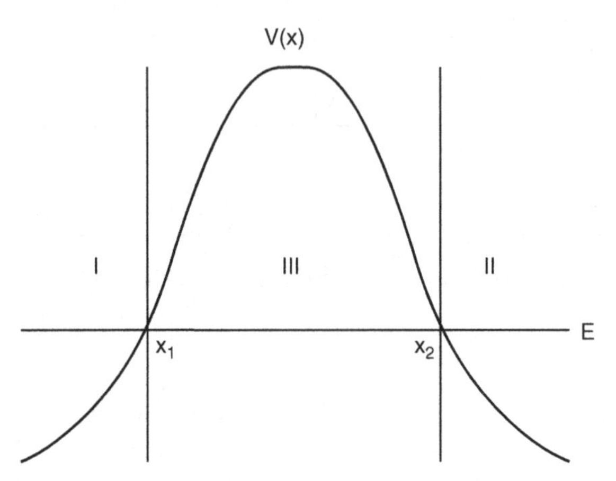

Figure 9.4 One dimensional potential barrier with classically allowed regions I, II and classically forbidden region III.

Another well-known example of the WKB method concerns penetration of a potential barrier of arbitrary shape (Figure 9.4). Here the only restriction is that the barrier must be sufficiently high and/or wide that the probability of penetration is small.

Let two waves u and v corresponding to the same energy E be incident from the left and right, respectively. Because each is reflected almost entirely by the barrier, we have a nearly perfect standing wave in region I ($x < x_1$), and the same is true in region II ($x > x_2$), where we write

$$v_{\text{II}} = \frac{1}{2i\sqrt{p}}\left[\exp\left(\frac{i}{\hbar}\int_{x_2}^{x>x_2} p\,dx + \frac{i\pi}{4}\right) - \exp\left(-\frac{i}{\hbar}\int_{x_2}^{x>x_2} p\,dx - \frac{i\pi}{4}\right)\right] \tag{9.22}$$

Now a small fraction of u is transmitted through the barrier, resulting in the following component traveling to the right in region II:

$$u_{\text{II}} = \frac{C}{\sqrt{p}}\exp\left(\frac{i}{\hbar}\int_{x_2}^{x>x_2} p\,dx - \frac{i\pi}{4}\right) \tag{9.23}$$

Inclusion of the phase factor $\exp(-i\pi/4)$ costs no loss of generality because we have not yet restricted the complex coefficient C. We know that $uv' - vu' = \text{const}$ and from (9.22) and (9.23) that the constant is $-iC/\hbar$. Now we continue the solutions u, v into the region III. From (9.22) and the connection between (9.15) and (9.16), we obtain

$$v_{\text{III}} = \frac{1}{2\sqrt{|p|}}\exp\left(-\frac{1}{\hbar}\int_{x<x_2}^{x_2} |p|\,dx\right) \tag{9.24}$$

Note that the magnitude of this solution decreases as x decreases. Meanwhile, the general form for u in region III is

$$u_{\text{III}} = \frac{C'}{\sqrt{|p|}}\exp\left(\frac{1}{\hbar}\int_{x<x_2}^{x_2} |p|\,dx\right) \tag{9.25}$$

the magnitude of which increases as x decreases. The coefficient C' is determined from the requirement $(uv' - vu')_{\text{III}} = -iC/\hbar$; we find that $C' = -iC$. As for u in region I, because it is a nearly perfect standing wave, we write

$$u_{\text{I}} = \frac{A}{\sqrt{p}} \sin\left(\int_{x<x_1}^{x_1} p\, dx + \frac{\pi}{4} \right) \tag{9.26}$$

Consequently, from (9.15) and (9.16),

$$u_{\text{III}} = \frac{A}{2\sqrt{|p|}} \exp\left(-\frac{1}{\hbar} \int_{x_1}^{x>x_1} |p|\, dx \right) \tag{9.27}$$

Comparing (9.25) with (9.27) and making use of $\int_{x_1}^{x>x_1} = \int_{x_1}^{x_2} - \int_{x<x_2}^{x_2}$, we obtain

$$C = i\frac{A}{2} \exp\left(-\frac{1}{\hbar} \int_{x_1}^{x_2} |p|\, dx \right)$$

Hence the transmission coefficient for barrier penetration in the WKB approximation is

$$T = \exp\left(-\frac{2}{\hbar} \int_{x_1}^{x_2} |p|\, dx \right) \tag{9.28}$$

9.3 Static perturbation theory

This is the most important general approximation method for time-independent problems. The basic approach was first introduced in the nineteenth century by Rayleigh for problems in classical physics and was developed for quantum mechanics by Schroedinger. We start with a zeroth-order Hamiltonian H_0 for which the eigenvalues E_{n0} and corresponding eigenstates $|n_0\rangle$ are known. The latter are assumed to form a complete orthonormal set. The Hamiltonian is then modified by adding a perturbing term $\lambda H'$, where λ is a small parameter. We wish to determine the eigenstates $|n\rangle$ and corresponding eigenvalues E_n of the new Hamiltonian $H = H_0 + \lambda H'$. To achieve this goal by successive approximations, we expand E_n and $|n\rangle$ in powers of λ; that is,

$$E_n = E_{n0} + \lambda E_{n1} + \lambda^2 E_{n2} + \cdots \tag{9.29}$$

$$|n\rangle = |n_0\rangle + \lambda |n_1\rangle + \lambda^2 |n_2\rangle + \cdots \tag{9.30}$$

This yields

$$(H_0 + \lambda H')(|n_0\rangle + \lambda|n_1\rangle + \lambda^2|n_2\rangle + \cdots) = (E_{n0} + \lambda E_{n1} + \lambda^2 E_{n2} + \cdots)(|n_0\rangle + \lambda|n_1\rangle + \lambda^2|n_2\rangle + \cdots) \tag{9.31}$$

Equating terms with equal powers of λ on both sides of (9.31), we obtain

$$(H_0 - E_{n0})|n_0\rangle = 0 \tag{9.32}$$

$$(H_0 - E_{n0})|n_1\rangle = (E_{n1} - H')|n_0\rangle \tag{9.33}$$

$$(H_0 - E_{n0})|n_2\rangle = (E_{n1} - H')|n_1\rangle + E_{n2}|n_0\rangle \tag{9.34}$$

In the kth equation of this system, $|n_k\rangle$, which always appears only on the left-hand side, is expressed in terms of the previous orders of approximation $|n_{k-1}\rangle, |n_{k-2}\rangle, \ldots$, which always appear only on the right-hand side. The left-hand side of the kth equation for $k > 0$ is not altered if we subtract any multiple of $|n_0\rangle$ from $|n_k\rangle$. Hence we can replace $|n_k\rangle$ by $|n_k'\rangle = |n_k\rangle - |n_0\rangle\langle n_0|n_k\rangle$, which gives $\langle n_k'|n_0\rangle = 0$ for all $k > 0$. From now on we adopt the $|n_k'\rangle$ as our corrections to $|n_0\rangle$ and drop the primes. Hence we have $\langle n_0|n_0\rangle = 1$, but $\langle n_0|n_1\rangle = \langle n_0|n_2\rangle = \cdots = 0$. To find E_{n1}, we first consider the case where the $|n_0\rangle$ are nondegenerate. Here multiplication of both sides of (9.33) on the left by $\langle n_0|$ immediately yields

$$E_{n1} = \langle n_0|H'|n_0\rangle \tag{9.35}$$

On the other hand, suppose that there exist s degenerate eigenvectors

$$|n_0^{(1)}\rangle, \quad |n_0^{(2)}\rangle, \quad \ldots, \quad |n_0^{(s)}\rangle$$

all of which correspond to the eigenvalue E_{n0} of H_0. In this case, the first-order energy shifts caused by H' are the eigenvalues of the $s \times s$ matrix $\langle n_0^{(i)}|H'|n_0^{(j)}\rangle$. In many cases, at least some of these eigenvalues are distinct; in other words, at least some of the degeneracy is lifted by imposition of the perturbation H' in first order. However, some degeneracy may remain and be removed only in second or higher order.

Once again assuming the nondegenerate case for simplicity, we multiply both sides of (9.33) on the left by a zero-order eigenstate $\langle m_0|$ that is distinct from $\langle n_0|$. Taking advantage of orthogonality, we obtain

$$\langle m_0|n_1\rangle = \frac{\langle m_0|H'|n_0\rangle}{E_{m0} - E_{n0}} \tag{9.36}$$

Now, multiplying both sides of (9.36) by $|m_0\rangle$, summing over all $|m_0\rangle \neq |n_0\rangle$, and using $\langle n_0|n_1\rangle = 0$, we arrive at

$$|n_1\rangle = \sum_{m_0 \neq n_0} |m_0\rangle\langle m_0|n_1\rangle = \sum_{m_0 \neq n_0} |m_0\rangle \frac{\langle m_0|H'|n_0\rangle}{E_{n0} - E_{m0}} \tag{9.37}$$

which gives the first-order correction to $|n_0\rangle$. Next we consider the second-order equation (9.34). Taking the scalar product of both sides with $\langle n_0|$, we obtain

$$E_{n2} = \langle n_0|H''|n_1\rangle. \tag{9.38}$$

Substitution of (9.37) into (9.38) then yields

$$E_{n2} = \sum_{m_0 \neq n_0} \frac{|\langle m_0|H'|n\rangle|^2}{E_{n0} - E_{m0}} \tag{9.39}$$

The general methods just described can be extended to third and higher order, but we have already obtained the formulae needed for most physical applications.

Problems for Chapter 9

9.1. A one-dimensional potential $V(x)$ satisfies the conditions

$$V(x) < 0 \qquad -\infty < x < \infty$$

and

$$\lim_{x\to\pm\infty} V(x) = 0$$

Use the variational method with a trial function of the form $\psi = N\exp(-\lambda x^2)$, where N is a normalization factor and λ is a real parameter, to show that $V(x)$, no matter how weak, always has at least one bound state.

It can be shown that any negative two-dimensional potential that vanishes at infinity also has at least one bound state, but the same statement does not hold in three dimensions.

9.2. Consider a spherically symmetric potential $V(r)$ that vanishes at infinity and for which there is at least one bound state. Use the variational method to show that the lowest bound state has no nodes and is therefore nondegenerate. You may assume that the lowest-bound-state wave function is spherically symmetric.

9.3. In Section 8.4 we derived the virial theorem from the equation

$$0 = \frac{d}{dt}\langle \boldsymbol{r}\cdot\boldsymbol{p}\rangle = \frac{1}{i\hbar}\langle[\boldsymbol{r}\cdot\boldsymbol{p}, H]\rangle$$

which holds for any bound state of the Hamiltonian H. An alternative and related derivation of the virial theorem proceeds from the following considerations: the functional

$$\langle H\rangle = \langle T\rangle + \langle V\rangle = \frac{\hbar^2}{2m}\int |\nabla\psi|^2\, d^3\boldsymbol{r} + \int |\psi|^2 V(r)\, d^3\boldsymbol{r}$$

is minimized by choosing $\psi(\boldsymbol{r})$ that satisfies Schroedinger's equation. Under the norm-preserving transformation $\psi(\boldsymbol{r}) \to a^{3/2}\psi(a\boldsymbol{r})$, $\langle H\rangle \to \langle H(a)\rangle$. However, the extremum property mentioned earlier requires that

$$\frac{d}{da}\langle H(a)\rangle\Big|_{a=1} = 0$$

Show that this yields the virial theorem. How is the scale change $\psi(\boldsymbol{r}) \to a^{3/2}\psi(a\boldsymbol{r})$ related to the operator $\boldsymbol{r}\cdot\boldsymbol{p}$?

9.4. A particle of unit mass is subject to the one-dimensional potential

$$V(x) = 4x^4$$

in units where $\hbar = 1$. It can be shown that the lowest bound state has energy $E_0 = 1.060362$ and the first excited state has energy $E_1 = 3.799673$ to eight significant figures.

Using the variational method, estimate E_0 with a trial function of the form

$$\psi_0 = N\exp\left(-\frac{\alpha^2 x^2}{2}\right)$$

and estimate E_1 with a trial function of the form

$$\psi_1 = N' x\exp\left(-\frac{\alpha^2 x^2}{2}\right)$$

9.5. This problem concerns a generalization of the usual variational method. It leads to a lower as well as an upper bound on the energy of a bound state of a system. Unfortunately, it is not easy to apply in practical situations. Consider a bound system with states u_i and energies E_i. Let v be a normalized trial function, and define

$$E = \int v^* H v\, d\tau$$
$$D = \int (Hv)^*(Hv)\, d\tau$$

Prove that there exists some bound state i such that

$$E + \sqrt{D - E^2} \geq E_i \geq E - \sqrt{D - E^2}$$

9.6. Just how good is the WKB approximation for bound states? This problem may give you some idea. Consider a bound state with energy E in the one-dimensional potential well $V(x)$ shown in Figure 9.5. We know from the WKB approximation that

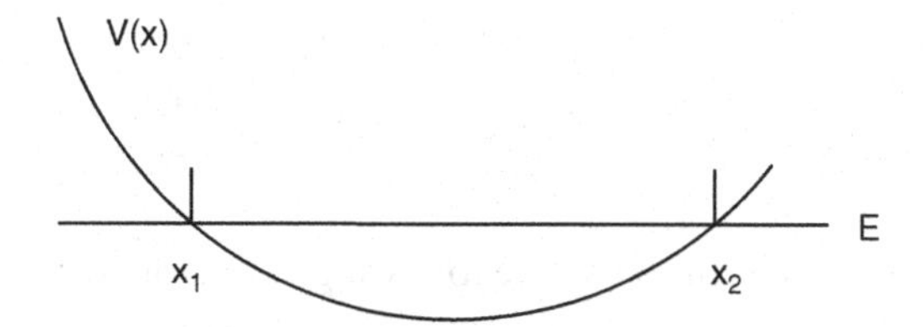

Figure 9.5 One dimensional potential well $V(x)$ with energy E such that classically allowed region is $x_1 \leq x \leq x_2$.

$$\int_{x_1}^{x_2} p(x)\,dx = \left(m+\frac{1}{2}\right)\pi\hbar \tag{1}$$

where $m = 0, 1, 2, \ldots$ and $p = \sqrt{2\mu\left[E - V(x)\right]}$.

(a) Let $V(x) = V(-x)$ and $V(x) = kx^a$ for $x \geq 0$, where k and a are positive real numbers. Show from (1) that

$$E_m = k^{2/(2+a)}\left(\frac{2\mu}{\hbar^2}\right)^{-a/(2+a)}\left[C(a)\left(\frac{m}{2}+\frac{1}{4}\right)\right]^{2a/(2+a)} \tag{2}$$

where

$$C(a) = \frac{2a\sqrt{\pi}\,\Gamma\left(\dfrac{3}{2}+\dfrac{1}{a}\right)}{\Gamma\left(\dfrac{1}{a}\right)} \tag{3}$$

(b) For $a = 1$, $2\mu = k^2\hbar^2$, the exact solutions are Airy functions. We tabulate a few of the corresponding exact eigenvalues for odd m below. Compare these values with the approximate WKB eigenvalues computed by you from (2).

m	*Exact eigenvalue*
1	2.33811
3	4.08795
5	5.52056
7	6.78671
9	7.94413
11	9.02265
13	10.04017

(c) Compare the approximate result (2) with the exact result for the one-dimensional simple harmonic oscillator ($a = 2$).

(d) Let $k = 4$, $a = 4$, $\mu = \hbar = 1$, as in Problem 2. Compare the approximate result (2) with the following very accurate eigenvalues resulting from numerical solution to the Schroedinger equation:

$$E_0 = 1.060362$$
$$E_1 = 3.799673$$
$$E_2 = 7.455703$$

Here you may wish to use the following values of the gamma function:

$$\Gamma\left(\frac{1}{4}\right) = 3.6256099$$
$$\Gamma\left(\frac{3}{4}\right) = 1.2254167$$

9.7. Consider the wave function for a bound state in a potential well with classical turning points $x_1 < x_2$. In this problem we use the WKB approximation and also ignore the small contribution to the normalization integral that comes from the regions $x < x_1$ and $x > x_2$. Thus, for a symmetric potential, we assume that

$$\int_0^{x_2} \psi^2 \, dx = \frac{1}{2}$$

Using the WKB approximation, treating the energy quantum number n as a continuous variable, and making a reasonable approximation in the evaluation of a certain integral, show that

$$|\psi(0)|^2 \approx \frac{1}{\pi\hbar}\sqrt{\frac{2\mu}{E_n}}\frac{\partial E_n}{\partial n}\cos^2\left(\frac{n\pi}{2}\right) \tag{1}$$

and

$$|\psi'(0)|^2 \approx \frac{(2\mu)^{3/2}}{\pi\hbar^3} E_n^{1/2}\frac{\partial E_n}{\partial n}\sin^2\left(\frac{n\pi}{2}\right) \tag{2}$$

Compare (1) or (2) with what is obtained in the exact solution for the one-dimensional simple harmonic oscillator for the states $n = 3$ and $n = 4$.

9.8. When a very intense electric field F in vacuum is applied normal to the surface of a metal and directed toward the surface, electrons are ejected from the metal (this is called *field emission*). In a simple but reasonably effective model of the phenomenon (Figure 9.6), conduction electrons in the metal are confined to a potential well of depth $\mu + \phi$ that exists for all $x < 0$ up to the metal-vacuum interface at $x = 0$. The conduction electrons form a Fermi gas at zero temperature, filling all levels up to the Fermi potential μ. The work function $\phi = e\phi_0$ of the metal is typically 4 to 5 eV. For all $x > 0$, there exists a potential energy $V = \mu + \phi - eFx$. Field emission occurs because conduction electrons can tunnel through the potential barrier and thus escape from the metal. Employing the WKB approximation to calculate the transmission coefficient T through this barrier and assuming that essentially all escaping electrons have energy $E \approx \mu$, show that

$$T \approx \exp\left(-6.8 \times 10^7 \frac{\phi_0^{3/2}}{F}\right)$$

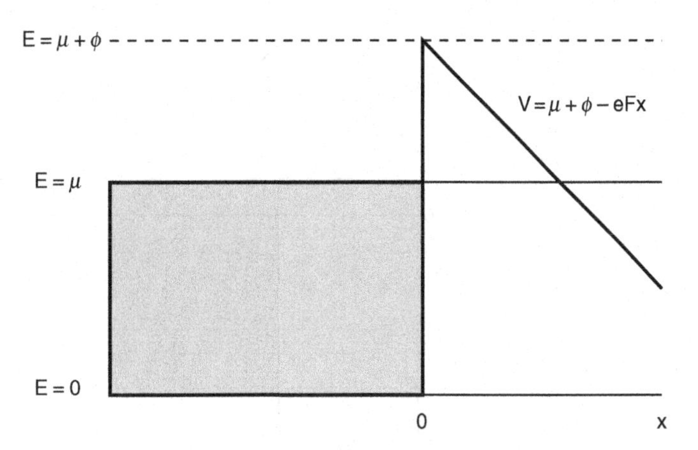

Figure 9.6 Schematic diagram of potential $V(x)$ and energy levels for calculation of field emission. Shaded region: filled electron levels.

where ϕ_0 is in electron volts and F is in volts per centimeter. The field emission current is proportional to T.

9.9. According to the WKB approximation (equation [9.28]), the transmission coefficient for barrier penetration in one spatial dimension is

$$T = e^{-G} = \exp\left[-\frac{2^{3/2} m^{1/2}}{\hbar} \int_{x_1}^{x_2} \sqrt{V(x) - E}\, dx\right] \tag{1}$$

This formula can be used to estimate the transition probability per unit time for alpha decay by a heavy nucleus. We picture the alpha particle as originally bound to the nucleus in a rectangular potential well of radius $R \approx 1.5 \times 10^{-13} A^{1/3}$ cm, where A is the nuclear mass number. Outside the well there is a positive potential $V(r) = 2Ze^2/4\pi r$ arising from the repulsive Coulomb force between the final nucleus of atomic number Z and the alpha particle (Figure 9.7). In (1), $m = 6.7 \times 10^{-24}$ g is the alpha-particle mass.

(a) Using equation (1) with $x_1 = R$, $x_2 = r_2 > R$ and ignoring the very minor contribution of orbital angular momentum of the alpha particle to the effective potential for $r > R$ so that $V_{\text{eff}}(r) = 2Ze^2/4\pi r$, evaluate G analytically in (1).

(b) The transition probability per unit time Γ for alpha decay can be estimated as the product of T and the rate ρ at which the alpha particle strikes the outer boundary of the nucleus because of its motion with velocity v_0 within the nucleus. Assuming that $v_0 \approx 10^9$ cm/s, estimate Γ and compare your results with the following experimental data for alpha-radioactive decays:

Initial nucleus	Final nucleus	Alpha energy	Trans. Prob./s
$^{232}_{90}Th$	$^{228}_{88}Ra$	4.011 MeV	$\Gamma = 1.73 \times 10^{-18}\ s^{-1}$
$^{230}_{90}Th$	$^{226}_{88}Ra$	4.687 MeV	$\Gamma = 3.01 \times 10^{-13}\ s^{-1}$
$^{228}_{90}Th$	$^{224}_{88}Ra$	5.427 MeV	$\Gamma = 1.178 \times 10^{-8}\ s^{-1}$
$^{226}_{90}Th$	$^{222}_{88}Ra$	6.34 MeV	$\Gamma = 4.26 \times 10^{-4}\ s^{-1}$

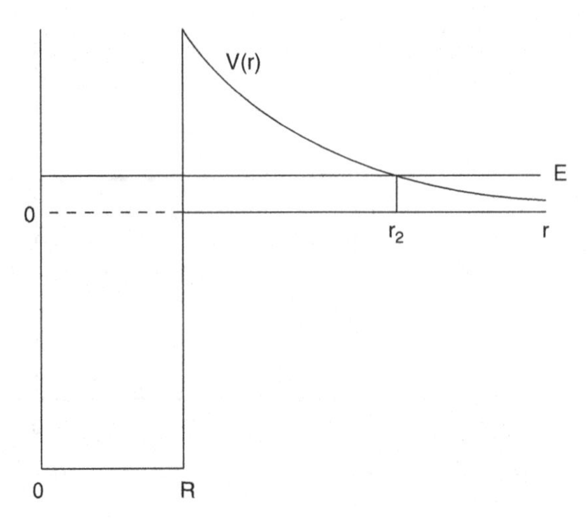

Figure 9.7 Energy and potential energy diagram for calculating alpha decay by means of the WKB approximation.

9.10. A three-dimensional isotropic harmonic oscillator of charge q is perturbed by an electric field of strength $\mathcal{E}$ in the positive z direction. Calculate the change in energy and the induced dipole moment for each energy level by solving the problem exactly. Show that if the polarizability $\underline{\alpha}$ of the oscillator is defined as the ratio of the induced electric dipole moment to $\mathcal{E}$, the change in energy is $-\underline{\alpha}\mathcal{E}^2/2$. Next, carry out the same calculations with second-order perturbation theory, and compare your result with the exact solution.

9.11. A system of three spin-1/2 particles has the Hamiltonian

$$H = AS_1 \cdot S_2 + BS_2 \cdot S_3 + CS_3 \cdot S_1$$

where A, B, and C are constants. Find the energies of the stationary states and their degeneracies.

9.12. In Section 9.3 we developed conventional Rayleigh-Schroedinger static perturbation theory (RSPT). In Chapter 6, Problem 6.7 we mention the Feynman-Hellmann theorem. Actually, a generalization of the latter theorem provides an alternative derivation of the formulas of RSPT, and this is the subject of this problem. Consider a Hamiltonian $H(\lambda)$ with eigenvectors $|n(\lambda)\rangle$ and associated eigenvalues $E_n(\lambda)$, all of which depend on a continuous real parameter λ; that is,

$$H(\lambda)|n(\lambda)\rangle = E_n(\lambda)|n(\lambda)\rangle \tag{1}$$

We assume for simplicity that the eigenstates $|n\rangle$ form a discrete orthonormal set

$$\langle n(\lambda)|m(\lambda)\rangle = \delta_{nm} \tag{2}$$

Hence

$$\langle n|H|m\rangle = \delta_{nm} E_m \tag{3}$$

(a) Differentiate both sides of (3) with respect to λ, and take into account (1) and (2) to show that

$$\delta_{nm}\frac{\partial E_m}{\partial \lambda}-\left\langle n\left|\frac{\partial H}{\partial \lambda}\right|m\right\rangle=(E_n-E_m)a_{nm} \tag{4}$$

where

$$a_{nm}=-a^*_{mn}=\left\langle n\left|\frac{\partial m}{\partial \lambda}\right.\right\rangle \tag{5}$$

(b) A special case of (4) is the original Feynman-Hellmann theorem; that is,

$$\left\langle n\left|\frac{\partial H}{\partial \lambda}\right|n\right\rangle=\frac{\partial E_n}{\partial \lambda} \tag{6}$$

Differentiate both sides of (6) with respect to λ, and use (4), completeness of which implies that

$$\left|\frac{\partial n}{\partial \lambda}\right\rangle=\sum_{m\neq n}a_{mn}|m\rangle \tag{7}$$

and Hermiticity of $\partial H/\partial\lambda$ (and higher derivatives of H) to show that

$$\frac{1}{2}\left(\frac{\partial^2 E_n}{\partial \lambda^2}-\left\langle n\left|\frac{\partial^2 H}{\partial \lambda^2}\right|n\right\rangle\right)=\sum_{m\neq n}\frac{|h_{nm}|^2}{E_n-E_m} \tag{8}$$

where

$$h_{nm}=\left\langle n\left|\frac{\partial H}{\partial \lambda}\right|m\right\rangle$$

(c) To apply (4), (5), and (8) to RSPT, assume that

$$H=H_0+\lambda H'$$
$$E_n=E_{n0}+\lambda E_{n1}+\lambda^2E_{n2}+\cdots$$
$$|n\rangle=|n_0\rangle+\lambda|n_1\rangle+\lambda^2|n_2\rangle+\cdots$$

and save only the leading terms to recover the first- and second-order energy shifts and the first-order change in the wave function in nondegenerate RSPT.

9.13. In Section 9.3 we introduced Rayleigh-Schroedinger static perturbation theory (RSPT). This problem concerns a somewhat different form of static perturbation theory developed by Brillouin and Wigner (BWPT). Although not used very often, BWPT does have advantages for certain problems.

(a) As in Section 9.3, let the unperturbed eigenvalue equation be $H_0|n_0\rangle = E_{n0}|n_0\rangle$. For each $|n_0\rangle$, define the operator

$$Q_{n_0} = I - |n_0\rangle\langle n_0| = \sum_{m_0 \neq n_0} |m_0\rangle\langle m_0| \tag{1}$$

Show that $\left[H_0, Q_{n_0}\right] = 0$.

(b) The perturbed eigenvalue equation is

$$(E_n - H_0)|n\rangle = H'|n\rangle \tag{2}$$

Using the result of part (a), show that (2) can be written as

$$Q_{n_0}|n\rangle = R_n H'|n\rangle \tag{3}$$

where

$$R_n = (E_n - H_0)^{-1} Q_{n_0} = \sum_{m_0 \neq n_0} \frac{|m_0\rangle\langle m_0|}{E_n - E_{m0}} \tag{4}$$

(c) We choose the normalization $\langle n_0|n\rangle = 1$. Thus, from (3), we obtain

$$|n\rangle = |n_0\rangle + Q_{n_0}|n\rangle = |n_0\rangle + R_n H'|n\rangle \tag{5}$$

Show that solution of (5) by iteration yields

$$|n\rangle = |n_0\rangle + \sum_{m_0 \neq n_0} \frac{|m_0\rangle\langle m_0|H'|n_0\rangle}{E_n - E_{m0}} + \sum_{\substack{m_0 \neq n_0 \\ k_0 \neq n_0}} \frac{|m_0\rangle\langle m_0|H'|k_0\rangle\langle k_0|H'|n_0\rangle}{(E_n - E_{m0})(E_n - E_{k0})} + \cdots \tag{6}$$

(d) To find the perturbed energies, we return to (2) and take the scalar product of both sides with $\langle n_0|$. Show that this yields

$$E_n = E_{n0} + \langle n_0|H'|n_0\rangle + \sum_{m_0 \neq n_0} \frac{|\langle n_0|H'|m_0\rangle|^2}{E_n - E_{m0}} + \cdots \tag{7}$$

Note that the denominator of the second term on the right-hand side of (7) is $E_n - E_{m0}$. Thus the perturbed energy E_n appears on both sides of (7). When the latter equation is truncated at any finite order, we then have a polynomial equation to solve for E_n.

10 Applications of Perturbation Theory: Bound States of Hydrogenic Atoms

In this chapter, we employ static perturbation theory to analyze the fine structure, hyperfine structure, Zeeman effect, and Stark effect in atomic hydrogen and the hydrogenic ions He^+, Li^{2+}, and so on. We also discuss the van der Waals interaction between hydrogen atoms. Calculations of these effects have major intrinsic importance, and they provide very good illustrations of perturbation theory. Moreover, the same or similar methods are used in the theory of many-electron atoms and in nuclear, elementary-particle, and condensed-matter physics.

Our starting point is the zero-order nonrelativistic Hamiltonian for an electron in the Coulomb field of a nucleus with infinite mass and atomic number Z. In atomic units,

$$H_0 = \frac{p^2}{2} - \frac{Z}{r} \tag{10.1}$$

As we know, the bound-state eigenvalues are given by the Balmer formula; that is,

$$E_{n\ell} = -\frac{Z^2}{2n^2} \tag{10.2}$$

Hence, as noted previously, the zero-order states with $\ell = 0,\ldots,n-1$ for given n (e.g., $2s$, $2p$ or $3s$, $3p$, $3d$) are degenerate.

10.1 Fine structure of hydrogenic atoms

The zero-order Hamiltonian (10.1), while a very good approximation for many purposes, does not account for a number of significant details. One very important feature that must be included is electron spin. The enlargement of Schroedinger's nonrelativistic wave mechanics by the inclusion of electron spin was largely due to Pauli (1927) and is known as the *Pauli-Schroedinger theory*. Here we introduce the electron-spin operator $\boldsymbol{S}$ ad hoc and define the total electronic angular momentum operator $\boldsymbol{J}$ by

$$\boldsymbol{J} = \boldsymbol{L} + \boldsymbol{S} \tag{10.3}$$

Because the electron-spin eigenvalue is $s = 1/2$, possible j eigenvalues are $j = \ell \pm 1/2$ for $\ell > 0$ and $j = 1/2$ for $\ell = 0$.[1] We label the energy eigenstates in standard spectroscopic notation by the expression

$$n^{2s+1}\ell_j$$

[1] It is customary to employ lowercase letters to describe hydrogenic quantum numbers but uppercase letters for total angular momenta in many-electron atoms.

Thus we have the states

$$\begin{array}{l} \cdots \\ (3^2 s_{1/2}),\quad (3^2 p_{1/2}, 3^2 p_{3/2}),\quad (3^2 d_{3/2}, 3^2 d_{5/2}) \\ (2^2 s_{1/2}),\quad (2^2 p_{1/2}, 2^2 p_{3/2}) \\ (1^2 s_{1/2}) \end{array}$$

In the absence of perturbations, all the n, ℓ, j, and m_j states of a given n would be degenerate. However, experiment shows that states of a given n and ℓ but different j are actually separated by fine-structure splittings. For example, Figure 10.1 shows the fine structure of the $n = 2$ levels of $^4He^+$.

Fine structure is a relativistic effect that can be treated in a consistent way only by means of the Dirac relativistic wave equation, to be discussed in later chapters. However, at this stage, we can give a partial explanation of fine structure by grafting two heuristically derived perturbations of comparable significance and relativistic origin onto the nonrelativistic zero-order Schroedinger theory.

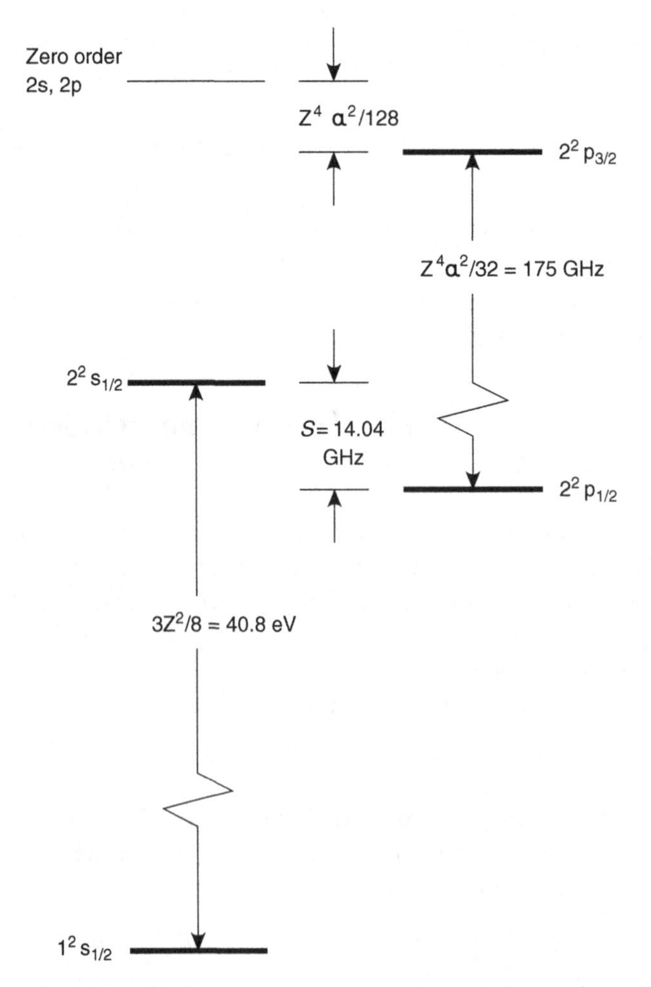

Figure 10.1 $n = 1$ and 2 energy levels of $^4He^+$ (not to scale). Note the fine-structure splitting between $2^2 p_{3/2}$ and $2^2 p_{1/2}$ states. S is the Lamb shift, a quantum-electrodynamic effect that requires quantum-field theory (and specifically renormalization) for its description.

The first perturbation, which does not involve electron spin, is a relativistic correction to the kinetic energy operator. In relativistic mechanics, the energy of a particle of rest mass m and momentum $\boldsymbol{p}$ is

$$E = \sqrt{p^2c^2 + m^2c^4} \tag{10.4}$$

Expanding this expression for $p \ll mc$, we obtain

$$E = mc^2 + \frac{p^2}{2m} - \frac{p^4}{8m^3c^2} + \cdots \tag{10.5}$$

The first term on the right-hand side of (10.5) is the rest energy, a constant. The second is the nonrelativistic kinetic energy

$$T = \frac{p^2}{2m}$$

and the third is the relativistic correction to the kinetic energy

$$-\frac{p^4}{8m^3c^2} = -\frac{T^2}{2mc^2} \tag{10.6}$$

With m chosen as the electron mass, (10.6) yields the following perturbation Hamiltonian in atomic units:

$$H_1 = -\frac{\alpha^2}{2}T^2 \tag{10.7}$$

where $\alpha = 1/c = (137.036)^{-1}$ is the fine-structure constant. From (9.35), the first-order energy shift due to H_1 is

$$\begin{aligned} \Delta E_1^{(1)} &= -\frac{\alpha^2}{2}\langle n\ell|T^2|n\ell\rangle = -\frac{\alpha^2}{2}\langle n\ell|(H_0 - V)^2|n\ell\rangle \\ &= -\frac{\alpha^2}{2}\left(E_{n\ell}^2 - 2E_{n\ell}\langle V\rangle + Z^2\left\langle\frac{1}{r^2}\right\rangle\right) \end{aligned} \tag{10.8}$$

Employing the virial theorem, which gives $\langle V\rangle = 2E_{n\ell}$, the Balmer formula (10.2), and

$$\left\langle\frac{1}{r^2}\right\rangle = \frac{Z^2}{n^3(\ell + 1/2)}$$

we see that (10.8) becomes

$$\Delta E_1^{(1)} = \frac{Z^4\alpha^2}{2n^4}\left(\frac{3}{4} - \frac{n}{\ell + 1/2}\right) \tag{10.9}$$

This formula reveals that the accidental degeneracy for states of the same n and different ℓ is lifted in first order by the perturbation H_1. An analogous effect occurs for a classical mass point describing an elliptical orbit in an attractive $1/r$ potential. There the Runge-Lenz vector is an invariant in the nonrelativistic limit, but when special relativity is taken into account, the semi-major axis precesses around the center of force in the orbit plane.

The second perturbation is the spin-orbit effect, which we introduce by the following heuristic semiclassical argument. Consider a laboratory reference frame in which there exists an electric field $\boldsymbol{\mathcal{E}}$ and in which an electron moves with velocity $\mathbf{v}$. In the electron rest frame there exists a magnetic field

$$\boldsymbol{B} = \gamma\frac{\boldsymbol{\mathcal{E}}\times\mathbf{v}}{c} = \gamma\alpha\boldsymbol{\mathcal{E}}\times\mathbf{v}$$

where $\gamma = (1 - \mathrm{v}^2/c^2)^{-1/2}$. In that rest frame, the equation of motion for the electron spin is

$$\frac{\partial \boldsymbol{S}}{\partial t} = \boldsymbol{\mu}_s \times \boldsymbol{B}$$

where $\boldsymbol{\mu}_s = -g_s\mu_B\boldsymbol{S} = -g_s\alpha\boldsymbol{S}/2$ is the spin magnetic moment. An energy of interaction $u = -\boldsymbol{\mu}_s \cdot \boldsymbol{B}$ corresponds to this equation of motion. Suppose further that the electron is in a hydrogenic atom so that $\boldsymbol{\mathcal{E}} = (Z/r^3)\boldsymbol{r}.$ is the Coulomb field of the nucleus in the laboratory frame. In such a field, the electron continually undergoes acceleration; hence the rest frame is not an inertial frame but instead rotates with a certain angular velocity $\boldsymbol{\omega}_T$ with respect to the laboratory frame. Thus the equation of motion for the electron spin in the laboratory frame is

$$\frac{d\boldsymbol{S}}{dt} = \frac{\partial \boldsymbol{S}}{\partial t} + \boldsymbol{\omega}_T \times \boldsymbol{S}$$

The interaction energy in the laboratory frame, corresponding to the latter equation of motion, is

$$U = u + \boldsymbol{S}\cdot\boldsymbol{\omega}_T \tag{10.10}$$

Now, because

$$\boldsymbol{B} = \frac{Z\alpha}{r^3}\boldsymbol{r}\times(\gamma\mathbf{v}) = \frac{Z\alpha}{r^3}\boldsymbol{r}\times\boldsymbol{p} = \frac{Z\alpha}{r^3}\boldsymbol{L}$$

we have

$$u = -\boldsymbol{\mu}_s\cdot\boldsymbol{B} = g_s\frac{Z\alpha^2}{2}\frac{1}{r^3}\boldsymbol{S}\cdot\boldsymbol{L} \tag{10.11}$$

This is called the *spin-orbit interaction* for the obvious reason that it contains the factor $\boldsymbol{S}\cdot\boldsymbol{L}$. The correction term $\boldsymbol{S}\cdot\boldsymbol{\omega}_T$ was first analyzed by L. H. Thomas (1926, 1927a; see also Jackson 1998). Thomas showed that (10.11) must be replaced with

$$H_{so} = U = u + \boldsymbol{S}\cdot\boldsymbol{\omega}_T = g_{\text{eff}}\frac{Z\alpha^2}{2}\frac{1}{r^3}\boldsymbol{S}\cdot\boldsymbol{L} \tag{10.12}$$

where

$$g_{\text{eff}} = g_s - \frac{2\gamma}{\gamma+1} \approx g_s - 1 = 1$$

Here $\gamma \approx 1$ because the nonrelativistic limit is a good approximation for the electron in a hydrogenic atom unless $Z \gg 1$. Thomas' derivation is somewhat subtle and lengthy, and because this entire subject is in any case treated more correctly by means of the Dirac equation, we do not present the derivation here. Rather, we assume (10.12) with $g_{\text{eff}} = 1$ and proceed with the consequences.

From $\boldsymbol{J} = \boldsymbol{L} + \boldsymbol{S}$, we have $\boldsymbol{J}^2 = \boldsymbol{L}^2 + \boldsymbol{S}^2 + 2\boldsymbol{L}\cdot\boldsymbol{S}$; thus

$$H_{so} = \frac{Z\alpha^2}{2}\frac{1}{r^3}\frac{\boldsymbol{J}^2 - \boldsymbol{L}^2 - \boldsymbol{S}^2}{2} \tag{10.13}$$

For a state with definite j, ℓ, and $s = 1/2$, the first-order energy shift due to H_{so} is

$$\Delta E_{so}^{(1)} = \frac{Z\alpha^2}{2}\left\langle\frac{1}{r^3}\right\rangle\frac{j(j+1) - \ell(\ell+1) - s(s+1)}{2} \tag{10.14}$$

We employ

$$\left\langle\frac{1}{r^3}\right\rangle = \frac{Z^3}{n^3\ell(\ell+1/2)(\ell+1)} \tag{10.15}$$

to obtain

$$\Delta E_{so}^{(1)} = \frac{Z^4\alpha^2}{4}\frac{j(j+1) - \ell(\ell+1) - s(s+1)}{n^3\ell(\ell+1/2)(\ell+1)} \tag{10.16}$$

This formula gives an accurate account of the fine-structure splitting between two states of the same n and $\ell \neq 0$ but different j, provided that Z is not too large. (If $Z \gg 1$, a modified formula derived from the Dirac equation must be used.) For illustration, we consider the states $2^2\,p_{3/2}$ and $2^2\,p_{1/2}$. According to (10.16), the difference in energies between these states is

$$\delta = \frac{Z^4\alpha^2}{4}\left[\frac{\left(\frac{3}{2}\frac{5}{2} - 1.2 - \frac{1}{2}\frac{3}{2}\right) - \left(\frac{1}{2}\frac{3}{2} - 1.2 - \frac{1}{2}\frac{3}{2}\right)}{8 \times 1 \times \frac{3}{2} \times 2}\right] = \frac{Z^4\alpha^2}{32} \tag{10.17}$$

which is 175 GHz in He^+. Recalling Figure 10.1, we see that result (10.17) agrees with experiment.

The situation is more subtle for s-states ($\ell = 0$). Here (10.16) appears ambiguous because the numerator and denominator on the right-hand side both vanish. Actually, (10.15) is strictly correct only in the limit of zero nuclear size. For a real nucleus, $\langle r^{-3}\rangle$ is finite for s-states, albeit very large, and thus $\Delta E_{so}^{(1)} = 0$. However, for s-states and only for s-states, there is an additional effect, first calculated from the Dirac equation by C. G. Darwin, grandson of the great nineteenth-century naturalist Charles Darwin. It is given by

$$\Delta E_{\text{Darwin}} = \frac{Z^4\alpha^2}{2n^3} \qquad \text{for } \ell = 0 \text{ only} \tag{10.18}$$

Unfortunately, there is no elementary intuitive explanation for the Darwin term; its origin must be deferred until we take up the Dirac equation. This effect, together with the contribution $\Delta E_1^{(1)}$ in (10.9), shifts the energy of $n^2s_{1/2}$ just enough to leave it degenerate with $n^2p_{1/2}$ in the present level of approximation. Indeed, combining (10.9) with (10.14) for $\ell \neq 0$ or (10.18) for $\ell = 0$, we obtain the energy shift

$$\Delta E = -\frac{Z^4\alpha^2}{2n^3}\left(\frac{1}{j+1/2} - \frac{3}{4n}\right) \tag{10.19}$$

which depends on j but does not contain ℓ explicitly. However, the degeneracy of $n^2s_{1/2}$ and $n^2p_{1/2}$ states is removed by an additional effect: the Lamb shift, first measured for $n = 2$ in H and He^+ by W. E. Lamb and coworkers in the years 1947–1951. Here the $2^2s_{1/2}$ state is displaced upward, and $2^2p_{1/2}$ is slightly shifted downward, resulting in splittings of 1.058 GHz in hydrogen and 14.04 GHz in He^+ (see Figure10.1). The Lamb shift actually affects all s-states (and to a much smaller extent p, d,...-states as well) and not just in hydrogenic atoms but also in other atoms. The phenomenon is not accounted for by the standard Dirac theory of the hydrogen atom; to calculate it, one needs the technique of renormalization in quantum electrodynamics. We postpone a discussion of the Lamb shift until Chapter 17.

Note that for hydrogenic atoms (H, He^+, Li^{++}, and so on), all the perturbations described in this section are proportional to $Z^4\alpha^2$ in leading order. Thus all these effects are of order $Z^2\alpha^2$ compared with the zero-order energy given by the Balmer formula, and while they are quite small for hydrogen, they grow rapidly with Z. Even for a neutral many-electron atom, the spin-orbit effect for a valence electron becomes very large and significant for high Z. This is discussed in more detail in Chapter 12.

10.2 Hyperfine structure of hydrogen

The most important interactions in an atom are the Coulomb interactions between the electrons and the nuclear charge. [The latter charge is sometimes called the nuclear monopole ($E0$) moment.] However, additional couplings exist between the electrons and higher magnetic and electric multipole moments of the nucleus. These couplings are responsible for hyperfine structure (hfs) observed in atomic spectra. If the nucleus has spin $I \geq 1/2$, it can possess a magnetic dipole moment $\boldsymbol{\mu}_I$. Also, if $J \geq 1/2$, a nonzero magnetic field $\boldsymbol{B}_e$ is generated at the nucleus from electron spin and orbital motion. The coupling of $\boldsymbol{\mu}_I$ to $\boldsymbol{B}_e$ causes magnetic dipole ($M1$) hfs, which is the most significant contribution to hfs. If $I \geq 1$, the nucleus can possess an electric quadrupole ($E2$) moment, and if $J \geq 1$, the electronic charge distribution generates a nonzero electric field gradient at the nucleus. The coupling of the quadrupole moment and the electric field gradient produces an additional ($E2$) hfs energy shift. Smaller still but observable nonetheless in a few cases is magnetic octupole ($M3$) hfs, which requires $I \geq 3/2$, $J \geq 3/2$. Note that static moments $E0$, $E2$, $E4$, … and $M1$, $M3$, … are allowed, but static moments $E1$, $E3$, … and $M2$, $M4$, … are forbidden by space-inversion and time-reversal symmetries, a point to be discussed later.

In this section we confine ourselves to a discussion of $M1$ hfs in the hydrogen atom. Here the perturbation Hamiltonian is

$$H_{hfs} = -\boldsymbol{\mu}_p \cdot \boldsymbol{B}_e \tag{10.20}$$

where $\boldsymbol{\mu}_p = g_p \mu_N \boldsymbol{I}$ is the proton spin magnetic moment operator. Let the position of the electron with respect to the nucleus be $\boldsymbol{r}_e$, and define $\boldsymbol{r} = -\boldsymbol{r}_e$. Now

$$\boldsymbol{B}_e = \nabla \times \boldsymbol{A}_e \tag{10.21}$$

where in atomic units

$$\boldsymbol{A}_e = \frac{\boldsymbol{\mu}_e \times \boldsymbol{r}}{r^3} = -\boldsymbol{\mu}_e \times \nabla\left(\frac{1}{r}\right) \tag{10.22}$$

In the analysis of hyperfine structure, there is a fundamental difference between s-states and states with $\ell > 0$. For s-states, the wave functions of which are spherically symmetric and nonzero at the origin, calculation of $\boldsymbol{B}_e$ requires special attention because $\boldsymbol{A}_e$ is singular at $r = 0$. No such difficulty occurs for states with $\ell > 0$ because these wave functions vanish at the origin. In the latter case, we can assume that $r \neq 0$ when using (10.21) and (10.22) to find $\boldsymbol{B}_e$. We start our analysis with the simpler case $\ell > 0$. The identity

$$\nabla \times (\boldsymbol{a} \times \boldsymbol{b}) = \boldsymbol{a}(\nabla \cdot \boldsymbol{b}) - \boldsymbol{b}(\nabla \cdot \boldsymbol{a}) + (\boldsymbol{b} \cdot \nabla)\boldsymbol{a} - (\boldsymbol{a} \cdot \nabla)\boldsymbol{b}$$

and the fact that the electron spin magnetic moment $\boldsymbol{\mu}_s$ is a fixed quantity independent of the coordinates imply that the magnetic field $\boldsymbol{B}_{es}$ generated by $\boldsymbol{\mu}_s$ is

$$\boldsymbol{B}_{es} = \nabla \times \boldsymbol{A}_{es} = \frac{1}{4\pi}\left[(\boldsymbol{\mu}_s \cdot \nabla)\nabla\left(\frac{1}{r}\right) - \boldsymbol{\mu}_s \nabla^2\left(\frac{1}{r}\right)\right] \tag{10.23}$$

The second term on the right-hand side of (10.23) vanishes for $r \neq 0$, and the first term is

$$\boldsymbol{B}_{e,s}^{\ell>0}(0) = \left(\frac{3\boldsymbol{\mu}_s \cdot \boldsymbol{r}\boldsymbol{r}}{r^5} - \frac{\boldsymbol{\mu}_s}{r^3}\right) \tag{10.24}$$

In addition, there exists a contribution $\boldsymbol{\mu}_\ell / r^3$ to $\boldsymbol{B}_e^{\ell>0}$ arising from orbital motion of the electron, where $\boldsymbol{\mu}_\ell = -\mu_B \boldsymbol{L} = -\mu_B (\boldsymbol{r}_e \times \boldsymbol{p})$. Altogether for $\ell > 0$ we have

$$\boldsymbol{B}_e^{\ell>0}(0) = \frac{3(\boldsymbol{\mu}_s \cdot \boldsymbol{r})\boldsymbol{r}}{r^5} - \frac{\boldsymbol{\mu}_s}{r^3} + \frac{\boldsymbol{\mu}_\ell}{r^3} \tag{10.25}$$

We now turn our attention to s-states and make use of the following picture: let an imaginary sphere be centered on the proton, with radius R larger than the proton radius but much smaller than the Bohr radius a_0. Because $R \ll a_0$,

$$\psi_{ns}(r_e) = \psi_{ns}(0)$$

is an excellent approximation for the spatial wave function at all points inside the sphere. Outside the sphere, $\psi_{ns}(r_e)$ does vary with r_e when the latter becomes comparable with a_0, but in any event, $\psi_{ns}(r_e)$ is spherically symmetric everywhere. Because the electron has a spin magnetic moment, there exists a magnetic moment density or magnetization

$$\boldsymbol{M}(r_e) = \boldsymbol{\mu}_s |\psi_{ns}(r_e)|^2$$

Using subscripts i and o to denote the interior and exterior of the sphere, respectively, we note that inside the sphere of radius R, the magnetization is uniform with value

$$\boldsymbol{M}_i = \boldsymbol{\mu}_s |\psi_{ns}(0)|^2$$

Outside the sphere, $\boldsymbol{M}_o$ is spherically symmetric everywhere, although it also varies with r_e when the latter becomes comparable with a_0. Thus $\boldsymbol{M}_o$ gives no contribution to $\boldsymbol{B}_e(0)$; however, from elementary magnetostatics, we have

$$\boldsymbol{B}_e^{\ell=0}(0) = \frac{8\pi}{3}\boldsymbol{M}_i = \frac{8\pi}{3}\boldsymbol{\mu}_s |\psi_{ns}(0)|^2 \tag{10.26}$$

Taking (10.25) and (10.26) into account, we write the hfs Hamiltonian (10.20) as

$$H_{\text{hfs}} = \left[-\frac{8\pi}{3}\boldsymbol{\mu}_p \cdot \boldsymbol{\mu}_s \delta^3(\boldsymbol{r})\right]_{\ell=0} + \left[\left(\frac{\boldsymbol{\mu}_p \cdot \boldsymbol{\mu}_s}{r^3} - \frac{3\boldsymbol{\mu}_p \cdot \boldsymbol{r}\,\boldsymbol{\mu}_s \cdot \boldsymbol{r}}{r^5}\right) - \frac{\boldsymbol{\mu}_p \cdot \boldsymbol{\mu}_\ell}{r^3}\right]_{\ell>0} \tag{10.27}$$

The first bracketed term on the right-hand side of (10.27) is sometimes called the *contact interaction*. In the Dirac theory, the derivation of (10.27) is actually more transparent, as we will see in Chapter 21. Equation (10.27) was first derived by E. Fermi in 1930 using the Dirac theory.

We now calculate the first-order hyperfine energies of s-states in hydrogen. Noting that

$$\boldsymbol{\mu}_p = g_p \frac{\alpha}{2m_p}\boldsymbol{I} = g_p \mu_B \frac{m_e}{m_p}\boldsymbol{I}$$

where $g_p = 5.58$ and $\boldsymbol{I}$ is the nuclear spin, and employing the contact interaction term in (10.27), we obtain

$$\Delta E_{\text{hfs}}^{(1)}(ns) = \frac{8\pi}{3} g_p g_s \mu_B^2 \frac{m_e}{m_p} \langle ns|\delta^3(\boldsymbol{r})|ns\rangle \langle \boldsymbol{I}\cdot\boldsymbol{S}\rangle \tag{10.28}$$

In atomic units, $\langle ns|\delta^3(\mathbf{r})|ns\rangle = 1/\pi n^3$ and $\mu_B = \alpha/2$. Thus (10.28) becomes

$$\Delta E^{(1)}_{\text{hfs}}(ns) = a\langle \mathbf{I}\cdot\mathbf{S}\rangle \tag{10.29}$$

where

$$a = \frac{2}{3n^3} g_p g_s \alpha^2 \frac{m_e}{m_p} \tag{10.30}$$

The total atomic angular-momentum operator, including nuclear spin, is defined as $\mathbf{F} = \mathbf{I} + \mathbf{J}$. Also, $i = 1/2$ for the proton, and $j = s = 1/2$ for the s-states; thus $f = 1$ or 0. The $f = 1$ multiplet contains three components: $m_F = 1$, 0, and -1, whereas the $f = 0$ state is a singlet with $m_F = 0$. In the absence of hfs interaction, all four components are degenerate. However, because

$$\langle \mathbf{I}\cdot\mathbf{S}\rangle = \frac{f(f+1) - i(i+1) - s(s+1)}{2} = \frac{f(f+1)}{2} - \frac{3}{4}$$

(10.29) yields

$$\begin{aligned} \Delta E^{(1)}_{\text{hfs}}(ns, f = 1) &= \frac{a}{4} \\ \Delta E^{(1)}_{\text{hfs}}(ns, f = 0) &= -\frac{3a}{4} \end{aligned} \tag{10.31}$$

for a hyperfine splitting between $f = 1$ and $f = 0$ of

$$\delta = a = \frac{2}{3n^3} g_p g_s \alpha^2 \frac{m_e}{m_p} \tag{10.32}$$

Note that even when H_{hfs} is included, the three m_F components of $f = 1$ remain degenerate in the absence of an external magnetic field. However, when such a field is imposed, the degeneracy is lifted by the Zeeman effect, to be discussed in the next section.

The hyperfine transition between $f = 1$ and $f = 0$ in the ground state of hydrogen is used in high-precision atomic clocks and is also extremely important in radioastronomy, where it is observed in absorption and emission. The quantity $a = 1.4204$ GHz (wavelength = 21 cm) has actually been measured to a precision of more than 12 significant figures using hydrogen maser techniques and is one of the most accurately determined physical quantities. For a precise comparison between theory and experiment, (10.32) with $g_s = 2$ is not accurate enough. The most important correction to a is the g-factor anomaly a_e in $g_s = 2(1 + a_e)$. In addition, there are small but important effects resulting from finite proton mass, proton recoil, relativistic electron motion, and so on. When all these corrections are included, theory and experiment agree, but the theoretical uncertainty, about 1 part per million (1 ppm), is much larger than the experimental uncertainty.

The hyperfine energies of states with $\ell > 0$ are derived from the second bracketed term on the right-hand side of (10.27). Assuming that $g_s = 2, g_\ell = 1$, it can be shown with the aid of the Wigner-Eckart theorem (see Section 7.12) that

$$\begin{aligned} \Delta E^{(1)}_{\text{hfs}}(n, \ell > 0, j, f) &= \alpha^2 \frac{m_e}{m_p} g_p \frac{\ell(\ell+1)[f(f+1) - i(i+1) - j(j+1)]}{4j(j+1)} \left\langle \frac{1}{r^3} \right\rangle \\ &= \alpha^2 \frac{m_e}{m_p} g_p \frac{[f(f+1) - i(i+1) - j(j+1)]}{2n^3(2\ell+1)j(j+1)} \end{aligned} \tag{10.33}$$

10.3 Zeeman effect

In the presence of an external magnetic field $\boldsymbol{B} = B_0\hat{z}$, the Hamiltonian for a hydrogenic atom is

$$H = \frac{1}{2\mu}\left(\boldsymbol{p} + \frac{e}{c}\boldsymbol{A}\right)^2 - \frac{Z}{r} + H_{fs} + H_{hfs} + g_s\mu_B\, \boldsymbol{S}\cdot\boldsymbol{B} - g_p\mu_N \boldsymbol{I}\cdot\boldsymbol{B} \tag{10.34}$$

where $H_{fs} = H_{\text{Darwin}} + H_1 + H_{so}$ and the external vector potential $\boldsymbol{A}$ can be chosen as

$$\boldsymbol{A} = \frac{B_0}{2}(x\hat{y} - y\hat{x})$$

to give $\boldsymbol{B} = \nabla \times \boldsymbol{A}$. Ignoring the distinction between the reduced mass μ and m_e, we expand the first term on the right-hand side of (10.34)

$$\frac{1}{2m_e}\left(\boldsymbol{p} + \frac{e}{c}\boldsymbol{A}\right)^2 = \frac{\boldsymbol{p}^2}{2m_e} + \frac{e}{2m_e c}(\boldsymbol{A}\cdot\boldsymbol{p} + \boldsymbol{p}\cdot\boldsymbol{A}) + \frac{e^2}{2m_e c^2}\boldsymbol{A}^2 \tag{10.35}$$

Because $\nabla\cdot\boldsymbol{A} = 0$, $\boldsymbol{p}\cdot\boldsymbol{A} = \boldsymbol{A}\cdot\boldsymbol{p} - i\nabla\cdot\boldsymbol{A} = \boldsymbol{A}\cdot\boldsymbol{p}$. Also,

$$\boldsymbol{A}\cdot\boldsymbol{p} = B_0\left(xp_y - yp_x\right)/2 = B_0 L_z/2$$

where $\boldsymbol{L} = (\boldsymbol{r}\times\boldsymbol{p})$. Thus (10.34) becomes

$$H = H_0 + H_{fs} + H_{hfs} + g_\ell\mu_B \boldsymbol{L}\cdot\boldsymbol{B} + g_s\mu_B \boldsymbol{S}\cdot\boldsymbol{B} - g_p\mu_N \boldsymbol{I}\cdot\boldsymbol{B} + \frac{\alpha^2}{2}A^2 \tag{10.36}$$

Here $H_0 = (\boldsymbol{p}^2/2m_e) + V$, and $g_\ell = 1$ is the orbital g-value. Later in this section we discuss the final term on the right-hand side of (10.36). For the present, we ignore it and concentrate on the remaining perturbing terms. In particular, for the ground state of hydrogen, we are concerned with the Zeeman effect of the hyperfine structure and thus with the third, fifth, and sixth terms of (10.36). Here the perturbing Hamiltonian is

$$H' = a\boldsymbol{I}\cdot\boldsymbol{S} + g_s\mu_B B_0 S_z - g_p\mu_N B_0 I_z \tag{10.37}$$

Defining the positive constants $k_1 = g_s \mu_B B_0$ and $k_2 = g_p \mu_N B_0$, where $k_1 \gg k_2$, we write (10.37) as

$$\begin{aligned} H' &= a\boldsymbol{I}\cdot\boldsymbol{S} + k_1 S_z - k_2 I_z \\ &= \frac{a}{2}[I_+ S_- + I_- S_+] + a I_z S_z + k_1 S_z - k_2 I_z \end{aligned} \tag{10.38}$$

In the absence of H', the ground-state components $f = 1$ ($m_F = 1, 0, -1$) and $f = 0$, $m_F = 0$ are degenerate. Hence we may choose any convenient orthonormal linear combinations of these components as a basis for the perturbation matrix of H'. We arbitrarily choose the four basis states:

$$u_{1s} \cdot \{\alpha_e \alpha_p,\ \beta_e \beta_p,\ \alpha_e \beta_p,\ \beta_e \alpha_p\} \tag{10.39}$$

where u_{1s} is the spatial wave function, and as usual, α (β) signifies spin up (down). With rows and columns in the same order as (10.39), the perturbation matrix is

$$\langle H' \rangle = \begin{pmatrix} \frac{a}{4} + \frac{k_1 - k_2}{2} & 0 & 0 & 0 \\ 0 & \frac{a}{4} - \frac{k_1 - k_2}{2} & 0 & 0 \\ 0 & 0 & -\frac{a}{4} + \frac{k_1 + k_2}{2} & \frac{a}{2} \\ 0 & 0 & \frac{a}{2} & -\frac{a}{4} - \frac{k_1 + k_2}{2} \end{pmatrix} \tag{10.40}$$

The eigenvalues

$$\lambda_1 = \frac{a}{4} + \frac{k_1 - k_2}{2} \tag{10.41}$$

and

$$\lambda_{-1} = \frac{a}{4} - \frac{k_1 - k_2}{2} \tag{10.42}$$

correspond to the $f = 1$ states $\alpha_e \alpha_p$ ($m_F = 1$) and $\beta_e \beta_p$ ($m_F = -1$), respectively, and are linear in B_0 with opposite slopes. To find the other two eigenvalues, we diagonalize the 2×2 submatrix in the lower right-hand corner of (10.40). Its secular determinant is

$$\begin{vmatrix} -\frac{a}{4} + \frac{k_1 + k_2}{2} - \lambda & \frac{a}{2} \\ \frac{a}{2} & -\frac{a}{4} - \frac{k_1 + k_2}{2} - \lambda \end{vmatrix} = 0$$

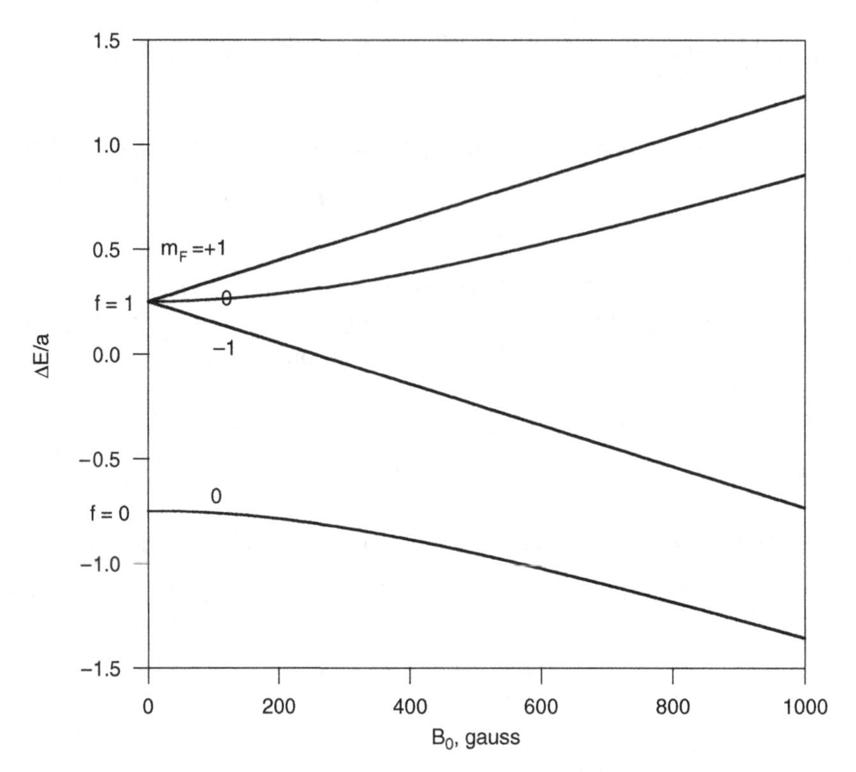

Figure 10.2 Zeeman effect of the hyperfine structure of the $1^2s_{1/2}$ state of hydrogen.

which yields

$$\left(\lambda+\frac{a}{4}\right)^2=\left(\frac{k_1+k_2}{2}\right)^2+\frac{a^2}{4} \tag{10.43}$$

We define $x=(k_1+k_2)/a$ and find from (10.43) that the two eigenvalues are

$$\lambda_\pm=-\frac{a}{4}\pm\frac{a}{2}\sqrt{1+x^2} \tag{10.44}$$

Because x is proportional to B_0, we have $\lambda_+(B_0=0)=a/4$, $\lambda_-(B_0=0)=-3a/4$. Thus $\lambda_\pm(B_0=0)$ obviously correspond to the states $f=1, m_F=0$, and $f=0, m_F=0$, respectively. When $0<x\ll 1$, $\sqrt{1+x^2}\approx 1+x^2/2$; hence, for small x, $\lambda_\pm$ vary quadratically with x. However, when $x\gg 1$, $\sqrt{1+x^2}\approx x$ and $\lambda_\pm$ are linear in x with opposite slopes. Figure 10.2 shows the four eigenvalues plotted as a function of B_0.

It remains to find the eigenvectors corresponding to $\lambda_\pm$. These can be expressed as

$$\psi_\pm=a_\pm(x)\alpha_e\beta_p+b_\pm(x)\beta_e\alpha_p \tag{10.45}$$

The coefficients $a_\pm(x), b_\pm(x)$ are found from the eigenvalue equation

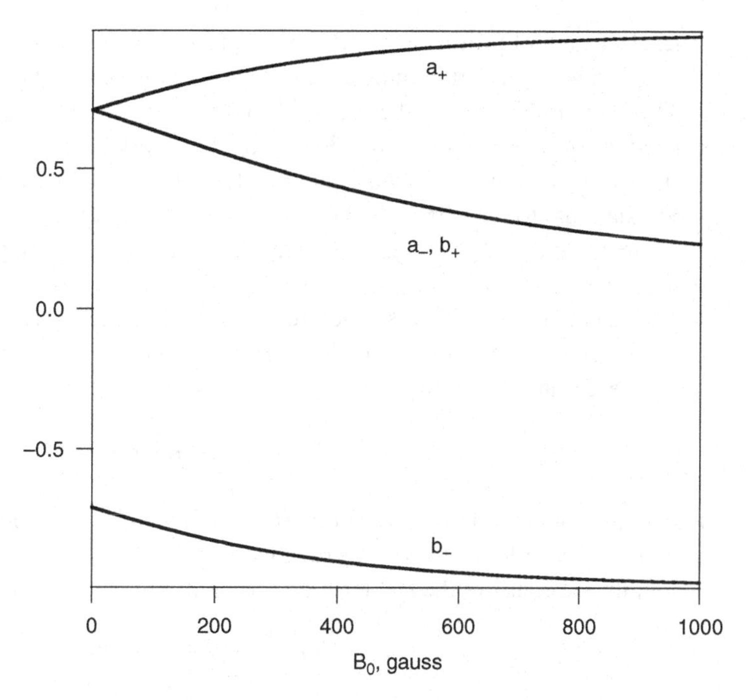

Figure 10.3 Coefficients $a_\pm$ and $b_\pm$ plotted versus B_0.

$$\begin{pmatrix} -\frac{a}{4}+\frac{ax}{2} & \frac{a}{2} \\ \frac{a}{2} & -\frac{a}{4}+\frac{ax}{2} \end{pmatrix}\begin{pmatrix} a_\pm \\ b_\pm \end{pmatrix} = \left(-\frac{a}{4}\pm\frac{a}{2}\sqrt{1+x^2}\right)\begin{pmatrix} a_\pm \\ b_\pm \end{pmatrix}$$

and from the normalization condition $a_\pm^2 + b_\pm^2 = 1$. We obtain

$$a_\pm = \frac{1}{\left[1+\left(x\mp\sqrt{1+x^2}\right)^2\right]^{1/2}} \tag{10.46a}$$

and

$$b_\pm = \frac{-\left(x\mp\sqrt{1+x^2}\right)}{\left[1+\left(x\mp\sqrt{1+x^2}\right)^2\right]^{1/2}} \tag{10.46b}$$

These coefficients are plotted versus B_0 in Figure 10.3.

Note that when $x = 0$, $a_+ = a_- = 2^{-1/2}$ and $b_+ = -b_- = 2^{-1/2}$, but when $x \gg 1$, $a_+ \to 1$ and $a_- \to 0$, whereas $b_+ \to 0$ and $b_- \to -1$. These relations have a simple physical meaning. When B_0 is small ($x \ll 1$), the electron and proton spins are much more tightly coupled with one another by the hyperfine interaction than they are with the weak external field. Thus neither m_I nor m_J is a good quantum number; only f and m_F are well defined. For large B_0 ($x \gg 1$), the reverse is true: electron and proton spins separately precess about the strong external field, so neither f nor m_F is a good quantum number, but m_I and m_J become well defined.

The Zeeman effect for other hyperfine levels in hydrogen or in a multielectron atom is calculated according to similar principles. The main task is to diagonalize a perturbation matrix such as (10.40). For $I = 1/2$ and any J or for $J = ½$ and any I, the matrix can be reduced to block diagonal form, where the submatrices are 2×2; hence only quadratic equations appear. In more complicated cases, the perturbation matrix is most easily diagonalized numerically by computer.

The Zeeman effect of fine-structure levels is calculated by the same method. For example, consider the $n = 2$, $\ell = 1$ states of $^4He^+$. Here the nuclear spin is zero, and there is no hyperfine structure. From (10.36), the relevant portion of the perturbing Hamiltonian is

$$H' = H_{fs} + g_\ell \mu_B \boldsymbol{L} \cdot \boldsymbol{B} + g_s \mu_B \boldsymbol{S} \cdot \boldsymbol{B} \tag{10.47}$$

and apart from an additive constant, $H_{fs} = (\alpha^2/3) \boldsymbol{L} \cdot \boldsymbol{S}$ in atomic units. The 6×6 perturbation matrix H' for $\ell = 1$, $s = 1/2$ is easily diagonalized.

Finally, we consider the last term on the right-hand side of (10.36); that is,

$$H_Q = \frac{\alpha^2}{2} A^2 = \frac{\alpha^2 B_0^2}{8} (x^2 + y^2) \tag{10.48}$$

This term is responsible for the quadratic Zeeman effect. It does not depend on $\boldsymbol{S}$, $\boldsymbol{L}$, $\boldsymbol{J}$, or $\boldsymbol{I}$ and therefore does not cause any splittings between different magnetic sublevels of a zero-order eigenstate with given values of n and ℓ. Because $\langle x^2 + y^2 \rangle = 2\langle r^2 \rangle / 3$ for s-states, the first-order energy shift in atomic units is

$$\Delta E_Q^{(1)} = \frac{\alpha^2 B_0^2}{12} \langle r^2 \rangle \tag{10.49}$$

Employing the formula

$$\langle r^2 \rangle = \frac{n^2}{2Z^2} \left[5n^2 + 1 - 3\ell(\ell + 1) \right] \tag{10.50}$$

we see that although $\Delta E_Q^{(1)}$ is very small for ordinary magnetic fields and low principal quantum numbers, it grows roughly in proportion to n^4 and thus becomes significant for $n \gg 1$. This is important in the case of a Rydberg atom, which is an atom (from almost anywhere in the periodic table) in which a valence electron is excited to a very high-lying state. The nucleus and remaining electrons form a compact core with effective charge $Z_{\text{eff}} = 1$, and the wave function of the valence electron is essentially hydrogenic with $n \gg 1$.

The positive-energy-shift quadratic in B_0 implies negative magnetic susceptibility; thus H_Q is directly related to diamagnetism. As is well known, this phenomenon arises from Lenz's law: if we apply an external magnetic field to an atom or a group of atoms by increasing the field from zero, the electron(s) experience a changing magnetic flux while the field is increasing. Thus, by Faraday's law, an electromagnetic force (emf) is generated that causes the electronic orbital currents to change. These incremental orbital currents generate an incremental magnetic field that is always opposite in direction to the applied field and generally much smaller in magnitude.

For illustration, we calculate the diamagnetic correction to the applied magnetic field at the origin of a hydrogen atom in its ground state. Such corrections are significant in high-precision nuclear magnetic resonance experiments and not merely for atomic hydrogen.

To start, we recall from equation (4.76) that in the presence of a vector potential $\boldsymbol{A}$, the probability current density in atomic units for an electron is

$$\boldsymbol{j} = \frac{1}{2i}(\psi^* \nabla \psi - \psi \nabla \psi^*) + \alpha \boldsymbol{A} \psi^* \psi$$

The electromagnetic current density is

$$\boldsymbol{j}_{\mathrm{EM}} = -e\boldsymbol{j} = \frac{-1}{2i}(\psi^* \nabla \psi - \psi \nabla \psi^*) - \alpha \boldsymbol{A} \psi^* \psi \tag{10.51}$$

In (10.51), only the second term on the right-hand side is important for diamagnetism; thus, in what follows, we ignore the first term. The electromagnetic current density generates a new vector potential according to the well-known formula

$$\boldsymbol{A}'(\boldsymbol{r}) = \frac{1}{c}\int \frac{\boldsymbol{j}_{\mathrm{EM}}(\boldsymbol{r}')}{|\boldsymbol{r}-\boldsymbol{r}'|}\, d^3\boldsymbol{r}' = -\alpha^2 \int \frac{\boldsymbol{A}(\boldsymbol{r}')\rho(\boldsymbol{r}')}{|\boldsymbol{r}-\boldsymbol{r}'|}\, d^3\boldsymbol{r}' \tag{10.52}$$

where $\rho = \psi^* \psi$. The magnetic field generated by this vector potential is

$$\boldsymbol{B}'(\boldsymbol{r}) = \nabla \times \boldsymbol{A}'(\boldsymbol{r}) = \alpha^2 \int \frac{\boldsymbol{r}-\boldsymbol{r}'}{|\boldsymbol{r}-\boldsymbol{r}'|^3} \times \boldsymbol{A}(\boldsymbol{r}')\rho(\boldsymbol{r}')\, d^3\boldsymbol{r}'$$

Hence, at the origin,

$$\boldsymbol{B}'(0) = -\alpha^2 \int \frac{\boldsymbol{r}' \times \boldsymbol{A}(\boldsymbol{r}')}{r'^3}\rho(\boldsymbol{r}')\, d^3\boldsymbol{r}' \tag{10.53}$$

Because $\boldsymbol{A}(\boldsymbol{r}') = (B_0/2)(x'\hat{y} - y'\hat{x})$,

$$\boldsymbol{r}' \times \boldsymbol{A}(\boldsymbol{r}') = \frac{B_0}{2}\left[-(x'z')\hat{x} - (y'z')\hat{y} + (x'^2 + y'^2)\hat{z}\right] \tag{10.54}$$

Only the third term on the right-hand side of (10.54) makes a nonzero contribution to the integral in (10.53). For the $1s$ state of hydrogen, it yields

$$\begin{aligned} \boldsymbol{B}'(0) &= -\frac{\alpha^2}{2} B_0 \hat{z} \int_0^{2\pi} d\phi \int_0^{\pi} \sin^3\theta\, d\theta \int_0^{\infty} r'\left(\frac{1}{\pi}\right)\exp(-2r')\, dr' \\ &= -\frac{\alpha^2}{3}\boldsymbol{B}(0) \end{aligned} \tag{10.55}$$

As anticipated, the diamagnetic correction to the applied magnetic field $\boldsymbol{B}$ is proportional to $\boldsymbol{B}$ but in the opposite direction, and because $\alpha^2/3 = 1.78 \times 10^{-5}$, it is much smaller than $\boldsymbol{B}$ itself.

10.4 Stark effect

We next consider an atom in a uniform external electric field $\boldsymbol{\mathcal{E}}$. The electrostatic potential due to $\boldsymbol{\mathcal{E}}$ is $\Phi = -\boldsymbol{\mathcal{E}}\cdot\boldsymbol{r}$, and the resulting perturbation Hamiltonian is

$$H_S = e\boldsymbol{\mathcal{E}}\cdot\boldsymbol{r} \tag{10.56}$$

Figure 10.4 is a plot of H_S plus the Coulomb potential energy along the direction of $\boldsymbol{\mathcal{E}}$. The plot reveals that, strictly speaking, no bound states can exist for the H atom in this potential (and more generally, no bound states can exist for any atom or molecule in an external electric field) because of the possibility of tunneling. To be sure, such tunneling is altogether negligible for low-lying states of typical atoms or molecules because in that case the internal electric fields experienced by the valence electrons are of the order 10^9 to 10^{10} V/cm, whereas laboratory electric fields do not exceed 10^5 to 10^6 V/cm. However, tunneling is observable in Rydberg atom experiments where $n \gg 1$.

The Stark effect in hydrogen can be analyzed in parabolic coordinates [see, e.g., Bethe (1957, pp. 228–240)], but for most purposes, especially where many-electron atoms are concerned, it is more practical to employ the usual spherical polar coordinates. In what follows we confine ourselves to spherical coordinates and to relatively weak electric fields where perturbation theory in lowest nonvanishing order is appropriate. Ignoring spin, we consider a nondegenerate zero-order state $|\psi\rangle$. Then, to first order, the energy shift due to (10.56) is

$$\Delta E_S^{(1)} = e\boldsymbol{\mathcal{E}}\cdot\langle \psi|\boldsymbol{r}|\psi\rangle$$

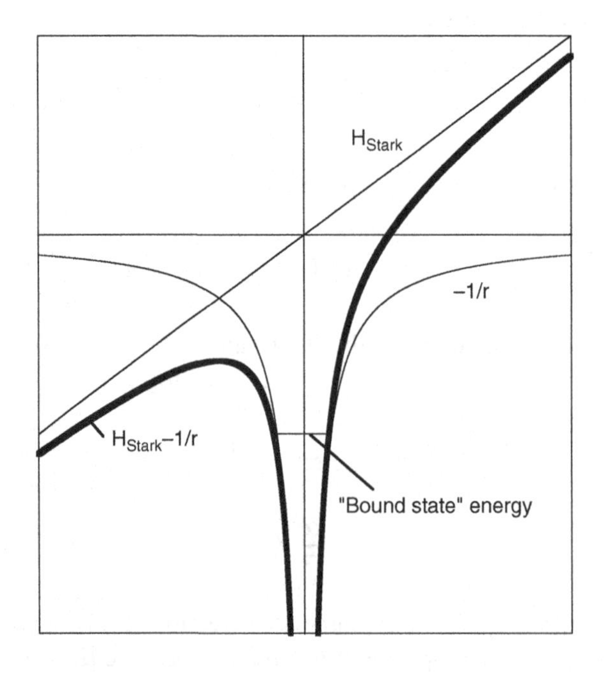

Figure 10.4 $H_s - r^{-1}$ plotted along the electric field direction. Bound states can tunnel to the left.

The quantity $-e\langle\psi|\mathbf{r}|\psi\rangle$ is the expectation value of the electric dipole moment operator and is frequently referred to as a *permanent electric dipole moment.* However, because the zero-order Hamiltonian is space-inversion symmetric, $|\psi\rangle$ has definite parity; that is, $P|\psi\rangle=\pm|\psi\rangle$, where P is the parity operator with $P^\dagger P = P^{-1}P = P^2 = I$ and $P\mathbf{r}P^{-1} = -\mathbf{r}$.
Thus

$$\langle\psi|\mathbf{r}|\psi\rangle=\langle\psi|PP\ \mathbf{r}\ PP|\psi\rangle=\langle\pm\psi|P\ \mathbf{r}\ P|\pm\psi\rangle=-\langle\psi|\mathbf{r}|\psi\rangle=0$$

Therefore, there is no first-order Stark energy shift for a nondegenerate state. In other words, a permanent electric dipole ($E1$) moment for a nondegenerate state is forbidden by space-inversion symmetry (and similar considerations rule out permanent $E3$, $E5$, ... and $M2$, $M4$, ... moments). Hence we must go to second order in perturbation theory to obtain a nonzero Stark energy shift

$$\Delta E_S^{(2)}=\sum_n\frac{|\langle n|e\boldsymbol{\mathcal{E}}\cdot\mathbf{r}|\psi\rangle|^2}{E_0-E_n} \tag{10.57}$$

where E_0 is the zero-order energy corresponding to $|\psi\rangle$, and the sum in (10.57) is taken over all zero-order states $|n\rangle$ except for $|\psi\rangle$. If $|\psi\rangle$ is the ground state, $E_0 < E_n$ for all n; hence the ground-state energy of any atom always *decreases* in proportion to $\mathcal{E}^2$. It is convenient to define the polarizability[2] $\underline{\alpha}$ by

$$\Delta E_S^{(2)}=-\frac{1}{2}\underline{\alpha}\mathcal{E}^2 \tag{10.58}$$

We now make a crude order-of-magnitude estimate of $\Delta E_S^{(2)}$ and thus of $\underline{\alpha}$ for a typical atom. If $|n\rangle$ and $|\psi\rangle$ are low-lying states of opposite parity that differ in ℓ by unity, $\langle n|\mathbf{r}|\psi\rangle\sim a_0$. Also, $|E_0-E_n|\sim e^2/a_0$. Hence

$$\Delta E_S^{(2)}\sim-\frac{e^2\mathcal{E}^2a_0^2}{e^2/a_0}\sim-a_0^3\,\mathcal{E}^2$$

Thus $\underline{\alpha}\sim a_0^3$ in magnitude ($\underline{\alpha}$ is roughly the atomic volume). Incidentally, the neutron, a neutral particle of nonzero spatial extent, also has polarizability roughly equal to its volume $\underline{\alpha}_n\sim(10^{-13}\text{ cm})^3$.This has been observed experimentally.

The first-order change in $|\psi\rangle$ induced by $H_S=e\boldsymbol{\mathcal{E}}\cdot\mathbf{r}$ is

$$\delta|\psi\rangle\equiv|\overline{\psi}\rangle-|\psi\rangle=e\boldsymbol{\mathcal{E}}\cdot\sum_n|n\rangle\frac{\langle n|\mathbf{r}|\psi\rangle}{E_0-E_n} \tag{10.59}$$

Henceforth, in employing atomic units and choosing the direction of $\boldsymbol{\mathcal{E}}$ to be the z-axis, we write (10.59) as

$$\delta|\psi\rangle=\mathcal{E}\sum_n\frac{|n\rangle\langle n|z|\psi\rangle}{E_0-E_n}=\mathcal{E}|\eta\rangle$$

[2] The symbol $\underline{\alpha}$ is used here to avoid confusion with the fine-structure constant α and the spinor $\alpha=\begin{pmatrix}1\\0\end{pmatrix}$.

where

$$|\eta\rangle = -\sum_n \frac{|n\rangle\langle n|z|\psi\rangle}{E_n - E_0}$$

The state $|\eta\rangle$ satisfies the equation

$$(H_0 - E_0)|\eta\rangle = -\sum_n |n\rangle\langle n|z|\psi\rangle \tag{10.60}$$

where the sum is over all states $|n\rangle$ except for $|\psi\rangle$. However, $\langle\psi|z|\psi\rangle = 0$; hence the sum can be taken over all the eigenstates of H_0. Because these states form a complete set, (10.60) reduces to the following equation in coordinate representation:

$$(H_0 - E_0)\eta = -z\psi \tag{10.61}$$

To illustrate the use of (10.61), we choose ψ to be the ground state of hydrogen. Then (10.61) becomes

$$-\frac{1}{2}\nabla^2\eta - \frac{1}{r}\eta + \frac{1}{2}\eta = -\frac{1}{\pi^{1/2}} r e^{-r}\cos\theta \tag{10.62}$$

It is clear from symmetry that η must be a superposition of states with $\ell = 1$ and $m_\ell = 0$ and therefore that the angular part of η must be Y_1^0. Substituting $\eta = f(r)Y_1^0$ into (10.62) and solving the resulting radial Schroedinger equation, we obtain

$$\eta = -\frac{1}{\pi^{1/2}}\left(\frac{r^2}{2} + r\right)e^{-r}\cos\theta \tag{10.63}$$

Therefore, from (10.57), the second-order energy shift in atomic units is

$$\begin{aligned}\Delta E_S^{(2)}(1s) &= \mathcal{E}^2\langle 1s|z|\eta\rangle \\ &= -\mathcal{E}^2\frac{1}{\pi}2\pi\int_0^\infty r^3\left(r + \frac{r^2}{2}\right)e^{-2r}\,dr\int_0^\pi \cos^2\theta\sin\theta\,d\theta \\ &= -\frac{9}{4}\mathcal{E}^2\end{aligned}$$

In cgs units, this result is

$$\Delta E_S^{(2)}(1s) = -\frac{9}{4}a_0^3\,\mathcal{E}^2 \tag{10.64}$$

which implies that the ground-state polarizability is $\underline{\alpha} = \frac{9}{2}a_0^3$. The $2p$ state makes the largest single contribution (~66 percent) to (10.64).

First-order Stark energy shifts do occur in certain circumstances. Consider a simple model in which there are two degenerate zero-order states $|u\rangle$ and $|v\rangle$ of opposite parity and where we impose two perturbations, $H_S = \mathcal{E}z$ and H''. In the $|u\rangle, |v\rangle$ basis, the diagonal matrix elements of H_S vanish, that is, $\mathcal{E}\langle u|z|u\rangle = \mathcal{E}\langle v|z|v\rangle = 0$, but we assume that $\mathcal{E}\langle v|z|u\rangle = \mathcal{E}\langle u|z|v\rangle \equiv w \neq 0$ We also assume that only the *diagonal* matrix elements of H'' are nonzero; that is, $\langle u|H''|u\rangle = \Delta/2$ and $\langle v|H''|v\rangle = -\Delta/2$, where Δ is very small compared with the energy spacings between $|u\rangle$ or $|v\rangle$ and all other states. Thus the perturbation matrix is

$$\begin{pmatrix} \Delta/2 & w \\ w & -\Delta/2 \end{pmatrix}$$

The eigenvalues of this matrix are

$$\lambda_{\pm} = \pm\frac{\Delta}{2}\sqrt{1+y^2} \tag{10.65}$$

where $y = 2w/\Delta$. When $y \ll 1$, these eigenvalues vary quadratically with $\mathcal{E}$, but if $y \gg 1$, they become linear in $\mathcal{E}$. The eigenvectors corresponding to $\lambda_{\pm}$ are

$$|\psi_{\pm}\rangle = a_{\pm}(y)|u\rangle + b_{\pm}(y)|v\rangle \tag{10.66}$$

where

$$a_{\pm} = \frac{1 \pm \sqrt{1+y^2}}{\sqrt{y^2 + \left(1 \pm \sqrt{1+y^2}\right)^2}} \tag{10.67}$$

and

$$b_{\pm} = \frac{y}{\sqrt{y^2 + \left(1 \pm \sqrt{1+y^2}\right)^2}} \tag{10.68}$$

When $y \ll 1$(weak electric field), $a_+ \approx 1$, $b_+ \approx 0$, and $a_- \approx 0$, $b_- \approx 1$; hence $|\psi_+\rangle \approx |u\rangle$ and $|\psi_-\rangle \approx |v\rangle$. On the other hand, when $y \gg 1$ (strong electric field) $a_{\pm} \to \pm 2^{-1/2}$, $b_{\pm} \to 2^{-1/2}$. In this linear Stark-effect limit, $|\psi_+\rangle \to 2^{-1/2}\left(|u\rangle + |v\rangle\right)$ and $|\psi_-\rangle \to 2^{-1/2}\left(|v\rangle - |u\rangle\right)$.

As an example of the foregoing, we let $|u\rangle$ and $|v\rangle$ be the $2s$ and $2p$, $m_\ell = 0$ states of hydrogen, respectively, and we ignore electron and nuclear spins, thus neglecting the complications resulting from fine and hyperfine structure. The $2s$ state is actually shifted upward slightly with respect to the $2p$ state by the Lamb shift (recall the quantity Δ in our model calculation), and it can be shown that $w = \mathcal{E}\langle 2p, m_\ell = 0|z|2s\rangle = -3\mathcal{E}$ in atomic units.

The foregoing two-level model is useful for understanding a phenomenon that occurs frequently in polar molecules and is often observed in chemistry and molecular spectroscopy. In such molecules, the energy splittings between adjacent spin-rotational states of opposite parity are often so small that a very modest external electric field is sufficient to cause a linear Stark effect. In such circumstances, it is often said (not altogether correctly) that the molecule possesses a "permanent" electric dipole moment.

10.5 Van der Waals interaction between two hydrogen atoms

It is known from countless observations that at distances $R \gg a_0$, two atoms experience a weak mutually attractive force. If $R < a_0/\alpha \approx 137a_0$, this force is derivable from a potential that varies as R^{-6} and is called the *van der Waals interaction*.[3] With almost no calculation, we can give a qualitative explanation of this interaction as follows: although in the absence of an external electric field the expectation value of the electric dipole moment of a hydrogen atom in its ground state is zero, the instantaneous electric dipole moment $\boldsymbol{p}_\varepsilon$ is not zero because the electron and proton are separated by a finite distance. This fluctuating dipole moment generates a fluctuating electric field that varies as $\mathcal{E} \sim p_\varepsilon/R^3$. A second hydrogen atom at distance R experiences this field, and from second-order perturbation theory, it suffers a change in energy

$$\Delta E = -\frac{1}{2}\alpha\mathcal{E}^2 \approx -a_0^3\frac{p_\varepsilon^2}{R^6}$$

Because $p_\varepsilon \approx -ea_0$, we have

$$\Delta E \approx -e^2\frac{a_0^5}{R^6} \qquad \text{(atom-atom interaction)} \tag{10.69}$$

In a similar way, consider the electric field $\mathcal{E} = -e/R^2$ at an atom due to an electron at distance $R \gg a_0$. The resulting energy shift is

$$\Delta E \approx -e^2\frac{a_0^3}{R^4} \qquad \text{(electron-atom interaction)} \tag{10.70}$$

Also, consider an atom at a distance $R \gg a_0$ from a perfectly conducting plane. In this case, the image charges are perfectly correlated with the atomic charges, and it can be shown that first-order perturbation theory is operative. Thus the interaction energy is

$$\Delta E \approx -\frac{p_\varepsilon^2}{R^3} \approx -e^2\frac{a_0^2}{R^3} \qquad \text{(atom-plane conductor interaction)} \tag{10.71}$$

Now we seek to understand the atom-atom van der Waals interaction more quantitatively. Our starting point is Figure 10.5, in which the coordinate labels of the electrons and protons in two H atoms are given.

Neglecting the kinetic energies of the two protons, we write the Hamiltonian of this two-atom system as

$$H = H_0 + H'$$

where the zero-order Hamiltonian is

[3] When $R \gg 137a_0$, the van der Waals interaction is attenuated by retardation.

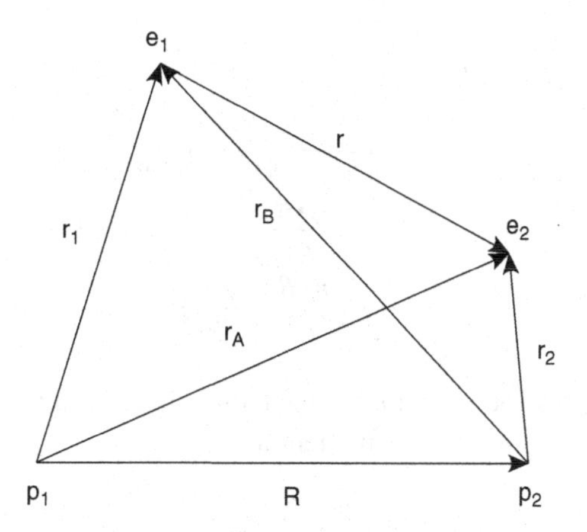

Figure 10.5 Diagram of coordinates for calculation of van der Waals interaction between two hydrogen atoms.

$$H_0 = \left(\frac{p_1^2}{2} + \frac{p_2^2}{2} - \frac{1}{r_1} - \frac{1}{r_2}\right) \tag{10.72}$$

and the perturbing Hamiltonian is

$$H' = \frac{1}{R} + \frac{1}{r} - \frac{1}{r_A} - \frac{1}{r_B} \tag{10.73}$$

From Figure 10.5 we see that

$$\begin{aligned} \boldsymbol{r}_A &= \boldsymbol{R} + \boldsymbol{r}_2 \\ \boldsymbol{r}_B &= -\boldsymbol{R} + \boldsymbol{r}_1 \\ \boldsymbol{r} &= \boldsymbol{R} + \boldsymbol{r}_2 - \boldsymbol{r}_1 \end{aligned}$$

Hence

$$H' = \frac{1}{R} + \frac{1}{|\boldsymbol{R}+\boldsymbol{r}_2-\boldsymbol{r}_1|} - \frac{1}{|\boldsymbol{R}+\boldsymbol{r}_2|} - \frac{1}{|\boldsymbol{R}-\boldsymbol{r}_1|} \tag{10.74}$$

Now we use the expansion $(1+x)^{-1/2} = 1 - x/2 + 3x^2/8 - \cdots$ to write

$$\begin{aligned} \frac{1}{|\boldsymbol{R}+\boldsymbol{r}_2|} &= \frac{1}{R\sqrt{1+\dfrac{2\boldsymbol{r}_2\cdot\boldsymbol{R}}{R^2}+\dfrac{r_2^2}{R^2}}} \\ &= \frac{1}{R} - \frac{\boldsymbol{r}_2\cdot\boldsymbol{R}}{R^3} + \frac{1}{2R^5}\left[3(\boldsymbol{r}_2\cdot\boldsymbol{R})^2 - R^2 r_2^2\right] - \cdots \end{aligned}$$

with similar expressions for $1/|\boldsymbol{R}+\boldsymbol{r}_2-\boldsymbol{r}_1|$ and $1/|\boldsymbol{R}-\boldsymbol{r}_1|$. Thus (10.74) becomes

$$\begin{aligned} H' = &\frac{1}{R} \\ &+\frac{1}{R}-\frac{(\boldsymbol{r}_2-\boldsymbol{r}_1)\boldsymbol{\cdot}\boldsymbol{R}}{R^3}+\frac{1}{2R^5}\left\{3\left[(\boldsymbol{r}_2-\boldsymbol{r}_1)\boldsymbol{\cdot}\boldsymbol{R}\right]^2-R^2(\boldsymbol{r}_2-\boldsymbol{r}_1)^2\right\} \\ &-\frac{1}{R}+\frac{\boldsymbol{r}_2\boldsymbol{\cdot}\boldsymbol{R}}{R^3}-\frac{1}{2R^5}\left[3(\boldsymbol{r}_2\boldsymbol{\cdot}\boldsymbol{R})^2-R^2r_2^2\right] \\ &-\frac{1}{R}-\frac{\boldsymbol{r}_1\boldsymbol{\cdot}\boldsymbol{R}}{R^3}-\frac{1}{2R^5}\left[3(\boldsymbol{r}_1\boldsymbol{\cdot}\boldsymbol{R})^2-R^2r_1^2\right] \end{aligned}$$

This expression simplifies considerably because a number of terms cancel. Choosing the z-axis along $\boldsymbol{R}$, we obtain to sufficient accuracy

$$H' = \frac{x_1x_2 + y_1y_2 - 2z_1z_2}{R^3} \tag{10.75}$$

Because the zero-order wave function is $\psi_0 = u_{1s}(r_1)u_{1s}(r_2)$, the first-order energy shift due to (10.75) is

$$\Delta E^{(1)} = \frac{1}{R^3}\left[\langle u_{1s}|x|u_{1s}\rangle^2 + \langle u_{1s}|y|u_{1s}\rangle^2 - 2\langle u_{1s}|z|u_{1s}\rangle^2\right] = 0$$

which vanishes because x, y, and z are odd-parity operators. Hence we must go to second order; that is,

$$\Delta E^{(2)} = \sum_n \frac{|\langle n|H'|\psi_0\rangle|^2}{E_0 - E_n} \tag{10.76}$$

In (10.76), the states $|n\rangle$ are all the bound and continuum states for which *each* H atom has $\ell = 1$. Thus, whereas $E_0 = -1/2 - 1/2 = -1$, $E_n = -1/8 - 1/8 = -1/4$ if both atoms are in the $2p$ state, $E_n = -1/8 - 1/18$ if one is in $2p$ and the other is in $3p$, and so forth. Hence $E_n \geq -1/4$ for all n in (10.76); consequently,

$$\Delta E^{(2)} \geq -\frac{4}{3}\sum_n |\langle n|H'|\psi_0\rangle|^2$$

Although this last sum is carried over all p states of both atoms, it can be extended to all eigenstates of H_0 because for states other than p states of both atoms, the matrix elements are zero. Thus, by completeness, we obtain

$$\Delta E^{(2)} \geq -\frac{4}{3}\langle \psi_0|H'^2|\psi_0\rangle \tag{10.77}$$

From (10.75), we have

$$H'^2 = \frac{1}{R^6}\left[x_1^2x_2^2 + y_1^2y_2^2 + 4z_1^2z_2^2 + \text{cross-terms}\right]$$

The cross-terms contribute nothing to the matrix element in (10.77). Also,

$$\langle x^2 \rangle = \langle y^2 \rangle = \langle z^2 \rangle = \frac{1}{3}\langle r^2 \rangle$$

Therefore, (10.77) yields

$$\Delta E^{(2)} \geq -\frac{4}{3R^6}\frac{6}{9}\langle r^2 \rangle^2 \tag{10.78}$$

For the 1s state of hydrogen, $\langle r^2 \rangle = 3$. Hence we arrive at

$$\Delta E^{(2)} \geq -\frac{8}{R^6}$$

or in hlu

$$\Delta E^{(2)} \geq -2\frac{e^2}{\pi}\frac{a_0^5}{R^6} \tag{10.79}$$

Also, it can be shown by a straightforward but tedious variational calculation that

$$\Delta E^{(2)} \leq -\frac{1.625e^2}{\pi}\frac{a_0^5}{R^6} \tag{10.80}$$

Finally, we remark that a person ignorant of quantum mechanics but with a knowledge of classical electricity and magnetism and acquainted with the fact that atoms experience a weak long-range van der Waals attraction might conclude that the latter is a new fundamental force of nature. Of course, we can see clearly with the aid of quantum mechanics that the van der Waals interaction is not really fundamental but is a residual higher-order manifestation of the Coulomb interaction. There is an analogous situation in nuclear physics. In 1935, long before the quark model and quantum chromodynamics appeared on the scene, H. Yukawa proposed that the exchange between nucleons of what we now know are pi mesons is the fundamental interaction that binds nucleons together in a nucleus. These days, however, we know that the Yukawa interaction is a higher-order manifestation of something more fundamental: the exchange of gluons between quarks.

Problems for Chapter 10

10.1. The deuterium atom ground state is $1^2S_{1/2}$. The nuclear spin is $I = 1$, and the nuclear magnetic moment is 0.857438 nuclear Bohr magnetons. Calculate the hyperfine splitting and Zeeman energies of all magnetic sublevels F and m_F in an external magnetic field of 200 G. Please give numerical results.

10.2. The sodium atom ground state is $3^2S_{1/2}$. The nuclear spin of the only stable isotope ($A = 23$) is $I = 3/2$, and the nuclear magnetic moment is 2.2175 nuclear Bohr magnetons. The hyperfine splitting in the ground state is 1,772 MHz. Calculate the Zeeman energies of all magnetic sublevels F and m_F in an external magnetic field of 500 G. Please give numerical results.

10.3. The bound system of one electron and one positron is called *positronium*. It is a hydrogen-like atom in which the positron takes the place of the proton. Positronium differs from normal hydrogenic atoms in several ways:

- The reduced mass is $\mu = m_e/2$.
- The spin magnetic moment of the positron is exactly equal in magnitude but opposite in sign to that of the electron, whereas the typical nuclear magnetic moment is approximately 1,000 times smaller.
- The electron and positron can annihilate each other, yielding two (singlet ground state) or three photons (triplet ground state). The lifetimes for these annihilation processes in zero external magnetic field are

$$\tau = 1.25 \times 10^{-10}\ \text{s} \qquad {}^1S_0\ \text{state}$$
$$\tau = 1.4 \times 10^{-7}\ \text{s} \qquad {}^3S_1\ \text{state}$$

The triplet and singlet states are separated by a fine-structure splitting

$$\Delta_{\text{expt}} = 2.034 \times 10^{11}\ \text{Hz} \tag{1}$$

(a) There are two contributions to the theoretical value of this splitting:

$$\Delta_{\text{theo}} = \Delta_1 + \Delta_2$$

Δ_1 arises from the usual "contact" interaction and is given by Fermi's formula. Δ_2 has no analogue in ordinary atomic physics. It arises from the fact that the electron and positron can annihilate one another, and its explanation thus requires quantum electrodynamics. It can be shown that the effective Hamiltonian yielding Δ_2 is

$$H_{\text{ann.}} = 2\pi\left(\frac{e\hbar}{m_e c}\right)^2 \delta_{S1}\delta^3(\mathbf{r}) \tag{2}$$

Here δ_{S1} is a Kronecker delta that vanishes unless the positronium is in the triplet state. Employ (2) and the Fermi formula to calculate Δ, and compare your result with Δ_{expt} in (1).

(b) Calculate the Zeeman effect of the fine structure, and find the spin function for each of the magnetic substates as a function of magnetic field B. Also show how the lifetime of each of the sublevels is modified at $B = 10^4$ G. Please give numerical results.

10.4. Figure 10.6 shows the ground $6^2P_{1/2}$ and excited $7^2S_{1/2}$ states of the ^{205}Tl atom (nuclear spin $I = 1/2$). The hyperfine splittings a and a' within these states are also shown. Figure 10.6 is not to scale: actually, $a/\Delta E = 2.6 \times 10^{-5}$ and $a'/\Delta E = 2.6 \times 10^{-6}$. Of course, the thallium atom has other levels as well, but in this problem we assume that the levels shown in Figure 10.6 are the only relevant ones.

(a) Suppose that an electric field $\boldsymbol{\mathcal{E}} = \mathcal{E}_0\hat{z}$ is applied to this atom in its ground state. Explain why the $6^2P_{1/2}, F = 1$ states $m_F = -1, 0$, and 1 undergo the Stark shifts indicated in Figure 10.7.

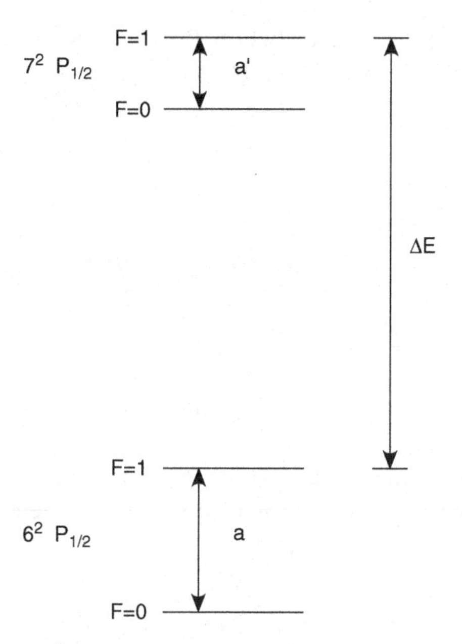

Figure 10.6 Energy level diagram for the thallium atom, not to scale.

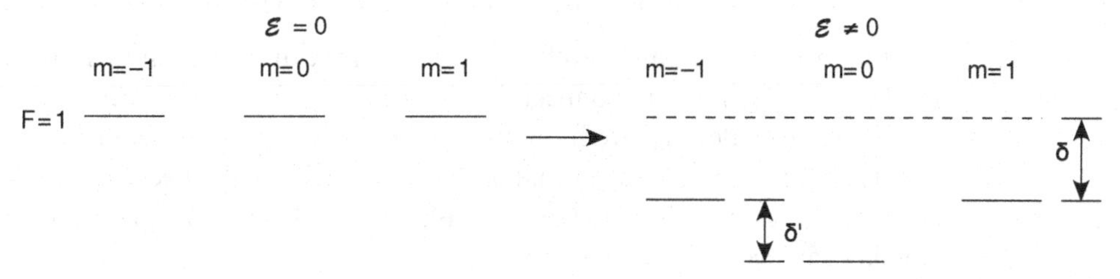

Figure 10.7 Quadratic Stark effect in the hfs of the ground state of thallium (not to scale).

In particular, give an estimate of the ratio of the "tensor" Stark shift to the "scalar" Stark shift, that is, the ratio δ'/δ.

(b) Now suppose that in addition to the electric field $\mathcal{E}$, a weak magnetic field $\boldsymbol{B} = B_z\hat{z} + B_y\hat{y}$ is applied. Assume that $B_y \ll B_z$ and that we can ignore the very small Zeeman shift in the $F = 1, m_F = 0$ level (which is proportional to B_z^2). Show that the difference between the energies of the $F = 1, m_F = \pm 1$ levels is

$$\lambda_+ - \lambda_- = 2k_1\left(1 + \frac{k_2^2}{k_1^2 - \delta'^2}\right) + \text{small terms} \tag{1}$$

where $k_1 = \mu_B B_z/3$, $k_2 = \mu_B B_y/3\sqrt{2}$, and we assume that $|k_2^2/k_1^2 - \delta'^2| \ll 1$.

10.5. This problem is concerned with the "no-crossing" theorem. Consider a zeroth-order Hamiltonian H_0 with orthonormal eigenstates $|u_1\rangle, |u_2\rangle$; that is,

$$H_0|u_1\rangle = E_1|u_1\rangle$$
$$H_0|u_2\rangle = E_2|u_2\rangle$$

We now impose a perturbation $V(x)$ that depends on a single continuous real parameter x. When V is present, the total Hamiltonian is

$$H = H_0 + V(x) \tag{1}$$

and

$$H|u\rangle = E|u\rangle \tag{2}$$

where

$$|u\rangle = c_1(x)|u_1\rangle + c_2(x)|u_2\rangle \tag{3}$$

(a) Take the scalar product of both sides of (2) with $\langle u_1|$ and with $\langle u_2|$. The resulting two linear equations possess a solution only if a certain determinant vanishes. From this, show that one obtains two possible energies

$$E_\pm = \frac{1}{2}(E_1 + E_2 + V_{11} + V_{22}) \pm \frac{1}{2}\sqrt{(E_1 - E_2 + V_{11} - V_{22})^2 + 4|V_{12}|^2} \tag{4}$$

where $V_{ij} = V_{ji}{}^* = \langle u_i|V|u_j\rangle$, with $i, j = 1, 2$. Equation (4) says that there are two separate energies that become the same (level crossing) when the discriminant vanishes. Now both the terms inside the square root are nonnegative; hence level crossing can occur only if both terms vanish simultaneously. Because we have only one parameter x, it is in general impossible to achieve this. However, it may happen that V_{12} vanishes identically because $|u_1\rangle$ and $|u_2\rangle$ have different symmetries. In this case, the discriminant in (4) can vanish for some value of x, and level crossing can occur.

(b) As an example of the foregoing, consider the Zeeman effect of the hyperfine structure of an atom with nuclear spin $I = 3/2$ and $J = 1$. Assume that the nuclear magnetic moment is positive and very small compared with the electronic magnetic moment. Sketch the Zeeman energy-level diagram schematically (not necessarily to scale) but show the correct ordering of the energy levels for the following cases:

- External magnetic field $B = 0$
- $|\mu_J B| \ll$ hyperfine energy
- $|\mu_I B| \ll$ hyperfine energy $\ll |\mu_J B|$
- $|\mu_I B| \gg$ hyperfine energy

What implications does the result of part (a) have for your diagram?

11 Identical Particles

11.1 Identical particles in classical and quantum mechanics

All electrons have the same intrinsic properties: mass m_e, charge $-e$, spin $s = 1/2$, and spin magnetic moment μ. Except for their trajectories and spin orientations, different electrons are identical and cannot be distinguished in any way. If classical mechanics governed the motion of electrons, the fact that they are identical would not be important because the trajectory of each and every electron (i.e., its position and momentum) could be specified perfectly, at least in principle. Thus, for example, in a collision between two electrons, we could follow each particle's trajectory and keep track of them separately.

However, in quantum mechanics, this is not always possible because there is no such thing as a precisely defined trajectory. Instead, we can only specify the wave function of a group of identical electrons. As a result, unless the electrons are well separated spatially, the fact that they are identical has very significant consequences. Of course, the same remarks hold for protons, neutrons, and any other type of identical particles.

11.2 Symmetric and antisymmetric wave functions

Consider two identical particles with coordinates $\boldsymbol{r}_1$, $\boldsymbol{r}_2$ (these coordinates could include spin) and wave function $\psi(\boldsymbol{r}_1, \boldsymbol{r}_2)$. Let us exchange the particles by employing a permutation operator P_{12}; that is,

$$P_{12}\psi(\boldsymbol{r}_1, \boldsymbol{r}_2) = \psi(\boldsymbol{r}_2, \boldsymbol{r}_1) \tag{11.1}$$

Because $P_{12}^2 = I$, the operator P_{12} has the following properties:

$$P_{12} = P_{12}^{\dagger} = P_{12}^{-1} \tag{11.2}$$

Any observable A associated with the two identical particles must be symmetric in the variables that describe these particles, and A therefore must be left invariant under the exchange of particles. Otherwise it would be possible to distinguish between the two particles, and they would not be identical. An example of a symmetric observable is the following Hamiltonian:

$$H_0 = \frac{\boldsymbol{p}_1^2}{2m} + \frac{\boldsymbol{p}_2^2}{2m} + V(|\boldsymbol{r}_1 - \boldsymbol{r}_2|) \tag{11.3}$$

A convenient way to state that any observable A is left invariant under the exchange is to say that A commutes with P_{12}; that is,

$$P_{12}A(1,2)P_{12}^{-1} = A(1,2) \tag{11.4}$$

Now suppose that $\psi(\boldsymbol{r}_1,\boldsymbol{r}_2)$ satisfies the Schroedinger equation

$$H(1,2)\psi(\boldsymbol{r}_1,\boldsymbol{r}_2) = E\psi(\boldsymbol{r}_1,\boldsymbol{r}_2) \tag{11.5}$$

where the Hamiltonian H is not necessarily H_0 in (11.3). Applying P_{12} on the left to both sides of (11.5) and using (11.4) with $A = H$, we have

$$H(1,2)\psi(\boldsymbol{r}_2,\boldsymbol{r}_1) = E\psi(\boldsymbol{r}_2,\boldsymbol{r}_1) \tag{11.6}$$

From (11.5) and (11.6), we see that $\psi(\boldsymbol{r}_1,\boldsymbol{r}_2)$ and $\psi(\boldsymbol{r}_2,\boldsymbol{r}_1)$ are both eigenfunctions of $H(1,2)$ corresponding to the same eigenvalue E. Hence these two wave functions are either linearly independent and thus degenerate or else $\psi(\boldsymbol{r}_1,\boldsymbol{r}_2) = c\psi(\boldsymbol{r}_2,\boldsymbol{r}_1)$, where c is a constant. In the latter case, we can exchange the two particles once again to obtain

$$\psi(\boldsymbol{r}_2,\boldsymbol{r}_1) = c\psi(\boldsymbol{r}_1,\boldsymbol{r}_2) = c^2\psi(\boldsymbol{r}_2,\boldsymbol{r}_1)$$

which implies that $c^2 = 1$ and thus $c = \pm 1$. Let $u(\boldsymbol{r}_1,\boldsymbol{r}_2)$ be an unsymmetrized solution to (11.5). When $c = +1$, we have the symmetric solution

$$\psi_S = \frac{1}{\sqrt{2}}\left[u(\boldsymbol{r}_1,\boldsymbol{r}_2) + u(\boldsymbol{r}_2,\boldsymbol{r}_1)\right] \tag{11.7}$$

and when $c = -1$, we have the antisymmetric solution

$$\psi_A = \frac{1}{\sqrt{2}}\left[u(\boldsymbol{r}_1,\boldsymbol{r}_2) - u(\boldsymbol{r}_2,\boldsymbol{r}_1)\right] \tag{11.8}$$

and these are the only nondegenerate solutions to (11.5).

Suppose instead that we have $N > 2$ identical particles. Once again, any observable A pertaining to these particles, including the Hamiltonian H, must be symmetric under any number of particle pair exchanges. Now N particles can be permuted in $N!$ ways, including the identity permutation, and each permutation is equivalent to a succession of pair exchanges, there being $N!/2$ even and $N!/2$ odd permutations corresponding to even and odd numbers of pair exchanges, respectively. From an unsymmetrized N-particle eigenfunction $u(\boldsymbol{r}_1,\boldsymbol{r}_2,\ldots,\boldsymbol{r}_N)$ we can construct the symmetric wave function

$$\psi_S = \frac{1}{\sqrt{N!}}\sum_{i=1}^{N!} u\left[P_i(\boldsymbol{r}_1,\boldsymbol{r}_2,\ldots,\boldsymbol{r}_N)\right] \tag{11.9}$$

and the antisymmetric wave function

$$\psi_A = \frac{1}{\sqrt{N!}}\sum_{i=1}^{N!} \varepsilon_i u\left[P_i(\boldsymbol{r}_1,\boldsymbol{r}_2,\ldots,\boldsymbol{r}_N)\right] \tag{11.10}$$

where $\varepsilon_i = \pm 1$ for even and odd permutations, respectively. Among the $N!$ different orthogonal linear combinations of the $u[P_i(r_1, r_2, ..., r_N)]$, only ψ_S and ψ_A are nondegenerate solutions.

It is an extremely important experimental fact, with no known exceptions, that if the identical particles have integral spin, the wave function is always symmetric with respect to exchange, whereas if the particles have half-integral spin, the wave function is always antisymmetric. In the former case, the particles obey Bose-Einstein statistics (and are called *bosons*); in the latter case, they obey Fermi-Dirac statistics and are called *fermions.*

The time-dependent many-particle Schroedinger equation

$$H\psi = i\hbar\dot{\psi} \tag{11.11}$$

ensures that a solution with one type of symmetry at some initial time preserves that symmetry at later times because, given that H is exchange symmetric, $\dot{\psi}$ has the same exchange symmetry as ψ.

The connection between spin and statistics is not just empirical. Pauli (1940) completed the work begun by others when he demonstrated that the spin-statistics connection must hold given certain very reasonable assumptions in conventional Lorentz-invariant relativistic quantum-field theory. (Actually, Pauli showed that given these assumptions, an exchange-symmetric wave function made up of identical fermions would lead to a contradiction, as would an exchange antisymmetric wave function composed of identical bosons.) No one has yet found a convincing argument based only on nonrelativistic quantum mechanics and simple reasoning to arrive at this profound result.

11.3 Composite bosons and composite fermions

Consider the ${}^4_2$He nucleus, which consists of two protons ($Z = 2$) and two neutrons ($N = 2$). Given two such nuclei, we can exchange them by exchanging one proton and one neutron at a time. Because this involves an even number of identical fermion exchanges, the wave function for two ${}^4_2$He nuclei is symmetric, and the ${}^4_2$He nucleus is thus a composite boson. On the other hand, an ${}^3_2$He nucleus is a composite fermion. The argument is easily extended to a neutral atom with atomic number Z and mass number A. If Z is even and A is odd or vice versa, the atom is a composite fermion, but if Z and A are both even or both odd, the atom is a composite boson. For example, the neutral atoms ${}^{1,3}_{1}$H, ${}^4_2$He, ${}^7_3$Li, ${}^{23}_{11}$Na, ${}^{85,87}_{37}$Rb, and ${}^{133}_{55}$Cs are bosons, whereas ${}^2_1$H, ${}^3_2$He, and ${}^6_3$Li, are fermions.

11.4 Pauli exclusion principle

Suppose that in a certain approximation we can ignore the interactions between N identical fermions. Then the Hamiltonian reduces to a sum of one-particle Hamiltonians of identical functional form; that is,

$$H = H_0(p_1, r_1) + H_0(p_2, r_2) + \cdots + H_0(p_N, r_N) \tag{11.12}$$

For example, if we could ignore the Coulomb interactions between N electrons in an ion of nuclear charge Ze, the Hamiltonian would be

$$H = \left(\frac{p_1^2}{2m_e} - \frac{Ze^2}{r_1} \right) + \left(\frac{p_2^2}{2m_e} - \frac{Ze^2}{r_2} \right) + \cdots + \left(\frac{p_N^2}{2m_e} - \frac{Ze^2}{r_N} \right)$$

When the Hamiltonian can be expressed as in (11.12), a particular solution to the Schroedinger equation before antisymmetrization can be written as a product of one-particle eigenstates of H_0; that is,

$$u = w_{q_1}(\boldsymbol{r}_1) w_{q_2}(\boldsymbol{r}_2) \cdots w_{q_N}(\boldsymbol{r}_N) \tag{11.13}$$

where $\boldsymbol{r}_i$ is a "coordinate" label (which can include spin) for the ith electron, and the q_i, $i = 1, \ldots, N$, label the quantum numbers of the one-particle wave function w (the latter is frequently called a *one-particle orbital*). Antisymmetrizing by inserting (11.13) in (11.10), we obtain

$$\psi = \frac{1}{\sqrt{N!}} \begin{vmatrix} w_{q_1}(\boldsymbol{r}_1) & \cdots & \cdots & w_{q_1}(\boldsymbol{r}_N) \\ w_{q_2}(\boldsymbol{r}_1) & \cdots & \cdots & w_{q_2}(\boldsymbol{r}_N) \\ \cdots & \cdots & \cdots & \cdots \\ w_{q_N}(\boldsymbol{r}_1) & \cdots & \cdots & w_{q_N}(\boldsymbol{r}_N) \end{vmatrix} \tag{11.14}$$

This is frequently called a *Slater determinant*, although Dirac first described the expression in (11.14). Because a determinant always vanishes if two rows or two columns are equal, we see that the orbitals must be distinct if ψ is to be nonzero; that is, no two single-partical orbitals w can have the same quantum numbers, including spin. This is the *Pauli exclusion principle* (which was actually discovered by W. Pauli at the end of 1924, one year before the invention of wave mechanics.) We emphasize that the exclusion principle only has meaning in the approximation where interactions between identical fermions can be neglected, whereas the antisymmetrization principle for identical fermions as embodied in (11.10) has a more general meaning. No restriction analogous to the Pauli principle applies for bosons. An unlimited number of bosons with the same single-particle quantum numbers are possible. For example, an arbitrarily large number of noninteracting photons can exist in the same electromagnetic field mode, so a monochromatic radio wave can have arbitrarily large intensity.

11.5 Example of atomic helium

We now consider some features of atomic helium that illustrate the antisymmetrization principle as well as other points of interest. Figure 11.1 shows several bound states of neutral helium (HeI) as well as the ground state and ionization limit of He^+ (HeII).

We now write the Hamiltonian for the two electrons in HeI in the approximation where the nucleus has infinite mass and we ignore relativistic effects and magnetic interactions between the electron spins; that is,

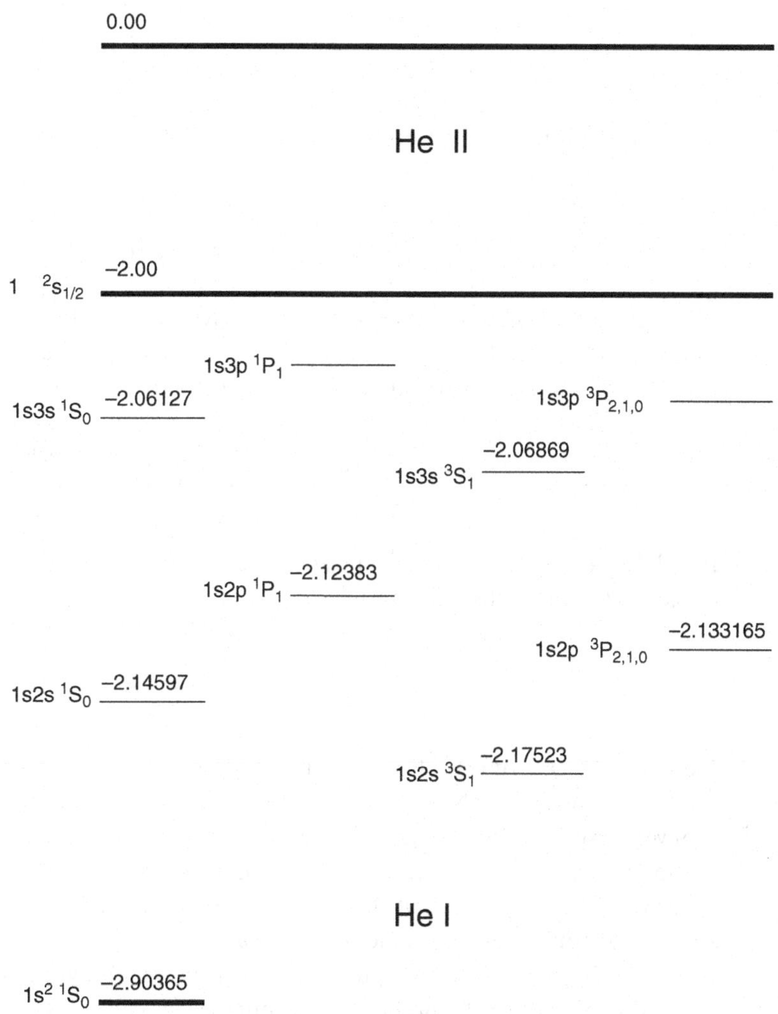

0.00

He II

$1\ ^2S_{1/2}$ −2.00

$1s3p\ ^1P_1$

$1s3s\ ^1S_0$ −2.06127

$1s3p\ ^3P_{2,1,0}$

$1s3s\ ^3S_1$ −2.06869

$1s2p\ ^1P_1$ −2.12383

$1s2p\ ^3P_{2,1,0}$ −2.133165

$1s2s\ ^1S_0$ −2.14597

$1s2s\ ^3S_1$ −2.17523

He I

$1s^2\ ^1S_0$ −2.90365

Figure 11.1 Some energy levels of atomic helium (not to scale). Following the customary notation, we write HeI for neutral helium and HeII for He^+. Energies are given in atomic units; thus the $1^2s_{1/2}$ ground state of HeII has energy $E = -Z^2/2 = -2.00$. The bound states of HeI form two independent systems, singlet and triplet, with no allowed intercombination transitions between them.

$$H = -\frac{1}{2}\nabla_1^2 - \frac{1}{2}\nabla_2^2 - \frac{Z}{r_1} - \frac{Z}{r_2} + \frac{1}{|\mathbf{r}_1 - \mathbf{r}_2|} \tag{11.15}$$

We regard

$$H_0 = -\frac{1}{2}\nabla_1^2 - \frac{1}{2}\nabla_2^2 - \frac{Z}{r_1} - \frac{Z}{r_2} \tag{11.16}$$

as a zeroth-order Hamiltonian and treat

$$H' = \frac{1}{|\boldsymbol{r}_1 - \boldsymbol{r}_2|} \tag{11.17}$$

as a perturbation. The zero-order unsymmetrized wave function for the ground state is

$$\begin{aligned} u_0(1s^2) &= w_1(\boldsymbol{r}_1)w_2(\boldsymbol{r}_2) \\ &= \phi_{1s}(\boldsymbol{r}_1)\chi_1 \cdot \phi_{1s}(\boldsymbol{r}_2)\chi_2 \end{aligned}$$

where $\chi_{1,2}$ are spinors for electrons 1 and 2, respectively. Because the spatial parts ϕ of the single-particle orbitals w are the same, the Pauli principle requires the spinors $\chi_{1,2}$ to be different. Thus the antisymmetrized ground-state zero-order wave function is

$$\psi_0 = u_0(1s^2) = \phi_{1s}(\boldsymbol{r}_1)\phi_{1s}(\boldsymbol{r}_2)\frac{\alpha_1\beta_2 - \alpha_2\beta_1}{\sqrt{2}} \tag{11.18}$$

In (11.18), the spin function is in fact the $J = 0$ combination of two electrons with spin-½; hence the ground state of HeI is a $^1S_0$ state. The zero-order energy of this state is

$$E_0(1s^2) = -\frac{Z^2}{2} - \frac{Z^2}{2} = -Z^2 = -4 \tag{11.19}$$

However, the experimentally determined energy is $E_{\text{expt}}(1s^2) = -2.90365$. We account for this large discrepancy shortly.

Several excited states of HeI are also shown in Figure 11.1. These are labeled by their configuration ($1s2s$, $1s2p$, $1s3s$, …) and by the usual spectroscopic notation for angular momentum ($^1S_0$, $^3S_1$, $^3P_{2,1,0}$, …). Note that these energy eigenstates of HeI form two independent systems: singlet and triplet. There are no allowed *intercombination* transitions between these systems. Note also that for a given configuration, a triplet level always lies lower in energy than the corresponding singlet level, and the energy difference is roughly three orders of magnitude larger than would be expected as a result of the magnetic interaction between electron spins. For example, $1s2s\ ^3S_1$ lies lower than $1s2s\ ^1S_0$ by 0.797 eV = 0.0293 atomic units. On the other hand, the spin-spin magnetic energy is crudely estimated as

$$\Delta E_{\text{mag}} \sim \mu_1 B_{21} \sim \frac{\mu_1\mu_2}{r^3} \sim \frac{\mu_B^2}{r^3}$$

In atomic units, $\mu_B^2 = \alpha^2/4$ and $r \sim 1$; hence $\Delta E_{\text{mag}} \sim \alpha^2/4 \sim 1.3\times10^{-5}$. The actual cause of the very large singlet-triplet splitting is an effect of crucial importance for all of atomic and molecular physics and for many aspects of condensed-matter physics as well. In what follows, we set ourselves two goals:

- Calculation of the ($1s^2\ ^1S_0$) ground-state energy with reasonable accuracy, and
- Explanation for the large singlet-triplet energy splittings.

11.6 Perturbation-variation calculation of the ground-state energy of helium

From (11.16), the first-order energy shift is

$$\Delta E^{(1)}(\text{HeI}, 1s^2) = \left\langle \phi_{1s}(\boldsymbol{r}_1)\phi_{1s}(\boldsymbol{r}_2) \left| \frac{1}{|\boldsymbol{r}_1 - \boldsymbol{r}_2|} \right| \phi_{1s}(\boldsymbol{r}_1)\phi_{1s}(\boldsymbol{r}_2) \right\rangle$$
$$= \frac{Z^6}{\pi^2} \iint e^{-2Zr_1} e^{-2Zr_2} \frac{1}{|\boldsymbol{r}_1 - \boldsymbol{r}_2|} d^3\boldsymbol{r}_1\, d^3\boldsymbol{r}_2$$

Now

$$\frac{1}{|\boldsymbol{r}_1 - \boldsymbol{r}_2|} = \frac{1}{r_>} \sum_{\ell=0}^{\infty} \left(\frac{r_<}{r_>}\right)^{\ell} P_\ell(\cos\theta)$$

where θ is the angle between $\boldsymbol{r}_1$ and $\boldsymbol{r}_2$. Thus

$$\Delta E^{(1)}(\text{HeI}, 1s^2) = (2\pi)^2 \frac{Z^6}{\pi^2} \int_0^\infty e^{-2Zr_2} r_2^2\, dr_2 \int_0^\pi \sin\theta_2\, d\theta_2$$
$$\left[\int_0^{r_2} dr_1 e^{-2Zr_1} \frac{r_1^2}{r_2} \sum_{\ell=0}^{\infty} \left(\frac{r_1}{r_2}\right)^{\ell} \int_0^\pi P_\ell(\cos\theta) \sin\theta\, d\theta \right.$$
$$\left. + \int_{r_2}^\infty dr_1 e^{-2Zr_1} \frac{r_1^2}{r_1} \sum_{\ell=0}^{\infty} \left(\frac{r_2}{r_1}\right)^{\ell} \int_0^\pi P_\ell(\cos\theta) \sin\theta\, d\theta \right] \tag{11.20}$$

Because $\int_0^\pi P_\ell(\cos\theta)\sin\theta\, d\theta = 2\delta_{\ell 0}$, (11.20) becomes

$$\Delta E^{(1)}(\text{HeI}, 1s^2) = 16Z^6 \left[\int_0^\infty r_2 e^{-2Zr_2}\, dr_2 \int_0^{r_2} r_1^2 e^{-2Zr_1} dr_1 + \int_0^\infty r_2^2 e^{-2Zr_2}\, dr_2 \int_{r_2}^\infty r_1 e^{-2Zr_1}\, dr_1 \right] \tag{11.21}$$

By exchanging the dummy variables $\boldsymbol{r}_1$ and $\boldsymbol{r}_2$, the second double integral on the right-hand side of (11.21) can be written

$$I_2 = 16Z^6 \int_0^\infty r_1^2 e^{-2Zr_1}\, dr_1 \int_{r_1}^\infty r_2 e^{-2Zr_2}\, dr_2 \tag{11.22}$$

It is clear from Figure 11.2a that I_2 in (11.22) is obtained by integrating along a given vertical strip and then adding up the strips, but this is equivalent to integrating along a horizontal strip as in Figure 11.2b and then adding up these strips as we do in the first double integral (I_1) on the right-hand side of (11.21). Therefore, $I_1 = I_2$, and

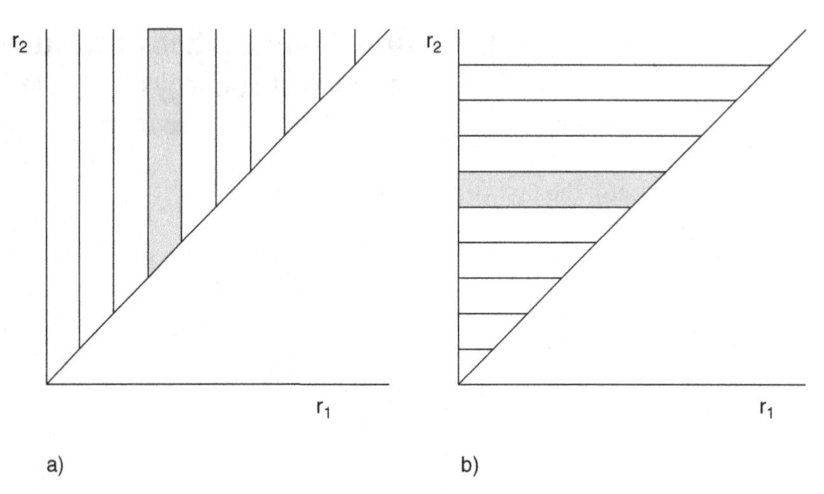

Figure 11.2 Diagrams (a) and (b) illustrate two different ways to evaluate the double integral *I2* in equation (11.22).

$$\begin{aligned}\Delta E^{(1)} &= 32Z^6\int_0^\infty r_2 e^{-2Zr_2}\,dr_2\int_0^{r_2} r_1^2 e^{-2Zr_1}\,dr_1\\ &= 32Z^6\left[\frac{1}{(2Z)^2}\frac{2}{(4Z)^3}+\frac{1}{2Z}\frac{3!}{(4Z)^4}\right]=\frac{5}{8}Z\end{aligned}\tag{11.23}$$

Combining (11.23) with (11.19) for Z = 2, we obtain

$$E_0+\Delta E^{(1)}=-4+\frac{5}{4}=-\frac{11}{4}=-2.75\tag{11.24}$$

Because $E_{\text{expt}} = -2.904$, $\Delta E^{(1)}$ is too large a correction. The result (11.24) is improved by using the variational method. For the ground state of helium, we choose the trial function

$$\psi=\frac{\lambda^3}{\pi}\exp\left[-\lambda(r_1+r_2)\right]\tag{11.25}$$

where λ, the effective nuclear charge, is now considered a variable parameter. This is physically reasonable because one electron experiences not only the nuclear charge but also the charge of the other electron. From (11.25), we find

$$\langle H\rangle=\langle H_0\rangle+\langle H'\rangle=\lambda^2-2\lambda Z+\frac{5}{8}\lambda\tag{11.26}$$

Hence

$$\frac{\partial\langle H\rangle}{\partial\lambda}=2\lambda-2Z+\frac{5}{8}$$

which vanishes for $\lambda = Z - 5/16$. Inserting this value in (11.26), we obtain

Table 11.1 Binding energy of the last electron in various helium like ions, calculated from equation (11.27).

Ion	Z	Binding Energy of Last Electron, eV	
		Calculated	Observed
He	2	23.3	24.5
Li^{+}	3	74.1	75.6
Be^{2+}	4	152.2	153.6
C^{4+}	6	390	393
O^{6+}	8	737	738

$$\langle H \rangle = -\left(Z - \frac{5}{16} \right)^2 \tag{11.27}$$

For helium ($Z = 2$), this result, namely,

$$\langle H \rangle_{\text{helium}} = -\left(2 - \frac{5}{16} \right)^2 = -2.8477 \tag{11.28}$$

is much closer to the experimental value than our original approximation (11.24). The variational method can be improved further by using a more complicated wave function with more parameters; for example, see Problem 11.1. The most sophisticated variational calculations of the helium ground-state energy yield results that differ from experiment by less than 1 ppm [see, e.g., Kinoshita (1957) and Peckeris (1958)].

The approximation (11.27) improves as Z increases for helium-like ions with $Z > 2$, as can be seen in Table 11.1.

In the case of the hydrogen negative ion H^- ($Z = 1$), our calculation (11.27) yields

$$\langle \mathrm{H} \rangle = -\left(1 - \frac{5}{16} \right)^2 = -0.4726$$

This is more positive than the ground-state energy of a neutral H atom and an electron at ∞, which is $-1/2$. Thus our simple variational calculation suggests that H^- has no bound state, but experiment shows that a bound state actually exists: $E = -0.528$. To reveal this, we need a better trial function with more parameters (Bethe and Salpeter 1957, p. 154). Also see Problem 11.1. Although the variational method yields the most accurate ground-state energy for helium, it does not provide us with the best ground-state wave function. Here it is better to use Hartree's self-consistent field method (see Section 12.2).

11.7 Excited states of helium: Exchange degeneracy

We now consider the $1s2s$ configuration, ignoring at first the Coulomb perturbation H' of (11.17). From the exclusion principle, the zeroth-order wave functions take the general form

$$\psi = \frac{1}{\sqrt{2}} \begin{vmatrix} \phi_{1s}(1)\chi_{11} & \phi_{1s}(2)\chi_{12} \\ \phi_{2s}(1)\chi_{21} & \phi_{2s}(2)\chi_{22} \end{vmatrix} \tag{11.29}$$

where χ_{12} means "spinor 1 for electron 2," and so forth. Because in each case the spinor can be α or β, there are four degenerate Slater determinants:

$$\begin{aligned} \psi_1 &= \frac{1}{\sqrt{2}} \begin{vmatrix} \phi_{1s}(1)\alpha_1 & \phi_{1s}(2)\alpha_2 \\ \phi_{2s}(1)\alpha_1 & \phi_{2s}(2)\alpha_2 \end{vmatrix} \\ &= \frac{1}{\sqrt{2}} [\phi_{1s}(1)\phi_{2s}(2) - \phi_{2s}(1)\phi_{1s}(2)]\alpha_1\alpha_2 \end{aligned} \tag{11.30}$$

$$\psi_2 = \frac{1}{\sqrt{2}} [\phi_{1s}(1)\phi_{2s}(2) - \phi_{2s}(1)\phi_{1s}(2)]\beta_1\beta_2 \tag{11.31}$$

$$\begin{aligned} \psi_A &= \frac{1}{\sqrt{2}} \begin{vmatrix} \phi_{1s}(1)\alpha_1 & \phi_{1s}(2)\alpha_2 \\ \phi_{2s}(1)\beta_1 & \phi_{2s}(2)\beta_2 \end{vmatrix} \\ &= \frac{1}{\sqrt{2}} [\phi_{1s}(1)\phi_{2s}(2)\alpha_1\beta_2 - \phi_{2s}(1)\phi_{1s}(2)\beta_1\alpha_2] \end{aligned} \tag{11.32}$$

$$\psi_B = \frac{1}{\sqrt{2}} [\phi_{1s}(1)\phi_{2s}(2)\beta_1\alpha_2 - \phi_{2s}(1)\phi_{1s}(2)\alpha_1\beta_2] \tag{11.33}$$

However, because any linear combination of degenerate eigenfunctions is another eigenfunction corresponding to the same eigenvalue, we are free to replace ψ_A and ψ_B by the linear combinations

$$\begin{aligned} \psi_3 &= \frac{1}{\sqrt{2}}(\psi_A + \psi_B) \\ &= \frac{[\phi_{1s}(1)\phi_{2s}(2) - \phi_{2s}(1)\phi_{1s}(2)]}{\sqrt{2}} \frac{[\alpha_1\beta_2 + \beta_1\alpha_2]}{\sqrt{2}} \end{aligned} \tag{11.34}$$

and

$$\begin{aligned} \psi_4 &= \frac{1}{\sqrt{2}}(\psi_A - \psi_B) \\ &= \frac{[\phi_{1s}(1)\phi_{2s}(2) + \phi_{2s}(1)\phi_{1s}(2)]}{\sqrt{2}} \frac{[\alpha_1\beta_2 - \beta_1\alpha_2]}{\sqrt{2}} \end{aligned} \tag{11.35}$$

We see from (11.30), (11.31), (11.34), and (11.35) that $\psi_{1,2,3}$ form three components of a $J = 1$ (triplet) system with antisymmetric spatial wave function and symmetric spin functions. On the other hand, ψ_4 is a $J = 0$ (singlet) state with antisymmetric spin function and symmetric spatial function. Of course, in each case, the total wave function including space and spin variables is antisymmetric.

We now include the electron-electron Coulomb interaction H', a perturbation that lifts the degeneracy. Our choice of $\psi_{1,2,3,4}$ as a basis is fortunate because the perturbation matrix is diagonal in this representation. In fact, $H' = r_{12}^{-1}$ does not affect spin variables, so the triplet terms 1, 2, 3, which all have the same spatial wave function, remain degenerate. We have

$$\langle \psi_{1,2,3} | r_{12}^{-1} | \psi_{1,2,3} \rangle = \frac{1}{2} \iint r_{12}^{-1} \left[\phi_{1s}(1)\phi_{2s}(2) - \phi_{1s}(2)\phi_{2s}(1) \right]^2 d^3\mathbf{r}_1\, d^3\mathbf{r}_2 \tag{11.36}$$

and

$$\langle \psi_4 | r_{12}^{-1} | \psi_4 \rangle = \frac{1}{2} \iint r_{12}^{-1} \left[\phi_{1s}(1)\phi_{2s}(2) + \phi_{1s}(2)\phi_{2s}(1) \right]^2 d^3\mathbf{r}_1\, d^3\mathbf{r}_2 \tag{11.37}$$

These can be written as

$$\langle \psi_{1,2,3} | r_{12}^{-1} | \psi_{1,2,3} \rangle = J - K \tag{11.38}$$

and

$$\langle \psi_4 | r_{12}^{-1} | \psi_4 \rangle = J + K \tag{11.39}$$

where J (the *Coulomb integral*) and K (the *exchange integral*) are given by

$$J = \iint \phi_{1s}^2(1)\phi_{2s}^2(2) r_{12}^{-1}\, d^3\mathbf{r}_1\, d^3\mathbf{r}_2 \tag{11.40}$$

and

$$K = \iint \phi_{1s}(1)\phi_{2s}(1) r_{12}^{-1} \phi_{1s}(2)\phi_{2s}(2) d^3\mathbf{r}_1 d^3\mathbf{r}_2 \tag{11.41}$$

Obviously, J is always positive, but K is also positive, as can be seen from the following: let $\rho = \phi_{1s}\phi_{2s}$. Then

$$K = \int d\tau_1 \rho(1) \int d\tau_2 \frac{\rho(2)}{r_{12}} \tag{11.42}$$

Now consider any potential function Φ that satisfies Poisson's equation, which in hlu is,

$$\nabla^2 \Phi = -\rho \tag{11.43}$$

Then, as we know from electrostatics, a formal solution to (11.43) is

$$\Phi(1) = \frac{1}{4\pi} \int d\tau_2 \frac{\rho(2)}{r_{12}}$$

Hence (11.42) can be written

$$K = \frac{1}{4\pi}\int \rho(1)\Phi(1)\, d\tau_1 = -\frac{1}{4\pi}\int \Phi\nabla^2\Phi\, d\tau_1$$

Integrating the last expression by parts, we obtain

$$K = \frac{1}{4\pi}\int \nabla\Phi\cdot\nabla\Phi\, d\tau_1 - \frac{1}{4\pi}\int \nabla\cdot(\Phi\nabla\Phi)\, d\tau_1 \tag{11.44}$$

The first integral on the right-hand side of (11.44) is obviously positive. The second integral may be converted to a surface integral by Gauss' theorem, and it vanishes because $\rho \to 0$ rapidly as $r_1 \to \infty$.

Evaluations of J (11.40) and K (12.41) proceed by techniques similar to those already employed to arrive at (11.23). A straightforward (and naive) calculation using hydrogenic orbitals for ϕ_{1s}, ϕ_{2s} yields

$$K = \left(\frac{4}{27}\right)^2 = 0.0219 \text{ atomic unit}$$

and thus a ${}^1S_0 - {}^3S_1$ splitting of $2K = 0.0438$ atomic unit As we have noted, the actual splitting is 0.0293 atomic unit. Thus our naive calculation is not very accurate; it overestimates the effect by a factor of approximately 1.5. However, it gives the correct order of magnitude, and this is the important point here: the calculation reveals the qualitative explanation for the large singlet-triplet splitting. The latter arises from the Coulomb repulsion between the two electrons. Because the singlet state has a symmetric spatial wave function, the electrons can get close together, whereas in the triplet case, the spatial wave function is antisymmetric, and the electrons are kept far apart. Thus the Coulomb repulsion energy is higher for the singlet state. In other words, although the actual spin-spin magnetic dipole interaction energy is very small, the Pauli principle generates a very strong *effective* spin-spin interaction through the Coulomb interaction, even though there are no spin operators in the Hamiltonian (11.15). Similar considerations are important for understanding complex atoms, covalent molecular bonding, the stability of metals, and ferromagnetism.

11.8 Matrix elements of determinantal wave functions

In previous sections we have used the conventional language of many-particle wave functions to introduce some very important concepts: the antisymmetrization principle for identical fermions, the related Pauli principle, and exchange degeneracy. Unfortunately, continued use of this conventional language for real calculations with many identical fermions can be very cumbersome, especially when we encounter a macroscopic number of fermions, as often occurs in astrophysics or condensed-matter physics. Moreover, in relativistic phenomena involving, for example, electrons and positrons, we frequently meet situations in which particles are created and destroyed. The conventional language of many-particle antisymmetrized wave functions is not well suited to such situations.

There exists an alternative language, called *second quantization*, that involves no new physical principles but *merely* requires a change in notation. This permits a radical simplification of many calculations and lends itself very well to situations where many particles are involved and/or where particles are created and destroyed. (We put the word *merely* in italics because notation is important: it can affect our thinking in profound ways.)

As a preparation for discussing second quantization, it is useful to obtain some general results concerning matrix elements of operators between Slater determinants. [We have already seen equations (11.36) and (11.37), which are specific examples.] Consider the general matrix element

$$M = \int \psi_B^\dagger F \psi_A \, d\mathbf{x}_1 \cdots d\mathbf{x}_N \tag{11.45}$$

where F is some operator, $\mathbf{x}_i$ denotes the space-spin coordinates of the ith fermion, and $\psi_{A,B}$ are two Slater determinants; that is,

$$\psi_A = \frac{1}{\sqrt{N!}} \sum_P \varepsilon_P \prod_{j=1}^{N} w_{Pj}(\mathbf{x}_j)$$
$$\psi_B = \frac{1}{\sqrt{N!}} \sum_Q \varepsilon_Q \prod_{i=1}^{N} u_i(Q\mathbf{x}_i).$$

Note that in ψ_A we permute the quantum numbers, whereas in ψ_B we permute the particle coordinates. We assume that the individual orbitals are orthonormal; that is,

$$\int u_i^\dagger(\mathbf{x}) u_j(\mathbf{x}) \, dx = \delta_{ij}$$

and similarly for the w_i. Insertion of the preceding expressions for $\psi_{A,B}$ in (11.45) yields

$$\begin{aligned} M &= \frac{1}{N!} \int d\mathbf{x}_1 \cdots d\mathbf{x}_N \sum_Q \varepsilon_Q \prod_{i=1}^{N} u_i^*(Q\mathbf{x}_i) F \sum_P \varepsilon_P \prod_{j=1}^{N} w_{Pj}(\mathbf{x}_j) \\ &= \frac{1}{N!} \int \sum_Q \varepsilon_Q \prod_{i=1}^{N} u_i^*(Q\mathbf{x}_i) F \sum_P \varepsilon_P \prod_{j=1}^{N} w_{Pj}(\mathbf{x}_j) \, d\mathbf{x}_j \end{aligned} \tag{11.46}$$

where each sum has $N!$ terms. On the right-hand side of the second line of (11.46) we set $j = Qi$ in the product over j. This is legitimate because Qi runs over all N when i does. Hence

$$\begin{aligned} M &= \frac{1}{N!} \int \sum_Q \varepsilon_Q \prod_{i=1}^{N} u_i^*(Q\mathbf{x}_i) F \sum_P \varepsilon_P w_{PQi}(Q\mathbf{x}_i) \, d\mathbf{x}_{Qi} \\ &= \frac{1}{N!} \int \sum_Q \sum_P \varepsilon_Q \varepsilon_P \prod_{i=1}^{N} u_i^*(Q\mathbf{x}_i) F w_{PQi}(Q\mathbf{x}_i) \, d\mathbf{x}_{Qi} \end{aligned} \tag{11.47}$$

Now $\varepsilon_Q \varepsilon_P = \varepsilon_{PQ}$, $Q\mathbf{x}_i$ is a dummy variable, and F, like all observables, is symmetric in the fermion coordinates. Hence we can label the coordinates as follows:

$$Q\mathbf{x}_i \to \mathbf{x}_i \tag{11.48}$$

Note that the quantum-number labels are not affected in (11.47) when we make the name change for the coordinate dummy variables. Also, we can sum over all permutations PQ for given Q, thus covering all permutations P. Therefore,

$$M = \frac{1}{N!}\sum_{Q}\left[\sum_{PQ}\varepsilon_{PQ}\int\prod_{i=1}^{N}u_i^*(\mathbf{x}_i)Fw_{PQi}(\mathbf{x}_i)\,d\mathbf{x}_i\right] \tag{11.49}$$

The quantity in square brackets on the right-hand side of (10.49) is independent of Q; hence we have $N!$ identical terms in the sum over Q. Therefore, M can be written

$$M = \sum_{P}\varepsilon_P\int\prod_{i=1}^{N}u_i^*(\mathbf{x}_i)Fw_{Pi}(\mathbf{x}_i)\,d\mathbf{x}_i \tag{11.50}$$

where we have relabeled PQ with P.

For almost all applications, F takes one of three possible forms:

1. $F = 1$. Here $\langle F\rangle = 0$ because of orthogonality of the single-particle orbitals unless there exists a permutation P such that $u_i = w_{Pi}$ for all i. This can happen for only one permutation P, and we can order the w's so that it is the identity permutation. In this case, $\langle F\rangle = 1$. We assume in what follows that the w's are ordered in this way.

2. $F = \sum_{j=1}^{N} f_j$, where f_j is a one-body operator acting on the fermion with coordinate $\mathbf{x}_j$. For example, f_j could be the kinetic energy operator $f_j = -\tfrac{1}{2}\nabla_j^2$. It is straightforward to obtain the following results:

   If $u_i \neq w_i$ for more than one i,

$$\langle F\rangle = 0 \tag{11.51}$$

   If $u_i \neq w_i$ for just one i but $u_j = w_j$ for all $j \neq i$, then

$$\langle F\rangle = \langle i|f|i\rangle = \int u_i^\dagger(\mathbf{x})fw_i(\mathbf{x})\,d\mathbf{x}. \tag{11.52}$$

   If $u_i = w_i$ for all i, then

$$\langle F\rangle = \sum_i \langle i|f|i\rangle \tag{11.53}$$

3. $F = \sum_{i>j} g(i,j)$ is a two-body operator, where, for example, $g(i,j) = e^2/(4\pi|\mathbf{x}_i - \mathbf{x}_j|)$ is the electrostatic interaction between electrons i and j. Because in all cases an observable is symmetric with respect to exchange of any pair of fermions, we always have $g(i,j) = g(j,i)$, whether or not F is the electrostatic interaction. The following results are obtained quite easily:

   If $u_i \neq w_i$ for more than two i,

$$\langle F \rangle = 0 \tag{11.54}$$

If $u_i \neq w_i$ and $u_j \neq w_j$ but $u_k = w_k$ for all k except i, j, then

$$\langle F \rangle = \langle ij|g|ij \rangle - \langle ij|g|ji \rangle \tag{11.55}$$

where

$$\langle ij|g|mn \rangle = \iint u_i^\dagger(x) u_j^\dagger(y) g(x,y) w_m(x) w_n(y)\, dx\, dy \tag{11.56}$$

If for a single i, $u_i \neq w_i$ but for all $j \neq i$, $u_j = w_j$, then

$$\langle F \rangle = \sum_{j \neq i} \left[\langle ij|g|ij \rangle - \langle ij|g|ji \rangle \right] \tag{11.57}$$

where the sum is over all values of j except $j = i$.
If $u_i = w_i$ for all i,

$$\langle F \rangle = \sum_{i<j} \left[\langle ij|g|ij \rangle - \langle ij|g|ji \rangle \right] \tag{11.58}$$

where the sum is over all pairs.

11.9 Second quantization for fermions

We begin a discussion of second quantization by recalling that the Hamiltonian for N identical noninteracting fermions is a sum of single-particle Hamiltonians identical in form, as in (11.12); that is,

$$H = \sum_{i=1}^{N} H_0(\boldsymbol{p}_i, \boldsymbol{x}_i) \tag{11.59}$$

A particular antisymmetrized solution to the many-particle Schroedinger equation with this Hamiltonian is the Slater determinant

$$\psi = \frac{1}{\sqrt{N!}} \begin{vmatrix} w_{q_1}(\boldsymbol{x}_1) & \dots & \dots & w_{q_1}(\boldsymbol{x}_N) \\ w_{q_2}(\boldsymbol{x}_1) & \dots & \dots & w_{q_2}(\boldsymbol{x}_N) \\ \dots & \dots & \dots & \dots \\ w_{q_N}(\boldsymbol{x}_1) & \dots & \dots & w_{q_N}(\boldsymbol{x}_N) \end{vmatrix} \tag{11.60}$$

where each of the single-particle orbitals w is normalized to unity. (Although we should be aware that a general antisymmetrized solution to the Schroedinger equation for N particles is a linear combination of $N \times N$ Slater determinants, in what follows we can make the essential points by assuming that ψ is just a single Slater determinant.)

Now consider all the possible distinct single-particle orbitals w (*distinct* meaning that no two of them have exactly the same quantum numbers). We can arrange these orbitals in a list with index $j = 1, 2, \ldots, K$, where each j corresponds to a different set of quantum numbers. Of course, this list can be infinitely long: $K = \infty$. Each Slater determinant ψ can be expressed as follows:

$$\psi = \psi(x_1, x_2, \ldots, x_N; n_1, n_2, \ldots, n_K) \tag{11.61}$$

where the numbers n_j (with $j = 1, \ldots, K$) are occupation numbers. Each n_j has only two possible values: $n_j = 1$ if the orbital is occupied, meaning that it appears in the Slater determinant, or $n_j = 0$ if the orbital is vacant, meaning that it does not appear in the Slater determinant. For example, suppose that the first, fifth, and sixth orbitals are occupied and all others are vacant. Then

$$\psi(x_1, x_2, x_3; 1,0,0,0,1,1,0,\ldots) = \frac{1}{\sqrt{3!}} \begin{vmatrix} w_1(1) & w_1(2) & w_1(3) \\ w_5(1) & w_5(2) & w_5(3) \\ w_6(1) & w_6(2) & w_6(3) \end{vmatrix}$$

Clearly, one Slater determinant is distinguished from another by specifying the occupation numbers for all the orbitals; we can describe any Slater determinant, no matter how large or small, by giving these numbers. In fact, we can think of a Slater determinant as the space-spin representative of an abstract vector $|\Psi\rangle$ in Fock space,[1] where $|\Psi\rangle$ is specified by the occupation numbers

$$|\Psi\rangle = |n_1, n_2, \ldots\rangle$$

Let us consider the manifold of such vectors in Fock space for all possible values of $n_1, n_2, \ldots, n_K$. One such vector is the vacuum state $|0\rangle$, where all the occupation numbers are 0. We assume that this state is normalized to unity; that is, $\langle 0|0\rangle = 1$. Let us also assume that there is a creation operator $b_j^\dagger$ corresponding to the jth orbital that, when applied to the vacuum state, results in the state where $n_j = 1$ and all other occupation numbers are zero; that is,

$$b_j^\dagger |0\rangle = |n_j = 1\rangle \tag{11.62}$$

Corresponding to this creation operator, we also assume a destruction or annihilation operator b_j that performs the reverse function; that is,

$$b_j |n_j = 1\rangle = |0\rangle \tag{11.63}$$

For comparison, we recall the simple harmonic oscillator creation and destruction operators $a^\dagger$ and a, respectively (see Section 6.13). Those operators satisfy the commutation relation

$$[a, a^\dagger] = 1 \tag{11.64}$$

For the harmonic oscillator, one defines a number operator $N = a^\dagger a$, and it follows from (11.64) that the eigenvalues of N are 0, 1, 2, 3, …. As we see in Chapter 15, the quantum theory of

[1] Named after V. Fock (USSR), who was one of the originators of these concepts in the early 1930s.

the simple harmonic oscillator is readily adapted to quantization of the various modes of the electromagnetic radiation field. Here we define operators a_j and $a_j^\dagger$, which are destruction and creation operators, respectively, for photons in the jth mode. Just as for the single oscillator, $N_j = a_j^\dagger a_j$ is the photon number operator for the jth mode, and from the relation

$$\left[a_j, a_k^\dagger\right] = \delta_{j,k} \tag{11.65}$$

it follows that the eigenvalues of N_j are 0, 1, 2, 3, ... Thus we can have 0, 1, 2, 3, ... photons in any given mode. In other words, photons are bosons.

Returning now to our theory of many identical fermions, the operator $N_j = b_j^\dagger b_j$ is also interpreted as a number operator, but its eigenvalues can only be 0 or 1. Therefore, the operators b_j and $b_k^\dagger$ cannot satisfy the algebra given by (11.65). What is their algebra? The solution is readily found when we consider the analogy between our present problem and the theory of angular momentum for a single particle of spin-1/2. In the latter case, there are two orthonormal states, spin up $|u\rangle$ and spin down $|d\rangle$, with respect to a z-axis. Also, there is a raising operator S_+ and a lowering operator S_- such that

$$\begin{aligned} S_+|d\rangle &= |u\rangle \\ S_-|u\rangle &= |d\rangle \\ S_+|u\rangle &= S_-|d\rangle = 0 \end{aligned}$$

In the representation where

$$|u\rangle = \alpha = \begin{pmatrix} 1 \\ 0 \end{pmatrix} \qquad |d\rangle = \beta = \begin{pmatrix} 0 \\ 1 \end{pmatrix}$$

we have

$$S_+ = \begin{pmatrix} 0 & 1 \\ 0 & 0 \end{pmatrix} \qquad S_- = \begin{pmatrix} 0 & 0 \\ 1 & 0 \end{pmatrix}$$

From this we see the obvious correspondence:

Fermion System		Spin-1/2 System
Occupied orbital	⇔	$\begin{pmatrix} 1 \\ 0 \end{pmatrix}$
Vacant orbital	⇔	$\begin{pmatrix} 0 \\ 1 \end{pmatrix}$
$b_j^\dagger$	⇔	$S_+ = \begin{pmatrix} 0 & 1 \\ 0 & 0 \end{pmatrix}$
b_j	⇔	$S_- = \begin{pmatrix} 0 & 0 \\ 1 & 0 \end{pmatrix}$

It is also easy to verify that

$$S_+S_- + S_-S_+ = I \tag{11.66}$$

where I is the 2×2 identity matrix. From (11.66) we can easily guess that the appropriate algebra for the jth orbital is

$$\{b_j, b_j^\dagger\} = 1 \tag{11.67}$$

where $\{\cdot\}$ stands for *anticommutator*. We now adopt the following *defined* rules governing the relations between creation and destruction operators for different orbitals:

$$\begin{aligned} \{b_j, b_k^\dagger\} &= \delta_{jk} \\ \{b_j, b_k\} &= 0 \\ \{b_j^\dagger, b_k^\dagger\} &= 0 \end{aligned} \tag{11.68}$$

With these definitions, it is possible to translate all the results that are contained in conventional wave-function language into Fock-space form. To this end, we define a fermion field operator as follows:

$$\Phi(x) = \sum_\alpha w_\alpha(x) b_\alpha \tag{11.69}$$

as well as the Hermitian conjugate field operator

$$\Phi^\dagger(x) = \sum_\beta w_\beta^\dagger(x) b_\beta^\dagger \tag{11.70}$$

Any Slater determinant can now be expressed as follows:

$$\psi(x_1, \ldots, x_N; n_1, \ldots, n_K) = \frac{1}{\sqrt{N!}} \langle 0 | \Phi(x_1) \cdots \Phi(x_N) | \Psi \rangle \tag{11.71}$$

Rather than give a general proof of this result, we demonstrate it for the special case of two particles where for example, one has the Slater determinant; that is,

$$\psi = \frac{1}{\sqrt{2!}} [w_1(x_1) w_2(x_2) - w_2(x_1) w_1(x_2)] \tag{11.72}$$

The corresponding Fock-space vector is $|\Psi\rangle = b_2^\dagger b_1^\dagger |0\rangle$. Thus, from (11.71), we get

$$\frac{1}{\sqrt{2!}} \langle 0 | \Phi(x_1) \Phi(x_2) | \Psi \rangle = \frac{1}{\sqrt{2!}} \sum_{\alpha,\beta} w_\alpha(x_1) w_\beta(x_2) \langle 0 | b_\alpha b_\beta b_2^\dagger b_1^\dagger | 0 \rangle \tag{11.73}$$

However,

$$\begin{aligned}\langle 0|b_\alpha b_\beta b_2^\dagger b_1^\dagger|0\rangle &= \langle 0|b_\alpha\left(\delta_{\beta 2} - b_2^\dagger b_\beta\right)b_1^\dagger|0\rangle \\ &= \langle 0|\delta_{\beta 2}\left(\delta_{\alpha 1} - b_1^\dagger b_\alpha\right) - \left(\delta_{\alpha 2} - b_2^\dagger b_\alpha\right)\left(\delta_{\beta 1} - b_1^\dagger b_\beta\right)|0\rangle \\ &= \delta_{\alpha 1}\delta_{\beta 2} - \delta_{\alpha 2}\delta_{\beta 1}\end{aligned}$$

Substituting the last expression into (11.73), we recover (11.72). The general proof of (11.71) proceeds in a similar way.

Next we consider the matrix element $\langle F\rangle = \int \psi^\dagger F \psi \, dx_1 \cdots dx_N$ of the operator $F = \sum_{i=1}^{N} f_i$, where the f_i are one-body operators. We have already seen in (11.53) that $\langle F\rangle = \sum_{i=1}^{N} \langle i|f|i\rangle$. We now show that in second-quantization form

$$\langle F\rangle = \langle \Psi|\int \Phi^\dagger(x) f(x)\Phi(x)\, dx|\Psi\rangle \tag{11.74}$$

To this end, we write (11.74) as

$$\langle F\rangle = \sum_{\alpha,\beta} \langle \Psi|b_\beta^\dagger b_\alpha|\Psi\rangle \int w_\beta^\dagger(x) f(x) w_\alpha(x)\, dx$$

The vectors $b_\alpha|\Psi\rangle$ and $b_\beta|\Psi\rangle$ are null vectors or mutually orthogonal unless $\alpha = \beta = i$, where i corresponds to an occupied orbital; hence (11.74) and (11.53) are equivalent. An important example of (11.74) is the total energy E of N noninteracting fermions; that is,

$$E = \langle \Psi|\int \Phi^\dagger(x) H_0(x)\Phi(x)\, dx|\Psi\rangle \tag{11.75}$$

where H_0 is the single-particle Hamiltonian as in (11.59).

Next, suppose that there are pairwise interactions between the N fermions, and we are concerned with the matrix element (11.58); that is,

$$\left\langle \sum_{i>j} g_{ij}\right\rangle = \sum_{i<j}\left[\langle ij|g|ij\rangle - \langle ij|g|ji\rangle\right] = \int \psi^\dagger \sum_{i>j} g_{ij}\psi \, dx_1 \cdots dx_N$$

We now show that this can be expressed in second-quantization form as

$$\left\langle \sum_{i>j} g_{ij}\right\rangle = \frac{1}{2}\langle \Psi|\iint \Phi^\dagger(x)\Phi^\dagger(x')g(x,x')\Phi(x')\Phi(x)\, d^3x\, d^3x'|\Psi\rangle \tag{11.76}$$

We start with

$$\begin{aligned}
\Phi(x')\Phi(x) &= \sum_{\alpha=0}^{\infty}\sum_{\beta=0}^{\infty} w_\alpha(x')w_\beta(x)b_\alpha b_\beta \\
&= \sum_{\alpha=0}^{\infty}\sum_{\beta=0}^{\alpha-1} w_\alpha(x')w_\beta(x)b_\alpha b_\beta + \sum_{\alpha=0}^{\infty}\sum_{\beta=\alpha+1}^{\infty} w_\alpha(x')w_\beta(x)b_\alpha b_\beta \\
&= \sum_{\alpha=0}^{\infty}\sum_{\beta=0}^{\alpha-1} w_\alpha(x')w_\beta(x)b_\alpha b_\beta - \sum_{\alpha=0}^{\infty}\sum_{\beta=\alpha+1}^{\infty} w_\alpha(x')w_\beta(x)b_\beta b_\alpha \\
&= \sum_{\alpha=0}^{\infty}\sum_{\beta=0}^{\alpha-1} w_\alpha(x')w_\beta(x)b_\alpha b_\beta - \sum_{\beta=0}^{\infty}\sum_{\alpha=0}^{\beta-1} w_\alpha(x')w_\beta(x)b_\beta b_\alpha \\
&= \sum_{\alpha=0}^{\infty}\sum_{\beta=0}^{\alpha-1}\left[w_\alpha(x')w_\beta(x) - w_\beta(x')w_\alpha(x)\right]b_\alpha b_\beta
\end{aligned} \tag{11.77}$$

Here we use $\{b_\alpha, b_\beta\} = 0$ in the second line for $\alpha = \beta$ and in the third line for $\alpha \neq \beta$, we change the order of summation over α, β for the second term in the fourth line, and we interchange the dummy variables α, β for the second term in the fifth line. Similarly,

$$\Phi^\dagger(x)\Phi^\dagger(x') = \sum_{\rho=0}^{\sigma-1}\sum_{\sigma=0}^{\infty}\left[w_\rho^\dagger(x)w_\sigma^\dagger(x') - w_\sigma^\dagger(x)w_\rho^\dagger(x')\right]b_\rho^\dagger b_\sigma^\dagger$$

Hence

$$\begin{aligned}
&\frac{1}{2}\left\langle\Psi\middle|\iint dx\,dx'\Phi^\dagger(x)\Phi^\dagger(x')g(x,x')\Phi(x')\Phi(x)\middle|\Psi\right\rangle \\
&= \frac{1}{2}\sum_{\alpha=0}^{\infty}\sum_{\beta=0}^{\alpha-1}\sum_{\rho=0}^{\sigma-1}\sum_{\sigma=0}^{\infty}\left\langle\Psi\middle|b_\rho^\dagger b_\sigma^\dagger b_\alpha b_\beta\middle|\Psi\right\rangle \\
&\times\iint dx\,dx'\left[w_\rho^\dagger(x)w_\sigma^\dagger(x') - w_\sigma^\dagger(x)w_\rho^\dagger(x')\right]g(x,x')\left[w_\alpha(x')w_\beta(x) - w_\beta(x')w_\alpha(x)\right]
\end{aligned} \tag{11.78}$$

Because of the restrictions on the sums and the orthogonality of the number eigenstates, only terms with $\sigma = \alpha$ and $\rho = \beta$ contribute to (11.78). Thus (11.76) is demonstrated. The Coulomb interaction between pairs of electrons, where $g(\mathbf{x} - \mathbf{x}') = e^2/(4\pi|\mathbf{x} - \mathbf{x}'|)$, is an instance where (11.76) can be used. The second-quantization formalism is applied to an important physical problem involving this interaction in Section 14.2.

11.10 Generalizations of exchange symmetrization and antisymmetrization

11.10.1 Isospin

There are many situations in elementary particle and nuclear physics where certain particles are so similar that they can usefully be treated as identical, even though they are not. For example, experiment shows that nuclear forces are charge independent; that is, the strong force between two neutrons of given relative momentum and spin orientation is the same as between a proton and a neutron or (if we neglect the Coulomb interaction) between two protons. Also, the proton and neutron differ in mass by hardly more than one-tenth of 1 percent. Therefore, in many circumstances, it is a good approximation to treat the proton and neutron as two different charge states of the same particle, the nucleon. The situation is analogous to the two distinct spin states of a particle of spin-½; that is,

$$\alpha = \begin{pmatrix} 1 \\ 0 \end{pmatrix} \text{ for spin up} \qquad \beta = \begin{pmatrix} 0 \\ 1 \end{pmatrix} \text{ for spin down}$$

To pursue the analogy, the concept of *isospin* has been invented, in which we form the following two-component isospinors in "charge space":

$$p = \begin{pmatrix} 1 \\ 0 \end{pmatrix} \qquad n = \begin{pmatrix} 0 \\ 1 \end{pmatrix}$$

Just as the angular-momentum spinors α, β can be transformed in various ways with the aid of the spin operators

$$\begin{aligned} S_z &= \sigma_z / 2 \\ S_+ &= \frac{1}{2}(\sigma_x + i\sigma_y) \\ S_- &= \frac{1}{2}(\sigma_x - i\sigma_y) \end{aligned}$$

so analogous isospin operators can be defined for transforming the isospinors p, n; that is,

$$\begin{aligned} t_z &= \tau_z / 2 \\ t_+ &= \frac{1}{2}(\tau_1 + i\tau_2) \\ t_- &= \frac{1}{2}(\tau_1 - i\tau_2) \end{aligned}$$

where τ_i is a 2×2 Pauli isospin matrix. Just as in the theory of angular momentum, we can build up a state with any nonnegative integral or half-integral value of J by combining states of spin-½ in various ways, so we can combine states of isospin-½ (nucleon states) in various ways to construct states with any nonnegative integral or half-integral value of isospin T (nuclear states).

If two or more nucleons are treated as identical particles, the wave function describing them in a compound nucleus must refer not only to space and spin variables but also to isospin, and the total wave function must be antisymmetric with respect to exchange. For example, consider the ^{3}He nucleus, which consists of two protons and one neutron (*ppn*). In a reasonably good approximation (the nuclear shell model), we may treat the motion of each nucleon as if it occurred in an effective central potential. Here the orbital angular momentum ℓ_i of each nucleon is a good quantum number. In fact, for the ground nuclear state of ^{3}He, it is known that $\ell_i = 0$ for each nucleon. Thus the nuclear spin I is entirely due to the spins of the constituent nucleons, and for ^{3}He it is $I = 1/2$. There are two independent ways to construct a state with $I = 1/2$, $m_I = 1/2$ from three nucleons with spin-½:

$$\chi_A = \frac{1}{\sqrt{2}}(\alpha_1\beta_2 - \alpha_2\beta_1)\alpha_3 \tag{11.79}$$

$$\chi_S = \sqrt{\frac{1}{3}}\left[\frac{1}{\sqrt{2}}(\alpha_1\beta_2 + \alpha_2\beta_1)\alpha_3\right] - \sqrt{\frac{2}{3}}\alpha_1\alpha_2\beta_3 \tag{11.80}$$

In (11.79), we form the $j = 0$ state of the first two nucleon spins and couple it to the third nucleon spin. This state is antisymmetric with respect to exchange of the first two spins. In (11.80), we form the $j = 1$, $m_j = 0,1$ combinations of the first two spins and couple them to α_3, β_3, respectively, as in (7.110). The state χ_S is symmetric with respect to exchange of the first two spins.

Now ^{3}He and ^{3}H, the latter of which consists of a proton and two neutrons, form an isospin doublet ($T = 1/2$) with $m_T = \pm 1/2$, respectively. In exact analogy to (11.79) and (11.80), there are two independent ways to construct a $T = 1/2$, $m_T = 1/2$ isospin state from three nucleons:

$$\eta_A = \frac{1}{\sqrt{2}}(p_1 n_2 - p_2 n_1)p_3 \tag{11.81}$$

$$\eta_S = \sqrt{\frac{1}{6}}(p_1 n_2 + p_2 n_1)p_3 - \sqrt{\frac{2}{3}}p_1 p_2 n_3 \tag{11.82}$$

The spin-isospin function for ^{3}He, which must be completely antisymmetric with respect to exchange, is easily constructed from (11.79) through (11.82). It is

$$\begin{aligned}\psi(^3\text{He}, m_I = 1/2) &= \frac{1}{\sqrt{2}}(\chi_S\eta_A - \chi_A\eta_S)\\ &= \frac{1}{\sqrt{6}}\left[p_1p_2n_3(\alpha_1\beta_2 - \alpha_2\beta_1)\alpha_3 + p_1n_2p_3(\alpha_3\beta_1 - \alpha_1\beta_3)\alpha_2 + n_1p_2p_3(\alpha_2\beta_3 - \alpha_3\beta_2)\alpha_1\right]\end{aligned} \tag{11.83}$$

A similar formula gives the spin-isospin function of ^{3}H. Now ^{3}H is beta-radioactive; that is,

$$^3\text{H} \rightarrow {}^3\text{He} + e^- + \bar{\nu}_e$$

To work out all the details of this beta transition theoretically, one needs (11.83) and its analogues for ^{3}H and $m_I = -1/2$.

11.10.1 Fermion-Antifermion States

The positronium atom provides an entirely different example. It consists of an electron and a positron in a bound state. The electron and positron can be treated as identical particles in states of equal and opposite electric charge. Hence the positronium wave function can be written as

$$\psi = \psi_{\text{space}} \psi_{\text{spin}} \psi_{\text{charge}}$$

Here ψ_{space} is the spatial wave function of relative motion of e^+ and e^-; it is symmetric with respect to exchange of these two particles if the orbital quantum number ℓ is even and antisymmetric if ℓ is odd. The spin function is symmetric if the total spin is $S = 1$ and antisymmetric if $S = 0$. The total wave function must be antisymmetric with respect to exchange; hence ψ_{charge} is symmetric if $\ell + S$ is even and antisymmetric if $\ell + S$ is odd. One usually says that the *charge parity* of positronium is $(-1)^{\ell+S}$. The ground state of positronium consists of two separate fine-structure components, ${}^1S_0$ with charge parity +1 and ${}^3S_1$ with charge parity −1. A positronium atom decays when the electron and positron annihilate to form two or more photons. In this electromagnetic interaction, charge parity is conserved. Also, it can be shown that the charge parity of a system of n photons is $(-1)^n$. Thus an ${}^1S_0$ positronium atom can only decay to an even number of photons (2, 4, …), whereas an ${}^3S_1$ positronium can only decay to an odd number of photons (3, 5, …).

11.10.2 Isospin of Pi Mesons

The pi mesons π^+, π^0, and π^- have zero spin and nearly equal masses [$m(\pi^\pm) = 139.6$ MeV/c^2; $m(\pi^0) = 135$ MeV/c^2]. The relatively small mass difference is thought to be due to electromagnetic interaction, in the absence of which the pions would form a perfect isospin triplet. In that approximation, the total wave function of two pions can be written as

$$\psi = \psi_{\text{space}} \psi_{\text{charge}}$$

and because pions are bosons, ψ must be symmetric with respect to exchange. As before, the spatial wave function is symmetric (antisymmetric) for even (odd) ℓ, respectively. Therefore, so is ψ_{charge}. Because each pion has isospin $T = 1$, the isospin of two pions can be $T = 2$, 1, or 0 a priori. However, $T = 2$ and $T = 0$ are symmetric with respect to exchange, and $T = 1$ is antisymmetric, just as in the theory of angular momentum. Hence a two-pion state with even ℓ must be a linear combination of $T = 2, 0$, whereas if ℓ is odd, $T = 1$.

Problems for Chapter 11

11.1. In Section 11.6 we calculate the ground-state energy of helium-like atoms by using the variational method with the simple trial function

$$\psi(r_1, r_2) = \frac{\lambda^3}{\pi} \exp\left[-\lambda(r_1 + r_2)\right] \tag{1}$$

This calculation can be improved substantially by using instead the following trial function with parameters s, λ:

$$\psi(\mathbf{r}_1,\mathbf{r}_2) = N\exp\left[-\lambda\left(sr_> + r_<\right)\right] \tag{2}$$

where $r_> = r_1$ if $r_1 > r_2$ and $r_> = r_2$ if $r_2 > r_1$. Note that

$$sr_> + r_< = \frac{s+1}{2}(r_1 + r_2) + \frac{s-1}{2}|r_1 - r_2|$$

(a) Show that N in (2) is given by

$$N = \sqrt{\frac{\lambda^6 s^3 (s+1)^5}{2\pi^2 (10s^2 + 5s + 1)}} \tag{3}$$

(b) Show that the expectation value of the kinetic energy of the two electrons is

$$\langle T_1\rangle + \langle T_2\rangle = \frac{s^2+1}{2}\lambda^2 \tag{4}$$

(c) Show that the expectation value of the potential energy of interaction of the electrons with the nucleus is

$$-Z\left\langle \frac{1}{r_1} + \frac{1}{r_2}\right\rangle = -Z\lambda(s+1) \tag{5}$$

(d) Show that the expectation value of the electron-electron interaction energy is

$$\left\langle \frac{1}{r_{12}}\right\rangle = \frac{s(s+1)(4s+1)}{10s^2 + 5s + 1}\lambda \tag{6}$$

To minimize the tedium of routine calculation for obtaining results (3) through (6), you may use the following relevant integrals:

$$\int_0^\infty r_1^2 \exp(-2\lambda s r_1)\, dr_1 \int_0^{r_1} r_2^2 \exp(-2\lambda r_2)\, dr_2 = \frac{1}{16\lambda^6}\left[\frac{1+5s+10s^2}{s^3(1+s)^5}\right] \tag{7}$$

$$\int_0^\infty r_1 \exp(-2\lambda s r_1)\, dr_1 \int_0^{r_1} r_2^2 \exp(-2\lambda r_2)\, dr_2 = \frac{1}{16\lambda^5}\frac{1+4s}{s^2(1+s)^4} \tag{8}$$

$$\int_0^\infty r_1^2 \exp(-2\lambda s r_1)\, dr_1 \int_0^{r_1} r_2 \exp(-2\lambda r_2)\, dr_2 = \frac{1}{16\lambda^5}\frac{1+4s+6s^2}{s^3(1+s)^4} \tag{9}$$

(e) From (4) through (6), the total energy as a function of λ and s is

$$E(\lambda,s)=\frac{s^2+1}{2}\lambda^2-Z\lambda(s+1)+\frac{s(s+1)(4s+1)}{10s^2+5s+1}\lambda \tag{10}$$

Minimize this function with respect to λ to show that the resulting value of λ is

$$\lambda_0=\frac{(s+1)\left[(10Z-4)s^2+(5Z-1)s+Z\right]}{(s^2+1)(10s^2+5s+1)} \tag{11}$$

and

$$E(\lambda_0,s)=-\frac{(s+1)^2\left[(10Z-4)s^2+(5Z-1)s+Z\right]^2}{2(s^2+1)(10s^2+5s+1)^2} \tag{12}$$

For Z = 1 (the hydrogen negative ion) and for Z = 2 (the helium atom), plot or tabulate $E(\lambda_0,s)$ versus s to find the minimum $E(\lambda_0,s_0)$ of $E(\lambda_0,s)$. You should find

$$\begin{array}{llll} Z=1: & \lambda_0=0.9144 & s_0=0.4598 & E(\lambda_0,s_0)=-0.506 \\ Z=2: & \lambda_0=1.85623 & s_0=0.817 & E(\lambda_0,s_0)=-2.87273 \end{array}$$

The Z = 1 result for E is significant: because it is less than –1/2, it shows that the H^- ion has a bound state. The most sophisticated variational calculations of H^- yield E = –0.528. For Z = 2, the value achieved for E is much closer to E_{obs} = –2.90365 than E = –2.8477 calculated with the simplest variational wave function given in equation (1).

11.2. The configurations of virtually all bound states of the helium atom are of the form $(1s\ n\ell)$. According to first-order perturbation theory, the energy difference between singlet and triplet terms of a given configuration is $\Delta E=2K$, where K is the exchange integral. Using hydrogenic orbitals with Z = 2 for 1*s* and Z = 1 for *np*, calculate the exchange integrals for the 1*s*2*p* and 1*s*3*p* configurations of atomic helium, and thus estimate the energy splitting between the ^1P and ^3P states for these configurations. Compare with the experimental values of these splittings, which in atomic units are

$$\Delta E(1s2p)_{\text{expt}}=0.00933$$
$$\Delta E(1s3p)_{\text{expt}}=0.00294$$

Explain why the agreement between calculation and experiment is better for $1s3p$ than for $1s2p$.

11.3. We have noted that the operator associated with any observable must be symmetric with respect to exchange of the coordinates of the particles it affects. This implies that for any observable A, all matrix elements of the form $\langle\Psi_S|A|\Psi_A\rangle$ must vanish, where $\Psi_{S,A}$ are symmetric (antisymmetric) wave functions, respectively. In Chapter 7 we introduced the concept of an irreducible spherical tensor operator T_L^M. Why must the rank L of any T_L^M always be integral and never half-integral?

11.4. (a) Equation (11.83) describes the wave function of the ^{3}He nuclear ground state with $m_I=1/2$. Write the analogous equation for the nuclear ground state of ^{3}H.

(b) One requires matrix elements of the operators $\sum_i t_{i+}$ and $\sqrt{3}\sum_1 \sigma_{iz} t_{i+}$ between ^{3}H and ^{3}He nuclear states to calculate the *vector* and *axial vector* amplitudes, respectively, for the beta decay $^3\mathrm{H} \rightarrow {}^3\mathrm{He} + e^- + \bar{\nu}_e$. Here the sums are over all nucleons in the nucleus. Using (11.83) and its analogue for ^{3}H, calculate these matrix elements.

(c) Consider the nuclear beta decay ${}^{19}_{10}\mathrm{Ne} \rightarrow {}^{19}_{9}\mathrm{F} + e^+ + \nu_e$. To a good approximation, the initial nucleus ^{19}Ne $(I = 1/2, T = 1/2, T_z = 1/2)$ consists of two protons and one neutron, each with zero orbital angular momentum, outside an inert ^{16}O core. When this nucleus decays to ^{19}F $(I = 1/2, T = 1/2, T_z = -1/2)$, one of the "valence" protons transforms into a neutron. To a good approximation, the final ^{19}F consists of two neutrons and a proton, each with zero orbital angular momentum, outside the ^{16}O core. Calculate the matrix elements of the operators $\sum_i t_{i-}$ and $\sqrt{3}\sum_i \sigma_{iz} t_{i-}$ for this beta decay, assuming that $m_I = 1/2$ for initial and final nuclei. Also estimate the nuclear magnetic moments of ^{19}Ne and ^{19}F

11.5. Consider three pions in a state with angular momentum $J = 0$. This state can be constructed from the orbital angular momentum $\vec{\ell}$ of relative motion of the first two pions and the orbital angular momentum $\boldsymbol{L}$ of the third pion about the center of mass of the three-pion system. Find the possible isospin multiplets of the three-pion system, and show that even values of $\ell = L$ must correspond to odd values of total isospin T, whereas odd values of $\ell = L$ must correspond to even values of T.

11.6. In a preliminary way by 1909 and more accurately by 1938 it was established that the ratio of charge to mass for beta rays and for ordinary atomic electrons is the same. Also, the charges of beta rays and atomic electrons were shown to be the same by 1940. Despite this, it was possible for a time to maintain the idea that beta rays and electrons are not identical particles but only very similar ones, until some decisive (and not very well-known) experiments were done in the period 1948–1950. Try to construct the principle of such an experiment using your own knowledge and intuition, and only then refer to Goldhaber (1948) and Davies (1951).

12 Atomic Structure

12.1 Central field approximation: General remarks

In principle, the many-particle Schroedinger equation with Coulomb interactions and spin should provide a good description of atomic structure. However, whereas exact solutions are known for hydrogenic atoms (one electron), and there exist excellent special approximation methods for two-electron systems (helium and helium-like ions), the situation is quite different for three or more electrons. Here we must resort to the *central field approximation* (CFA), where it is assumed that each electron in an N-electron atom moves in an effective central potential $\bar{V}$ that arises from the nuclear charge and a suitable effective charge distribution due to the other $N-1$ electrons.

The CFA and the antisymmetrization postulate are the two cornerstones of the theory of atomic structure, which gives a remarkably accurate quantitative account of thousands of precise experimental results and is one of the most important triumphs of quantum mechanics. Our discussion of the CFA is as follows: in this section we confine ourselves to an overall survey and qualitative remarks. Then we briefly sketch how the CFA is carried out in practice by means of the Hartree, Hartree-Fock, Thomas-Fermi, and related methods. Finally, we consider the principal, and very important, corrections to the approximation.

If each electron in an N-electron atom moves in an effective central potential, then taking into account the exclusion principle, the N-electron wave function may be written as a Slater determinant[1]; that is,

$$\psi = \frac{1}{\sqrt{N!}} \begin{vmatrix} \phi_{n_1\ell_1 m_{\ell 1}}(\boldsymbol{r}_1)\chi_{11} & \cdots & \cdots & \phi_{n_1\ell_1 m_{\ell 1}}(\boldsymbol{r}_N)\chi_{1N} \\ \cdots & \cdots & \cdots & \cdots \\ \cdots & \cdots & \cdots & \cdots \\ \phi_{n_N\ell_N m_{\ell N}}(\boldsymbol{r}_1)\chi_{N1} & \cdots & \cdots & \phi_{n_N\ell_N m_{\ell N}}(\boldsymbol{r}_N)\chi_{NN} \end{vmatrix} \tag{12.1}$$

Because $\bar{V}$ is central, the spatial part ϕ of each one-electron orbital can be expressed as a product of a radial function and a spherical harmonic

$$\phi_{n_i\ell_i m_{\ell i}}(\boldsymbol{r}_j) = R_{n_i\ell_i}(r_j)\, Y_{\ell_i}^{m_{\ell i}}(\hat{r}_j) \tag{12.2}$$

[1] Strictly speaking, in the CFA, one does not require the same central potential for all electrons but merely for all electrons in the same shell.

Table 12.1 Ground-state configurations of the first transition elements

Z	Element	Ground-state configuration
21	Sc	$1s^2 \ldots 4s^2 3d$
22	Ti	$1s^2 \ldots 4s^2 3d^2$
23	V	$1s^2 \ldots 4s^2 3d^3$
24	Cr	$1s^2 \ldots 4s 3d^5$
25	Mn	$1s^2 \ldots 4s^2 3d^5$
26	Fe	$1s^2 \ldots 4s^2 3d^6$
27	Co	$1s^2 \ldots 4s^2 3d^7$
28	Ni	$1s^2 \ldots 4s^2 3d^8$
29	Cu	$1s^2 \ldots 4s 3d^{10}$
30	Zn	$1s^2 \ldots 4s^2 3d^{10}$

The principal quantum number n_i is defined so that the number of radial nodes in R for finite r_j is $n_i - \ell_i - 1$, as in hydrogen. For very small r_j,

$$\bar{V}(r_j) \to -\frac{Z}{r_j} \tag{12.3}$$

whereas for very large r_j,

$$\bar{V}(r_j) \to -\frac{(Z - N + 1)}{r_j} \tag{12.4}$$

because a given electron is exposed to the full nuclear charge when it is close to the nucleus, but the nuclear charge is screened by the remaining $N - 1$ electrons when the given electron is far from the origin. Thus, as r_j increases, $|\bar{V}|$ falls to zero more rapidly than the Coulomb potential arising from the unscreened nuclear charge. Hence there is no accidental degeneracy, and single-electron orbitals with the same n have energies that increase with ℓ.

Even from this very elementary consideration, we can gain a rudimentary understanding of the ground-state electronic configurations of atoms (the periodic table). Adding electrons one by one, we first fill the $1s$ shell with two electrons, then the $2s$ shell with two more, then the $2p$ shell with six, then the $3s$ shell with two more, and then the $3p$ shell with six more. Thus we arrive at argon ($Z = 18$). Two additional electrons go into the $4s$ shell to yield potassium ($Z = 19$) and then calcium ($Z = 20$). However, beyond this point, experiment shows and CFA calculations predict that less energy is required to fill the $3d$ shell than the $4p$ shell. In fact, there is a competition between $4s$ and $3d$, as can be seen in Table 12.1 from the ground-state configurations of the first transition group of elements. Notice in particular that in Cr and Cu there is only one $4s$ electron, whereas the other elements in this group have a complete $4s$ shell.

Starting with $Z = 31$, we fill the $4p$ shell with six electrons and then the $5s$ shell with two more. At this point we again encounter competition between ns and $(n - 1)d$ shells in the second transition group, followed by np. To summarize, the shells are filled in the following order, with $\ell \leq 3$ in all cases:

1s, 2s, 2p, 3s, 3p, (4s, 3d), 4p, (5s, 4d), 5p, (6s, 4f, 5d), 6p, (7s, 5f, [6d])
It is well known that the chemical properties of a neutral atom depend on the electrons in the outer shell(s); elements with similar outer shells are similar physically and chemically. They also have similar patterns of excited-state energy levels. Here are just a few examples:

- The noble gases with complete p shells are very similar:

$$\mathrm{Ne}(2p^6),\ \mathrm{Ar}(3p^6),\ \mathrm{Kr}(4p^6),\ \mathrm{Xe}(5p^6),\ \mathrm{Ra}(6p^6)$$

  These also resemble helium ($1s^2$), which has no p electrons but only a complete 1s shell.

- The alkali metals with a single $(n + 1)s$ electron after a completed np shell are very similar:

$$\mathrm{Na}(2p^63s),\ \mathrm{K}(3p^64s),\ \mathrm{Rb}(4p^65s),\ \mathrm{Cs}(5p^66s),\ \mathrm{Fr}(6p^67s)$$

  These also resemble Li($2s^23s$) and even in some respects H(1s).
- The halogens each lack one electron to complete a p shell:

$$\mathrm{F}(2p^5),\ \mathrm{Cl}(3p^5),\ \mathrm{Br}(4p^5),\ \mathrm{I}(5p^5),\ \mathrm{At}(6p^5)$$

- Group II elements

$$\mathrm{Be}(1s^22s^2) \quad \text{and} \quad \mathrm{Mg}(1s^2 \ldots 3s^2)$$

  are very similar. The next electron would be np. Ca(… $4s^2$), Sr(… $5s^2$), and Ba(… $6s^2$) are very similar. The next electron would be $(n-1)d$ or, in the case of Ba, $(n-2)f$. Zn(… $3d^{10}4s^2$), Cd(… $4d^{10}5s^2$), and Hg(… $5d^{10}6s^2$) are very similar.

- The noble metals Cu, Ag, and Au all have the structure [$\ldots(n-1)d^{10}ns$] and are similar.
- The metals with one p electron outside a completed d shell are similar:

$$\mathrm{Ga}(\ldots 3d^{10}4p),\ \mathrm{In}(\ldots 4d^{10}5p),\ \mathrm{Tl}(\ldots 5d^{10}6p)$$

12.2 Hartree's self-consistent field method

We now turn to actual implementation of the CFA and start with the method developed by D. R. Hartree (1928). Here the many-electron wave function is assumed to be

$$\psi = \phi_1(r_1)\cdots\phi_N(r_N) \tag{12.5}$$

Thus exchange antisymmetry is ignored, and the Pauli principle is taken into account merely by restricting the number of electrons that can have a spatial orbital associated with a given shell (two in 1s, two in 2s, six in 2p, etc). A Hartree calculation starts with an assumed set of zero-order spatial orbitals $\phi_i^{(0)}(\boldsymbol{r}_i)$, $i = 1, \ldots, N$. These are obtained from intelligent guesswork or from the Thomas-Fermi model (described later). The following potential is computed from these orbitals:

$$V_i = -\frac{Z}{r_i} + \sum_{k \neq i} \int \frac{\left|\phi_k^{(0)}(\boldsymbol{r}_k)\right|^2}{r_{ik}} d\tau_k \qquad i = 1, \ldots, N \tag{12.6}$$

Here the first term on the right-hand side is obviously the potential energy of interaction of the ith electron with the nucleus, whereas the sum represents an electrostatic interaction of the ith electron with the remaining $N - 1$ electrons, where the charge density of the kth electron is assumed to be $-\left|\phi_k^{(0)}(\boldsymbol{r}_k)\right|^2$ in atomic units. The next step is to replace V_i by its average over angular coordinates; that is,

$$V_i \rightarrow \bar{V}_i = \frac{1}{4\pi} \int V_i \, d\Omega_i \tag{12.7}$$

This is where the CFA is made. We now have a set of N differential equations for the next approximation to the ϕ_i; that is,

$$-\frac{1}{2} \nabla_i^2 \phi_i^{(1)} + \bar{V}_i \phi_i^{(1)} = \varepsilon_i^{(1)} \phi_i^{(1)} \qquad i = 1, \ldots, N \tag{12.8}$$

These are solved numerically to find the $\phi_i^{(1)}$ and their corresponding eigenvalues $\varepsilon_i^{(1)}$. The entire process is then repeated with as many iterations as necessary to yield a potential, orbitals, and eigenvalues that are self-consistent with the desired precision. To explain the meaning of the eigenvalues ε_i, we multiply (12.8) on the left by ϕ_i^* and integrate over τ_i using the normalization condition

$$\int |\phi_i|^2 \, d\tau_i = 1$$

This yields

$$T_i + V_{i,\text{nuc}} + \overline{\sum_{k \neq i} \iint \frac{|\phi_k|^2 |\phi_i|^2}{r_{ik}} \, d\tau_i \, d\tau_k} = \varepsilon_i \tag{12.9}$$

Here T_i is the kinetic energy of the ith electron, and the bar indicates that a spherically symmetric average has been performed, as described earlier. Thus $-\varepsilon_i$ is approximately the energy required to remove the ith electron from the atom in question to infinity. (It is not exactly the removal energy because the self-consistent field of the resulting ion is slightly different from that of the original atom.) Summing (12.9) over all N electrons, we get

$$\sum_i \varepsilon_i = T + V_{\text{nuc}} + 2\overline{\sum_{\text{pairs}} \iint \frac{|\phi_k|^2 |\phi_i|^2}{r_{ik}} \, d\tau_i \, d\tau_k} \tag{12.10}$$

Meanwhile, the total electron energy in the Hartree central field approximation is

$$E = T + V_{\text{nuc}} + \sum_{\text{pairs}} \iint \frac{|\phi_k|^2 |\phi_i|^2}{r_{ik}} d\tau_i \, d\tau_k \tag{12.11}$$

Notice the factor of 2 that is present in (12.10): it means that we have counted the energy of the electron-electron Coulomb interaction twice in summing over all eigenvalues ε_i.

The simplest application of the Hartree method is to the ground state of helium. Here we only need a single Hartree equation

$$-\frac{1}{2}\nabla^2\phi(r) - \frac{Z}{r}\phi(r) + \left[\int \frac{|\phi(r')|}{|\boldsymbol{r}-\boldsymbol{r}'|} d^3\boldsymbol{r}'\right]\phi(r) = \varepsilon\phi(r) \tag{12.12}$$

because both spatial orbitals pertain to the $1s$ shell, and each may be written as $\phi = \phi_{1s}$. Also, in the present case, V is already spherically symmetric, and there is no need to make an angular average.

When (12.12) is solved numerically by iteration, we find $\langle H \rangle = -2.86168$ atomic units, which is somewhat better than the result obtained by the simplest variational method [see equation (11.28)] but not as good as that from the most sophisticated variational treatments. The main virtue of Hartree's method for the helium ground state is not that it gives a good value of the energy but rather that it gives quite a good wave function.

12.3 Hartree-Fock method

Although Hartree's equations were originally obtained intuitively, they can be derived using the variational method. A substantial improvement is made when, in addition, determinantal wave functions are employed. This is the *Hartree-Fock method*, actually developed by V. Fock (1930) and J. Slater (1930). We start with the Hamiltonian

$$H = \sum_i \left(-\frac{1}{2}\nabla_i^2 - \frac{Z}{r_i} \right) + \sum_{i<j} \frac{1}{r_{ij}}$$

Here the first sum on the right-hand side is a sum of one-body operators $\sum_i f_i$, whereas the second sum is a sum of two-body operators $\sum_{i<j} g_{ij}$. Using the determinantal wave function

$$\psi = \frac{1}{\sqrt{N!}} \begin{vmatrix} w_1(\boldsymbol{x}_1) & \dots & w_1(\boldsymbol{x}_N) \\ \dots & \dots & \dots \\ w_N(\boldsymbol{x}_1) & \dots & w_N(\boldsymbol{x}_N) \end{vmatrix}$$

where $\boldsymbol{x}_i$ denotes the space-spin coordinates of the ith electron, we construct $\langle \psi | H | \psi \rangle$ with the aid of equations (11.53) and (11.58); that is,

$$E = \langle \psi | H | \psi \rangle = \sum_i \langle i | f | i \rangle + \sum_{i<j} \left[\langle ij | g | ij \rangle - \langle ij | g | ji \rangle \right]$$

This is an energy functional that depends on the values of the orbitals $w(\boldsymbol{x})$. We vary each of these orbitals and thus vary E subject to the constraints

$$\int w_i^\dagger (\boldsymbol{x}) w_j (\boldsymbol{x})\, d\boldsymbol{x} = \delta_{ij}$$

These constraints are incorporated into the calculation in the usual way by employing a set of Lagrange multipliers. By demanding that E reach an extreme value, which can be shown to be a minimum, we thus obtain a set of N Hartree-Fock equations

$$\begin{aligned} &-\frac{1}{2}\nabla_1^2 w_i(\boldsymbol{r}_1) - \frac{Z}{r_1} w_i(\boldsymbol{r}_1) + \left[\sum_j \int d\tau_2 \frac{1}{r_{12}} |w_j(\boldsymbol{r}_2)|^2 \right] w_i(\boldsymbol{r}_1) \\ &- \sum_j \delta(m_{si}, m_{sj}) \left[\int d\tau_2 \frac{1}{r_{12}} w_j{}^*(\boldsymbol{r}_2) w_i(\boldsymbol{r}_2) \right] w_j(\boldsymbol{r}_1) = \varepsilon_i w_i(\boldsymbol{r}_1) \end{aligned} \tag{12.13}$$

These equations differ from the corresponding Hartree equations because they contain an additional term on the left-hand side; that is,

$$-\sum_{j \neq i} \delta(m_{si}, m_{sj}) \left[\int d\tau_2 \frac{1}{r_{12}} w_j{}^*(\boldsymbol{r}_2) w_i(\boldsymbol{r}_2) \right] w_j(\boldsymbol{r}_1)$$

Note in particular how the indices i, j appear in this last expression. In the Hartree-Fock method, the effective potential energy of a given electron is generated by interaction with

- The nucleus;
- All the electrons having spin opposite to that of the given electron; and
- A charge distribution of electrons having the same spin as the given electron.

This last charge distribution adds up to one less than the total number of electrons in this spin state. Effectively, it is as if the given electron carried a "hole" with it. The potential energy in the Hartree-Fock method is systematically lower than that in the Hartree method because of exchange.

The procedure for solving the Hartree-Fock equations is similar to that for the Hartree case. One chooses an initial Slater determinant, calculates the effective potential, makes a spherically symmetric average (which for $\ell \leq 3$ is a very mild approximation), solves the resulting differential equations numerically, and thus obtains a new Slater determinant. The process is repeated until the results are self-consistent to desired precision. This method and modern variants of it employing more than one configuration and/or relativistic wave equations are very powerful, and they usually yield impressively accurate numerical results. See, for example, the excellent monograph by Johnson (2007).

12.4 Thomas-Fermi model

This is a relatively simple approach to the CFA that makes use of easily comprehended physical ideas, provides valuable intuitive insights about atomic structure, and leads directly to potentials that can be used in Hartree-Fock calculations or are even employed effectively in their own right (Thomas 1927b; Fermi 1928). The basic idea is that if the number of electrons in a given atom is sufficiently large, we can divide the atom into various volume elements, where in each volume element the electrostatic potential varies rather slowly and thus can be approximated as a constant, whereas the electrons in that volume element are sufficient in number to be in statistical equilibrium because of their Coulomb interactions. Hence they can be treated as a degenerate Fermi gas. This approximation obviously fails in the immediate vicinity of the nucleus, where the electrostatic potential varies rapidly, and it also fails at very large distances from the nucleus, where the electron density approaches zero. Nevertheless, it is remarkably effective even for atomic numbers Z as small as 10 or so.

We start with a gas of "free" electrons in the absence of a potential. (In fact, they must interact with one another to some extent so that equilibrium can be established, but we assume that this interaction is very feeble.) According to Fermi-Dirac statistics, the number of free electrons per unit volume with momentum $\boldsymbol{p}$ in range $d^3\boldsymbol{p}$ at temperature T is

$$dn = \frac{1}{\pi^2\hbar^3}\frac{p^2 dp}{\exp\left[\frac{(E-\mu)}{k_B T}\right]+1} \tag{12.14}$$

where E is the electron energy, and μ is the chemical potential. Thus the electron number density is

$$n = \frac{1}{\pi^2\hbar^3}\int_0^\infty \frac{p^2 dp}{\exp\left[\frac{(E-\mu)}{k_B T}\right]+1} \tag{12.15}$$

and the kinetic energy per unit volume (assumed here to be nonrelativistic) is

$$\varepsilon = \frac{1}{2m_e\pi^2\hbar^3}\int_0^\infty \frac{p^4 dp}{\exp\left[\frac{(E-\mu)}{k_B T}\right]+1} \tag{12.16}$$

For an atom in laboratory conditions, the electrons are highly degenerate, and the actual temperature of the surroundings is orders of magnitude less than the Fermi temperature. Thus we can make the approximation $T \to 0$, in which case the Fermi distribution

$$f = \left\{\exp\left[\frac{(E-\mu)}{k_B T}\right]+1\right\}^{-1}$$

becomes

$$f = 1 \quad \text{for } E < \mu_F$$
$$f = 0 \quad \text{for } E > \mu_F$$

and where $\mu_F = \mu(T = 0)$. Hence, at $T = 0$, we have

$$n = \frac{1}{\pi^2 \hbar^3} \int_0^{p_F} p^2 \, dp = \frac{p_F^3}{3\pi^2 \hbar^3} \tag{12.17}$$

and

$$\varepsilon = \frac{1}{2m_e \pi^2 \hbar^3} \int_0^{p_F} p^4 \, dp = \frac{p_F^5}{10 m_e \pi^2 \hbar^3} = \frac{3^{5/3} \pi^{4/3}}{10} \frac{\hbar^2}{m_e} n^{5/3} \tag{12.18}$$

where $p_F = \sqrt{2m_e \mu_F}$ is the Fermi momentum (the maximum momentum of an electron in the zero-temperature gas).

Now suppose that a given electron is no longer free but has potential energy $V(r) = -e\Phi$, where Φ, assumed to be central, is the electrostatic potential due to the nucleus and all the electrons. Then the energy of an electron is

$$E = \frac{p^2}{2m_e} - e\Phi$$

and the maximum value of this energy is

$$E_{\max} = \mu_F = \frac{p_F^2}{2m_e} - e\Phi \tag{12.19}$$

In an atom, Φ, n, and hence $p_{\rm F}$ all depend on r. However, μ_F must be independent of r for statistical equilibrium. This is so because it is a general principle of statistical mechanics that for two systems to be in equilibrium, not only their temperatures but also their chemical potentials must be equal. We can apply this principle to two successive annular shells of electrons, one between $r - dr$ and r, the other between r and $r + dr$. Employing (12.17) in (12.19), we obtain

$$\frac{\left(3\pi^2 \hbar^3 n\right)^{2/3}}{2m_e} = \mu_F + e\Phi \equiv e\Phi_0$$

or

$$n = \frac{\left(2m_e e\right)^{3/2}}{3\pi^2 \hbar^3} \Phi_0^{3/2} \tag{12.20}$$

Poisson's equation gives an additional relation between n and Φ_0; that is,

$$\nabla^2 \Phi = \nabla^2 \Phi_0 = ne - Ze\delta^3(\boldsymbol{r}) \tag{12.21}$$

On the right-hand side of (12.21), the first and second terms are due to the electron charge distribution and the nucleus, respectively. For $r > 0$, we combine (12.20) and (12.21) to obtain

$$\nabla^2 \Phi_0 = e\frac{(2m_e e)^{3/2}}{3\pi^2\hbar^3}\Phi_0^{3/2} \tag{12.22}$$

For very small r, $\Phi_0(r) \to Ze/4\pi r$. Thus it is convenient to define a new function $\phi(r)$ by

$$\Phi_0(r) = \frac{Ze}{4\pi r}\phi(r) \tag{12.23}$$

where $\phi(0) = 1$. Substitution of (12.23) into (12.22) yields

$$\frac{Ze}{r}\frac{d^2\phi}{dr^2} = 4\pi e\frac{(2m_e e)^{3/2}}{3\pi^2\hbar^3}\frac{(Ze)^{3/2}}{r^{3/2}}\phi^{3/2}$$

or

$$\frac{d^2\phi}{dr^2} = \left(\frac{2^{7/2}Z^{1/2}}{3\pi a_0^{3/2}}\right)\frac{\phi^{3/2}}{r^{1/2}} \tag{12.24}$$

This equation is simplified by writing $r = ax$, where x is a dimensionless parameter, and a is a constant with dimension of length, defined by

$$a = \frac{(3\pi)^{2/3}a_0}{2^{7/3}Z^{1/3}} = 0.885Z^{-1/3}a_0 \tag{12.25}$$

Thus (12.24) becomes

$$\frac{d^2\phi}{dx^2} = \frac{\phi^{3/2}}{x^{1/2}} \tag{12.26}$$

with the boundary condition $\phi(0) = 1$. This is the *Thomas-Fermi differential equation.* It is nonlinear, and as a practical matter, it must be solved numerically. [Not so many years ago a rather complicated analytic solution was discovered (Esposito 2002) in the papers of E. Majorana, the brilliant and reclusive Italian physicist who disappeared under mysterious circumstances in 1938. Majorana had found the solution while he was Fermi's student in the late 1920s.] Solutions are conveniently divided into three classes, as shown in Figure 12.1. Note that in all cases, $d^2\phi / dx^2 > 0$ for $\phi > 0$, so all three types of solutions curve upward.

We now show that solutions of types II and III correspond to a neutral atom and a positive ion, respectively. Suppose that an atom with N electrons that contains a nucleus with atomic number Z has a finite radius R_0 within which all the charge is contained. Then $\phi(R_0) = 0$, and

$$N = 4\pi\int_0^{R_0} nr^2\, dr \tag{12.27}$$

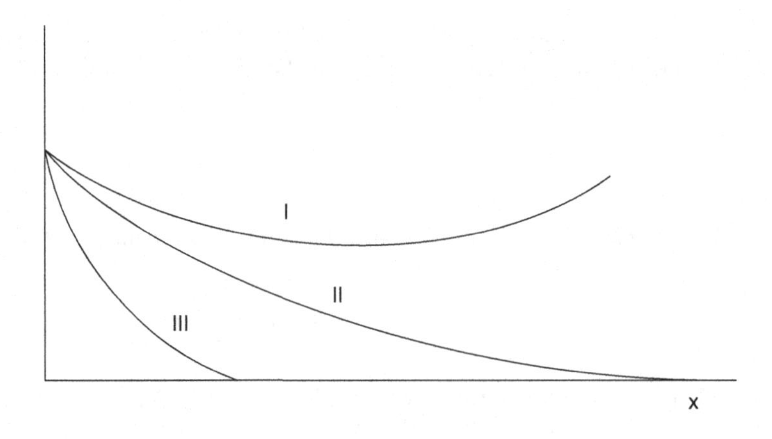

Figure 12.1 The three types of solutions to the Thomas-Fermi equation: (1) $\varphi(x) > 0$ for all x (an atom under pressure as in a Coulomb lattice), (2) asymptotic to the x-axis at $x = \infty$ [a neutral atom $\phi'(0) = -1.588071$], and (3) intersects x-axis for finite positive x (a positive ion).

Using (12.20), (12.23), and (12.25), we rewrite (12.27) as

$$N = Z\int_0^{x_0} x^{1/2}\phi^{3/2}\,dx \tag{12.28}$$

where $R_0 = ax_0$. Employing (12.26) in (12.28), we obtain

$$\begin{aligned}\frac{N}{Z} &= \int_0^{x_0} x\phi''\,dx\\ &= x_0\phi'(x_0) - \phi(x_0) + 1\end{aligned}$$

or, because $\phi(x_0) = 0$,

$$\frac{N}{Z} - 1 = x_0\phi'(x_0) \tag{12.29}$$

Hence, if $\phi'(x_0) < 0$ (solution type III), we have $N < Z$, a positive ion. If $N = Z$ (neutral atom), the solution must go asymptotically to zero as $x_0 \to \infty$, for if, on the contrary, $x_0\phi'(x_0) = 0$ and $\phi(x_0) = 0$ at finite x_0, the solution would be identically zero, as can be seen by making a Taylor expansion of ϕ about x_0 and employing (12.26). Because for a neutral atom $n(r)$ and $\Phi(r)$ both vanish at infinity, we must have $\mu_F = 0$. However, in the case of a positive ion, $n(R_0) = 0$, but $\Phi(R_0) > 0$, so $\mu_F < 0$.

Although negative ions do exist in nature, the binding of the last electron(s) to a neutral atom for such ions is due to subtle polarization effects, and these cannot be accommodated in the Thomas-Fermi model, which does not describe negative ions.

It is easy to verify that $144/x^3$ is a solution to (12.26), although it does not satisfy the boundary condition $\phi(0) = 1$. However, it can be shown that the neutral-atom solution is asymptotic to $144/x^3$ for very large x. Because $n = \text{const}\cdot(\phi/x)^{3/2}$, we see that for very large x, this implies that $n \propto r^{-6}$ in the Thomas-Fermi model. This is a shortcoming of the model because in a neutral atom the electron density at very large distances from the nucleus actually drops exponentially to zero. Also, the relation $n = \text{const}\cdot(\phi/x)^{3/2}$ implies that n diverges at the origin, but in reality, the electron density at the nucleus is finite.

Let us consider how various quantities of physical interest scale with Z in the Thomas-Fermi model. From (12.25), we see that the length scale varies as $Z^{-1/3}$. This means that although the charge distribution goes gradually to zero as $r \to \infty$, we may define a radius R within which exists a certain fixed fraction of the charge, and R is proportional to $Z^{-1/3}$. From this we deduce the following:

$$\begin{array}{lll} Z \propto nR^3 & \text{thus} & n \propto Z^2 \\ p_F \propto n^{1/3} & \text{thus} & p_F \propto Z^{2/3} \\ \varepsilon \propto p_F^2 n & \text{thus} & E_{\text{kin,total}} \propto R^3\varepsilon \propto Z^{-1}Z^{4/3}Z^2 = Z^{7/3} \end{array}$$

$$\begin{array}{llll} \text{Electrostatic potential:} & \Phi \propto Z/R & \text{thus} & \Phi \propto Z^{4/3} \\ \text{Electric field:} & \mathcal{E} \propto \Phi/R & \text{thus} & \mathcal{E} \propto Z^{5/3} \end{array}$$

Also, because the wavelength λ of an electron with maximum kinetic energy is inversely proportional to p_F, $\lambda \propto Z^{-2/3}$, whereas $R \propto Z^{-1/3}$. Thus $\lambda/R \propto Z^{-1/3}$, and this gives an intuitive explanation for why the Thomas-Fermi approximation improves with increasing Z. It can be shown that the approximation becomes exact in the limit $Z \to \infty$.

We now give an important application of the Thomas-Fermi model, saving other examples for later discussion and for the problems. Let us consider the total energy of all the electrons in a neutral atom of atomic number Z. First, we calculate the kinetic energy E_1, starting with the kinetic energy density. In atomic units,

$$\varepsilon = \frac{3^{5/3}\pi^{4/3}}{10} n^{5/3}$$

Because

$$n = \frac{Z}{4\pi a^3} \frac{\phi^{3/2}}{x^{3/2}} \tag{12.30}$$

we have

$$E_1 = 4\pi \int_0^\infty \varepsilon r^2 \, dr = \frac{3^{5/3}\pi^{2/3}}{5\cdot 2^{7/3}} \frac{Z^{5/3}}{a^2} \int_0^\infty \frac{\phi^{5/2}}{x^{1/2}} \, dx \tag{12.31}$$

The integral in (12.31) can be evaluated in two different ways as follows:

$$\begin{aligned} I &= \int_0^\infty \phi \frac{\phi^{3/2}}{x^{1/2}} \, dx = \int_0^\infty \phi\phi'' \, dx \\ &= -\phi'(0) - \int_0^\infty \phi'^2 \, dx \end{aligned} \tag{12.32}$$

and

$$
\begin{aligned}
I = \int_0^\infty \frac{\phi^{5/2}}{x^{1/2}}\,dx &= 2x^{1/2}\phi^{5/2}\Big|_0^\infty - 5\int_0^\infty x^{1/2}\phi^{3/2}\phi'\,dx \\
&= -5\int_0^\infty x\phi'\phi''\,dx \\
&= -\frac{5}{2}x\phi'^2\Big|_0^\infty + \frac{5}{2}\int_0^\infty \phi'^2\,dx \\
&= \frac{5}{2}\int_0^\infty \phi'^2\,dx.
\end{aligned}
\tag{12.33}
$$

Comparing the last lines of (12.32) and (12.33), we obtain

$$
J \equiv \int_0^\infty \phi'^2\,dx = -\frac{2}{7}\int_0^\infty = -\frac{2}{7}\phi'(0) = \frac{2}{7}1.58807 = 0.4537 \tag{12.34}
$$

Hence (12.31) becomes

$$
E_1 = \frac{3}{2}\frac{Z^2}{a}J \tag{12.35}
$$

Next we consider the potential energy of interaction between the electrons and the nucleus; that is,

$$
\begin{aligned}
E_2 &= -\int_0^\infty \frac{Z}{r}nr^2\,dr \\
&= -\frac{Z^2}{a}\int_0^\infty \frac{\phi^{3/2}}{x^{3/2}}x\,dx \\
&= -\frac{Z^2}{a}\int_0^\infty \phi''\,dx \\
&= \frac{Z^2}{a}\phi'(0) \\
&= -\frac{7}{2}\frac{Z^2}{a}J
\end{aligned}
\tag{12.36}
$$

Finally, we consider the potential energy of interaction of the electrons with one another; that is,

$$
E_3 = \frac{1}{2}\iint \frac{n(\boldsymbol{r})n(\boldsymbol{r}')}{|\boldsymbol{r}-\boldsymbol{r}'|}d^3\boldsymbol{r}\,d^3\boldsymbol{r}' \tag{12.37}
$$

This may be evaluated by the same method employed to calculate the expectation value of $|\boldsymbol{r}_1 - \boldsymbol{r}_2|^{-1}$ for the ground state of atomic helium (recall Section 11.6). After straightforward manipulations, one finds

$$
E_3 = \frac{Z^2}{2a}J \tag{12.38}
$$

Therefore, from (12.35), (12.37), and (12.38), the total energy is

$$\begin{aligned} E &= E_1 + E_2 + E_3 \\ &= \left(\frac{3}{2} - \frac{7}{2} + \frac{1}{2}\right)\frac{Z^2}{a}J \\ &= -\frac{3}{2}\frac{Z^2}{a}J = -0.7687 Z^{7/3}\frac{e^2}{a_0} = -20.93 Z^{7/3}\ \text{eV} \end{aligned} \tag{12.39}$$

The result $E = -E_1$ expresses the virial theorem for the Thomas-Fermi model. The numerical value given in (12.39) is in fair but not perfect agreement with experimental determinations of the total electronic energy of atoms. The discrepancy is due mainly to the breakdown of the Thomas-Fermi approximation in the immediate vicinity of the origin. At $r = 0$, the Thomas-Fermi electron number density is infinite, whereas in reality, n is finite. A correction to E for this effect is proportional to $Z^{6/3} = Z^2$ (Scott 1952; Schwinger 1980). Another correction, proportional to $Z^{5/3}$, arises from two distinct physical effects. One is exchange (Dirac 1930), and the other is that even for electrons that are neither very close to the nucleus nor very far distant from it, the variation in potential energy with r is not totally negligible in any given small volume (Schwinger 1981). When these corrections are included, the agreement between calculated and experimental values of the total electronic energy is improved considerably.

The Thomas-Fermi method can be applied not only to atoms but also to nuclei, molecules, solids, and matter at extremely high densities (as in white dwarf and neutron stars). However, one can show that within the Thomas-Fermi approximation, the energy of two isolated atoms or a separated atom and molecule is always less than the energy of a molecule formed from these entities (Teller 1962; Lieb 1973), so molecular binding cannot be explained by the approximation. This is not so surprising: the Thomas-Fermi model works well for the great bulk of electrons in the core of the atom but not for the outermost electrons, which are the ones that participate in the formation of chemical bonds.

12.5 Corrections to the central field approximation: Introduction

The theory as outlined so far suffers from two major deficiencies. First, we have averaged

$$\sum_{\text{pairs}} \frac{1}{r_{ij}}$$

so as to obtain a central field, but this so-called electrostatic interaction term in the Hamiltonian is clearly not central in general. Second, we have ignored the spin-orbit interaction. However, the latter grows in significance roughly in proportion to Z^2 for atoms in a given periodic table group (e.g., the alkali atoms or the halogen atoms; see Figure 12.3). For large Z, the spin-orbit effect is very significant.

For the moment, we confine ourselves to fairly light atoms ($Z < 40$ or so), where the spin-orbit effect is still small enough that we can concentrate on the electrostatic interaction and treat the spin-orbit effect as a minor perturbation. (This is called the *Russell-Saunders* or *L-S coupling scheme*). The zero-order Hamiltonian is

$$H = \sum_i \frac{p_i^2}{2} - Z\sum_i \frac{1}{r_i} + \sum_{i>j} \frac{1}{r_{ij}} \tag{12.40}$$

Let us consider which operators of physical significance commute with this Hamiltonian. First, $[\boldsymbol{J}, H] = 0$ even if the spin-orbit interaction is included because $\boldsymbol{J}$ is the total electronic angular momentum, and H is rotationally invariant. Next, consider the orbital angular momentum operator for the ith electron $\boldsymbol{L}_i$. We now show that this operator does not commute with H but that the total electronic orbital angular momentum $\boldsymbol{L}$ does commute with H. For simplicity, we give the proof for a two-electron atom; the extension to more than two electrons is shown similarly. We write $\boldsymbol{L} = \boldsymbol{L}_1 + \boldsymbol{L}_2$. Then

$$[\boldsymbol{L}_1, H] = \left[\boldsymbol{L}_1, \frac{1}{r_{12}}\right] = \left[\boldsymbol{r}_1 \times \boldsymbol{p}_1, \frac{1}{r_{12}}\right]$$
$$= i\boldsymbol{r}_1 \times \frac{\boldsymbol{r}_1 - \boldsymbol{r}_2}{|\boldsymbol{r}_1 - \boldsymbol{r}_2|^3} = -i\frac{\boldsymbol{r}_1 \times \boldsymbol{r}_2}{|\boldsymbol{r}_1 - \boldsymbol{r}_2|^3}$$

Similarly,

$$[\boldsymbol{L}_2, H] = -i\frac{\boldsymbol{r}_2 \times \boldsymbol{r}_1}{|\boldsymbol{r}_1 - \boldsymbol{r}_2|^3}$$

Thus

$$[\boldsymbol{L}, H] = 0 \tag{12.41}$$

Furthermore, $[\boldsymbol{S}, H] = 0$, which follows from the fact that $\boldsymbol{J}$ and $\boldsymbol{L}$ separately commute with H. One might be tempted to assert that the individual electron-spin operators also commute with H because the latter does not contain the operators $\boldsymbol{S}_i$ explicitly. However, we must take into account the antisymmetrization principle: although it does not appear in the Hamiltonian, it gives rise to an effective spin-spin coupling from the Coulomb interaction, as was mentioned in Chapter 11.

To summarize, if we ignore the spin-orbit interaction, $\boldsymbol{S} = \sum_i \boldsymbol{S}_i$ and $\boldsymbol{L} = \sum_i \boldsymbol{L}_i$ separately commute with H, as does $\boldsymbol{J}$. Therefore, energy eigenstates simultaneously have definite values of J, L, and S. Our next task is to determine the possible values of J, L, and S for a given configuration and establish reasons for energy ordering of states of given J, L, and S.

12.6 Theory of multiplets in the Russell-Saunders scheme

In the CFA, the energy of an atom is determined solely by the configuration (assignment of n, ℓ values to individual orbitals). Thus there is in general a great deal of degeneracy in each configuration. To see this, consider a given ℓ shell that can contain up to $N_0 = 2(2\ell + 1)$ electrons. Suppose that in fact it contains $n_0 < N_0$ electrons. For example, if $\ell = 1$ (p shell), $N_0 = 6$. For carbon, we have two equivalent p electrons, so $n_0 = 2$. For nitrogen, $n_0 = 3$; for oxygen, $n_0 = 4$; and so on. The first electron may be placed in any one of the N_0 orbitals, the second in $N_0 - 1$, the third in $N_0 - 2$, and so forth. Therefore, a priori, there are

$$\frac{N_0!}{(N_0 - n_0)!}$$

possibilities. However, some of these are equivalent because the electrons are indistinguishable. Thus we must divide by the number of ways $n_0!$ that the electrons can be permuted among themselves, so the degeneracy is

$$g = \frac{N_0!}{n_0!(N_0 - n_0)!} = \binom{N_0}{n_0} \tag{12.42}$$

For example,

$$g(\text{carbon}) = \binom{6}{2} = 15$$
$$g(\text{nitrogen}) = \binom{6}{3} = 20$$
$$g(\text{oxygen}) = \binom{6}{4} = 15$$
$$g(\text{iron}) = \binom{10}{6} = 210$$
$$g(\text{gadolinum}) = \binom{14}{8} = 3{,}003$$

Thus, in the CFA, there are 15 degenerate states associated with the ground configuration of carbon, 210 for iron, and so on. However, because of the electrostatic interaction, the potential is not truly central, and much of this degeneracy is lifted. Because L, S, and J are still good quantum numbers, all states with given L, S, and J remain degenerate (they form a *term*), but terms with distinct values of L, S, and J in general have distinct energies. (The word *multiplet* is reserved for states with given L and S. Here there can sometimes be several values of J. For example, the ground multiplet of oxygen is ^3P, which consists of the terms $J = 2$, 1, and 0.) We have already seen the simplest example in Chapter 11: in helium, the $1s2s$ configuration has a degeneracy of 4, and it breaks up into two distinct terms

$$^1S_0\colon\ g = 1$$
$$^3S_1\colon\ g = 3$$

We now use the Pauli principle to determine which multiplets correspond to a given configuration.

1. If an atom contains only closed shells (e.g., the rare gas atoms and the Group II elements Be, Mg, Ca, Sr, Ba, Zn, Cd, and Hg), the Pauli principle requires a pairing off of spins and orbital angular momenta to give $S = 0$, $L = 0$. Thus we must have a $^1S_0$ ground state.

2. If there is one electron outside closed shells, the total spin S (=1/2) and the total orbital angular momentum L are that of the valence electron. Thus we obtain the following ground states:

   H, Li, Na, K, Rb, Cs, Fr: $^2S_{1/2}$
   Ag, Au: $^2S_{1/2}$
   Ga, In, Tl: $^2P_{1/2}$

   Note that for Ga, In, and Tl we have written $J = 1/2$ for the ground state. In fact, we need some knowledge of the spin-orbit interaction to determine that $J = 1/2$ has lower energy than $J = 3/2$.
3. Atoms that lack just one electron to complete a shell (e.g., the halogens, which have five equivalent p electrons) may be treated as having one "hole," and the total spin and orbital angular momentum are that of the hole. Thus the halogen ground states are $^2P_{3/2}$ (once again, J is determined from knowledge of the spin-orbit interaction).
4. If there are two electrons in the outermost incomplete shell, we employ the following theorem: the wave function describing these electrons, which must be antisymmetric with respect to exchange, can always be written as a product of a spatial part and a spin part. If the spatial part is symmetric, the spin part is antisymmetric, and vice versa. The proof of this theorem is exactly the same as for the $1s2s$ configuration in helium (see Section 11.7), and we do not repeat it. For example, consider carbon, which has two equivalent p $(\ell = 1)$ electrons. The possible values of S are 1 and 0, whereas the possible values of L are 2,1, and 0. A priori, we may form the multiplets ^3D, ^3P, ^3S, ^1D, ^1P, and ^1S. However, because of the symmetry properties of vector coupling coefficients, $L = 0, 2$ are symmetric spatial states, whereas $L = 1$ is antisymmetric. Also, $S = 1$ is a symmetric spin state, whereas $S = 0$ is antisymmetric. Therefore, the multiplets ^3D, ^3S, and ^1P are excluded, and we are left with the multiplets

   $$^3P \text{ (9 states)} \qquad ^1D \text{ (5 states)} \qquad ^1S \text{ (1 state)}$$

   for a total of 15 states. The theorem also may be used if we have two equivalent holes in an incomplete shell. Thus the conclusions we have just arrived at for carbon also apply for oxygen, which has two equivalent $\ell = 1$ holes.
5. If there are more than two equivalent electrons, it is no longer possible to factor the wave function into space and spin parts with opposite exchange symmetry. However the multiplets can be enumerated in a systematic if somewhat laborious way by constructing a table. We illustrate with the example of three equivalent p electrons in Table 12.2.

The columns of the table are labeled

$$m_{\ell 1},\ m_{\ell 2},\ m_{\ell 3},\ m_{s1},\ m_{s2},\ m_{s3},\ M_L,\ M_S,\ L, \text{ and } S$$

We start by entering in the first row the maximum possible values of $m_{\ell i}$, $i = 1, 2, 3$. For three equivalent p electrons, this would be +1, +1, +1, yielding $M_L = 3$. Now each of the m_{si} can only be ±½. However, because $m_{\ell 1} = m_{\ell 2} = m_{\ell 3}$, no two of the m_{si} can be the same. Therefore, we cannot have a state with $M_L = 3$. Hence we cannot form an $L = 3$ multiplet; the largest possible

Table 12.2 Multiplet table for p^3

m_{l1}	m_{l2}	m_{l3}	m_{s1}	m_{s2}	m_{s3}	M_L	M_S	L	S
1	1	1							
1	1	0	+	–	+	2	1/2	2	1/2
1	1	–1	+	–	+	1	1/2	2	1/2
1	0	0	+	–	+	1	1/2	1	1/2
1	0	–1	+	+	+	0	3/2	0	3/2

value is $L = 2$. Accordingly, we try to construct a state with $M_L = 2$ (the next row in the table). As can be seen, this is possible provided that not all values of m_s are the same; hence for this particular state, we have $M_S = ½$. Therefore, this line of the table corresponds to a ^2D multiplet, which has altogether 10 distinct states. Because the total degeneracy of a p^3 configuration is 20, we must identify 10 more states. In the next lines of the table we write out those configurations associated with $M_L = 1$. It can be seen that there are two independent possibilities. One linear combination must correspond to the ^2D, $M_L = 1$ state. The orthogonal combination must be the $M_L = 1$ component of a new multiplet with $M_S = ½$ and hence $S = ½$. This is obviously ^2P and contains altogether 6 states. It remains to identify 4 of the 20 states. To this end, we start to write out the lines of the table corresponding to $M_L = 0$. It can be seen that there is one state with $M_S = 3/2$. Because $M_S = 3/2$ did not occur in any of the previous lines, it must be associated with $L = 0$, $S = 3/2$ (a $^4S_{3/2}$ term with four states). We have now identified all 20 states, and it is unnecessary to fill in any more lines of the table. To recapitulate, a p^3 configuration contains the multiplets ^2D, ^2P, and ^4S. This method can be extended in a routine way to more complicated cases.

Now that we have a procedure for constructing the possible multiplets of a given configuration, how are they to be ordered in energy? There are several rules, named after the German spectroscopist F. Hund, who worked in the early decades of the twentieth century and arrived at the rules empirically.

Hund's first rule: Terms with the highest spin multiplicity lie lowest in energy.

This owes its origin to the same phenomenon that causes $^3S_1$ to lie lower than $^1S_0$ in helium. Recall that the spatial wave function of the two electrons in $^3S_1$ is antisymmetric; thus the probability that both electrons are found in the same small region of space is vanishingly small. Hence the average value of the repulsive interaction $1/r_{12}$ is much smaller than for $^1S_0$, where the spatial wave function is symmetric. More generally, given a many-electron configuration, a multiplet with large S has a more antisymmetric spatial wave function than a multiplet with small S.

Hund's second rule: For multiplets of the same S, those with higher L lie lower in energy.

For given spin, multiplets with larger L tend to have electrons farther apart than those with smaller L. Several simple examples of these rules are provided by carbon, oxygen, and nitrogen. In carbon and oxygen, ^3P lies lowest, followed by ^1D and then ^1S. In nitrogen, the ground multiplet is ^4S, followed by ^2D and then ^2P. However, whereas Hund's rules provide a useful general guide, they are not always valid. Configuration mixing and other effects do alter the energy ordering of multiplets in many cases.

12.7 Calculation of multiplet energies in the *L-S* coupling scheme

Quantitative calculations of multiplet splittings are not trivial, even for relatively simple atoms. The basic approach is to start with the CFA, assume only a single configuration, and use first-order perturbation theory with the Slater sum rule. We illustrate with the $1s^2 2s^2 2p^2$ configuration, for which we have already shown that the distinct multiplets are ^3P, ^1D, and ^1S. Our goal is to calculate the first-order energy shifts; that is,

$$\begin{aligned}
\Delta E^{(1)}(^3P) &= \langle LM_LSM_S|H'|LM_LSM_S\rangle \qquad L=1, S=1 \\
\Delta E^{(1)}(^1D) &= \langle LM_LSM_S|H'|LM_LSM_S\rangle \qquad L=2, S=0 \\
\Delta E^{(1)}(^1S) &= \langle LM_LSM_S|H'|LM_LSM_S\rangle \qquad L=0, S=0
\end{aligned}$$

where H' is the electrostatic interaction. In general, a multiplet component $|LM_LSM_S\rangle$ is constructed as a linear combination of Slater determinants containing electrons with different $m_{\ell i}, m_{si}$ values such that $\sum_i m_{\ell i} = M_L$ and $\sum_i m_{si} = M_S$. For example, in the case of two equivalent p electrons, consider the state $|L=2, M_L=2, S=0, M_S=0\rangle = |2,2,0,0\rangle$, which is one component of the ^1D multiplet. In this particular case, there is only one possible Slater determinant, which corresponds to the values

$$m_{\ell 1}=1 \quad m_{\ell 2}=1 \quad m_{s1}=+\tfrac{1}{2} \quad m_{s2}=-\tfrac{1}{2}$$

This Slater determinant is conveniently denoted by the symbol (1+, 1–). Thus we have

$$|2,2,0,0\rangle = (1+,1-) \tag{12.43}$$

The remaining components of the ^1D multiplet are easily found by applying the lowering operator $L_- = L_{1-} + L_{2-}$ to both sides of (12.43)

$$\begin{aligned}
|2,2,0,0\rangle &= (1+,1-) \\
|2,1,0,0\rangle &= \frac{1}{\sqrt{2}}(0+,1-) + \frac{1}{\sqrt{2}}(1+,0-) \\
|2,0,0,0\rangle &= \frac{1}{\sqrt{6}}(-1+,1-) + \frac{2}{\sqrt{6}}(0+,0-) + \frac{1}{\sqrt{6}}(1+,-1-) \\
|2,-1,0,0\rangle &= \frac{1}{\sqrt{2}}(-1+,0-) + \frac{1}{\sqrt{2}}(0+,-1-) \\
|2,-2,0,0\rangle &= (-1+,-1-)
\end{aligned} \tag{12.44}$$

We see that the coefficients in the transformation from the $(m_{\ell i}, m_{si})$ basis to the (LM_LSM_S) basis are vector coupling coefficients. Similarly, we could express the components of the ^3P multiplet, as well as the ^1S state, as linear combinations of Slater determinants. However, it is not necessary to make all this effort because there are several features that simplify the problem. The first is that in the absence of external magnetic and electric fields, all M_L, M_S components of a given multiplet have the same energy. Thus, to calculate $\Delta E^{(1)}$ for a given multiplet, it is sufficient to calculate it for just one choice of M_L, M_S. Now, as we have just seen, the ^1D component $|2,2,0,0\rangle$ is expressed as the single Slater determinant (1+, 1–). Similarly, the ^3P component $|1,1,1,1\rangle = (1+,0+)$ is just a single Slater determinant. Hence we have

$$\begin{aligned}\Delta E^{(1)}(^1D) &= \langle(1+,1-)|H'|(1+,1-)\rangle \\ \Delta E^{(1)}(^3P) &= \langle(1+,0+)|H'|(1+,0+)\rangle\end{aligned} \tag{12.45}$$

The $^1S_0$ state is not a single Slater determinant; it is a linear combination of the determinants (1+, –1–), (1–, –1+), and (0+, 0–). However, for given M_L, M_S, we can use the fact that the trace of the H' submatrix is invariant under the unitary transformation from the $(m_{\ell i}, m_{si})$ basis to the (LM_LSM_S) basis. In particular, for $M_L = M_S = 0$, we have

$$\Delta E^{(1)}(^1S) + \Delta E^{(1)}(^1D) + \Delta E^{(1)}(^3P) = \langle(1+,-1-)\rangle + \langle(1-,-1+)\rangle + \langle(0+,0-)\rangle \tag{12.46}$$

where we employ the shorthand

$$\langle(0+,0-)\rangle \equiv \langle(0+,0-)|H'|(0+,0-)\rangle$$

and similarly for the other Slater determinants. Substituting (12.45) in (12.46) and transposing, we obtain

$$\Delta E^{(1)}(^1S) = \langle(1+,-1-)\rangle + \langle(1-,-1+)\rangle + \langle(0+,0-)\rangle - \langle(1+,1-)\rangle - \langle(1+,0+)\rangle \tag{12.47}$$

We have just employed a simple application of the Slater sum rule. Equations (12.45) and (12.47) reveal that our problem is reduced to calculation of the diagonal matrix elements of the electrostatic interaction for the Slater determinants: (1+, 1–), (1+, 0+), (1+, –1,–), (1–, –1+), and (0+, 0–).

We now discuss the procedure for calculation of such matrix elements. First of all, it can be shown from (11.58) that, in general, the matrix element can be expressed as

$$\sum_{i<j}\left[\langle ij|r_{12}^{-1}|ij\rangle - \langle ij|r_{12}^{-1}|ji\rangle\right] \tag{12.48}$$

where the sum is over all pairs of electrons in the incomplete shell. (Note that we are ignoring the contribution of the complete shells to the electrostatic energy; it is an additive constant common to all the multiplets in question.) In (12.48), the first and second terms in square brackets are the direct and exchange contributions, respectively. A general form for each matrix element in (12.48) is

$$\langle 12|r_{12}^{-1}|34\rangle = \langle n_1\ell_1 m_{\ell 1} m_{s1}, n_2\ell_2 m_{\ell 2} m_{s2}|r_{12}^{-1}|n_3\ell_3 m_{\ell 3} m_{s3}, n_4\ell_4 m_{\ell 4} m_{s4}\rangle \tag{12.49}$$

The angular and spin parts of this expression are

$$\begin{aligned}&\delta(m_{s1}, m_{s3})\delta(m_{s2}, m_{s4})\times \\ &\iint d\Omega_1 d\Omega_2 Y_{\ell 1}^{m\ell 1*}(\Omega_1) Y_{\ell 2}^{m\ell 2*}(\Omega_2) r_{12}^{-1} Y_{\ell 3}^{m\ell 3}(\Omega_1) Y_{\ell 4}^{m\ell 4}(\Omega_2)\end{aligned} \tag{12.50}$$

Now

$$\frac{1}{r_{12}} = 4\pi \sum_{k=0}^{\infty} \frac{1}{2k+1} \frac{r_<^k}{r_>^{k+1}} \sum_{\mu=-k}^{k} Y_k^{\mu *}(\Omega_1) Y_k^{\mu}(\Omega_2) \tag{12.51}$$

Inserting this in (12.50), we obtain

$$\begin{gathered} \delta(m_{s1}, m_{s3}) \delta(m_{s2}, m_{s4}) \sum_{k=0}^{\infty} \frac{4\pi}{2k+1} \frac{r_<^k}{r_>^k} \times \\ \sum_{\mu=-k}^{k} \int d\Omega_1 Y_{\ell 1}^{m\ell 1 *} Y_k^{\mu *} Y_{\ell 3}^{m\ell 3} \times \int d\Omega_2 Y_{\ell 2}^{m\ell 2 *} Y_k^{\mu} Y_{\ell 4}^{m\ell 4} \end{gathered} \tag{12.52}$$

Each integral of three spherical harmonics is usually expressed in a standard way as follows:

$$\int Y_{\ell'}^{m*} Y_k^{\mu} Y_{\ell}^{m} d\Omega = \sqrt{\frac{2k+1}{4\pi}} \delta(\mu, m' - m) c^k(\ell', m', \ell, m) \tag{12.53}$$

where the c^k are tabulated coefficients defined by (12.53), and ℓ, ℓ', and k are constrained by the triangle rule. Taking all this into account and including the radial portion, we find that (12.49) becomes

$$\begin{aligned} \langle 12 | r_{12}^{-1} | 34 \rangle = {} & \delta(m_{s1}, m_{s3}) \delta(m_{s2}, m_{s4}) \delta(m_{\ell 1} + m_{\ell 2}, m_{\ell 3} + m_{\ell 4}) \times \\ & \sum_k c^k(\ell_3 m_{\ell 3} \ell_1 m_{\ell 1})\, c^k(\ell_2 m_{\ell 2} \ell_4 m_{\ell 4}) R^k(12, 34) \end{aligned} \tag{12.54}$$

where

$$R^k(12, 34) = \int_0^{\infty} r_1^2\, dr_1 \int_0^{\infty} r_2^2\, dr_2 R_{n1\ell 1}(r_1) R_{n2\ell 2}(r_2) R_{n3\ell 3}(r_1) R_{n4\ell 4}(r_2) \frac{r_<^k}{r_>^{k+1}} \tag{12.55}$$

Further standard notations are employed when the initial and final states are the same, as is the case for our diagonal matrix elements

$$\begin{aligned} F^k(n_i \ell_i, n_j \ell_j) &\equiv R^k(ij, ij) \\ G^k(n_i \ell_i, n_j \ell_j) &\equiv R^k(ij, ji) \\ a^k(\ell_i m_{\ell i}, \ell_j m_{\ell j}) &\equiv c^k(\ell_i m_{\ell i}, \ell_i m_{\ell i}) c^k(\ell_j m_{\ell j}, \ell_j m_{\ell j}) \\ b^k(\ell_i m_{\ell i}, \ell_j m_{\ell j}) &\equiv \left[c^k(\ell_i m_{\ell i}, \ell_j m_{\ell j})\right]^2 \end{aligned}$$

Thus, in (12.48),

$$\begin{aligned} & \langle ij | r_{12}^{-1} | ij \rangle - \langle ij | r_{12}^{-1} | ji \rangle = \\ & \sum_k \left[a^k(\ell_i m_{\ell i} \ell_j m_{\ell j}) F^k(n_i \ell_i n_j \ell_j) - \delta(m_{si}, m_{sj}) b^k(\ell_i m_{\ell i} \ell_j m_{\ell j}) G^k(n_i \ell_i n_j \ell_j) \right] \end{aligned} \tag{12.56}$$

Table 12.3 Partial list of coefficients c^k

	$m_{\ell i}$	$m_{\ell j}$	k = 0	1	2
ss	0	0	1	0	0
sp	0	±1	0	$-\sqrt{1/3}$	0
	0	0	0	$\sqrt{1/3}$	0
sd	0	±2	0	0	$\sqrt{1/5}$
	0	±1	0	0	$-\sqrt{1/5}$
pp	±1	±1	1	0	$-\sqrt{1/25}$
	±1	0	0	0	$\sqrt{3/25}$
	±1	∓1	0	0	$-\sqrt{6/25}$
	0	0	1	0	$\sqrt{4/25}$

Note: When there are two ± signs, the two upper or the two lower signs must be employed together.

We now return to the specific problem of evaluating the matrix elements in (12.45) and (12.47). Making use of Table 12.3 for coefficients $c^k\left(\ell_i m_{\ell i}\ell_j m_{\ell j}\right)$ and carrying out straightforward algebra, we obtain

$$\begin{aligned}\Delta E^{(1)}\left({}^1D\right) &= F^0 + \frac{1}{25}F^2\\ \Delta E^{(1)}\left({}^3P\right) &= F^0 - \frac{1}{5}F^2\\ \Delta E^{(1)}\left({}^1S\right) &= F^0 + \frac{2}{5}F^2\end{aligned} \tag{12.57}$$

Only F^2 contributes to the splittings between multiplets. To evaluate F^2, we need radial wave functions, which can be obtained from the Hartree-Fock model. However, even without knowledge of radial functions, (12.57) enables us to predict the ratio of the multiplet splittings

$$R \equiv \frac{{}^1S - {}^1D}{{}^1D - {}^3P} = 1.5 \tag{12.58}$$

This result does not agree very well with experiment. The reason is that splittings due to the electrostatic interaction are comparable in magnitude to the splittings between different configurations (Figure 12.2). Thus the assumption of a single configuration in the Slater sum rule calculation is not realistic; to improve the agreement between theory and experiment, one must take into account configuration interaction.

12.8 Spin-orbit interaction

In the CFA, description of the spin-orbit interaction in a complex atom is quite similar to that in hydrogen (recall Section 10.2). Suppose that the ith electron in an N-electron atom moves

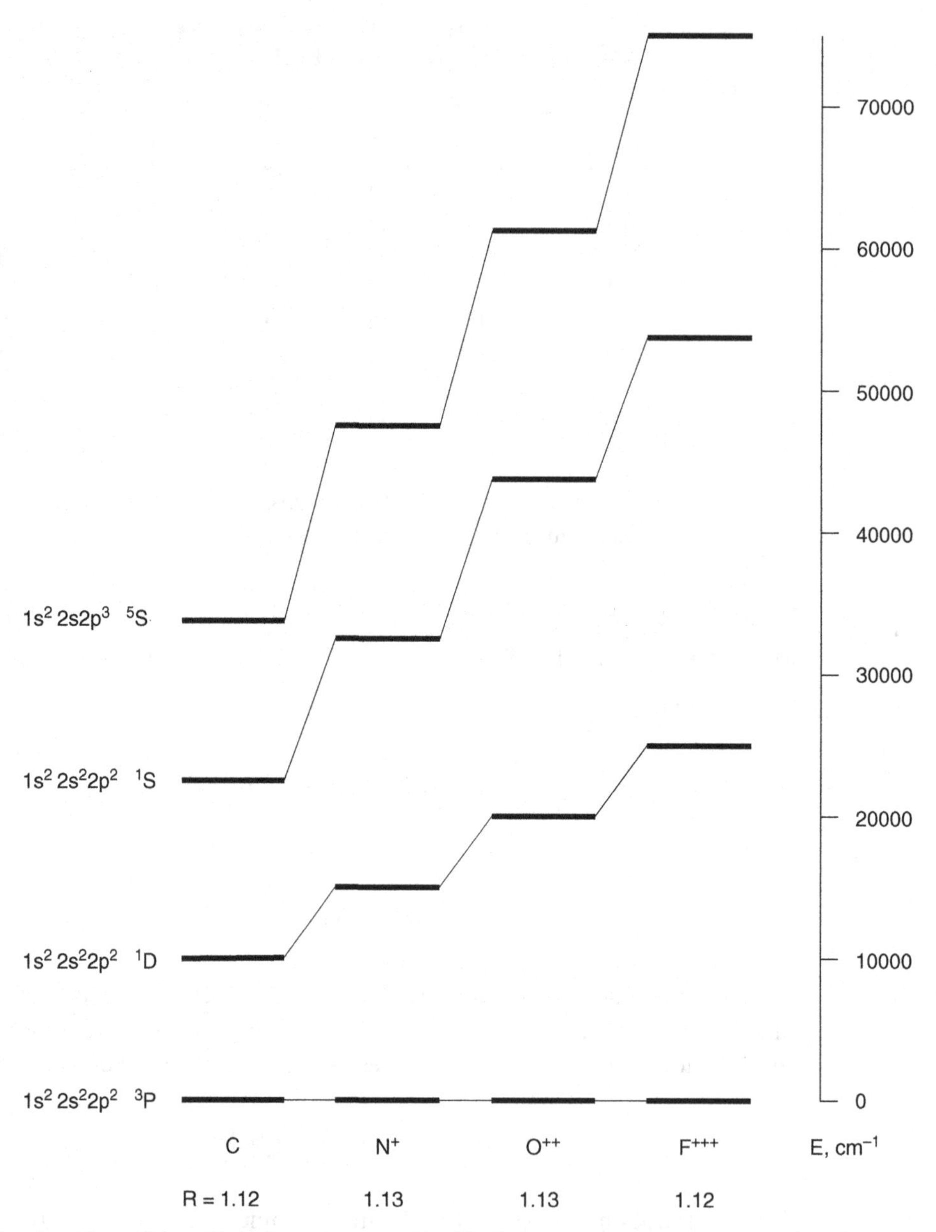

Figure 12.2 Energies of low-lying multiplets of the isoelectronic sequence C, N^+, O^{2+}, F^{2+}. Assuming the single configuration $1s^22s^22p^2$ and calculating the multiplet splittings by first-order perturbation theory using the Slater sum rule, one predicts that $R = [E(^1S) - E(^1D)]/[E(^1D) - E(^3P)] = 1.50$. The observed values of R for C, N^+, O^{2+}, F^{2+} are shown in the figure. The discrepancies between theory and experiment arise from neglect of neighboring configurations, which are in fact quite close in energy to the multiplets of interest. To show this, we include the $1s^22s2p^3\,{}^5S$ level for each member of the isoelectronic sequence.

in a central electrostatic potential $\Phi(r_i)$. Then, if this electron described uniform motion, the magnetic field in its rest frame would be

$$\boldsymbol{B}_i = \frac{1}{c}\boldsymbol{\mathcal{E}}_i \times \mathbf{v}_i = -\frac{\alpha}{r_i}\frac{\partial \Phi(r_i)}{\partial r_i}\boldsymbol{L}_i \tag{12.59}$$

The spin-orbit interaction of the ith electron is given by the phenomenological expression

$$H_i^{(so)} = -g_{\text{eff}} \frac{\alpha^2}{2} \frac{1}{r_i} \frac{\partial \Phi}{\partial r_i} \boldsymbol{S}_i \cdot \boldsymbol{L}_i$$

where, following Thomas' analysis, the noninertial nature of the ith electron rest frame is compensated by replacing $g_s \approx 2$ with $g_{eff} = g_s - 1 \approx 1$. The total spin-orbit interaction Hamiltonian is

$$H^{(so)} = \sum_{i=1}^{N} \xi_i \boldsymbol{S}_i \cdot \boldsymbol{L}_i \tag{12.60}$$

where

$$\xi_i = -\frac{g_{\text{eff}} \alpha^2}{2} \frac{1}{r_i} \frac{\partial \Phi}{\partial r_i} \tag{12.61}$$

It is intuitively obvious that for electrons in closed shells, the individual terms in the sum in (12.60) cancel in pairs, leaving no net contribution. However, there is a residual spin-orbit effect for electrons in unfilled shells, and for a given periodic table group, it increases approximately as Z^2. This dependence may be understood from (12.61) and the Thomas Fermi model, which tells us that because r scales as $Z^{-1/3}$ and Φ as $Z^{4/3}$, ξ_i scales as Z^2. In Figure 12.3, we plot the fine-structure splittings (spin-orbit effect) versus Z on a log-log scale for several groups of atoms. These data show that for a given periodic table group, fine-structure splittings are indeed approximately proportional to Z^2.

We first confine ourselves to fairly light atoms, where L-S coupling is a good approximation, and the spin-orbit effect can be treated as a perturbation. Using the Wigner-Eckart theorem, we now show that diagonal matrix elements of $H^{(so)}$ in the $LSJM_J$ representation are proportional to those of the operator $\boldsymbol{L} \cdot \boldsymbol{S}$. We write

$$\begin{aligned} &\langle LSJM_J | H^{(so)} | LSJM_J \rangle = \\ &\sum_{M_L, M_L'} \langle LM_L' SM_S' | JM_J \rangle \langle LM_L' SM_S' | H^{(so)} | LM_L SM_S \rangle \langle LM_L SM_S | JM_J \rangle \end{aligned} \tag{12.62}$$

Now

$$\langle LM_L' SM_S' | H^{(so)} | LM_L SM_S \rangle = \sum_{i=1}^{N} \xi_i \langle LM_L' SM_S' | \boldsymbol{L}_i \cdot \boldsymbol{S}_i | LM_L SM_S \rangle \tag{12.63}$$

However,

$$\langle LM_L' SM_S' | \boldsymbol{L}_i \cdot \boldsymbol{S}_i | LM_L SM_S \rangle = \langle LM_L' | \boldsymbol{L}_i | LM_L \rangle \cdot \langle SM_S' | \boldsymbol{S}_i | SM_S \rangle$$

and also $\boldsymbol{L}_i$ and $\boldsymbol{S}_i$ are first-rank tensor operators. Thus we have

$$\langle LM_L' | L_i^{(m)} | LM_L \rangle = \langle LM_L 1m | LM_L' \rangle \langle L \| L_i \| L \rangle \tag{12.64}$$

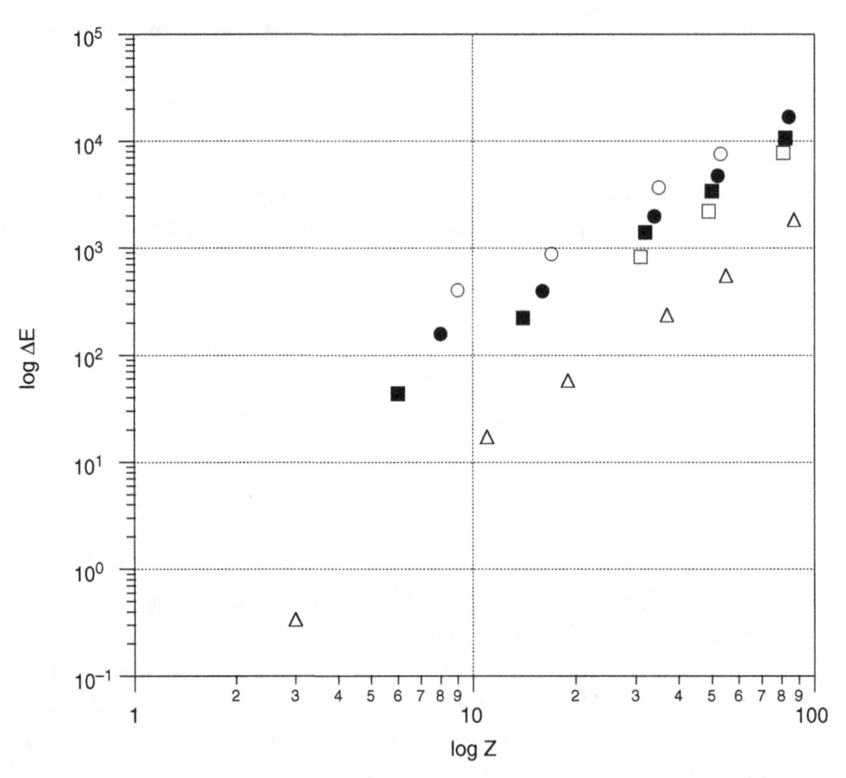

Figure 12.3 Fine-structure splittings. Plot of log ΔE versus. Log Z for the following cases:

○: F, Cl, Br, I np^5 $\Delta E : ({}^2P_{1/2} - {}^2P_{3/2})$
●: O, S, Se, Te, Po np^4 $\Delta E : ({}^3P_1 - {}^3P_2)$
□: Ga, In, Tl np $\Delta E : ({}^2P_{3/2} - {}^2P_{1/2})$
■: C, Si, Ge, Sn, Pb np^2 $\Delta E : ({}^3P_2 - {}^3P_0)$
△: Li, Na, K, Rb, Cs, Fr [closed shell + $(n+1)p*$] $\Delta E : ({}^2P_{3/2} - {}^2P_{1/2})$

and

$$\left\langle SM_S' \middle| S_i^{(m)} \middle| SM_S \right\rangle = \left\langle SM_S 1m \middle| SM_S' \right\rangle \left\langle S \middle\| S_i \middle\| S \right\rangle \tag{12.65}$$

where $m = -1, 0$, or 1. We also know that in general

$$\left\langle jM' \middle| J^{(m)} \middle| jM \right\rangle = \sqrt{j(j+1)} \left\langle jM1m \middle| jM' \right\rangle$$

which implies that

$$\left\langle LM_L' \middle| L_i^{(m)} \middle| LM_L \right\rangle = \left\langle LM_L' \middle| L^{(m)} \middle| LM_L \right\rangle \frac{\left\langle L \middle\| L_i \middle\| L \right\rangle}{\sqrt{L(L+1)}} \tag{12.66}$$

and

$$\langle SM'_S | S_i^{(m)} | SM_S \rangle = \langle SM'_S | S^{(m)} | SM_S \rangle \frac{\langle S \| S_i \| S \rangle}{\sqrt{S(S+1)}} \tag{12.67}$$

Thus (12.62) becomes

$$\langle LSJM_J | H^{(so)} | LSJM_J \rangle = \left[\sum_{M_L, M'_L} \langle LM'_L SM'_S | JM_J \rangle \langle LM'_L SM'_S | \mathbf{L} \cdot \mathbf{S} | LM_L SM_S \rangle \langle LM_L SM_S | JM_J \rangle \times \sum_{i=1}^{N} \xi_i \frac{\langle L \| L_i \| L \rangle \langle S \| S_i \| S \rangle}{\sqrt{L(L+1)}\sqrt{S(S+1)}} \right] \tag{12.68}$$

The sum over i is a numerical constant that we call A. Thus (12.68) becomes

$$\begin{aligned} \langle LSJM_J | H^{(so)} | LSJM_J \rangle &= A \langle LSJM_J | \mathbf{L} \cdot \mathbf{S} | LSJM_J \rangle \\ &= \frac{A}{2} \left[J(J+1) - L(L+1) - S(S+1) \right] \end{aligned} \tag{12.69}$$

This result implies the *Lande interval rule*, which states that the splitting between terms with J and $J-1$ within a given multiplet is proportional to J. For example, in a ^3P multiplet, the possible values of J are 2, 1, and 0. The interval rule predicts that

$$R = \frac{E(J=2) - E(J=1)}{E(J=1) - E(J=0)} = 2 \tag{12.70}$$

How well is the interval rule obeyed? Consider Table 12.4, which lists atoms having two equivalent p electrons in the outer shell and a ^3P ground multiplet.

Only in silicon is R in moderately good agreement with the simple prediction of (12.70). In the very light carbon atom, the discrepancy is due to relativistic effects. In the heavier atoms, and especially in Sn and Pb, L-S coupling is no longer a good approximation.

Another example is shown in Table 12.5. The ground configuration of iron is $1s^2 \cdots 4s^2 3d^6$, and the ground multiplet is ^5D ($L = S = 2$). The various terms in this multiplet, in order of increasing energy, are $J = 4, 3, 2, 1, 0$. (This ordering is inverted because the incomplete shell is more than half full – a general rule for fine-structure splittings.) The energies are given in cm^{-1}.

If the interval rule were obeyed exactly, all the numbers in the last column would be the same. Here agreement with the interval rule is good but not perfect. Calculations of the factor A are facilitated by use of the Slater sum-rule technique. We refer the reader to Condon and Shortley (1953) for a detailed discussion of this question.

We now turn to consideration of the spin-orbit effect in heavy atoms, where it is no longer a minor perturbation but a very major effect (see Figure 12.4).

In heavy atoms, neither L nor S is a good quantum number, but only J. For illustration, we consider mercury (Hg), where $Z = 80$. Within the CFA, the ground configuration is $1s^2 \cdots 5d^{10} 6s^2$, and most of the low-lying excited states have configurations in which one of the

Table 12.4 Fine structure splittings for atoms with 2 equivalent p electrons, (^3P ground multiplet)

Atom	Z	$E(^3P_0)$	$E(^3P_1)$	$E(^3P_2)$	R
C	6	0.00	16.4 cm^{-1}	43.5	1.65
Si	14	0.00	77.15	223.31	1.89
Ge	32	0.00	557	1410	1.53
Sn	50	0.00	1692	3428	1.03
Pb	82	0.00	7819	10650	0.36

Table 12.5 Spin-orbit splittings in the ground 5D multiplet of iron

J	$E(J)$	$E(J) - E(J-1)$	$[E(J) - E(J-1)]J^{-1}$
4	0.000	—	—
3	415.934	−415.934	−104.0
2	704.001	−288.067	−96.0
1	888.126	−184.125	−92.1
0	978.068	−89.942	−89.9

$6s$ electrons is promoted to a higher orbital ($6p$, $7s$, $5d$, etc.). If L-S coupling were valid, we could describe the ground state as $6s^2$ ($^1S_0$) and the terms with the $6s6p$ configuration as $6s6p$ ($^1P_1$) and $6s6p$ ($^3P_{2,1,0}$). These states are in fact labeled this way, but it is misleading. Let us focus our attention on the $^3P_1$ level. Once produced, an atom in this state decays to the ground state by emission of a photon at 254 nm, with a mean lifetime $\tau = 10^{-7}$ s. The transition probability per unit time for this decay ought to be much smaller (for reasons that will be explained in a later chapter); hence τ should be orders of magnitude longer. The answer to the puzzle is that L-S coupling is not legitimate because of the large spin-orbit interaction. Thus what we label as a $^3P_1$ state is really a $J = 1$ state with the following major components: ($S = L = 1$: ^3P), ($S = 0$, $L = 1$, ^1P), and ($S = 1$, $L = 0$,^3S). In particular, the ^1P component, which enters into the wave function with a coefficient of several percent, has a very large matrix element for decay to the ^1S ground state.

In the limit where the spin-orbit effect is large and the electrostatic interaction can be ignored to zeroth order, it is most convenient to employ the representation in which each individual $\boldsymbol{L}_i$ and $\boldsymbol{S}_i$ couple to form a $\boldsymbol{J}_i$, and these are added together to give the total $\boldsymbol{J}$. This is called the *j-j coupling scheme*. The procedure for finding the possible j-j multiplets, given a specific configuration, is described in chapter 10 of Condon and Shortley (1953). The L-S and j-j schemes are, of course, both limiting cases; more generally, it would be desirable to find a representation that diagonalizes the Hamiltonian and contains both electrostatic and spin-orbit effects. Unfortunately, this intermediate coupling scheme, discussed in chapter 11 of Condon and Shortley (1953), is quite complicated in most cases.

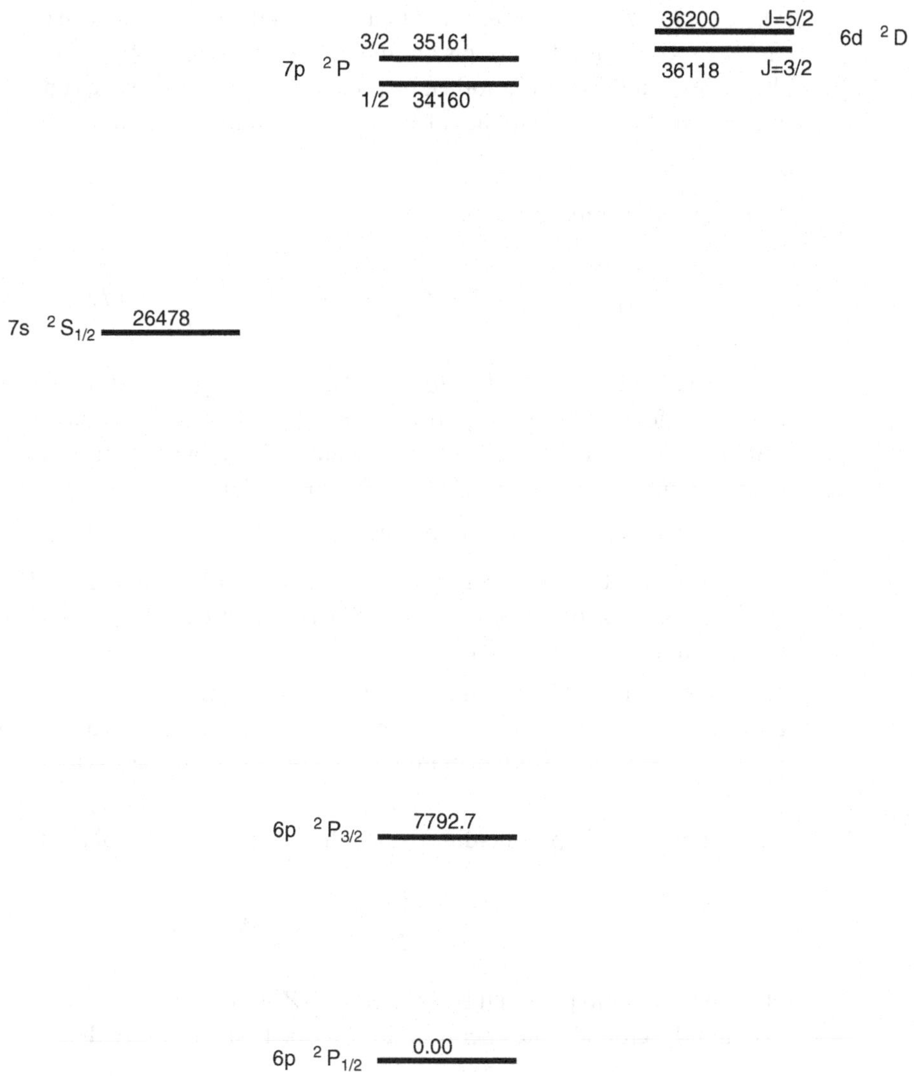

3/2 35161
7p ^{2}P
1/2 34160

36200 J=5/2
6d ^{2}D
36118 J=3/2

7s $^2S_{1/2}$ 26478

6p $^2P_{3/2}$ 7792.7

6p $^2P_{1/2}$ 0.00

Figure 12.4 Low-lying energy levels of atomic thallium (Tl), $Z = 81$, approximately to scale. Note the very large fine-structure splitting between the two lowest levels.

Problems for Chapter 12

12.1. The Thomas-Fermi differential equation (12.26) and the Thomas-Fermi formula (12.40) for the total energy of a neutral atom can be derived by starting with the *energy functional*

$$E[n] = c\int n^{5/3}\, d\tau - Ze^2\int \frac{n}{r}\, d\tau + \frac{e^2}{2}\iint \frac{n(\boldsymbol{r})n(\boldsymbol{r}')}{|\boldsymbol{r}-\boldsymbol{r}'|}\, d\tau d\tau' \tag{1}$$

where $c = (3^{5/3}\pi^{4/3})/10(\hbar^2/m_e)$, the first term on the right-hand side of (1) is the kinetic energy, and the second and third terms are the potential energies of interaction of the electrons with the nucleus and with one another, respectively. Here n is an as-yet-undetermined function of r. We want to find n by using the calculus of variations to minimize $E[n]$ subject to the constraint $Z = \int n\, d\tau$.

(a) Consider the auxiliary functional

$$h[n] = c\int n^{5/3}\, d\tau - Ze^2\int \frac{n}{r}\, d\tau + \frac{e^2}{2}\iint \frac{n(\mathbf{r})n(\mathbf{r}')}{|\mathbf{r}-\mathbf{r}'|}\, d\tau d\tau' - \lambda\left(\int n\, d\tau\right) \tag{2}$$

where λ is an undetermined multiplier. Vary h by making arbitrary small variations in n. Show that the condition $\delta h = 0$ yields the Thomas-Fermi differential equation with $\mu_F = 0$ for a neutral atom. It can be shown that the extremum of h found this way is in fact a minimum. Hence the minimum energy with respect to arbitrary variations in n is given by (12.40).

(b) The condition $Z = \int n\, d\tau$ is left invariant if we replace $n(r)$ by $\beta^3 n(\beta r)$, where β is a positive real parameter. Show that when this replacement is made, $E \to E(\beta) = \beta^2 E_1 + \beta E_2 + \beta E_3$. Using this expression and the fact that $E(\beta)$ is minimized when $\beta \to 1$, obtain the viral theorem for the Thomas-Fermi model.

12.2. In Chapter 10 in an analysis leading to equation (10.55), we discuss the diamagnetic correction to the magnetic field at the nucleus of a hydrogen atom. Use the Thomas-Fermi model to calculate a similar diamagnetic correction at the nucleus for a neutral atom with $Z \gg 1$.

12.3. Consider a simple one-dimensional hydrogen atom that obeys the equation

$$-\frac{1}{2}\frac{\partial^2\psi}{\partial x^2} - Z\delta(x)\psi = E\psi \tag{1}$$

Here the Coulomb potential is replaced by $-Z\delta(x)$.

(a) Find the ground-state energy and wave function and verify that

$$\langle T\rangle = -E = -\frac{1}{2}\langle V\rangle$$

(b) Now consider the one-dimensional helium atom, which obeys

$$H\psi(x_1,x_2) = -\frac{1}{2}\frac{\partial^2\psi}{\partial x_1^2} - \frac{1}{2}\frac{\partial^2\psi}{\partial x_2^2} - Z\delta(x_1)\psi - Z\delta(x_2)\psi + \delta(x_1 - x_2)\psi = E\psi \tag{2}$$

where x_1 and x_2 are the coordinates of the two electrons. First, treat the $\delta(x_1 - x_2)$ term as a perturbation, and find the ground-state energy to first order. Compare with three-dimensional helium.

(c) Employ the variational method to improve the result of (b). Use a trial function analogous to that employed in the text for three-dimensional helium (recall Section 11.6). Find the best value of $\langle H\rangle$, and compare it with three-dimensional helium.

(d) This is the main part of the problem. Still referring to the Hamiltonian of (2), consider the Hartree self-consistent field solution $\Psi(x_1,x_2)=\Phi(x_1)\Phi(x_2)$ for the ground state. Show that

$$-\frac{1}{2}\frac{\partial^2\Phi}{\partial x^2}-Z\delta(x)\Phi(x)+\Phi^3=\varepsilon\Phi \tag{3}$$

where $\langle H\rangle=2\varepsilon-\langle\delta(x_1-x_2)\rangle$. Find the normalized analytic solution to (3) and thus show that

$$\langle H\rangle_{\text{Hartree}}=-\left(Z-\frac{1}{4}\right)^2-\frac{1}{48} \tag{4}$$

Compare this result with the results obtained in (b) and (c). This model is interesting because it yields an exact analytic solution to the Hartree problem, with results that are very similar to those obtained numerically for three-dimansional helium.

12.4. (a) List the multiplets that can be constructed from the configuration $(ns)(n'd)^2$, and order them according to Hund's rules.
(b) Find the possible multiplets for the configuration $(nd)^7$.

12.5. Find the linear combinations of Slater determinants that make up the ${}^2P_{3/2}$ and ${}^2P_{1/2}$ states constructed from three equivalent $\ell=1(p)$ electrons.

12.6. In very heavy atoms, the spin-orbit interaction is so strong that the *L-S* coupling scheme breaks down. Here it is more appropriate to use the *j-j* coupling scheme, in which the orbital and spin angular momenta of a given electron couple to form the angular momentum ***j*** of that electron, and the various ***j***'s are coupled together to form the total ***J***. A similar situation exists in the nuclear shell model, where spin-orbit coupling can be so strong that the *j-j* coupling scheme is necessary. Suppose that n equivalent electrons in an atom, each with the same value of j, are in a partially filled shell and that all the other electrons are in closed shells with zero resultant angular momentum. Taking into account the Pauli principle, what are the possible values of J that can result? This question can be answered methodically by constructing a table that is somewhat analogous to the multiplet tables in the *L-S* coupling scheme (see Section 12.6). In the present case, we label the table columns as

$$m_{j1}\quad m_{j2}\quad\cdots\quad m_{jn}\quad M_J\quad J$$

To satisfy the Pauli principle, no two values of m_j can be the same.
(a) Find the possible values of J for $n=3$, $j=5/2$.
(b) Find the possible values of J for $n=4$, $j=7/2$. In this case, you should find that two values of J can each be realized in two independent ways.

13 Molecules

13.1 The Born-Oppenheimer approximation

Even the simplest diatomic molecule is difficult to analyze quantitatively because there is more than one force center, so we cannot use the central field approximation (CFA). However, one essential feature does exist that permits considerable simplification, as was recognized by M. Born and J. R. Oppenheimer (Born and Oppenheimer 1927): the nuclei are much more massive than the electrons and therefore move much more slowly. To see how the Born-Oppenheimer approximation is implemented, we consider the Schroedinger equation

$$\Delta_e \psi + \Delta_N \psi + V_e(R)\psi + \frac{Z_1 Z_2}{R}\psi = E\psi \tag{13.1}$$

where $\Delta_e \psi$ and $\Delta_N \psi$ are the electronic and nuclear kinetic energy terms (written schematically here), R is the internuclear separation, and $V_e(R)$ is the potential energy of interaction of the electrons with one another and with the nuclei. The last term on the left-hand side of (13.1) is the Coulomb repulsion energy of the two nuclei (where $Z_{1,2}$ are the nuclear electric charges in atomic units), and E is the total energy of the molecule. The first step in the Born-Oppenheimer approximation is to ignore the nuclear kinetic energy. Then $Z_1 Z_2 / R$ is a constant, whereas the electronic wave function $\psi_e(R)$ depends on R as a parameter and satisfies the Schroedinger equation

$$\Delta_e \psi_e(R) + V_e(R)\psi_e(R) = E_e(R)\psi_e(R) \tag{13.2}$$

where the electronic energy $E_e(R)$ is not a real number but rather a real function of the parameter R. In principle, we can solve (13.2) to find $E_e(R)$ and $\psi_e(R)$ for each value of R. Once this is done, we form the quantity

$$V(R) = E_e(R) + Z_1 Z_2 / R \tag{13.3}$$

which is called the *molecular potential energy.* In the next level of approximation, we go back to (13.1) and include the nuclear kinetic energy

$$\Delta_N \psi + \left[\Delta_e \psi + V_e(R)\psi + \frac{Z_1 Z_2}{R}\psi\right] = E\psi \tag{13.4}$$

We also assume that $\psi = \psi_e \psi_N$ is a product wave function. Then (13.4) becomes

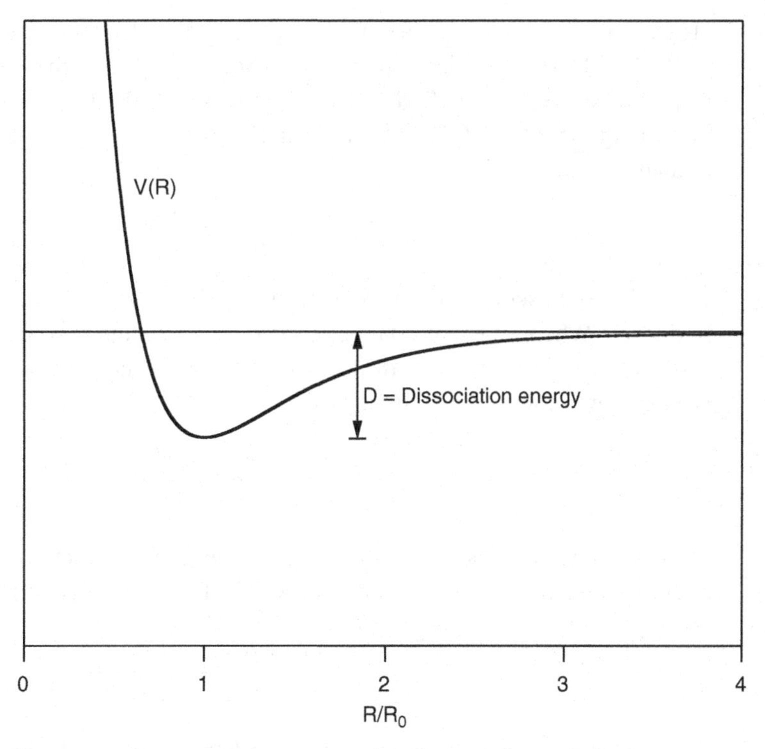

Figure 13.1 Typical shape of a molecular potential energy curve corresponding to a bound state. R_0, the value of R at which $V(R)$ is minimum, is also the equilibrium separation between the nuclei to a good approximation. D is the dissociation energy.

$$\psi_e \Delta_N \psi_N + \psi_e V(R) \psi_N = \psi_e E \psi_N$$

Dividing both sides of this equation by ψ_e, we obtain

$$\Delta_N \psi_N + V(R) \psi_N = E \psi_N \tag{13.5}$$

This equation reveals that the nuclear motion is determined by the molecular potential energy $V(R)$. For bound molecular states, $V(R)$ generally has the shape shown in Figure 13.1.

In first approximation, the nuclei describe two distinct types of motion in $V(R)$

- Vibration about the equilibrium position (stretching mode along the internuclear axis)
- Rigid-body (dumbbell) rotations with two degrees of freedom

We now make some rough but useful estimates of the electronic, vibrational, and rotational energies. A molecule, like an atom, has a linear size of approximately 1 Bohr radius (1 atomic unit). Thus, from the uncertainty principle,

$$\Delta p \Delta x \approx \hbar = 1$$

$$\Delta p \approx \frac{\hbar}{a_0} = 1$$

$$\therefore E_e \approx \frac{\Delta p^2}{2m_e} \approx 1$$

Hence a typical molecule, like a typical atom, has electronic energies of order 1 atomic unit ($\approx$ 10 eV). If the molecular diameter is of order unity, then the width of $V(R)$ and the equilibrium nuclear separation R_0 are also of order unity. Because V(R) is approximately parabolic in the neighborhood of R_0, the vibrational motion is approximately that of a simple harmonic oscillator with energies

$$E_{\text{vib}} \approx M\omega^2 x^2 \tag{13.6}$$

where x is the oscillator displacement, M is the nuclear mass, and ω is the oscillator angular frequency. When x becomes large $(x \to 1)$, the molecule becomes sufficiently distorted that it dissociates, and this requires an electronic excitation energy $E_e \approx 1$. Thus we have $M\omega^2 \approx 1$; that is, vibrational energy splittings are of order

$$\hbar\omega = \omega = \frac{1}{M^{1/2}} \tag{13.7}$$

Next we consider rigid-body rotations of the nuclei, which are characterized by an orbital angular momentum quantum number $\boldsymbol{K}$. The kinetic energy of rotation is

$$E_{\text{rot}} = \frac{\boldsymbol{K}^2}{2I} \approx \frac{\hbar^2 K(K+1)}{MR_0^2}$$

where I is the moment of inertia. Because $\hbar$ = 1 and $R_0 \approx 1$, the separation between adjacent rotational states for small K is

$$E_{\text{rot}} \approx M^{-1} \tag{13.8}$$

Thus, to summarize,

$$E_e : E_{\text{vib}} : E_{\text{rot}} \approx 1 : M^{-1/2} : M^{-1} \tag{13.9}$$

Even for the lightest diatomic molecules (H_2^+, H_2), $M \approx 10^3$. This justifies a posteriori the approximation scheme of Born and Oppenheimer.

13.2 Classification of diatomic molecular states

Because the force on a molecular electron is noncentral, the electronic orbital angular momentum operator $\boldsymbol{L}$ does not commute with the Hamiltonian; consequently, there is no definite orbital angular momentum eigenvalue L. However, a diatomic molecule is axially symmetric about the internuclear axis z; therefore, L_z commutes with the Hamiltonian, and M, the eigenvalue of L_z, is well defined. Electronic states of the molecule are classified by $\Lambda = |M|$ as

$$\Sigma, \Pi, \Delta, \ldots$$

according to whether Λ = 0, 1, 2, …, respectively. The capital Greek letters Σ, Π, Δ, and so on are used in place of the Latin letters S, P, D, and so on that we employed to describe orbital angular momenta in atoms.

Consider any plane that contains the internuclear axis, and reflect the electronic wave function through this plane. If we ignore possible parity violating effects due to the weak interaction (and these are exceedingly small in any case), this reflection does not affect any physically observable property of the molecule. On the other hand, under the reflection, $M \to -M$ in the case of Π, Δ, … states. Thus each Π, Δ, ... state is twofold degenerate. However, Σ states are not degenerate; reflection through the plane either leaves the electronic wave function unchanged (an Σ^+ state) or changes the sign of the wave function (an Σ^- state). There exists an additional symmetry for homonuclear diatomic molecules: the electronic Hamiltonian is symmetric with respect to inversion about the midpoint between the two nuclei. Hence, according to the parity theorem of Chapter 6, nondegenerate solutions ψ are either even (*g* for *gerade* in German) or odd (*u* for *ungerade*) under reflection about the midpoint. Thus we have the states

$$\begin{array}{l} 1\Sigma_g^{\pm}, 2\Sigma_g^{\pm}, \ldots \\ 1\Sigma_u^{\pm}, 2\Sigma_u^{\pm}, \ldots \\ 1\Pi_g, 2\Pi_g, \ldots \\ 1\Pi_u, 2\Pi_u, \ldots \end{array}$$

where here the numbers 1, 2, … simply label the states in order of increasing energy.

13.3 Analysis of electronic motion in the hydrogen molecular ion

We now consider in some detail the electronic wave functions and energies for the simplest of all molecules: H_2^+. This is important because some of the features revealed here can be generalized to yield an understanding of the covalent bond in H_2 and more complicated diatomic molecules.[1] Figure 13.2 shows the prolate spheroidal coordinates that are convenient for describing H_2^+. The Schroedinger equation can be separated in these coordinates and solved exactly for each fixed value of R (Bates et al. 1953). The resulting eigenfunctions are expressed as

$$\psi(\lambda, \mu, \phi) = L(\lambda) M(\mu) e^{im\phi} \tag{13.10}$$

where

$$\lambda = \frac{r_a + r_b}{R} \qquad \mu = \frac{r_a - r_b}{R}$$

and ϕ is an angle of rotation about the internuclear axis.

In Figure 13.3 we show the exact energies obtained in the calculation by Bates, Ledsham, and Stewart for the lowest σ and π states as a function of R. (Note that lowercase Greek letters are used to denote H_2^+ states.)

We consider the lowest σ solutions at the two extremes $R = 0$ and $R = \infty$, as well as at finite R. For $R = \infty$, we obviously have a proton and a hydrogen atom separated by infinite distance. The electron could be centered on the first proton with wave function

[1] In our discussions of binding forces in diatomic molecules in this chapter, we ignore the very feeble van der Waals interaction (discussed in Section 10.6).

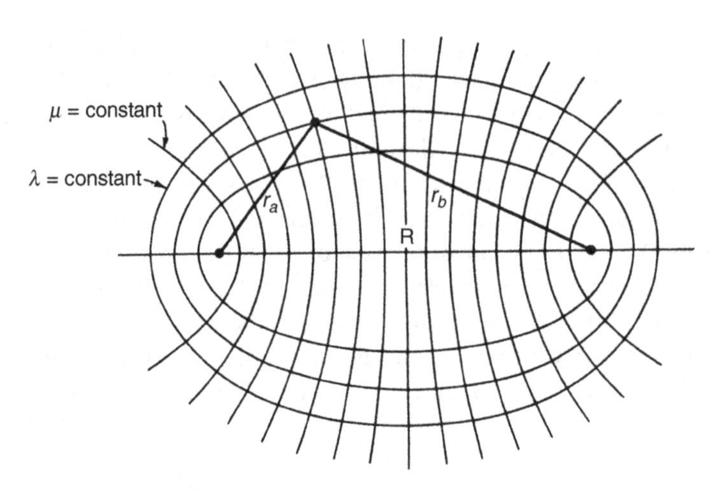

Figure 13.2 Prolate spheroidal coordinates for the hydrogen molecular ion: λ is constant on ellipsoids of revolution with the protons at the two foci, whereas μ is constant on the conjugate hyperboloids of revolution. (Reprinted with permission from *Quantum Theory of Matter*, 2nd ed., by John C. Slater; copyright 1968, The McGraw-Hill Companies, Inc.)

$$\psi_1 = \frac{1}{\pi^{1/2}} e^{-r_a} \tag{13.11}$$

or it could be centered on the second proton with wave function

$$\psi_2 = \frac{1}{\pi^{1/2}} e^{-r_b} \tag{13.12}$$

Because ψ_1 and ψ_2 are degenerate at $R = \infty$ with energy $E_e = -\frac{1}{2}$, any linear combination of these solutions is also a solution with the same energy. Thus we can form the solutions

$$\psi(1\sigma_g, R = \infty) = \frac{1}{\sqrt{2}}(\psi_1 + \psi_2) = \frac{1}{\sqrt{2\pi}}(e^{-r_a} + e^{-r_b}) \tag{13.13}$$

and

$$\psi(1\sigma_u, R = \infty) = \frac{1}{\sqrt{2}}(\psi_1 - \psi_2) = \frac{1}{\sqrt{2\pi}}(e^{-r_a} - e^{-r_b}) \tag{13.14}$$

It can be seen from Figure 13.3 that at any finite R, the $1\sigma_g$ solution has lower energy than the $1\sigma_u$ solution. In Figure 13.4 we plot the $1\sigma_g$ and $1\sigma_u$ solutions along the internuclear axis for various values of R. These plots clearly reveal how $1\sigma_g$ gradually transforms from (13.13) at $R = \infty$ to the $1s$ state of He^+ with energy $E = -2$ at $R = 0$. Meanwhile, $1\sigma_u$ changes from (13.14) to the $2p$ state of He^+ (with energy $E = -\frac{1}{2}$) at $R = 0$.

When the Coulomb repulsion energy $1/R$ of the two protons is added to the energies of Figure 13.3, we obtain the molecular potential energy curves of Figure 13.5. From this figure it is clear that H_2^+ is stable in the $1\sigma_g$ state with a dissociation energy $D = 0.1$ atomic unit.

Qualitatively, binding occurs for $1\sigma_g$ because the probability is high to find the electron in the region between the two protons. Here the potential energy of interaction between the electron and the two protons is large and negative. This can be seen in Figure 13.6, where we plot the electronic potential energy along the internuclear axis for various values of R. This phenomenon is

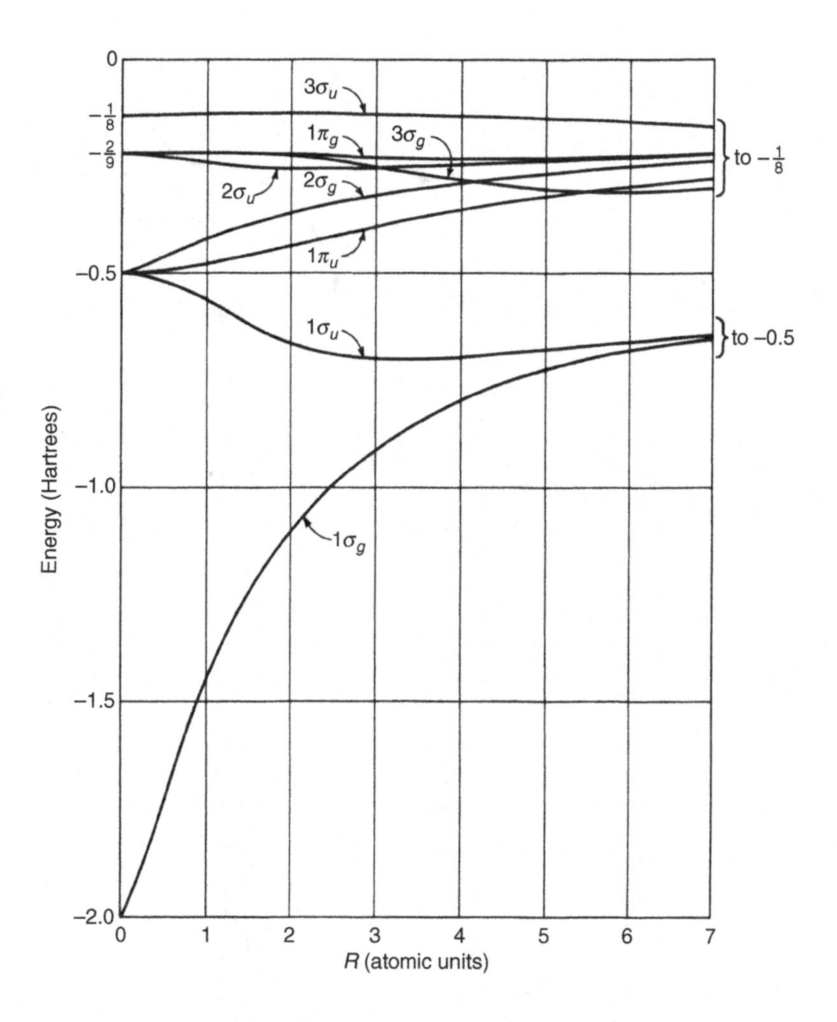

Figure 13.3 Lowest electronic energies (in atomic units) of H_2^+ as a function of internuclear distance *R* from the exact calculation for fixed nuclei by Bates, Ledsham, and Stewart. (Reprinted with permission from *Quantum Theory of Matter*, 2nd ed., by John C. Slater; copyright 1968, The McGraw-Hill Companies, Inc.)

very general and occurs for more complex diatomic molecules and in particular for H_2, where low energy is associated with large probability for both electrons to be located in the region between the nuclei (i.e., in an exchange-symmetric spatial state). Because of the Pauli principle, this requires the two-electron spin function to be antisymmetric: a singlet. Thus, although in the $1s2s$ configuration of atomic helium the triplet state lies lower, for H_2 and indeed for most (but not all) diatomic molecules, the lowest electronic state is $^1\Sigma$.

13.4 Variational method for the hydrogen molecular ion

Although the electronic Schroedinger equation for H_2^+ has been solved analytically for any fixed R, it would be extremely difficult and impractical to generalize this method to the case of H_2 or more complex molecules. Therefore, we seek a reasonably accurate but simpler approach that

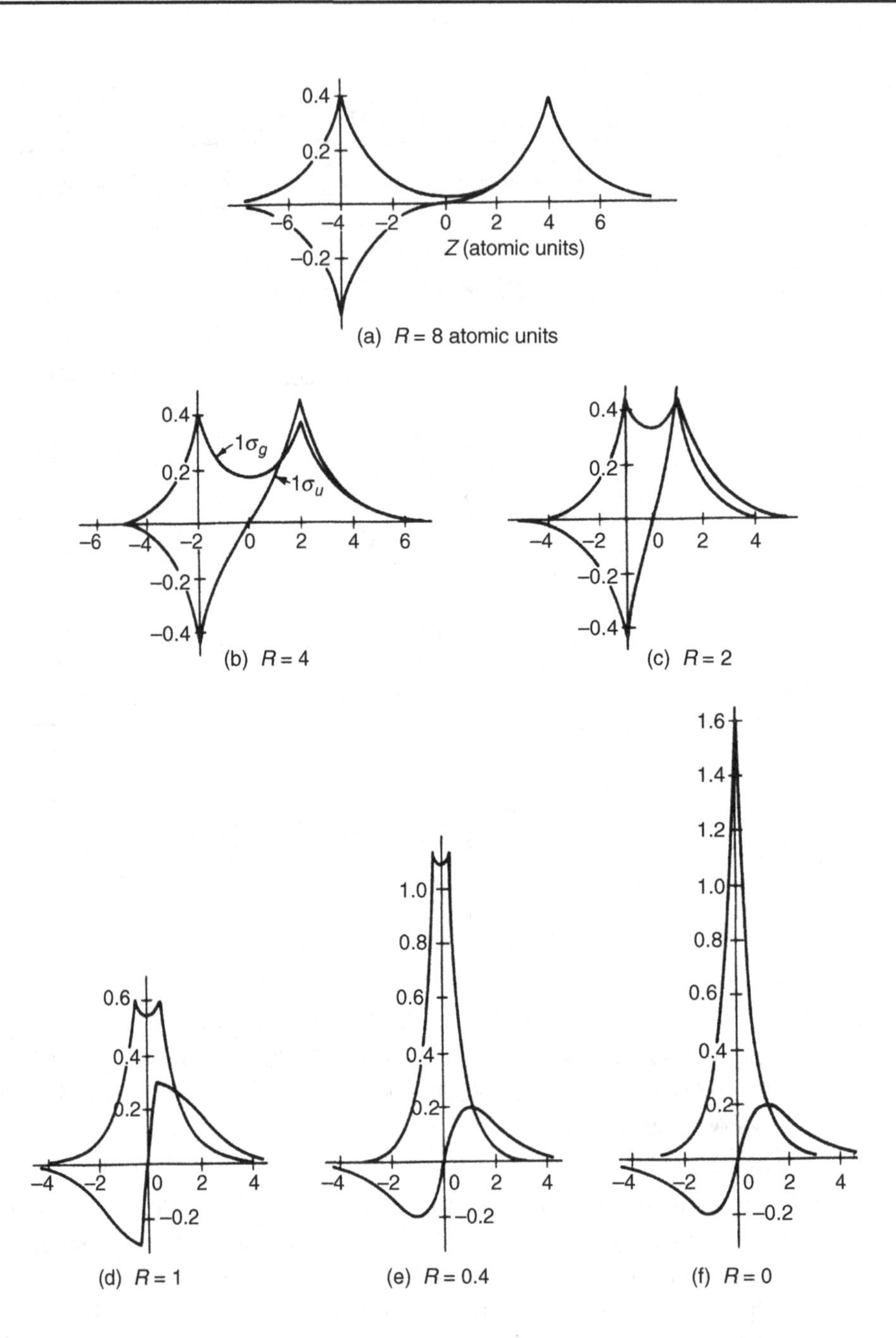

Figure 13.4 The $1\sigma_g$ and $1\sigma_u$ wave functions of H_2^+ plotted along the internuclear axis for various nuclear separations. (Reprinted with permission from *Quantum Theory of Matter*, 2nd ed., by John C. Slater; copyright 1968, The McGraw-Hill Companies, Inc.)

can be generalized. To this end, we recall (13.13), which describes the $1\sigma_g$ state at $R = \infty$. This solution is a linear combination of atomic orbitals (LCAO) with effective nuclear charge $\alpha = 1$ in

$$\psi = N(\infty)\left[e^{-\alpha r_a} + e^{-\alpha r_b}\right] \tag{13.15}$$

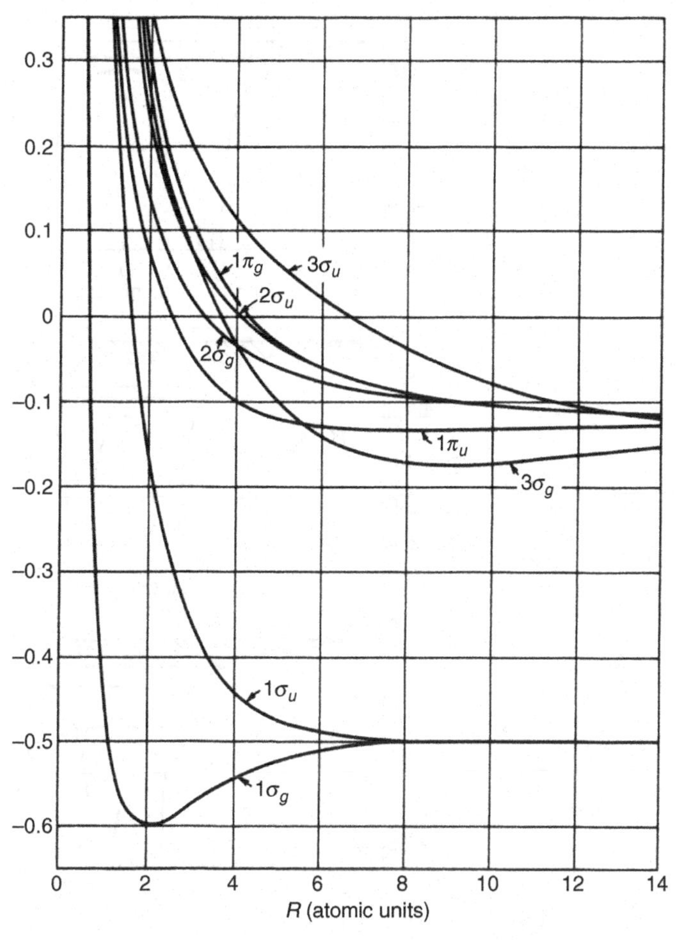

Figure 13.5 Molecular potential energy curves for H_2^+ obtained by adding the nuclear repulsion energy $1/R$ to the exact solution curves of Figure 13.3. (Reprinted with permission from *Quantum Theory of Matter*, 2nd ed., by John C. Slater; copyright 1968, The McGraw-Hill Companies, Inc.)

where N is a normalization factor. (Do not confuse this α with the fine-structure constant or the spin-½ spinor!) When $R \to 0$ and hence $r_{a,b} \to r$, we have a solution with effective nuclear charge $\alpha = 2$

$$\psi = N(0)e^{-2r} \tag{13.16}$$

This suggests that we try a solution of the form

$$\psi(1\sigma_g) = N(R)\left[e^{-\alpha r_a} + e^{-\alpha r_b}\right] \tag{13.17}$$

where α is chosen at each R to minimize the energy by means of the variational method. We first consider the normalization constant $N(R)$. Here we must evaluate the integral

$$I = N^{-2} = \int \left[\exp(-2\alpha r_a) + \exp(-2\alpha r_b) + 2\exp(-\alpha r_a - \alpha r_b)\right] d\tau \tag{13.18}$$

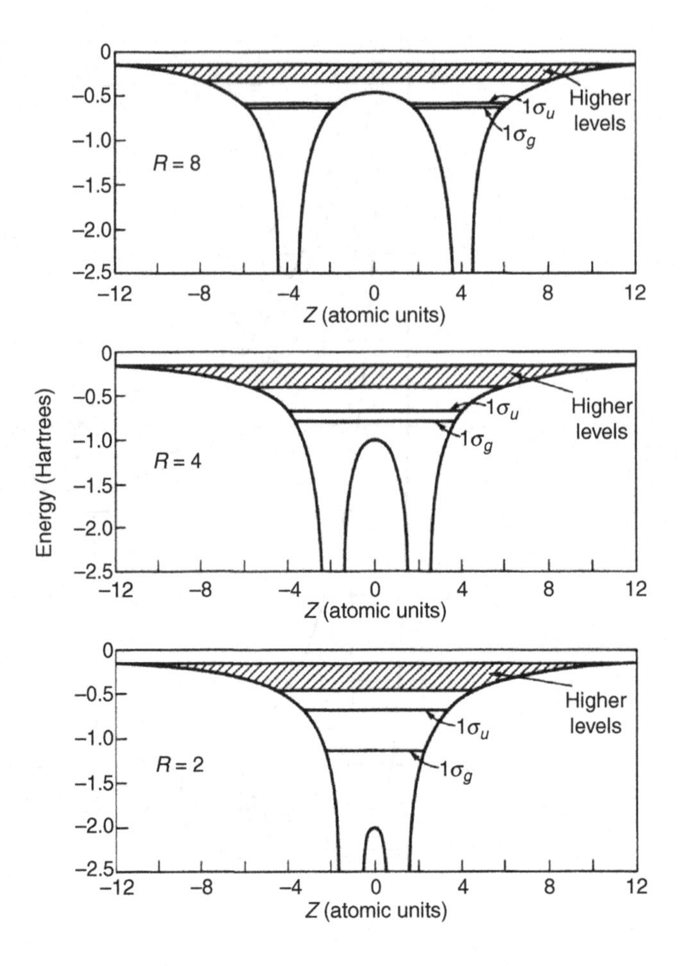

Figure 13.6 Potential energy of electron along internuclear axis in H_2^+ for various internuclear separations. Energy levels of $1\sigma_{g,u}$ are also shown. The shaded band indicates the range of energies of the higher levels shown in Figure 13.3. (Reprinted with permission from *Quantum Theory of Matter*, 2nd ed., by John C. Slater; copyright 1968, The McGraw-Hill Companies, Inc.)

The first and second integrals are easily done in spherical polar coordinates centered on protons a and b, respectively. The third integral must be calculated in prolate spheroidal coordinates. The result is

$$I = \frac{2\pi}{\alpha^3}\left[1 + e^{-\alpha r}\left(1 + \alpha R + \frac{\alpha^2 R^2}{3}\right)\right] \equiv \frac{2\pi}{\alpha^3}(1+S) \tag{13.19}$$

Hence

$$\psi(1\sigma_g) = \sqrt{\frac{\alpha^3}{2\pi(1+S)}}\left[e^{-\alpha r_a} + e^{-\alpha r_b}\right] \tag{13.20}$$

We now consider $E_e(R) = \langle\psi|H|\psi\rangle$, where

$$H = -\frac{1}{2}\nabla^2 - \frac{1}{r_a} - \frac{1}{r_b}$$

It can be shown that

$$-\frac{1}{2}\int \psi^* \nabla^2 \psi \, d\tau = \alpha^2 F_1(\alpha R) \tag{13.21}$$

where

$$F_1(\alpha R) = \frac{1}{2}\frac{1 + e^{-\alpha R}\left(1 + \alpha R - \dfrac{\alpha^2 R^2}{3}\right)}{1 + e^{-\alpha R}\left(1 + \alpha R + \dfrac{\alpha^2 R^2}{3}\right)} \tag{13.22}$$

and that

$$\left\langle -\frac{1}{r_a} - \frac{1}{r_b} \right\rangle = \alpha F_2(\alpha R) \tag{13.23}$$

where

$$F_2(\alpha R) = -\frac{1 + 2e^{-\alpha R}(1 + \alpha R) + \dfrac{1}{\alpha R} - \left(1 + \dfrac{1}{\alpha R}\right)e^{-2\alpha R}}{1 + e^{-\alpha R}\left(1 + \alpha R + \dfrac{\alpha^2 R^2}{3}\right)} \tag{13.24}$$

It is easy to verify that $F_1(0) = 1/2$ and $F_2(0) = -2$. Therefore, because $\alpha(0) = 2$,

$$\langle T(0) \rangle = 2 \qquad \langle V(0) \rangle = -4 \qquad E_e(0) = -2$$

Also, $F_1(\infty) = 1/2$ and $F_2(\infty) = -1$. Because $\alpha(\infty) = 1$, this yields

$$\langle T(\infty) \rangle = 1/2 \qquad \langle V(\infty) \rangle = -1 \qquad E_e(\infty) = -1/2$$

These results agree with the exact solution. At finite $R \neq 0$, we have

$$E_e = \alpha^2 F_1(\alpha R) + \alpha F_2(\alpha R) \tag{13.25}$$

It is convenient to employ the new variable $w = \alpha R$, in terms of which (13.25) is written

$$E_e = \alpha^2 F_1[w(\alpha)] + \alpha F_2[w(\alpha)] \tag{13.26}$$

Now, calculating $\partial E_e / \partial \alpha$ and setting it equal to zero, we obtain after simple algebra

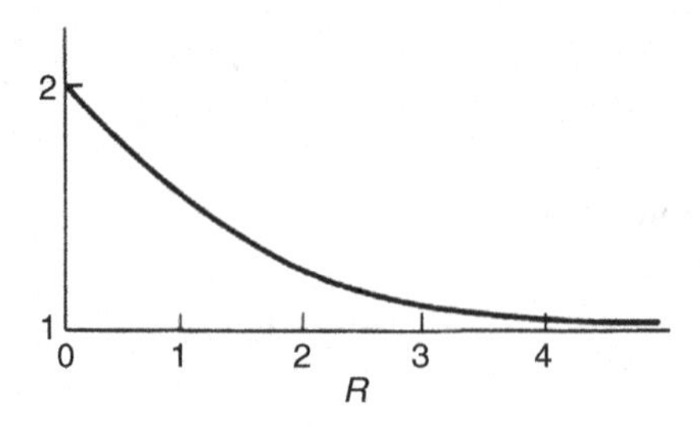

Figure 13.7 The variational parameter α plotted versus R for H_2^+. (Reprinted with permission from *Quantum Theory of Matter,* 2nd ed., by John C. Slater; copyright 1968, The McGraw-Hill Companies, Inc.)

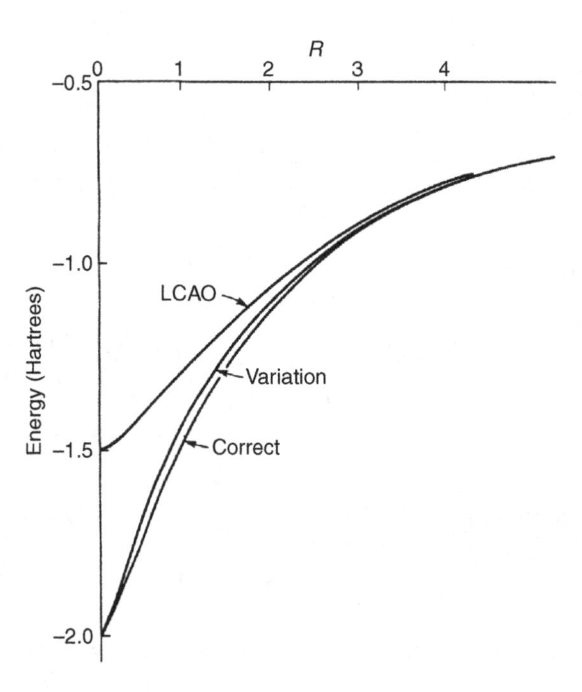

Figure 13.8 Comparison of variational, exact, and LCAO energies for the $1\sigma_g$ state of H_2^+. (Reprinted with permission from *Quantum Theory of Matter,* 2nd ed., by John C. Slater; copyright 1968, The McGraw-Hill Companies, Inc.)

$$\alpha = -\frac{F_2 + w\dfrac{\partial F_2}{\partial w}}{2F_1 + w\dfrac{\partial F_1}{\partial w}} \tag{13.27}$$

This yields α as a function of R, which is plotted in Figure 13.7.

With α determined, we calculate $E_e(R)$ from (13.25). The result is shown in Figure 13.8 together with the exact solution and with the case where $\alpha = 1$ is held fixed (LCAO). Clearly, the variational energy curve is very close to the exact solution.

13.5 Molecular orbital and Heitler-London methods for H_2

Because the variational method for H_2^+ is successful, it seems natural to try to extend it to H_2. Let us assume that a given electron has a spatial wave function

$$g = \sqrt{\frac{1}{2(1+S)}}(a+b) \tag{13.28}$$

where

$$a = \left(\frac{\alpha^3}{\pi}\right)^{1/2} e^{-\alpha r_a} \qquad b = \left(\frac{\alpha^3}{\pi}\right)^{1/2} e^{-\alpha r_b} \tag{13.29}$$

It therefore seems reasonable to construct the following wave function for two identical electrons:

$$\psi(1,2) = \frac{1}{\sqrt{2}}\begin{vmatrix} g(1)\alpha_1 & g(2)\alpha_2 \\ g(1)\beta_1 & g(2)\beta_2 \end{vmatrix} = g(1)g(2)\frac{\alpha_1\beta_2 - \alpha_2\beta_1}{\sqrt{2}} \tag{13.30}$$

This is obviously an $S = 0$ (singlet) combination of the two electron spins. It is easy to see that in the limit $R \to 0$, (13.30) becomes the $1s^2\ ^1S_0$ wave function of atomic helium in the approximation where electron-electron correlations are neglected. The next step is to calculate $E(\alpha) = \langle\psi|H|\psi\rangle$ using (13.30) and

$$H = -\frac{1}{2}\nabla_1^2 - \frac{1}{2}\nabla_2^2 - \frac{1}{r_{1a}} - \frac{1}{r_{2a}} - \frac{1}{r_{1b}} - \frac{1}{r_{2b}} + \frac{1}{r_{12}} \tag{13.31}$$

where the coordinates employed in (13.31) are as shown in Figure 13.9.

Calculation of $E(\alpha)$ using (13.30) and (13.31) and minimization for each value of R yield the result shown in Figure 13.10, in which the $1/R$ Coulomb repulsion of the protons is included. The figure shows a serious disagreement between the experimentally determined molecular potential energy and that calculated by this variational molecular orbital method. The disagreement is particularly severe for large R.

Why have we failed badly here when essentially the same method worked so well for H_2^+? We can readily see the answer by considering (13.30) in the limit of large R. In that case, $S \to 0$, and we have

$$g(1)g(2) = \frac{1}{2}\left[a(1)a(2) + b(1)b(2) + a(1)b(2) + a(2)b(1)\right] \tag{13.32}$$

In (13.32), $a(1)a(2)$ corresponds to both electrons centered on proton a and none on proton b (i.e., a hydrogen negative ion H^- and a proton), whereas $b(1)b(2)$ corresponds to both electrons centered on proton b and none on a (also a hydrogen negative ion and a proton). Only the third and fourth terms on the right-hand side of (13.32) correspond to two neutral hydrogen atoms.

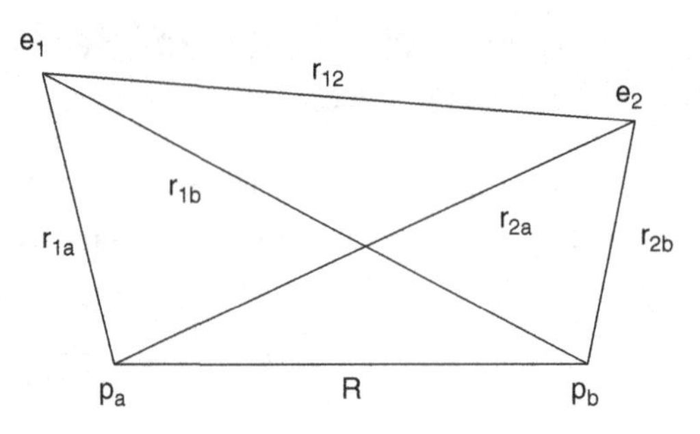

Figure 13.9 Coordinates for description of the electrons and protons in the H_2 Molecule.

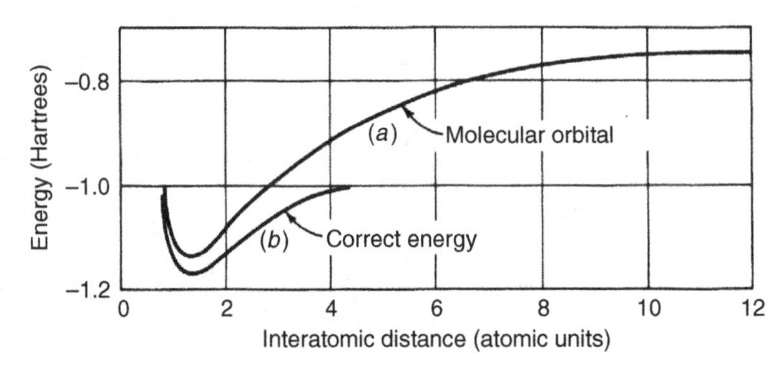

Figure 13.10 Ground-state energy of H_2 as a function of R: (a) molecular orbital calculation based on (13.30); (b) experimentally determined energy curve. (Reprinted with permission from *Quantum Theory of Matter*, 2nd ed., by John C. Slater; copyright 1968, The McGraw-Hill Companies, Inc.)

However, a ground-state hydrogen molecule always dissociates into two H atoms and never into a hydrogen negative ion and a proton: the latter would cost too much energy.

This problem is remedied by eliminating the terms $a(1)a(2)$ and $b(1)b(2)$ so that we obtain the two-electron spatial wave function

$$\psi_{\text{space}}(1,2) = \text{const}\left[a(1)b(2)+b(1)a(2)\right] \tag{13.33}$$

which was originally proposed by W. Heitler and F. London (Heitler and London 1927). As before, we can fix $\alpha = 1$ as in the LCAO solution for H_2^+, or we can vary α to minimize E at each R. The latter was first done by Wang (1928) and by Rosen (1931), with further improvements by many authors [see, e.g., James (1933, 1935)]. Figure 13.11 shows the results obtained by Rosen for the energy.

The variational parameter α of Wang and Rosen is plotted versus R in Figure 13.12. Note that at $R = 0$, $\alpha = 1.6875 = 27/16$. This is not difficult to understand, for when $R \to 0$, the H_2 molecule becomes a helium atom as far as the electrons are concerned. We have only to recall the simple variational calculation that we did on the ground state of helium [see Section 11.6 and equation (11.27)] to obtain the factor 27/16.

We now return to the original molecular orbital spatial function that appears in (13.30). It is

$$g(1)g(2) = \frac{1}{2(1+S)}\left[a(1)a(2)+b(1)b(2)+a(1)b(2)+a(2)b(1)\right] \tag{13.34}$$

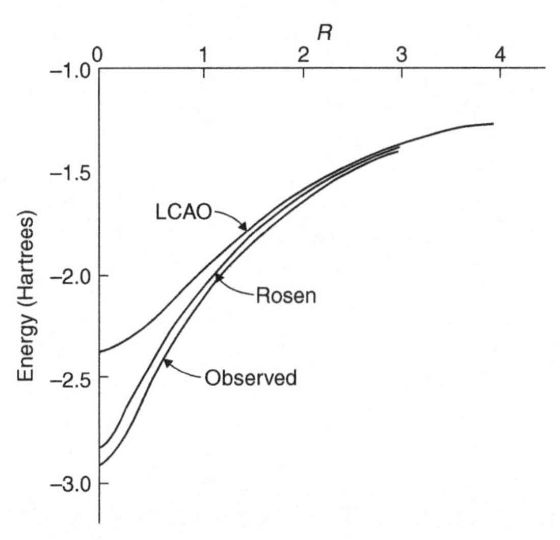

Figure 13.11 Electronic energy in H_2: Heitler-London method, as calculated for $\alpha = 1$ (LCAO), and by variational approach (Rosen). (Reprinted with permission from *Quantum Theory of Matter*, 2nd ed., by John C. Slater; copyright 1968, The McGraw-Hill Companies, Inc.)

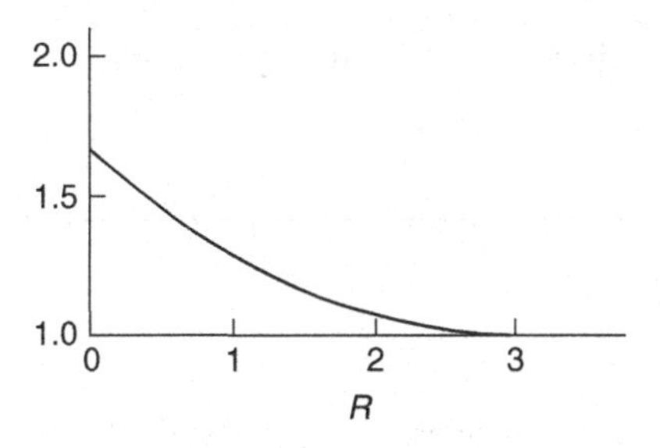

Figure 13.12 Variational parameter α plotted versus R from Wang and Rosen calculation of H_2. (Reprinted with permission from *Quantum Theory of Matter*, 2nd ed., by John C. Slater; copyright 1968, The McGraw-Hill Companies, Inc.)

It is also of interest to consider the function $u(1)u(2)$, where

$$u = \sqrt{\frac{1}{2(1+S)}}(a-b)$$

We have

$$u(1)u(2) = \frac{1}{2(1+S)}\left[a(1)a(2)+b(1)b(2)-a(1)b(2)-b(1)a(2)\right] \tag{13.35}$$

From (13.34) and (13.35) we can form the linear combinations

$$g(1)g(2)-u(1)u(2) = \text{const}\left[a(1)b(2)+b(1)a(2)\right] \quad \text{(Heitler-London)} \tag{13.36}$$

and

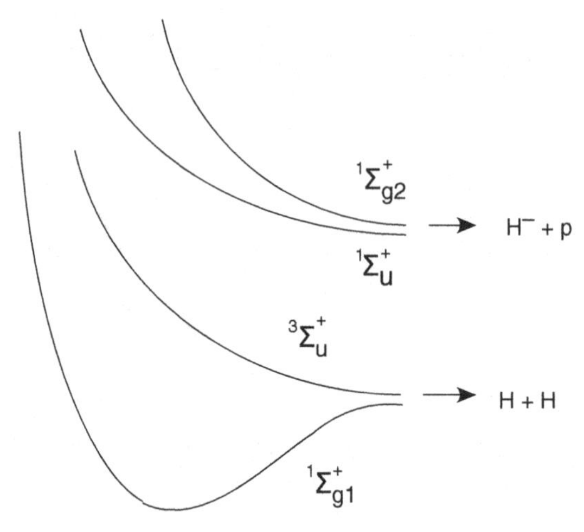

Figure 13.13 Schematic diagram (not to scale) of low-lying electronic states of H_2 described in text.

$$g(1)g(2)+u(1)u(2)=\text{const}\left[a(1)a(2)+b(1)b(21)\right] \quad \text{(ionic)} \tag{13.37}$$

both of which correspond to spin singlets. In addition, we can construct the following combinations:

$$1\sigma_g 1\sigma_u \text{ triplet:} \quad g(1)u(2)-u(1)g(2)=\text{const}\left[a(1)b(2)-b(1)a(2)\right] \tag{13.38}$$

and

$$1\sigma_g 1\sigma_u \text{ singlet:} \quad g(1)u(2)+u(1)g(2)=\text{const}\left[-a(1)a(2)+b(1)b(2)\right] \tag{13.39}$$

We have already seen that the Heitler-London function (13.36) corresponds to the ground $\left(^1\Sigma^+_{g1}\right)$ electronic state, which dissociates to two $1s$ H atoms with electron spins opposed. The triplet function (13.38) corresponds to an unbound electronic state $\left(^3\Sigma^+_u\right)$, which dissociates to two $1s$ hydrogen atoms with parallel electron spins. The two remaining singlet functions are associated with excited (unbound) electronic states, each of which dissociates to an H^- ion and a proton (Figure 13.13).

13.6 Valency: An elementary and qualitative discussion of the chemical bond

We have indicated that in the hydrogen molecule, binding can only occur when both electrons have a high probability of being located in the region between the nuclei. This requires the two electrons to be in an exchange-symmetric spatial state. Because of the Pauli principle, this, in turn, implies that the two-electron spin function must be antisymmetric: the spins must be opposed. The following general principle is suggested:

Atoms can only combine to form stable molecules if the spins of their shared valence electrons are opposed.

This is perhaps the single most powerful principle in all of chemistry, and it explains a great deal about the nature of the chemical bond. In what follows, we give a number of elementary examples.

Normally, hydrogen gas is in the form of molecular hydrogen, not atomic hydrogen. However, in several relatively recent experiments, a gas of electron-spin-polarized atomic hydrogen has been created. These atoms do not combine to form molecular hydrogen because the $^3\Sigma_u^+$ state is not bound (recall Figure 13.13). However, if either inadvertently or deliberately a spin-polarization relaxation mechanism is introduced, the atoms do combine, sometimes with explosively destructive results. Consider next the interaction between a hydrogen atom and a helium atom in their ground states. The spins of the two electrons in the helium atom are opposed to one another. Thus, regardless of whether the electron spin in the hydrogen atom is up or down, it must encounter an electron with parallel spin in helium. Therefore, there is no bound state of the HHe system. Similarly, two normal helium atoms cannot form a bound state. More generally, because all the inert gas atoms in their ground states have closed electron shells with spins paired off, they do not show any chemical activity.

It is convenient to define the *valency* as an integer equal to twice the net electronic spin of an atom. Obviously, the valency of each rare gas atom is zero, whereas the valency of a hydrogen atom, and of the alkali atoms Li, Na, K, Rb, Cs, and Fr in Group I of the periodic table, is unity.

Group II atoms (Be, Mg, Ca, Sr, Ba, etc.) have an s^2 ground configuration, with zero valency. However, there exists an sp configuration with valency equal to 2, relatively close to the ground configuration. The energy difference between sp and s^2 is more than repaid when an alkaline earth atom combines with another atom or atoms: thus the principal valency of alkaline earth elements is 2.

Elements of Group III have ground configuration s^2p, which would give a valency of unity. However, the configuration sp^2 is relatively close, and here it is possible for the total spin to be $S = 3/2$, which gives a valency of 3. The lighter members of this group (B and Al) are only tervalent, whereas the heavier members are both univalent and tervalent (e.g., consider the compounds TlCl and $TlCl_3$).

Group IV atoms (C, Si, etc.) have ground states $s^2p^2\,{}^3P$ with spin of unity and hence a valency of 2. However, for the light atoms in this group, and especially for carbon, the state $sp^3\,{}^5S$ with spin of 2 and hence valency of 4 lies relatively close to the ground state (see Figure 12.2). Thus carbon is mainly quadrivalent, which enables it to form an enormous number of compounds. The heavier members of Group IV are increasingly bivalent.

Atoms of Group V, for example, nitrogen, have the ground state $s^2p^3\,{}^4S_{3/2}$ with a total spin of 3/2; these are therefore tervalent (e.g., consider ammonia NH_3). However, there exists the configuration s^2p^2s', where one of the electrons is promoted to a higher s orbital (s'). Although the energy of this orbital is quite large, it is possible for the valency of nitrogen to be 5 in some cases, for example, in HNO_3.

Group VI elements have the ground state $s^2p^4\,{}^3P$ with total spin of unity and hence a valency of 2. The lightest member of the group, oxygen, is solely bivalent. However, for the heavier members, the excited configurations s^2p^3s' and $s^2p^2s'p'$ yield valencies of 4 and 6, respectively.

Thus, for example, one has the compounds H_2S, SO_2, and SO_3, where sulfur has valencies 2, 4, and 6.

The halogens (Group VII) have s^2p^5 ground states with spin ½, which yields a valency of unity, and this is the only valency exhibited by fluorine. However, excited configurations of the heavier halogens yield valencies of 3, 5, and 7. For example, one has HCl (valency of 1), $HClO_2$ (valency of 3), $HClO_3$ (valency of 5), and $HClO_4$ (valency of 7).

Similar qualitative remarks are useful for understanding a great deal about chemical compounds of transition metals (with incomplete d shells) and rare earths (with incomplete f shells).

It is well known to every student of elementary chemistry that molecules have a definite and sometimes very complex spatial structure. This arises from the fact that valencies are often associated with definite directions in space. We begin with a simple example: the tervalent nitrogen atom. As we know, the ground state is $1s^22s^22p^3\,{}^4S_{3/2}$. Because the total orbital angular momentum is $L = 0$, we must have $M_L = 0$, which means that each of the p electrons must have a different value of m_ℓ. We thus have the states

$$\begin{aligned} |p_z\rangle &= Y_1^0 = \sqrt{\frac{3}{4\pi}}\cos\theta \\ |p_x\rangle &= \frac{1}{\sqrt{2}}(-Y_1^1 + Y_1^{-1}) = \sqrt{\frac{3}{4\pi}}\sin\theta\cos\phi \\ |p_y\rangle &= \frac{i}{\sqrt{2}}(Y_1^1 + Y_1^{-1}) = \sqrt{\frac{3}{4\pi}}\sin\theta\sin\phi \end{aligned} \tag{13.40}$$

where the last two states are independent linear combinations of the degenerate $m_\ell = \pm 1$ states. The probability distributions for the first two functions in (13.40) are indicated schematically in Figure 13.14.

Clearly, the directions where these three electronic wave functions yield maximum probabilities are mutually orthogonal. We are therefore led to expect that the chemical bonds involving tervalent nitrogen should be at right angles to one another. In fact, the NH_3 molecule forms a tetrahedron, where the angles between the NH bonds are 107° × 3. This is larger than 90° because of the mutual repulsion of the hydrogen atoms.

Phosphorus, arsenic, antimony, and bismuth are similar to nitrogen: they all have ground states with p^3 configurations and $L = 0$. Thus one expects that their bonds are also mutually orthogonal. Indeed, in PH_3, the angles are 93° × 3.

More than one bond can exist between two given atoms. For example, consider the two carbon atoms in ethylene (C_2H_4), with one carbon displaced relative to the other along the z-axis (Figure 13.15).

Here each carbon is quadrivalent. Two of the electrons from each carbon are shared with two H atoms. This leaves two more to be shared with the other carbon. One from each carbon can have an orbital of the form $|p_z\rangle$ in (13.40). These form a σ bond, as shown in Figure 13.16a. Here the charge overlap is substantial, as indicated by the shaded region. The other pair of electrons is oriented along x or y, with symmetry axes in planes perpendicular to z. These electrons form a π bond (Figure 13.16b), which is considerably weaker than a σ bond because the charge overlap is much smaller.

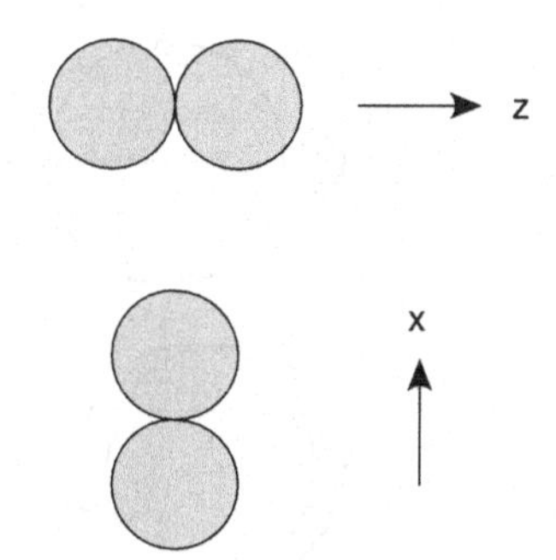

Figure 13.14 Schematic diagrams showing orientation of *pz*, *px* orbitals.

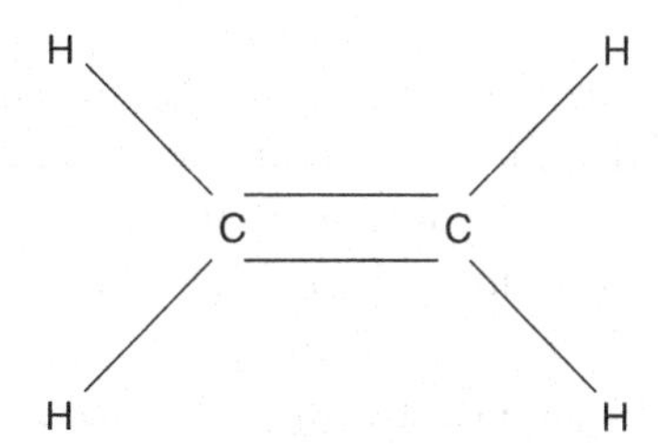

Figure 13.15 Diagram of ethylene: C_2H_4. Each carbon atom is quadrivalent and shares single bonds with two hydrogen atoms and a double bond with the other carbon.

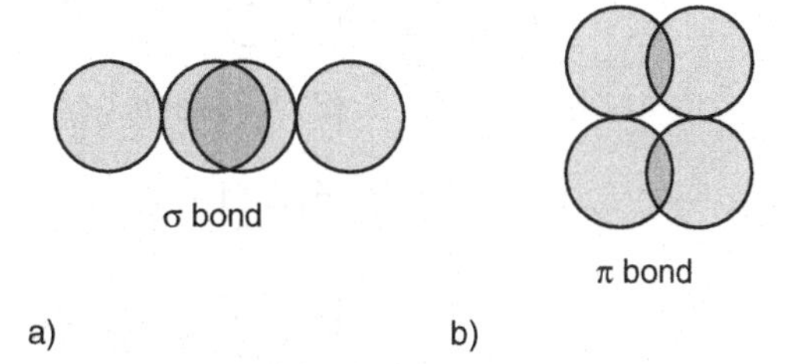

Figure 13.16 Schematic diagrams showing charge overlap: a) σ bond, b) π bond.

Two ground-state nitrogen atoms are joined together with a *triple* bond (one σ bond, a π_x bond, and a π_y bond). Hence N_2 is one of the most strongly bound diatomic molecules.

Atomic oxygen has the ground configuration np^4. To form the molecule O_2, we employ three of the p electrons from each atom to form a σ bond and $\pi_{x,y}$ bonds as in N_2. What should we do with the fourth p electron in each atom? The smallest cost in energy (minimum Coulomb repulsion) occurs if we keep them as far apart as possible. This occurs in a spatially antisymmetric combination of a $|p_x\rangle$ and a $|p_y\rangle$ orbital. The corresponding spin function is symmetric; hence the O_2 electronic ground state is $^3\Sigma$, contrary to the general rule that almost all diatomic molecular electronic ground states are $^1\Sigma$.

Let us return to Figure 13.16a, which is a schematic representation of a σ bond. It is evident that we could improve the charge overlap and consequently the strength of this bond if we could make each orbital asymmetric, as shown in Figure 13.17.

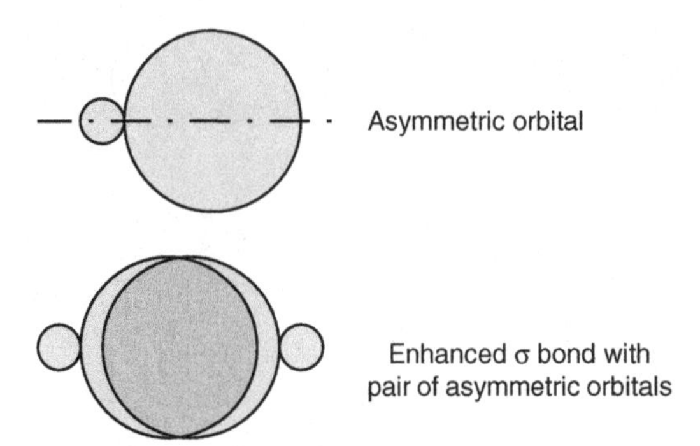

Figure 13.17 Schematic diagrams illustrating asymmetric (hybrid) orbitals.

Such an asymmetry occurs if, for example, we mix an *s* orbital with a *p* orbital. Consider the lithium atom, which has the ground configuration $1s^2 2s$. A *2p* orbital has higher energy than *2s*, but the cost in energy is more than regained in better charge overlap in the σ bond between two lithium atoms to form the Li_2 molecule. The orbital in question is $|2s + \lambda \bullet 2p_z\rangle$, where the optimal value of λ is approximately 0.3. This mixing of orbitals is called *hybridization*.

For our final example, we consider H_2O. Although the ground configuration of oxygen is $1s^2 2s^2 2p^4$, considerable hybridization occurs in the H_2O molecule for the six electrons with $n = 2$. Let us assume that this hybridization is perfect and form the four orthogonal combinations

$$\begin{aligned}
|1\rangle &= \frac{1}{2}|s + p_x + p_y + p_z\rangle \\
|2\rangle &= \frac{1}{2}|s + p_x - p_y - p_z\rangle \\
|3\rangle &= \frac{1}{2}|s - p_x + p_y - p_z\rangle \\
|4\rangle &= \frac{1}{2}|s - p_x - p_y + p_z\rangle
\end{aligned} \tag{13.41}$$

where $|s\rangle = (4\pi)^{-1/2}$, and the *p* functions are given by (13.40). It is easy to show that the absolute squares of the four functions $|1\rangle, |2\rangle, |3\rangle$, and $|4\rangle$ are maximum along the directions defined by the vectors

$$\begin{aligned}
&\hat{\imath} + \hat{\jmath} + \hat{k} \\
&\hat{\imath} - \hat{\jmath} - \hat{k} \\
&-\hat{\imath} + \hat{\jmath} - \hat{k} \\
&-\hat{\imath} - \hat{\jmath} + \hat{k}
\end{aligned}$$

respectively, and that the angle between any two of these vectors is $\Theta = \cos^{-1}(-1/3) = 109.47°$. Given the six electrons, we can place two along any one of these directions, two along any other, and one each along the third and fourth directions. Each of these last two electrons forms a bond with a hydrogen atom. Therefore, according to this picture, the angle between the two hydrogen bonds in H_2O should be 109.47°. In fact, this angle is 104.5°. The agreement is reasonably good, and the residual discrepancy arises from the fact that the hybridization between $2s$ and $2p$ orbitals is not perfect. Note that the two "arms" of H_2O that do not have hydrogen bonds still have two electrons each, which are capable of attracting a proton from another H_2O molecule. Thus any given water molecule can be linked to four others in a chain (a polymer). However, the links in this chain are weak and easily disrupted by thermal motion. Hence the physical properties of liquid water are strongly temperature dependent, especially near 0°C. For example, the density of liquid water reaches a maximum at 4°C.

13.7 Nuclear vibration and rotation

Let the nuclei have masses M_1 and M_2, and let their relative position vector $\boldsymbol{R}$ have polar coordinates R, θ, and ϕ. The Schroedinger equation of relative motion of the two nuclei is

$$-\frac{\hbar^2}{2\mu}\nabla^2\psi_N + V(R)\psi_N = E\psi_N \tag{13.42}$$

where $\mu = M_1M_2/(M_1+M_2)$. As a practical matter, one finds that the potential energy curves for low-lying bound electronic states can be described quite accurately by the Morse function

$$V(R) = V_0\left(e^{-2(R-R_0)/a} - 2e^{e^{-(R-R_0)/a}}\right) \tag{13.43}$$

which has three independent parameters V_0, R_0, and a. The shape of the Morse function is shown in Figure 13.1. Equation (13.42) can be separated in spherical coordinates to yield

$$\psi_N = \frac{\chi(R)}{R}Y_K^{M_K}(\theta,\phi)$$

where

$$-\frac{\hbar^2}{2\mu}\frac{d^2\chi}{dR^2} + W(R)\chi = E\chi \tag{13.44}$$

with

$$W(R) = V(R) + \frac{\hbar^2 K(K+1)}{2\mu R^2} \tag{13.45}$$

and $K = 0, 1, 2, \ldots$.. As usual, we require $\chi(0) = 0$. If K is not too large, the shape of $W(R)$ also resembles the curve of Figure 13.1. Because we are interested in small oscillations about the minimum of this curve, we expand W about its minimum R_1

$$W(R) = W_0 + \frac{1}{2}K_0(R-R_1)^2 + b(R-R_1)^3 + c(R-R_1)^4 + \cdots \tag{13.46}$$

(Note that $R_1 = R_0$ only when $K = 0$.) If the terms in b and c are neglected and the range of R is extended to $-\infty$, (13.44) becomes the Schroedinger equation for a one-dimensional harmonic oscillator. In a better approximation, the b and c terms are treated as perturbations on the oscillator. A straightforward but tedious calculation shows that to lowest order in b and c,

$$E = W_0 + \hbar\sqrt{\frac{K_0}{\mu}}\left(v+\frac{1}{2}\right) - \frac{\hbar^2 b^2}{\mu K_0^2}\left[\frac{15}{4}\left(v+\frac{1}{2}\right)+\frac{7}{16}\right] + \frac{3\hbar^2 c}{2\mu K_0}\left[\left(v+\frac{1}{2}\right)^2+\frac{1}{4}\right] \tag{13.47}$$

where $v = 0, 1, 2, \ldots$. Assuming $V(R)$ to have the form (12.43), it can be shown that to second order in $v + 1/2$ and $K(K+1)$,

$$R_1 = R_0 + \frac{\hbar^2 K(K+1)a^2}{2\mu R_0^3 V_0} \tag{13.48}$$

$$W_0 = -V_0 + \frac{\hbar^2 K(K+1)}{2\mu R_0^2} - \frac{\hbar^4 \left[K(K+1)\right]^2 a^2}{4\mu^2 R_0^6 V_0} \tag{13.49}$$

$$K_0 = \frac{2V_0}{a^2} - \frac{3\hbar^2 K(K+1)}{\mu R_0^2 a^2}\frac{a}{R_0}\left(1-\frac{a}{R_0}\right) \tag{13.50}$$

$$b = -\frac{V_0}{a^2} \tag{13.51}$$

$$c = \frac{7V_0}{12a^4} \tag{13.52}$$

The second term on the right-hand side of (13.48) describes centrifugal stretching of the equilibrium position. The right-hand side of (13.49) contains the equilibrium energy $-V_0$, the first-order rotational energy, and the second-order rotational energy. Equation (13.50) describes the change in stiffness due to stretching. If we limit ourselves to the first two terms on the right-hand side of (13.47), neglect the second-order rotational energy in (13.49), and keep only the first term on the right-hand side of (13.50), we obtain

$$E = -V_0 + \hbar\omega\left(v+\frac{1}{2}\right) + \frac{\hbar^2 K(K+1)}{2\mu R_0^2} \tag{13.53}$$

where $\omega = \sqrt{2V_0/\mu a^2}$. Equation (13.53), which is a good approximation for small to moderate values of v and K, tells us that the energy shift with respect to $-V_0$ associated with nuclear motion contains two parts: the vibrational part and the rigid-body rotational part. Their magnitudes are roughly as described in (13.9).

13.8 Quantum statistics of homonuclear diatomic molecules

By definition, the two nuclei in a homonuclear diatomic molecule are identical; hence the two-nucleus wave function must be exchange symmetric if the nuclei are bosons or exchange antisymmetric if they are fermions. Now, in the approximation that leads to (13.53), the two-nucleus wave function including spin can be written as

$$\psi_N = \psi_{\text{vib}}\psi_{\text{rot}}\psi_{\text{spin}} \tag{13.54}$$

The function ψ_{vib} is always symmetric with respect to exchange. The exchange symmetry of the rotational function is $(-1)^K$; hence, if the nuclei are bosons, the spin function is antisymmetric if K is odd and symmetric if K is even. If the nuclei are fermions, the spin function is antisymmetric if K is even and symmetric if K is odd. To illustrate, we consider H_2, which has two protons (fermions), and D_2, which has two deuterons (bosons). First, with regard to H_2, each proton has spin-½. Thus the total nuclear spin can be $I = 1$ (exchange-symmetric triplet) or $I = 0$ (exchange-antisymmetric singlet). Therefore, if $I = 1$, K must be odd, whereas if $I = 0$, K must be even. At low temperatures, H_2 gas can be formed in the singlet state (*para*-hydrogen) or in the triplet state (*ortho*-hydrogen). Either of these gases can persist for a long time in a container before relaxing to thermal equilibrium (a mixture of the two). Disregarding nuclear spin for the moment, one can show that the rotational partition function would be

$$Z_{\text{rot}} = \sum_{K=0}^{\infty}(2K+1)e^{-\Theta K(K+1)} = 1+3e^{-2\Theta}+5e^{-6\Theta}+7e^{-12\Theta}+\cdots \tag{13.55}$$

where $\Theta = \hbar^2/2\mu R_0^2 k_B T$. At very low temperatures, $\Theta \gg 1$, and the sum on the right-hand side of (13.55) converges rapidly. If one has *ortho*-hydrogen, the terms in (13.55) with even K are missing, and the partition function including nuclear spin is

$$Z(ortho) = 3\left(3e^{-2\Theta}+7e^{-12\Theta}+\cdots\right) \tag{13.56}$$

where the extra factor of 3 arises from the spin statistical weight. If the gas is *para*-hydrogen,

$$Z(para) = 1+5e^{-6\Theta}+\cdots \tag{13.57}$$

These distinct partition functions yield very different specific heats at low temperature for the two gases. Interpretation of the measured specific heats by D. Dennison (1927) was important because it led to the discovery of proton spin.

In the case of D_2, each nucleus has spin 1, so the total nuclear spin can be 2 (symmetric), 1 (antisymmetric), or 0 (symmetric). Thus, for $I = 2$ or 0, K must be even, whereas if $I = 1$, K must be odd. Finally, for the molecule HD, there are no restrictions on K because the proton and the deuteron are not identical particles.

Problems for Chapter 13

13.1. This problem illustrates the Born-Oppenheimer approximation. Consider a-one dimensional "H_2^+ molecule" containing one light "electron" of mass m connected to two heavy "protons" each with mass M by two identical springs, each of unstretched length a and spring constant k. The Hamiltonian is

$$H = \frac{p_1^2}{2M} + \frac{p_3^2}{2M} + \frac{p_2^2}{2m} + \frac{k}{2}\left[(x_2 - x_1 - a)^2 + (x_3 - x_2 - a)^2\right] \tag{1}$$

(a) Find the normal modes of these coupled oscillators and thus obtain the exact quantized energy levels of this system.
(b) Starting once again from (1), fix the distance between the protons at R, and find the energy spectrum $E_n(R)$ of the electron as a function of R.
(c) Now employ $E_n(R)$ as a potential energy of interaction between the protons, and solve for the proton energies by doing the calculation in the proton center of mass frame. Compare your result for the total energy to that of part (a). How does the discrepancy depend on m/M?

13.2. (a) Derive (13.19) from (13. 18).
(b) Fill in the steps leading from (13.20) to (13.21) with (13.22).
(c) Show how (13.20) yields (13.23) with (13.24). It may be helpful to make use of the following integrals:

$$\int_x^\infty y^n e^{-ay}\, dy = \frac{n!e^{-ax}}{a^{n+1}}\left[1 + ax + \cdots + \frac{(ax)^n}{n!}\right]$$

$$\int f(\lambda,\mu,\phi)\, d\tau = \frac{R^3}{8}\int_0^{2\pi} d\phi \int_{-1}^{1} d\mu \int_1^\infty f(\lambda,\mu,\phi)(\lambda^2 - \mu^2)\, d\lambda$$

where λ, μ, and ϕ are prolate spheroidal coordinates defined in Section 13.3.

13.3. Starting from (13.26) for E_e expressed in terms of the functions F_1 and F_2 as given by (13.22) and (13.24), respectively, derive the result (13.27).

13.4. We have seen that when two H atoms in their ground states combine to form a hydrogen molecule, the possible electronic states of that molecule are $^1\Sigma$ and $^3\Sigma$. What are the possible electronic states of the following diatomic molecules formed from separated atoms in their ground states?

N_2 O_2 Cl_2 HCl CN TlF

Confine your solution to statements concerning possible values of Λ, that is, Σ, Π, and so on;, possible spin multiplicities, that is, singlet, triplet, and so on; and the number of independent states of each type. Do not concern yourself with which of these states are even or odd with respect to reflection through a plane containing the internuclear axis or which homonuclear molecular states are of the form g or u. (The arguments concerning these latter points are somewhat subtle, and for lack of space, we do not discuss them in this chapter.)

14 The Stability of Matter

14.1 Stabilities of the first and second kind: The thermodynamic limit

Ordinary matter is stable, a remarkable fact of nature that often goes unappreciated. Why don't the electrons, which are attracted to the nuclei by Coulomb forces, collapse into those nuclei, yielding matter with densities many orders of magnitude greater than are actually observed? Classical physics gives no answer to this question. Quantum mechanics does provide a satisfactory answer, but when we go beyond the one-electron atom and consider many-electron atoms or many atoms in bulk matter, the analysis leading to the answer is not easy. The first such analysis, given by F. Dyson and A. Lenard (Dyson and Lenard 1967; Lenard and Dyson 1968), was extremely lengthy and difficult. A simpler and more effective analysis was constructed by E. Lieb and W. Thirring in the 1970s, and this was improved in recent decades by Lieb and collaborators, as well as by other investigators [see, e.g., Lieb (1976) and Lieb and Seiringer (2010)]. However, even these efforts are far too lengthy, intricate, and subtle for a detailed exposition in this book. Hence, in what follows, we limit ourselves to a brief and superficial summary of their achievements.

The stability-of-matter problem encompasses three different concepts:

1. Stability of the first kind
2. Stability of the second kind
3. Existence of a thermodynamic limit for the total energy of a system of particles

We consider each of these in turn.

A system of electrons and nuclei interacting by means of Coulomb forces has stability of the first kind if the lowest bound state has finite rather than infinite negative energy. The simplest such system is a hydrogenic atom, consisting of a single electron and a single nucleus with atomic number Z. In classical physics, a mass point bound to an inverse square force center can have an orbit arbitrarily close to the origin and thus an energy that is arbitrarily large and negative. Hence there is no finite lower bound to the energy. According to quantum mechanics, there is a finite lower bound given exactly for the nonrelativistic hydrogenic atom by the Balmer formula: $E_{1s} = -Z^2/2$ in atomic units. If we did not know the Balmer formula, we could try to estimate the lower bound from the Heisenberg uncertainty principle using a very well-known argument as follows: suppose that the electron in the atom is limited to a spatial region of linear size Δx so that the potential energy is of order $-Ze^2/\Delta x$. The Heisenberg uncertainty principle states that

$$\Delta p \Delta x \geq \frac{\hbar}{2}$$

Therefore, the electron kinetic energy must be at least of order $\hbar^2/m_e\Delta x^2$, so the total energy is

$$E \approx \frac{\hbar^2}{m_e\Delta x^2} - \frac{Ze^2}{\Delta x}$$

This is minimized when $\Delta x \approx \hbar^2/Zm_e e^2 \approx a_0/Z$, and the corresponding minimum energy is $E \approx -Z^2e^2/a_0 = -Z^2$ in atomic units.

However, this argument is flawed because it fails to account for the possibility that the wave function might have two or more widely separated lobes. Furthermore, if we consider a many-electron atom or a system of many electrons and nuclei, we find that the Heisenberg uncertainty principle (which is derived from the Cauchy-Schwarz inequality) is not an effective tool. Instead, it is far better to use an uncertainty principle obtained from two other mathematical inequalities, one due to Hölder and the other due to Sobolev. These are derived in Appendix A. Hölder's integral inequality is

$$\left(\int f^p\, d\tau\right)^{1/p}\left(\int g^q\, d\tau\right)^{1/q} \geq \int fg\, d\tau \tag{14.1}$$

where f and g are nonnegative real functions of x, y, z, $d\tau = dxdydz$, the integrations are over all space, and p and q are two positive real numbers satisfying the condition

$$\frac{1}{p} + \frac{1}{q} = 1$$

Note that when $p = q = 2$, Hölder's integral inequality reduces to the Cauchy-Schwarz integral inequality

$$\int f^2\, d\tau \int g^2\, d\tau \geq \left(\int fg\, d\tau\right)^2$$

Sobolev's integral inequality in three dimensions is

$$\int (\nabla F)^2\, d\tau \geq C\left(\int |F|^6\, d\tau\right)^{1/3} \tag{14.2}$$

where F is a real differentiable function of x, y, z, C is a positive constant, $d\tau = dxdydz$, and both integrals are carried out over all space. As shown in Appendix A, (14.2) yields the following lower bound on the electron kinetic energy in the hydrogen atom:

$$\langle T\rangle \geq D_C \int \rho_1^{5/3}\, d\tau \tag{14.3}$$

where $D_C = 3/10\left(6\pi^2\right)^{2/3}\left(\hbar^2/m_e\right)$, and $\rho_1 = |\psi|^2$. This leads to the following lower bound on the energy in atomic units: $E \geq -3^{1/3}Z^2/2$. The precise value of the latter bound is not important. Instead, the main point is that (14.3), in which the inequality for the kinetic energy is expressed in terms of an integral of the 5/3 power of the probability density, is of a form that can be generalized to apply to an N-electron atom.

The electronic wave function for such an atom is antisymmetric with respect to exchange of any pair of electrons. It can be written

$$\psi(\boldsymbol{x}_1, \ldots, \boldsymbol{x}_N; \sigma_1, \ldots, \sigma_N)$$

where $\boldsymbol{x}_j$ and σ_j, $j = 1, \ldots, N$, are the spatial and spin coordinates, respectively, for the jth electron. Assuming that ψ is normalized to unity, we write the expectation value of the kinetic energy as

$$\langle T \rangle = -\frac{N}{2}\int \psi^\dagger \nabla_1^2 \psi \, d\tau_1 \ldots d\tau_N = \frac{N}{2}\int |\nabla_1 \psi|^2 \, d\tau_1 \ldots d\tau_N$$

In addition, it is convenient to define

$$\rho_N(\boldsymbol{x}) = N\int \left|\psi(\boldsymbol{x}, \boldsymbol{x}_2, \ldots, \boldsymbol{x}_N; \sigma_1, \ldots, \sigma_N)\right|^2 d\tau_2 \ldots d\tau_N \tag{14.4}$$

where the integral on the right-hand side is taken over all spatial coordinates except $\boldsymbol{x}$. The quantity $\rho_N(\boldsymbol{x})$ is a single-particle electron density function and is analogous to the electron density n in the Thomas-Fermi model. Integration of both sides of (14.4) with respect to $\boldsymbol{x}$ yields

$$N = \int \rho_N \, d\tau$$

[For a hydrogenic atom, $N = 1$, and (14.4) reduces to $\rho_1 = |\psi|^2$.]

In 1975, Lieb and Thirring showed that for an N-electron atom,

$$\langle T \rangle \geq \left(\frac{1}{4\pi}\right)^{2/3} \frac{D_C}{q^{2/3}} \int \rho_N^{5/3} \, d\tau \tag{14.5}$$

where q is the spin statistical weight (in other words, the maximum number of fermions that can have the same single-particle spatial orbital). Of course, for electrons, $q = 2$. [The factor of $(4\pi)^{-2/3}$ on the right-hand side of (14.5) has no fundamental significance but only resulted from a number of steps in the original derivation of Lieb and Thirring. Subsequent mathematical improvements have brought this factor much closer to unity.]

The significance of (14.5) becomes clear if we assume that $\rho_N(\boldsymbol{x})$ is zero outside a certain finite volume V. Although this is not strictly true in any real situation because ρ_N goes to zero exponentially for large x, it is an excellent approximation if V is sufficiently large Then, by Hölder's inequality,

$$\int_V \rho_N^{5/3} \, d\tau \geq \left(\int_V \rho_N \, d\tau\right)^{5/3} \left(\int_V d\tau\right)^{-2/3} = \frac{N^{5/3}}{V^{2/3}}$$

which implies that $\langle T \rangle$ increases with N at least as fast as $N^{5/3}$. This result, essential for what follows, is valid because the electronic wave function is antisymmetric. It would not be true for bosons.

We now address the questions: what is stability of the second kind, and what does it mean to speak of a thermodynamic limit? Consider an ordinary macroscopic object of a size found in a laboratory, such as a lump of copper. Its energy E and its volume V are extensive quantities. That is, if we were to double the number of particles $N \to 2N$, then to very high precision, the volume and energy also would double: $V \to 2V$ and $E \to 2E$. In the absence of such extensivity, ordinary matter could not have any of the commonplace properties that are familiar from elementary physics and chemistry. For example, if the energy were not extensive, the specific heat of the lump of copper would depend on its size. More dramatically, an enormous release of energy would result whenever we tried to join two macroscopic samples of matter together, or it would be impossible to have any bound macroscopic objects in the first place.

Energy and volume are generally extensive if we can ignore gravitational forces within an object. When gravitational forces become important, as in a star, extensivity fails. In a typical star, a good fraction of the energy is gravitational potential energy, which is of order $-GM^2/R$, where G is Newton's constant, M is the mass, and R is the radius. In this case, neither the energy nor the volume is proportional to the total number of particles in the star. Although inverse square (Coulomb) forces govern the interaction of electrons and nuclei in ordinary objects of laboratory size, there are two signs of electric charge. Because ordinary macroscopic objects are always very nearly electrically neutral, Coulomb forces are very effectively screened, and thus extensivity of V and E holds to an excellent approximation when gravity can be neglected.

If the energy of a very large number of N electrons and M nuclei is extensive, that is, if $E \propto (N + cM)$ *in the limit of very large* $N + cM$*, where c is a constant, then one says that the energy is extensive in the thermodynamic limit. The system has stability of the second kind if a lower bound exists on the energy of the form*

$$E_{\min} = -f(Z)(N + cM)$$

where f(Z) is a positive function that may depend on the atomic number Z of the nuclei but does not depend on N or M.

Stability of the second kind and existence of the thermodynamic limit for the energy have in fact been derived from quantum mechanics by the authors mentioned earlier. A remarkable result emerges in the course of this derivation: for a system of electrons and nuclei, stability of the second kind and existence of the thermodynamic limit only hold because electrons obey Fermi-Dirac statistics. A macroscopic object consisting solely of bosons with equal and opposite charges, with Coulomb interaction, cannot exist.

There are several ways to derive stability of the second kind. One route starts from the kinetic-energy bound (14.5) for an N-electron atom and makes use of the Thomas-Fermi model to extend it. Here one employs a result we have already mentioned in Section 12.4: within the Thomas-Fermi approximation, the energy of any two isolated atoms or isolated atom and molecule is always less than the energy of a molecule formed from these constituents. The result of the derivation is a bound on the nonrelativistic energy of N electrons and M nuclei of charges $Z_j e$, where $j = 1, \ldots, M$, all interacting by Coulomb forces. In atomic units, one obtains

$$E \geq -0.231 q^{2/3} N \left(1 + 1.77 \sqrt{\frac{1}{N} \sum_{j=1}^{M} Z_j^{7/3}} \right)^2 \tag{14.6}$$

Recall that q is the spin statistical weight ($q = 2$ for electrons and for protons). When (14.6) is generalized to the case $q = N$, it applies to bosonic matter, and we thus obtain a lower bound to the energy, which is

$$E_{\text{boson}} \geq -CN^{5/3} \tag{14.7}$$

where C is a constant. However, although (14.7) is a lower bound, in recent years it has been shown that for the special case where the bosons all have equal masses and positive and negative charges of equal magnitude, $E_{\text{boson}} = -CN^{7/5}$ in the limit of large N. Hence, for bosons, one has neither stability of the second kind nor a thermodynamic limit for the energy.

In this brief summary we have confined ourselves to nonrelativistic particles interacting solely by means of Coulomb forces. However, the analyses of stability carried out by Lieb and others have been extended in recent years to include relativistic motion, the effects of magnetic fields, and coupling of charges to the radiation field (Lieb and Seiringer 2010).

14.2 An application of second quantization: The stability of a metal

A typical metal consists of a crystal lattice with positive-ion cores at the vertices and a Fermi sea of conduction electrons. We discuss it here for several reasons: First, it provides a practical illustration of the stability of matter. Second, the analysis that follows affords a good exercise in second quantization. Finally, allowing our sample of metal to become astronomical in size so that gravitational forces become important, we see very easily (in Section 14.3) what happens when extensivity fails. We shall employ the *jellium model*, in which the positive-ion cores are replaced by a smooth continuum of inert positive charge, just enough to balance the negative charge of the conduction electrons. This approximation is simple enough so that our calculation is relatively straightforward, but it retains the essential features of a real metal. In the jellium model we imagine that our sample of metal occupies a large cube of edge L and volume $v = L^3$, and it has N conduction electrons with total charge $-Ne$. We assume that there exists in addition a smooth positive continuum of total charge Ne, so that the total charge is zero. Thus the positive charge density is

$$\rho_+ = \frac{Ne}{\text{v}}. \tag{14.8}$$

The total energy of the metal is:

$$E = E_+ + E_- + E_{\text{int}} \tag{14.9}$$

where E_+ is the energy of the positive charges, E_- is the energy of the electrons, and E_{int} is the energy of interaction of the positive charges with the electrons. As we shall see, E_+ and E_{int} are easy to calculate. E_- consists of two parts:

$$E_- = \langle T \rangle + \langle V \rangle$$

where $E_0 \equiv \langle T \rangle$ is the expectation value of the electron kinetic energy and $\Delta E \equiv \langle V \rangle$ is the expectation value of the potential energy, which arises from electron-electron Coulomb interactions. From the viewpoint of Fermi-Dirac statistics an ordinary metal is essentially at zero temperature; (in other words, the actual temperature is orders of magnitude less than the Fermi temperature). Thus to a good approximation:

$$E_0 = \frac{3}{5} N \mu_F \tag{14.10}$$

where μ_F, the "Fermi energy," is proportional to the 5/3 power of the electron density. ΔE is the only portion of E that requires some effort to calculate. We shall find that ΔE consists of two parts: $\Delta E = \Delta E_1 + \Delta E_2$. The first part ΔE_1 cancels $E_+ + E_{\text{int}}$. The remaining part ΔE_2 is the "exchange" contribution. Its form, not easily guessed intuitively, results from a subtle interplay of quantum mechanics and the anti-symmetrization principle. We shall find that ΔE_2 per electron is negative and that it varies as the 4/3 power of the electron density. When E_0 and ΔE_2 are combined we obtain stable equilibrium at reasonably realistic values of energy E and electron number density.

Let us start with E_+. It is:

$$\begin{aligned} E_+ &= \frac{1}{2}\int d^3\boldsymbol{x}\int d^3\boldsymbol{x}' \frac{\rho_+(\boldsymbol{x})\rho_+(\boldsymbol{x}')}{4\pi|\boldsymbol{x}-\boldsymbol{x}'|} \\ &= \frac{1}{2}\left(\frac{Ne}{\mathrm{v}}\right)^2 \int d^3\boldsymbol{x}\int d^3\boldsymbol{x}' \frac{1}{4\pi|\boldsymbol{x}-\boldsymbol{x}'|}. \end{aligned} \tag{14.11}$$

To deal with this integral, we insert an integrating factor $\exp(-\eta|\boldsymbol{x}-\boldsymbol{x}'|)$ in the integrand, carry out the integration, then take the limit as $\eta \to 0$. While this appears illegitimate on mathematical grounds, there is good physical justification for it: in real matter the Coulomb potential of any ion core or electron is always screened to some extent by the surrounding charges. Thus (14.11) becomes:

$$\begin{aligned} E_+ &= \frac{1}{2}\left(\frac{Ne}{\mathrm{v}}\right)^2 \int d^3\boldsymbol{x}\int d^3\boldsymbol{x}' \frac{e^{-\eta|\boldsymbol{x}-\boldsymbol{x}'|}}{|\boldsymbol{x}-\boldsymbol{x}'|} \\ &= \frac{1}{8\pi}\left(\frac{Ne}{\mathrm{v}}\right)^2 \int d^3\boldsymbol{x}\int_0^\infty 4\pi y \cdot \mathrm{e}^{-\eta y} dy \\ &= \frac{N^2e^2}{2\mathrm{v}\eta^2}. \end{aligned} \tag{14.12}$$

Next we calculate E_{int} in a similar way:

$$\begin{aligned} E_{\text{int}} &= -e\sum_{i=1}^{N}\int \frac{\rho_+(\boldsymbol{x})}{4\pi|\boldsymbol{x}-\boldsymbol{x}_i|} d^3\boldsymbol{x} \\ &= -\frac{Ne^2}{4\pi\mathrm{v}}\sum_{i=1}^{N}\int \frac{e^{-\eta|\boldsymbol{x}-\boldsymbol{x}_i|}}{|\boldsymbol{x}-\boldsymbol{x}_i|} d^3\boldsymbol{x} \\ &= -\frac{N^2e^2}{\mathrm{v}\eta^2}. \end{aligned} \tag{14.13}$$

Combining (14.12) and (14.13) we have:

$$E_{+} + E_{\text{int}} = -\frac{N^2 e^2}{2v\eta^2}. \tag{14.14}$$

We do not worry at this stage about the dependence of this expression on η; it will become clear later. For now, we proceed to calculate ΔE. The formula we need for this purpose is (11.76). In the present application, the Fock state is the Fermion ground state $|F\rangle$, in which all occupation numbers are unity up to the Fermi energy μ_F; beyond this, all occupation numbers are zero. Also, in each of the four field operators, the single-particle orbital w has the form

$$w_{k,\alpha} = \frac{1}{v^{1/2}} e^{ik\cdot x} \chi_\alpha \tag{14.15}$$

which is a product of a plane wave and a spin-½ spinor χ. Here $\boldsymbol{k}$ can take any value allowed by the boundary conditions (periodic boundary conditions at the cube surfaces), whereas α can have only two values (1 and 2 for spin up and spin down, respectively). Also, for ease of integration, we make the usual replacement

$$\frac{1}{|\boldsymbol{x}-\boldsymbol{x}'|} \to \frac{\exp(-\eta|\boldsymbol{x}-\boldsymbol{x}'|)}{|\boldsymbol{x}-\boldsymbol{x}'|}$$

Taking all this into account in (11.76), we obtain

$$\Delta E = \frac{e^2}{2v^2} \sum_{\substack{k1,k2,k3,k4 \\ \alpha1,\alpha2,\alpha3,\alpha4}} \int d^3\boldsymbol{x} \int d^3\boldsymbol{x}' e^{i(\boldsymbol{k}1-\boldsymbol{k}4)\cdot\boldsymbol{x}} e^{i(\boldsymbol{k}2-\boldsymbol{k}3)\cdot\boldsymbol{x}'} \frac{\exp(-\eta|\boldsymbol{x}-\boldsymbol{x}'|)}{4\pi|\boldsymbol{x}-\boldsymbol{x}'|} \times \\ \left\langle F \left| b^\dagger_{k4,\alpha4} b^\dagger_{k3,\alpha3} b_{k2,\alpha2} b_{k1,\alpha1} \right| F \right\rangle \chi^\dagger_{\alpha4} \chi^\dagger_{\alpha3} \chi_{\alpha2} \chi_{\alpha1} \tag{14.16}$$

It is convenient to rewrite the integral in (14.16) as

$$\int d^3\boldsymbol{x} \int d^3\boldsymbol{x}' e^{i(\boldsymbol{k}1-\boldsymbol{k}4)\cdot\boldsymbol{x}} e^{i(\boldsymbol{k}2-\boldsymbol{k}3)\cdot\boldsymbol{x}'} \frac{\exp(-\eta|\boldsymbol{x}-\boldsymbol{x}'|)}{4\pi|\boldsymbol{x}-\boldsymbol{x}'|} = \\ \frac{1}{4\pi} \int d^3\boldsymbol{x} e^{i(\boldsymbol{k}1+\boldsymbol{k}2-\boldsymbol{k}3-\boldsymbol{k}4)\cdot\boldsymbol{x}} \int d^3\boldsymbol{y} e^{i(\boldsymbol{k}2-\boldsymbol{k}3)\cdot\boldsymbol{y}} \frac{e^{-\eta y}}{y} \tag{14.17}$$

where $\boldsymbol{y} = \boldsymbol{x}' - \boldsymbol{x}$. The first integral on the right-hand side of (14.17) is

$$\int d^3\boldsymbol{x} e^{i(\boldsymbol{k}1+\boldsymbol{k}2-\boldsymbol{k}3-\boldsymbol{k}4)\cdot\boldsymbol{x}} = \delta(\boldsymbol{k}1+\boldsymbol{k}2, \boldsymbol{k}3+\boldsymbol{k}4)v \tag{14.18}$$

Because this requires that

$$\boldsymbol{k}1 + \boldsymbol{k}2 = \boldsymbol{k}3 + \boldsymbol{k}4 \tag{14.19}$$

the sum in (14.16) is really carried out over only three vectors. It is convenient to change our notation slightly and to represent the various $\boldsymbol{k}$ vectors on a diagram (Figure 14.1).

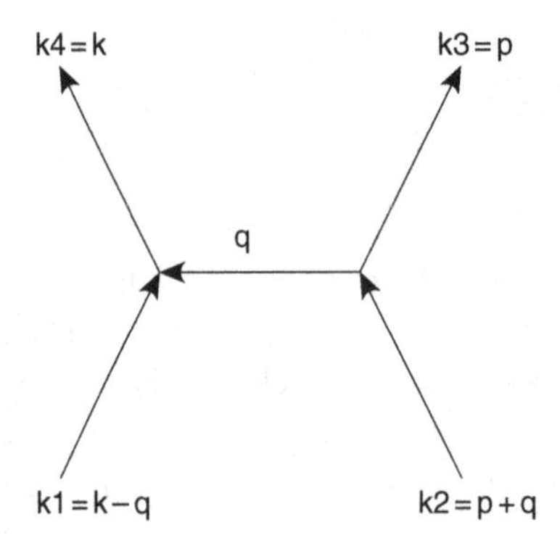

Figure 14.1 Diagram of the momenta appearing in equation (14.16).

Here we define

$$\begin{aligned} \boldsymbol{k}1 &= \boldsymbol{k} - \boldsymbol{q} \\ \boldsymbol{k}2 &= \boldsymbol{p} + \boldsymbol{q} \\ \boldsymbol{k}3 &= \boldsymbol{p} \end{aligned}$$

and thus from (14.19), $\boldsymbol{k}4 = \boldsymbol{k}$. The second integral I on the right-hand side of (14.17) is most easily evaluated by employing spherical polar coordinates with the polar axis along $\boldsymbol{q} = \boldsymbol{k}2 - \boldsymbol{k}3$; that is,

$$\begin{aligned} I &= 2\pi \int_0^\infty y\, dy e^{-\eta y} \int_0^\pi e^{iqy\cos\theta} \sin\theta\, d\theta \\ &= 2\pi \int_0^\infty y\, dy e^{-\eta y} \frac{1}{iqy}\left(e^{iqy} - e^{-iqy}\right) \\ &= \frac{2\pi}{iq}\left(\frac{1}{\eta - iq} - \frac{1}{\eta + iq}\right) = \frac{4\pi}{\eta^2 + q^2} \end{aligned}$$

Thus (14.16) becomes

$$\Delta E = \frac{e^2}{2\mathrm{v}} \sum_{\substack{\boldsymbol{k},\boldsymbol{p},\boldsymbol{q} \\ \alpha1,\alpha2,\alpha3,\alpha4}} \frac{1}{\eta^2 + q^2} \bullet \left\langle F \middle| b^\dagger_{\boldsymbol{k},\alpha4} b^\dagger_{\boldsymbol{p},\alpha3} b_{\boldsymbol{p}+\boldsymbol{q},\alpha2} b_{\boldsymbol{k}-\boldsymbol{q},\alpha1} \middle| F \right\rangle \chi^\dagger_{\alpha4} \chi^\dagger_{\alpha3} \chi_{\alpha2} \chi_{\alpha1} \tag{14.20}$$

This can be simplified further by noting that

$$\begin{aligned} \chi^\dagger_{\alpha3}\chi_{\alpha2} &= \delta_{\alpha2,\alpha3} \\ \chi^\dagger_{\alpha4}\chi_{\alpha1} &= \delta_{\alpha1,\alpha4} \end{aligned}$$

Thus

$$\Delta E = \frac{e^2}{2\mathrm{v}} \sum_{\substack{\boldsymbol{k},\boldsymbol{p},\boldsymbol{q} \\ \alpha1,\alpha2}} \frac{1}{\eta^2 + q^2} \bullet \left\langle F \middle| b^\dagger_{\boldsymbol{k},\alpha1} b^\dagger_{\boldsymbol{p},\alpha2} b_{\boldsymbol{p}+\boldsymbol{q},\alpha2} b_{\boldsymbol{k}-\boldsymbol{q},\alpha1} \middle| F \right\rangle \tag{14.21}$$

At this point it is convenient to consider separately the contributions to ΔE that arise from $\boldsymbol{q}=0$ and $\boldsymbol{q}\neq 0$. Starting with $\boldsymbol{q}=0$, we have

$$\begin{aligned}\Delta E(\boldsymbol{q}=0) &= \frac{e^2}{2v\eta^2}\sum_{\substack{\boldsymbol{k},\boldsymbol{p}\\ \alpha1,\alpha2}}\left\langle F\left|b^\dagger_{\boldsymbol{k},\alpha1}b^\dagger_{\boldsymbol{p},\alpha2}b_{\boldsymbol{p},\alpha2}b_{\boldsymbol{k},\alpha1}\right|F\right\rangle \\ &= -\frac{e^2}{2v\eta^2}\sum_{\substack{\boldsymbol{k},\boldsymbol{p}\\ \alpha1,\alpha2}}\left\langle F\left|b^\dagger_{\boldsymbol{k},\alpha1}b^\dagger_{\boldsymbol{p},\alpha2}b_{\boldsymbol{k},\alpha1}b_{\boldsymbol{p},\alpha2}\right|F\right\rangle \\ &= -\frac{e^2}{2v\eta^2}\sum_{\substack{\boldsymbol{k},\boldsymbol{p}\\ \alpha1,\alpha2}}\left\langle F\left|b^\dagger_{\boldsymbol{k},\alpha1}\left(\delta_{\boldsymbol{k},\boldsymbol{p}}\delta_{\alpha1,\alpha2}-b_{\boldsymbol{k},\alpha1}b^\dagger_{\boldsymbol{p},\alpha2}\right)b_{\boldsymbol{p},\alpha2}\right|F\right\rangle\end{aligned} \tag{14.22}$$

where in the last line we have used anticommutation relations. Recalling that

$$N_{\boldsymbol{k},\alpha1}=b^\dagger_{\boldsymbol{k},\alpha1}b_{\boldsymbol{k},\alpha1}\quad\text{and}\quad N_{\boldsymbol{k},\alpha2}=b^\dagger_{\boldsymbol{k},\alpha2}b_{\boldsymbol{k},\alpha2}$$

we see that (14.22) becomes

$$\begin{aligned}\Delta E(\boldsymbol{q}=0) &= \frac{e^2}{2v\eta^2}\left[\sum_{\substack{\boldsymbol{k},\boldsymbol{p}\\ \alpha1,\alpha2}}\left\langle F\left|N_{\boldsymbol{k},\alpha1}N_{\boldsymbol{p},\alpha2}\right|F\right\rangle-\sum_{\boldsymbol{k},\alpha1}\left\langle F\left|N_{\boldsymbol{k},\alpha1}\right|F\right\rangle\right] \\ &= \frac{e^2}{2v\eta^2}\left(N^2-N\right)\end{aligned} \tag{14.23}$$

Combining this with (14.14), we obtain

$$E_+ + E_{\text{int}} + \Delta E(\boldsymbol{q}=0) = -\frac{Ne^2}{2L^3\eta^2} \tag{14.24}$$

This represents an energy per electron of $-e^2/2L^3\eta^2$. When we take the limit as $L\to\infty$ and $\eta\to 0$ in such a way that $L\eta$ is kept constant, this residual energy per electron vanishes. Thus we conclude that

$$E_+ + E_{\text{int}} + \Delta E(\boldsymbol{q}=0) = 0 \tag{14.25}$$

We now return to (14.21) and consider the contribution $\Delta E(\boldsymbol{q}\neq 0)$. Here we can set $\eta=0$; that is,

$$\Delta E(\boldsymbol{q}\neq 0) = \frac{e^2}{2v}\sum_{\substack{\boldsymbol{k},\boldsymbol{p},\boldsymbol{q}\\ \alpha1,\alpha2}}\frac{1}{q^2}\left\langle F\left|b^\dagger_{\boldsymbol{k},\alpha1}b^\dagger_{\boldsymbol{p},\alpha2}b_{\boldsymbol{p}+\boldsymbol{q},\alpha2}b_{\boldsymbol{k}-\boldsymbol{q},\alpha1}\right|F\right\rangle \tag{14.26}$$

The only nonzero contributions to the sum occur for $\boldsymbol{p}=\boldsymbol{k}-\boldsymbol{q}$. From Figure 14.1 we see that this implies exchange: $\boldsymbol{k}1=\boldsymbol{k}3$ and $\boldsymbol{k}2=\boldsymbol{k}4$. Also, the sum in (14.26) is carried over only two vectors (e.g., $\boldsymbol{k}$ and $\boldsymbol{q}$), and the sum over two spin indices reduces to a sum over just one:

$$\Delta E(\boldsymbol{q}\neq 0)=\frac{e^2}{2\mathrm{v}}\sum_{\substack{\boldsymbol{k},\boldsymbol{q}\neq 0\\ \alpha 1}}\frac{1}{q^2}\left\langle F\left|b^{\dagger}_{\boldsymbol{k},\alpha 1}b^{\dagger}_{\boldsymbol{k}-\boldsymbol{q},\alpha 1}b_{\boldsymbol{k},\alpha 1}b_{\boldsymbol{k}-\boldsymbol{q},\alpha 1}\right|F\right\rangle$$
$$=-\frac{e^2}{2\mathrm{v}}\sum_{\substack{\boldsymbol{k},\boldsymbol{q}\neq 0\\ \alpha 1}}\frac{1}{q^2}\left\langle F\left|b^{\dagger}_{\boldsymbol{k},\alpha 1}b_{\boldsymbol{k},\alpha 1}b^{\dagger}_{\boldsymbol{k}-\boldsymbol{q},\alpha 1}b_{\boldsymbol{k}-\boldsymbol{q},\alpha 1}\right|F\right\rangle \tag{14.27}$$
$$=-\frac{e^2}{2\mathrm{v}}\sum_{\substack{\boldsymbol{k},\boldsymbol{q}\neq 0\\ \alpha 1}}\frac{1}{q^2}\left\langle F\left|N_{\boldsymbol{k},\alpha 1}N_{\boldsymbol{k}-\boldsymbol{q},\alpha 1}\right|F\right\rangle$$

This may be rewritten as

$$\Delta E(\boldsymbol{q}\neq 0)=-\frac{e^2}{v}\sum_{\boldsymbol{k},\boldsymbol{q}\neq 0}\frac{1}{q^2}\theta\left(k_F-|\boldsymbol{k}|\right)\theta\left(k_F-|\boldsymbol{k}-\boldsymbol{q}|\right) \tag{14.28}$$

where $k_F = p_F/\hbar$, where p_F is the Fermi momentum, and $\theta(x)=1$ if $x>0$, whereas $\theta(x)=0$ if $x<0$. We now replace the sums in (14.28) by integrals using the substitutions

$$\frac{1}{\mathrm{v}}\sum_{\boldsymbol{k}}\Rightarrow\frac{1}{(2\pi)^3}\int d^3\boldsymbol{k},$$
$$\sum_{\boldsymbol{q}}\Rightarrow\frac{v}{(2\pi)^3}\int d^3\boldsymbol{q}.$$

Thus (14.28) becomes

$$\Delta E(\boldsymbol{q}\neq 0)=-\frac{e^2 v}{(2\pi)^6}\int\frac{d^3\boldsymbol{q}}{q^2}\int d^3\boldsymbol{k}\,\theta\left(k_F-|\boldsymbol{k}|\right)\theta\left(k_F-|\boldsymbol{k}-\boldsymbol{q}|\right) \tag{14.29}$$

The integrals in (14.29) are easily evaluated by means of the simple geometric construction shown in Figure 14.2. Two spheres, each of radius k_F, are separated by a distance $|\boldsymbol{q}|$. The vectors $\boldsymbol{k}$ and $\boldsymbol{k}-\boldsymbol{q}$ are constrained by the theta functions to meet in the region where the two spheres overlap. Therefore, the integral over $\boldsymbol{k}$ in (14.29) is just the volume of this overlap region; that is,

$$\int d^3\boldsymbol{k}\,\theta\left(k_F-|\boldsymbol{k}|\right)\theta\left(k_F-|\boldsymbol{k}-\boldsymbol{q}|\right)=2\pi\int_{q/2}^{k_F}\left(k_F^2-x^2\right)dx$$
$$=2\pi k_F^3\left(\frac{2}{3}-\frac{q}{2k_F}+\frac{q^3}{24k_F^3}\right) \tag{14.30}$$

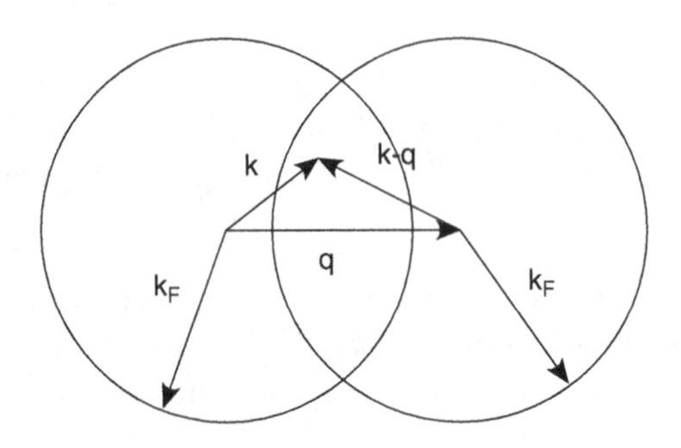

Figure 14.2 Geometric construction for evaluation of the integrals in equation (14.29)

Inserting (14.30) in (14.29), we obtain

$$\Delta E_2 \equiv \Delta E(\boldsymbol{q} \neq 0) = -\frac{e^2 v}{16\pi^4} k_F^4 \tag{14.31}$$

From Fermi-Dirac statistics, we have

$$k_F = (3\pi^2)^{1/3} \left(\frac{N}{v}\right)^{1/3} \tag{14.32}$$

Thus our final result for ΔE_2 is

$$\Delta E_2 = -\frac{3^{4/3}}{16\pi^{4/3}} \frac{e^2 N^{4/3}}{v^{1/3}} \tag{14.33}$$

Including the electron kinetic energy at zero temperature from Fermi-Dirac statistics, we obtain the total energy of the metal

$$E = E_0 + \Delta E_2 = \frac{3^{5/3}\pi^{4/3}}{10} \frac{\hbar^2}{m_e} \frac{N^{5/3}}{v^{2/3}} - \frac{3^{4/3}}{16\pi^{4/3}} \frac{e^2 N^{4/3}}{v^{1/3}} \tag{14.34}$$

It is convenient to write the electron number density as

$$n_e = \frac{N}{v} = \frac{3}{4\pi a_0^3} \frac{1}{r_s^3} \tag{14.35}$$

where $a_0 = 4\pi\hbar^2/m_e e^2$ is the Bohr radius, and r_s is a dimensionless parameter. Thus (14.34) becomes

$$E = \left(\frac{1.105}{r_s^2} - \frac{0.458}{r_s}\right) N \frac{e^2}{4\pi a_0} \tag{14.36}$$

where the first and second terms on the right-hand side correspond to E_0 and ΔE_2, respectively. We take the derivative of E with respect to r_s and set it equal to zero to find the position of stable equilibrium; that is,

$$r_s = 4.83 \tag{14.37}$$

Substitution of (14.37) into (14.35) and (14.36) yields n_e and E at equilibrium; that is,

$$n_e = 1.43 \times 10^{22}\ \text{cm}^{-3} \tag{14.38}$$

and

$$E = -0.047 N \frac{e^2}{4\pi a_0} = -1.28\ \text{eV} \bullet N \tag{14.39}$$

First, note that (14.39) is consistent with the general bound (14.6). Second, considering the crude simplicity of the jellium model, the results in (14.38) and (14.39) agree reasonably well with measured parameters for some metals (e.g., the alkali metals; see Problem 14.3). That said, there are two basic limitations to the calculation we have just completed. First, the discrete ion cores were replaced by a smooth continuum of positive charge. This defect is remedied in more sophisticated calculations. Next, we employed first-order perturbation theory to calculate ΔE. This procedure is usually reliable when the first-order correction is very small compared with the zeroth-order energy. However, in the present case ΔE and E_0 have comparable magnitudes in the vicinity of equilibrium. Thus substantial corrections are expected, and they do occur when the next order of perturbation is included. Nevertheless, the foregoing calculation does give a reasonably good description of the electron density and binding energy of a metal.

14.3 Some astrophysical consequences

For very hot stars, radiation pressure is the dominant mechanism resisting gravitational crush. In stars with moderate temperatures and densities, such as the Sun, ideal gas pressure is most important. However, for white dwarf stars, electron degeneracy pressure dominates, and as an application of the results of the previous section, it is degeneracy pressure that we now want to consider. To make our main point in the simplest possible fashion, we consider the following thought experiment: let us start with a lump of iron the size of a cannonball and at absolute zero temperature, and let us add atoms one by one, always maintaining mechanical (i.e., hydrostatic) equilibrium. Does the lump grow in size indefinitely, or does it reach a maximum radius and then shrink as we gradually add mass? In fact, we shall find that at first the density of the lump remains constant, so the radius R grows in proportion to $M^{1/3}$. However, eventually, our lump of metal reaches the size of an astronomical object, and gravitational forces become appreciable. At this point, *pressure ionization* begins to liberate electrons that were bound in the ion cores; eventually, all the electrons are ionized. Also, gravitational forces squeeze the central portion of the sample to higher and higher densities until the radius begins to decrease. As we see momentarily, for a considerable range of masses in this regime, $R \propto M^{-1/3}$, but eventually R decreases much more rapidly, going to zero at a certain critical mass M_C. These stages are illustrated schematically in Figure 14.3.

To reduce the problem to its essentials, we make three simplifications. First, we idealize the metal as a lump of jellium. Second, we replace all factors of order unity (such as $3^{5/3}$, $\pi^{4/3}$, etc.) by unity. Third, we replace differential equations by difference equations. It turns out that these simplifications do not seriously invalidate our argument.

Before we begin, we need to establish the condition for hydrostatic equilibrium in the presence of Newtonian gravity. This is a simple exercise in classical mechanics. Consider a spherical object with radius R, total mass M, and mass density $\rho(r)$ (Figure 14.4). We focus on a small cylinder of cross-sectional area A located between r and $r + dr$. Its volume is $dV = drdA$, and its mass is $dm = \rho(r)drdA$. The radial force on the cylinder consists of three parts: the pressure forces on the two ends and the gravitational force on dm; that is,

$$F_r = P(r)dA - P(r+dr)dA - \frac{GM(r)\rho(r)}{r^2}drdA \tag{14.40}$$

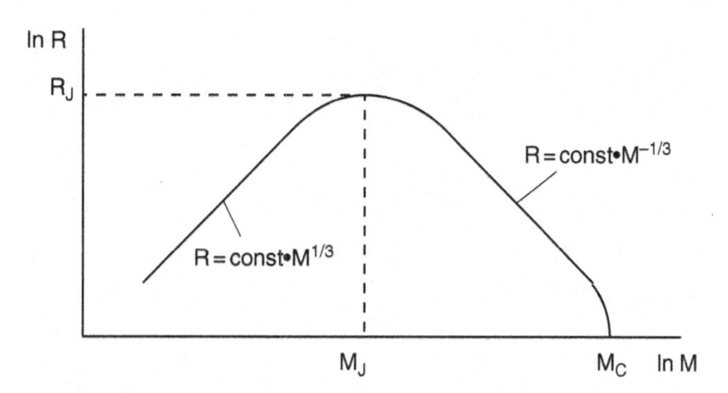

Figure 14.3 Schematic diagram (not to scale) showing the dependence of radius on mass for a cold lump of matter. R_J is the maximum radius, and M_J is the corresponding mass.

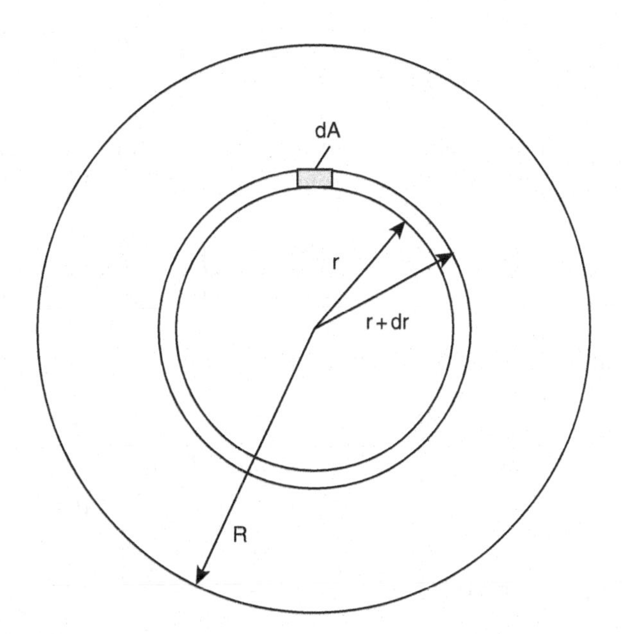

Figure 14.4 Diagram of a star, indicating various quantities that appear in equation (14.40).

where $M(r)$, the mass interior to r, is related to $\rho(r)$ by

$$\frac{\partial M(r)}{\partial r} = 4\pi r^2 \rho(r)$$

and G is Newton's constant. At hydrostatic equilibrium, $F_r = 0$; hence (14.40) yields

$$\frac{\partial P(r)}{\partial r} = -G\frac{M(r)\rho(r)}{r^2} \tag{14.41}$$

This is the equation of hydrostatic equilibrium for a star.

Because we have agreed to replace differential equations by difference equations, we use $\rho \approx M/R^3$, and rewrite (14.41) as

$$P \approx G\frac{M^2}{R^4} \tag{14.42}$$

From (14.34), the total kinetic and electrostatic energy in a lump of metal at zero temperature is

$$E \approx \frac{e^2}{4\pi}\left(a_0\frac{N^{5/3}}{v^{2/3}} - \frac{N^{4/3}}{v^{1/3}}\right) \tag{14.43}$$

The pressure is

$$P = -\frac{\partial E}{\partial v} \approx \frac{e^2}{4\pi}\left[a_0\left(\frac{N}{v}\right)^{5/3} - \left(\frac{N}{v}\right)^{4/3}\right] \tag{14.44}$$

Now $N/v = n_e = Z\rho/Am_p \approx \rho/m_p$, where A is the mass number, Z is the atomic number, and m_p is the proton mass. Hence, from (14.42) and (14.44), we have

$$\frac{GM^2}{R^4} \approx \frac{e^2}{4\pi}\left(a_0\frac{M^{5/3}}{m_p^{5/3}R^5} - \frac{M^{4/3}}{m_p^{4/3}R^4}\right) \tag{14.45}$$

Rearranging this equation, we obtain

$$R \approx \frac{e^2 a_0}{4\pi G m_p^{5/3} M^{1/3}} \frac{1}{\left(1 + \dfrac{e^2}{4\pi G m_p^{4/3} M^{2/3}}\right)} \tag{14.46}$$

If $e^2/\left(4\pi G m_p^{4/3} M^{2/3}\right) \gg 1$ $\left[\text{i.e., if } M \ll \left(e^2/4\pi G\right)^{3/2}\left(1/m_p^2\right)\right]$, then

$$R \approx \frac{a_0 M^{1/3}}{m_p^{1/3}} \tag{14.47}$$

Of course, this means that the density $\rho \approx m_p/a_0^3$ is constant: the volume is extensive. Here our sample is relatively small in mass, and Coulomb forces are much more important than gravity. On the other hand, if $M \gg \left(e^2/4\pi G\right)^{3/2}\left(1/m_p^2\right)$, then (14.46) implies that

$$R \approx \frac{e^2 a_0}{4\pi G m_p^{5/3} M^{1/3}} \tag{14.48}$$

Here gravity is much more important than Coulomb forces, and the pressure is supplied almost entirely by nonrelativistic electron degeneracy. [In other words, the kinetic-energy term on the right-hand side of (14.45) is much more important than the exchange-energy term.] Obviously,

(14.47) and (14.48) apply to the two portions of the curve that have constant positive and negative slopes, respectively, in Figure 14.3. Also, it is clear from the preceding remarks that M_J in Figure 14.3 must be approximately

$$M_J \approx \left(\frac{e^2}{4\pi G}\right)^{3/2} \frac{1}{m_p^2} \approx 10^{30}\, g \tag{14.49}$$

and from (14.46), $R_J \approx 5 \times 10^9$ cm. These numerical values correspond roughly to the mass and radius, respectively, of the planet Jupiter: $M_{\text{Jupiter}} = 1.90 \times 10^{30}$g, and $R_{\text{Jupiter}} = 7.14 \times 10^9$ cm.

We now return to the case $M \gg M_J$, where (14.48) applies. As we add atoms one by one in this regime, the density increases in proportion to M^2, and as it does, the electron Fermi momentum p_F also increases. Eventually, $p_F \approx m_e c$, in which case the electrons are no longer nonrelativistic. Let us carry this situation to the limit where the electrons are ultrarelativistic. Then, according to zero-temperature Fermi-Dirac statistics, the electron kinetic-energy density is

$$\begin{aligned}\varepsilon = \frac{1}{\pi^2\hbar^3}\int_0^{p_F} pc \cdot p^2\, dp &= \frac{c}{4\pi^2\hbar^3} p_F^4 \\ &= \frac{3^{4/3}\pi^{2/3}}{4}\hbar c n^{4/3}\end{aligned}$$

Thus the pressure is

$$P \approx \hbar c n_e^{4/3} \approx \frac{\hbar c}{m_p^{4/3}} \frac{M^{4/3}}{R^4} \tag{14.50}$$

Again employing (14.42), we obtain

$$\frac{GM^2}{R^4} \approx \frac{\hbar c}{m_p^{4/3}} \frac{M^{4/3}}{R^4}$$

Because we have R^{-4} on both sides of this equation, a solution can be found for only one value of the mass; that is,

$$M = M_C \approx \left(\frac{\hbar c}{G}\right)^{3/2} \frac{1}{m_p^2} \approx 1 \text{ solar mass} \tag{14.51}$$

We have just given a crude derivation of the *Chandrasekhar limit* on the mass of cold white dwarf stars [see, e.g., Chandrasekhar (1939)]. Finally, it is interesting to compare (14.49) with (14.51). We see that the ratio of the mass of "Jupiter" to the Chandrasekhar mass is

$$\frac{M_J}{M_C} \approx \alpha^{3/2} \tag{14.52}$$

where $\alpha = e^2/4\pi\hbar c = 1/137.036$ is the fine-structure constant.

Problems for Chapter 14

14.1. Work through the steps between equations (A.30) and (A.33) of Appendix A.

14.2. Suppose that bulk matter consisted entirely of bosons with energy

$$E = -\frac{e^2}{4\pi a_0} N^{7/5}$$

where N is the number of particles, with equal numbers of positive and negative charges. Consider two separate samples of bulk matter, where N is Avogadro's number for each sample. Estimate the energy released when these two samples are brought together, and compare it with the energy released when an equal number of ^{235}U nuclei undergo fission.

14.3. In this problem we are concerned with the simple model of a metal discussed in Section 14.2.

(a) Compare the result (14.38) for electron-number density with experimental data for the following metals in solid form:

Li, Na, K, Rb, Cs, Cu, Au

(b) Use results (14.38) and (14.39) to calculate the metal bulk modulus at the equilibrium value $r_s = 4.83$. A numerical result is requested.

14.4. In this problem you are asked to make simple estimates of the type described in Section 14.3.

(a) Estimate the radius R of a star of one solar mass ($M = 2 \times 10^{33} g$) that has a mass density of ordinary nuclear matter. It may be useful to employ the following formula for the radius of a nucleus:

$$R_N = 1.4 \times 10^{-13} A^{1/3} \text{ cm}$$

where A is the nuclear mass number (the number of neutrons plus protons in the nucleus).

(b) A star with this large density usually consists for the most part of free neutrons. For the purposes of this part of the problem, assume that the star in question consists entirely of free neutrons. Using the equation of hydrostatic equilibrium, and assuming that the pressure is entirely nonrelativistic neutron degeneracy pressure, estimate the equilibrium radius of this model star. How does your answer compare with that in part (a)?

14.5. In Section 14.3 we give a crude argument resulting in an approximate expression for the Chandrasekhar mass M_C [equation (14.51)]. This problem concerns a proper derivation of Chandrasekhar's equation, which yields M_C.

(a) From kinetic theory, we know that the pressure in any gas, relativistic or nonrelativistic, is $P = \frac{1}{3}\langle vp \rangle$, where v and p are the speed and momentum of a particle, respectively, and $\langle \ldots \rangle$ denotes an average over all the particles. For a completely degenerate electron gas at zero temperature, show from Fermi-Dirac statistics that this yields

$$P = \frac{m_e^4 c^5}{3\pi^2 \hbar^3} \int_0^x \frac{z^4 dz}{\sqrt{1+z^2}} = Af(x) \tag{1}$$

where $x = p_F / m_e c$, p_F is the Fermi momentum,

$$A = \frac{m_e^4 c^5}{24\pi^2 c^3} = 6.01 \times 10^{22}$$

and

$$f(x) = x(2x^2 - 3)(x^2 + 1)^{1/2} + 3\sinh^{-1}(x) \tag{2}$$

(b) Let n_e be the electron-number density, and write the mass density as

$$\rho = \mu_e m_p n_e \tag{3}$$

where m_p is the proton mass, and μ_e is equal to unity for hydrogen but otherwise approximately equal to 2. From Fermi-Dirac statistics, show that

$$\rho = Bx^3 \tag{4}$$

where

$$B = \frac{m_e^3 c^3 \mu_e m_p}{3\pi^2 \hbar^3} = 9.82 \times 10^5 \mu_e$$

Employing the equation of hydrostatic equilibrium

$$\frac{\partial P}{\partial r} = -G \frac{M(r)\rho}{r^2} \tag{5}$$

as well as (1) and (4), show that

$$\frac{1}{r^2} \frac{\partial}{\partial r}\left(r^2 \frac{\partial y}{\partial r}\right) = -\frac{\pi G B^2}{2A}(y^2 - 1)^{3/2} \tag{6}$$

where $y = \sqrt{1 + x^2}$.

(c) Let $y = y_0 \phi$, where $y_0 = \sqrt{1 + x_0^2}$, $x_0 = x(r = 0)$, and $\phi(r = 0) = 1$. Also, let $r = a\eta$, where η is a dimensionless parameter, and

$$a = \left(\frac{2A}{\pi G}\right)^{1/2} \frac{1}{B y_0} \tag{7}$$

With these variable changes, show that (6) becomes

$$\frac{1}{\eta^2} \frac{\partial}{\partial \eta}\left(\eta^2 \frac{\partial \phi}{\partial \eta}\right) = -\left(\phi^2 - \frac{1}{y_0^2}\right)^{3/2} \tag{8}$$

which is Chandrasekhar's equation. The function $\phi(\eta)$ runs from $\phi(0) = 1$ to $\phi(\eta_1) = 1/y_0$. The radius of the star corresponds to η_1, where $x = 0$.

(d) Show that the mass of the star is

$$M = -4\pi \left(\frac{2A}{\pi G} \right)^{3/2} \frac{1}{B^2} \left(\eta^2 \frac{\partial \phi}{\partial \eta} \right)_{\eta_1} \tag{9}$$

In (9) y_0 does not appear explicitly, but M is a function of y_0 because of the dependence of ϕ on that quantity, as seen in (8). Also, $M > 0$ because $\partial \phi / \partial \eta < 0$ at η_1. The limiting mass M_C occurs for $y_0 \to \infty$. Numerical integration of (8) in this case yields

$$M_C = 5.75 \mu_e^{-2} M_\odot$$

Because $\mu_e = 2$ is a good approximation, we finally have

$$M_C = 1.44 M_\odot$$

Chandrasekhar's theory, as outlined in this problem, is based solely on Newtonian gravitation and zero-temperature Fermi-Dirac statistics for the electron gas. To make the theory more realistic, corrections must be applied to account for the Coulomb exchange interaction and general relativistic, finite temperature, inverse beta decay, and nonuniform electron-density effects.

15 Photons

15.1 Hamiltonian form of the classical radiation field

We begin this chapter with the classical theory of radiation, which is derived from Maxwell's equations in vacuum

$$\nabla \cdot \boldsymbol{\mathcal{E}} = \rho \tag{15.1a}$$

$$\nabla \cdot \boldsymbol{B} = 0 \tag{15.1b}$$

$$\nabla \times \boldsymbol{B} = \frac{1}{c}\boldsymbol{j} + \frac{1}{c}\frac{\partial \boldsymbol{\mathcal{E}}}{\partial t} \tag{15.2a}$$

$$\nabla \times \boldsymbol{\mathcal{E}} = -\frac{1}{c}\frac{\partial \boldsymbol{B}}{\partial t} \tag{15.2b}$$

where ρ and $\boldsymbol{j}$ are the free charge and current densities, respectively, and we employ Heaviside-Lorentz units (hlus). In general, we can express $\boldsymbol{\mathcal{E}}$ and $\boldsymbol{B}$ in terms of scalar and vector potentials; that is,

$$\boldsymbol{\mathcal{E}} = -\nabla\Phi - \frac{1}{c}\frac{\partial \boldsymbol{A}}{\partial t} \tag{15.3}$$

$$\boldsymbol{B} = \nabla \times \boldsymbol{A} \tag{15.4}$$

The potentials are not unique because we can always make a gauge transformation; that is,

$$\Phi' = \Phi + \frac{1}{c}\frac{\partial \chi}{\partial t}$$

$$\boldsymbol{A}' = \boldsymbol{A} - \nabla\chi$$

where χ is an arbitrary real scalar function. Under this transformation, the electric and magnetic fields remain invariant; that is,

$$\boldsymbol{\mathcal{E}}' = -\nabla\Phi' - \frac{1}{c}\frac{\partial \boldsymbol{A}'}{\partial t} = -\nabla\Phi - \frac{1}{c}\frac{\partial \boldsymbol{A}}{\partial t} = \boldsymbol{\mathcal{E}}$$
$$\boldsymbol{B}' = \nabla \times \boldsymbol{A}' = \nabla \times \boldsymbol{A} = \boldsymbol{B}$$

We rewrite (15.1a) and (15.2a) in terms of the potentials Φ and $\boldsymbol{A}$. Using (15.3) and (15.4), we obtain

$$\nabla^2\Phi - \frac{1}{c^2}\ddot{\Phi} = -\rho - \frac{1}{c}\frac{\partial}{\partial t}\left(\nabla\boldsymbol{\cdot}\boldsymbol{A} + \frac{1}{c}\dot{\Phi}\right) \tag{15.5}$$

and

$$\nabla^2\boldsymbol{A} - \frac{1}{c^2}\ddot{\boldsymbol{A}} = -\frac{1}{c}\boldsymbol{j} + \nabla\left(\nabla\boldsymbol{\cdot}\boldsymbol{A} + \frac{1}{c}\dot{\Phi}\right) \tag{15.6}$$

In a class of gauges collectively called *Lorenz gauge*,[1] we choose the function χ so that

$$\nabla\boldsymbol{\cdot}\boldsymbol{A} + \frac{1}{c}\dot{\Phi} = 0 \tag{15.7}$$

In Lorenz gauge, (15.5) and (15.6) become

$$\nabla^2\Phi - \frac{1}{c^2}\ddot{\Phi} = -\rho \tag{15.8}$$

and

$$\nabla^2\boldsymbol{A} - \frac{1}{c^2}\ddot{\boldsymbol{A}} = -\frac{1}{c}\boldsymbol{j} \tag{15.9}$$

respectively. As is well known, the following retarded potentials are solutions to the inhomogeneous wave equations [(15.8) and (15.9)]:

$$\Phi(\boldsymbol{r},t) = \frac{1}{4\pi}\int \frac{\rho(\boldsymbol{r}',t')}{|\boldsymbol{r}-\boldsymbol{r}'|}\, d^3\boldsymbol{r}'$$

$$\boldsymbol{A}(\boldsymbol{r},t) = \frac{1}{4\pi c}\int \frac{\boldsymbol{j}(\boldsymbol{r}',t')}{|\boldsymbol{r}-\boldsymbol{r}'|}\, d^3\boldsymbol{r}'$$

where $t' = t - (1/c)|\boldsymbol{r}-\boldsymbol{r}'|$ is the retarded time. One important advantage of Lorenz gauge is that the gauge condition (15.7) is invariant under Lorentz transformations. That is, if we find a set of potentials that satisfy (15.7) in one inertial frame and then go to another inertial frame by means of a Lorentz transformation, the potentials in the new frame also satisfy (15.7).

We might instead choose to employ Coulomb gauge, defined by the condition $\nabla\boldsymbol{\cdot}\boldsymbol{A} = 0$. In Coulomb gauge, (15.5) becomes Poisson's equation $\nabla^2\Phi = -\rho$. It has the solution

[1] Named for the Danish physicist L. V. Lorenz (1829–1891), not to be confused with the Dutch physicist H. A. Lorentz (1853–1928).

$$\Phi(\boldsymbol{r},t) = \frac{1}{4\pi}\int \frac{\rho(\boldsymbol{r}',t)}{|\boldsymbol{r}-\boldsymbol{r}'|}\,d^3\boldsymbol{r}' \tag{15.10}$$

If the charge distribution is bounded, then at large distances from this distribution, Φ in (15.10) goes to zero at least as fast as r^{-1}. At present, we are interested in electromagnetic radiation fields far from their charge-current sources. For example, this could be the electromagnetic waves radiated by the Sun or a more distant star, the radio waves miles from a transmitting antenna, or the thermal radiation inside a cavity with reflecting walls. In all such circumstances, we can ignore Φ in Coulomb gauge, in which case (15.6) with $\boldsymbol{j} = 0$ becomes

$$\nabla^2 \boldsymbol{A} - \frac{1}{c^2}\ddot{\boldsymbol{A}} = 0 \tag{15.11}$$

with the supplementary condition $\nabla\boldsymbol{\cdot}\boldsymbol{A} = 0$. Equation (15.11) is the homogeneous vector wave equation. An arbitrary solution to this linear partial differential equation satisfying specific boundary conditions can be built up as a superposition of certain fundamental solutions satisfying the same boundary conditions by Fourier synthesis. This is analogous to the synthesis of a complicated standing wave on a violin string from the superposition of the various harmonics. We write the fundamental solutions to (15.11) as

$$\boldsymbol{A}_{k,\alpha}(\boldsymbol{r},t) = \hat{\varepsilon}_{k,\alpha} c_{k,\alpha}(t) A_k^0(\boldsymbol{r}) \tag{15.12}$$

where $\boldsymbol{k}$ is the wave vector, and $\hat{\varepsilon}_{k,\alpha}$ is the polarization vector, with $\boldsymbol{k}\boldsymbol{\cdot}\hat{\varepsilon}_{k,\alpha} = 0$ because $\nabla\boldsymbol{\cdot}\boldsymbol{A} = 0$. Also, $\alpha = 1,2$ because there are two independent polarization states in the plane perpendicular to $\boldsymbol{k}$ (these could be linear polarizations or circular polarizations). Inserting (15.12) into (15.11), we obtain

$$c_{k,\alpha}\nabla^2 A_k^0 - \frac{1}{c^2}\ddot{c}_{k,\alpha} A_k^0 = 0$$

or

$$\frac{\nabla^2 A_k^0}{A_k^0} = \frac{1}{c^2}\frac{\ddot{c}_{k,\alpha}}{c_{k,\alpha}} \tag{15.13}$$

Because the left-hand side of (15.13) is a function of $\boldsymbol{r}$ only, whereas the right-hand side is a function of t only, both must be equal to a constant. Hence

$$\nabla^2 A_k^0 + k^2 A_k^0 = 0 \tag{15.14}$$

and

$$\ddot{c}_{k,\alpha} + \omega^2 c_{k,\alpha} = 0 \tag{15.15}$$

where $k = \omega/c$. Equation (15.14), also called the *wave equation*, has fundamental solutions in Cartesian coordinates of the form

$$A_k^0 = e^{\pm i k \cdot r}$$

or

$$A_k^0 = \cos \boldsymbol{k} \cdot \boldsymbol{r} \qquad \sin \boldsymbol{k} \cdot \boldsymbol{r}$$

for traveling waves or standing waves, respectively. Equation (15.15) is the equation of a classical harmonic oscillator. We see presently that this harmonic oscillator plays an important role in quantization of the electromagnetic field.

Let us divide all space into a lattice of cubes of edge L and require that A satisfy periodic boundary conditions on the surfaces of each cube. We do this so that we can deal with a denumerable infinity of electromagnetic modes rather than a continuum of such modes. (We shall see that all results of physical interest are independent of L, which is chosen just for convenience.) Now we express a general solution of (15.11) as a superposition of fundamental solutions of the traveling-wave type as follows:

$$\boldsymbol{A}(\boldsymbol{r},t) = \frac{1}{L^{3/2}} \sum_{k,\alpha} \hat{\varepsilon}_{k,\alpha} \left[c_{k,\alpha}(t) e^{i k \cdot r} + c_{k,\alpha}^*(t) e^{-i k \cdot r} \right] \tag{15.16}$$

The periodic boundary conditions restrict $\boldsymbol{k} = k_x \hat{i} + k_y \hat{j} + k_z \hat{k}$ to the values

$$k_{x,y,z} = \frac{2\pi n_{1,2,3}}{L}$$

where $n_1, n_2, n_3 = 0, \pm 1, \pm 2, \ldots$, but not all $n_{1,2,3}$ are simultaneously zero. The distinct solutions are orthogonal in the following sense: integrating over the volume of one cube, we have

$$\iiint \frac{1}{L^3} \hat{\varepsilon}_{k,\alpha} \cdot \hat{\varepsilon}_{k',\alpha'} e^{i k \cdot r} e^{-i k' \cdot r} \, dx dy dz = \delta_{k,k'} \delta_{\alpha,\alpha'} \tag{15.17}$$

In (15.16) we have expanded an arbitrary vector potential in plane traveling waves. Alternatively, we could have expanded A in plane standing waves (appropriate for a rectangular cavity with perfectly reflecting walls) or in spherical waves and so on. In Chapter 16 we see why and how one would employ spherical waves (also see Appendix C).

Now consider the energy stored in the field in volume $V = L^3$. From Maxwell's theory and in hlus, this is

$$H = \frac{1}{2} \int_V (\boldsymbol{\mathcal{E}} \cdot \boldsymbol{\mathcal{E}} + \boldsymbol{B} \cdot \boldsymbol{B}) \, d\tau \tag{15.18}$$

Let us express (15.18) in terms of the coefficients $c_{k,\alpha}(t)$. We have

$$\boldsymbol{\mathcal{E}} = -\frac{1}{c} \frac{\partial A}{\partial t} = \frac{i}{L^{3/2}} \sum_{k,\alpha} k \hat{\varepsilon}_{k,\alpha} \left(c_{k,\alpha} e^{i k \cdot r} - c_{k,\alpha}^* e^{-i k \cdot r} \right) \tag{15.19}$$

Therefore,

$$\mathcal{E}\cdot\mathcal{E} = -\frac{1}{L^3}\sum_{k,\alpha}\sum_{k',\alpha'} kk'\hat{\varepsilon}_{k,\alpha}\cdot\hat{\varepsilon}_{k',\alpha'}\left(c_{k,\alpha}e^{ik\cdot r} - c^*_{k,\alpha}e^{-ik\cdot r}\right)\left(c_{k',\alpha'}e^{ik'\cdot r} - c^*_{k',\alpha'}e^{-ik'\cdot r}\right)$$

From this expression, and making use of (15.17), we find

$$\int \mathcal{E}\cdot\mathcal{E}\, d\tau = \sum_{k,\alpha} k^2\left[2\left|c_{k,\alpha}\right|^2 - \hat{\varepsilon}_{k,\alpha}\cdot\hat{\varepsilon}_{-k,\alpha}\left(c_{k,\alpha}c_{-k,\alpha} + c^*_{k,\alpha}c^*_{-k,\alpha}\right)\right] \tag{15.20}$$

Similarly,

$$\boldsymbol{B} = \nabla\times\boldsymbol{A} = \frac{i}{L^{3/2}}\sum_{k,\alpha} k\hat{\boldsymbol{k}}\times\hat{\varepsilon}_{k,\alpha}\left(c_{k,\alpha}e^{ik\cdot r} - c^*_{k,\alpha}e^{-ik\cdot r}\right) \tag{15.21}$$

and after straightforward manipulations, this yields

$$\int \boldsymbol{B}\cdot\boldsymbol{B}\, d\tau = \sum_{k,\alpha} k^2\left[2\left|c_{k,\alpha}\right|^2 + \hat{\varepsilon}_{k,\alpha}\cdot\hat{\varepsilon}_{-k,\alpha}\left(c_{k,\alpha}c_{-k,\alpha} + c^*_{k,\alpha}c^*_{-k,\alpha}\right)\right] \tag{15.22}$$

Combining (15.20) and (15.22), we see that (15.18) becomes

$$H = 2\sum_{k,\alpha}\frac{\omega^2}{c^2}\left|c_{k,\alpha}\right|^2 \tag{15.23}$$

At this stage it is convenient to make the following change of variables:

$$c_{k,\alpha} = \frac{c}{2}\left(Q^0_{k,\alpha} + \frac{i}{\omega}P^0_{k,\alpha}\right)$$

$$c^*_{k,\alpha} = \frac{c}{2}\left(Q^0_{k,\alpha} - \frac{i}{\omega}P^0_{k,\alpha}\right)$$

In terms of the new variables, (15.23) is

$$H = \frac{1}{2}\sum_{k,\alpha}\left[\left(P^0_{k,\alpha}\right)^2 + \omega^2\left(Q^0_{k,\alpha}\right)^2\right] \tag{15.24}$$

By expressing the Hamiltonian for the classical radiation field in this form, we see that it is a sum of harmonic oscillator Hamiltonians, one for each mode characterized by $\boldsymbol{k}$ and α. We can now guess how to quantize the radiation field: simply quantize each harmonic oscillator.

15.2 Quantization of the radiation field in Coulomb gauge

Recalling our treatment of the simple harmonic oscillator in Section 6.13, we make the following substitutions:

$$Q^0_{k,\alpha} = \left(\frac{\hbar}{\omega}\right)^{1/2} Q_{k,\alpha}$$

$$P^0_{k,\alpha} = (\hbar\omega)^{1/2} P_{k,\alpha}$$

Then (15.24) becomes

$$H = \sum_{k,\alpha} \frac{\hbar\omega}{2}\left(P^2_{k,\alpha} + Q^2_{k,\alpha}\right)$$

The transition from classical to quantum theory is accomplished by replacing the classical variables $Q_{k,\alpha}$ and $P_{k,\alpha}$ by quantum-mechanical operators with the same names that satisfy the commutation relations

$$\left[Q_{k,\alpha}, P_{k',\alpha'}\right] = i\delta_{k,k'}\delta_{\alpha,\alpha'}$$

It is also convenient to define the annihilation (destruction) operators

$$a_{k,\alpha} = \frac{1}{\sqrt{2}}\left(Q_{k,\alpha} + iP_{k,\alpha}\right)$$

and the corresponding creation operators

$$a^\dagger_{k,\alpha} = \frac{1}{\sqrt{2}}\left(Q_{k,\alpha} - iP_{k,\alpha}\right)$$

These satisfy the commutation relations

$$\left[a_{k,\alpha}, a^\dagger_{k',\alpha'}\right] = \delta_{k,k'}\delta_{\alpha,\alpha'} \tag{15.25}$$

Furthermore, following exactly the same path as we did for the simple harmonic oscillator, we define a number operator for each mode; that is,

$$N_{k,\alpha} = a^\dagger_{k,\alpha} a_{k,\alpha}$$

These operators have the following properties:

1. The eigenstates of $N_{k,\alpha}$ are the $\left|n_{k,\alpha}\right\rangle$ with eigenvalues $n_{k,\alpha} = 0, 1, 2, 3, \ldots$.
2. A state $\left|n_{k,\alpha}\right\rangle$ is said to consist of $n_{k,\alpha}$ photons, all of which have the same wave vector $\boldsymbol{k}$ and polarization denoted by α. The state $\left|n_{k,\alpha}\right\rangle$ can be generated from the vacuum state $|0\rangle$ (which contains no photons) by repeated application of the creation operator $a^\dagger_{k,\alpha}$; that is,

$$\left|n_{k,\alpha}\right\rangle = \frac{1}{\sqrt{n_{k,\alpha}!}}\left(a^\dagger_{k,\alpha}\right)^{n_{k,\alpha}} |0\rangle$$

   Similarly, photons are destroyed by application of $a_{k,\alpha}$; that is,

$$a_{k,\alpha}\left|n_{k,\alpha}\right\rangle = \sqrt{n_{k,\alpha}}\left|n_{n_{k,\alpha}} - 1\right\rangle$$

3. The Hamiltonian can be expressed as follows:

$$H = \sum_{k,\alpha} \hbar\omega \left(N_{k,\alpha} + \frac{1}{2} \right)$$

Thus the energy stored in the radiation field is distributed in discrete parcels in the various modes: if $n_{k,\alpha}$ photons exist in the $\boldsymbol{k}$, α mode, the energy in that mode is

$$\left(n_{k,\alpha} + \frac{1}{2} \right) \hbar\omega_k$$

Even when no photons are present, there still exists the zero-point energy $\hbar\omega_{k,\alpha}/2$. Because there are infinitely many modes, and the total energy is obtained by summing over all modes, the total zero-point energy in the radiation field is infinite. This is the first of a number of problems that we encounter in formulating a quantum-field theory. We return to the zero-point energy for a more detailed discussion in the next section.

4. The operators $a_{k,\alpha}$, $a^\dagger_{k,\alpha}$ are obviously the quantum-mechanical counterparts of $c_{k,\alpha}$, $c^*_{k,\alpha}$, respectively. It is easy to see that when we make the transition from classical to quantum theory,

$$c_{k,\alpha} \rightarrow \sqrt{\frac{\hbar c^2}{2\omega}} a_{k,\alpha} \qquad c^*_{k,\alpha} \rightarrow \sqrt{\frac{\hbar c^2}{2\omega}} a^\dagger_{k,\alpha}$$

Thus the classical expressions (15.16), (15.19), and (15.21) for $\boldsymbol{A}$, $\boldsymbol{\mathcal{E}}$, and $\boldsymbol{B}$, respectively, go over to the following operators in quantum theory:

$$\boldsymbol{A}(\boldsymbol{r}) = \frac{1}{L^{3/2}} \sum_{k,\alpha} \sqrt{\frac{\hbar c^2}{2\omega}} \hat{\varepsilon}_{k,\alpha} \left(a_{k,\alpha} e^{i\boldsymbol{k}\cdot\boldsymbol{r}} + a^\dagger_{k,\alpha} e^{-i\boldsymbol{k}\cdot\boldsymbol{r}} \right) \tag{15.26}$$

$$\boldsymbol{\mathcal{E}} = \frac{i}{L^{3/2}} \sum_{k,\alpha} \sqrt{\frac{\hbar\omega}{2}} \hat{\varepsilon}_{k,\alpha} \left(a_{k,\alpha} e^{i\boldsymbol{k}\cdot\boldsymbol{r}} - a^\dagger_{k,\alpha} e^{-i\boldsymbol{k}\cdot\boldsymbol{r}} \right) \tag{15.27}$$

$$\boldsymbol{B} = \frac{i}{L^{3/2}} \sum_{k,\alpha} \sqrt{\frac{\hbar\omega}{2}} \hat{k} \times \hat{\varepsilon}_{k,\alpha} \left(a_{k,\alpha} e^{i\boldsymbol{k}\cdot\boldsymbol{r}} - a^\dagger_{k,\alpha} e^{-i\boldsymbol{k}\cdot\boldsymbol{r}} \right) \tag{15.28}$$

Note that each of these operators is time independent (the Schroedinger picture). However, we can go to the Heisenberg picture, where

$$a_{k,\alpha} \rightarrow a_{k,\alpha} e^{-i\omega t} \qquad a^\dagger_{k,\alpha} \rightarrow a^\dagger_{k,\alpha} e^{i\omega t}$$

and where the operators $\boldsymbol{A}$, $\boldsymbol{\mathcal{E}}$, and $\boldsymbol{B}$ are time dependent.

5. The momentum in the field can be interpreted in terms of the momenta of the field quanta (photons). In classical electrodynamics, the momentum in the field is

$$P = \frac{1}{c}\int \mathcal{E} \times B \, d\tau \tag{15.29}$$

Substituting (15.27) and (15.28) into (15.29), integrating over V and making use of orthogonality as in (15.17), and doing some routine algebra, we obtain

$$P = \sum_{k,\alpha} \frac{\hbar k}{2}\left(a_{k,\alpha}a_{k,\alpha}^{\dagger} + a_{k,\alpha}^{\dagger}a_{k,\alpha} + a_{k,\alpha}a_{-k,\alpha} + a_{k,\alpha}^{\dagger}a_{-k,\alpha}^{\dagger}\right) \tag{15.30}$$

In this last expression, we must sum over all possible values of k; thus, if a given vector k_0 is included in the sum, the opposite vector $-k_0$ is also included. By adding up these contributions in pairs and taking into account the commutation rules, we see that the third and fourth types of terms on the right-hand side of (15.30) give no net contribution. Thus, using the commutation rule (15.25), we obtain

$$P = \sum_{k,\alpha} \frac{\hbar k}{2}\left(2N_{k,\alpha} + 1\right) = \sum_{k,\alpha} \hbar k N_{k,\alpha}$$

This means that each photon in mode k, α has momentum $\hbar k$ as well as energy $\hbar kc$. From the relation

$$E^2 = p^2c^2 + m_\gamma^2 c^4$$

we conclude that in the present theory, photons must have zero rest mass. In fact, if photons did possess nonzero rest mass, it would be necessary to modify Maxwell's equations in a fundamental way (see Section 15.8).

15.3 Zero-point energy and fluctuations in the field

We have noted that a state of the radiation field can be specified by giving the number of photons in each k, α mode; that is,

$$|\psi\rangle = |n_1, n_2, n_3, \ldots\rangle$$

The vacuum state is that state where all the occupation numbers are zero. Assuming that we can normalize this state to unity, that is, $\langle 0|0\rangle = 1$, we now calculate the expectation value of various physical quantities in the vacuum state, starting with the electric field $\mathcal{E}$. From (15.27), we have

$$\langle 0|\mathcal{E}|0\rangle = \frac{i}{L^{3/2}}\sum_{k,\alpha}\sqrt{\frac{\hbar\omega}{2}}\,\hat{\varepsilon}_{k,\alpha}\langle 0|\left(a_{k,\alpha}e^{ik\cdot r} - a_{k,\alpha}^{\dagger}e^{-ik\cdot r}\right)|0\rangle$$

Because $\langle 0|a_{k,\alpha}|0\rangle = \langle 0|a^\dagger_{k,\alpha}|0\rangle = 0$, we obtain $\langle 0|\mathcal{E}|0\rangle = 0$. This is a very reasonable result because there is no preferred direction in three-space associated with the vacuum state. In fact, the existence of such a direction would constitute a violation of spatial isotropy. Next, we consider

$$\langle 0|\mathcal{E}\cdot\mathcal{E}|0\rangle = -\frac{1}{2L^3}\sum_{\substack{k,\alpha,\\ k',\alpha'}} \hbar\sqrt{\omega\omega'}\hat{\varepsilon}_{k,\alpha}\cdot\hat{\varepsilon}_{k',\alpha'}$$
$$\langle 0|\left(a_{k,\alpha}e^{ik\cdot r} - a^\dagger_{k,\alpha}e^{-ik\cdot r}\right)\left(a_{k',\alpha'}e^{ik'\cdot r} - a^\dagger_{k',\alpha'}e^{-ik'\cdot r}\right)|0\rangle$$

In this expression, only terms of the form $\langle 0|a_{k,\alpha}a^\dagger_{k,\alpha}|0\rangle$ make a nonzero contribution. Hence

$$\langle 0|\mathcal{E}\cdot\mathcal{E}|0\rangle = \frac{1}{2L^3}\sum_{k,\alpha}\hbar\omega\langle 0|a^\dagger_{k,\alpha}a_{k,\alpha}+1|0\rangle = \frac{1}{2L^3}\sum_{k,\alpha}\hbar\omega_k \tag{15.31}$$

A similar result $\langle 0|\boldsymbol{B}\cdot\boldsymbol{B}|0\rangle = (1/2L^3)\sum_{k,\alpha}\hbar\omega_k$ is obtained for the magnetic field. Because the sum on the right-hand side of (15.31) must be carried out over an infinite number of modes, that sum diverges, so we have

$$\langle 0|\mathcal{E}\cdot\mathcal{E}|0\rangle = \langle 0|\boldsymbol{B}\cdot\boldsymbol{B}|0\rangle = \infty$$

Thus the mean square dispersion of the electric or magnetic field in the vacuum state is

$$\begin{aligned}(\Delta\mathcal{E})^2 &= \langle 0|\mathcal{E}\cdot\mathcal{E}|0\rangle - \langle 0|\mathcal{E}|0\rangle\cdot\langle 0|\mathcal{E}|0\rangle \\ &= \langle 0|\mathcal{E}\cdot\mathcal{E}|0\rangle = \infty\end{aligned}$$

This result should not be surprising because we know that

$$H = \frac{1}{2}\int[\mathcal{E}\cdot\mathcal{E} + \boldsymbol{B}\cdot\boldsymbol{B}]\,d\tau = \sum_{k,\alpha}\hbar\omega\left(N_{k,\alpha}+\frac{1}{2}\right)$$

Thus

$$\langle 0|H|0\rangle = \frac{1}{2}\int\left[\langle 0|\mathcal{E}\cdot\mathcal{E} + \boldsymbol{B}\cdot\boldsymbol{B}|0\rangle\right]d\tau = \frac{1}{2}\sum_{k,\alpha}\hbar\omega_k = \infty$$

In other words, the infinite zero-point energy and infinite fluctuations of $\mathcal{E}$ and $\boldsymbol{B}$ are equivalent.

We next consider how rapidly the zero-point energy sum diverges. To this end, we first determine the number of modes with angular frequency between ω and $\omega + d\omega$. This is actually a quantity of major importance that we shall refer to repeatedly. First, consider the number Z of modes with k less than or equal to a fixed value k_0. Because in general for a mode with specified $\boldsymbol{k}$ we have

$$k_x = \frac{2\pi n_1}{L} \qquad k_y = \frac{2\pi n_2}{L} \qquad k_z = \frac{2\pi n_3}{L}$$

and for each choice of the triplet of numbers $n_{1,2,3}$ there are two independent polarizations, Z is twice the number of ways that we can form integers $n_{1,2,3}$ such that

$$k_0^2 \geq \left(\frac{2\pi}{L}\right)^2 \left(n_1^2 + n_2^2 + n_3^2\right) = \frac{4\pi^2}{L^2} n^2$$

where $n^2 \equiv n_1^2 + n_2^2 + n_3^2$. For sufficiently large n, Z is just twice the volume of a sphere of radius n; that is,

$$Z = \frac{8\pi}{3} n^3 = \frac{8\pi}{3} \frac{L^3 k_0^3}{\left(4\pi^2\right)^{3/2}}$$
$$= \frac{L^3 \omega^3}{3\pi^2 c^3}$$

where $\omega = k_0 c$. Hence the number of modes between ω and $\omega + d\omega$ is

$$\frac{dZ}{d\omega} d\omega = \frac{V \omega^2 d\omega}{\pi^2 c^3} \tag{15.32}$$

Thus the zero-point energy summed over all modes is

$$\sum \frac{\hbar\omega}{2} \approx \frac{\hbar V}{2\pi^2 c^3} \int \omega^3 \, d\omega \tag{15.33}$$

which diverges in proportion to ω^4. Let's now imagine that instead of carrying the integral to arbitrarily high frequency, we cut it off at some upper limit $\omega_{max} = 2\pi c/\lambda_{min}$, where λ_{min} is a "shortest length." Then the total zero-point energy per unit volume $\varepsilon_{EM} = (1/2V)\sum \hbar\omega$ is

$$\varepsilon_{EM} \approx \frac{2\pi^2 \hbar c}{\lambda_{min}^4}$$

Is there a reasonable physical choice for λ_{min}? The *classical radius* of the electron

$$r_0 = \frac{e^2}{4\pi m_e c^2} \approx 2.8 \times 10^{-13} \text{ cm}$$

is far too large because we know from many experiments that quantum electrodynamics is valid to distances smaller by orders of magnitude than r_0. Many physicists believe that λ_{min} is the Planck length R, defined as the Compton wavelength of a particle of mass M such that its gravitational self-energy and relativistic rest energy are comparable. Employing Newton's constant G, we have

$$\frac{GM^2}{R} \approx Mc^2$$

and thus

$$M \approx \frac{Rc^2}{G} = \frac{\hbar}{Mc}\frac{c^2}{G}$$

which yields

$$R \approx \frac{(\hbar G)^{1/2}}{c^{3/2}} \approx 2\times10^{-33}\ \text{cm}$$

In this case, ε_{EM} would be

$$\varepsilon_{EM} \approx \frac{c^7}{\hbar G^2} \approx 10^{114}\ \text{erg/cm}^3 \tag{15.34}$$

If such an enormous energy density actually existed, it would be necessary to include it in the Friedmann equation of general relativity that describes the expansion of the universe according to the big bang cosmological model. However, this would yield a description of the Hubble expansion that is grossly in contradiction with known facts. Indeed, observations made since 1998 of the luminosity of type Ia supernovas with known redshifts lead to the conclusion that the Hubble expansion is accelerating, and this is consistent with the existence of an all-pervasive dark energy density ε_D that is approximately 120 orders of magnitude smaller than ε_{EM} in (15.34).

This enormous discrepancy suggests several possibilities. Perhaps ε_{EM} is very nearly canceled by zero-point energies of the opposite sign arising from other fields, as proposed in speculative supersymmetric models. But why should the cancellation be so nearly perfect yet not complete? Alternatively, the whole concept of zero-point energy might be wrong: it might be a mathematical artifice without physical reality – a manifestation of something deeply wrong with quantum-field theory, and ε_D might be due to something else entirely. However, the possibility that there is no zero-point energy does not seem reasonable because we have persuasive experimental evidence that the zero-point energy of the EM field actually exists, and this evidence is from the Casimir-Polder effect.

15.4 The Casimir-Polder effect

In Chapter 10 we learned that there exists an attractive van der Waals interaction between two ordinary nonrelativistic polarizable atoms separated by a distance R (e.g., two H atoms in the ground state). Essentially this occurs as follows: an isolated atom has, on average, no electric dipole moment. However, at any given instant, either one of the atoms (e.g., atom 1) has a dipole moment, which gives rise to a fluctuating dipolar electric field. Such a field induces a dipole moment on atom 2, and the interaction of dipoles 1 and 2 is the van der Waals potential. For $a_0 \ll R \ll \alpha^{-1}a_0 = 137a_0$, we found from second-order perturbation theory that the van der Waals potential energy is

$$U^0_{a-a} \approx -\frac{e^2 a_0^5}{R^6} \tag{15.35}$$

The restriction $R < \alpha^{-1}a_0$ occurs because the quasi-static electric field generated by the fluctuating dipole 1 takes time $T = R/c$ to propagate from atom 1 to atom 2, whereas the "natural time" for motion of an electron in an atom is $\tau \approx a_0/ac$. To avoid retardation effects that wash out the van der Waals potential, we require $T \ll \tau$.

We also learned that there is a van der Waals interaction between an atom and a perfectly grounded conducting plane. Here the instantaneous atomic dipole sets up an image dipole; this time the two dipoles are perfectly correlated, and the interaction energy (now nonvanishing in first-order perturbation theory) is

$$U^0_{a-w} \approx \frac{-e^2 a_0^2}{R^3} \tag{15.36}$$

where R is now the distance between the atom and the plane. Here again we require $R < \alpha^{-1}a_0$ to avoid retardation effects.

However, in addition to the ordinary van der Waals effect, there is an entirely new feature arising from quantization of the radiation field. According to the theory just developed in preceding sections, the atoms (and the conducting planes) should exist in a space pervaded by fluctuating electric fields associated with zero-point energy. A fluctuating zero-point electric field at atom 1 must induce a fluctuating electric dipole moment in atom 1. Of course, we can Fourier analyze these fluctuations into sinusoidal oscillations. Furthermore, we know from elementary electricity and magnetism that an oscillating dipole generates its own radiation field. This radiation field will interact with atom 2, and the energy of interaction should be described by a new potential (originating entirely from zero-point energy) that yields a correction to the ordinary van der Waals potential.

For various practical reasons, this correction to the van der Waals interaction for two isolated atoms is very difficult to observe. However, the same kind of argument ought to apply to the interaction between an atom and a perfectly conducting grounded plane or to the interaction between two uncharged macroscopic conductors. In fact, it turns out that the force of attraction between an atom and a plane grounded conductor can be measured experimentally, as can the force of attraction between two uncharged plane parallel conductors. These extensions of the van der Waals interaction, known as the *Casimir-Polder effect* (Casimir 1948) are demonstrations of the reality of zero-point energy.

Let's now see how to calculate the Casimir-Polder effect [see, e.g., Spruch (1978, 1996)]. In order to bring out the main points as simply and clearly as possible, we give only order-of-magnitude estimates rather than precise calculations. The one exception is precise calculation of the force of attraction between two plane parallel uncharged conductors, which appears as Problem 15.3 at the end of this chapter. Imagine two isolated objects 1 and 2 separated by a distance r. Each object is polarizable: if an electric field exists, each object acquires an induced electric dipole moment $\boldsymbol{p}$; that is,

$$\boldsymbol{p}_1 = \underline{\alpha}_1 \boldsymbol{\mathcal{E}}(\boldsymbol{r}_1, t) \qquad \boldsymbol{p}_2 = \underline{\alpha}_2 \boldsymbol{\mathcal{E}}(\boldsymbol{r}_2, t)$$

where $\underline{\alpha}_{1,2}$ are the polarizabilities of objects 1 and 2 respectively, and $\boldsymbol{\mathcal{E}}$ is the fluctuating electric field that exists even when no photons are present. Now consider the energy U of dipole 1 due to the radiation field $\boldsymbol{\mathcal{E}}_{21}$ generated by fluctuating dipole 2. It is

$$U \approx -p_1 \mathcal{E}_{21}$$

However, from elementary radiation theory, we know that the radiation-field component with frequency ω generated by an oscillating dipole is

$$\mathcal{E}_{21} \approx \frac{\omega^2}{c^2 r} p_2$$

Hence the contribution to the energy from this frequency component is

$$U \approx -p_1 p_2 \frac{\omega^2}{c^2 r} = -\underline{\alpha}_1 \underline{\alpha}_2 \mathcal{E}(\boldsymbol{r}_1, t) \mathcal{E}(\boldsymbol{r}_2, t) \frac{\omega^2}{c^2 r}$$

This expression describes the contribution of one mode with frequency ω, but we are interested in the total contribution to the energy from all modes; that is,

$$U \approx L^3 \int \underline{\alpha}_1 \underline{\alpha}_2 \mathcal{E}_\omega(\boldsymbol{r}_1, t) \mathcal{E}_\omega(\boldsymbol{r}_2, t) \frac{\omega^2}{c^2 r} \frac{\omega^2 d\omega}{c^3} \tag{15.37}$$

The integral is very difficult to evaluate if we include arbitrarily high frequencies, but on physical grounds, it is reasonable to cut off the integration at a maximum frequency $\omega_{\max} \approx c/r$. The reason is that for much higher frequencies, $\mathcal{E}_\omega(\boldsymbol{r}_1, t) \mathcal{E}_\omega(\boldsymbol{r}_2, t)$ oscillates rapidly and does not contribute effectively to the integral in (15.37). For frequencies much lower than $\omega_{\max}$, $L^3 \mathcal{E}_\omega(\boldsymbol{r}_1, t) \mathcal{E}_\omega(\boldsymbol{r}_2, t) = V \mathcal{E}^2 \approx \hbar\omega$, so (15.37) becomes

$$U \approx \int^{c/r} \underline{\alpha}_1 \underline{\alpha}_2 \frac{\hbar \omega^5}{c^5 r} d\omega \tag{15.38}$$

This formula will now be employed to draw conclusions concerning the following interactions:

1. Electron-electron
2. Electron-atom
3. Atom-atom
4. Atom-wall
5. Wall-wall

These are distinguished from one another by different products of polarizabilities $\underline{\alpha}_1 \underline{\alpha}_2$. We obtain a rough estimate of the polarizability in each case by resorting to a classical harmonic oscillator model of an electron in an atom. Here the electron is imagined to be bound to the nucleus with a restoring force $F = -m_e \omega_0^2 x$. If an external oscillating electric field is applied to the atom with applied frequency ω, the electron motion is described by

$$m_e \ddot{x} + m_e \omega_0^2 x = -e \mathcal{E}_0 e^{i\omega t}$$

We try a solution of the form $x = x_0 e^{i\omega t}$ and obtain

$$x_0 = \frac{-e}{m_e \left(\omega_0^2 - \omega^2\right)} \mathcal{E}_0$$

The electric dipole moment is

$$p = -ex_0 = \frac{e^2}{m_e\left(\omega_0^2 - \omega^2\right)}\mathcal{E}_0$$

and the polarizability at frequency ω is

$$\underline{\alpha}(\omega) = \frac{e^2}{m_e\left(\omega_0^2 - \omega^2\right)} \tag{15.39}$$

If $\omega_0 \gg \omega$, which is an appropriate approximation for an ordinary atom when $\omega < c/r$, we have

$$\underline{\alpha} = \frac{e^2}{m_e\omega_0^2}$$

Now ω_0 can be determined from the requirement that $\hbar\omega_0 \approx e^2/a_0$, which is a typical atomic energy. Hence

$$\underline{\alpha} = \frac{e^2}{m_e\left(e^2/\hbar a_0\right)^2} = a_0^3 \tag{15.40}$$

This agrees with what we learned in Chapter 10 when studying the Stark effect: typical atomic polarizabilities are indeed $\underline{\alpha} \approx a_0^3$. We now proceed to consider the various interactions listed earlier.

Electron-Electron

The dominant interaction between two electrons is the Coulomb interaction

$$U_{\text{Coulomb}} = \frac{e^2}{4\pi r}$$

However, there exists a small "retarded potential" correction, given by the integral in (15.38), with $\underline{\alpha}_{1,2}$ determined by setting $\omega_0 = 0$ in (15.39); that is,

$$\underline{\alpha}_1 = \underline{\alpha}_2 = -\frac{e^2}{m_e\omega^2} \tag{15.41}$$

Inserting these in (15.38) and integrating, we obtain

$$U_{e-e} \approx -\frac{e^4\hbar}{m_e^2c^3}\frac{1}{r^3} = -\left(\frac{\alpha^3 a_0^2}{r^2}\right)\frac{e^2}{r}$$

Of course, this is only a crude estimate of an effect that would require considerable effort to calculate accurately, but it is basically correct, and the same goes for the estimates that follow.

Electron-Atom

For $r \ll \alpha^{-1}a_0$, the interaction energy of an electron and a polarizable atom is given by the following ordinary van der Waals result:

$$U_{e-a}^{0} \approx -\underline{\alpha}_{\text{atom}}\mathcal{E}^2 \approx -\frac{e^2 a_0^3}{r^4} \tag{15.42}$$

For $r > \alpha^{-1}a_0$, we employ (15.40), and for the atom and electron polarizabilities, we employ (15.41). Then (15.38) yields

$$U_{e-a} \approx -\alpha\frac{e^2 a_0^4}{r^5} \tag{15.43}$$

It may be possible to distinguish between (15.42) and (15.43) in future experiments with Rydberg atoms.

Atom-Atom

In this case, as we have noted, the ordinary van der Waals interaction is described by the approximate formula (15.35). For $r > \alpha^{-1}a_0$, we employ (15.40) for both atomic polarizabilities to calculate the integral in (15.38). This yields the Casimir-Polder contribution

$$U_{a-a} \approx -\hbar c\frac{a_0^6}{r^7} = -\left(\frac{a_0}{\alpha r}\right)\frac{e^2 a_0^5}{r^6}$$

Atom-Wall

Here the ordinary van der Waals interaction is described by the approximate formula (15.36). To estimate the Casimir-Polder contribution, we replace the conducting wall with a large conducting sphere of radius z and place the atom at a distance $2z$ from the center of the sphere. The polarizability of the atom is given by (15.42); that of the sphere is z^3 from elementary electricity and magnetism. Inserting these in (15.38) and evaluating the integral, we obtain

$$U_{a-w} \approx -\hbar c\frac{a_0^3}{z^4} = -\left(\frac{a_0}{\alpha z}\right)\frac{e^2 a_0^2}{z^3}$$

Wall-Wall

To make an estimate in this case, we imagine two identical conducting spherical shells of radius z with centers separated by $3z$. Then, because the polarizability of each is z^3, (15.38) yields

$$U_{w-w} \approx -\frac{\hbar c}{z}$$

Now the surface area of each sphere is $4\pi z^2$. Thus the energy per unit area is

$$u \approx -\frac{\hbar c}{z^3} \tag{15.44}$$

This expression should also be valid for the energy per unit area associated with a pair of uncharged parallel plane conductors separated by a distance z. Therefore, from (15.44), we obtain an estimate of the force of attraction per unit area between these conductors; that is,

$$F = -\frac{\partial u}{\partial z} \approx -\frac{\hbar c}{z^4}$$

Let us try to understand why such a force exists from a slightly different but equivalent viewpoint. We know that by choosing L^3 sufficiently large, we can have modes of arbitrarily long wavelength (low frequency) in free space. This is true whether we expand the vector potential in plane traveling waves, in plane standing waves, or in terms of any other complete set of orthogonal vector waves. However, if we now introduce a pair of plane parallel conductors separated by a finite distance z, it is no longer true that modes of arbitrarily long wavelength can exist. In fact, it is impossible to have any standing waves with half-wavelengths larger than z in the direction normal to the plates. Thus the zero-point energy sums (15.33) for the two cases – free space and space occupied by the pair of uncharged plates – are both infinite but differ by a perfectly definite and calculable finite amount because of the difference in low-frequency modes. A straightforward calculation (see Problem 15.3) yields the following refinement of (15.44) for the energy per unit area:

$$u = -\frac{\pi^2}{720}\frac{\hbar c}{z^3}$$

The first accurate measurements of the effect were reported in 1997 by Lamoreaux (Lamoreaux 1997, 2012). Since then, it has also been measured accurately by several other groups of experimenters [see, e.g., Bressi (2002)].

15.5 Blackbody radiation and Planck's law

Consider electromagnetic radiation in thermal equilibrium at temperature T. What is the average number of photons in a given mode at frequency ω? We recall that a given mode is described as if it were a one-dimensional harmonic oscillator. Thus we are led to consider the average excitation of such an oscillator at temperature T. According to statistical mechanics, its partition function is

$$\begin{aligned} Z &= \sum_{n=0}^{\infty} \exp\left(-\frac{E_n}{k_B T}\right) \\ &= \exp\left(-\frac{\hbar\omega}{2k_B T}\right)\sum_{n=0}^{\infty} \exp\left(-\frac{n\hbar\omega}{k_B T}\right) \end{aligned}$$

Thus the average occupation number is

$$\langle n \rangle = \frac{\sum_{n=0}^{\infty} n x^n}{\sum_{n=0}^{\infty} x^n} \tag{15.45}$$

where $x = \exp(-\hbar\omega/k_B T)$. Equation (15.45) can be written

$$\langle n \rangle = \frac{x \dfrac{\partial}{\partial x} \sum x^n}{\sum x^n} = -x \frac{\partial}{\partial x} \ln(1-x) = \frac{x}{1-x}$$

Thus

$$\langle n \rangle = \frac{1}{\exp\left(\dfrac{\hbar\omega}{k_B T}\right) - 1} \tag{15.46}$$

From this expression it is easy to find the energy per unit volume in the field between frequencies ω and $\omega + d\omega$. We multiply $\langle n \rangle$ from (15.46) by the energy $\hbar\omega$ of each photon and by the number of modes per unit volume between ω and $\omega + d\omega$; that is,

$$\varepsilon(\omega) d\omega = \frac{1}{\exp\left(\dfrac{\hbar\omega}{k_B T}\right) - 1} \bullet \hbar\omega \bullet \frac{\omega^2 d\omega}{\pi^2 c^3}$$

This gives Planck's law for the spectral energy density $\varepsilon(\omega)$ as

$$\varepsilon(\omega) = \frac{\hbar\omega^3}{\pi^2 c^3} \frac{1}{\exp\left(\dfrac{\hbar\omega}{k_B T}\right) - 1}$$

The total energy per unit volume in all frequencies is

$$\begin{aligned} \varepsilon &= \int_0^\infty \varepsilon(\omega)\, d\omega \\ &= \frac{\hbar}{\pi^2 c^3} \int_0^\infty \omega^3 \left[\exp\left(\frac{\hbar\omega}{k_B T}\right) - 1\right]^{-1} d\omega \\ &= \frac{(k_B T)^4}{\pi^2 (\hbar c)^3} \sum_{n=1}^{\infty} \int_0^\infty x^3 e^{-nx}\, dx \\ &= \frac{(k_B T)^4}{\pi^2 (\hbar c)^3} \sum_{n=1}^{\infty} \frac{3!}{n^4} \\ &= aT^4 \end{aligned} \tag{15.47}$$

where

$$a = \frac{\pi^2 k_B^4}{15 \hbar^3 c^3} = 7.56 \times 10^{-15} \text{ erg cm}^{-3} \text{ deg}^{-4}$$

is the Stefan-Boltzmann constant. The blackbody radiation exerts a pressure p that is obtained from elementary kinetic theory by thinking of the radiation as a relativistic photon gas in which each photon carries momentum $\hbar k$ as well as energy $\hbar\omega$. For any ultrarelativistic gas, one has $p = \varepsilon/3$; thus, for blackbody radiation,

$$p = \frac{1}{3}aT^4 \tag{15.48}$$

Next, we calculate the entropy of blackbody radiation from the thermodynamic identity

$$TdS = dU + pdV$$

where $U = \varepsilon V$ is the energy, and V is the volume. From (15.47) and (15.48) we obtain

$$\begin{aligned} TdS &= d(VaT^4) + \frac{aT^4}{3}dV \\ &= \frac{4}{3}aT^4 dV + 4aVT^3 dT \end{aligned}$$

or

$$dS = \frac{4}{3}aT^3 dV + 4aVT^2 dT$$

Comparing this expression with

$$dS = \left(\frac{\partial S}{\partial V}\right)_T dV + \left(\frac{\partial S}{\partial T}\right)_V dT$$

we obtain

$$\begin{aligned} \left(\frac{\partial S}{\partial V}\right)_T &= \frac{4}{3}aT^3 \\ \left(\frac{\partial S}{\partial T}\right)_V &= 4aVT^2 \end{aligned}$$

Integration of either of these expression yields

$$S = \frac{4}{3}aT^3 V$$

Therefore, in an isentropic (constant-entropy) expansion, sometimes called an *adiabatic expansion*, VT^3 remains constant. This requires that

$$\varepsilon = aT^4 = \text{const} \bullet V^{-4/3}$$

15.6 Classical limit of the quantized radiation field

Because the quantized radiation field is described as a collection of quantized harmonic oscillators, one for each mode, we can make the transition to the classical limit for the radiation field by recalling how this is done for the simple harmonic oscillator (SHO). In Chapter 6 we discussed the coherent state $|\psi\rangle$, which is obtained by displacing the SHO ground state from the origin by amount s. We found that $|\psi\rangle$ is a superposition of number n eigenstates, that the probability of obtaining a definite n value in state $|\psi\rangle$ is given by a Poisson distribution, and that the mean value of n is proportional to s^2. Also, $|\psi\rangle$ oscillates coherently about the origin in such a way that for sufficiently large values of s, it displays all the properties we normally associate with a classical harmonic oscillator.

It is intuitively clear that we should follow a similar path for the radiation field. To make the writing as simple as possible, we consider just one radiation field mode so that it is not necessary to write sums over $\boldsymbol{k}$, α or exhibit the latter quantities repeatedly. We also recall from Section 6.13.5 that

$$|\psi(t)\rangle = e^{-s^2/4}e^{-i\omega t/2}\sum_{n=0}^{\infty}\left(\frac{s}{\sqrt{2}}\right)^n \frac{1}{n!}e^{-in\omega t}\left(a^\dagger\right)^n|0\rangle$$

We now assume that the classical vector potential for the mode in question is

$$\boldsymbol{A}_c \equiv \langle\psi|\boldsymbol{A}|\psi\rangle = \frac{\hat{\varepsilon}}{L^{3/2}}\sqrt{\frac{\hbar c^2}{2\omega}}\langle\psi|ae^{i\boldsymbol{k}\cdot\boldsymbol{r}} + a^\dagger e^{-i\boldsymbol{k}\cdot\boldsymbol{r}}|\psi\rangle \tag{15.49}$$

To evaluate this expression, we must calculate the matrix elements $\langle\psi|a|\psi\rangle$ and $\langle\psi|a^\dagger|\psi\rangle$. We have

$$a|\psi\rangle = e^{-s^2/4}e^{-i\omega t/2}\sum_{n=0}^{\infty}\left(\frac{s}{\sqrt{2}}\right)^n \frac{1}{n!}e^{-in\omega t}a\left(a^\dagger\right)^n|0\rangle \tag{15.50}$$

Now

$$\begin{aligned} a\left(a^\dagger\right)^n|0\rangle &= \left[a,\left(a^\dagger\right)^n\right]|0\rangle + \left(a^\dagger\right)^n a|0\rangle \\ &= \left[a,\left(a^\dagger\right)^n\right]|0\rangle \\ &= n\left(a^\dagger\right)^{n-1}|0\rangle \end{aligned}$$

Thus (15.50) is

$$a|\psi\rangle = \frac{s}{\sqrt{2}}e^{-i\omega t}|\psi\rangle$$

This can be generalized slightly by noting that it's not necessary to start the coherent state with a maximum displacement at $t = 0$. Instead, we can include a phase factor

$$a|\psi\rangle = \frac{s}{\sqrt{2}} e^{-i(\omega t+\phi)} |\psi\rangle$$

Then, because $|\psi\rangle$ is normalized to unity, we have

$$\langle\psi|a|\psi\rangle = \frac{s}{\sqrt{2}} e^{-i(\omega t+\phi)} \tag{15.51}$$

Similarly,

$$\langle\psi|a^\dagger|\psi\rangle = \frac{s}{\sqrt{2}} e^{i(\omega t+\phi)} \tag{15.52}$$

Furthermore, $\langle n\rangle = s^2/2$; hence (15.51) and (15.52) can be written

$$\langle\psi|a|\psi\rangle = \langle n\rangle^{1/2} e^{-i(\omega t+\phi)} \tag{15.53}$$

and

$$\langle\psi|a^\dagger|\psi\rangle = \langle n\rangle^{1/2} e^{i(\omega t+\phi)} \tag{15.54}$$

Substitution of (15.53) and (15.54) into (15.49) yields

$$\boldsymbol{A}_c = \frac{\hat{\varepsilon}}{L^{3/2}} \langle n\rangle^{1/2} \sqrt{\frac{\hbar c^2}{\omega}} \cos(\boldsymbol{k}\cdot\boldsymbol{r} - \omega t - \phi) \tag{15.55}$$

From (15.55), we obtain the following expressions for $\boldsymbol{\mathcal{E}}_c$ and $\boldsymbol{B}_c$:

$$\boldsymbol{\mathcal{E}}_c = -\frac{1}{c}\frac{\partial \boldsymbol{A}_c}{\partial t} = -\frac{\hat{\varepsilon}}{L^{3/2}} \langle n\rangle^{1/2} \sqrt{\hbar\omega} \sin(\boldsymbol{k}\cdot\boldsymbol{r} - \omega t - \phi) \tag{15.56}$$

and

$$\boldsymbol{B}_c = -\frac{\hat{k}\times\hat{\varepsilon}}{L^{3/2}} \langle n\rangle^{1/2} \sqrt{\hbar\omega} \sin(\boldsymbol{k}\cdot\boldsymbol{r} - \omega t - \phi) \tag{15.57}$$

We emphasize that the phase ϕ in these expressions is associated with oscillation of the coherent state. We cannot associate a definite phase with a photon number eigenstate.

When is classical electromagnetic theory legitimate; that is, when are we permitted to use (15.55)–(15.57)? A sensible criterion for a given mode is that the time average of $\boldsymbol{\mathcal{E}}_c \cdot \boldsymbol{\mathcal{E}}_c$ over one period of oscillation, which we denote by $\overline{\boldsymbol{\mathcal{E}}_c \cdot \boldsymbol{\mathcal{E}}_c}$, should be much greater than the contribution of this mode to $\langle 0|\boldsymbol{\mathcal{E}} \cdot \boldsymbol{\mathcal{E}}|0\rangle$. From (15.56), we have

$$\overline{\boldsymbol{\mathcal{E}}_c \cdot \boldsymbol{\mathcal{E}}_c} = \frac{\hbar\omega}{2L^3} \langle n\rangle$$

whereas from (15.31), we have

$$\langle 0|\boldsymbol{\mathcal{E}}\cdot\boldsymbol{\mathcal{E}}|0\rangle_{\text{single mode}} = \frac{\hbar\omega}{2L^3}$$

Thus, for the classical description to be valid for any given mode, the mean number of photons in that mode must be much greater than unity.

In most real observations and/or experiments, we usually deal with N photons distributed over a range of modes. Let us suppose that these modes are in the frequency range ω to $\omega + \delta\omega$ so that the relevant number of modes is $L^3\omega^2\delta\omega/\pi^2c^3$. In this case, the classical description is valid when

$$\frac{N}{L^3} >> \frac{\omega^2\delta\omega}{\pi^2c^3}$$

In the limit $\omega \to 0$ (static electric and/or magnetic field), the classical description is always valid.

15.7 Digression on special relativity: Covariant description of the radiation field

We now provide a short summary and review of special relativity so that we can write the equations of electrodynamics in covariant form. According to special relativity, all inertial frames are equivalent, and the laws of nature take the same form no matter which inertial frame they are expressed in. Maxwell's theory is assumed to be the correct description of electromagnetic wave propagation, and the velocity of light c is the same in all inertial frames.

In special relativity, an *event* (the emission or absorption of an infinitely short light pulse) occurs at a single space-time point and is defined by the spatial coordinates $\boldsymbol{x} = x, y, z$ and the time t or $x^0 = ct$. The contravariant coordinate 4-vector of this event with respect to a particular inertial frame is defined as

$$x^\mu = (x^0, \boldsymbol{x}) = (x^0, x^1, x^2, x^3)$$

It is also useful to define the covariant coordinate 4-vector x_μ by

$$x_0 = x^0 \qquad x_i = -x^i \qquad i = 1,2,3$$

The metric tensor is

$$g_{\mu\nu} = g^{\mu\nu} = \begin{pmatrix} 1 & 0 & 0 & 0 \\ 0 & -1 & 0 & 0 \\ 0 & 0 & -1 & 0 \\ 0 & 0 & 0 & -1 \end{pmatrix}$$

We convert a contravariant 4-vector to its covariant counterpart, or vice versa, by means of the transformations

$$x^{\mu} = g^{\mu\nu} x_{\nu} \qquad x_{\rho} = g_{\rho\sigma} x^{\sigma}$$

where the repeated index summation convention is assumed here and in what follows.

The interval between two events x^{μ} and X^{μ} is defined as s, with

$$\begin{aligned} s^2 &= \left(x^0 - X^0\right)^2 - \left(x^1 - X^1\right)^2 - \left(x^2 - X^2\right)^2 - \left(x^3 - X^3\right)^2 \\ &= \left(x^{\mu} - X^{\mu}\right)\left(x_{\mu} - X_{\mu}\right) \end{aligned}$$

Choosing event X^{μ} to be at the origin, we have

$$s^2 = x^{\mu} x_{\mu} \tag{15.58}$$

It follows from the invariance of the velocity of light and the homogeneity and isotropy of space-time that the quadratic form (15.58) is invariant under linear transformations from one inertial frame to another (Lorentz transformations). These are classified as inhomogeneous or homogeneous according to whether or not a shift in the origin of 4-space accompanies the transformation. For our purposes, only the homogeneous transformations are necessary. Each homogeneous Lorentz transformation is described by an orthogonal matrix $a^{\mu}{}_{\nu}$ with

$$x^{\mu\prime} = a^{\mu}{}_{\nu} x^{\nu}$$

The orthogonality of the transformation matrix is expressed by the relation

$$a^{\mu}{}_{\nu} a^{\nu}{}_{\sigma} = \delta^{\mu}{}_{\sigma}$$

The matrix of the transformation inverse to $a^{\mu}{}_{\nu}$ is

$$g_{\rho\mu} g^{\sigma\nu} a^{\mu}{}_{\nu} = a_{\rho}{}^{\sigma}$$

The homogeneous Lorentz transformations are further classified as proper ($\det a = +1$) or improper ($\det a = -1$). The proper transformations include spatial rotations, for example,

$$\begin{aligned} x^{0\prime} &= x^0 \\ x^{1\prime} &= x^1 \cos\theta - x^2 \sin\theta \\ x^{2\prime} &= x^1 \sin\theta + x^2 \cos\theta \\ x^{3\prime} &= x^3 \end{aligned}$$

Thus

$$a = \begin{pmatrix} 1 & 0 & 0 & 0 \\ 0 & \cos\theta & -\sin\theta & 0 \\ 0 & \sin\theta & \cos\theta & 0 \\ 0 & 0 & 0 & 1 \end{pmatrix}$$

They also include Lorentz boosts, for example,

$$t' = \gamma\left(t - \frac{\mathrm{v}}{c^2}x^1\right)$$
$$x^{1\prime} = \gamma(x^1 - \mathrm{v}t)$$
$$x^{2\prime} = x^2$$
$$x^{3\prime} = x^3$$

where $\gamma = \left(1-\beta^2\right)^{-1/2}$ and $\beta = \mathrm{v}/c$. Thus

$$a = \begin{pmatrix} \gamma & -\beta\gamma & 0 & 0 \\ -\beta\gamma & \gamma & 0 & 0 \\ 0 & 0 & 1 & 0 \\ 0 & 0 & 0 & 1 \end{pmatrix} \qquad a^{-1} = \begin{pmatrix} \gamma & \beta\gamma & 0 & 0 \\ \beta\gamma & \gamma & 0 & 0 \\ 0 & 0 & 1 & 0 \\ 0 & 0 & 0 & 1 \end{pmatrix}$$

The improper Lorentz transformations include, for example, spatial inversion, with

$$a = \begin{pmatrix} 1 & 0 & 0 & 0 \\ 0 & -1 & 0 & 0 \\ 0 & 0 & -1 & 0 \\ 0 & 0 & 0 & -1 \end{pmatrix}$$

The totality of homogeneous Lorentz transformations forms a group. Any set of four quantities V^μ is called a *4-vector* if it transforms like x^μ under a Lorentz transformation; that is,

$$V^{\mu\prime} = a^\mu{}_\nu V^\nu$$

Examples include

$j^\mu = c\rho, \boldsymbol{j}$ — *Electromagnetic current density*

$A^\mu = \Phi, \boldsymbol{A}$ — 4-vector potential

$p^\mu = \dfrac{E}{c}, \boldsymbol{p}$ — 4-momentum of a particle

The scalar product of two 4-vectors A, B is

$$A \bullet B = A^\mu B_\mu = A_\mu B^\mu = A^0 B^0 - \boldsymbol{A} \bullet \boldsymbol{B}$$

This is invariant under a Lorentz transformation. Thus, for example,

$$p^\mu p_\mu = \frac{E^2}{c^2} - \boldsymbol{p} \bullet \boldsymbol{p} = -(m_0 c)^2$$

and

$$j^\mu A_\mu = c\left(\rho\Phi - \frac{1}{c}\boldsymbol{j} \bullet \boldsymbol{A}\right)$$

are invariants. Also,

$$\partial_\mu \equiv \frac{\partial}{\partial x^\mu} = \left(\frac{1}{c}\frac{\partial}{\partial t}, \nabla\right)$$

whereas

$$\partial^\mu \equiv \frac{\partial}{\partial x_\mu} = \left(\frac{1}{c}\frac{\partial}{\partial t}, -\nabla\right)$$

Thus

$$\partial_\mu j^\mu = \frac{\partial \rho}{\partial t} + \nabla \cdot \boldsymbol{j} = 0 \quad \text{(Equation of continuity)}$$

and

$$\partial^\mu A_\mu = \frac{1}{c}\frac{\partial \Phi}{\partial t} + \nabla \cdot \boldsymbol{A} = 0 \quad \text{(Lorenz gauge condition)}$$

are invariant conditions. Also,

$$\partial^\mu \partial_\mu = \partial_\mu \partial^\mu = \frac{1}{c^2}\frac{\partial^2}{\partial t^2} - \nabla^2$$

is an invariant operator. A set of 16 quantities that transform in the following way is called a *contravariant second-rank tensor*:

$$T'^{\rho\sigma} = a^\rho{}_\mu a^\sigma{}_\nu T^{\mu\nu}$$

We can also form covariant tensors and mixed tensors; that is,

$$T_{\rho\sigma} = g_{\rho\mu} g_{\sigma\nu} T^{\mu\nu} \qquad T^\rho{}_\sigma = g_{\sigma\nu} T^{\rho\nu}$$

The electromagnetic field tensor

$$F_{\mu\nu} = \partial_\mu A_\nu - \partial_\nu A_\mu$$

is an important antisymmetric second-rank tensor. It can be written as a 4×4 matrix

$$F_{\mu\nu} = \begin{pmatrix} 0 & \mathcal{E}_x & \mathcal{E}_y & \mathcal{E}_z \\ -\mathcal{E}_x & 0 & -B_z & B_y \\ -\mathcal{E}_y & B_z & 0 & -B_x \\ -\mathcal{E}_z & -B_y & B_x & 0 \end{pmatrix}$$

We now recall the first and third Maxwell equations expressed in terms of the potentials

$$\nabla^2\Phi - \frac{1}{c^2}\ddot{\Phi} = -\rho - \frac{1}{c}\frac{\partial}{\partial t}\left(\nabla \bullet A + \frac{1}{c}\dot{\Phi}\right) \tag{15.59}$$

$$\nabla^2 A - \frac{1}{c^2}\ddot{A} = -\frac{1}{c}j + \nabla\left(\nabla \bullet A + \frac{1}{c}\dot{\Phi}\right) \tag{15.60}$$

Equations (15.59) and (15.60) are written as follows in covariant notation:

$$\partial^\mu \partial_\mu A_\nu - \partial_\nu\left(\partial^\mu A_\mu\right) = \frac{1}{c}j_\nu$$

Reversal of the order of partial differentiations in the second term on the left-hand side yields the important relation

$$\partial^\mu F_{\mu\nu} = \frac{1}{c}j_\nu \tag{15.61}$$

The 4-vector potential A_μ becomes an operator analogous to A in Coulomb gauge [recall equation (15.26)] when the radiation field is quantized. We write this new operator in Lorenz gauge and in the Heisenberg picture as follows:

$$A_\mu = \frac{1}{L^{3/2}}\sum_{k,\varepsilon}\sqrt{\frac{\hbar c^2}{2\omega}}\varepsilon_\mu\left(a_{k,\varepsilon}e^{-ik\cdot x} + a^\dagger_{k,\varepsilon}e^{ik.x}\right) \tag{15.62}$$

where ε_μ is a polarization 4-vector, $k \cdot x = \omega t - \boldsymbol{k}\bullet\boldsymbol{r}$, and the Lorenz gauge condition implies that $k \cdot \varepsilon \equiv k^\mu \varepsilon_\mu = 0$. Note that while we have until now employed a vector potential A with two independent polarization components for each mode, both transverse to the direction of propagation, there are four components of the 4-vector ε_μ in (15.62). Two of the latter can be considered orthogonal to $\boldsymbol{k}$; one is longitudinal (collinear with $\boldsymbol{k}$), and one is timelike. However, when we deal with the emission and absorption of real photons, it turns out that the effects due to the longitudinal and timelike components cancel one another, and thus the two vector potentials A and A_μ are consistent.

Finally, in relativistic quantum mechanics, it is convenient to employ natural units,, where $\hbar = c = m_e = 1$. These units, which will be useful in Chapters 19–24, imply that

Unit of length = electron Compton wavelength $= \lambda\!\!\!^{-}_e = \frac{\hbar}{m_e c} = 3.86\times10^{-11}$ cm.

Unit of mass $= m_e = 9.11\times10^{-28}$ g.
Unit of velocity $= c = 2.998\times10^{10}$ cm/s.

Unit of time $= \frac{\hbar}{m_e c^2} = 1.287\times10^{-21}$ s.

Unit of energy $= m_e c^2 = 0.511$ MeV.

In the Heaviside-Lorentz system, $\alpha = e^2/4\pi\hbar c = (137.036)^{-1}$. Thus $e = \sqrt{4\pi\alpha} = 0.303$. Note that in natural units, because $c = 1$, energy has the same dimension as mass and as inverse length and inverse time.

15.8 The possibility of nonzero photon rest mass

It is known from observations that the mass of the photon m_γ is less than 1.5×10^{-51} g, which is approximately 24 orders of magnitude smaller than the electron mass. This is consistent with quantization of the electromagnetic field starting from Maxwell's equations, a procedure that must yield $m_\gamma = 0$. However, we cannot exclude the possibility that some future experiment might reveal $m_\gamma \neq 0$, in which case it would be necessary to modify Maxwell's equations. How would this be done? We start with the first and third of Maxwell's equations expressed in terms of $\boldsymbol{A}$ and Φ in Lorenz gauge; that is,

$$\nabla^2\Phi - \frac{1}{c^2}\ddot{\Phi} = -\rho \tag{15.63}$$

$$\nabla^2\boldsymbol{A} - \frac{1}{c^2}\ddot{\boldsymbol{A}} = -\frac{1}{c}\boldsymbol{j} \tag{15.64}$$

A plane-wave solution to (15.63) or (15.64) with $\rho = \boldsymbol{j} = 0$ is proportional to $e^{i(\boldsymbol{k}\cdot\boldsymbol{r}-\omega t)}$, where $k = |\boldsymbol{k}| = \omega/c$. The latter expression is a relation between energy and momentum; that is,

$$E = \hbar\omega = \hbar kc = pc \tag{15.65}$$

valid for $m_\gamma = 0$. If we had $m_\gamma \neq 0$, it would be necessary to amend (15.65) as follows:

$$E^2 = p^2c^2 + m_\gamma^2c^4$$

or, equivalently,

$$\hbar^2\omega^2 = m_\gamma^2c^4 + \hbar^2c^2k^2$$

To accommodate this new relation, we revise (15.63) and (15.64) to read

$$\begin{aligned} \nabla^2\Phi - \frac{1}{c^2}\ddot{\Phi} &= -\rho + \left(\frac{m_\gamma c}{\hbar}\right)^2 \Phi \\ \nabla^2\boldsymbol{A} - \frac{1}{c^2}\ddot{\boldsymbol{A}} &= -\frac{1}{c}\boldsymbol{j} + \left(\frac{m_\gamma c}{\hbar}\right)^2 \boldsymbol{A} \end{aligned} \tag{15.66}$$

which are equivalent to the following covariant equation, first described by Proca in 1930:

$$\partial^\mu\partial_\mu A_\nu + \kappa^2 A_\nu = \frac{1}{c} j_\nu \tag{15.67}$$

where $\kappa = m_\gamma c/\hbar$. To see if this modification holds in another gauge, we go back to the more general equation (15.61) and append the mass term, just as in (15.67); that is,

$$\partial^{\mu}\partial_{\mu}A_{\nu}-\partial^{\mu}\partial_{\nu}A_{\mu}+\kappa^{2}A_{\nu}=\frac{1}{c}j_{\nu} \tag{15.68}$$

Applying ∂^{ν} on the left to both sides of (15.68), we obtain

$$\partial^{\nu}\partial^{\mu}\partial_{\mu}A_{\nu}-\partial^{\nu}\partial^{\mu}\partial_{\nu}A_{\mu}+\kappa^{2}\partial^{\nu}A_{\nu}=\frac{1}{c}\partial^{\nu}j_{\nu}$$

Because the order of partial differentiations in the first two terms on the left-hand side of this equation may be rearranged at will, and because μ and ν are both repeated indices and therefore dummy indices in these two terms, the latter terms cancel one another, and we are left with the condition

$$\partial^{\nu}A_{\nu}=\frac{1}{c\kappa^{2}}\partial^{\nu}j_{\nu}$$

Now the electromagnetic current density satisfies the equation of continuity $\partial^{\nu}j_{\nu}=0$; hence we also require the Lorenz gauge condition $\partial^{\nu}A_{\nu}=0$. This means the following: if the photon mass is strictly zero, we have the freedom to choose any gauge, but if the photon mass is nonzero, no matter how small, that gauge freedom is lost because of current conservation, and we are restricted to Lorenz gauge.

Finally, we consider (15.66) in the static limit; that is,

$$\nabla^{2}\Phi-\kappa^{2}\Phi=-\rho \tag{15.69}$$

Suppose that we have a point charge q at the origin. Then $\rho=q\delta^{3}(\boldsymbol{r})$. Thus, for $r\neq 0$, (15.69) becomes $\nabla^{2}\Phi-\kappa^{2}\Phi=0$. The solution to this equation is obviously spherically symmetric and is easily seen to be

$$\Phi=\frac{q}{4\pi}\frac{\exp(-\kappa r)}{r}$$

Thus the Coulomb potential of a point charge is replaced by a Yukawa potential. The quantity κ^{-1} has the dimension of length: it is the Compton wavelength of the massive photon. We have mentioned that the experimental upper limit on the photon mass is about 10^{-24} m_e. The corresponding value of κ^{-1} is 1.5×10^{13} cm, about the distance from the Earth to the Sun. In the physics of weak interactions, there exist massive vector bosons ($W^{\pm}$, Z^{o}) that play roles somewhat analogous to the role of the photon in electromagnetic interactions. The masses of these bosons are

$$m_W=80.3\ \text{GeV}/c^2 \qquad m_Z=91.2\ \text{GeV}/c^2$$

The corresponding values of κ^{-1} are

$$\kappa^{-1}(W^{\pm})=2.47\times10^{-16}\ \text{cm} \qquad \kappa^{-1}(Z^{o})=2.17\times10^{-16}\ \text{cm}$$

These lengths are much smaller than the proton radius, which is approximately 10^{-13} cm.

Problems for Chapter 15

15.1. Consider the expressions for the operators $\boldsymbol{\mathcal{E}}$ and $\boldsymbol{B}$ of the quantized electromagnetic radiation field in the Heisenberg picture.

(a) Calculate the commutators

$$\left[\mathcal{E}_i(\boldsymbol{r},t), B_j(\boldsymbol{r}',t)\right] \tag{1}$$

$$\left[\mathcal{E}_i(\boldsymbol{r},t), \mathcal{E}_j(\boldsymbol{r}',t)\right] \tag{2}$$

Hint: At a certain stage in your calculation, convert a sum to an integral, and make use of the delta function.

(b) Using the expression for the Hamiltonian of the radiation field

$$H = \frac{1}{2}\int\left(\boldsymbol{\mathcal{E}}^2 + \boldsymbol{B}^2\right)d\tau$$

and the result of part (a), calculate the commutators of $\boldsymbol{\mathcal{E}}$ and $\boldsymbol{B}$ with H, and employ Heisenberg's equation to obtain Maxwell's third and fourth equations in the absence of charge and current densities.

15.2. In how many ways N can we choose the integers n_1, n_2, n_3 such that $n_1^2 + n_2^2 + n_3^2 \leq n^2$ for given n, when $n_{1,2,3}$ can be positive or negative integers, excluding $n_1 = n_2 = n_3 = 0$? To derive the important formula (15.33) for the number of modes in the radiation field, we used the following approximation:

$$N = \frac{4\pi}{3}n^3 \tag{1}$$

How good is this approximation? To find out, write a short computer program to calculate N exactly, and compare your result with (1) for $n = 1$ to 50.

15.3. In this problem, we calculate the Casimir force of attraction between two uncharged parallel conducting plates separated by distance d. Consider a very large cubical box of volume $V = L^3$ with perfectly conducting walls. Place a plane conducting plate P parallel to one wall at a distance $d \ll L$ from that wall. The total zero-point energy in the box is the sum of the contributions from the volume of width d and the remaining volume of width $L - d$; that is,

$$U = U_d + U_{L-d}$$

We are interested in the difference Δ between U and the zeropoint energy U_0 that would have existed if the conducting plate had not been introduced. U_0 is, of course, the zero-point energy of the original cubical box. We have

$$\Delta = U_d - \left(U_0 - U_{L-d}\right) \tag{1}$$

To calculate U_{L-d} and U_0, it is sufficient to employ the usual formula for the number of modes between ω and $\omega + d\omega$ and replace each sum over all modes by an integral. Thus the contribution $(U_0 - U_{L-d})$ is equal to an integral over ω (called J for short) multiplied by $L^2 d$. If we could carry out the same procedure for U_d, it would also be equal to $JL^2 d$, and then Δ would be zero. However, in U_d, we cannot replace the sum Σ over all modes by J because modes with a sufficiently small component of $\boldsymbol{k}$ in the direction of normal to plate P are cut off. Hence we have

$$\Delta = (\Sigma - J) L^2 d \tag{2}$$

(a) Employ the Euler-Maclaurin sum formula

$$\begin{aligned}\sum_{0}^{N-1} f(n) = \int_0^N f(x)\,dx - \frac{1}{2}\left[f(N) - f(0)\right] + \frac{1}{12}\left[f^{(1)}(N) - f^{(1)}(0)\right] \\ - \frac{1}{720}\left[f^{(3)}(N) - f^{(3)}(0)\right] + \frac{1}{30{,}240}\left[f^{(5)}(N) - f^{(5)}(0)\right] - \cdots\end{aligned} \tag{3}$$

where $f^{(j)} \equiv d^j f / dx^j$, to evaluate $\Sigma - J$ and thus derive the result

$$\Delta = -\frac{\pi^2}{720}\frac{\hbar c}{d^3} L^2 \tag{4}$$

This yields an energy per unit area of

$$\frac{\Delta}{L^2} = -\frac{\pi^2}{720}\frac{\hbar c}{d^3} \tag{5}$$

and from this one immediately obtains the force of attraction per unit area. Here are several hints:

- To calculate $\Sigma - J$, one should employ standing waves, appropriate in the presence of conducting walls.
- Are there two possible polarization states for all allowed values of $\boldsymbol{k}$, or is there one special mode or class of modes for which only one polarization contributes?
- It appears at first that Σ and J are both infinite because of the divergent contribution of high-frequency (short wavelength) modes. However, this is not a real problem because for sufficiently high frequencies, plates constructed from any real conducting material become transparent. How would this be incorporated into the calculation?

(b) Consider two large uncharged parallel plane sheets of copper, each of thickness 1 mm and initially at rest and separated by a gap of 1 mm. These sheets are attracted to one another because of the Casimir force, the value of which can be found from equation (5). Estimate the time it takes for the two sheets to come in contact.

15.4. Consider thermal radiation at absolute temperature T, which is described by Planck's law. If $k_B T \approx m_e c^2$ or greater (where m_e is the electron rest mass), then electron-positron pairs can be created from two or more photons in the radiation field. Also, e^+ and e^- can annihilate to form two or more photons. Thus, at sufficiently high temperature, we can have an electron-positron gas in thermal equilibrium with the radiation field.

(a) Using Fermi-Dirac statistics, find the number densities of electrons and positrons as a function of T in the two limits $k_B T \gg m_e c^2$ and $k_B T \ll m_e c^2$, assuming thermal equilibrium in each case.

(b) Again assuming thermal equilibrium, find the entropy of the system photons plus electron-positron pairs when $k_B T \gg m_e c^2$.

15.5. In Section 15.8 we discuss the modifications of Maxwell's equations that would be necessary if photons had nonzero rest mass. In particular, we show that current conservation requires that A and Φ satisfy the Lorenz gauge condition

$$\nabla \bullet A + \frac{1}{c}\frac{\partial \Phi}{\partial t} = 0 \tag{1}$$

In covariant form, this is

$$\partial^\mu A_\mu = 0 \tag{2}$$

The radiation field A_μ can be expressed as a superposition of plane-wave solutions in the usual way; that is,

$$A_\mu = \frac{1}{L^{3/2}} \sum \sqrt{\frac{\hbar c^2}{2\omega}} \varepsilon_\mu \left(a_{k\varepsilon} e^{-ik\cdot x} + a_{k\varepsilon}^\dagger e^{ik\cdot x} \right) \tag{3}$$

where ε_μ is a polarization 4-vector satisfying $\varepsilon^\mu \varepsilon_\mu{}^* = 1$, and also $k\cdot x = \omega t - \boldsymbol{k}\bullet\boldsymbol{x}$, whereas $a_{k\varepsilon}$ and $a_{k\varepsilon}^\dagger$ are destruction and creation operators, respectively, in the Heisenberg picture.

(a) Consider a massive photon in its rest frame. Let the spin of this particle be along the z-direction. Find the four components of ε_μ. What if the spin lies along $-z$? What if the particle is "linearly" polarized along the z-direction?

(b) Consider the case where the spin lies along $\pm z$. Now make a Lorentz transformation to a frame in which the photon moves along z with speed v. What are the four components of ε_μ now? Do the same thing for the case where the particle is linearly polarized along z. Pay special attention to the latter case in the limit where the kinetic energy of the massive photon is much larger than its rest energy.

15.6. We show in Section 15.8 that if the photon has rest mass, the static Coulomb potential in hlus of a point charge q is replaced by the Yukawa potential

$$\Phi(r) = \frac{q}{4\pi}\frac{e^{-\lambda r}}{r}$$

where $\lambda = m_{\text{photon}} c/\hbar$. In such a world, conducting bodies could still exist, within which the electrostatic field $\boldsymbol{\mathcal{E}} = -\nabla\Phi$ must vanish. However, Poisson's equation and Gauss's law are modified, and if a charge Q is placed on a conductor, only a portion of it remains on the surface, and the remainder is distributed throughout the volume.

(a) Find the modifications of Poisson's equation and Gauss's law, and show that the volume charge density in a conductor must be the same from one spatial point to another regardless of the shape of the conductor.

(b) Consider a solid conducting sphere of radius a with total charge Q. Find the potential everywhere, and find the ratio of volume charge to surface charge.

16 Interaction of Nonrelativistic Charged Particles and Radiation

16.1 General form of the Hamiltonian in Coulomb gauge

It is intuitively clear that the Hamiltonian for a system of charged particles and radiation must consist of three parts: one for the radiation, another for the particles, and a third for the interactions between the radiation and the particles; that is,

$$H = H_{\text{radiation}} + H_{\text{particle}} + H_{\text{interaction}} \tag{16.1}$$

If no charged particles are present, the electromagnetic field consists solely of transverse waves (the pure radiation case). However, if charges and currents are present, the electromagnetic field also has longitudinal (static and quasi-static) components. We begin our discussion by considering the energy U_{EM} of the electromagnetic field in this more general case; that is,

$$U_{EM} = \frac{1}{2}\int (\boldsymbol{\mathcal{E}}^2 + \boldsymbol{B}^2)\, d^3x$$

In Coulomb gauge, we first consider the term

$$\begin{aligned} \frac{1}{2}\int \boldsymbol{\mathcal{E}}^2\, d^3x &= \frac{1}{2}\int \left(-\nabla\Phi - \frac{1}{c}\frac{\partial \boldsymbol{A}}{\partial t}\right)^2 d^3x \\ &= \frac{1}{2}\int \left[\nabla\Phi\boldsymbol{\cdot}\nabla\Phi + \frac{2}{c}\nabla\Phi\boldsymbol{\cdot}\frac{\partial \boldsymbol{A}}{\partial t} + \frac{1}{c^2}\left(\frac{\partial \boldsymbol{A}}{\partial t}\right)^2\right] d^3x \end{aligned} \tag{16.2}$$

Because in Coulomb gauge $\nabla^2\Phi = -\rho$, we have

$$\begin{aligned} \frac{1}{2}\int \nabla\Phi\boldsymbol{\cdot}\nabla\Phi\, d^3x &= \frac{1}{2}\int \nabla\boldsymbol{\cdot}(\Phi\nabla\Phi)\, d^3x - \frac{1}{2}\int \Phi\nabla^2\Phi\, d^3x \\ &= \frac{1}{2}\int \nabla\boldsymbol{\cdot}(\Phi\nabla\Phi)\, d^3x + \frac{1}{2}\int \rho\phi\, d^3x \end{aligned} \tag{16.3}$$

The first term on the far right side of (16.3) can be converted to a surface integral by Gauss's theorem, and assuming that the charge distribution is bounded, this surface integral vanishes. Next consider

$$\frac{1}{c}\int \nabla\Phi\boldsymbol{\cdot}\frac{\partial \boldsymbol{A}}{\partial t}\, d^3x = \frac{1}{c}\int \nabla\boldsymbol{\cdot}\left(\Phi\frac{\partial \boldsymbol{A}}{\partial t}\right) d^3x - \frac{1}{c}\int \Phi\frac{\partial \nabla\boldsymbol{\cdot}\boldsymbol{A}}{\partial t}\, d^3x \tag{16.4}$$

In this chapter we use Heaviside-Lorentz units unless otherwise noted.

Both terms on the right-hand side of this equation vanish, the first by virtue of Gauss's theorem and the second because $\nabla \cdot \boldsymbol{A} = 0$ in Coulomb gauge. Thus, defining

$$\boldsymbol{\mathcal{E}}_\perp = -\frac{1}{c}\frac{\partial \boldsymbol{A}}{\partial t}$$

which is the transverse part of $\boldsymbol{\mathcal{E}}$, we see that (16.2) reduces to

$$\frac{1}{2}\int \boldsymbol{\mathcal{E}}^2\, d^3x = \frac{1}{2}\int \boldsymbol{\mathcal{E}}_\perp^2\, d^3x + \frac{1}{2}\int \rho\Phi\, d^3x \tag{16.5}$$

The radiation Hamiltonian is

$$H_{\text{radiation}} = \frac{1}{2}\int (\boldsymbol{\mathcal{E}}_\perp^2 + \boldsymbol{B}^2)\, d^3x \tag{16.6}$$

with $\boldsymbol{\mathcal{E}}_\perp = -(1/c)(\partial \boldsymbol{A}/\partial t)$ and $\boldsymbol{B} = \nabla \times \boldsymbol{A}$. Because in Coulomb gauge

$$\Phi(x) = \frac{1}{4\pi}\int \frac{\rho(\boldsymbol{x}')}{|\boldsymbol{x}-\boldsymbol{x}'|}\, d^3\boldsymbol{x}'$$

we can write the second term on the right-hand side of (16.5) as

$$V = \frac{1}{2}\iint \frac{\rho(\boldsymbol{x})\rho(\boldsymbol{x}')}{4\pi|\boldsymbol{x}-\boldsymbol{x}'|}\, d^3\boldsymbol{x} d^3\boldsymbol{x}' \tag{16.7}$$

V is often called the *instantaneous Coulomb interaction*, and because it is expressed entirely in terms of the particles and does not refer explicitly to the field, we place it in H_{particle} rather than in $H_{\text{radiation}}$. If we have N-point particles at locations $\boldsymbol{x}_1, \ldots, \boldsymbol{x}_N$, V becomes

$$V = \frac{1}{4\pi}\sum_{i>j} \frac{q_i q_j}{|\boldsymbol{x}_i - \boldsymbol{x}_j|} \tag{16.8}$$

which is the familiar sum over all pairwise Coulomb interactions. In particular, for just two particles,

$$V = \frac{q_1 q_2}{4\pi|\boldsymbol{x}_1 - \boldsymbol{x}_2|} \tag{16.9}$$

Finally, we consider the effect of the vector potential $\boldsymbol{A}$ on the charged-particle Hamiltonian. We learned in Chapter 4 that in the presence of $\boldsymbol{A}$, the following modification is required:

$$\begin{aligned}
\sum_i \frac{\boldsymbol{p}_i^2}{2m_i} + V &\to \sum_i \frac{1}{2m_i}\left[\boldsymbol{p}_i - \frac{q}{c}\boldsymbol{A}(\boldsymbol{x}_i)\right]^2 + V \\
&= \sum_i \left(\frac{\boldsymbol{p}_i^2}{2m_i} - \frac{q}{2m_i c}\boldsymbol{p}_i \cdot \boldsymbol{A} - \frac{q}{2m_i c}\boldsymbol{A}\cdot\boldsymbol{p}_i + \frac{q^2}{2m_i c^2}A^2\right) + V \\
&= \sum_i \left(\frac{\boldsymbol{p}_i^2}{2m_i} + V\right) - \sum_i \left(\frac{q}{m_i c}\boldsymbol{A}\cdot\boldsymbol{p}_i - \frac{q^2}{2m_i c^2}A^2\right)
\end{aligned}$$

where the $\boldsymbol{p}_i$ are canonical momenta, and we have used $\nabla\boldsymbol{\cdot}\boldsymbol{A} = 0$ to obtain the last line. Taking all the foregoing into account, we have

$$H_{\text{particle}} = \sum_i \frac{p_i^2}{2m_i} + V \tag{16.10}$$

$$H_{\text{interaction}} = -\sum_i \left(\frac{q}{m_i c} \boldsymbol{A}\boldsymbol{\cdot}\boldsymbol{p}_i - \frac{q^2}{2m_i c^2} A^2 \right) \tag{16.11}$$

$$H_{\text{radiation}} = \frac{1}{2} \int (\boldsymbol{\mathcal{E}}_\perp^2 + \boldsymbol{B}^2)\, d^3x \tag{16.12}$$

In Chapter 15 we already learned how to quantize $H_{\text{radiation}}$ in Coulomb gauge. If in addition there are external static or quasi-static electric and/or magnetic fields, these can be treated classically, and the interactions of the particles with these fields are described according to rules we have learned previously and are easily included. If the particles have spin and associated spin magnetic moments, these interact with the magnetic field (radiation and/or static and quasi-static) by means of the additional phenomenologic Hamiltonian term; that is,

$$H_{\text{spin}} = \sum_i g_i \frac{\mp\mu_B}{\hbar} \boldsymbol{S}_i\boldsymbol{\cdot}\boldsymbol{B}$$

where $\mp$ applies for a particle with positive (negative) magnetic moment. When we discuss relativistic wave equations, in particular, the Dirac equation, it will be seen that the spin contribution arises naturally.

Equation (16.11) should be used with caution because in it the interaction Hamiltonian is expressed in terms of the vector potential $\boldsymbol{A}$, which is not unique because it can be altered by a gauge transformation, even within the Coulomb gauge condition. Although for the calculations of various physical phenomena considered in this and following chapters (16.11) and its relativistic generalization are legitimate, uncritical use of (16.11) without due regard for the requirements of gauge invariance yields incorrect answers in certain situations.

For the moment, let us ignore the interaction term (16.11). In this case, because there is no coupling of particles to the fields, eigenstates of the Hamiltonian are product eigenstates; that is,

$$|u\rangle = |u_{\text{particle}}\rangle |u_{\text{radiation}}\rangle \exp\left(-i\frac{E}{\hbar}t\right) \tag{16.13}$$

where E is the total energy of the particles plus the radiation field, $|u_{\text{particle}}\rangle$ is a particle eigenstate, and $|u_{\text{radiation}}\rangle$ is a photon number eigenstate. Such a product eigenstate is not very interesting because in the absence of interaction, neither the particle quantum numbers nor the photon occupation numbers can change with time.

16.2 Time-dependent perturbation theory

When H_{int} (we call it H' from now on) is included in the Hamiltonian, the resulting effects are usually too complicated to calculate exactly. Instead, we have to employ a method of successive approximations called *time-dependent perturbation theory*, which is formulated as follows: consider a Hamiltonian consisting of two parts

$$H = H_0 + H' \tag{16.14}$$

and suppose we know the eigenstates $|u_n\rangle$ and corresponding eigenvalues E_n of H_0 exactly. [In our discussion of electrodynamics, the $|u_n\rangle$ would be the product eigenstates in (16.13).] We want to study the time-dependent Schroedinger equation

$$H|\psi\rangle = i\hbar|\dot{\psi}\rangle \tag{16.15}$$

where H is the complete Hamiltonian, given by (16.14). We first express $|\psi\rangle$ as a superposition of the eigenstates of H_0 with time-dependent coefficients; that is,

$$|\psi\rangle = \sum_n a_n(t)|u_n\rangle \exp(-i\omega_n t) \tag{16.16}$$

where $\omega_n = E_n/\hbar$. Substituting (16.16) into (16.15), we obtain

$$(H_0 + H')\sum_n a_n(t)|u_n\rangle \exp(-i\omega_n t) = i\hbar\sum_n (\dot{a}_n - i\omega_n a_n)|u_n\rangle \exp(-i\omega_n t)$$

which yields

$$\sum_n a_n(t)H'|u_n\rangle \exp(-i\omega_n t) = i\hbar\sum_n \dot{a}_n(t)|u_n\rangle \exp(-i\omega_n t) \tag{16.17}$$

The eigenstates $|u_n\rangle$ form a complete orthonormal set. For convenience, we assume that they are discrete so that we can write

$$\langle u_m|u_n\rangle = \delta_{nm}$$

Multiplying (16.17) on the left by $\langle u_m|$, we obtain

$$\dot{a}_m = -\frac{i}{\hbar}\sum_n a_n H'_{mn} e^{i\omega_{mn}t} \tag{16.18}$$

where $H'_{mn} \equiv \langle u_m|H'|u_n\rangle$ and $\omega_{mn} = \omega_m - \omega_n$. Equations (16.18) (one for each value of m) form a system of coupled differential equations that are equivalent to (16.15), and so far, no approximations have been made.

Except for a small number of very simple systems, equations (16.18) are too complicated to solve exactly. Thus, assuming that at $t = 0$, $a_i(t = 0) = 1$ for some i and $a_{n\neq i}(0) = 0$, we try to solve (16.18) by approximating the $a_n(t)$ on the right-hand side by their initial values. Thus we replace (16.18) by

$$\dot{a}_m = -\frac{i}{\hbar} H'_{mi} e^{i\omega_{mi}t} \tag{16.19}$$

In many cases of interest, H' is independent of time. [This includes the interaction of a charge with the radiation field in the Schroedinger picture, as given in (16.11).] Then (16.19) can be integrated immediately to yield

$$a_m = -\frac{i}{\hbar} H'_{mi} \frac{e^{i\omega_{mi}t} - 1}{i\omega_{mi}} = H'_{mi} \frac{1 - e^{i\omega_{mi}t}}{\hbar\omega_{mi}} \qquad \text{(First order)} \tag{16.20}$$

It may happen that for given m, $H'_{mi} = 0$; nevertheless, it may be possible to obtain $a_m \neq 0$ in second order. To see this, replace m by n in (16.20), and use (16.20) in (16.18) to obtain

$$\begin{aligned} \dot{a}_m &= -\frac{i}{\hbar} \sum_n H'_{mn} e^{i\omega_{mn}t} \frac{H'_{ni}}{\hbar\omega_{ni}} \left(1 - e^{i\omega_{ni}t}\right) \\ &= -\frac{i}{\hbar} \sum_n \frac{H'_{mn} H'_{ni}}{\hbar\omega_{ni}} \left(e^{i\omega_{mn}t} - e^{i\omega_{mi}t}\right) \end{aligned} \tag{16.21}$$

where $\omega_{mi} = \omega_{mn} + \omega_{ni} = (\omega_m - \omega_n) + (\omega_n - \omega_i) = \omega_m - \omega_i$. Integrating (16.21) with the initial condition $a_m(0) = 0$, we have

$$a_m = \sum_n \frac{H'_{mn} H'_{ni}}{\hbar\omega_{ni}} \left(\frac{1 - e^{i\omega_{mn}t}}{\hbar\omega_{mn}} - \frac{1 - e^{i\omega_{mi}t}}{\hbar\omega_{mi}} \right) \qquad \text{(Second order)} \tag{16.22}$$

In many practical applications, the only significant terms in (16.22) are those in which energy is conserved between the initial state i and the final state m. In these cases, it is usually possible to ignore the first set of terms on the right-hand side of (16.22), which then becomes

$$a_m = \sum_n \frac{H'_{mn} H'_{ni}}{\hbar\omega_{in}} \left(\frac{1 - e^{i\omega_{mi}t}}{\hbar\omega_{mi}} \right) \qquad \text{(Second order)} \tag{16.23}$$

Comparing (16.20) with (16.23), we see that both expressions contain the factor

$$\frac{1 - e^{i\omega_{mi}t}}{\hbar\omega_{mi}}$$

However, in (16.20), the first-order matrix element appears; that is,

$$M_1 = H'_{mi} \tag{16.24}$$

whereas in (16.23) there is the second-order matrix element; that is,

$$M_2 = \sum_n \frac{H'_{mn} H'_{ni}}{\hbar\omega_{in}} \tag{16.25}$$

The analogy with first- and second-order static perturbation theory is obvious.

From (16.20) or (16.23), we obtain the transition probability $|a_m(t)|^2$; that is,

$$|a_m(t)|^2 = \frac{|M|^2}{\hbar^2 \omega_{mi}^2} 4\sin^2\left(\frac{\omega_{mi}t}{2}\right) \tag{16.26}$$

Although this formula is useful in some circumstances, we are most frequently interested in the transition probability to a group of states m with a density (number of states per unit energy) $\rho(E)$. For example, consider an atom in an excited state that decays by spontaneous emission of a photon. The final state consists of the atom in its ground state and the emitted photon. By conservation of energy (and we see presently that this is not assumed a priori but emerges from the formalism we have constructed), the photon energy must be equal to the energy difference between initial and final atomic states. However, even with this restriction, also restricting the photon polarization to a definite value, and fixing the direction of photon emission within a very small solid angular uncertainty, the number of possible photon modes in most practical cases turns out to be extremely large. In this situation, we are interested in the transition probability summed over all relevant modes.

From (16.26), the transition probability to a band of states between energy E and $E + dE$ is

$$dP = \frac{|M|^2}{\hbar^2\left(\frac{\omega_{mi}}{2}\right)^2}\sin^2\left(\frac{\omega_{mi}t}{2}\right)\rho(E)dE \tag{16.27}$$

However, $dE = 2\hbar d(\omega_{mi}/2)$. Therefore, (16.27) can be written

$$dP = \frac{2\pi t|M|^2\rho(E)}{\hbar}\left[\frac{\sin^2\left(\frac{\omega_{mi}t}{2}\right)}{\pi t\left(\frac{\omega_{mi}}{2}\right)^2}\right]d\left(\frac{\omega_{mi}}{2}\right) \tag{16.28}$$

Consider the factor f in square brackets on the right-hand side of (16.28). Writing $x = \omega_{mi}t/2$, we have

$$f = \frac{t}{\pi}\frac{\sin^2 x}{x^2} \tag{16.29}$$

The function $\sin^2 x/x^2$ equals unity for $x = 0$ but drops rapidly to zero for $|x| \gg 1$. Hence, for any value of t, $f = t/\pi$ when $\omega_{mi} = 0$, but for any fixed $\omega_{mi} \neq 0$, $f \to 0$ as $t \to \infty$. In fact, from equation (2.46),

$$\lim_{t\to\infty} f = \lim_{t\to\infty}\left[\frac{\sin^2\left(\frac{\omega_{mi}t}{2}\right)}{\pi t\left(\frac{\omega_{mi}}{2}\right)^2}\right] = \delta\left(\frac{\omega_{mi}}{2}\right) \tag{16.30}$$

Thus, for sufficiently large t, we can write (16.28) as

$$dP = \frac{2\pi t |M|^2 \rho(E)}{\hbar} \delta\left(\frac{\omega_{mi}}{2}\right) d\left(\frac{\omega_{mi}}{2}\right) \tag{16.31}$$

Instead of dP, it is more useful to discuss the differential transition probability per unit time $dW = dP/dt$. Making use of the formula $\int g(x)\delta(ax)\,dx = (1/a)\int g(x)\,dx$, we thus obtain

$$dW = \frac{2\pi |M|^2 \rho(E)}{\hbar} \delta(E_{\text{initial}} - E_{\text{final}})\,dE \tag{16.32}$$

The total transition probability per unit time is obtained by integrating dW over all E; that is,

$$W = \frac{2\pi}{\hbar} |M|^2 \rho(E) \tag{16.33}$$

Formulas (16.32) and (16.33) are known as *Fermi's golden rule*, although they were first described by Dirac. Of course, we must be careful to apply these well-known and very useful formulas only within their domain of validity. Because $x = \omega_{mi}t/2$, it is clear that for extremely short times we need a large ω_{mi} to generate a moderate value of x. In this case, replacement of f by the delta function is not always a good approximation. When t is extremely long, replacement of the a_n in (16.18) by their initial values appears very suspicious because the a_n surely must evolve from their initial values as time elapses.

However, we shall see that the latter problem can be dealt with effectively in almost all practical cases.

16.3 Single-photon emission and absorption processes

16.3.1 Electric Dipole Transitions

We now apply the results of the preceding sections to calculate single-photon emission and absorption processes in atoms. Our starting point is Fermi's golden rule, with M the first-order matrix element of

$$H' = \frac{e}{m_e c} \boldsymbol{A} \cdot \boldsymbol{p} + \frac{e^2}{2m_e c^2} A^2 + g_s \frac{\mu_B}{\hbar} \boldsymbol{S} \cdot \boldsymbol{B} \tag{16.34}$$

from (16.11) with $q = -e$. Actually, the term in A^2 cannot contribute to a single-photon process because the latter requires a matrix element linear in a photon creation or destruction operator, whereas A^2 contains terms with pairs of creation and/or destruction operators. Also, we ignore the spin term in (16.34) for the moment. Thus we have

$$H' = \frac{e}{m_e c} \boldsymbol{A} \cdot \boldsymbol{p} \tag{16.35}$$

with

$$A = \frac{1}{L^{3/2}} \sum_{k'\alpha'} \sqrt{\frac{\hbar c^2}{2\omega'}} \hat{\varepsilon}_{k'\alpha'} \left(a_{k'\alpha'} e^{i\boldsymbol{k}'\cdot\boldsymbol{r}} + a^{\dagger}_{k'\alpha'} e^{-i\boldsymbol{k}'\cdot\boldsymbol{r}} \right) \tag{16.36}$$

Consider emission of a single photon in mode $\boldsymbol{k}\alpha$ by an atom that is initially in the state $|i_a\rangle$ and finally in the state $|f_a\rangle$. Clearly, only one term in (16.36) can yield a nonzero matrix element for such a process; it is the term in $a^{\dagger}_{k\alpha}$. If there are n photons in the mode $\boldsymbol{k}\alpha$ in the initial state, the matrix element is

$$\begin{aligned} M &= \frac{e}{m_e c} \langle f | \boldsymbol{A}\cdot\boldsymbol{p} | i \rangle = \frac{e}{m_e c} \langle (n+1)_{k\alpha}, f_a | \boldsymbol{A}\cdot\boldsymbol{p} | n_{k\alpha}, i_a \rangle \\ &= \frac{e}{m_e c} \frac{1}{L^{3/2}} \sqrt{\frac{\hbar c^2}{2\omega}} \langle (n+1)_{k\alpha} | a^{\dagger}_{k\alpha} | n_{k\alpha} \rangle \langle f_a | e^{-i\boldsymbol{k}\cdot\boldsymbol{r}} \hat{\varepsilon}_{k\alpha} \cdot \boldsymbol{p} | i_a \rangle \end{aligned} \tag{16.37}$$

Now

$$\langle (n+1)_{k\alpha} | a^{\dagger}_{k\alpha} | n_{k\alpha} \rangle = \sqrt{(n+1)_{k\alpha}}$$

Thus (16.37) becomes

$$M_{\text{emission}} = \frac{e}{m_e c} \frac{1}{L^{3/2}} \sqrt{\frac{\hbar c^2}{2\omega}} (n+1)^{1/2} \langle f_a | e^{-i\boldsymbol{k}\cdot\boldsymbol{r}} \hat{\varepsilon}\cdot\boldsymbol{p} | i_a \rangle \tag{16.38}$$

where we have dropped the subscript $\boldsymbol{k}\alpha$. The quantity $n + 1$ has the following significance: n refers to stimulated emission, but even when $n = 0$, we still have a factor of unity, which accounts for spontaneous emission.

By a similar argument, we find the matrix element for stimulated absorption; that is,

$$M_{\text{absorption}} = \frac{e}{m_e c} \frac{1}{L^{3/2}} \sqrt{\frac{\hbar c^2}{2\omega}} (n)^{1/2} \langle f_a | e^{i\boldsymbol{k}\cdot\boldsymbol{r}} \hat{\varepsilon}\cdot\boldsymbol{p} | i_a \rangle \tag{16.39}$$

Note that in the matrix element for absorption there is a factor $(n)^{1/2}$ instead of $(n+1)^{1/2}$.

The factor $\langle f_a | e^{-i\boldsymbol{k}\cdot\boldsymbol{r}} \hat{\varepsilon}\cdot\boldsymbol{p} | i_a \rangle$ in (16.38) is an integral over the atomic coordinates of the form

$$\int \psi^*_{fa} e^{-i\boldsymbol{k}\cdot\boldsymbol{r}} \hat{\varepsilon}\cdot\boldsymbol{p} \psi_{ia} \, d^3\boldsymbol{r}$$

Let R be the radius of the region over which the electronic wave function is appreciably different from zero, and let λ be the photon wavelength. If $\lambda \gg R$, then $\boldsymbol{k}\cdot\boldsymbol{r} \ll 1$ for all $|\boldsymbol{r}| \le R$; hence we can replace $\exp(-i\boldsymbol{k}\cdot\boldsymbol{r})$ by unity (the long-wavelength approximation).

The factor $\exp(-i\boldsymbol{k}\cdot\boldsymbol{r})$ arose in the first place because we chose to Fourier analyze the radiation field in Cartesian coordinates, which yields plane waves. For atomic, nuclear, or elementary particle emission and absorption processes, it would obviously be more natural to solve the vector wave equation

$$\nabla^2 \boldsymbol{A} - \frac{1}{c^2}\ddot{\boldsymbol{A}} = 0$$

in spherical polar coordinates centered on the radiating system. Although it is simpler to retain Cartesian coordinates, let us summarize very briefly the main features of vector spherical wave solutions, leaving details for Appendix C. Divergence-free vector spherical waves are constructed from solutions to the scalar wave equation in spherical coordinates. As is well known, the latter solutions are of the form $g_\ell(kr)Y_\ell^m(\theta,\phi)$, where the g_ℓ are appropriate linear combinations of the spherical Bessel functions $j_\ell(kr)$ and $n_\ell(kr)$. Such combinations can be constructed to describe outgoing waves (the spherical Hankel functions $h_\ell^{(1)} = j_\ell + in_\ell$), incoming waves (the spherical Hankel functions $h_\ell^{(2)} = j_\ell - in_\ell$), or standing waves described entirely by the j_ℓ. In any of these cases, the vector spherical waves consist of two distinct types: *magnetic* (m) and *electric* (e) *multipoles.* Each multipole mode is characterized by two distinct indices L and M (in addition to the wave number k). The permissible values of L are 1, 2, 3, …, and the possible values of M for a given L are $-L$, …, L. For any given L, M, the electric field $\boldsymbol{\mathcal{E}}_{LM}^{(m)}$ of a magnetic multipole is purely transverse (it has no radial component), but the magnetic field $\boldsymbol{B}_{LM}^{(m)}$ does have a radial component. In fact, in the near-field zone (where $kr \ll 1$), $\boldsymbol{B}_{LM}^{(m)}$ becomes like a magnetostatic field of multipolarity L, M, and in this region it greatly dominates over $\boldsymbol{\mathcal{E}}_{LM}^{(m)}$. For electric ($e$) multipoles, the situation is opposite: there the magnetic field $\boldsymbol{B}_{LM}^{(e)}$ is transverse everywhere, but $\boldsymbol{\mathcal{E}}_{LM}^{(e)}$ does have a radial component, and in the near-field zone it becomes like an electrostatic L,M–pole field that greatly dominates over $\boldsymbol{B}_{LM}^{(e)}$.

The plane running waves that result from Fourier analysis of the radiation field in rectangular coordinates form a complete set for given k: any arbitrary divergence-free solution to the vector wave equation can be expressed as a superposition of such plane running waves. The same remarks hold for the e and m vector spherical waves: together they form a complete set. It follows that we can express a plane running electromagnetic wave as a superposition of m and e vector spherical waves summed over L and M. In this "multipole" expansion, the leading term is electric dipole ($E1$); it dominates over all others when $kr \ll 1$. Thus the expressions *long-wavelength approximation* and *electric dipole approximation* are equivalent.

Now, returning to the matrix element $\langle f_a|\hat{\varepsilon}\cdot\boldsymbol{p}|i_a\rangle$ in the $E1$ approximation, we note that

$$[\boldsymbol{r}, H_{\text{atom}}] = \left[\boldsymbol{r}, \frac{\boldsymbol{p}^2}{2m_e}\right] = \frac{i\hbar}{m_e}\boldsymbol{p} \tag{16.40}$$

Hence

$$\langle f_a|\hat{\varepsilon}\cdot\boldsymbol{p}|i_a\rangle = \frac{m_e}{i}(\omega_{ia} - \omega_{fa})\langle f_a|\hat{\varepsilon}\cdot\boldsymbol{r}|i_a\rangle$$

Consequently, in the long-wavelength approximation, (16.38) becomes

$$\begin{aligned} M_{\text{emission}} &= -\frac{e}{m_e c}\frac{i}{L^{3/2}}\sqrt{\frac{\hbar c^2}{2\omega}}m_e\omega(n+1)^{1/2}\langle f_a|\hat{\varepsilon}\cdot\boldsymbol{r}|i_a\rangle \\ &= \frac{-i}{L^{3/2}}e(\hbar/2)^{1/2}\,\omega^{1/2}(n+1)^{1/2}\langle f_a|\hat{\varepsilon}\cdot\boldsymbol{r}|i_a\rangle \end{aligned} \tag{16.41}$$

where we have used $\omega_{if}^a = \omega$, which follows from conservation of energy. Thus, from (16.32), the differential transition probability per unit time for spontaneous emission is

$$dW = \frac{\pi e^2}{2L^3}\omega\left|\langle f_a|\hat{\varepsilon}\cdot\boldsymbol{r}|i_a\rangle\right|^2 d\rho \tag{16.42}$$

Now the number of modes between ω and $\omega + d\omega$ for photons radiated into solid angle $d\Omega$ with given polarization is

$$dZ = \frac{L^3}{2}\frac{\omega^2 d\omega}{\pi^2 c^3}\frac{d\Omega}{4\pi} = \frac{L^3\omega^2 d\omega d\Omega}{8\pi^3 c^3}$$

Thus

$$d\rho = \frac{dZ}{dE} = \frac{dZ}{d\hbar\omega} = \frac{L^3\omega^2 d\Omega}{8\pi^3\hbar c^3} \tag{16.43}$$

Hence (16.42) becomes

$$dW = \frac{e^2}{4\pi\hbar c}\frac{\omega^3}{2\pi c^2}\left|\langle\hat{\varepsilon}\cdot\boldsymbol{r}\rangle\right|^2 d\Omega = \frac{\alpha}{2\pi}\frac{\omega^3}{c^2}\left|\langle\hat{\varepsilon}\cdot\boldsymbol{r}\rangle\right|^2 d\Omega \tag{16.44}$$

The quantity

$$\begin{aligned}\hat{\varepsilon}\cdot\boldsymbol{r} &= \left(\frac{\hat{\varepsilon}_x - i\hat{\varepsilon}_y}{\sqrt{2}}\right)\left(\frac{x+iy}{\sqrt{2}}\right) + \left(\frac{\hat{\varepsilon}_x + i\hat{\varepsilon}_y}{\sqrt{2}}\right)\left(\frac{x-iy}{\sqrt{2}}\right) + \hat{\varepsilon}_z z \\ &= -\hat{\varepsilon}^{(-1)}r^{(1)} - \hat{\varepsilon}^{(1)}r^{(-1)} + \hat{\varepsilon}^{(0)}r^{(0)}\end{aligned}$$

is the scalar product of two first-rank tensors, where $\hat{\varepsilon}$ refers to the radiation field, and $\boldsymbol{r}$ refers to the atom. Applying the Wigner-Eckart theorem to $\boldsymbol{r}$, we have

$$\langle j_f^a, m_f^a|r^{(M)}|j_i^a, m_i^a\rangle = \langle j_f, m_f, 1, M|j_i, m_i\rangle\langle j_f\|r\|j_i\rangle \tag{16.45}$$

The vector coupling coefficient on the right-hand side of (16.45) implies that $m_f + M = m_i$ and that $j_f = j_i$ or $j_f = j_i \pm 1$, but $j_i = 0$ cannot go to $j_f = 0$. The last conditions are summarized by stating that these quantum numbers must obey the *triangle rule*: $\Delta(j_i 1 j_f)$. Also, $\boldsymbol{r}$ is an odd-parity operator, which implies that the parities of the initial and final atomic states must be opposite; that is,

$$\pi_i\pi_f = -1 \tag{16.46}$$

As a specific example of an $E1$ transition, we calculate the transition probability per unit time for spontaneous decay of the $2p$ state of a hydrogenic atom of nuclear charge Ze to the $1s$ state. Here we include electron spin but ignore nuclear spin. The relevant wave functions are

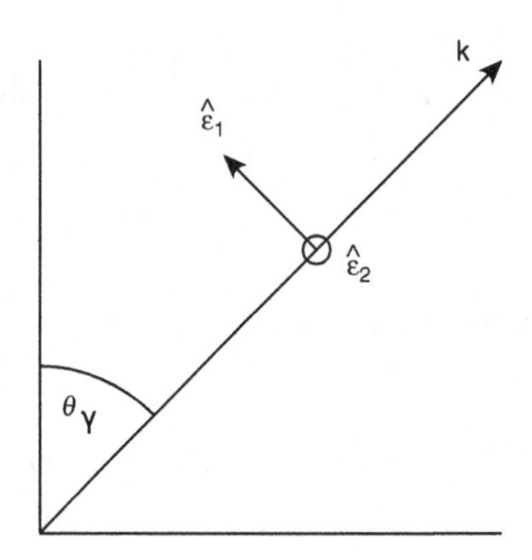

Figure 16.1 Vectors $k, \hat{\varepsilon}_1, \hat{\varepsilon}_2$ that describe an emitted photon.

$$\begin{aligned}
&\psi\left(2^2 P_{1/2}, m_j = 1/2\right) = \sqrt{\frac{1}{3}} R_{2p} Y_1^0 \alpha - \sqrt{\frac{2}{3}} R_{2p} Y_1^1 \beta \\
&\psi\left(2^2 P_{1/2}, -1/2\right) = -\sqrt{\frac{1}{3}} R_{2p} Y_1^0 \beta + \sqrt{\frac{2}{3}} R_{2p} Y_1^{-1} \alpha \\
&\psi(1^2 S_{1/2}, 1/2) = R_{1s} Y_0^0 \alpha \\
&\psi(1^2 S_{1/2}, -1/2) = R_{1s} Y_0^0 \beta
\end{aligned} \tag{16.47}$$

where, as usual, α and β are up, down spin-½ spinors, respectively. For the transition $m_{ji} = 1/2 \rightarrow m_{jf} = 1/2$, we have

$$\begin{aligned}
\langle \hat{\varepsilon} \cdot \boldsymbol{r} \rangle_{1/2 \rightarrow 1/2} &= \varepsilon^{(0)} \sqrt{\frac{1}{3}} \int R_{1s} R_{2p} r^3 \, dr \int Y_0^0 \sqrt{\frac{4\pi}{3}} Y_0^0 Y_1^0 \, d\Omega \\
&= \frac{\varepsilon^{(0)}}{3} \int R_{1s} R_{2p} r^3 \, dr
\end{aligned} \tag{16.48}$$

In the case $m_{ji} = 1/2 \rightarrow m_{jf} = -1/2$,

$$\begin{aligned}
\langle \hat{\varepsilon} \cdot \boldsymbol{r} \rangle_{1/2 \rightarrow -1/2} &= -\varepsilon^{(+1)} \sqrt{\frac{2}{3}} \int R_{1s} R_{2p} r^3 \, dr \int Y_0^0 \sqrt{\frac{4\pi}{3}} Y_1^{-1} Y_1^1 \, d\Omega \\
&= \frac{\varepsilon^{(+1)} \sqrt{2}}{3} \int R_{1s} R_{2p} r^3 \, dr
\end{aligned} \tag{16.49}$$

In each transition, there are two possible polarization states a priori. These are shown in Figure 16.1.

We have

$$\begin{aligned}
\hat{\varepsilon}_{1x} &= \cos\theta_\gamma & \hat{\varepsilon}_{2x} &= 0 \\
\hat{\varepsilon}_{1y} &= 0 & \hat{\varepsilon}_{2y} &= 1 \\
\hat{\varepsilon}_{1z} &= \sin\theta_\gamma & \hat{\varepsilon}_{2z} &= 0
\end{aligned}$$

Thus $\hat{\varepsilon}_1^{(0)} = \sin\theta_\gamma$, $\hat{\varepsilon}_2^{(0)} = 0$, $\hat{\varepsilon}_1^{(+1)} = -\left(1/\sqrt{2}\right)\cos\theta_\gamma$, and $\hat{\varepsilon}_2^{(+1)} = -\left(i/\sqrt{2}\right)$. Therefore,

$$\sum_{\text{pols}} \left| \langle \hat{\varepsilon} \cdot \boldsymbol{r} \rangle_{1/2 \to 1/2} \right|^2 = \frac{1}{9} \sin^2 \theta_\gamma \left(\int R_{1s} R_{2s} r^3 \, dr \right)^2 \tag{16.50}$$

$$\sum_{\text{pols}} \left| \langle \hat{\varepsilon} \cdot \boldsymbol{r} \rangle_{1/2 \to -1/2} \right|^2 = \frac{1}{9} \left(1 + \cos^2 \theta_\gamma \right) \left(\int R_{1s} R_{2s} r^3 \, dr \right)^2 \tag{16.51}$$

Hence

$$\begin{aligned} \int \sum_{\text{pols}} \left| \langle \hat{\varepsilon} \cdot \boldsymbol{r} \rangle_{1/2 \to 1/2} \right|^2 d\Omega_\gamma &= \frac{1}{9} \left(\int R_{1s} R_{2s} r^3 \, dr \right)^2 2\pi \int_0^\pi \sin^3 \theta_\gamma \, d\theta_\gamma \\ &= \frac{8\pi}{27} \left(\int R_{1s} R_{2s} r^3 \, dr \right)^2 \end{aligned} \tag{16.52}$$

and

$$\begin{aligned} \int \sum_{\text{pols}} \left| \langle \hat{\varepsilon} \cdot \boldsymbol{r} \rangle_{1/2 \to -1/2} \right|^2 d\Omega_\gamma &= \frac{1}{9} \left(\int R_{1s} R_{2s} r^3 \, dr \right)^2 2\pi \int_0^\pi (1 + \cos^2 \theta_\gamma) \sin \theta_\gamma \, d\theta_\gamma \\ &= \frac{16\pi}{27} \left(\int R_{1s} R_{2s} r^3 \, dr \right)^2 \end{aligned} \tag{16.53}$$

In atomic units,

$$\begin{aligned} R_{1s} &= 2Z^{3/.2} e^{-Zr} \\ R_{2p} &= \frac{Z^{5/2}}{2\sqrt{6}} r e^{-Zr/2} \end{aligned}$$

Therefore,

$$\int_0^\infty R_{1s} R_{2p} r^3 \, dr = \frac{Z^4}{\sqrt{6}} \int_0^\infty r^4 e^{-3Zr/2} \, dr = \sqrt{\frac{2^{15}}{3^9}} \frac{1}{Z} \tag{16.54}$$

In atomic units, the total decay rate from $2^2 P_{1/2}, m_j = 1/2$ is obtained from (16.54) as

$$W = \frac{\alpha^3 \omega^3}{2\pi} \sum_{\text{pols}} \left(\int \left| \langle \hat{\varepsilon} \cdot \boldsymbol{r} \rangle_{1/2 \to 1/2} \right|^2 d\Omega_\gamma + \int \left| \langle \hat{\varepsilon} \cdot \boldsymbol{r} \rangle_{1/2 \to -1/2} \right|^2 d\Omega_\gamma \right) \tag{16.55}$$

Employing (16.52) through (16.54) and $\omega = \frac{3}{8} Z^2$ in (16.55), we obtain

$$W = \left(\frac{2}{3} \right)^8 Z^4 \alpha^3 \tag{16.56}$$

Because W is a transition probability per unit time, it has dimension (time)$^{-1}$. Therefore, to find W in ordinary units, we must multiply the right-hand side of (16.56) by $\alpha c / a_0$. The result is called the *spontaneous-emission decay rate A*; that is,

$$A(2p \to 1s) = \left(\frac{2}{3} \right)^8 Z^4 \frac{\alpha^4 c}{a_0} = 6.25 \times 10^8 Z^4 \text{ s}^{-1} \tag{16.57}$$

Table 16.1 *E*1 Spontaneous emission probabilities in hydrogen in 10^8 s^{-1}

Initial	Final	$n = 1$	2	3
2*s*	*np*	—	—	—
2*p*	*ns*	6.25	—	—
3*s*	*np*	—	0.063	—
3*p*	*ns*	1.64	0.22	—
3*d*	*np*	—	0.64	—
4*s*	*np*	—	0.025	0.018
4*p*	*ns*	0.68	0.095	0.030
4*p*	*nd*	—	—	0.003
4*d*	*np*	—	0.204	0.070
4*f*	*nd*	—	—	0.137

Note the Z^4 dependence, which arises from a factor Z^{-2} in the square of the matrix element and a factor Z^6 in ω^3. Result (16.57) agrees with experiment, as do A coefficients for other allowed transitions in hydrogenic atoms (all of which can be calculated exactly). Table 16.1 lists some of them.

The same principles are involved in calculating allowed $E1$ transition rates for more complex atoms, and results of the same order of magnitude are expected: $A \approx \alpha^3\omega^3$ in atomic units. However, because radial-wave functions are known only approximately for complex atoms, the A coefficients are correspondingly uncertain. For the alkali atoms and a few others, one can achieve precisions of 1 to 5 percent with considerable computational effort using the Hartree-Fock and related methods.

A beam of light incident on an atom can cause stimulated emission and/or absorption. We now calculate the rate for this process in the allowed approximation, starting from (16.38) for stimulated emission and (16.39) for stimulated absorption. Replacing the factors $\exp(\pm i\boldsymbol{k}\cdot\boldsymbol{r})$ by unity, writing $n + 1 \approx n$ for $n \gg 1$, and following the same line of argument as given for spontaneous emission in earlier paragraphs, we obtain for the differential transition probability per unit time

$$dW = \frac{n}{2\pi}\alpha\frac{\omega^3}{c^2}\left|\langle\hat{\varepsilon}\cdot\boldsymbol{r}\rangle\right|^2 d\Omega \tag{16.58}$$

Suppose that we have a beam of light with intensity $I_{00}d\omega d\Omega$, where I_{00} is the intensity per unit angular frequency per steradian, with dimensions erg cm^{-2} s^{-1} rad^{-1} s sterad^{-1}. Clearly,

$$\begin{aligned} I_{00}d\omega d\Omega &= \frac{\text{no. of photons}}{\text{mode}\times\text{cm}^3}\frac{\text{no. of modes}}{\text{sterad}\times\text{rad/s}}\frac{\text{energy}}{\text{photon}}c\,d\omega d\Omega \\ &= \frac{n}{L^3}\frac{L^3\omega^2}{8\pi^3c^3}\hbar\omega c\,d\omega d\Omega \\ &= \frac{n\hbar\omega^3}{(2\pi)^3c^2}d\omega d\Omega \end{aligned} \tag{16.59}$$

We employ (16.59) to rewrite (16.58) as

$$dW = I_{00}d\Omega\frac{(2\pi)^2\alpha}{\hbar}\left|\langle\hat{\varepsilon}\cdot\boldsymbol{r}\rangle\right|^2$$

The total intensity per unit angular frequency in the beam of light is $I_0 \equiv I_{00} d\Omega$. Thus we obtain the following useful formula for the transition probability per unit time for stimulated emission or absorption:

$$W = (2\pi)^2 I_0 \frac{\alpha}{\hbar} |\langle \hat{\varepsilon} \bullet \boldsymbol{r} \rangle|^2 \tag{16.60}$$

16.3.2 Magnetic dipole and electric quadrupole transitions

Now consider the effect of the next term in the expansion of $\exp(\pm i\boldsymbol{k}\bullet\boldsymbol{r})$ in the matrix elements (16.38) and (16.39). The spontaneous emission matrix element is

$$H'_{fi} = \frac{e}{m_e c} \frac{1}{L^{3/2}} \sqrt{\frac{\hbar c^2}{2\omega}} \left\langle \psi_f \left| -i\boldsymbol{k}\bullet\boldsymbol{r}\hat{\varepsilon}\bullet\boldsymbol{p} \right| \psi_i \right\rangle \tag{16.61}$$

We write

$$\begin{aligned} \boldsymbol{k}\bullet\boldsymbol{r}\hat{\varepsilon}\bullet\boldsymbol{p} &= k_i \hat{\varepsilon}_j x_i p_j \\ &= \frac{1}{2} k_i \hat{\varepsilon}_j \left[(x_i p_j + p_i x_j) + (x_i p_j - p_i x_j) \right]. \end{aligned} \tag{16.62}$$

Now $p_i x_j = x_j p_i - i\hbar\delta_{ij}$, and in Coulomb gauge, $k_i \hat{\varepsilon}_j \delta_{ij} = 0$. Thus (16.62) can be written

$$\boldsymbol{k}\bullet\boldsymbol{r}\hat{\varepsilon}\bullet\boldsymbol{p} = \frac{1}{2} k_i \hat{\varepsilon}_j \left[(x_i p_j + p_i x_j) + (x_i p_j - x_j p_i) \right] \tag{16.63}$$

For the moment, we confine our attention to the second term on the right-hand side of (16.63):

$$\frac{1}{2} k_i \hat{\varepsilon}_j (x_i p_j - x_j p_i) = \frac{\omega}{2c} \hat{k} \times \hat{\varepsilon}\bullet\boldsymbol{L} \tag{16.64}$$

where $\boldsymbol{L}$ is the electron orbital angular momentum operator. Inserting this into (16.61), we obtain

$$H'^{(L)}_{fi} = \frac{-ie}{2m_e c} \frac{1}{L^{3/2}} \sqrt{\frac{\hbar\omega}{2}} \left\langle \psi_f \left| \hat{k} \times \hat{\varepsilon}\bullet\boldsymbol{L} \right| \psi_i \right\rangle \tag{16.65}$$

At this point we recall the spin term in (16.52), which we have neglected until now; that is,

$$H'^{(S)} = +\frac{g_s \mu_B}{\hbar} \boldsymbol{S}\bullet\boldsymbol{B} \tag{16.66}$$

Noting that

$$\boldsymbol{B} = \frac{i}{L^{3/2}} \sum_{k\alpha} \sqrt{\frac{\hbar\omega}{2}} (\hat{k} \times \hat{\varepsilon}) \left(a_{k\alpha} e^{i\boldsymbol{k}\bullet\boldsymbol{r}} - a^\dagger_{k\alpha} e^{-i\boldsymbol{k}\bullet\boldsymbol{r}} \right)$$

we calculate the matrix element of $H'^{(S)}$ in the long-wavelength approximation (i.e., we replace the exponential factors in $\boldsymbol{B}$ by unity). Thus

$$H'^{(S)}_{fi} = -ig_s \frac{e}{2m_e c} \frac{1}{L^{3/2}} \sqrt{\frac{\hbar\omega}{2}} \left\langle \psi_f \left| \hat{k} \times \hat{\varepsilon} \cdot \boldsymbol{S} \right| \psi_i \right\rangle \tag{16.67}$$

Combining (16.65) and (16.67), we obtain the magnetic dipole ($M1$) matrix element

$$H'_{fi}(M1) = -i\frac{\mu_B}{\hbar} \frac{1}{L^{3/2}} \sqrt{\frac{\hbar\omega}{2}} \left\langle \psi_f \left| \hat{k} \times \hat{\varepsilon} \cdot (\boldsymbol{L} + g_s \boldsymbol{S}) \right| \psi_i \right\rangle \tag{16.68}$$

The quantity $\hat{k} \times \hat{\varepsilon} \cdot (\boldsymbol{L} + g_s \boldsymbol{S})$ is a scalar product of two first-rank tensors, where $\hat{k} \times \hat{\varepsilon}$ refers to the radiation field and $\boldsymbol{L} + g_s \boldsymbol{S}$ refers to the atom. From this we immediately conclude that $M1$ transitions, like $E1$ transitions, satisfy the selection rules $m_f + M = m_i$ and $\Delta(j_i 1 j_f)$. However, $\boldsymbol{L} + g_s \boldsymbol{S}$, unlike **r**, is an even-parity operator ($\boldsymbol{L} + g_s \boldsymbol{S}$ is an axial vector, whereas $\boldsymbol{r}$ is a polar vector). Thus an $M1$ transition connects initial and final atomic states of the same parity; that is,

$$\pi_i \pi_f = +1 \tag{16.69}$$

For example, allowed $M1$ transitions occur in atoms between hyperfine components of a given state or between fine-structure components (e.g., between $n^2 p_{3/2}$ and $n^2 p_{1/2}$ levels in hydrogen), or between m_J components of a state with given J. The selection rules just stated also allow an $M1$ transition between $2^2 s_{1/2}$ and $1^2 s_{1/2}$ states in hydrogen. In the nonrelativistic limit, the $M1$ matrix element in this case is zero because the radial-wave functions are orthogonal. The matrix element is actually nonzero because of relativistic corrections, although it is exceedingly small. (The chief mechanism for decay of the $2s$ state of hydrogen in the absence of perturbing electric fields is emission of *two* electric dipole photons, with a spontaneous emission mean lifetime of 1/8 s.) *Forbidden* $M1$ transitions similar to that just mentioned for $2s$-$1s$ in hydrogen do occur and have been observed in the alkali atoms (e.g., $7s$-$6s$ in cesium) and in thallium.

Let us compare the matrix elements for allowed $M1$ and $E1$ transitions of similar frequencies. From (16.68) and (16.41), we have

$$\left| \frac{H'(ml)}{H'(el)} \right| \approx \left| \frac{\mu_B}{e a_0} \right| \approx \frac{\alpha}{2} \approx 10^{-2} \tag{16.70}$$

Next, we return to (16.63) and consider the term

$$\frac{1}{2} k_i \hat{\varepsilon}_j \left(x_i p_j + p_i x_j \right)$$

which is responsible for electric quadrupole ($E2$) transitions. Employing the identity

$$\left[x_k, H_{\text{atom}} \right] = \frac{i\hbar}{m_e} p_k$$

we have

$$\frac{1}{2}k_i\hat{\varepsilon}_j\left(x_ip_j+p_ix_j\right)=\frac{m_e}{2i\hbar}k_i\hat{\varepsilon}_j\left(x_i\left[x_j,H_a\right]+\left[x_i,H_a\right]x_j\right)$$
$$=\frac{m_e}{2i\hbar}k_i\hat{\varepsilon}_j\left(x_ix_jH_a-x_iH_ax_j+x_iH_ax_j-H_ax_ix_j\right)$$
$$=\frac{m_e}{2i\hbar}k_i\hat{\varepsilon}_j\left(x_ix_jH_a-H_ax_ix_j\right)$$

Thus the matrix element for spontaneous emission of $E2$ radiation is

$$\begin{aligned}H'_{fi}(e2)&=-i\frac{e}{m_ec}\frac{m_e}{2i\hbar}\frac{1}{L^{3/2}}\sqrt{\frac{\hbar c^2}{2\omega}}\hbar\omega\left\langle\psi_f\left|k_i\hat{\varepsilon}_j\left(x_ix_j-\frac{1}{3}\delta_{ij}r^2\right)\right|\psi_i\right\rangle\\&=-\frac{1}{L^{3/2}}\frac{e\omega}{2c}\sqrt{\frac{\hbar\omega}{2}}\left\langle\psi_f\left|\hat{k}_i\hat{\varepsilon}_j\left(x_ix_j-\frac{1}{3}\delta_{ij}r^2\right)\right|\psi_i\right\rangle\end{aligned}\tag{16.71}$$

In (16.71) we have included a term $\frac{1}{3}\hat{k}_i\hat{\varepsilon}_j\delta_{ij}r^2$ that gives no contribution to the matrix element because $\boldsymbol{k}\cdot\hat{\varepsilon}=0$. This is done to make the quadrupole tensor $x_ix_j-\frac{1}{3}\delta_{ij}r^2$ an irreducible second-rank tensor. Because this tensor operator has even parity, the selection rules for $E2$ radiation are as follows: $\Delta(j_i2j_f)$ and $\pi_i\pi_f=+1$. Comparing (16.71) with an $E1$ matrix element (16.41) of comparable frequency, we have

$$\left|\frac{H'(e2)}{H'(e1)}\right|\approx\frac{\omega}{c}a_0$$

Taking $\omega\approx1$, $a_0=1$, and $c=\alpha^{-1}$ in atomic units, we see that $E2$ amplitudes, like $M1$ amplitudes, are typically of order α compared with $E1$ amplitudes in atomic transitions. The same scaling does not hold for gamma-ray transitions in nuclei.

16.4 Damping and natural linewidth

We now recall our original formulation of time-dependent perturbation theory, where

$$H=H_0+H'$$

$$H_0\left|u_n\right\rangle=\hbar\omega_n\left|u_n\right\rangle$$

$$H\left|\psi\right\rangle=i\hbar\frac{\partial\left|\psi\right\rangle}{\partial t}\tag{16.72}$$

and

$$\left|\psi\right\rangle=\sum_n a_n(t)\left|u_n\right\rangle e^{-i\omega_nt}$$

We have shown that the system of differential equations

$$\dot{a}_m = -\frac{i}{\hbar}\sum_n a_n H'_{mn} e^{i\omega_{mn}t} \tag{16.73}$$

is equivalent to (16.72). Suppose that, as before, the initial conditions are

$$a_i(0) = 1 \qquad a_{n\neq i}(0) = 0$$

We have described how the first-order approximation consists of replacing all the a_n in (16.73) by their initial values, which results in the equations

$$\dot{a}_m = -\frac{i}{\hbar} H'_{mi} e^{i\omega_{mi}t} \qquad m \neq i$$
$$\dot{a}_i = 0$$

Clearly, this approximation must fail for sufficiently long times because $|a_i|$ decreases from its initial value of unity, and the other $|a_n|$ increase from zero. We now see how to improve the approximation substantially by assuming that $a_i(t)$ takes the form

$$a_i(t) = \exp\left(-\frac{\gamma}{2}t\right) \tag{16.74}$$

where $\gamma = \gamma_1 + i\gamma_2$ and $\gamma_{1,2}$ are real numbers (Weisskopf and Wigner 1930). With this substitution, the amended equations are

$$\dot{a}_n = -\frac{i}{\hbar} H'_{ni} e^{i\omega_{mi}t} \exp\left(-\frac{\gamma}{2}t\right) \qquad n \neq i \tag{16.75}$$

$$\dot{a}_i = -\frac{\gamma}{2}\exp\left(-\frac{\gamma}{2}t\right) = -\frac{i}{\hbar}\sum_n H'_{in} a_n e^{-i\omega_{ni}t} \tag{16.76}$$

We integrate (16.75) subject to $a_n(0) = 0$ to obtain

$$a_n = H'_{ni}\frac{\exp\left[-i\left(\omega_{in} - i\frac{\gamma}{2}\right)t\right] - 1}{\hbar\left(\omega_{in} - i\frac{\gamma}{2}\right)} \tag{16.77}$$

and insert (16.77) into (16.76) to obtain

$$\gamma = \frac{2i}{\hbar}\sum_n \frac{|H'_{in}|^2\left\{1 - \exp\left[i\left(\omega_{in} - i\frac{\gamma}{2}\right)t\right]\right\}}{\hbar\left(\omega_{in} - \frac{i\gamma}{2}\right)} \tag{16.78}$$

Typically, we might have an atom in some initial excited state i_a that decays by spontaneous emission to a final state n consisting of the atom in final state f_a and a photon with energy ε

in range $d\varepsilon$ and with density of states $\rho(\varepsilon)$. In this case we replace the sum in (16.78) by an integral

$$\gamma = \frac{2i}{\hbar}\int \rho(\varepsilon)\,d\varepsilon \frac{|H'_{in}|^2\left\{1-\exp\left[\frac{i}{\hbar}\left(E_{if}^a - \varepsilon - i\frac{\hbar\gamma}{2}\right)t\right]\right\}}{E_{if}^a - \varepsilon - i\frac{\hbar\gamma}{2}} \tag{16.79}$$

It is impossible to solve this integral equation exactly because γ appears not only on the left-hand side but also in a complicated integrand on the right-hand side. However, we can find an approximate solution by ignoring γ in the integrand. Then

$$\gamma = \gamma_1 + i\gamma_2 \approx \frac{2i}{\hbar}\int \rho(\varepsilon)\,d\varepsilon \frac{|H'_{in}|^2\left\{1-\exp\left[\frac{i}{\hbar}\left(E_{if}^a - \varepsilon\right)t\right]\right\}}{E_{if}^a - \varepsilon} \tag{16.80}$$

In the factor

$$\frac{1-\exp\left[\frac{i}{\hbar}\left(E_{if}^a - \varepsilon\right)t\right]}{E_{if}^a - \varepsilon} = \frac{1-\cos\left[\frac{1}{\hbar}\left(E_{if}^a - \varepsilon\right)t\right]}{E_{if}^a - \varepsilon} - i\frac{\sin\left[\frac{1}{\hbar}\left(E_{if}^a - \varepsilon\right)t\right]}{E_{if}^a - \varepsilon} \tag{16.81}$$

the first term on the right-hand side of (16.81) is associated with γ_2. One finds that $\hbar\gamma_2$ is a small energy shift in level i_a due to its coupling with the radiation field. We return to this for a more detailed discussion in Chapter 17. However, for the moment, we ignore this term and concentrate on the second term on the right-hand side of (16.81), which is associated with γ_1. We make use of the following relations:

$$\begin{aligned}\delta(x) &= \lim_{k\to\infty}\frac{1}{2\pi}\int_{-k}^{k} e^{i\omega x}\,d\omega \\ &= \lim_{k\to\infty}\frac{1}{\pi}\int_{0}^{k}\cos\omega x\,d\omega \\ &= \lim_{k\to\infty}\frac{1}{\pi}\frac{\sin kx}{x}\end{aligned}$$

we see from (16.80) that in the limit as $t \to \infty$,

$$\begin{aligned}\gamma_1 &= \frac{2\pi}{\hbar}\int |H'_{ni}|^2 \delta\left(E_{if}^a - \varepsilon\right)\rho(\varepsilon)\,d\varepsilon \\ &= \frac{2\pi}{\hbar}|H'_{ni}|^2 \rho\left(E_{if}^a\right)\end{aligned} \tag{16.82}$$

Thus γ_1 is just the total transition probability per unit time as given by the golden rule. This is reasonable because given the expression

$$|a_i|^2 = e^{-\gamma_1 t}$$

we have

$$\frac{d}{dt}|a_i|^2 = -\gamma_1 e^{-\gamma_1 t}$$

or

$$\gamma_1 = -\frac{1}{|a_i|^2}\frac{d|a_i|^2}{dt}$$

which is a very plausible expression for the transition probability per unit time out of state i_a. The existence of "damping," as expressed by the factor $\exp(-\gamma_1 t)$, implies that in the transition between states i_a and f_a with energies E_i^a and E_f^a, the emitted photon energy is somewhat uncertain. To see this, we go back to (16.77) and ask for the transition probability (not the transition probability per unit time) to a state n consisting of an atom in state f_a and a photon with frequency ω in range $d\omega$. This is

$$\begin{aligned} dp &= |H'_{ni}|^2 \left| \frac{\exp\left[-i\left(\omega_{in} - i\frac{\gamma}{2}\right)t\right] - 1}{\hbar\left(\omega_{in} - i\frac{\gamma}{2}\right)} \right|^2 \rho(\omega)d\omega \\ &= |H'_{ni}|^2 \frac{1 - 2e^{-\gamma t/2}\cos\omega_{in}t + e^{-\gamma t}}{\hbar^2\left[\left(\omega_{if}^a - \omega\right)^2 + \frac{\gamma^2}{4}\right]}\rho(\omega)d\omega \end{aligned}$$

In the limit of large t, this becomes

$$dp = |H'_{ni}|^2 \frac{1}{\hbar^2\left[\left(\omega_{if}^a - \omega\right)^2 + \frac{\gamma^2}{4}\right]}\rho(\omega)d\omega \tag{16.83}$$

Because in most applications $|H'_{ni}|^2$ and $\rho(\omega)$ are slowly varying functions in the neighborhood of $\omega = \omega_{if}^a$, we conclude that the photon frequency is distributed about ω_{if}^a in a Lorentz distribution; that is,

$$\frac{1}{\left(\omega_{if}^a - \omega\right)^2 + \frac{\gamma^2}{4}} \tag{16.84}$$

with full width at half maximum (FWHM) equal to γ. In other words, state i_a does not have a sharp energy E_i^a but rather a natural energy width $\Gamma = \hbar\gamma_1 = \hbar A$, where A is the total spontaneous emission decay coefficient for i_a. For example, $A = 6.25\times10^8$ s^{-1} for the $2p \rightarrow 1s$ transition in hydrogen. Thus $\Gamma/\hbar = 6.25\times10^8$ rad/s $= 100$ MHz is the frequency uncertainty of the $2p$ state.

16.5 Approximate character of the exponential law of decay

In the preceding section we discussed the "exponential" law of spontaneous decay, which appears in so many contexts (e.g., decay of radioactive nuclei, spontaneous decay of elementary particles, and spontaneous emission of photons by atoms, molecules, and nuclei in excited states) that it appears to be a fundamental law. However, we have seen that in fact it is not fundamental but rather emerges from a succession of subtle approximations in time-dependent perturbation theory. The approximate nature of the exponential decay law also can be seen from the following general argument: suppose that at time $t = 0$ a system is prepared in state $|w\rangle$. For example, $|w\rangle$ could describe a nucleus that has not decayed. At time t, the state vector is $U(t,0)|w\rangle = \exp(-iHt/\hbar)|w\rangle$ in the Schroedinger picture. The probability that the nucleus has not decayed at time t is

$$P(t) = |A(t)|^2 \tag{16.85}$$

where $A(t) = \langle w|e^{-iHt/\hbar}|w\rangle$. Let us express $|w\rangle$ in terms of the eigenstates of the Hamiltonian H; that is,

$$|w\rangle = \sum |u_n\rangle\langle u_n|w\rangle$$

where $H|u_n\rangle = E_n|u_n\rangle$ (in fact, the eigenvalue spectrum could be continuous, and the last sum could be an integral). Then

$$\begin{aligned} A(t) &= \sum_{n,m} \langle w|u_m\rangle\langle u_m|u_n\rangle\langle u_n|w\rangle e^{-iE_nt/\hbar} \\ &= \sum_n |\langle u_n|w\rangle|^2 e^{-iE_nt/\hbar} \\ &= \int \eta(E) e^{-iEt/\hbar}\, dE \end{aligned} \tag{16.86}$$

where $\eta(E) = \sum_n |\langle u_n|w\rangle|^2 \delta(E - E_n)$ is called the *spectral function*. (If the energy spectrum is continuous, the last sum becomes an integral, and η can be a continuous function of E.) We see from (16.86) that $A(t)$ is the Fourier transform of the spectral function. In the special case where the spectral function is Lorentzian, that is,

$$\eta(E) = \frac{\Gamma}{2\pi} \frac{1}{(E - E_0)^2 + \frac{\Gamma^2}{4}} \tag{16.87}$$

we find from a simple contour integration that

$$A(t) = \frac{\gamma}{2\pi} \int_{-\infty}^{\infty} \frac{e^{-i\omega t}}{(\omega - \omega_0)^2 + \frac{\gamma^2}{4}}\, d\omega = e^{-i(\omega_0 - i\gamma/2)t} \tag{16.88}$$

where $E_0 = \hbar\omega_0$ and $\Gamma = \hbar\gamma$. However, if η deviates from Lorentzian shape, $A(t)$ is not exponential. In particular, because the energy spectrum of any real physical system must have a finite lower bound $E_{\min}$, we require $\eta = 0$ for $E < E_{\min}$, and this implies that there must be some deviation from the exponential decay law. At sufficiently long times, the decay is slower than exponential (Khalfin 1958). Experimental evidence for this deviation has been found (Rothe et al. 2006).

16.6 Second-order processes: Scattering of light by an atomic electron

We now consider processes in which two photons are involved. These include two-photon absorption, two-photon emission as in the decay $2s$-$1s$ in hydrogen, and absorption of one photon with emission of another (e.g., Raman effect, photon scattering). We shall discuss photon scattering in some detail, but before considering the quantum-mechanical aspects, it is worth our while to analyze it classically by assuming that the atom consists of a classical electron bound harmonically to a fixed nucleus. Here electron motion is described by the equation

$$m_e\ddot{\boldsymbol{x}} + m_e\gamma_0\dot{\boldsymbol{x}} + m_e\omega_0^2\boldsymbol{x} = \boldsymbol{F} \tag{16.89}$$

where $-m_e\omega_0^2\boldsymbol{x}$ is the harmonic restoring force, $-m_e\gamma_0\dot{\boldsymbol{x}}$ is a damping force, and $\boldsymbol{F}$ is an external force. Suppose that an atom at the origin is exposed to a plane electromagnetic wave of frequency ω. Assuming the electron motion to be nonrelativistic so that the force term $-e(\mathbf{v}\times\boldsymbol{B})/c$ can be neglected and writing $\boldsymbol{\mathcal{E}} = \mathcal{E}_0\hat{z}e^{-i\omega t}$, we have

$$m_e\ddot{z} + m_e\gamma_0\dot{z} + m_e\omega_0^2 z = -e\mathcal{E}_0 e^{-i\omega t} \tag{16.90}$$

We try a solution of the form $z = z_0 e^{-i\omega t}$, which yields

$$z_0 = -\frac{e\mathcal{E}_0}{m_e}\frac{1}{\left(\omega_0^2 - \omega^2\right) - i\omega\gamma_0} \tag{16.91}$$

A classical accelerated charge radiates energy at the instantaneous rate

$$S_0 = \frac{2}{3}\frac{e^2a^2}{4\pi c^3}$$

where $a = -\omega^2 z$ is the acceleration. The average of S_0 over any number of complete periods is

$$\overline{S_0} = \frac{1}{3}\frac{e^2\omega^4}{4\pi c^3}|z_0|^2 \text{ erg s}^{-1} \tag{16.92}$$

Meanwhile, the intensity of radiation in the incoming plane wave is

$$P = \frac{c}{2}\mathcal{E}_0^2 \text{ erg cm}^{-2}\text{ s}^{-1} \tag{16.93}$$

The ratio $\sigma \equiv \overline{S_0}/P$ has dimension of length squared and is called the *scattering cross section.* From (16.91) through (16.93), we obtain

$$\sigma = \frac{8\pi}{3} r_0^2 \frac{\omega^4}{\left(\omega_0^2 - \omega^2\right)^2 + \omega^2 \gamma_0^2} \tag{16.94}$$

where $r_0 = e^2/4\pi m_e c^2 \approx 2.8 \times 10^{-13}$ cm is the *classical electron radius.* We now consider three separate cases of (16.94):

1. Rayleigh scattering: $\omega \ll \omega_0$.
2. Resonance fluorescence: $\omega \approx \omega_0$.
3. "Free" electron (Thomson) scattering: $\omega \gg \omega_0$.

Rayleigh Scattering

When $\omega \ll \omega_0$,

$$\sigma = \frac{8\pi}{3} r_0^2 \frac{\omega^4}{\left(\omega_0^2 - \omega^2\right)^2 + \omega^2 \gamma_0^2} \approx \frac{8\pi}{3} r_0^2 \frac{\omega^4}{\omega_0^4} \tag{16.95}$$

The well-known ω^4 dependence is characteristic of scattering of sunlight by dust grains in the atmosphere and is responsible for the blueness of the sky and the redness of sunsets.

Resonance fluorescence: $\omega \approx \omega_0$

Here $\omega_0^2 - \omega^2 = (\omega_0 + \omega)(\omega_0 - \omega) \approx 2\omega_0(\omega_0 - \omega)$. Hence

$$\begin{aligned} \sigma &= \frac{8\pi}{3} r_0^2 \frac{\omega^4}{\left(\omega_0^2 - \omega^2\right)^2 + \omega^2 \gamma_0^2} \approx \frac{8\pi}{3} r_0^2 \frac{\omega_0^4}{4\omega_0^2 (\omega_0 - \omega)^2 + \omega_0^2 \gamma_0^2} \\ &= \frac{2\pi}{3} r_0^2 \frac{\omega_0^2}{(\omega_0 - \omega)^2 + \frac{\gamma_0^2}{4}} \end{aligned} \tag{16.96}$$

The cross section exhibits a characteristic Lorentz shape in the neighborhood of ω_0.

Free Electron Scattering: $\omega \gg \omega_0$

Here $\sigma \approx (8\pi/3) r_0^2$, which is a constant, the *Thomson scattering cross section.* In fact, it is the nonrelativistic limit of the Compton scattering cross section.

Now we study the same problem in quantum mechanics. Imagine an incoming photon beam with n photons in mode $\boldsymbol{k}_i \alpha_i$ and an initial atomic state i_a. We want to calculate the probability per unit time that one of the $\boldsymbol{k}_i \alpha_i$ photons is scattered into the mode $\boldsymbol{k}_f \alpha_f$ (i.e., one $\boldsymbol{k}_i \alpha_i$ photon is destroyed and one $\boldsymbol{k}_f \alpha_f$ photon is created) while the atom simultaneously ends up in the state f_a. We ignore the spin term and write the perturbation Hamiltonian as

$$H' = \frac{e}{m_e c} \boldsymbol{A} \cdot \boldsymbol{p} + \frac{e^2}{2 m_e c^2} A^2 \tag{16.97}$$

Because we want to create one photon and destroy another, we employ the $\boldsymbol{A}\bullet\boldsymbol{p}$ term in second order and the $\boldsymbol{A}^2$ term in first order. Let us begin with the latter. The first-order matrix element is

$$\begin{aligned}
H_{fi}^{\prime(1)} &= \frac{e^2}{2m_e c^2}\left\langle f_a;(n-1)\boldsymbol{k}_i\alpha_i;\boldsymbol{k}_f\alpha_f\left|\boldsymbol{A}\bullet\boldsymbol{A}\right|i_a;n\boldsymbol{k}_i\alpha_i\right\rangle \\
&= \frac{e^2}{2m_e c^2}\frac{1}{L^3}\sum_{\substack{k\alpha\\k'\alpha'}}\left(\frac{\hbar c^2}{2}\right)\sqrt{\frac{1}{\omega\omega'}}\hat{\varepsilon}\bullet\hat{\varepsilon}' \\
&\left\langle f_a;(n-1)\boldsymbol{k}_i\alpha_i;\boldsymbol{k}_f\alpha_f\left|a_{k\alpha}a^{\dagger}_{k'\alpha'}e^{i(\boldsymbol{k}-\boldsymbol{k}')\bullet\boldsymbol{r}}+a^{\dagger}_{k\alpha}a_{k'\alpha'}e^{-i(\boldsymbol{k}-\boldsymbol{k}')\bullet\boldsymbol{r}}\right|i_a;n\boldsymbol{k}_i\alpha_i\right\rangle \\
&= \frac{e^2}{m_e c^2}\frac{n^{1/2}}{L^3}\left(\frac{\hbar c^2}{2}\right)\sqrt{\frac{1}{\omega_i\omega_f}}\hat{\varepsilon}_i\bullet\hat{\varepsilon}_f\left\langle f_a\left|e^{i(\boldsymbol{k}_i-\boldsymbol{k}_f)\bullet\boldsymbol{r}}\right|i_a\right\rangle
\end{aligned} \tag{16.98}$$

In the long-wavelength approximation,

$$\left\langle f_a\left|e^{i(\boldsymbol{k}_i-\boldsymbol{k}_f)\bullet\boldsymbol{r}}\right|i_a\right\rangle = e^{i(\boldsymbol{k}_i-\boldsymbol{k}_f)\bullet\boldsymbol{R}}\left\langle f_a|i_a\right\rangle = e^{i(\boldsymbol{k}_i-\boldsymbol{k}_f)\bullet\boldsymbol{R}}\delta_{f_a,i_a}$$

where $\boldsymbol{R}$ is the location of the atomic center of mass. Therefore, in this limit, the $\boldsymbol{A}^2$ term contributes only when $|i_a\rangle = |f_a\rangle$, in which case the initial and final photons have the same energy but not necessarily the same direction. The change in photon momentum is permitted by atomic recoil, which is not taken into account explicitly in this calculation because we assume that the atomic nucleus has infinite mass.

It is convenient in what follows to employ a different form of (16.98) that has identical content. Assuming $|i_a\rangle = |f_a\rangle$, $\omega_i = \omega_f \equiv \omega$, and $\boldsymbol{R} = 0$, we have

$$H_{fi}^{\prime(1)} = \frac{\hbar e^2}{2m_e\omega}\frac{n^{1/2}}{L^3}\hat{\varepsilon}_i\bullet\hat{\varepsilon}_f \tag{16.99}$$

Now

$$\begin{aligned}
\hat{\varepsilon}_i\bullet\hat{\varepsilon}_f &= \hat{\varepsilon}_i\bullet\frac{[\boldsymbol{x},\boldsymbol{p}]}{i\hbar}\bullet\hat{\varepsilon}_f = \frac{1}{i\hbar}\sum_{n_a}\left(\hat{\varepsilon}_i\bullet\boldsymbol{x}_{in}\boldsymbol{p}_{ni}\bullet\hat{\varepsilon}_f - \hat{\varepsilon}_i\bullet\boldsymbol{p}_{in}\boldsymbol{x}_{ni}\bullet\hat{\varepsilon}_f\right) \\
&= \hat{\varepsilon}_f\bullet\frac{[\boldsymbol{x},\boldsymbol{p}]}{i\hbar}\bullet\hat{\varepsilon}_i = \frac{1}{i\hbar}\sum_{n_a}\left(\hat{\varepsilon}_f\bullet\boldsymbol{x}_{in}\boldsymbol{p}_{ni}\bullet\hat{\varepsilon}_i - \hat{\varepsilon}_f\bullet\boldsymbol{p}_{in}\boldsymbol{x}_{ni}\bullet\hat{\varepsilon}_i\right)
\end{aligned} \tag{16.100}$$

In addition, we have

$$\boldsymbol{p} = \frac{m_e}{i\hbar}[\boldsymbol{x},H_a] \tag{16.101}$$

Substitution of the last relation into (16.100) and some rearrangement yield

$$H_{fi}^{\prime(1)} = \frac{e^2}{2\hbar\omega}\frac{n^{1/2}}{L^3}\sum_{n_a}\left(\hat{\varepsilon}_i\bullet\boldsymbol{x}_{in}\boldsymbol{x}_{ni}\bullet\hat{\varepsilon}_f + \hat{\varepsilon}_f\bullet\boldsymbol{x}_{in}\boldsymbol{x}_{ni}\bullet\hat{\varepsilon}_i\right)E^a_{ni} \tag{16.102}$$

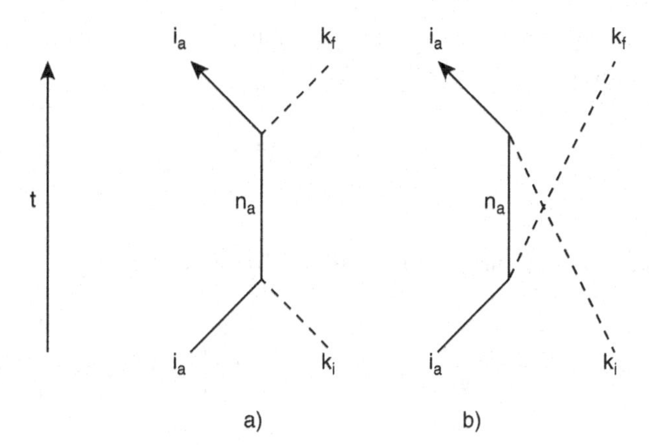

Figure 16.2 Second order diagrams showing 2 types of intermediate states.

where $E_{ni}^{a} = E_{n}^{a} - E_{i}^{a}$. We use (16.102) momentarily, but for now, we consider the contribution of the term $(e/m_e c)\boldsymbol{A}\bullet\boldsymbol{p}$ in second order. In general, the second-order matrix element is

$$M_2 = \sum_n \frac{H_{fn} H'_{ni}}{E_{in}}$$

In the present case, we have two types of intermediate states, associated with Figure 16.2a and b.

In Figure 16.2a, the intermediate state consists of the atom in state $|n_a\rangle$, $n-1$ photons in mode $\boldsymbol{k}_i\alpha_i$, and no photons in mode $\boldsymbol{k}_f\alpha_f$. However, in Figure 16.2b, there are n photons in mode $\boldsymbol{k}_i\alpha_i$ and one photon in mode $\boldsymbol{k}_f\alpha_f$ in the intermediate state. Denoting the photon energy by ε, we have the following contribution from the $\boldsymbol{A}\bullet\boldsymbol{p}$ term in second order:

$$H_{fi}^{\prime(2)} = \sum \frac{H_{fn}^{\prime\mu} H_{ni}^{\prime\mu}}{E_{in}^{a} + \varepsilon} + \sum \frac{H_{fn}^{\prime\nu} H_{ni}^{\prime\nu}}{E_{in}^{a} - \varepsilon} \tag{16.103}$$

where

$$H_{fn}^{\prime\mu} = \frac{e}{m_e c} \frac{1}{L^{3/2}} \sqrt{\frac{\hbar c^2}{2\omega}} \left\langle i_a \middle| \hat{\varepsilon}_f \bullet \boldsymbol{p} \middle| n_a \right\rangle \tag{16.104}$$

$$H_{ni}^{\prime\mu} = \frac{e}{m_e c} \frac{n^{1/2}}{L^{3/2}} \sqrt{\frac{\hbar c^2}{2\omega}} \left\langle n_a \middle| \hat{\varepsilon}_i \bullet \boldsymbol{p} \middle| i_a \right\rangle \tag{16.105}$$

$$H_{fn}^{\prime\nu} = \frac{e}{m_e c} \frac{n^{1/2}}{L^{3/2}} \sqrt{\frac{\hbar c^2}{2\omega}} \left\langle i_a \middle| \hat{\varepsilon}_i \bullet \boldsymbol{p} \middle| n_a \right\rangle \tag{16.106}$$

$$H_{fn}^{\prime\nu} = \frac{e}{m_e c} \frac{1}{L^{3/2}} \sqrt{\frac{\hbar c^2}{2\omega}} \left\langle n_a \middle| \hat{\varepsilon}_f \bullet \boldsymbol{p} \middle| i_a \right\rangle \tag{16.107}$$

Inserting (16.104) through (16.107) into (16.103) and employing the identity (16.101), we find, after some algebra, that

$$H'^{(2)}_{fi} = \frac{e^2}{2\hbar\omega}\frac{n^{1/2}}{L^3}\sum_{n_a}\left[\frac{\left(E^a_{ni}\right)^2}{E^a_{in}+\varepsilon}\hat{\varepsilon}_f \bullet \boldsymbol{x}_{in}\boldsymbol{x}_{ni} \bullet \hat{\varepsilon}_i + \frac{\left(E^a_{ni}\right)^2}{E^a_{in}-\varepsilon}\hat{\varepsilon}_i \bullet \boldsymbol{x}_{in}\boldsymbol{x}_{ni} \bullet \hat{\varepsilon}_f\right] \tag{16.108}$$

We now combine (16.108) with (16.102) to obtain

$$M_{\text{total}} = \frac{e^2}{2\hbar\omega}\frac{n^{1/2}}{L^3}\sum_{n_a}\left\{\left[E^a_{ni} - \frac{\left(E^a_{ni}\right)^2}{E^a_{ni}-\varepsilon}\right]\hat{\varepsilon}_f \bullet \boldsymbol{x}_{in}\boldsymbol{x}_{ni} \bullet \hat{\varepsilon}_i + \left[E^a_{ni} - \frac{\left(E^a_{ni}\right)^2}{E^a_{ni}+\varepsilon}\right]\hat{\varepsilon}_i \bullet \boldsymbol{x}_{in}\boldsymbol{x}_{ni} \bullet \hat{\varepsilon}_f\right\} \tag{16.109}$$

In the case of Rayleigh scattering, $\varepsilon \ll E^a_{ni}$ for any n. Thus we can employ the expansions

$$E^a_{ni} - \frac{\left(E^a_{ni}\right)^2}{E^a_{ni} \pm \varepsilon} \approx E^a_{ni}\left[1 - \left(1 \mp \frac{\varepsilon}{E^a_{ni}} + \left(\frac{\varepsilon}{E^a_{ni}}\right)^2 \mp \cdots\right)\right]$$

to write (16.109) as follows:

$$M_{\text{total}} = -\frac{e^2}{2\hbar\omega}\frac{n^{1/2}}{L^3}\sum\left[\varepsilon\left(\hat{\varepsilon}_f \bullet \boldsymbol{x}_{in}\boldsymbol{x}_{ni} \bullet \hat{\varepsilon}_i - \hat{\varepsilon}_i \bullet \boldsymbol{x}_{in}\boldsymbol{x}_{ni} \bullet \hat{\varepsilon}_f\right) + \frac{\varepsilon^2}{E^a_{ni}}\left(\hat{\varepsilon}_f \bullet \boldsymbol{x}_{in}\boldsymbol{x}_{ni} \bullet \hat{\varepsilon}_i + \hat{\varepsilon}_i \bullet \boldsymbol{x}_{in}\boldsymbol{x}_{ni} \bullet \hat{\varepsilon}_f\right)\right]$$

However, because the atomic states form a complete set, that part of the sum proportional to ε on the right-hand side of this expression vanishes. Thus

$$M^{\text{Rayleigh}}_{\text{total}} = -\frac{e^2}{2}\varepsilon\frac{n^{1/2}}{L^3}\sum\frac{1}{E^a_{ni}}\left(\hat{\varepsilon}_f \bullet \boldsymbol{x}_{in}\boldsymbol{x}_{ni} \bullet \hat{\varepsilon}_i + \hat{\varepsilon}_i \bullet \boldsymbol{x}_{in}\boldsymbol{x}_{ni} \bullet \hat{\varepsilon}_f\right) \tag{16.110}$$

The transition probability per unit time is given by the golden rule; that is,

$$dW = \frac{2\pi}{\hbar}|M|^2\, d\rho$$

where

$$d\rho = \frac{L^3\omega^2}{8\pi^3\hbar c^3}d\Omega$$

for radiation of a final photon of given polarization into solid angle $d\Omega$. Thus, from (16.110), we have

$$dW = \frac{\alpha^2 n}{cL^3}\omega^4\left|\sum\frac{1}{\omega^a_{ni}}\left(\hat{\varepsilon}_f \bullet \boldsymbol{x}_{in}\boldsymbol{x}_{ni} \bullet \hat{\varepsilon}_i + \hat{\varepsilon}_i \bullet \boldsymbol{x}_{in}\boldsymbol{x}_{ni} \bullet \hat{\varepsilon}_f\right)\right|^2 d\Omega \tag{16.111}$$

The incoming photon flux is $J = cn/L^3$, and the differential cross section is

$$d\sigma = \frac{dW}{J} = \frac{\alpha^2}{c^2}\omega^4 \left| \sum \frac{1}{\omega_{ni}^a} \left(\hat{\varepsilon}_f \bullet \boldsymbol{x}_{in} \boldsymbol{x}_{ni} \bullet \hat{\varepsilon}_i + \hat{\varepsilon}_i \bullet \boldsymbol{x}_{in} \boldsymbol{x}_{ni} \bullet \hat{\varepsilon}_f \right) \right|^2 d\Omega \tag{16.112}$$

Let us make an order-of-magnitude estimate of the sum in (16.112). It is approximately a_0^2/ω_0, where $\hbar\omega_0 \approx e^2/4\pi a_0$. Therefore,

$$\begin{aligned} \sigma &\approx \frac{\alpha^2}{c^2}\frac{\omega^4}{\omega_0^4}\omega_0^2 a_0^4 \\ &\approx \frac{\alpha^2}{c^2}\frac{\omega^4}{\omega_0^4}\frac{e^4}{\hbar^2 a_0^2}a_0^4 \\ &= \alpha^4 \frac{\omega^4}{\omega_0^4}a_0^2 \\ &= r_0^2 \frac{\omega^4}{\omega_0^4} \end{aligned} \tag{16.113}$$

This is consistent with the classical result (16.95) for Rayleigh scattering. We next turn to Thomson scattering, where $\varepsilon \gg E_{ni}^a$. Here the second-order contribution (16.108) is very small, so we ignore it and consider only the first-order contribution (16.99). We have

$$dW = r_0^2 \frac{nc}{L^3} \left| \hat{\varepsilon}_i \bullet \hat{\varepsilon}_f \right|^2 d\Omega$$

Hence

$$d\sigma = r_0^2 \left| \hat{\varepsilon}_i \bullet \hat{\varepsilon}_f \right|^2 d\Omega. \tag{16.114}$$

It is easy to show that when this expression is summed over final polarizations, the quantity $\left| \hat{\varepsilon}_i \bullet \hat{\varepsilon}_f \right|^2$ contributes the factor $\cos^2\theta\cos^2\phi + \sin^2\phi$, where θ and ϕ are the scattering angles (between $\boldsymbol{k}_i$ and $\boldsymbol{k}_f$). Integrating over all angles, we obtain the total cross section

$$\sigma = \frac{8\pi}{3} r_0^2 \tag{16.115}$$

which agrees with the classical result.

Finally, we consider the case of resonance fluorescence, where the energy of the incoming photon is very nearly equal to the energy difference between a particular excited atomic state $|n_a\rangle$ and the ground state $|i_a\rangle$. Clearly, the A^2 contribution can be neglected here, and in (16.108), we need to consider only a single one of the second type of terms on the right-hand side. Hence

$$M \Rightarrow -\frac{e^2}{2\varepsilon}\frac{n^{1/2}}{L^3}\hat{\varepsilon}_f \bullet \boldsymbol{x}_{in}\boldsymbol{x}_{ni} \bullet \hat{\varepsilon}_i \frac{\left(E_{ni}^a\right)^2}{E_{ni}^a - \varepsilon} \tag{16.116}$$

Actually, in the vicinity of resonance, it is important to take into account the natural width of the state n_a, which is done by making the following replacement in the denominator of the last factor in (16.116):

$$E_{ni}^{a}-\varepsilon \rightarrow E_{ni}^{a}-\varepsilon-\frac{i\Gamma}{2}$$

From this we obtain the differential cross section using methods that are now familiar; that is,

$$d\sigma=\frac{\alpha^2}{c^2}\left|\hat{\varepsilon}_f \cdot \boldsymbol{x}_{in}\right|^2\left|\hat{\varepsilon}_i \cdot \boldsymbol{x}_{in}\right|^2 \frac{\omega_0^4}{(\omega-\omega_0)^2+\frac{\gamma^2}{4}} d\Omega \tag{16.117}$$

It is easy to show that this resembles the classical result (16.96).

Problems for Chapter 16

16.1. In Chapter 7 we discussed magnetic resonance and derived Rabi's formula [equation (7.71)] for the transition probability from the state $m = +½$ to the state $m = -½$ for a particle (e.g., a proton) of spin-½ and magnetic moment $\mu = g\mu_B \boldsymbol{S}$ in the magnetic field $\boldsymbol{B} = B_1\left[(\cos \omega t)\hat{i}-(\sin \omega t)\hat{j}\right]+B_0\hat{k}$. Treat the same problem by first-order time-dependent perturbation theory, where the interaction of the magnetic moment with the static magnetic field $B_0\hat{z}$ is considered to be the zeroth-order Hamiltonian, whereas the interaction with the rotating field $\boldsymbol{B}_1$ is the perturbation. Compare your result with the exact solution.

16.2. Consider two particles of spin-½ coupled by an interaction of the form

$$H' = a(t)\,\boldsymbol{S}_1 \cdot \boldsymbol{S}_2$$

where $a(t)$ is a smooth function of time that approaches zero as $|t|$ becomes very large and has appreciable size a_0 only for $|t| < T/2$.

(a) Assume that at large negative t the system is in the initial state (1: spin up; 2: spin down). Calculate without approximations the state at very large positive t. Show that the probability of finding the system in the state (1: spin down; 2: spin up) depends only on the integral $\int_{-\infty}^{\infty} a(t)dt$.

(b) Calculate the transition probability $P(+,- \rightarrow -,+)$ by first-order time-dependent perturbation theory, and compare your result with that of part (a).

(c) Now assume that in addition, the two spins are coupled to an external magnetic field B in the z-direction by the Zeeman interaction

$$H_{\text{Zeeman}} = -B(\gamma_1 S_{1z} + \gamma_2 S_{2z})$$

where $\gamma_{1,2}$ are two distinct constants. Also assume that $a = a_0 \exp(-t^2/T^2)$. Again, use first-order time-dependent perturbation theory to calculate $P(+,- \rightarrow -,+)$. How does the result depend on B for fixed a_0 and T?

16.3. Calculate the free-space spontaneous emission decay rate for the $1^2 s_{1/2}\left(F=1 \rightarrow F=0\right)$ magnetic dipole transition (1,420 MHz) in atomic hydrogen.

16.4. Ignoring electron spin, calculate the free-space spontaneous emission decay rate for the single photon $3d \rightarrow 2s$ electric quadrupole transition in atomic hydrogen. Compare your result with the rate for the single photon $3d \rightarrow 2p$ electric dipole transition. What is the angular distribution of emitted photons in the $3d \rightarrow 2s$ transition if the initial state is $\ell = 2, m_\ell = 1$?

16.5. The following interesting phenomenon occurs with electrons orbiting in a storage ring. Even though the electrons are initially unpolarized, they gradually acquire a polarization P perpendicular to the plane of the ring. It can be shown that

$$P(t) = P_0\left(1 - e^{-t/t_0}\right)$$

where $P_0 = 8/5\sqrt{3} = 0.9238$ is the maximum polarization, and

$$t_0 = \left(\frac{5\sqrt{3}}{32\pi}\frac{e^2\hbar\gamma^5}{m_e^2c^2R^3}\right)^{-1}$$

in Heaviside-Lorentz units, where $\gamma = \left(1 - v^2/c^2\right)^{-1/2}$ and R is the orbit radius. The exact derivation of these results is lengthy and difficult, but one can give a simple and intuitive approximate derivation that forms the subject of this problem.

Here we imagine an electron describing a classical planar orbit in a uniform magnetic field $\boldsymbol{B}$ with speed v in the laboratory frame. We transform to the instantaneous rest frame of the electron, where the electron experiences a magnetic field $\boldsymbol{B}'$ and also an electric field $\boldsymbol{\mathcal{E}}'$. Because the electron possesses a spin magnetic moment, there exist two magnetic energy states in field $\boldsymbol{B}'$ that are separated by a frequency difference ω. An electron in the upper state can spontaneously emit a magnetic dipole photon and end in the lower state. The transition probability per unit time for this process can be calculated from the formalism we have developed in this chapter. When we transform back to the laboratory frame, this transition probability per unit time is altered by a well-known factor familiar from special relativity. Fill in the details to show that according to this approximate treatment, the maximum polarization is 100 percent and

$$t_0 = \left(\frac{g_s^5}{192\pi}\frac{e^2\hbar\gamma^5}{m_e^2c^2R^3}\right)^{-1} \tag{1}$$

where $g_s \approx 2$ is the electron spin g value, and we assume that the electron is ultrarelativistic. According to (1), what is the electron energy in the laboratory frame if $t_0 = 1$ hour and $|\boldsymbol{B}| = 10^4$ G?

In this approximate treatment, we have combined a classical description of the electron orbit with a quantum-mechanical description of spontaneous emission. This inconsistency is what causes the discrepancy between the exact and approximate results.

16.6. In this problem we consider possible mechanisms for spontaneous decay of the $2^2 s_{1/2}$ state in a hydrogenic atom (H, He$^+$, Li^{2+}, etc.). Recall that the $2^2 s_{1/2}$ state lies above the $2^2 p_{1/2}$

state by the Lamb shift, which in hydrogen is $S = 1{,}058$ MHz. The Lamb shift scales roughly as Z^4.

(a) Although angular momentum and parity selection rules do not prohibit a magnetic dipole transition $2^2 s_{1/2} \to 1^2 s_{1/2} + M1$ photon, the matrix element for this process vanishes in the nonrelativistic limit because of the orthogonality of the nonrelativistic $2s$ and $1s$ radial-wave functions. When relativity is taken into account (via the Dirac equation), the $M1$ matrix element no longer vanishes but is now proportional to $Z^2\alpha^2$. How does the $M1$ transition rate Γ_{M1} scale with Z?

(b) The $2^2 s_{1/2}$ state can decay by electric dipole emission to the $2^2 p_{1/2}$ state, but the transition rate Γ_{E1} is negligible because the Lamb shift is so small. Calculate Γ_{E1} explicitly, including its Z dependence and assuming that the Lamb shift scales as Z^4.

(c) The principal mode of decay of $2s$ in the absence of external electric fields is by two-photon emission to the ground $1s$ state. Each photon is $E1$, and the intermediate atomic states in this second-order process are all the discrete and continuum p states. The $2^2 p_{1/2}$ state gives only a very minor contribution compared with that of all the other p states combined. The rate for this process for hydrogenic atoms is

$$\Gamma_{2E1} \approx 8Z^6 \text{ s}^{-1}$$

Explain the Z^6 dependence. Note that for hydrogen, the $M1$ rate is much smaller; that is,

$$\Gamma_{M1}(H) \approx (2 \text{ days})^{-1}$$

(d) In the presence of a static external electric field $\boldsymbol{\mathcal{E}} = \mathcal{E}\hat{z}$, the lifetime of the $2s$ state is shortened by admixing of p states because of the Stark effect. Here the $2^2 p_{1/2}$ state plays a *dominant* role. Why? Neglecting other p states, let the atomic wave function for $m_j = 1/2$ be written

$$\psi(t) = a_{2s}(t)u_{2s}(\boldsymbol{r})\exp\left(\frac{-iE_{2s}t}{\hbar}\right) + a_{2p}(t)u_{2p}(\boldsymbol{r})\exp\left(\frac{-iE_{2p}t}{\hbar}\right)$$

where the coefficients a_{2s} and a_{2p} satisfy the coupled differential equations

$$\begin{aligned}\dot{a}_{2s} &= -\frac{i}{\hbar}a_{2p}\langle 2s|H'|2p\rangle\exp(i\omega_{sp}t)\\ \dot{a}_{2p} &= -\frac{i}{\hbar}a_{2s}\langle 2p|H'|2s\rangle\exp(-i\omega_{sp}t) - \frac{\gamma}{2}a_{2p}\end{aligned} \tag{1}$$

Here $\gamma = 6\times 10^8$ s^{-1} is the A-coefficient for decay of $2^2 p_{1/2}$, and we assume that $2s$ would have an infinitely long lifetime if the Stark coupling Hamiltonian H' were zero. Solve these equations for a weak electric field to show that the $2s$ state decays with the rate

$$\Gamma_{\text{Stark}} = \frac{3\gamma e^2 a_0^2 \mathcal{E}^2}{S^2 + \dfrac{\hbar^2\gamma^2}{4}} \tag{2}$$

What external electric field in volts per centimeter is required to give a decay rate by this mechanism that is comparable with the decay rate by two-photon emission? Your answer should reveal that the metastability (long lifetime) of the $2s$ state of hydrogen is severely compromised even by very weak external electric fields.

16.7. This problem concerns sum rules, which are very useful in the general analysis of radiation by atoms, molecules, nuclei, and so on. Before we proceed with the problem, we give an example of a sum rule. Define the x oscillator strength of a transition between two atomic states n and k as

$$f_{kn}^x = \frac{2m_e\omega_{kn}}{\hbar}\left|\langle k|x|n\rangle\right|^2$$

and similarly for y and z. Also define $f_{kn} = f_{kn}^x + f_{kn}^y + f_{kn}^z$. Then the single-electron oscillator strength sum rule (otherwise known as the *Thomas-Reich-Kuhn sum rule*) is

$$\sum_k f_{kn}^x = \sum_k f_{kn}^y = \sum_k f_{kn}^z = 1$$

where the sum is taken over all states k including the continuum, and n is a given state. For example, the $1s$-$2p$ x-oscillator strength for hydrogen is 0.4162 by explicit computation. Thus the sum of the x-oscillator strengths for transitions from the ground state to all other states including the continuum must be $1 - 0.4162 = 0.5838$. Now for the problem. Show that

(a) $\sum_k |x_{kn}|^2 = \langle n|x^2|n\rangle$

(b) $\sum_k f_{kn}^x = 1$

(c) $\sum_k \omega_{kn}^2 |x_{kn}|^2 = \frac{1}{m_e^2}\sum_k |p_{kn}^x|^2$

where p^x is the linear momentum along x.

(d) Show that if state n is spherically symmetric, (c) becomes

$$\sum_k \omega_{kn}^2 |x_{kn}|^2 = \frac{2}{3m_e}\left(E_n - \langle n|V|n\rangle\right)$$

(e) $\sum_k \omega_{kn}^3 |x_{kn}|^2 = \frac{\hbar}{2m_e^2}\left\langle n\left|\frac{\partial^2 V}{\partial x^2}\right|n\right\rangle$

(f) Show that if state n is spherically symmetric, (e) becomes

$$\sum_k \omega_{kn}^3 |x_{kn}|^2 = \frac{2\pi\hbar e^2}{3m_e^2}\langle n|\rho(r)|n\rangle$$

where ρ is the probability density.

16.8. Consider the experimental arrangement shown in Figure 16.3, where an ensemble of atoms with ${}^1S_0$ ground state is located in a small volume centered at the origin and is exposed to a beam of circularly polarized photons. These are absorbed in a transition to an excited ${}^1P_1$

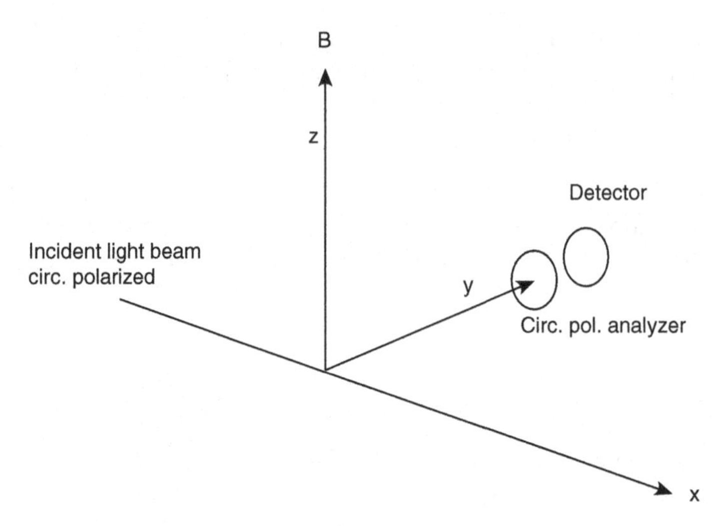

Figure 16.3 Experimental arrangement for observing the Hanle effect.

state. The entire experiment takes place in a magnetic field $\boldsymbol{B}$ in the z-direction. Because the light is circularly polarized, the excited atoms are initially polarized along the $\pm x$-direction. However, the polarization vector precesses with frequency ω (proportional to B) in the xy-plane because the excited atom has a magnetic moment. The excited atom also has a finite mean lifetime $\tau = \gamma^{-1}$ for spontaneous decay.
When the atom decays, it may emit a photon into the very small solid angle subtended by a detector that lies along the y-axis. We arrange the detector to contain an ideal circular polarization analyzer so that only positive-helicity photons are detected. Show that the detector registers a signal proportional to

$$I = \frac{3}{2\gamma} \pm \frac{2\omega}{\gamma^2 + \omega^2} - \frac{\gamma}{2(\gamma^2 + 4\omega^2)}$$

where we employ $\pm$ for positive (negative) helicity of incoming photons. This practical method for measuring the lifetimes of a considerable number of excited atomic states is called the *Hanle effect.* Analogous methods are used to study the decays of polarized muons in $g - 2$ experiments, the Garwin-Lederman experiment, and so on.

16.9. Consider an intense laser beam of very narrow bandwidth tuned to one-half the frequency corresponding to the separation between the $1^2 s_{1/2}$ and $2^2 s_{1/2}$ hyperfine components of atomic hydrogen. An experiment is carried out to excite the $2^2 s_{1/2}$ state by two-photon absorption starting from the $1^2 s_{1/2}$ state. Of the possible transitions

$$F = 0 \to F = 0$$
$$F = 0 \to F = 1$$
$$F = 1 \to F = 0$$
$$F = 1 \to F = 1$$

which are forbidden, and why?

17 Further Topics in Perturbation Theory

17.1 Lamb shift

Recall the Hamiltonian for charges, radiation, and their interaction; that is,

$$H = H_{\text{particle}} + H_{\text{radiation}} + H' \tag{17.1}$$

If we could somehow turn off H', the bound-state energies of a hydrogen atom would be given by the following Balmer formula:

$$E_{n\ell} = -\frac{1}{2n^2}\frac{e_0^4 m_0}{(4\pi)^2 \hbar^2} \tag{17.2}$$

where m_0 and e_0 are the bare mass and bare charge, respectively, of the electron in the make-believe world where there is no H'. Of course, there is no way to determine m_0 or e_0 because H' is always present. However, this does not mean that there are no observable manifestations of the shifts in electron mass and charge caused by H'.

Let us consider the *relative* energy shift of the $2^2 s_{1/2}$ and $2^2 p_{1/2}$ states, which are degenerated in the Pauli-Schroedinger theory and in the Dirac theory, prior to the introduction of H'. When H' is included, the degeneracy is removed: the $2^2 s_{1/2}$ level is shifted upward, the $2^2 p_{1/2}$ is shifted very slightly downward, and the measured splitting between them is the Lamb shift $S = 1{,}058$ MHz. The experimental discovery and careful measurement of this shift by W. E. Lamb and coworkers (Lamb and Retherford 1947; Lamb 1951) and theoretical efforts to explain it by H. Bethe, V. Weisskopf, R. Feynman, F. Dyson, J. Schwinger, and others in the years 1947–1951 were extremely important steps in the development of modern quantum electrodynamics. We now try to understand this shift in the simplest possible way. First of all, we assume something that is very nearly true: that the splitting S arises *entirely* from the upward shift of the $2^2 s_{1/2}$ state. This assumption means that if H' did not exist, the zeroth-order energy of $2s$ would be that of the physical $2p_{1/2}$ state. (By physical, we mean the energy of $2p_{1/2}$ calculated using the experimental values of m_e and e). Thus

$$\langle 2p_{1/2}|H_p(e,m_e)|2p_{1/2}\rangle = \langle 2s_{1/2}|H_p(e,m_e)|2s_{1/2}\rangle \tag{17.3}$$

where the subscript p is shorthand for particle. Thus, given our assumption, the Lamb shift is

$$\begin{aligned} S &= E_{2s}^{\text{obs}} - \langle 2s|H_p(e,m_e)|2s\rangle \\ &= \left[E_{2s}^{\text{obs}} - \langle 2s|H_p(e_0,m_0)|2s\rangle\right] - \left[\langle 2s|H_p(e,m_e)|2s\rangle - \langle 2s|H_p(e_0,m_0)|2s\rangle\right] \end{aligned} \tag{17.4}$$

Now it can be shown that only a very small fraction of the Lamb shift arises from the difference between e and e_0 (charge renormalization). The great bulk of the effect is due to the difference between m and m_0. Thus, ignoring the effects of charge renormalization, we simplify our calculation by replacing (17.4) with

$$S \approx \left[E_{2s}^{\text{obs}} - \langle 2s|H_p(e,m_0)|2s\rangle\right] - \left[\langle 2s|H_p(e,m_e)|2s\rangle - \langle 2s|H_p(e,m_0)|2s\rangle\right] \tag{17.5}$$

However,

$$\langle 2s|H_p(e,m_e)|2s\rangle - \langle 2s|H_p(e,m_0)|2s\rangle \approx \left\langle 2s\left|\frac{p^2}{2m} - \frac{p^2}{2m_0}\right|2s\right\rangle \tag{17.6}$$

because the potential energy $V = -e^2/4\pi r$ does not contain the mass, and the remaining small contributions to H_p (spin-orbit effect, etc) are of negligible significance in the present problem. The right-hand side of (17.6) is the energy shift ΔE^{free} of a nonrelativistic free electron averaged over the $2s$ state. Therefore, (17.5) can be written

$$S = \Delta E_{2s} - \Delta E_{2s}^{\text{free}} \tag{17.7}$$

where $\Delta E_{2s} = E_{2s}^{\text{obs}} - \langle 2s|H_p(e,m_0)|2s\rangle$.

Each of the energy shifts ΔE_{2s} and $\Delta E_{2s}^{\text{free}}$ is obtained in lowest approximation by employing the $\boldsymbol{A}\bullet\boldsymbol{p}$ term in second-order of static perturbation theory. Such an approach would appear to be valid only if the energy shift in each case is quite small. When we calculate these shifts using a nonrelativistic description of the electron and with the long-wavelength approximation, we find that each shift is extremely large. Indeed, each diverges linearly when we integrate over all possible intermediate photon energies. However, as we shall see, in the difference $\Delta E_{2s} - \Delta E_{2s}^{\text{free}}$, the linear divergences cancel, and only a much milder logarithmic divergence remains. If the nonrelativistic approximation is replaced by a proper relativistic analysis, each of the shifts ΔE_{2s} and $\Delta E_{2s}^{\text{free}}$ diverges only logarithmically, and $\Delta E_{2s} - \Delta E_{2s}^{\text{free}}$ is finite. Unfortunately, the relativistic treatment is complicated and beyond the scope of this book. Thus we content ourselves with the nonrelativistic calculation (Bethe 1947). It is much simpler and reveals the main features of the effect.

We begin by calculating ΔE_{2s}. According to second-order perturbation theory, it is given by

$$\Delta E_{2s} = \sum_n \frac{H'_{in}H'_{ni}}{E_i - E_n} \tag{17.8}$$

and this is represented by the diagram of Figure 17.1. Here the state $|i\rangle$ is the $2s$ state of hydrogen with no photons present, whereas the states $|n\rangle$ are all hydrogenic states that can be connected to the $2s$ state by the operator $\boldsymbol{p}$ [all discrete and continuum $\ell = 1$ (p) states] and all states of the radiation field with one photon. Thus (17.8) can be written

$$\Delta E_{2s} = \left(\frac{e}{m_e c}\right)^2 \frac{\hbar c^2}{2L^3} \sum_{n_a} \int \frac{1}{\omega} \frac{L^3\omega^2}{8\pi^3\hbar c^3} \frac{\left|\langle n_a|e^{-i\boldsymbol{k}\cdot\boldsymbol{r}}\hat{\boldsymbol{\varepsilon}}\cdot\boldsymbol{p}|i_a\rangle\right|^2}{E_i^a - E_n^a - \varepsilon}\, d\varepsilon\, d\Omega \tag{17.9}$$

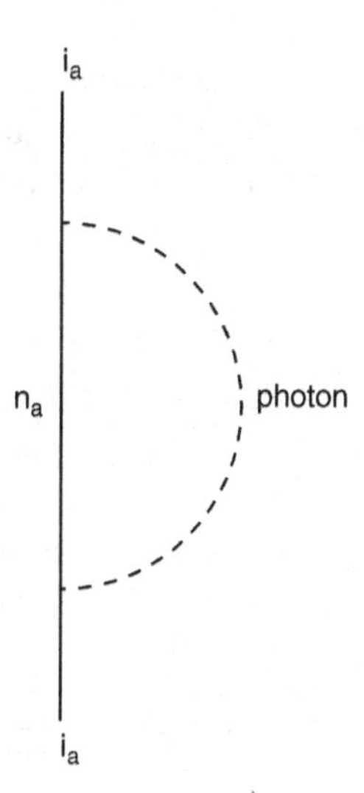

Figure 17.1 Lowest order self-energy diagram for calculation of Lamb shift

where $\varepsilon = \hbar\omega$ is the energy of the "virtual" photon, and $\hat{\varepsilon}$ is its polarization. If we replace $\exp(-i\boldsymbol{k}\cdot\boldsymbol{r})$ by unity (the long-wavelength approximation), then (17.9) becomes

$$\Delta E_{2s} = \frac{1}{4\pi^2}\frac{\alpha}{(m_e c)^2}\sum_{n_a}\int \frac{\varepsilon}{E_i^a - E_n^a - \varepsilon}\left|\langle n_a|\hat{\varepsilon}\cdot\boldsymbol{p}|i_a\rangle\right|^2 d\varepsilon d\Omega \tag{17.10}$$

For any given n_a, we can integrate over all photon solid angles by choosing the vector $\boldsymbol{p}_{ni} \equiv \langle n_a|\boldsymbol{p}|i_a\rangle$ to lie along a polar axis, in which case

$$\int |\hat{\varepsilon}_z|^2\, d\Omega = 2\pi\int_0^\pi \sin^3\theta\, d\theta = \frac{8\pi}{3}$$

Hence (17.10) becomes

$$\Delta E_{2s} = \frac{2}{3\pi}\frac{\alpha}{(m_e c)^2}\sum_{n_a}\int \frac{\varepsilon}{E_i^a - E_n^a - \varepsilon}\boldsymbol{p}_{in}\cdot\boldsymbol{p}_{ni}\, d\varepsilon \tag{17.11}$$

It is easy to see that the integral in (17.11) diverges linearly with ε. However, although we have assumed the long-wavelength approximation in the interest of simplicity, it breaks down at energies of order $\varepsilon \approx \alpha m_e c^2 \approx 3{,}700$ eV, where the photon wavelength is comparable with a_0. For larger photon energies, the factor $\exp(-i\boldsymbol{k}\cdot\boldsymbol{r})$ oscillates very rapidly, and $\langle n_a|e^{-i\boldsymbol{k}\cdot\boldsymbol{r}}\hat{\varepsilon}\cdot\boldsymbol{p}|i\rangle$ goes to zero. Therefore, it is very sensible to cut off the divergent integral in (17.11) by assuming a superior limit of

$$\varepsilon_{\max} = m_e c^2 = 5\times 10^5 \text{ eV}$$

We now want to turn our attention to ΔE^{free}, but first let us consider the energy shift of a free nonrelativistic electron of fixed momentum $\boldsymbol{p}$ due to H'. Once again, we use static perturbation theory with the $\boldsymbol{A}\bullet\boldsymbol{p}$ term in second order. The matrix element H'_{ni} is

$$\begin{aligned} H'_{ni} &= \frac{1}{L^{3/2}}\frac{e}{m_e c}\sqrt{\frac{\hbar c^2}{2\omega}}\frac{1}{L^3}\int e^{-i\boldsymbol{p}'\bullet\boldsymbol{r}/\hbar}e^{-i\boldsymbol{k}\bullet\boldsymbol{r}}\,\hat{\boldsymbol{\varepsilon}}\bullet\boldsymbol{p}e^{i\boldsymbol{p}\bullet\boldsymbol{r}/\hbar}d^3\boldsymbol{r} \\ &= \frac{1}{L^{3/2}}\frac{e}{m_e c}\sqrt{\frac{\hbar c^2}{2\omega}}\delta_{\boldsymbol{p}',\boldsymbol{p}-\hbar\boldsymbol{k}}\,\hat{\boldsymbol{\varepsilon}}\bullet\boldsymbol{p} \end{aligned} \tag{17.12}$$

where $\boldsymbol{p}'$ is the momentum of the electron in the intermediate state. Employing steps similar to those used in going from (17.9) to (17.11), we obtain

$$\Delta E(\text{free},\boldsymbol{p}) = \frac{2}{3\pi}\frac{\alpha}{(m_e c)^2}p^2\int\frac{\varepsilon\, d\varepsilon}{\dfrac{p^2}{2m_e}-\dfrac{(\boldsymbol{p}-\hbar\boldsymbol{k})^2}{2m_e}-\varepsilon} \tag{17.13}$$

Now

$$\frac{p^2}{2m_e}-\frac{(\boldsymbol{p}-\hbar\boldsymbol{k})^2}{2m_e}-\varepsilon = \frac{\hbar}{m_e}\boldsymbol{p}\bullet\boldsymbol{k}-\frac{\hbar^2k^2}{2m_e}-\varepsilon$$

The first term on the right-hand side of this equation is approximately $\varepsilon(p/m_e c)$, which is very small compared with ε because the momentum of a nonrelativistic electron is small compared with $m_e c$. The second term on the right-hand side has magnitude $\varepsilon(\varepsilon/m_e c^2)$, which is much smaller than ε if $\varepsilon \ll m_e c^2$. Hence it is a good approximation to replace the entire denominator in the integrand of (17.13) by $-\varepsilon$, in which case we obtain

$$\Delta E(\mathit{free},\boldsymbol{p}) = -\frac{2}{3\pi}\frac{\alpha}{(m_e c)^2}p^2\int d\varepsilon = -\frac{2}{3\pi}\frac{\alpha}{(m_e c)^2}p^2\varepsilon_{\max} \tag{17.14}$$

Next, we average $\Delta E(\text{free},\boldsymbol{p})$ over the 2s state, thereby obtaining

$$\Delta E^{\text{free}} = -\frac{2}{3\pi}\frac{\alpha}{(m_e c)^2}\sum_{n_a}\boldsymbol{p}_{in}\bullet\boldsymbol{p}_{ni}\int d\varepsilon \tag{17.15}$$

Now, taking the difference between (17.11) and (17.15), we obtain

$$\begin{aligned} S &= \frac{2\alpha}{3\pi}\frac{1}{(m_e c)^2}\sum_{n_a}\int\left(\frac{\varepsilon}{E^a_{in}-\varepsilon}+\frac{E^a_{in}-\varepsilon}{E^a_{in}-\varepsilon}\right)(\boldsymbol{p}_{in}\bullet\boldsymbol{p}_{ni})\,d\varepsilon \\ &= \frac{2\alpha}{3\pi}\frac{1}{(m_e c)^2}\sum_{n_a}\int\left(\frac{E^a_{in}}{E^a_{in}-\varepsilon}\right)(\boldsymbol{p}_{in}\bullet\boldsymbol{p}_{ni})\,d\varepsilon \\ &= \frac{2\alpha}{3\pi}\frac{1}{(m_e c)^2}\sum_{n_a}E^a_{ni}(\boldsymbol{p}_{in}\bullet\boldsymbol{p}_{ni})\,\ell n\left(\frac{E^a_{ni}+\varepsilon_{\max}}{E^a_{ni}}\right) \end{aligned} \tag{17.16}$$

Because $\varepsilon_{\max} >> E^a_{ni}$ for all p states that contribute anything appreciable, we replace (17.16) with

$$S = \frac{2\alpha}{3\pi}\frac{1}{(m_e c)^2}\sum_{n_a} E^a_{ni}\left(\boldsymbol{p}_{in}\bullet\boldsymbol{p}_{ni}\right)\ell n\left(\frac{\varepsilon_{\max}}{E^a_{ni}}\right) \tag{17.17}$$

The logarithmic factor in the sum of (17.17) varies extremely slowly with n. Hence it is convenient to define

$$\ell n\left\langle E^a_{ni}\right\rangle \equiv \frac{\sum E^a_{ni}\ell n E^a_{ni}\left(\boldsymbol{p}_{in}\bullet\boldsymbol{p}_{ni}\right)}{\sum E^a_{ni}\left(\boldsymbol{p}_{in}\bullet\boldsymbol{p}_{ni}\right)} \tag{17.18}$$

The quantity $\left\langle E^a_{ni}\right\rangle$ is evaluated numerically by explicit calculation from hydrogenic wave functions, and it turns out to be $8.32Z^2e^2/(4\pi a_0)$ virtually independent of whether i refers to the $1s$ state, the $2s$ state, and so on. Accepting this, we concentrate on the sum in the denominator of the right-hand side of (17.18). For the $2s$ state, it is

$$\begin{aligned}\sum\left\langle 2s\right|\left[\boldsymbol{p},H_a\right]\left|n_a\right\rangle\bullet\left\langle n_a\right|\boldsymbol{p}\left|2s\right\rangle &= -\sum\left\langle 2s\right|\boldsymbol{p}\left|n_a\right\rangle\bullet\left\langle n_a\right|\left[\boldsymbol{p},H_a\right]\left|2s\right\rangle \\ &= -\frac{1}{2}\left\langle 2s\right|\left[\boldsymbol{p}\bullet\left[\boldsymbol{p},H_a\right]\right]\left|2s\right\rangle\end{aligned} \tag{17.19}$$

Employing $[\boldsymbol{p},H_a] = -i\hbar\nabla V$, we find for any hydrogenic state $|\psi\rangle$

$$S_\psi = \frac{\alpha}{3\pi}\frac{\hbar^2}{(m_e c)^2}\left\langle\psi\right|\nabla^2 V\left|\psi\right\rangle \ell n\frac{\varepsilon_{\max}}{\left\langle E^a_{ni}\right\rangle}$$

Now $\nabla^2 V = Ze^2\delta^3(\boldsymbol{r})$ for a hydrogenic atom of nuclear charge Ze. Thus

$$S_\psi = \frac{\alpha}{3\pi}\frac{\hbar^2}{(m_e c)^2}Ze^2\left|\psi(0)\right|^2 \ell n\frac{\varepsilon_{\max}}{\left\langle E^a_{ni}\right\rangle} \tag{17.20}$$

The wave function vanishes at the origin unless ψ is an s state, which explains why there is no shift for the $2p_{1/2}$ state in the nonrelativistic approximation. For the ns state,

$$\left|\psi_{ns}(0)\right|^2 = \frac{Z^3}{\pi n^3 a_0^3} \tag{17.21}$$

Therefore, we obtain

$$S_{ns} = \frac{4\alpha^3}{3\pi}\frac{Z^4}{n^3}\frac{e^2}{4\pi a_0}\ell n\frac{\varepsilon_{\max}}{\left\langle E^a_{ni}\right\rangle} \tag{17.22}$$

As mentioned previously, it is plausible to assume that $\varepsilon_{\max} = m_e c^2$, and this is verified by the complete relativistic calculation. Employing this and the numerical value of $\left\langle E^a_{ni}\right\rangle$ already mentioned and inserting $n = 2$, one obtains from (17.22)

$$S_{2s}(H) = 1{,}040 \text{ MHz} \tag{17.23}$$

which is to be compared with the experimental value

$$S_{2s}(H, \text{obs}) = 1{,}058 \text{ MHz} \tag{17.24}$$

The discrepancy between (17.23) and (17.24) is removed when higher-order perturbations, relativistic corrections, the small shift of the $2p_{1/2}$ state, the vacuum polarization (charge renormalization) effect, and other small effects are taken into account. In the years since Lamb's historic achievement, the Lamb shift has been measured accurately in the ground state of hydrogen, in the $n = 2$ states not only of hydrogen and deuterium but also of He^+ and hydrogenic ions with large Z, and in several other systems and states. Agreement with sophisticated theoretical calculations is excellent.

We have sketched the main features of Bethe's nonrelativistic calculation of the Lamb shift. Whereas neglect of retardation and the apparently rather arbitrary choice of $\varepsilon_{\max} = m_e c^2$ are questionable features, this calculation is basically correct, and it shows that the Lamb shift is essentially a low-energy nonrelativistic phenomenon.

17.2 Adiabatic approximation: The geometric phase

Consider a physical system in initial state $|n\rangle$ where the spacing between energy levels n and k is $E_n - E_k = \hbar\omega_{nk}$. Suppose that this system is exposed to a perturbing Hamiltonian $H'(t)$ that is a function of t. Also suppose that H' varies slowly with t, by which we mean that H' contains only those Fourier components with frequencies ω such that $\omega \ll \omega_{nk}$ for given n and all possible $k \neq n$. Then we expect intuitively that H' will not induce transitions between states $|n\rangle$ and $|k\rangle$. Instead, the system should remain in initial state $|n\rangle$, which is now a function of t, as is its corresponding energy $E_n(t)$. For example, consider a particle with spin and an associated spin magnetic moment in an external magnetic field. We know that if the field varies rapidly, transitions can occur between various magnetic quantum-number (m) levels. However, if the magnetic field varies slowly (in magnitude and/or direction), then we expect that transitions between various m states should not occur. Instead, the spin direction should follow the direction of the magnetic field adiabatically. [Here *adiabatic* simply means that transitions out of state $|n(t)\rangle$ do not occur.]

We now try to make these ideas more precise by starting with the time-dependent Schroedinger equation; that is,

$$H(t)|\Psi(t)\rangle = i\hbar\frac{\partial}{\partial t}|\Psi(t)\rangle \tag{17.25}$$

with

$$|\Psi(t)\rangle = \sum_r a_r(t)|r(t)\rangle \exp\left[-\frac{i}{\hbar}\int_0^t E_r(t')\,dt'\right] \tag{17.26}$$

where the $|r(t)\rangle$, which satisfy

$$H(t)|r(t)\rangle = E_r(t)|r(t)\rangle \tag{17.27}$$

are assumed to be orthonormal and, for simplicity, nondegenerate as well as discrete. Substitution of (17.26) into (17.25) with use of (17.27) yields the result

$$\sum_r (\dot{a}_r|r\rangle + a_r|\dot{r}\rangle)\exp\left[-\frac{i}{\hbar}\int_0^t E_r(t')\,dt'\right] = 0$$

We multiply this equation on the left by $\langle k|$ to obtain

$$\dot{a}_k = -\sum_r a_r(t)\langle k|\dot{r}\rangle \exp\left[-\frac{i}{\hbar}\int_0^t (E_r - E_k)\,dt'\right] \tag{17.28}$$

To obtain a useful expression for $\langle k|\dot{r}\rangle$ with $k \neq r$, we differentiate both sides of (17.27) with respect to t

$$\frac{\partial H}{\partial t}|r\rangle + H(t)|\dot{r}\rangle = \dot{E}_r|r\rangle + E_r|\dot{r}\rangle$$

and multiply this equation on the left by $\langle k|$. Thus we obtain

$$\langle k|\dot{r}\rangle = \frac{\left\langle k\left|\frac{\partial H}{\partial t}\right|r\right\rangle}{E_r - E_k} \qquad k \neq r \tag{17.29}$$

and therefore (17.28) becomes

$$\dot{a}_k = -a_k\langle k|\dot{k}\rangle - \sum_{r\neq k} a_r(t)\frac{\left\langle k\left|\frac{\partial H}{\partial t}\right|r\right\rangle}{E_r - E_k}\exp\left[-\frac{i}{\hbar}\int_0^t (E_r - E_k)\,dt'\right] \tag{17.30}$$

We now obtain an estimate of $\dot{a}_k$ by assuming that all quantities on the right-hand side of (17.30) (which have already been assumed to vary only slowly with time) are now assumed to be constant in time. (This is the adiabatic approximation.) It includes the replacement of all the a_r including a_k by their initial values

$$a_r \to \delta_{rn}$$

Hence we obtain

$$\dot{a}_k \approx -\frac{\left\langle k\left|\frac{\partial H}{\partial t}\right|n\right\rangle}{\hbar\omega_{nk}}\exp(-i\omega_{nk}t) \qquad k \neq n$$

Integration of both sides of this equation yields

$$a_k \approx \frac{e^{i\omega_{kn}t}-1}{i\omega_{kn}}\left(\frac{\left\langle k\left|\frac{\partial H}{\partial t}\right|n\right\rangle}{\hbar\omega_{kn}}\right)$$

$$\approx e^{i\omega_{kn}t/2}\,\frac{e^{i\omega_{kn}t/2}-e^{-i\omega_{kn}t/2}}{i\omega_{kn}}\left(\frac{\left\langle k\left|\frac{\partial H}{\partial t}\right|n\right\rangle}{\hbar\omega_{kn}}\right)$$

$$\approx e^{i\omega_{kn}t/2}\,\frac{\sin\left(\frac{\omega_{kn}t}{2}\right)}{\left(\frac{\omega_{kn}t}{2}\right)}\left\{\frac{\langle k|[H(t)-H(0)]|n\rangle}{\hbar\omega_{kn}}\right\}$$

We see that in the adiabatic approximation, a_k oscillates with an amplitude roughly equal to the change in the matrix element of the Hamiltonian over time interval $\Delta t \approx \pi/\omega_{kn}$ divided by the energy difference $E_k - E_n$. For a sufficiently slowly varying Hamiltonian, the amplitude of a_k does indeed become extremely small.

Now consider once again the example of a particle with a spin magnetic moment in a slowly varying magnetic field. Suppose that the tip of the magnetic field vector traces a closed path in space in time T subtending a certain solid angle Ω with respect to its origin. We now show that the quantum state describing the particle (which has magnetic quantum number m) acquires a corresponding phase increment; that is,

$$\gamma = -m\Omega \tag{17.31}$$

This is called the *geometric phase* or *Berry's phase* after the British physicist Michael Berry, who made substantial contributions to this topic several decades ago. The geometric phase is wholly distinct from the usual dynamical phase.

Actually, the geometric phase appears in many and diverse physical situations and is not limited to a spin magnetic moment in a magnetic field, nor is it even limited to quantum mechanics. (For example, it can be shown that the slow precession of the plane of oscillation of a Foucault pendulum in classical mechanics involves a geometric phase.) Hence, in the discussion that follows, we take a general point of view and consider a Hamiltonian that depends on a number of parameters $R_1(t), R_2(t), \ldots$. It is convenient to think of these parameters as defining a vector $\boldsymbol{R}(t)$ in parameter space. Next we consider the eigenstates of H and label them with the symbol m; that is,

$$H[\boldsymbol{R}(t)]|m\rangle = E_m|m\rangle \tag{17.32}$$

Of course, the eigenstates $|m\rangle$ and associated eigenvalues E_m as well as the Hamiltonian are functions of $\boldsymbol{R}(t)$. Suppose that at $t = 0$ the system of interest is in one such state $|m\rangle$. Then the state at some later time can be expressed as

$$|\psi\rangle = |m\rangle \exp\left(-\frac{i}{\hbar}\int_0^t E_m\, dt\right) e^{i\gamma} \tag{17.33}$$

where the second and third factors on the right-hand side contain the dynamical and geometric phases, respectively. To find γ, we once again make use of the time-dependent Schroedinger equation as follows:

$$H|\psi\rangle = i\hbar|\dot{\psi}\rangle$$

or

$$E_m|m\rangle \exp\left(-\frac{i}{\hbar}\int_0^t E_m dt\right) e^{i\gamma} = i\hbar\left(\frac{\partial|m\rangle}{\partial t} - \frac{i}{\hbar}E_m|m\rangle + i\dot{\gamma}|m\rangle\right)\exp\left(-\frac{i}{\hbar}\int_0^t E_m\, dt\right) e^{i\gamma}$$

which yields

$$\dot{\gamma}|m\rangle = i|\dot{m}\rangle \tag{17.34}$$

We take the scalar product of both sides of (17.34) with $\langle m|$ and use $\langle m|m\rangle = 1$ to obtain

$$\dot{\gamma} = i\langle m|\dot{m}\rangle \tag{17.35}$$

Note that because $\langle m|m\rangle = 1$, $\langle m|\dot{m}\rangle = \langle \dot{m}|m\rangle^* = -\langle \dot{m}|m\rangle$; hence $\langle m|\dot{m}\rangle$ is imaginary and therefore γ is real. Also, because $|m\rangle$ depends on time through its dependence on $\boldsymbol{R}(t)$, we can write (17.35) as

$$\dot{\gamma} = -\mathrm{Im}\langle m|\nabla m\rangle \cdot \dot{\boldsymbol{R}} \tag{17.36}$$

where by $|\nabla m\rangle$ we mean the vector in parameter space with components

$$\frac{\partial|m\rangle}{\partial R_1}, \frac{\partial|m\rangle}{\partial R_2}, \ldots$$

Now suppose that $\boldsymbol{R}$ traces a closed curve in parameter space, arriving back at its original value at time T. Then, from (17.36), the increment in γ is

$$\gamma = -\mathrm{Im}\oint \langle m|\nabla m\rangle \cdot d\boldsymbol{R} \tag{17.37}$$

In three-dimensional parameter space, this loop integral is converted into a surface integral by Stokes' theorem; that is,

$$\gamma = -\mathrm{Im}\int \nabla \times \langle m|\nabla m\rangle \cdot d\boldsymbol{\sigma} \tag{17.38}$$

where, as usual, the latter integral is taken over a surface bounded by the closed curve traced by $\boldsymbol{R}$. Because

$$\nabla \times \langle m|\nabla m\rangle = \langle \nabla m|\times|\nabla m\rangle + \langle m|\nabla \times \nabla m\rangle$$
$$= \langle \nabla m|\times|\nabla m\rangle$$

(17.36) becomes:

$$\gamma = -\operatorname{Im}\int \langle \nabla m|\times|\nabla m\rangle \cdot d\boldsymbol{\sigma} \tag{17.39}$$

We now use the completeness relation for the eigenstates of the Hamiltonian to write

$$\gamma = -\operatorname{Im}\sum_{M}\int \left(\langle \nabla m|M\rangle \times \langle M|\nabla m\rangle\right)\cdot d\boldsymbol{\sigma} \tag{17.40}$$

We remind ourselves that M refers to any of the eigenstates of the Hamiltonian, whereas m refers to a particular eigenstate. When $M = m$, the integrand on the right-hand side of (17.40) vanishes because $\langle \nabla m|m\rangle = -\langle m|\nabla m\rangle$. Therefore, (17.40) can be written

$$\gamma = -\operatorname{Im}\sum_{M\neq m}\int \left(\langle \nabla m|M\rangle \times \langle M|\nabla m\rangle\right)\cdot d\boldsymbol{\sigma} \tag{17.41}$$

Now we develop a useful alternative form for the integrand in (17.41). Let us take the gradient of both sides of $H|m\rangle = E_m|m\rangle$. This yields

$$\nabla H|m\rangle + H|\nabla m\rangle = \nabla E_m|m\rangle + E_m|\nabla m\rangle \tag{17.42}$$

We take the scalar product of both sides of (17.42) with $\langle M| \neq \langle m|$; that is,

$$\langle M|\nabla H|m\rangle + \langle M|H|\nabla m\rangle = \nabla E_m\langle M|m\rangle + E_m\langle M|\nabla m\rangle$$

or, because $\langle M|m\rangle = 0$,

$$\langle M|\nabla m\rangle = \frac{\langle M|\nabla H|m\rangle}{E_m - E_M} \tag{17.43}$$

Therefore, (17.41) can be written as

$$\gamma = -\operatorname{Im}\sum_{M\neq m}\int \frac{\langle m|\nabla H|M\rangle \times \langle M|\nabla H|m\rangle}{(E_m - E_M)^2}\cdot d\boldsymbol{\sigma} \tag{17.44}$$

At this point we return to the case of a particle with spin magnetic moment $k\boldsymbol{S}$ (where k is a constant and $\boldsymbol{S}$ is the spin operator) in a magnetic field $\boldsymbol{B}$. Here the Hamiltonian is

$$H = -k\boldsymbol{S}\cdot\boldsymbol{B}$$

hence $\nabla H = -k\boldsymbol{S}$. Now, because $\boldsymbol{S}$ is a first-rank tensor, the sum in (17.44) is restricted to terms for which $M = m \pm 1$, and in each such term the denominator of the integrand in (17.44) is

$$\left(E_m - E_M\right)^2 = k^2 B^2$$

Thus, contracting the sum over states in (17.44), we obtain

$$\gamma = -\operatorname{Im}\int \frac{k^2 \langle m|\boldsymbol{S}\times\boldsymbol{S}|m\rangle}{k^2 B^2} \cdot d\boldsymbol{\sigma} \tag{17.45}$$

We recall from the commutation relations for angular momentum operators that $\boldsymbol{S}\times\boldsymbol{S}$ is not zero; in fact,

$$S_x S_y - S_y S_x = i S_z$$

Thus, choosing a local coordinate system so that the magnetic field is oriented along the local z-axis at each point in space, which is also normal to $d\boldsymbol{\sigma}$, and noting that $S_z|m\rangle = m|m\rangle$, we see that (17.45) yields

$$\gamma = -m\int \frac{d\sigma_z}{B^2} = -m\Omega \tag{17.46}$$

Equations (17.44) and (17.46) are the main results of interest. Referring to (17.46), let the magnetic field remain constant in magnitude, but let it describe a cone with opening half-angle θ. Then the solid angle subtended after one cycle is $\Omega = 2\pi(1-\cos\theta)$, and the corresponding increment in geometric phase is

$$\Delta\gamma = -2\pi m\left(1-\cos\theta\right) \tag{17.47}$$

17.3 Sudden approximation

In the preceding section we considered Hamiltonians that depend on external parameters that vary very slowly. By this we mean that if the time dependence of the Hamiltonian is Fourier analyzed, the component frequencies are all much smaller than the frequencies corresponding to energy differences between eigenstates of the Hamiltonian. In this adiabatic limit, no transitions occur between the various eigenstates.

Now we consider the opposite extreme, where the Hamiltonian H changes rapidly over a very short period of time. During this small time interval, the wave function ψ cannot change very much, for otherwise its time derivative would be enormous in magnitude; hence, from the time-dependent Schroedinger equation, the same would be true of $H\psi$. In fact, in the limit of a discontinuous change in H at time $t = 0$, the wave function immediately before $t = 0$ must be equal to the wave function immediately after $t = 0$: $\psi(0-) = \psi(0+)$.

Suppose that the initial Hamiltonian is H with eigenstates $|u_n\rangle$ and corresponding energies E_n, and let us assume that prior to $t = 0$, the system of interest is in one such state $|u_n\rangle$. After $t = 0$, let the Hamiltonian be H', with eigenstates $|w_n\rangle$ and corresponding eigenvalues F_n. Then, immediately after $t = 0$, the wave function is still $|u_n\rangle$, but this is not an eigenstate of H'. Instead, it is a superposition of the eigenstates of H'; that is,

$$|\psi(0+)\rangle = |u_n\rangle = \sum_m |w_m\rangle\langle w_m|u_n\rangle \tag{17.48}$$

Thus the probability p_m of finding the system in the state $|w_m\rangle$ at time $t = 0+$ is

$$p_m = |\langle w_m|u_n\rangle|^2 \tag{17.49}$$

For example, consider an atom with a beta-radioactive nucleus of atomic number Z. Suppose that beta decay occurs at $t = 0$. The beta ray (electron) and antineutrino depart rapidly from the atom in a time that is very short compared with the periods of atomic electron motion. Thus, from the point of view of an atomic electron, all that happens is a sudden change of the nuclear charge from Ze to $(Z+1)e$. In the new nuclear Coulomb field, the atomic electron wave function is the same immediately after the decay as it was before, but, of course, the new Hamiltonian is different.

17.4 Time-dependent perturbation theory and elementary theory of beta decay

Nuclear beta decay was the first observed example of a weak interaction, and it has a long and interesting history. The first serious and successful attempt to describe beta decay theoretically was made by E. Fermi in 1934, and his work is the basis of all further theoretical developments in the study of weak interactions. For the moment, we consider beta decay as an elementary application of the golden rule of time-dependent perturbation theory.

The simplest beta decay is that of the free neutron

$$n \to pe^-\bar{\nu}_e$$

Here a neutron at rest decays to a proton, an electron, and an antineutrino. The final state consists of three particles, but because the proton is very massive compared with the electron and antineutrino, the latter two particles carry off essentially all the released energy. This can be seen as follows: by conservation of linear momentum,

$$\mathbf{p}_p + \mathbf{p}_e + \mathbf{p}_{\bar{\nu}} = 0$$

in the neutron rest frame. Hence the proton linear momentum must be comparable with that of the electron. Thus the proton kinetic energy is approximately

$$\mathrm{KE}_p \approx \frac{p_e^2}{2m_p} \tag{17.50}$$

However, in neutron decay, as in most other beta decays, the electron kinetic energy is of order 1 MeV, comparable in order of magnitude with the electron rest energy $m_e c^2$. Thus $p_e \approx m_e c$, and (17.50) becomes

$$\mathrm{KE}_p \approx \frac{(m_e c)^2}{2m_p} \approx \frac{m_e}{m_p} m_e c^2 \tag{17.51}$$

This shows that the proton kinetic energy is of order $m_e/m_p \approx 10^{-3}$ compared with the electron (or antineutrino) kinetic energy, which are themselves roughly comparable. Therefore, it is a good approximation to ignore the proton kinetic energy entirely, in which case we need only consider the electron and neutrino motions in the final state. From the golden rule, the differential transition probability per unit time is

$$dW = \frac{2\pi}{\hbar}\left|H'_{fi}\right|^2 \left[\frac{d^3 \boldsymbol{p}_e}{(2\pi\hbar)^3}\frac{d^3 \boldsymbol{p}_{\bar{\nu}}}{(2\pi\hbar)^3}\delta(\Delta - E_e - E_{\bar{\nu}})\right] \tag{17.52}$$

Here the first two factors in the square brackets refer to the electron and antineutrino densities of states, whereas the delta function accounts for conservation of energy. Also, $\Delta = 1.3$ MeV is the difference in rest energy between neutron and proton. To a good approximation, the matrix element is just a constant. Hence, apart from constant factors,

$$dW = p_e^2 dp_e p_{\bar{\nu}}^2 dp_{\bar{\nu}} \delta(\Delta - E_e - E_{\bar{\nu}}) \tag{17.53}$$

Furthermore, because in units where $c = 1$,

$$p_e^2 dp_e = \sqrt{E_e^2 - m_e^2}\, E_e dE_e$$

$$p_{\bar{\nu}}^2 dp_{\bar{\nu}} = E_{\bar{\nu}}^2 dE_{\bar{\nu}}$$

where in the last expression we ignore a very minute neutrino mass, we have

$$dW = \sqrt{E_e^2 - m_e^2}\, E_e dE_e E_{\bar{\nu}}^2 dE_{\bar{\nu}} \delta(\Delta - E_e - E_{\bar{\nu}}) \tag{17.54}$$

In almost all beta decay experiments, one does not observe the outgoing neutrino. Thus, to obtain an expression that can be compared with experiment, we integrate (17.54) over neutrino energies, thereby eliminating the delta function to obtain

$$dW = \left[\sqrt{E_e^2 - m_e^2}\, E_e (\Delta - E_e)^2\right] dE_e \tag{17.55}$$

The factor in square brackets in (17.55) is the electron energy spectrum function (apart from a constant factor and uncorrected for "final state" Coulomb interaction with the proton).

Problems for Chapter 17

17.1. In this problem we calculate the geometric phase γ for a simple harmonic oscillator with the Hamiltonian

$$H = \frac{\hbar\omega}{2}\left[\left(P - \sqrt{2}x_2\right)^2 + \left(Q - \sqrt{2}x_1\right)^2\right] \tag{1}$$

where Q and P are the operators for position and momentum satisfying $[Q, P] = i$, and x_1 and x_2 are slowly varying real parameters that, over the course of a cycle, specify a closed curve in the x_1x_2 plane.

(a) Making use of the identity

$$e^A B e^{-A} = B + [A, B] + \frac{1}{2!}[A,[A, B]] + \cdots$$

show that

$$H = D(\alpha) H_0 D^\dagger(\alpha) = \hbar\omega\left[(a^\dagger - \alpha^*)(a - \alpha) + \frac{1}{2}\right] \tag{2}$$

where

$$D(\alpha) = \exp[\alpha a^\dagger - \alpha^* a] \tag{3}$$

$$\begin{aligned} &a = \frac{1}{\sqrt{2}}(Q + iP) \qquad a^\dagger = \frac{1}{\sqrt{2}}(Q - iP) \qquad \alpha = (x_1 + ix_2) \\ &H_0 = \hbar\omega\left(a^\dagger a + \frac{1}{2}\right) \end{aligned} \tag{4}$$

The eigenvalues and eigenstates of H_0 are, of course, $\hbar\omega(n + \frac{1}{2})$ and $|n\rangle$, respectively. The corresponding eigenstates of H are the coherent states $|n,\alpha\rangle = D(\alpha)|n\rangle$.

(b) According to (17.37), the geometric phase for coherent state $|n,\alpha\rangle$ is given by the formula

$$\begin{aligned} \gamma_n &= -\operatorname{Im}\oint \langle n,\alpha|\nabla|n,\alpha\rangle \cdot d\mathbf{R} \\ &= -\operatorname{Im}\oint \langle n|D^\dagger(\alpha)\nabla D(\alpha)|n\rangle \cdot d\mathbf{R} \end{aligned} \tag{5}$$

Show that

$$\begin{aligned} D^\dagger \frac{\partial D}{\partial x_1} &= -ix_2 + (a^\dagger - a) \\ D^\dagger \frac{\partial D}{\partial x_2} &= ix_1 + i(a^\dagger + a) \end{aligned} \tag{6}$$

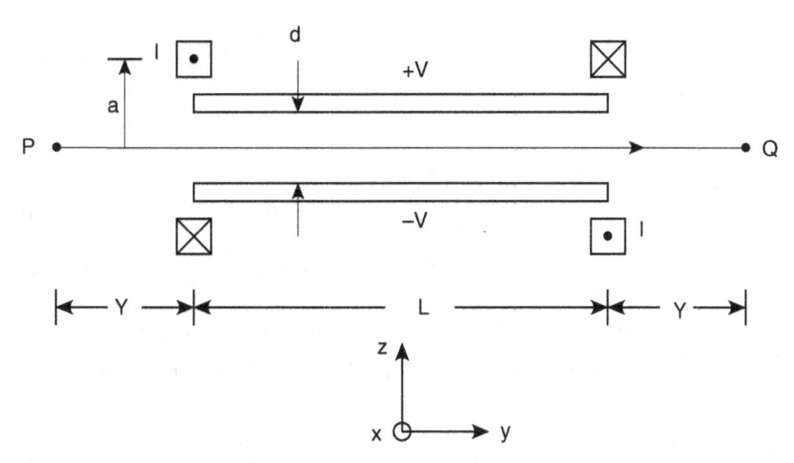

Figure 17.2 Schematic diagram of experimental setup described in Problem 17.2

Thus show that for any n, (5) yields

$$\gamma = \oint (x_2 dx_1 - x_1 dx_2) \quad (7)$$

What is the geometric interpretation of the integral on the right-hand side of (7)?

17.2. Figure 17.2 is a schematic diagram (not to scale) of an experimental setup. A well-collimated atomic beam is directed along the y-axis from point P to point Q. Each atom has total angular momentum $J = ½$, and the atomic magnetic moment in the $J = \frac{1}{2}, m_J = \frac{1}{2}$ state is $-\mu_B$. The beam velocity is $v = 10^4$ cm/s. Along the entire beam length there is a homogeneous magnetic field in the z-direction $B_z = 0.2$ G. Two single-turn circular coils, each of radius $a =$ 10 cm, generate a magnetic field along the y-axis. The currents in the two coils are opposed, and the magnitude of each current is $I = 0.016$ A. Two parallel conducting plates, separated by distance $d = 0.2$ cm, are charged to potentials $\pm V$, where $V = 10{,}000$ V. The length of the plates is $L = 100$ cm. Distance Y is much larger than a, so the magnetic fields due to the circular coils at P and Q are negligible.

At point P, each atom is in the following state in its own rest frame:

$$|\psi_P\rangle = \frac{1}{\sqrt{2}}\left(|J = 1/2, m_J = 1/2\rangle + |1/2, -1/2\rangle\right)$$

At point Q, the rest frame state is

$$|\psi_Q\rangle = \frac{1}{\sqrt{2}}\left(|1/2, 1/2\rangle e^{i\phi} + |1/2, -1/2\rangle e^{-i\phi}\right)$$

Estimate the value of that part of the phase ϕ that reverses with the electric field between the plates. If you use the adiabatic approximation, justify it.

17.3. In this problem we consider a physical example that can be solved analytically and that illustrates the adiabatic limit, the sudden approximation limit, and the entire range between. A beam of neutral atoms with $J = ½$ moves along the positive z-axis with velocity v, and for all

$z < 0$, each atom is in the state $m_J = +\frac{1}{2}$. A magnetic field $\boldsymbol{B}$ is applied, which takes the following values along the z-axis:

$$\begin{aligned} z < 0: &\quad \boldsymbol{B} = B_0 \hat{z}, \\ 0 \le z \le L: &\quad \boldsymbol{B} = B_0 \left(\hat{z}\cos\frac{\pi z}{L} + \hat{x}\sin\frac{\pi z}{L} \right), \\ z > L: &\quad \boldsymbol{B} = -B_0 \hat{z}. \end{aligned}$$

Each atom possesses a magnetic moment; hence the Hamiltonian of interaction with the magnetic field is $H = -\boldsymbol{\mu}\boldsymbol{\cdot}\boldsymbol{B} = -\mu\boldsymbol{\sigma}\boldsymbol{\cdot}\boldsymbol{B}$. Let $\omega = \pi v/L$, $\omega_0 = \mu B_0/\hbar$, and $\Omega = \sqrt{\omega_0^2 + (\omega^2/4)}$. Also, let the spin wave function be

$$\psi = \begin{bmatrix} a_+(z) \\ a_-(z) \end{bmatrix}$$

Assume that at $z = 0$, $a_+ = 1$, and $a_- = 0$. Show that

$$a_+(z = L) = \frac{\omega}{2\Omega}\sin\frac{\Omega L}{v} \tag{1}$$

$$a_-(z = L) = \cos\frac{\Omega L}{v} + i\frac{\omega_0}{\Omega}\sin\frac{\Omega L}{v} \tag{2}$$

Discuss the behavior of the solutions (1) and (2) in the adiabatic limit and in the sudden-approximation limit. Also, plot $|a_+(z = L)|^2$ for the range

$$0.1 \le \frac{\omega}{\omega_0} \le 10$$

17.4. The tritium nucleus undergoes beta decay

$$^3\mathrm{H} \to {}^3\mathrm{He} + e^- + \bar{\nu}_e$$

The maximum electron kinetic energy in this allowed beta decay is 18.6 keV.
(a) Calculate the probability that the resulting $^3\mathrm{He}^+$ ion is in each of the $1s$, $2s$, $2p$ states. Use the sudden approximation, and justify it.
(b) What modification in the shape of the beta decay spectrum occurs near the end point (place of maximum electron kinetic energy) if the antineutrino is assumed to have a mass of 2 eV/c^2?
(c) To see how difficult it would be to observe such a mass from an experiment using tritium decay, calculate the fraction of decays that would occur for which the electron kinetic energy is within 5 eV of the end point, assuming that the neutrino rest mass is actually zero. (In this problem, use the analysis of the beta decay spectrum discussed in Section17.4, which ignores the Coulomb interaction of the outgoing beta particle with the final nucleus. In tritium beta decay, this Coulomb effect actually distorts the beta spectrum considerably.)

18 Scattering

18.1 Typical scattering experiment

Scattering experiments and the information gained from them are of very great importance in many areas of physics, especially elementary particle and nuclear physics. The literature devoted to the theoretical and experimental aspects of scattering is enormous, and there are a number of monographs that give a very detailed account of scattering theory [see, e.g., Goldberger and Watson (1964) and Newton (1982)]. Our intention in this chapter is to summarize some of the most important topics in scattering theory as simply as possible. Only very few scattering problems permit exact quantum-mechanical analysis; therefore, approximation methods are necessary. The most important of these for both nonrelativistic and relativistic collisions is the Born approximation, which receives considerable attention in this and later chapters.

Scattering experiments can be done with colliding beams of particles or with a beam of projectiles incident on a fixed target. The main features of a typical fixed-target experiment are shown schematically in Figure 18.1. The projectile beam is usually well collimated, and the projectiles are usually prepared with reasonably well-defined momentum P_z along the beam axis (z) so that $\Delta P_z \ll P_z$ and $|P_{x,y}| \ll P_z$. Interaction between the projectile and target wave packets generates an outgoing scattered wave, which is observed with a detector that subtends a small solid angle $d\Omega$ at the target, is oriented at angles θ and ϕ with respect to the z-axis, and is at a macroscopic distance r from the target. The distance r is many orders of magnitude greater than the range of the projectile-target interaction, and r is usually at least several orders of magnitude greater than the collimated transverse width of the projectile beam.

The experimenter measures the number of particles dN_D scattered into the detector per unit time. This quantity can be expressed in terms of the flux F_0 of projectiles, the number density n_0 of target particles, the effective volume V_T of the target, the detection efficiency ε_D, the solid angle $d\Omega$ subtended by the detector at the target, and the differential scattering cross section $\sigma(\theta,\phi)$; that is,

$$dN_D = F_0 \cdot n_0 \cdot V_T \cdot \varepsilon_D d\Omega \cdot \sigma(\theta,\phi) \tag{18.1}$$

It is clear from (18.1) that $\sigma(\theta,\phi)$, which has dimension (length squared), is the radial probability current of scattered particle per unit solid angle and per target particle divided by the probability current of the incident particle. The total cross section is

$$\sigma_T = \int \sigma(\theta,\phi)\, d\Omega$$

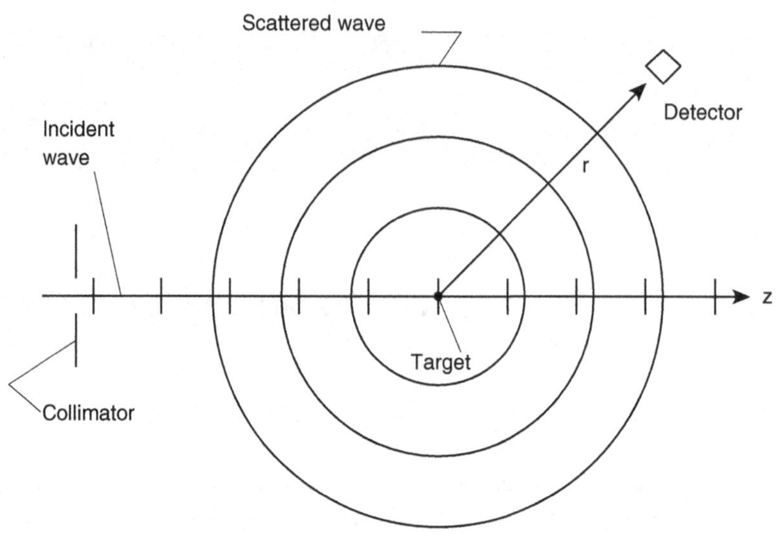

Figure 18.1 Schematic diagram of a typical scattering experiment.

The differential and total cross sections are the fundamental quantities that characterize the interaction between projectile and target, and it is the job of scattering theory to calculate $\sigma(\theta,\phi)$ and σ_T from assumptions about this interaction.

There are two distinct approaches to analysis of a scattering event: time dependent and time independent. In the former, each projectile is treated as a localized wave packet, the time evolution of which is calculated with the time-dependent Schroedinger equation and, in particular, with the aid of a time-dependent Green function. In Section 6.1 we constructed the free-particle Green function, and in Section 6.14 we saw that a more general time-dependent Green function, which includes the effect of a potential, can be constructed with the path integral method. The latter is developed further for analysis of scattering in Section 18.6.

The time-independent approach is simpler and easier to employ, is adequate for most applications, and for the most part is the method we use in this chapter. It is based on the fact, already mentioned, that the incoming projectiles are prepared in a well-collimated beam with reasonably well-defined momentum P_z along the beam axis so that $\Delta P_z \ll P_z$ and $|P_{x,y}| \ll P_z$. Hence the spatial wave function of the incident projectile can be approximated as a plane wave. The effects of scattering on this plane wave are calculated with the time-independent Schroedinger equation.

Scattering can be elastic or inelastic. In elastic scattering, the target and projectile have no internal degrees of freedom, or if they do, these degrees of freedom are not excited by the collision. Thus, in elastic scattering, there is no loss or gain of kinetic energy E. Inelastic scattering does involve excitation of internal degrees of freedom of the target and/or projectile, and thus the kinetic energy of relative motion is generally altered. Inelastic scattering includes the possibility (sometimes called *reaction scattering*) that the very identity of the reactants is transformed by the collision; for example,

$$\pi^0 + p \to n + \pi^+$$

$$e^+ + e^- \rightarrow \mu^+ + \mu^-$$

$$d + {}^3\text{H} \rightarrow n + {}^4\text{He}$$

We show later that whereas a collision can be purely elastic, it cannot be purely inelastic. In other words, if the cross section for an inelastic scattering channel is nonzero, the cross section for elastic scattering also must be nonzero.

18.2 Amplitude for elastic potential scattering

We start with elastic potential scattering of spinless particles, which is the simplest case, and we use the time-independent method. Here we assume that the interaction between a projectile of mass m_1 and a target particle of mass m_2 is described by a potential $V(\boldsymbol{r})$ that falls off sufficiently rapidly as r increases. It is convenient to work in the center-of-mass frame, where we consider the scattering of a particle with reduced mass μ by a fixed-force center. Because the relative momentum is $p = \hbar k$ and the energy of relative motion is $E = \hbar^2 k^2/(2\mu)$, the Schroedinger equation is

$$(\nabla^2 + k^2)\psi(\boldsymbol{r}) = \frac{2\mu}{\hbar^2}V(\boldsymbol{r})\psi = U(\boldsymbol{r})\psi \tag{18.2}$$

We are interested in the asymptotic solution to this equation at a distance from the origin so large that $U(\boldsymbol{r})$ is negligible. This solution can be expressed as

$$\psi = e^{ikz} + f(\theta,\phi)\frac{e^{ikr}}{r} \tag{18.3}$$

where the first term on the right-hand side represents the incident wave, whereas the second term describes the outgoing scattered wave. It is easy to see that (18.3) is in fact a solution to (18.2) for $U = 0$ in the limit of large r:

$$\nabla^2\left[f(\theta,\phi)\frac{e^{ikr}}{r}\right] = -k^2 f\frac{e^{ikr}}{r} + \frac{1}{r^3}e^{ikr} \times (\text{angular factors})$$

Note before we go further that the extension of the incoming "plane wave" e^{ikz} in the transverse (x,y)-directions is assumed to be large compared with the range of the potential but small compared with $\theta_{\min}r$, where $\theta_{\min}$ is the smallest polar angle that can be measured in a practical apparatus. This requirement is always easily achieved in real experiments.

All interesting information concerning the scattering is contained in the scattering amplitude $f(\theta,\phi)$. Let us calculate the outgoing radial probability current density j_r at the detector and compare it with the incident probability current density j_{inc}. Ignoring interference terms between the incident and scattered wave, which is legitimate if we choose $r\theta \gg \delta$, where δ is the collimated radius of the incident beam, we have

$$j_r = \frac{\hbar}{2\mu i}|f|^2\left[\frac{e^{-ikr}}{r}\frac{\partial}{\partial r}\left(\frac{e^{ikr}}{r}\right) - \frac{e^{ikr}}{r}\frac{\partial}{\partial r}\left(\frac{e^{-ikr}}{r}\right)\right]$$
$$= \frac{\hbar k}{\mu r^2}|f|^2 = \frac{|f|^2}{r^2}v$$

where $v = \hbar k / \mu$. Meanwhile,

$$j_{\text{inc}} = \frac{\hbar}{2\mu i}\left[e^{-ikz}\frac{\partial}{\partial z}(e^{ikz}) - e^{ikz}\frac{\partial}{\partial z}(e^{-ikz})\right] = v$$

Thus

$$\frac{j_r}{j_{\text{inc}}} = \frac{|f(\theta,\phi)|^2}{r^2}$$

The total probability current radiated into solid angle $d\Omega$ per unit incident current and for an individual scattering center is thus

$$\frac{|f|^2}{r^2}r^2 d\Omega = |f|^2\, d\Omega$$

Therefore,

$$\sigma(\theta,\phi) = |f(\theta,\phi)|^2 \tag{18.4}$$

Hence the cross section is completely determined from the scattering amplitude.

18.3 Partial wave expansion of the scattering amplitude for a central potential

If $V = V(r)$ is a central potential, we have axial symmetry about z, and thus ψ can be expressed as a sum of products of radial functions and Legendre polynomials; that is,

$$\begin{aligned}\psi &= \sum_{\ell=0}^{\infty}(2\ell+1)i^\ell R_\ell(r)P_\ell(\cos\theta)\\ &= \sum_{\ell=0}^{\infty}(2\ell+1)i^\ell\frac{\chi_\ell(r)}{r}P_\ell(\cos\theta)\end{aligned} \tag{18.5}$$

where

$$\chi_\ell'' + \left[k^2 - U(r) - \frac{\ell(\ell+1)}{r^2}\right]\chi_\ell = 0 \tag{18.6}$$

Because $U(r)$ and r^{-2} approach zero as $r \to \infty$, we try a solution to (18.6) of the form

$$\chi_\ell = \alpha e^{\pm ikr} \exp\left[\int_{r_0}^{r} g(r)\, dr\right] \tag{18.7}$$

where α is a constant. Substitution of (18.7) into (18.6) results in the equation

$$g^2 \pm 2ikg + g' - U(r) - \frac{\ell(\ell+1)}{r^2} = 0 \tag{18.8}$$

It is useful to distinguish between the Coulomb potential $U(r) = c/r$, where c is a constant, and all other potentials of physical interest, which approach zero more rapidly than $1/r$ as $r \to \infty$. For the Coulomb potential, (18.8) yields

$$g \to \frac{c}{2ikr} + O(r^{-2})$$

Hence

$$\chi_\ell = \alpha e^{\pm ikr} \exp\left(\mp \frac{ic}{2k} \ln r\right)$$

Thus, for Coulomb scattering, to be discussed in detail in Section 18.5, the phase of χ_ℓ varies logarithmically with r at large distances. For potentials that approach zero more rapidly than $1/r$ as $r \to \infty$, $\int_{r_0}^{r} g(r)\, dr$ converges and $\chi_\ell \to \text{const} \exp(\pm ikr)$ for large r. Considering only the latter potentials for the moment, we see that the asymptotic form of χ_ℓ can be written quite generally as

$$\chi_\ell \to \frac{A_\ell}{k} \sin\left(kr + \delta_\ell - \frac{\ell\pi}{2}\right) \tag{18.9}$$

where A_ℓ and δ_ℓ are constants, and δ_ℓ is real. Thus, from (18.5), we have

$$\begin{aligned} \lim_{r\to\infty} \psi &= e^{ikz} + f(\theta)\frac{e^{ikr}}{r} \\ &= \sum_{\ell=0}^{\infty} A_\ell (2\ell+1) i^\ell \frac{\sin(kr + \delta_\ell - \ell\pi/2)}{kr} P_\ell(\cos\theta) \end{aligned} \tag{18.10}$$

Our next step is to expand the free-particle wave function $\exp(ikz)$ in a sum of products of radial functions and Legendre polynomials. With the substitutions $k^2 = 2\mu E^2/\hbar^2$ and $\rho = kr$, the radial Schroedinger equation for a free particle is

$$\frac{d^2R}{d\rho^2} + \frac{2}{\rho}\frac{dR}{d\rho} + \left[1 - \frac{\ell(\ell+1)}{\rho^2}\right] R = 0 \tag{18.11}$$

This second-order differential equation has two linearly independent solutions: the spherical Bessel functions $j_\ell(\rho)$ and $n_\ell(\rho)$. The first few of these functions are

$$
\begin{aligned}
j_0(\rho) &= \frac{\sin\rho}{\rho} \\
j_1(\rho) &= \frac{\sin\rho}{\rho^2} - \frac{\cos\rho}{\rho} \\
j_2(\rho) &= \left(\frac{3}{\rho^3} - \frac{1}{\rho}\right)\sin\rho - \frac{3}{\rho^2}\cos\rho
\end{aligned}
\tag{18.12}
$$

and

$$
\begin{aligned}
n_0(\rho) &= -\frac{\cos\rho}{\rho} \\
n_1(\rho) &= -\frac{\cos\rho}{\rho^2} - \frac{\sin\rho}{\rho} \\
n_2(\rho) &= -\left(\frac{3}{\rho^3} - \frac{1}{\rho}\right)\cos\rho - \frac{3}{\rho^2}\sin\rho
\end{aligned}
\tag{18.13}
$$

These functions take the following limiting forms:

$$
\lim_{\rho\to 0} j_\ell(\rho) \to \frac{\rho^\ell}{(2\ell+1)!!} \qquad \lim_{\rho\to 0} n_\ell(\rho) \to -\frac{(2\ell-1)!!}{\rho^{\ell+1}} \tag{18.14}
$$

$$
\lim_{\rho\to\infty} j_\ell(\rho) \to \frac{\sin\left(\rho - \frac{\ell\pi}{2}\right)}{\rho} \qquad \lim_{\rho\to\infty} n_\ell(\rho) \to -\frac{\cos\left(\rho - \frac{\ell\pi}{2}\right)}{\rho} \tag{18.15}
$$

Note that while each j_ℓ is regular at the origin, each n_ℓ diverges there. Because the wave function of a free particle with definite energy and orbital angular momentum is regular at the origin, it must take the form

$$
\psi_{\text{free}}(r,\theta,\phi) = \text{const } j_\ell(kr) Y_\ell^m(\theta,\phi) \tag{18.16}
$$

For fixed k, these functions form a complete set, and therefore, it is possible to express a free-particle plane wave exp $(i\boldsymbol{k}\cdot\boldsymbol{r})$ as a linear combination of spherical waves (18.16). To find the coefficients in this linear combination, we first choose a coordinate system where $\boldsymbol{k}$ is parallel to the z-axis. Then exp $(i\boldsymbol{k}\cdot\boldsymbol{r})$ becomes $\exp(ikz)$, which is independent of ϕ and therefore may be expressed as

$$
e^{ikz} = e^{i\rho\cos\theta} = \sum_{\ell=0}^{\infty} (2\ell+1) i^\ell B_\ell j_\ell(\rho) P_\ell(\cos\theta) \tag{18.17}
$$

To determine the B_ℓ, we expand both sides of (18.17) in powers of $\rho\cos\theta$; that is,

$$
\exp(i\rho\cos\theta) = \sum_{n=0}^{\infty} \frac{i^n}{n!}\rho^n \cos^n\theta \tag{18.18}
$$

and

$$\sum_{\ell=0}^{\infty}(2\ell+1)i^{\ell}B_{\ell}j_{\ell}(\rho)P_{\ell}(\cos\theta)$$
$$=\sum_{\ell=0}^{\infty}(2\ell+1)i^{\ell}B_{\ell}\left[\frac{\rho^{\ell}}{(2\ell+1)!!}+\cdots\right]\left\{\frac{(2\ell)!}{2^{\ell}(\ell!)^{2}}\left[\cos^{\ell}\theta-\frac{\ell(\ell+1)}{2(2\ell+1)!!}\cos^{\ell-2}\theta+\cdots\right]\right\} \tag{18.19}$$

For given $\ell < n$, the highest power of $\cos\theta$ in (18.19) is $\ell < n$. For given $\ell > n$, the lowest power of ρ in (18.19) is $\ell > n$. Therefore, in (18.19), only $\ell = n$ contributes to $(\rho\cos\theta)^n$ in (18.18). Hence

$$B_{\ell}=\frac{(2\ell+1)!!2^{\ell}(\ell!)^{2}}{(2\ell+1)(2\ell)!\ell!}=1$$

and thus

$$e^{ikz}=e^{i\rho\cos\theta}=\sum_{\ell=0}^{\infty}(2\ell+1)i^{\ell}j_{\ell}(\rho)P_{\ell}(\cos\theta) \tag{18.20}$$

Note that it is sometimes convenient to use the spherical harmonic addition theorem

$$P_{\ell}(\cos\theta)=\frac{4\pi}{2\ell+1}\sum_{m=-\ell}^{\ell}Y_{\ell}^{m*}(\hat{k})Y_{\ell}^{m}(\hat{r})$$

where θ is the angle between $\boldsymbol{k}$ and $\boldsymbol{r}$, to rewrite (18.20) as

$$e^{i\boldsymbol{k}\cdot\boldsymbol{r}}=4\pi\sum_{\ell=0}^{\infty}i^{\ell}j_{\ell}(kr)\sum_{m=-\ell}^{\ell}Y_{\ell}^{m*}(\hat{k})Y_{\ell}^{m}(\hat{r})$$

Now employing

$$\lim_{\rho\to\infty}j_{\ell}(\rho)\to\frac{\sin\left(\rho-\frac{\ell\pi}{2}\right)}{\rho}$$

in (18.20) evaluated at large r, inserting the latter in (18.10), and transposing, we obtain

$$f(\theta)\frac{e^{ikr}}{r}=\sum_{\ell=0}^{\infty}\frac{(2\ell+1)i^{\ell}}{2ikr}\left[i^{-\ell}e^{ikr}\left(A_{\ell}e^{i\delta_{\ell}}-1\right)-i^{\ell}e^{-ikr}\left(A_{\ell}e^{-i\delta_{\ell}}-1\right)\right]P_{\ell}(\cos\theta) \tag{18.21}$$

The left-hand side of (18.21) contains only an outgoing wave (proportional to e^{ikr}); hence the coefficient of e^{-ikr} must vanish for each ℓ on the right-hand side. This implies that $A_{\ell}=e^{i\delta_{\ell}}$ and therefore that

$$f(\theta)=\frac{1}{2ik}\sum_{\ell=0}^{\infty}(2\ell+1)\left(e^{2i\delta_{\ell}}-1\right)P_{\ell}(\cos\theta) \tag{18.22a}$$

or, alternatively,

$$f(\theta)=\frac{1}{k}\sum_{\ell=0}^{\infty}(2\ell+1)\ e^{i\delta_\ell}\sin\delta_\ell P_\ell(\cos\theta) \tag{18.22b}$$

In this important result, called the *partial wave expansion of the scattering amplitude*, the latter is completely characterized by the real phase shifts δ_ℓ. A number of significant consequences follow immediately from (18.22a) or (18.22b); namely,

1. The asymptotic wave function is

$$\psi=e^{ikz}+f(\theta)\frac{e^{ikr}}{r}=\frac{1}{2ikr}\sum_\ell\left[e^{2i\delta_\ell}e^{ikr}-(-1)^\ell e^{-ikr}\right](2\ell+1)P_\ell(\cos\theta) \tag{18.23}$$

   This formula reveals that in elastic scattering, the presence of the potential U has no effect on the incoming partial waves and only causes a phase shift but no change in the magnitude of each outgoing partial wave. Therefore, for each ℓ, the incoming radial probability current density is just balanced by the outgoing radial probability current density: the net radial probability current density is zero.
2. The differential cross section is

$$\begin{aligned}\sigma(\theta)&=|f(\theta)|^2\\&=\frac{1}{k^2}\sum_{\ell,\ell'}(2\ell+1)(2\ell'+1)\exp\left[i(\delta_\ell-\delta'_\ell)\right]\sin\delta_\ell\sin\delta'_\ell P_\ell(\cos\theta)P_{\ell'}(\cos\theta)\end{aligned}$$

   The total cross section is obtained by integrating over the entire solid angle; that is,

$$\sigma_T=2\pi\int_0^\pi\sigma(\theta)\sin\theta\,d\theta$$

   In this integration,

$$\int_0^\pi P_\ell(\cos\theta)P_{\ell'}(\cos\theta)\sin\theta\,d\theta=\frac{2}{2\ell+1}\delta_{\ell,\ell'}$$

   Hence

$$\sigma_T=\frac{4\pi}{k^2}\sum_{\ell=0}^{\infty}(2\ell+1)\sin^2\delta_\ell \tag{18.24}$$

   Comparison of (18.24) with (18.22) for $\theta=0$ yields the optical theorem

$$\sigma_T=\frac{4\pi}{k}\operatorname{Im}f(0)$$

3. In (18.24), we may think of each partial wave ℓ as contributing its own partial cross section; that is,

$$\sigma_T = \sum_{\ell=0}^{\infty} \sigma_\ell$$

where

$$\sigma_\ell = \frac{4\pi}{k^2}(2\ell+1)\sin^2\delta_\ell \tag{18.25}$$

Obviously,

$$\sigma_\ell \le \frac{4\pi}{k^2}(2\ell+1) \tag{18.26}$$

no matter how strong the potential is.

4. Recalling the radial equation

$$\chi_\ell'' + \left[k^2 - U(r) - \frac{\ell(\ell+1)}{r^2}\right]\chi_\ell = 0$$

we see that

$$\frac{\chi_\ell''}{\chi_\ell} = U + \frac{\ell(\ell+1)}{r^2} - k^2 = U + \frac{s_\ell''}{s_\ell} \tag{18.27}$$

where s_ℓ is the function to which χ_ℓ reduces when there is no scattering (when $U = 0$). For given ℓ and sufficiently large r, the inequalities

$$k^2 > \frac{\ell(\ell+1)}{r^2} + U$$

and

$$k^2 > \frac{\ell(\ell+1)}{r^2}$$

are always satisfied. Thus for sufficiently large r, $\chi_\ell''/\chi_\ell < 0$ and $s_\ell''/s_\ell < 0$, and from (18.27), if $U < 0$, $\chi_\ell''/\chi_\ell < s_\ell''/s_\ell < 0$. Thus, in the case of an attractive potential, χ_ℓ is "pulled in" toward the origin relative to s_ℓ; in other words, δ_ℓ is positive. Conversely, if the potential is repulsive, the phase shifts are negative.

We now generalize this formalism to include the possibility of inelastic scattering. First, we note that the coefficients of the outgoing waves in (18.23) must be reduced in magnitude to account for diversion of scattered probability current from the elastic channel to one or more inelastic channels. Thus

$$\lim_{r\to\infty} \psi = \frac{1}{2ikr}\sum_{\ell=0}^{\infty}(2\ell+1)\left[e^{2i\delta_\ell}e^{ikr} - (-1)^\ell e^{-ikr}\right]P_\ell(\cos\theta)$$

must be replaced by

$$\lim_{r\to\infty} \psi = \frac{1}{2ikr}\sum_{\ell=0}^{\infty}(2\ell+1)\left[\eta_\ell e^{ikr} - (-1)^\ell e^{-ikr}\right]P_\ell(\cos\theta) \tag{18.28}$$

where $|\eta_\ell| < 1$ in the presence of inelastic scattering. To obtain the scattered wave, we subtract the incoming wave

$$\lim_{r\to\infty} e^{ikz} = \sum_{\ell=0}^{\infty}(2\ell+1)\left[\frac{e^{ikr} - (-1)^\ell e^{-ikr}}{2ikr}\right]P_\ell(\cos\theta)$$

from (18.28), which yields

$$f_{\text{el}}(\theta) = \frac{1}{2ik}\sum_{\ell=0}^{\infty}(2\ell+1)(\eta_\ell - 1)P_\ell(\cos\theta) \tag{18.29}$$

Now, forming $|f_{\text{el}}(\theta)|^2$, integrating over the solid angle, and making use of the orthogonality of Legendre polynomials, we arrive at the revised elastic scattering cross section

$$\sigma_{\text{el}} = \frac{\pi}{k^2}\sum_{\ell=0}^{\infty}(2\ell+1)|1-\eta_\ell|^2 \tag{18.30}$$

The reaction cross section (i.e., the inelastic cross section for a particular channel) is determined by the number of particles removed from the incoming beam per second; in other words, it is obtained from the net inward probability current density through a sphere of radius r centered on the target calculated from (18.23). We thus obtain

$$\sigma_{\text{inel},\ell} = \frac{\pi}{k^2}(2\ell+1)\left(1-|\eta_\ell|^2\right) \tag{18.31}$$

From (18.30) and (18.31), we see that if $\sigma_{\text{el},\ell} = 0$, we must have $\eta_\ell = 1$, which implies $\sigma_{\text{inel},\ell} = 0$ as well. However, if $\eta_\ell = -1$, $\sigma_{inel,\ell} = 0$, but $\sigma_{\text{el},\ell} = (4\pi/k^2)(2\ell+1) \neq 0$. When $\eta_\ell = 0$, $\sigma_{\text{inel},\ell}$ reaches its maximum for given k, and $\sigma_{\text{el},\ell} = \sigma_{\text{inel},\ell} = (\pi/k^2)(2\ell+1)$.

18.4 *s*-Wave scattering at very low energies: Resonance scattering

The partial wave expansion is especially useful in cases where only a few ℓ values contribute. In particular, if $V(r)$ has very short range, only the $\ell = 0$ partial wave is significant (*s*-wave scattering). This is so because only $j_0(kr)$, among all the j_ℓ, is nonzero at the origin

$$\lim_{r\to 0} j_\ell(kr) \to \frac{(kr)^\ell}{(2\ell+1)!!} \tag{18.32}$$

It is easy to see from (18.32) that for fixed k and a potential of finite range b, the contribution of partial waves with $\ell > 0$ must be negligible if $kb \ll 1$.

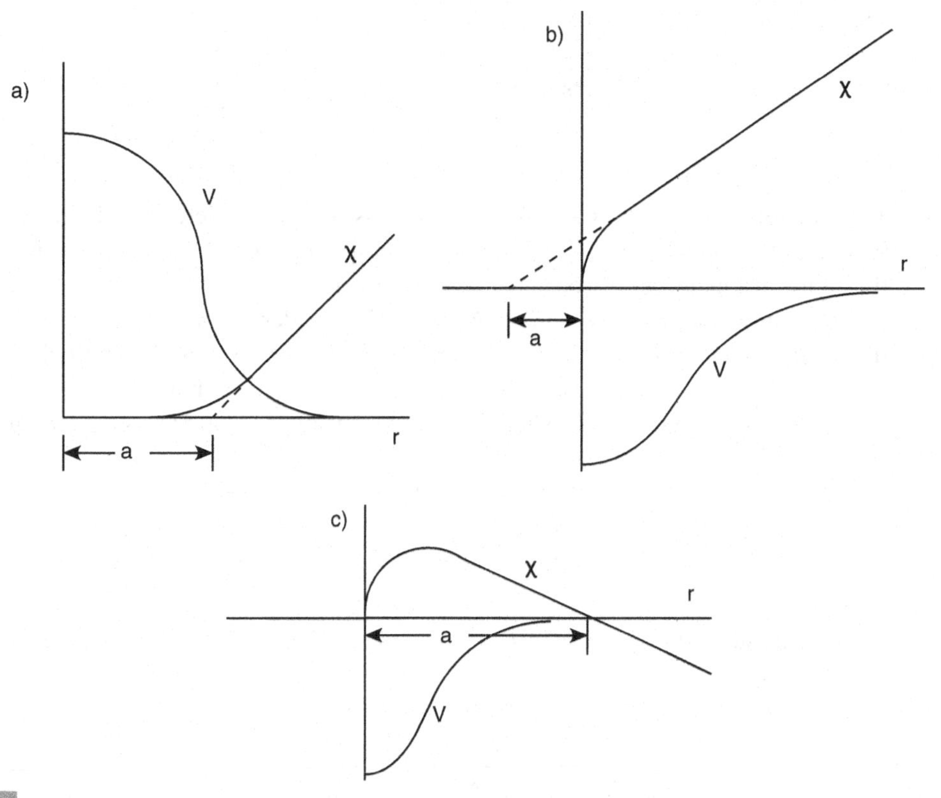

Figure 18.2 (a) Repulsive potential V: $a(k) > 0$. (b) Attractive potential V: $a(k) < 0$: unbound state. (c) Attractive potential V: $a(k) > 0$: bound state.

For s-wave scattering, the total cross section is

$$\sigma = \frac{4\pi}{k^2}\sin^2\delta_0 \tag{18.33}$$

Because σ remains finite when $k \to 0$, we have

$$\frac{\sin^2\delta_0}{k^2} \to \frac{\delta_0^2}{k^2} = [a(k)]^2$$

The quantity $a(k) = -\delta_0/k$ obviously has the dimension of length and is called the *scattering length*. At very small k, the radial wave function $\chi = rR$ immediately outside the range of force is proportional to $\sin(kr + \delta_0) \simeq kr + \delta_0$. If the force is repulsive, the phase shift is negative; hence $a(k) > 0$. This is illustrated in Figure 18.2a. If the potential is attractive, the scattering length can be either negative or positive. These possibilities are illustrated in Figure 18.2b and c.

Although we have defined $a(k) = -\delta_0/k$ for very small k, it is convenient to define the scattering length for arbitrary k as

$$a(k) = -\frac{1}{k}\tan\delta_0 \tag{18.34}$$

The total cross section is expressed in terms of $a(k)$ as

$$\sigma = \frac{4\pi \sin^2 \delta_0}{k^2} = \frac{4\pi}{k^2}\left(\frac{1}{1+\cot^2 \delta_0}\right) = \frac{4\pi}{k^2 + \dfrac{1}{a^2(k)}} \tag{18.35}$$

Thus the total cross section as a function of k^2 is completely determined if we know $a(k)$ as a function of k. Before proceeding with this general discussion of s-wave scattering, we illustrate the points just made with two elementary examples.

First, consider a particle of energy $E = \hbar^2 k^2/2m$ incident on an attractive square-well potential, where $V = -V_0$ when $r < b$ and $V = 0$ for $r > b$. We assume that the de Broglie wavelength of the projectile is much larger than the range of the potential b. Hence $kb \ll 1$, so we can confine ourselves to s-wave scattering. Then the one-dimensional radial Schroedinger equation is

$$\chi'' + \frac{2m}{\hbar^2}(E + V_0)\chi = 0 \qquad r < b$$

$$\chi'' + \frac{2m}{\hbar^2}E\chi = 0 \qquad r > b$$

Making the substitutions $k^2 = 2mE/\hbar^2$, $U_0 = 2mV_0/\hbar^2$, and $\kappa^2 = k^2 + U_0$, we have

$$\chi'' + \kappa^2 \chi = 0 \qquad r < b \tag{18.36}$$

$$\chi'' + k^2 \chi = 0 \qquad r > b \tag{18.37}$$

Because $\chi(0) = 0$, the solution to (18.36) is $\chi_i = A \sin \kappa r$, where A is a constant and subscript i means "interior." The solution to (18.37) is $\chi_e = B \sin(kr + \delta_0)$, where B is a constant, δ_0 is the s-wave phase shift, and subscript e means "exterior." Because the radial-wave function χ and its derivative must be continuous at $r = b$, we have

$$A \sin \kappa b = B \sin(kb + \delta_0) \tag{18.38}$$

and

$$\kappa A \cos \kappa b = kB \cos(kb + \delta_0) \tag{18.39}$$

Division of (18.38) by (18.39) yields

$$\tan(kb + \delta_0) = \frac{k}{\kappa}\tan \kappa b$$

We solve for $\tan \delta_0$ to obtain

$$\tan \delta_0 = \frac{kb \tan \kappa b - \kappa b \tan kb}{\kappa b + kb \tan \kappa b \tan kb} \tag{18.40}$$

Because $kb \ll 1$,

$$\tan \delta_0 \approx \frac{kb(\tan \kappa b - \kappa b)}{\kappa b + (kb)^2 \tan \kappa b} \tag{18.41}$$

If $\kappa b \ll \pi/2$, we can use the approximation $\tan \kappa a \approx \kappa b + (\kappa b)^3/3$. Then, neglecting the second term in the denominator on the right-hand side of (18.41), we have

$$\tan \delta_0 \approx kb(kb)^2/3 \approx \sin \delta_0$$

in which case the total cross section for s-wave scattering is

$$\begin{aligned} \sigma &= \frac{4\pi}{k^2} \sin^2 \delta_0 \\ &\approx \frac{4\pi b^2}{9} (\kappa b)^4 \end{aligned}$$

If $\kappa b \ll \pi/2$ but $U_0 \gg k^2$, then $\kappa^2 \approx U_0$, in which case

$$\sigma = \frac{4\pi b^2}{9} (U_0 b^2)^2$$

Hence, in this regime, the cross section is proportional to U_0^2. If $\kappa b \approx \pi/2, 3\pi/2, 5\pi/2$, and so on, the first term in the numerator and the second term in the denominator of the right-hand side of (18.40) become dominant. In this case,

$$\tan \delta_0 \approx \frac{1}{kb} \gg 1$$

so

$$\sin^2 \delta_0 \approx 1$$

and

$$\sigma \approx \frac{4\pi}{k^2}$$

These maxima (*resonances*) in σ for $U_0 b^2 \approx \pi^2/4, 9\pi^2/4, \ldots$ correspond to the appearance of successive $\ell = 0$ bound states.

Equation (18.41) reveals that $\sigma \to 0$ when $\tan \kappa b \approx \kappa b$. This explains the *Ramsauer-Townsend effect* in low-energy-electron rare gas atom scattering. We summarize this example by plotting the variation of σ versus κb over a large range in Figure 18.3. Here, in order to visualize various features clearly, we have chosen $kb = 0.2$ and calculated $\sigma = (4\pi/k^2)\sin^2 \delta_0$ from (18.40).

Second, consider scattering by a hard sphere. Let $V = \infty$ for $r \le b$ and $V = 0$ for $r > b$. Then $\chi_\ell = 0$ for $r \le b$, and χ_ℓ / r is a linear combination of $j_\ell(kr)$ and $n_\ell(kr)$ for $r > b$. Because χ_ℓ must be continuous at $r = b$, we have

$$\chi_\ell(r) = C_\ell r\left[n_\ell(kb) j_\ell(kr) - j_\ell(kb) n_\ell(kr)\right] \qquad r \ge b \tag{18.42}$$

where C_ℓ is a constant. Using (18.15), we write the asymptotic form of (18.42), namely,

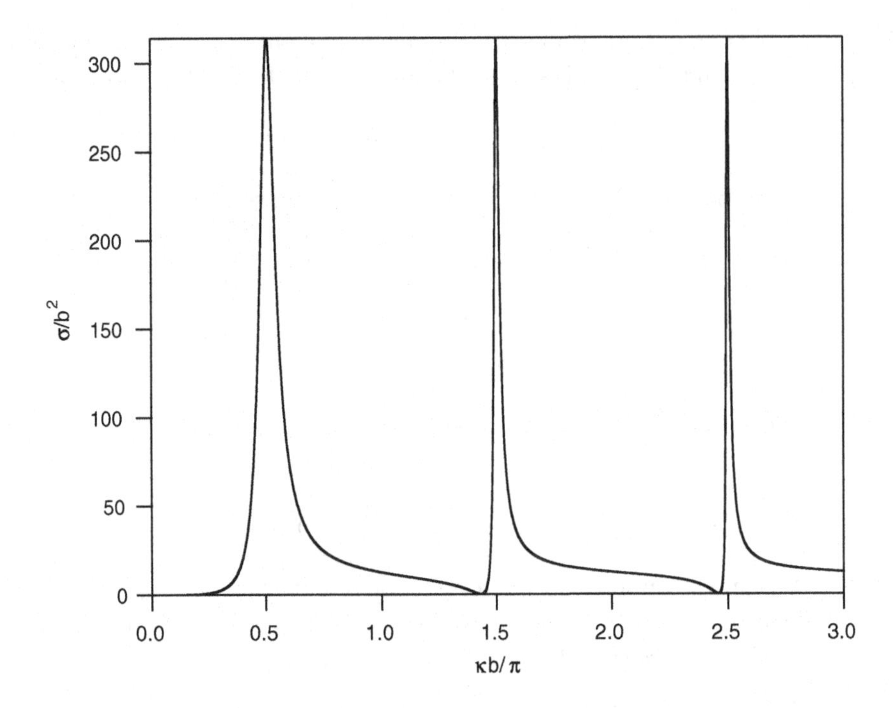

Figure 18.3 σ/b^2 versus $\kappa b/\pi$ for s-wave scattering from an attractive square well potential with $kb = 0.2$. Note the resonances at $\kappa b = \pi/2, 3\pi/2, \ldots$, which correspond to the appearance of successive s-wave bound states. Also note that the cross section vanishes at $\kappa b \approx 1.45\pi, \ldots$ (the Ramsauer-Townsend effect).

$$\chi_\ell \to \frac{C_\ell}{k}\left[n_\ell(kb)\sin\left(kr - \frac{\ell\pi}{2}\right) + j_\ell(kb)\cos\left(kr - \frac{\ell\pi}{2}\right)\right] \tag{18.43}$$

and we compare this with the general asymptotic form

$$\chi_\ell \to \frac{e^{i\delta_\ell}}{k}\sin\left(kr + \delta_\ell - \frac{\ell\pi}{2}\right)$$

to obtain

$$\tan\delta_\ell = \frac{j_\ell(kb)}{n_\ell(kb)}$$

Thus the total scattering cross section is

$$\sigma_T = \frac{4\pi}{k^2}\sum_{\ell=0}^{\infty}(2\ell+1)\left[\frac{j_\ell^2(kb)}{j_\ell^2(kb)+n_\ell^2(kb)}\right] \tag{18.44}$$

If $kb \ll 1$, the wavelength $\lambda = 2\pi/k$ of the incident and scattered waves is much larger than b, and then only the first few terms of the sum in (18.44) are significant. Using (18.14), it is easy to evaluate the factors in square brackets in (18.44) and thus to show that when $kb \ll 1$,

$$\begin{aligned}\sigma_{\ell=0} &\simeq 4\pi b^2\left(1-\frac{k^2b^2}{3}\right)\\ \sigma_{\ell=1} &\simeq 4\pi b^2\left(\frac{k^4b^4}{3}+\cdots\right) \ll \sigma_{\ell=0}\end{aligned} \tag{18.45}$$

For comparison, the cross section for scattering of a classical pointlike projectile by a hard sphere of radius b is $\sigma_{\text{classical}} = \pi b^2$. The extra factor of approximately 4 in $\sigma_T = \sigma_0 + \sigma_1$ revealed by (18.45) occurs because of the wave nature of quantum-mechanical scattering, in which there are diffraction effects.

We now continue with our general discussion of s-wave scattering. If inelastic as well as elastic scattering occurs, we must employ the parameter $\eta_0 = e^{2i\delta_0}$. This can be written as

$$\eta_0 = e^{2i\delta_0} = \frac{\dfrac{1}{a(k)} - ik}{\dfrac{1}{a(k)} + ik}$$

Hence

$$\eta_0 = \frac{1 - ika(k)}{1 + ika(k)}$$

Thus the cross section for elastic scattering is

$$\begin{aligned}\sigma_{\text{el}} &= \frac{\pi}{k^2}\left|1-\eta_0\right|^2 = \frac{\pi}{k^2}\left|1 - \frac{1-ika(k)}{1+ika(k)}\right|^2\\ &= \frac{4\pi}{\left|ik + \dfrac{1}{a(k)}\right|^2}\end{aligned} \tag{18.46}$$

whereas the inelastic cross section is

$$\sigma_{\text{inel}} = \frac{\pi}{k^2}\left(1 - \left|\eta_0\right|^2\right) = \frac{\pi}{k^2}\left[1 - \left|\frac{\dfrac{1}{a(k)} - ik}{\dfrac{1}{a(k)} + ik}\right|^2\right] \tag{18.47}$$

If $a(k)$ is real, we have only elastic scattering. Then it is obvious from (18.46) that σ_{el} reaches a maximum when $1/a(k)$ goes to zero. Suppose that the energy at which this happens is E_0. Then we can write the Taylor expansion

$$\begin{aligned}\frac{1}{a(E)} &= \frac{1}{a(E_0)} + (E - E_0)\frac{d}{dE}\left(\frac{1}{a}\right)\bigg|_{E=E_0} + \cdots\\ &= (E - E_0)\frac{2k}{\Gamma_s} + \cdots\end{aligned}$$

where Γ_s is defined by

$$\frac{2k}{\Gamma_s} = \frac{d}{dE}\left(\frac{1}{a}\right)\bigg|_{E=E_0} \tag{18.48}$$

Note that because the right-hand side of (18.48) is a constant, Γ_s is proportional to k. If $(E - E_0)$ is sufficiently small, we may neglect higher-order terms in the Taylor expansion, in which case (18.46) becomes

$$\sigma_{\text{el}} = \frac{4\pi}{k^2 + (E - E_0)^2 \dfrac{4k^2}{\Gamma_s^2}} = \frac{\pi}{k^2}\frac{\Gamma_s^2}{(E - E_0)^2 + \dfrac{\Gamma_s^2}{4}} \tag{18.49}$$

Suppose instead that $a(k)$ is complex. We write $1/a(k) = \alpha + i\beta$, where α and β are both real. Then (18.47) becomes

$$\sigma_{\text{inel}} = \frac{4\pi}{k}\frac{\text{Im}\left[\dfrac{1}{a(k)}\right]}{\left|\dfrac{1}{a(k)} + ik\right|^2} \tag{18.50}$$

In addition, now let the energy E_0 have a small imaginary part; that is,

$$E_0 = \varepsilon_0 - i\frac{\Gamma_r}{2}$$

where ε_0 and Γ_r are both real. (This would happen if spontaneous decay could occur from the level with energy E_0 to some lower state.) In this case, the first-order Taylor expansion becomes

$$\frac{1}{a(E)} = \frac{1}{a(E_0)} + \left(E - \varepsilon_0 + \frac{i\Gamma_r}{2}\right)\frac{d}{dE}\left(\frac{1}{a}\right)\bigg|_{E_0} \tag{18.51}$$

In (18.51), we must take into account the fact that

$$\frac{d}{dE}\left(\frac{1}{a}\right)\bigg|_{E_0} = \frac{2k}{\Gamma_s} + ik\alpha \tag{18.52}$$

is also complex (here Γ_s and α are both real). From (18.51) and (18.52), we have

$$\begin{aligned} \text{Re}\left[\frac{1}{a(E)}\right] &= (E - \varepsilon_0)\frac{2k}{\Gamma_s} - \frac{\Gamma_r k\alpha}{2} \\ &= \frac{2k}{\Gamma_s}\left(E - \varepsilon_0 - \frac{\Gamma_r\Gamma_s\alpha}{4}\right) \end{aligned} \tag{18.53}$$

and

$$\operatorname{Im}\left[\frac{1}{a(E)}\right] = k\alpha(E-\varepsilon_0) + \frac{2k\Gamma_r}{2\Gamma_s} = k\left[\frac{\Gamma_r}{\Gamma_s} + \alpha(E-\varepsilon_0)\right] \tag{18.54}$$

Thus the inelastic cross section becomes

$$\sigma_{\text{inel}} = \frac{4\pi\left[\frac{\Gamma_r}{\Gamma_s} + \alpha(E-\varepsilon_0)\right]}{\left|\frac{2k}{\Gamma_s}\left(E-\varepsilon_0-\frac{\Gamma_r\Gamma_s\alpha}{4}\right) + ik\left[1+\frac{\Gamma_r}{\Gamma_s}+\alpha(E-\varepsilon_0)\right]\right|^2} \tag{18.55}$$

In many applications it turns out that α is a very small quantity. Then we may ignore the term $\alpha(E-\varepsilon_0)$ in (18.54) in the neighborhood of resonance, in which case (18.55) simplifies to

$$\sigma_{\text{inel}} = \frac{\pi}{k^2}\frac{\Gamma_r\Gamma_s}{(E-E_r)^2 + \frac{(\Gamma_r+\Gamma_s)^2}{4}} \tag{18.56}$$

where $E_r = \varepsilon_0 + \alpha(\Gamma_r\Gamma_s/4)$ is the shifted resonance energy. Also, the elastic scattering cross section becomes

$$\sigma_{\text{el}} = \frac{\pi}{k^2}\frac{\Gamma_s^2}{(E-E_r)^2 + \frac{(\Gamma_s+\Gamma_r)^2}{4}} \tag{18.57}$$

Equations (18.56) and (18.57) are very useful in a wide variety of situations. They are usually called the *Breit-Wigner resonance* relations.

18.5 Coulomb scattering

We now consider scattering by the potential $V(r) = Z_1Z_2(e^2/4\pi r)$, where Z_1e and Z_2e are the electric charges of the projectile and target, respectively. This problem is most conveniently analyzed in parabolic coordinates (recall Section 8.6) where, as usual, the z-axis is the direction of the incoming projectile. Schroedinger's equation is

$$-\frac{4}{\xi+\eta}\left[\frac{\partial}{\partial\xi}\left(\xi\frac{\partial\psi}{\partial\xi}\right) + \frac{\partial}{\partial\eta}\left(\eta\frac{\partial\psi}{\partial\eta}\right)\right] - \frac{1}{\xi\eta}\frac{\partial^2\psi}{\partial\phi^2} + \frac{4nk}{\hbar(\xi+\eta)}\psi = k^2\psi \tag{18.58}$$

where

$$n = \frac{Z_1 Z_2 e^2 \mu}{4\pi\hbar k} = \frac{Z_1 Z_2 e^2}{4\pi\hbar v} \tag{18.59}$$

If the potential were absent, the solution to (18.58) would be exp(ikz). With the potential present, we try to construct a solution that differs from exp(ikz) by an outgoing wave. Thus we write

$$\psi = e^{ikz} f(\eta) \tag{18.60}$$

Substitution of (18.60) into (18.58) reveals that f indeed depends only on η and satisfies the ordinary differential equation

$$\eta f'' + (1 - ik\eta) f' - nkf = 0 \tag{18.61}$$

This second-order equation has two linearly independent solutions. We choose the solution that is regular at $\eta = 0$; that is,

$$f = cF(-in, 1, ik\eta) \tag{18.62}$$

where c is a constant, and F is the confluent hypergeometric function. For large x, the asymptotic form of $F(a,b,x)$ is

$$\begin{aligned}\lim_{x\to\infty} F(a,b,x) = & \frac{\Gamma(b)}{\Gamma(b-a)} e^{-a\ln(-x)} \left[1 + \frac{a(a+1-b)}{x} + \cdots\right] \\ & + \frac{\Gamma(b)}{\Gamma(a)} e^{x+(a-b)\ln(x)} \left[1 + \frac{(1-a)(b-a)}{x} + \cdots\right]\end{aligned}$$

Thus the asymptotic form of (18.60), normalized to unit incident flux, is

$$\begin{aligned}\psi \to \frac{1}{\sqrt{v}} \Bigg\{ & e^{i[kz + n\ln k(r-z)]} \left[1 - \frac{n^2}{ik(r-z)} + \cdots\right] \\ & + A_C(\theta) \frac{e^{i(kr - n\ln 2kr)}}{r} \left[1 - \frac{(1+in)^2}{ik(r-z)} + \cdots\right]\Bigg\}\end{aligned} \tag{18.63}$$

Here

$$A_C(\theta) = \frac{n}{2k\sin^2\frac{\theta}{2}} \exp\left[ni\ln\left(\sin^2\frac{\theta}{2}\right) + 2i\eta_0 + i\pi\right] \tag{18.64}$$

is the scattering amplitude, and

$$\eta_0 = Arg\Gamma(1 + in) \tag{18.65}$$

We now consider the most important features of (18.63).

1. In order that the leading terms of both square brackets dominate, we require that

$$\left|\frac{n^2}{k(r-z)}\right| \ll 1$$

This, in turn, requires that kr and $k(r - z)$ be large (i.e., θ must not be too small). In any real experiment, this restriction is not of practical significance because θ is in any case large enough that the detector is not sensitive to the incoming beam.

2. The differential scattering cross section is

$$\sigma_C(\theta) = |A_C(\theta)|^2 = \frac{n^2}{4k^2 \sin^4 \frac{\theta}{2}} = \frac{Z_1^2 Z_2^2}{16E^2} \frac{e^4}{(4\pi)^2} \frac{1}{\sin^4 \frac{\theta}{2}} \tag{18.66}$$

This is the famous Rutherford formula, originally derived from classical mechanics. One can also calculate $\sigma_C(\theta)$ and obtain result (18.66) by means of the Born approximation, to be discussed later.

3. Next, we examine the behavior of ψ at the origin. For any r, we have

$$\psi = \frac{1}{\sqrt{V}} \Gamma(1+in) e^{-n\pi/2} e^{ikr\cos\theta} F\left(-in, 1, 2ikr \sin^2 \frac{\theta}{2}\right)$$

Because $F(-in,1,0) = 1$ and $\Gamma^*(1+in) = \Gamma(1-in)$, this yields

$$|\psi(0)|^2 = \frac{1}{V} \Gamma(1+in)\Gamma(1-in) e^{-n\pi}$$

Now

$$\Gamma(1+in)\Gamma(1-in) = in\Gamma(in)\Gamma(1-in)$$
$$= \frac{in\pi}{\sin(in\pi)} = \frac{2\pi n}{e^{\pi n} - e^{-\pi n}}$$

Therefore,

$$|\psi(0)|^2 = \frac{2\pi n}{V} \frac{1}{e^{2\pi n} - 1} \tag{18.67}$$

There are several limiting cases of interest for this important formula:

a. $Z_1 Z_2 > 0$ (repulsive force), and $n \gg 1$ (low velocity). Here,

$$|\psi(0)|^2 \approx \frac{2\pi n}{V} e^{-2\pi n} \tag{18.68}$$

This case is of particular interest for nuclear reactions between one nuclide (Z_1) and another (Z_2) at low relative kinetic energy and where the reaction can only occur by the short-range

weak or strong interaction when the nuclei come in contact with one another. Here the reaction cross section is proportional to $|\psi(0)|^2$ as given by (18.68), and it is therefore inhibited by the Coulomb barrier, expressed by the factor $\exp(-2\pi n)$. This situation occurs in stellar interiors, where the kinetic energies of relative motion of nuclides engaged in thermonuclear reactions are frequently in the range 1–10 keV.
b. $Z_1Z_2 > 0$ (repulsive force), and $n \ll 1$ (high velocity)

$$|\psi(0)|^2 \approx \frac{1}{v}$$

c. $Z_1Z_2 < 0$ (attractive force), and $|n| \gg 1$ (low velocity)

$$|\psi(0)|^2 \approx \frac{2\pi|n|}{v}$$

d. $Z_1Z_2 < 0$ (attractive force), and $|n| \ll 1$ (high velocity)

$$|\psi(0)|^2 \approx \frac{1}{v}$$

4. The Coulomb scattering wave function can be expressed as a sum of spherical partial waves; that is,

$$\psi = \frac{1}{\sqrt{v}} e^{-n\pi} \sum_{\ell=0}^{\infty} \frac{\Gamma(\ell+1+in)}{(2\ell)!} (2ikr)^{\ell} e^{ikr} F(\ell+1+in, 2\ell+2, -2ikr) P_\ell(\cos\theta)$$

One can also write the scattering amplitude $A_C(\theta)$ as a sum over ℓ values; that is,

$$A_C(\theta) = \frac{1}{2ik} \sum_{\ell=0}^{\infty} (2\ell+1) e^{2i\eta_\ell} P_\ell(\cos\theta) \tag{18.69}$$

where $\eta_\ell = \arg\Gamma(1+\ell+in)$.

5. Ordinarily we think of k as a real quantity, but for some purposes it is interesting to consider it as a complex variable. In particular, let $k = iZ/N$, where N is a positive integer; choose $Z_1 = -1$, $Z_2 = Z$; and employ atomic units with $\mu = 1$. In this case, $n = -Z/k = iN$, and $E = -Z^2/2N^2$, which is just the Balmer formula once again, where N denotes the principal quantum number. In (18.69), we have

$$e^{2i\eta_\ell} = \frac{\Gamma(1+\ell-N)}{\Gamma(1+\ell+N)} \tag{18.70}$$

Because $\Gamma(z)$ is an analytic function of z with poles at $z = 0, -1, -2, -3, \ldots$, we see that A_C has poles at $\ell = 0, 1, \ldots, N-1$, which correspond to the possible bound states with given N.

18.6 Green functions: The path integral method and Lippmann-Schwinger equation

We noted in Chapter 6 that because the time-dependent Schroedinger equation is first order in t, it is possible to obtain a wave function $\psi(\boldsymbol{r},t)$ at a given $\boldsymbol{r}$ and time $t \geq t_0$ from the values of $\psi(\boldsymbol{r}_0,t_0)$ at all $\boldsymbol{r}_0$ and at time t_0 by means of the integral equation

$$\psi(\boldsymbol{r},t) = \int G(\boldsymbol{r},t;\boldsymbol{r}_0,t_0)\psi(\boldsymbol{r}_0,t_0)\, d^3\boldsymbol{r}_0$$

where $G(\boldsymbol{r},t;\boldsymbol{r}_0,t_0)$ is the *Green function.* In Section 6.14 we discussed the path integral method for obtaining this Green function. There, working in one spatial dimension for simplicity, we found that

$$G(x,t;x_0,t_0) = \int [dx]\exp\left[\frac{i}{\hbar}\int_{t_0}^{t}\left(\frac{m}{2}\dot{x}^2 - V\right)d\tau\right]$$

where $\int_{t_0}^{t}\left[(m/2)\dot{x}^2 - V\right]d\tau$ is the action S, and $\int[dx]$ denotes a sum over all paths with end points (x_0,t_0) and (x,t). We were able to find G by explicit evaluation of the path integral in a few simple cases: the free particle, a particle in a uniform gravitational field, and a simple harmonic oscillator. The free-particle Green function G_0 that we found by the path integral method is, of course, identical to G_0 as calculated in Section 6.1.

Now we want to develop a method of successive approximations for determining the Green function in the case of an arbitrary potential, and we wish to apply this method to scattering problems. This is the time-dependent approach to scattering. Once again, we find it convenient to employ the path integral technique, and we initially use one spatial dimension for simplicity.

We start by focusing on the term $\int_{t_0}^{t} V\, ds$ in the action, where s is a timelike dummy variable of integration. The expansion

$$\exp\left(\frac{-i}{\hbar}\int_{t_0}^{t} V\, ds\right) = 1 - \frac{i}{\hbar}\int_{t_0}^{t} V\, ds + \frac{1}{2!}\left(-\frac{i}{\hbar}\int_{t_0}^{t} V\, ds\right)^2 + \cdots$$

enables us to separate the Green function into a series of terms

$$G = G_0 + G_1 + G_2 + \cdots$$

where

$$G_0(x,t;x_0,t_0) = \int [dx]\exp\left(\frac{i}{\hbar}\int_{t_0}^{t}\frac{m}{2}\dot{x}^2\, d\tau\right) \tag{18.71}$$

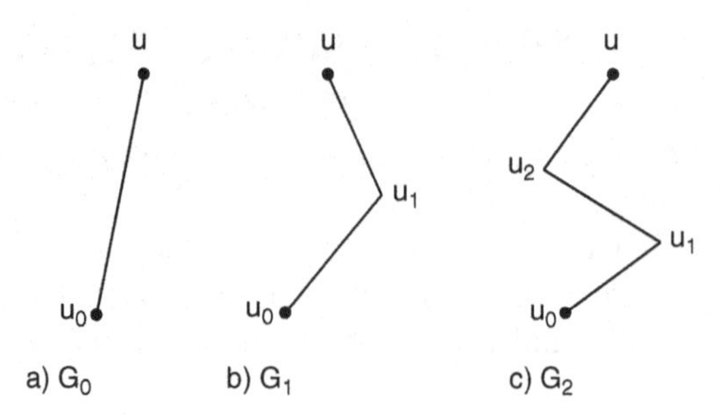

Figure 18.4 Schematic diagrams for the Green functions G_0, G_1, G_2. Time advances in the upward direction.

is the free-particle Green function,

$$G_1(x,t;x_0,t_0) = -\frac{i}{\hbar}\int ds \int [dx] \exp\left(\frac{i}{\hbar}\int_{t_0}^{t} \frac{m}{2}\dot{x}^2\, d\tau\right) V(x,s) \tag{18.72a}$$

is a first-order correction, and so on. We evaluate G_1 by writing out the path integral on the right-hand side of (18.72) explicitly. Dividing the time interval $t - t_0$ into the usual $(n+1)\varepsilon$ slices as in Section 6.14 and writing $s = t_j$, we have

$$\begin{aligned}
&\int [dx] \exp\left\{\frac{i}{\hbar}\int_{t_0}^{t} \frac{m}{2}\dot{x}^2 d\tau\right\} V(x,s) \\
&= \left(\frac{m}{2\pi i\hbar\varepsilon}\right)^{\frac{n+1}{2}} \int dx_n \cdots \int dx_{j+1} \int dx_j \int dx_{j-4} \cdots \int dx_1 \\
&\exp\left\{\frac{im}{2\hbar\varepsilon}\left[(x_{n+1}-x_n)^2 + \ldots + (x_{j+1}-x_j)^2\right]\right\} V(x_i) \exp\left\{\frac{im}{2\hbar\varepsilon}\left[(x_j - x_{j-1})^2 + \ldots + (x_1 - x_0)^2\right]\right\} \\
&= \int_{-\infty}^{\infty} dx' G_0(x,t;x',s) V(x',s) G_0(x',s;x_0,t_0)
\end{aligned} \tag{18.72b}$$

where in the last expression on the right-hand side we have changed the label x_j to x'. This yields

$$G_1(x,t;x_0,t_0) = -\frac{i}{\hbar}\int_{t_0}^{t} ds \int_{-\infty}^{\infty} dx' G_0(x,t;x',s) V(x',s) G_0(x',s;x_0,t_0) \tag{18.73}$$

A simple graphic interpretation can be given to G_0 and to G_1. G_0 is the Green function (*kernel*) that allows a particle to propagate freely from $u_0 = x_0, t_0$ to $u = x, t$, and this is illustrated in Figure 18.4a. From (18.73), we interpret G_1 as follows:

- A particle propagates freely from u_0 to $u_1 = (x', s)$: [the kernel $G_0(x', s; x_0, t_0)$].
- The particle is scattered by potential $V(x', s)$.
- The particle propagates freely from u_1 to u: [the kernel $G_0(x, t; x', s)$].

These three steps are illustrated in Figure 18.4b. The product of the three associated factors, integrated over all possible values x' and s and multiplied by $-i/\hbar$ is G_1.

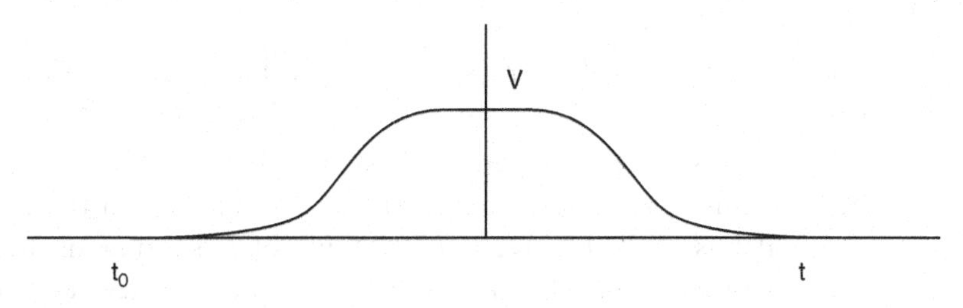

Figure 18.5 Diagram illustrating gradual turn-on and turn-off of scattering potential.

Similarly, the following expression is obtained for G_2:

$$G_2(x,t;x_0,t_0)=\left(\frac{-i}{\hbar}\right)^2\int_{-\infty}^{\infty}dx_2\int_{t_0}^{t}ds_2\int_{-\infty}^{\infty}dx_1\int_{t_0}^{t}ds_1 G_0(u,u_2)V(u_2)G_0(u_2,u_1)V(u_1)G_0(u_1,u_0)$$
$$=\left(\frac{-i}{\hbar}\right)^2\int du_2\int du_1 G_0(u,u_2)V(u_2)G_0(u_2,u_1)V(u_1)G_0(u_1,u_0) \quad (18.74)$$

The form of this expression should be intuitively clear from the preceding discussion. The only subtle point is the absence of a factor of $(2!)^{-1}$ on the right-hand side of (18.74). This is so because we require $G_0(u_2,u_1)=0$ if $t_1>t_2$. The interpretation of G_2 is illustrated in Figure 18.4c. Extensions to G_3, G_4, and so on follow a similar pattern and are straightforward.

Because the symbols u_1 and u_2 represent dummy variables of integration, the expansion

$$G(u,u_0)=G_0(u,u_0)+\left(\frac{-i}{\hbar}\right)\int du_1 G_0(u,u_1)V(u_1)G_0(u_1,u_0)$$
$$+\left(\frac{-i}{\hbar}\right)^2\int du_2\int du_1 G_0(u,u_2)V(u_2)G_0(u_2,u_1)V(u_1)G_0(u_1,u_0)+\cdots$$

can be written as

$$G(u,u_0)=G_0(u,u_0)+\left(\frac{-i}{\hbar}\right)\int du_1 G_0(u,u_1)V(u_1)\bullet$$
$$\left[G_0(u_1,u_0)-\frac{i}{\hbar}\int du_2 G_0(u_1,u_2)V(u_2)G_0(u_2,u_0)+\cdots\right] \quad (18.75)$$

Now the factor in square brackets on the right-hand side of (18.75) is $G(u,u_0)$. Thus (18.75) is the integral equation

$$G(u,u_0)=G_0(u,u_0)-\frac{i}{\hbar}\int G_0(u,u_1)V(u_1)G(u_1,u_0)\,du_1 \quad (18.76)$$

Also, because

$$\psi(u)=\int G(u,u_0)\psi(u_0)\,dx_0$$

we have

$$\psi(u) = \int G_0(u,u_0)\psi(u_0)\,dx_0 - \frac{i}{\hbar}\int du_1 \int G_0(u,u_1)V(u_1)G(u_1,u_0)\psi(u_0)\,dx_0 \tag{18.77}$$

Now, in any scattering experiment, we can think of t_0 as a time in the remote past when the incoming projectile wave packet is still distant from the target, and the effect of the potential V is negligible. We can also think of t as a time in the far future when the interaction between projectile and target has been completed, and the potential is once again negligible. Thus we can imagine a graph of the potential as in Figure 18.5, where V turns on gradually at some time between t_0 and 0 and then turns off gradually at some time between 0 and t.

If the scattering is elastic, then the energy E associated with $\psi(u)$ is the same as that associated with $\psi(u_0)$. We can thus write

$$\psi(u_0) = \psi(x_0,t_0) = \phi_0(x_0)e^{-iEt_0/\hbar}$$
$$\psi(u) = \psi(x,t) = \phi(x)e^{-iEt/\hbar}$$

where ϕ_0 and ϕ are called the *in* and *out* spatial wave functions, respectively.

Inserting these expressions in (18.77), we obtain

$$\begin{aligned}
\phi(x) &= \phi_0(x) - \frac{i}{\hbar}\int dx_1 \int_{t_0}^{t} dt_1 \int e^{iEt/\hbar} G_0(u,u_1)V(u_1)G(u_1,u_0)\psi(u_0)\,dx_0 \\
&= \phi_0(x) - \frac{i}{\hbar}\int dx_1 \int_{t_0}^{t} e^{iEt/\hbar} dt_1 G_0(u,u_1)V(u_1)\int G(u_1,u_0)\psi(u_0)\,dx_0 \\
&= \phi_0(x) - \frac{i}{\hbar}\int dx_1 \int_{t_0}^{t} e^{iEt/\hbar} dt_1 G_0(u,u_1)V(u_1)\psi(u_1) \\
&= \phi_0(x) - \frac{i}{\hbar}\int dx_1 \int_{t_0}^{t} e^{iE(t-t_1)/\hbar} dt_1 G_0(u,u_1)V(u_1)\phi(x_1)
\end{aligned} \tag{18.78}$$

Now $\phi(x) = \langle x|\Psi\rangle$ and $\phi_0(x) = \langle x|\Psi_0\rangle$, where $|\Psi_0\rangle$ and $|\Psi\rangle$ are in, out state vectors, respectively. Also, $G_0(u,u_1) = \langle x|e^{-iH_0(t-t_1)/\hbar}|x_1\rangle$, where $H_0 = \hat{p}^2/2\mu$ is the free-particle Hamiltonian. Thus (18.78) can be written

$$\langle x|\Psi\rangle = \langle x|\Psi_0\rangle - \frac{i}{\hbar}\int dx_1 \int_{t_0}^{t} e^{iE(t-t_1)/\hbar} dt_1 \langle x|e^{-iH_0(t-t_1)/\hbar}|x_1\rangle V(x_1,t_1)\langle x_1|\Psi\rangle$$

which implies that

$$|\Psi\rangle = |\Psi_0\rangle - \frac{i}{\hbar}\left\{\int_{-\infty}^{t} \exp\left[i(E-H_0)(t-t_1)/\hbar\right]\hat{V}_0(t_1)\,dt_1\right\}|\Psi\rangle \tag{18.79}$$

where $\hat{V}_0$ is the potential written in operator form, and we have chosen $t_0 = -\infty$. In (18.79) we have freed ourselves from the coordinate representation. In the integral on the right-hand side of (18.79), let us make the change of variable $T = t - t_1$. Then the integral becomes

$$I_1 = -\frac{i}{\hbar}\int_0^\infty \exp\left[\frac{i}{\hbar}(E-H_0)T\right]\hat{V}_0(t-T)\,dT$$

To account for the slow time variation of $\hat{V}_0$, we write $\hat{V}_0(t-T)=\hat{V}e^{-\varepsilon T/\hbar}$, where $\hat{V}$ is the potential operator evaluated at times when the projectile and target are in full interaction, and ε is a positive real infinitesimal. Then I_1 is easily evaluated,[1] and (18.79) becomes

$$|\Psi\rangle = |\Psi_0\rangle + \frac{1}{E-H_0+i\varepsilon}\hat{V}|\Psi\rangle \tag{18.80}$$

This is the important Lippmann-Schwinger equation, which contains the operator

$$(E-H_0+i\varepsilon)^{-1}$$

We now employ the Lippmann-Schwinger equation to establish the connection between the time-dependent view of scattering, which is characterized by (18.75) through (18.77), and the time-independent view. Returning to three spatial dimensions, we multiply (18.80) on the left by $\langle \boldsymbol{r}|$ and use the completeness relation on the right-hand side to obtain

$$\langle \boldsymbol{r}|\Psi\rangle = \langle \boldsymbol{r}|\Psi_0\rangle + \int\left(\frac{\hbar^2}{2\mu}\langle \boldsymbol{r}|\frac{1}{E-H_0+i\varepsilon}|\boldsymbol{r}'\rangle\right)\langle \boldsymbol{r}'|\hat{U}|\Psi\rangle d^3\boldsymbol{r}' \tag{18.81}$$

Now consider the factor in parentheses in the integrand on the right-hand side of (18.81). It is

$$\begin{aligned}\frac{\hbar^2}{2\mu}\langle \boldsymbol{r}|\frac{1}{E-H_0+i\varepsilon}|\boldsymbol{r}'\rangle &= \frac{\hbar^2}{2\mu}\iint\langle \boldsymbol{r}|\boldsymbol{p}''\rangle\frac{\langle \boldsymbol{p}''|\boldsymbol{p}'\rangle}{\frac{p^2}{2\mu}-\frac{p'^2}{2\mu}+i\varepsilon}\langle \boldsymbol{p}'|\boldsymbol{r}'\rangle\, d^3\boldsymbol{p}''d^3\boldsymbol{p}' \\ &= \frac{\hbar^2}{2\mu}\left(\frac{1}{2\pi\hbar}\right)^3\iint e^{i\boldsymbol{p}''\cdot\boldsymbol{r}/\hbar}e^{-i\boldsymbol{p}'\cdot\boldsymbol{r}'/\hbar}\frac{\delta^3(\boldsymbol{p}'-\boldsymbol{p}'')}{\frac{p^2}{2\mu}-\frac{p'^2}{2\mu}+i\varepsilon}\,d^3\boldsymbol{p}''d^3\boldsymbol{p}' \\ &= \left(\frac{1}{2\pi}\right)^3\int\frac{e^{i\boldsymbol{k}'\cdot(\boldsymbol{r}-\boldsymbol{r}')}}{k^2-k'^2+i\varepsilon'}\,d^3\boldsymbol{k}'\end{aligned}$$

where $\varepsilon' = (2\mu/\hbar^2)\varepsilon$ is also a positive infinitesimal. The integral

$$G(\boldsymbol{r},\boldsymbol{r}') = \frac{1}{(2\pi)^3}\int\frac{1}{k^2-k'^2+i\varepsilon'}e^{i\boldsymbol{k}'\cdot(\boldsymbol{r}-\boldsymbol{r}')}\,d^3\boldsymbol{k}' \tag{18.82}$$

is evaluated by choosing polar coordinates in $\boldsymbol{k}'$ space with $\boldsymbol{R}=\boldsymbol{r}-\boldsymbol{r}'$ along the polar axis; that is,

[1] The choice $\hat{V}_0(t-T)=\hat{V}e^{-\varepsilon T/\hbar}$ makes evaluation of I_1 easy, but it appears to be quite arbitrary. However, it can be shown that the choice of any function of T that goes very gradually to zero as $T\to\infty$ yields the same result for I_1.

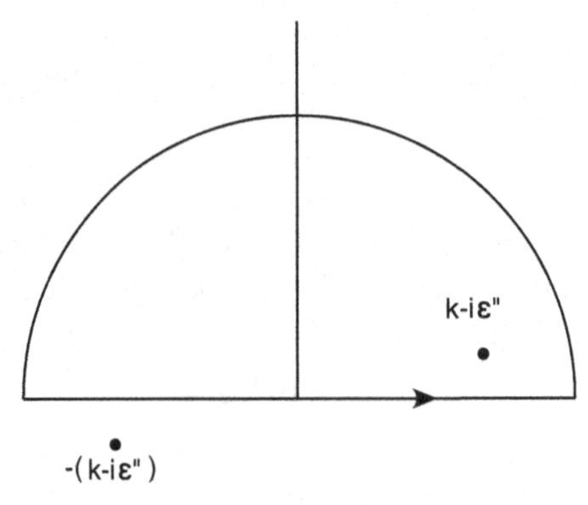

Figure 18.6 Contour in the complex k' plane for the evaluation of the integral in (18.83).

$$
\begin{aligned}
G(\boldsymbol{r},\boldsymbol{r}') &= \frac{2\pi}{(2\pi)^3}\int_0^\infty \frac{k'^2 dk'}{k^2-k'^2+i\varepsilon'}\int_0^\pi e^{ik'R\cos\theta}\sin\theta\, d\theta \\
&= \frac{1}{(2\pi)^2}\int_0^\infty \frac{k'^2 dk'}{k^2-k'^2+i\varepsilon'}\int_{-1}^1 e^{ik'Rx}\, dx \\
&= \frac{1}{(2\pi)^2}\int_0^\infty \frac{k'^2}{k^2-k'^2+i\varepsilon'}\frac{e^{ik'R}-e^{-ik'R}}{ik'R}\, dk' \\
&= \frac{i}{(2\pi)^2 R}\int_{-\infty}^\infty \frac{k' e^{ik'R}}{(k'+k+i\varepsilon'')(k'-k-i\varepsilon'')}\, dk'
\end{aligned} \tag{18.83}
$$

where $\varepsilon'' = \varepsilon'/2k$. The integrand in this last integral has a simple pole at $k' = k + i\varepsilon''$. Contour integration in the complex k' plane with the contour of Figure 18.6 and use of Cauchy's theorem yields

$$G(\boldsymbol{r},\boldsymbol{r}') = -\frac{1}{4\pi}\frac{e^{ik|\boldsymbol{r}-\boldsymbol{r}'|}}{|\boldsymbol{r}-\boldsymbol{r}'|} \tag{18.84}$$

which is frequently called the *time-independent Green function*. From (18.81), we have

$$\phi(\boldsymbol{r}) = \phi_0(\boldsymbol{r}) + \int G(\boldsymbol{r},\boldsymbol{r}')\, U(\boldsymbol{r}')\phi(\boldsymbol{r}')\, d^3\boldsymbol{r}' \tag{18.85a}$$

At this point, we change the names of the wave functions of interest to conform with our earlier notation: $\phi_0(\boldsymbol{r}) \to \psi_0(\boldsymbol{r})$, $\phi(\boldsymbol{r}) \to \psi(\boldsymbol{r})$. Then (18.70) is rewritten as

$$\psi(\boldsymbol{r}) = \psi_0(\boldsymbol{r}) + \int G(\boldsymbol{r},\boldsymbol{r}')\, U(\boldsymbol{r}')\psi(\boldsymbol{r}')\, d^3\boldsymbol{r}' \tag{18.85b}$$

Now ψ_0 describes a free particle and therefore satisfies the equation $(\nabla^2 + k^2)\psi_0 = 0$, whereas $(\nabla^2 + k^2)\psi(\boldsymbol{r}) = U(\boldsymbol{r})\psi(\boldsymbol{r})$. Applying the operator $(\nabla_r^2 + k^2)$ on the left to both sides of (18.85b), we have

$$U(\boldsymbol{r})\psi(\boldsymbol{r}) = \int (\nabla_r^2 + k^2) G(\boldsymbol{r},\boldsymbol{r}')\, U(\boldsymbol{r}')\psi(\boldsymbol{r}')\, d^3\boldsymbol{r}'$$

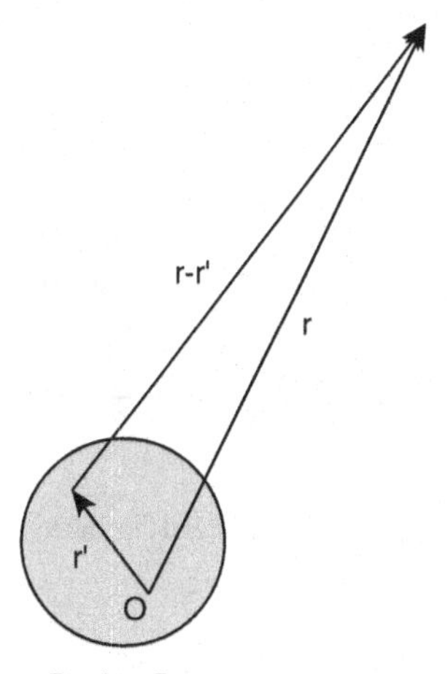

Figure 18.7 Vectors r, r' in calculation of scattering amplitude in first Born approximation.

which implies that

$$(\nabla_r^2 + k^2)G(\boldsymbol{r},\boldsymbol{r}') = \delta^3(\boldsymbol{r}-\boldsymbol{r}') \tag{18.86}$$

18.7 Potential scattering in the Born approximation

The starting point for developing the Born approximation in time-independent scattering theory is the integral equation (18.85b). It is solved by iteration: we repeatedly substitute for $\psi(\boldsymbol{r}')$ in the integrand of (18.85b), its value given by the left-hand side; that is,

$$\psi(\boldsymbol{r}) = \psi_0(\boldsymbol{r}) + \int d^3\boldsymbol{r}' G(\boldsymbol{r},\boldsymbol{r}')U(\boldsymbol{r}')\left[\psi_0(\boldsymbol{r}') + \int d^3\boldsymbol{r}'' G(\boldsymbol{r}',\boldsymbol{r}'')U(\boldsymbol{r}'')\psi(\boldsymbol{r}'')\right] \tag{18.87}$$

or

$$\begin{aligned}\psi(\boldsymbol{r}) = \psi_0(\boldsymbol{r}) &+ \int d^3\boldsymbol{r}' G(\boldsymbol{r},\boldsymbol{r}')U(\boldsymbol{r}')\psi_0(\boldsymbol{r}') \\ &+ \iint d^3\boldsymbol{r}' d^3\boldsymbol{r}'' G(\boldsymbol{r},\boldsymbol{r}')U(\boldsymbol{r}')G(\boldsymbol{r}',\boldsymbol{r}'')U(\boldsymbol{r}'')\psi_0(\boldsymbol{r}'') + \cdots\end{aligned} \tag{18.88}$$

In (18.88), the leading term on the right-hand side is the free wave (zeroth Born approximation), the first two terms together give the first Born approximation, the first three terms the second Born approximation, and so on. The right-hand side of (18.88) is frequently called the *Born series* or the *Neumann series*.

In the first Born approximation, we have

$$\psi(\boldsymbol{r}) \cong \psi_0(\boldsymbol{r}) - \frac{1}{4\pi}\int \frac{e^{ik|\boldsymbol{r}-\boldsymbol{r}'|}}{|\boldsymbol{r}-\boldsymbol{r}'|} U(\boldsymbol{r}')\,\psi_0(\boldsymbol{r}')\,d^3\boldsymbol{r}' \tag{18.89}$$

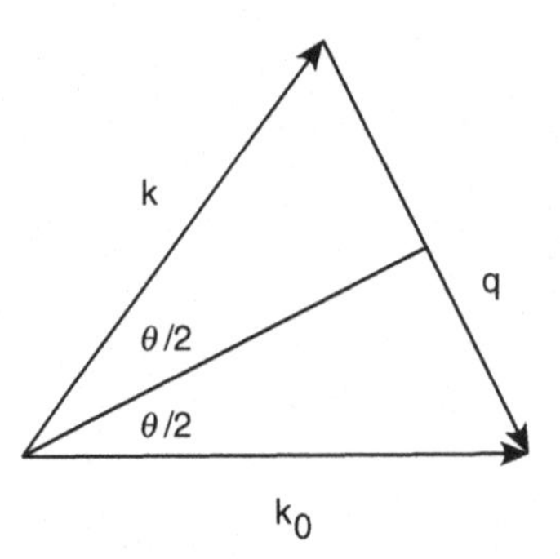

Figure 18.8 Diagram showing vectors k_0, k, and $q = k_0 - k$ and the scattering angle θ.

We are interested in asymptotic values of $\psi(\boldsymbol{r})$ far from the region D of interaction between projectile and target (as in Figure 18.7).

Thus $|\boldsymbol{r}| \gg |\boldsymbol{r}'|$; hence

$$|\boldsymbol{r}-\boldsymbol{r}'| = r\sqrt{1-2\frac{\boldsymbol{r}\cdot\boldsymbol{r}'}{r^2}+\frac{r'^2}{r^2}} \approx r-\hat{\boldsymbol{r}}\cdot\boldsymbol{r}'$$

where $\hat{\boldsymbol{r}}$ is a unit vector along $\boldsymbol{r}$. We use this approximation in the exponential factor in the integrand of (18.89), but it is sufficient to approximate the denominator of the integrand by r^{-1}. Thus we obtain

$$\psi(\boldsymbol{r}) \cong e^{ikz} - \frac{1}{4\pi}\frac{e^{ikr}}{r}\int e^{ikz'}e^{-ik\boldsymbol{r}'\cdot\hat{\boldsymbol{r}}}U(\boldsymbol{r}')\,d^3\boldsymbol{r}' \tag{18.90}$$

Let $\boldsymbol{k}_0$ and $\boldsymbol{k}$ be the incident and scattered wave vectors, respectively. Then $kz' = \boldsymbol{k}_0\cdot\boldsymbol{r}'$ and $k\boldsymbol{r}'\cdot\hat{\boldsymbol{r}} = \boldsymbol{k}'\cdot\boldsymbol{r}'$. Hence (18.90) becomes

$$\psi(\boldsymbol{r}) \cong e^{ikz} - \frac{1}{4\pi}\frac{e^{ikr}}{r}\int \exp(i\boldsymbol{q}\cdot\boldsymbol{r}')U(\boldsymbol{r}')\,d^3\boldsymbol{r}' \tag{18.91}$$

where $\hbar\boldsymbol{q} = \hbar(\boldsymbol{k}_0-\boldsymbol{k})$ is the momentum transfer. For elastic scattering, $|\boldsymbol{k}| = |\boldsymbol{k}_0|$, and as Figure 18.8 shows, $|\boldsymbol{q}| = 2k_0\sin(\theta/2)$.

From (18.91), the scattering amplitude in the first Born approximation is

$$\begin{aligned} f(\theta) &= -\frac{1}{4\pi}\int e^{i\boldsymbol{q}\cdot\boldsymbol{r}'}U(\boldsymbol{r}')\,d^3\boldsymbol{r}' \\ &= -\frac{\mu}{2\pi\hbar^2}\int e^{i\boldsymbol{q}\cdot\boldsymbol{r}'}V(\boldsymbol{r}')\,d^3\boldsymbol{r}' \end{aligned} \tag{18.92}$$

Apart from a constant factor, $f(\theta)$ is just the Fourier transform of the potential V with respect to $\boldsymbol{q}$.

18.8 Criterion for the validity of the Born approximation

We write the first Born approximation (18.89) in the form

$$\psi = e^{ikz} + \psi_1$$

Clearly, this approximation is valid only if the second term on the right-hand side is much smaller in magnitude than the first, that is, if $|\psi_1| \ll 1$. Now

$$\nabla^2 \psi_1 + k^2 \psi_1 = U e^{ikz} \tag{18.93}$$

We try to find a solution to this equation of the form $\psi_1 = g e^{ikz}$. Substitution of the latter expression into (18.93) yields

$$e^{ikz} \nabla^2 g + 2ik e^{ikz} \frac{\partial g}{\partial z} = U e^{ikz}$$

If k is sufficiently large, the second term on the left-hand side greatly dominates over the first, and we can neglect the latter. In this case, we obtain

$$\frac{\partial g}{\partial z} = \frac{U}{2ik} = -i \frac{\mu V}{\hbar^2 k}$$

which yields

$$\psi_1 = -\frac{i}{\hbar \mathrm{v}} e^{ikz} \int V dz \approx -\frac{i}{\hbar \mathrm{v}} e^{ikz} \bar{V} d$$

where d is the range of the potential, $\bar{V}$ is a suitable average of V over this range, and $\mathrm{v} = \hbar k / \mu$ is the projectile velocity. Then the requirement $|\psi_1| \ll 1$ implies

$$\bar{V} \ll \frac{\hbar \mathrm{v}}{d} \tag{18.94}$$

For scattering of a fast electron by an atom with atomic number Z, we have $d \approx a_0$ and $\bar{V} \approx Ze^2 / 4\pi a_0$. Thus the criterion (18.94) becomes $v/c \gg Z\alpha$. This criterion is easily satisfied for low and moderate values of Z but obviously becomes impossible to satisfy when $Z\alpha$ is comparable to unity. The criterion is obviously more easily satisfied for scattering of a massive projectile (e.g., muon, proton, etc.) by an atom.

18.9 Coulomb scattering in the first Born approximation

Let us consider the Yukawa potential

$$V(r) = \frac{Z_1 Z_2 e^2}{4\pi r} e^{-\lambda r}$$

where λ is a real parameter, and $Z_1 e, Z_2 e$ are the charges of the projectile and target, respectively. The scattering amplitude for this potential is

$$\begin{aligned} f(\theta) &= -\frac{\mu}{2\pi\hbar^2} Z_1 Z_2 \frac{e^2}{4\pi} \int_0^\infty r e^{-\lambda r}\, dr \bullet 2\pi \int_0^\pi e^{iqr\cos\theta} \sin\theta\, d\theta \\ &= -\frac{\mu}{\hbar^2} Z_1 Z_2 \frac{e^2}{4\pi} \int_0^\infty r e^{-\lambda r} \frac{e^{iqr} - e^{-iqr}}{iqr}\, dr \\ &= \frac{i\mu}{\hbar^2 q} Z_1 Z_2 \frac{e^2}{4\pi} \int_0^\infty \left(e^{-(\lambda - iq)r} - e^{-(\lambda + iq)r} \right) dr \\ &= \frac{i\mu}{\hbar^2 q} Z_1 Z_2 \frac{e^2}{4\pi} \left[\frac{1}{(\lambda - iq)} - \frac{1}{(\lambda + iq)} \right] \\ &= -\frac{2\mu}{\hbar^2} Z_1 Z_2 \frac{e^2}{4\pi} \frac{1}{\lambda^2 + q^2} = -\frac{2 Z_1 Z_2 \mu}{\hbar^2} \frac{e^2}{4\pi} \frac{1}{\lambda^2 + 4k^2 \sin^2(\theta/2)} \end{aligned} \tag{18.95}$$

In the limit of vanishingly small λ, (18.95) yields the scattering amplitude for the Coulomb potential; that is,

$$f(\theta) = -\frac{2 Z_1 Z_2 \mu}{\hbar^2} \frac{e^2}{4\pi} \frac{1}{4k^2 \sin^2(\theta/2)} \tag{18.96}$$

The foregoing procedure is flawed mathematically because the radial integral in (18.95) is not defined when $\lambda = 0$. However, as we mentioned in Chapter 14, the calculation is reasonable on physical grounds. In any real Coulomb scattering experiment, the Coulomb potential generated by a target (e.g., a nucleus) is screened by the charges of surrounding matter to some extent; hence a "Coulomb" potential actually can be approximated by the Yukawa form with some extremely small and undetectable but nevertheless nonzero λ.

Equation (18.96) yields the differential cross section for Rutherford scattering; that is,

$$\sigma(\theta) = |f(\theta)|^2 = \left(\frac{Z_1 Z_2}{4E} \right)^2 \frac{e^4}{(4\pi)^2} \frac{1}{\sin^4(\theta/2)} \tag{18.97}$$

where $E = \hbar^2 k^2 / 2\mu$ for a nonrelativistic projectile. Note that (18.97) is exactly the same result that was obtained from classical mechanics by Rutherford and is also identical to the exact cross section derived in equation (18.66). How can the first Born approximation yield the same cross section for Coulomb scattering as the *exact* solution to Schroedinger's equation? The explanation is quite subtle and requires a careful definition of the scattering amplitude when higher-order terms are included in the Born approximation (Holstein 2007).

18.10 Elastic scattering of fast electrons by atoms in the first Born approximation

The scattering of a fast electron from an atom of atomic number Z is an interesting and instructive example that can be treated in the first Born approximation. We start with the scattering amplitude

$$f(\theta) = -\frac{m_e}{2\pi\hbar^2}\int e^{i\boldsymbol{q}\cdot\boldsymbol{r}} V(\boldsymbol{r})\, d^3\boldsymbol{r} \tag{18.98}$$

where V is now the potential energy of interaction of the projectile electron with the nucleus and the atomic electrons. Because

$$e^{i\boldsymbol{q}\cdot\boldsymbol{r}} = -\frac{1}{q^2}\nabla^2 e^{i\boldsymbol{q}\cdot\boldsymbol{r}}$$

we can write

$$f(\theta) = \frac{m_e}{2\pi\hbar^2 q^2}\int V(\boldsymbol{r})\left(\nabla^2 e^{i\boldsymbol{q}\cdot\boldsymbol{r}}\right) d^3\boldsymbol{r}$$

Now

$$\int V(\boldsymbol{r})\left(\nabla^2 e^{i\boldsymbol{q}\cdot\boldsymbol{r}}\right) d^3\boldsymbol{r} = \int \nabla\cdot\left(V\nabla e^{i\boldsymbol{q}\cdot\boldsymbol{r}}\right) d^3\boldsymbol{r} - \int \nabla V\cdot\nabla e^{i\boldsymbol{q}\cdot\boldsymbol{r}}\, d^3\boldsymbol{r} \tag{18.99}$$

By Gauss's theorem, the first integral on the right-hand side of (18.99) can be converted to a surface integral that vanishes if the volume this surface encloses has sufficiently large radius. Also,

$$-\int \nabla V\cdot\nabla e^{i\boldsymbol{q}\cdot\boldsymbol{r}}\, d^3\boldsymbol{r} = -\int \nabla\cdot\left(e^{i\boldsymbol{q}\cdot\boldsymbol{r}}\nabla V\right) d^3\boldsymbol{r} + \int e^{i\boldsymbol{q}\cdot\boldsymbol{r}}\nabla^2 V\, d^3\boldsymbol{r} \tag{18.100}$$

and once again, the first integral on the right-hand side of (18.100) vanishes by Gauss's theorem. Hence

$$f(\theta) = \frac{\mu}{2\pi\hbar^2 q^2}\int e^{i\boldsymbol{q}\cdot\boldsymbol{r}}\left(\nabla^2 V\right) d^3\boldsymbol{r} \tag{18.101}$$

However, from Poisson's equation,

$$\nabla^2 V = e^2\left[Z\delta^3(\boldsymbol{r}) - \rho(\boldsymbol{r})\right] \tag{18.102}$$

where the first term on the right-hand side refers to the nucleus, whereas ρ is the atomic electron number density, with

$$\int \rho\, d^3\boldsymbol{r} = Z \tag{18.103}$$

for a neutral atom. Substitution of (18.102) into (18.101) yields

$$f(\theta) = \frac{2m_e}{\hbar^2 q^2}\frac{e^2}{4\pi}\left[Z - F(\boldsymbol{q})\right] \tag{18.104}$$

where

$$F(\boldsymbol{q}) = \int e^{i\boldsymbol{q}\cdot\boldsymbol{r}}\rho(\boldsymbol{r})\, d^3\boldsymbol{r} \tag{18.105}$$

is called the *form factor* of the atomic electron charge distribution. The differential cross section is

$$\begin{aligned}\frac{d\sigma}{d\Omega} = |f(\theta)|^2 &= \left(\frac{2m_e}{\hbar^2 q^2}\right)^2 \frac{e^4}{(4\pi)^2}[Z - F(\boldsymbol{q})]^2 \\ &= \frac{4}{(a_0 q^2)^2}[Z - F(\boldsymbol{q})]^2\end{aligned} \tag{18.106}$$

Let R be the radius over which ρ is significantly greater than zero. When $qR \ll 1$, it is useful to expand $Z - F(\boldsymbol{q})$ in a power series as follows:

$$Z - F(\boldsymbol{q}) = \left(Z - \int \rho\, d^3\boldsymbol{r}\right) - \left(i\boldsymbol{q}\cdot\int \boldsymbol{r}\rho\, d^3\boldsymbol{r}\right) + \left[\frac{1}{2!}\int (\boldsymbol{q}\cdot\boldsymbol{r})^2 \rho\, d^3\boldsymbol{r}\right] - \cdots \tag{18.107}$$

The first term in parentheses on the right-hand side of (18.107) vanishes because of (18.103). The second term also vanishes because in $\rho = |\psi|^2$ the atomic wave function ψ has definite parity. [Indeed, every term in the series on the right-hand side of (18.107) with an odd power of q vanishes for the same reason.] The term in square brackets is $Zq^2\langle r^2\rangle/6$, which yields

$$f(\theta) = Z\frac{\langle r^2\rangle}{3a_0} \tag{18.108}$$

which is independent of the scattering angle. Thus the singularity in Rutherford scattering at $q = 0$ [recall (18.97)] does not appear in the first Born approximation when a fast electron scatters from a neutral atom. The physical explanation for this is very simple. The singularity in Rutherford scattering occurs at $\theta = 0$, which is equivalent to an extremely large impact parameter. However, when an electron passes a neutral atom at a very large distance, it sees no potential whatever in first approximation because the nuclear potential is canceled by that of the atomic electrons.

The form factor F can be determined experimentally from detailed measurements of the scattering cross section as a function of q. From (18.107), this yields

$$-\frac{\partial F}{\partial q^2}\bigg|_{q^2=0} = \frac{Z}{6}\langle r^2\rangle \tag{18.109}$$

Hence one can determine the root-mean-square (rms) charge radius of the atomic electron distribution from experimental scattering data. A similar method has been used in high-energy physics with relativistic electrons as projectiles to determine the mean-square charge radius of the proton (0.8×10^{-13} cm), as well as the mean-square charge radius of other nuclei.

The total cross section for electron-atom scattering in first Born approximation is obtained from (18.106) by integrating over all angles. Making the change of variable $q = 2k\sin(\theta/2)$, we easily obtain

$$\sigma = \frac{8\pi}{a_0^2 k^2}\int_0^{2k} \frac{(Z - F)^2}{q^3}\, dq \tag{18.110}$$

Because for very small q, $(Z-F)^2$ is of order q^4, whereas for very large q $F \to 0$, the integral $\int_0^\infty \left[(Z-F)^2/q^3\right] dq$ is finite; hence, at large k, σ is proportional to $1/E$.

For atomic hydrogen, the density ρ and the form factor F are easily calculated exactly from the ground state wave function; that is,

$$\psi = \frac{1}{\sqrt{\pi a_0^3}} e^{-r/a_0}$$

We obtain

$$F(q) = \frac{1}{\left(1+\dfrac{a_0^2 q^2}{4}\right)^2} \tag{18.111}$$

Substitution of (18.111) into (18.106) with $Z = 1$ yields

$$\frac{d\sigma}{d\Omega} = \frac{4a_0^2\left(8+a_0^2q^2\right)^2}{\left(4+a_0^2q^2\right)^4} \tag{18.112}$$

For atomic helium, use of the simplest ground-state variational wave function[2] yields a cross section that agrees reasonably well with experiment. In the case of more complex atoms, the Thomas-Fermi and Hartree methods are useful, and more precise results are obtained at the expense of additional computational complexity by including exchange (i.e., the Hartree-Fock method).

18.11 Connection between the Born approximation and time-dependent perturbation theory

It is intuitively clear that the Born approximation is closely related to perturbation theory. We now show that these two approaches are equivalent by demonstrating that the Born approximation and the golden rule of time-dependent perturbation theory lead to exactly the same results for nonrelativistic elastic scattering. The first Born approximation gives

$$\begin{aligned} d\sigma &= \sigma(\theta)d\Omega = |f(\theta)|^2\, d\Omega \\ &= \frac{\mu^2}{4\pi^2\hbar^4}\left|\int e^{i(\mathbf{k}_i-\mathbf{k}_f)\cdot\mathbf{r}} V(\mathbf{r})\, d^3\mathbf{r}\right|^2 d\Omega \end{aligned} \tag{18.113}$$

On the other hand, the golden rule yields the following transition probability per unit time:

[2] Recall Section 11.6.

$$dW = \frac{2\pi}{\hbar}|\langle V\rangle|^2 \rho(E_f) \tag{18.114}$$

where the matrix element is $\langle V\rangle = \int \psi_f^*(\boldsymbol{r})V\psi_i(\boldsymbol{r})d^3\boldsymbol{r}$, and

$$\psi_i = \frac{1}{L^{3/2}}e^{i\boldsymbol{k}_i\cdot\boldsymbol{r}} \quad \text{and} \quad \psi_f = \frac{1}{L^{3/2}}e^{i\boldsymbol{k}_f\cdot\boldsymbol{r}}$$

Also, for a nonrelativistic final projectile,

$$\rho = L^3\frac{d^3\boldsymbol{p}_f}{(2\pi\hbar)^3\,dE_f} = L^3\frac{\mu p_f^2 dp_f d\Omega}{(2\pi\hbar)^3\,p_f dp_f} = L^3\frac{\mu p_f d\Omega}{(2\pi\hbar)^3}$$

Hence (18.114) becomes

$$dW = \frac{2\pi}{\hbar}\frac{1}{L^6}\left|\int e^{i(\boldsymbol{k}_i-\boldsymbol{k}_f)\cdot\boldsymbol{r}}V(\boldsymbol{r})\,d^3\boldsymbol{r}\right|^2\frac{L^3\mu p_f d\Omega}{8\pi^3\hbar^3}$$

The differential cross section is $d\sigma = dW/j$, where $j = p_i/(\mu L^3)$ is the incident probability current density. Thus we obtain

$$d\sigma = \frac{\mu^2}{4\pi^2\hbar^4}\left|\int e^{i(\boldsymbol{k}_i-\boldsymbol{k}_f)\cdot\boldsymbol{r}}V(\boldsymbol{r})\,d^3\boldsymbol{r}\right|^2\frac{p_f}{p_i}d\Omega \tag{18.115}$$

Because $p_i = p_f$ for elastic scattering, (18.115) and (18.113) are indeed identical.

18.12 Inelastic scattering in the Born approximation

The equivalence of (18.115) and (18.113) for elastic scattering suggests that the Born approximation can be extended to describe inelastic scattering. Here we limit ourselves to a target, but not a projectile, with internal degrees of freedom. The target can be excited by interaction with the passing projectile, and the final energy of the projectile thus is less than its initial energy. Let E_0 and $E_n > E_0$ be the internal energies of the target before and after the scattering event, respectively, and let u_0 and u_n be the initial and final wave functions of the target corresponding to these energies, respectively. Then conservation of energy requires

$$E_n - E_0 = \frac{\hbar^2}{2\mu}\left(k_i^2 - k_f^2\right) \tag{18.116}$$

Also, for inelastic scattering, the integral in (18.115) is replaced by

$$T = \iint e^{i\boldsymbol{q}\cdot\boldsymbol{r}}u_n^* V u_0\, d^3\boldsymbol{r}\,d\tau \tag{18.117}$$

where

$$q = k_i - k_f \quad q^2 = k_i^2 + k_f^2 - 2k_i k_f \cos\theta \quad \text{and} \quad d\tau = \prod_{i=1}^{N} d\tau_i \tag{18.118}$$

where N is the number of particles in the target, $d\tau_i$ is the volume element corresponding to the ith particle, and V is the potential energy of interaction of the projectile with each target particle. For inelastic scattering of a fast electron from a neutral many-electron atom with atomic number Z,

$$V = -\frac{Ze^2}{4\pi r} + \sum_{i=1}^{Z} \frac{e^2}{4\pi|r - r_i|} \tag{18.119}$$

Here the first term on the right-hand side describes interaction between the projectile electron and the nucleus (the latter located at the origin), whereas the remaining terms refer to the interaction of the projectile with Z atomic electrons. In fact, because the many-electron wave functions u_0 and u_n are orthogonal, the first term on the right-hand side of (18.119) does not contribute to (18.117).

The general procedure for analyzing fast electron-atom inelastic scattering is illustrated by consideration of electron scattering from a hydrogen atom, where, ignoring the first term on the right-hand side of (18.119), we have from (18.117)

$$\begin{aligned} T &= \frac{e^2}{4\pi} \iint e^{iq\cdot r} u_n^* \frac{1}{|r - r_1|} u_0 \, d^3r d^3r_1 \\ &= -\frac{e^2}{4\pi q^2} \iint e^{iq\cdot r} u_n^*(r_1) \left(\nabla_r^2 \frac{1}{|r - r_1|} \right) u_0(r_1) \, d^3r d^3r_1 \\ &= \frac{e^2}{q^2} \iint e^{iq\cdot r} u_n^*(r_1) \delta^3(r - r) u_0(r_1) \, d^3r d^3r_1 \\ &= \frac{e^2}{q^2} \int e^{iq\cdot r} u_n^*(r) u_0(r) \, d^3r \end{aligned} \tag{18.120}$$

The integral on the right-hand side of the last line is frequently called the *transition form factor* $F_n(q)$, and from (18.115), the differential cross section is

$$\left(\frac{d\sigma}{d\Omega} \right)_{0\to n} = \left(\frac{2}{a_0 q^2} \right)^2 \frac{p_f}{p_i} [F_n(q)]^2 \tag{18.121}$$

When $qR \ll 1$, where R is the atomic radius, we can expand the exponential in the integrand of (18.120); that is,

$$e^{iq\cdot r} \approx 1 + iq\cdot r$$

The leading term in this expansion contributes nothing to F_n because of the orthogonality of u_n and u_0. The next term does contribute if the selection rules for an allowed electric dipole transition are obeyed for a transition between u_0 and u_n. If $qR \approx 1$ or $qR \gg 1$, this restriction on u_n is relaxed. For example, consider the transition form factors for a hydrogen atom excited by a fast electron in the following two cases:

$$1s \to 2p, m_\ell = 0: \qquad F_{2p}(q) = \frac{15}{4\sqrt{2}} \frac{iqa_0}{\left(\frac{9}{4} + q^2 a_0^2\right)^3}$$

$$1s \to 2s: \qquad F_{2s}(q) = \frac{8}{\sqrt{2}} \frac{(qa_0)^2}{\left(\frac{9}{4} + q^2 a_0^2\right)^3}$$

These formulas show that when $qa_0 \ll 1$, $|F_{2s}| \ll |F_{2p}|$; however, when $qa_0 \approx 1$, $|F_{2s}| \approx |F_{2p}|$.

We return to the Born approximation in later chapters.

Problems for Chapter 18

18.1. This problem concerns some general properties of the phase shifts appearing in (18.22a) for scattering from a central potential $V(r)$.

(a) What is the largest possible total cross section if all phase shifts vanish for $\ell > L$?

(b) Let $U(r) = [2\mu V(r)]/\hbar^2$ and assume that $U(r) \to \text{const} \cdot r^{-s}$ for large r. Use the WKB approximation to show that if $s > 1$ and $\ell \gg 1$,

$$\delta_\ell \simeq -\frac{\mu}{\hbar^2} \frac{\ell}{k^2} V\left(\frac{\ell}{k}\right)$$

Thus show that σ_{total} is finite if $s > 2$. If $2 > s > 1$, the total cross section is infinite. Is this due to large or small impact parameter scattering? In the real world, how are such infinities avoided?

18.2. Determine the phase shift δ_0 for scattering from the potential $V(r) = -V_0 \exp(-r/a)$. What is the connection between δ_0 and the spectrum of bound states with $\ell = 0$?

18.3. Consider the Coulomb scattering of two identical fermions of charge Ze, for example, electrons. Show that in the CM frame the cross section to obtain either particle scattered into angle θ within small solid angle $d\Omega$ is

$$d\sigma(\theta) = \frac{Z^4}{16E^2} \frac{e^4}{(4\pi)^2} \frac{1}{\sin^4 \frac{\theta}{2}} \left(1 + \tan^4 \frac{\theta}{2} \pm 2Q \tan^2 \frac{\theta}{2}\right) d\Omega$$

where $Q = \cos\{n\, \ell n[\tan^2(\theta/2)]\}$, $\pm$ refers to total spin 1 (0), and $n = Z^2 e^2 / 4\pi\hbar v$. How is this differential cross section reconciled with the Rutherford formula for scattering of two identical electrons in the classical limit? What modifications would be required if we were describing the scattering of two identical bosons?

18.4. For Coulomb scattering of two positive charges $Z_1 e, Z_2 e$, we show [equation (18.67)] that the square of the wave function for relative motion takes the following value at the origin:

$$|\psi(0)|^2 = \frac{2\pi n}{v}\exp(-2\pi n) \qquad \text{(Gamow's formula)}$$

where $n = Z_1Z_2e^2/4\pi\hbar v > 0$, and where the relative velocity v is assumed to be nonrelativistic: $v = c$. This formula has an important application in astrophysics – to thermonuclear reactions in stellar interiors. Suppose that one has an ionized gas with bare nuclei of atomic numbers Z_1 and Z_2 and number densities n_1 and n_2, respectively, at temperature T. The rate λ of thermonuclear reactions per cubic centimeter between these species is proportional to $|\psi(0)|^2\, n_1 n_2 v$. Show that to a good approximation

$$\lambda = \frac{\text{const}}{T^{2/3}}\exp\left(-b\frac{Z_1^{2/3}Z_2^{2/3}}{T^{1/3}}\right)$$

where

$$b = \frac{3}{2^{5/3}}\left(\frac{\mu e^4}{\hbar^2 k_B}\right)^{1/3}$$

and k_B is Boltzmann's constant, whereas μ is the reduced mass of the two reactants. Also show that practically all the reactions occur in a narrow band of energies at

$$E = E_0 = \left(\frac{\pi}{\sqrt{2}}Z_1Z_2\alpha\sqrt{\mu c^2}\,k_B T\right)^{2/3}$$

What is E_0/k_BT for the proton-proton reaction at the center of the Sun, where $T = 1.2\times10^7$ K? Your result should reveal that almost all *pp* reactions in the Sun are generated with protons that are far out on the high-energy tail of the relative-velocity Maxwell distribution.

18.5. In this problem we consider the photoelectric effect in hydrogen. A photon with frequency ω has $\boldsymbol{k}$ along z and linear polarization along x. The photon is absorbed by a hydrogen atom at the origin, initially in the $1s$ ground state. The final state consists of a proton (the recoil of which we neglect) and an ionized electron with energy

$$E = \frac{\hbar^2 k_f^2}{2m_e} = \frac{m_e v^2}{2}$$

and final state wave function u_f. We assume that

$$m_e c^2\alpha^2 \ll \hbar\omega \ll m_e c^2$$

(That is, the photon energy is much larger than the binding energy of the ground-state H atom but much less than the electron rest energy.) Use the Born approximation to show that

$$\frac{d\sigma}{d\Omega} = 2^5\alpha\left(\frac{\hbar}{m_e c}\right)\frac{c}{\omega}\left(\frac{1}{k_f a_0}\right)^5\frac{\sin^2\theta\cos^2\phi}{\left(1-\frac{v}{c}\cos\theta\right)^4}$$

and that the total cross section is

$$\sigma = \frac{2^8 \pi}{3} \alpha a_0^2 \left(\frac{e^2}{8\pi a_0} \frac{1}{\hbar \omega} \right)^{7/2}$$

18.6. In this lengthy and very important problem we analyze the connection between scattering in classical mechanics and quantum mechanics.

(a) A classical particle in a central potential $V(r)$ moves in a plane, and its motion is described by the usual plane polar coordinates r, ϕ. Let J be the angular momentum with respect to the force center at the origin, and let r_i, ϕ_i be the initial values of r, ϕ. Show that

$$\phi = \phi_i - \int_{r_i}^{r} \frac{\partial}{\partial J} \sqrt{2\mu[E - V(r')] - \frac{J^2}{r'^2}}\, dr' \tag{1}$$

and that the scattering angle is given by

$$\pm\theta = \pi + 2\int_{r_0}^{\infty} \frac{\partial}{\partial J} \sqrt{2\mu[E - V(r')] - \frac{J^2}{r'^2}}\, dr' \tag{2}$$

where $\pm$ refers to a repulsive (attractive) potential, and r_0 is the largest root of $\sqrt{2\mu[E - V(r)] - (J^2/r^2)}$. Thus show that the differential cross section is

$$\frac{d\sigma}{d\Omega} = -\frac{b}{\sin\theta} \frac{db}{d\theta} \tag{3}$$

$$\frac{d\sigma}{d\Omega} = -\frac{J}{p^2 \sin\theta} \frac{dJ}{d\theta} \tag{4}$$

where $b = J/\mu v_0 = J/p$ is the impact parameter, and v_0 is the initial velocity. Note that $d\theta/db < 0$, so the differential cross section in [(3) and (4)] is positive.

(b) In quantum mechanics, we seek δ_ℓ in the asymptotic solution

$$\lim_{r \to \infty} u = \sin\left(kr - \frac{\ell\pi}{2} + \delta_\ell \right)$$

of the equation

$$\frac{d^2u}{dr^2} + \frac{2\mu}{\hbar^2}[E - V(r)] - \frac{\ell(\ell+1)}{r^2} = 0$$

Show that in the WKB approximation,

$$\delta_\ell = \lim_{r \to \infty} \left[\int_0^r F(r')\, dr' - kr + \frac{\pi}{2}\left(\ell + \frac{1}{2} \right) \right] \tag{5}$$

where

$$F(r) = \sqrt{2\mu[E - V(r)] - \frac{\left(\ell + \frac{1}{2}\right)^2}{r^2}}$$

and r_0 is the largest root of F^2.

(c) The goal of this part of the problem is to show that for large ℓ and in the WKB approximation, the scattering amplitude for nonforward scattering $(\theta \neq 0)$ is

$$f(\theta) = -\frac{1}{k\sqrt{2\pi \sin\theta}} \int_0^\infty \sqrt{\ell} \left\{ \exp i\left[2\delta_\ell + \left(\ell + \frac{1}{2}\right)\theta + \frac{\pi}{4}\right] - \exp i\left[2\delta_\ell - \left(\ell + \frac{1}{2}\right)\theta - \frac{\pi}{4}\right] \right\} d\ell \tag{6}$$

To arrive at (6), first prove that

$$\sum_\ell (2\ell + 1) P_\ell(\cos\theta) = 2\delta(1 - \cos\theta) \tag{6a}$$

which is infinite at $\theta = 0$ but zero for $\theta \neq 0$. Thus, for $\theta \neq 0$, the usual partial-wave expansion of the elastic scattering amplitude yields

$$f(\theta) = \frac{1}{2ik} \sum_\ell (2\ell + 1) \exp(2i\delta_\ell) P_\ell(\cos\theta) \tag{6b}$$

Next, show that for $\ell \gg 1$, $\ell\theta \gg 1$, and $\ell(\pi - \theta) \gg 1$,

$$P_\ell(\cos\theta) \to \sqrt{\frac{2}{\pi \ell \sin\theta}} \sin\left[\left(\ell + \frac{1}{2}\right)\theta + \frac{\pi}{4}\right] \tag{6c}$$

Hint: Start with the Legendre differential equation and make the substitution $\chi(\theta) = \sqrt{\sin\theta}\, P_\ell(\cos\theta)$. This should lead you to the result

$$P_\ell(\cos\theta) \to A \frac{\sin\left[\left(\ell + \frac{1}{2}\right)\theta + \alpha\right]}{\sqrt{\sin\theta}}$$

where A and α are coefficients. To determine these coefficients, consider the connection between the Legendre polynomial $P_\ell(\cos\theta)$ and the Bessel function $J_0\left[(\ell + 1/2)\theta\right]$ that exists for large ℓ and small θ. Use of (6c) in (6b) and conversion of the sum to an integral yield (6).

(d) The exponentials in the integrand of (6) oscillate rapidly except near the extremum of the exponents as a function of ℓ. The main contribution to the integral thus comes from the neighborhood of this extremum (the stationary-phase approximation). Show that the extremum occurs when the following condition is fulfilled:

$$2\hbar \frac{\partial \delta_\ell}{\partial J} = \pm\theta \qquad \text{for } J = \hbar\ell \text{ and } \ell \gg 1 \tag{7}$$

Show that (7) is the same condition as (2). Thus, assuming the WKB approximation and $\ell \gg 1$, the classical and quantum scattering angles are the same.

(e) The classical and quantum-mechanical differential cross sections are also the same. To see this, assume that (7) holds for $\ell = \ell_0$. Then, making a Taylor expansion, we have

$$2\delta_\ell + \left(\ell + \frac{1}{2}\right)\theta + \frac{\pi}{4} = 2\delta_{\ell_0} + \left(\ell_0 + \frac{1}{2}\right)\theta + \frac{\pi}{4} + \left.\frac{\partial^2 \delta_\ell}{\partial \ell^2}\right|_{\ell=\ell_0} (\ell - \ell_0)^2 \tag{8}$$

Thus show that if $A(\ell_0) = 2\delta_{\ell_0} + (\ell_0 + ½)\theta + \pi/4$,

$$f(\theta) = \exp i\left[A(\ell_0) + \frac{3\pi}{4}\right] \cdot \sqrt{\frac{-J}{\sin\theta}\frac{\partial J}{\partial \theta}} \tag{9}$$

The absolute square of (9) gives the differential cross section, which is the same as in (4).

(f) What is the difference between quantum-mechanical scattering for large ℓ in the WKB approximation and classical scattering? The essential difference appears in forward scattering. Let's assume that the phase shifts are large for $\ell < \ell_1$ but small for $\ell \gg \ell_1$, where $\ell_1 = ka$ and a is the radius of some sphere. Then

$$f(\theta) \approx \frac{1}{2ik}\int_0^{\ell_1} 2\ell\left[\exp(2i\delta_\ell) - 1\right] P_\ell(\cos\theta)\, d\ell \tag{10}$$

For small angles, show that the P_ℓ vary only slowly with ℓ. Thus destructive interference eliminates the contribution of the $\exp(2i\delta_\ell)$ to the integral in (10). From this, show that

$$f(\theta) \approx \frac{ia}{\theta} J_1(ka\theta) \tag{11}$$

where J_1 is the first Bessel function. It follows that

$$\frac{d\sigma}{d\Omega} = \frac{a^2}{\theta^2}\left[J_1(ka\theta)\right]^2 \tag{12}$$

Compare this to Fraunhofer diffraction of light around a black sphere of radius a [see, e.g., Jackson (1998)].

19 Special Relativity and Quantum Mechanics: The Klein-Gordon Equation

19.1 General remarks on relativistic wave equations

We now attempt to extend quantum mechanics to the relativistic domain. It would seem only natural to start by searching for an appropriate relativistic single-particle wave equation (or equations) to replace the Schroedinger equation. In fact, one finds that the form of the relativistic equation depends on the spin of the particle; that is,

Spin-0: Klein-Gordon equation
Spin-½: Dirac equation
Massive spin-1: Proca equation

It is important to study these one-particle equations and their solutions in detail because by doing so we gain insight into many significant physical phenomena. However, no matter what the spin, if we pursue a single-particle relativistic quantum theory far enough, we encounter fundamental inconsistencies, and the theory breaks down. The essential reason for this failure is that whereas energy is conserved in special relativity, mass is not conserved, and particles can be created and destroyed in real physical processes; for example,

$$
\begin{aligned}
&e^+ + e^- \to 2\gamma, 3\gamma, \ldots \\
&\mu^- \to e^- \bar{\nu}_e \nu_\mu \\
&\pi^- + p \to \pi^0 + n
\end{aligned}
$$

and so forth. Obviously, no single-particle theory is capable of accounting for such phenomena. Therefore, at a certain stage, we are forced to abandon the single-particle approach and go over to a many-particle relativistic quantum theory in which particles can be created and destroyed. This is relativistic quantum field theory, a discipline that we can only begin to study in this book.

19.2 The Klein-Gordon equation

In nonrelativistic mechanics, the energy of a free particle is $E = p^2/2m$. To obtain the Schroedinger equation for a free spinless particle, we make the well-known substitutions

$$E \to i\hbar \frac{\partial}{\partial t} \quad \text{and} \quad \boldsymbol{p} \to -i\hbar\nabla \tag{19.1}$$

which give

$$-\frac{\hbar^2}{2m}\nabla^2\psi = i\hbar\frac{\partial\psi}{\partial t}$$

In relativistic mechanics, the energy of a free particle is

$$E = \sqrt{p^2c^2 + m^2c^4} \tag{19.2}$$

Using the substitutions of (19.1) in (19.2), we obtain

$$i\hbar\dot{\psi} = \sqrt{-\hbar^2c^2\nabla^2 + m^2c^4}\,\psi \tag{19.3}$$

In coordinate representation, it is difficult to interpret the operator on the right-hand side of (19.3). Thus, instead of (19.2), we try

$$E^2 = p^2c^2 + m^2c^4 \tag{19.4}$$

Again employing (19.1), we obtain

$$\left(i\hbar\frac{\partial}{\partial t}\right)^2\psi = -\hbar^2c^2\nabla^2\psi + m^2c^4\psi$$

or

$$\left(\nabla^2 - \frac{1}{c^2}\frac{\partial^2}{\partial t^2}\right)\psi = \left(\frac{mc}{\hbar}\right)^2\psi \tag{19.5}$$

This is the Klein-Gordon equation for a free spinless particle. Actually, it was first obtained by Schroedinger in 1926 before he developed the nonrelativistic equation that bears his name. In natural units (recall Section 15.7), (19.5) is written as

$$\left(\partial^\mu\partial_\mu + m^2\right)\psi = 0 \tag{19.6}$$

Plane-wave solutions to (19.5) are readily found:

$$\psi = \exp\left[\frac{i}{\hbar}(\boldsymbol{p}\cdot\boldsymbol{r} - Et)\right] = \exp\left(-\frac{i}{\hbar}p\cdot x\right) \tag{19.7}$$

where

$$E^2 = \boldsymbol{p}^2c^2 + m^2c^4$$

and thus

$$E = \pm\sqrt{\boldsymbol{p}^2c^2 + m^2c^4} \tag{19.8}$$

Thus there is a negative-energy solution as well as a positive-energy solution for each $\boldsymbol{p}$. It would seem at first that we should simply discard the negative-energy solution as unphysical.

Indeed, as long as we consider a free particle, defined initially to be in a positive-energy state, this would be perfectly legitimate because there is no mechanism by which a transition could occur to a negative-energy state. However, suppose that the particle of interest has electric charge q and we impose an external electromagnetic potential. The Klein-Gordon equation then must be altered by the usual replacements

$$E \to E - q\Phi \quad \text{and} \quad \boldsymbol{p} \to \boldsymbol{p} - \frac{q}{c}\boldsymbol{A}$$

to become

$$\left(i\hbar\partial_t - q\Phi\right)^2 \psi = c^2\left(-i\hbar\nabla - \frac{q}{c}\boldsymbol{A}\right)^2 \psi + m^2c^4\psi \tag{19.9}$$

or in covariant notation and natural units

$$\left(\partial^\mu + iqA^\mu\right)\left(\partial_\mu + iqA_\mu\right)\psi + m^2\psi = 0 \tag{19.10}$$

The solution to this equation can always be expressed as a superposition of free-particle solutions, but only if the latter form a complete set, and this requires inclusion of the negative-energy solutions, which therefore cannot be discarded.

For the nonrelativistic time-dependent Schroedinger equation, we were able to define a non-negative probability density $\rho_S = \psi^*\psi$ and a probability current density $\boldsymbol{j}_S$ that satisfy the equation of continuity

$$\nabla \cdot \boldsymbol{j}_S + \frac{\partial \rho_S}{\partial t} = 0$$

We attempt to follow the same procedure for the Klein-Gordon equation. First, multiply (19.10) on the left by ψ^*; that is,

$$\psi^*\left(\partial^\mu + iqA^\mu\right)\left(\partial_\mu + iqA_\mu\right)\psi + m^2\psi^*\psi = 0 \tag{19.11}$$

Next, take the complex conjugate of (19.10) and multiply by ψ; that is,

$$\psi\left(\partial^\mu - iqA^\mu\right)\left(\partial_\mu - iqA_\mu\right)\psi^* + m^2\psi^*\psi = 0 \tag{19.12}$$

We subtract (19.12) from (19.11) to obtain

$$\partial^\mu\left(\psi^*\partial_\mu\psi - \psi\partial_\mu\psi^* + 2iqA_\mu\psi^*\psi\right) = 0 \tag{19.13}$$

This may be interpreted as an equation of continuity for the probability four-current density j_μ; that is,

$$j_\mu = \frac{\hbar}{2mi}\left(\psi^*\partial_\mu\psi - \psi\partial_\mu\psi^* + \frac{2iq}{\hbar c}A_\mu\psi^*\psi\right) \tag{19.14}$$

In (19.14) we exhibit $\hbar$ and c explicitly and choose an overall multiplicative factor so that the present definition of $\boldsymbol{j}$ agrees with that obtained earlier in the nonrelativistic Schroedinger theory. Equation (19.14) implies that for the Klein-Gordon equation, the probability density is

$$\rho = \frac{i\hbar}{2mc^2}\left[\psi^*\dot{\psi} - \psi\dot{\psi}^*\right] - \frac{q\Phi}{mc^2}\psi^*\psi \tag{19.15}$$

Of course, for a free particle, $\Phi = 0$, in which case

$$\rho_{\text{free}} = \frac{i\hbar}{2mc^2}\left[\psi^*\dot{\psi} - \psi\dot{\psi}^*\right] \tag{19.16}$$

It can be shown that in the nonrelativistic limit, where all significant momentum components in a free-particle wave packet satisfy $|\boldsymbol{p}| \ll mc$, (19.16) reduces to $\rho_{\text{free}} = \psi_+^*\psi_+ - \psi_-^*\psi_-^*$, where the plus and minus signs refer to positive- and negative-energy contributions, respectively. If we ignore the negative-energy contribution, then in this nonrelativistic limit ρ_{free} reduces to the familiar probability density associated with the Schroedinger equation. However, in general, ρ_{free} as given by (19.16) is not necessarily positive, and thus it makes no sense to employ it as a probability density for a single particle. This fundamental difficulty arises because the Klein-Gordon equation is second order in the time, so ρ_{free} contains first-order time derivatives. The problem does not arise for the nonrelativistic Schroedinger equation, which is first order in the time.

Because of this difficulty, the Klein-Gordon equation was discarded for some years, during which the Dirac equation (Dirac 1928 a, b) was thought to be the only valid single-particle relativistic wave equation. However, it was ultimately discovered that there are also profound difficulties associated with the Dirac equation when it is employed in a single-particle theory. Such difficulties are overcome only by constructing a many-particle *Dirac field*, the quanta of which have spin-½. In 1934, W. Pauli and V. Weisskopf showed that the Klein-Gordon equation can also be used in a field theory, in which the quanta are massive spin-0 (scalar) particles. In this theory, when ρ and $\boldsymbol{j}$ are multiplied by e, they become the charge and current densities, respectively, of the field quanta. The fact that ρ can be negative then causes no difficulty: scalar particles with both signs of charge appear.

Problems for Chapter 19

19.1. The probability density for the free-particle Klein-Gordon equation is

$$\rho_{\text{KG}} = \frac{i\hbar}{2mc^2}\left(\psi^*\frac{\partial\psi}{\partial t} - \psi\frac{\partial\psi^*}{\partial t}\right)$$

Show that in the nonrelativistic limit (and where we neglect negative-energy solutions), $\rho_{\text{KG}} \to \psi^*\psi$.

19.2. Consider the lowest bound state of a particle of mass m that satisfies the Klein-Gordon equation for a spherically symmetric square well of radius a and well depth V_0. Employ units where $\hbar = c = 1$, and assume that the orbital angular momentum of the particle is $\ell = 0$. Find V_0 in terms of a and m for each of the following two cases:

(a) The particle energy is $E = m$ (the bound state just appears).

(b) The particle energy is $E = -m$.

Can a well-behaved bound state exist with $E < -m$?

19.3. Consider the one-dimensional potential step

$$\begin{aligned} V &= 0 \qquad && x < 0 \\ V &= +V_0 \qquad && x \geq 0 \end{aligned}$$

A plane-wave solution to the Klein-Gordon equation $\psi = \exp\left[i(kx - \omega t)\right]$ is incident from the left. Find the reflection and transmission coefficients for the three cases

$$\begin{aligned} &V_0 < E - mc^2 \\ &E - mc^2 < V_0 < E + mc^2 \\ &E + mc^2 < V_0 \end{aligned}$$

What pathologic features are associated with case 3?

20 The Dirac Equation

20.1 Derivation of the Dirac equation

In 1928, P. A. M. Dirac obtained the very important equation that bears his name (Dirac 1928a, b). His original line of argument, though outmoded, is very instructive, so we temporarily suspend our skepticism about the feasibility of a single-particle theory and retrace Dirac's steps as follows: the Klein-Gordon equation for a single particle presents fundamental difficulties because it is second order in the time. This suggests that we should find a new equation that is first order in the time. In special relativity, the spatial coordinates and the time should appear in a symmetric way in any physical equation, so we require that the new equation should be first order in the spatial derivatives as well. Furthermore, the wave function should satisfy not only this first-order equation but also the Klein-Gordon equation because the latter merely expresses the relation

$$E^2 = \boldsymbol{p}^2 c^2 + m^2 c^4$$

An analogous situation appears in classical electrodynamics, where Maxwell's equations are first order in the spatial and time derivatives, and in addition, each component of $\boldsymbol{\mathcal{E}}$ and $\boldsymbol{B}$ satisfies a second-order wave equation. This can happen because the various components of $\boldsymbol{\mathcal{E}}$ and $\boldsymbol{B}$ are coupled together in Maxwell's equations. This suggests that the wave function ψ has several (N) components and that we are really dealing with a set of N coupled first-order relativistic wave equations. For a free particle, these can be written quite generally as

$$\frac{1}{c}\frac{\partial \psi_m}{\partial t} + \sum_{k=1}^{3}\sum_{n=1}^{N} \alpha_k^{mn} \frac{\partial \psi_n}{\partial x_k} + \frac{imc}{\hbar}\sum_{n=1}^{N} \beta^{mn}\psi_n = 0 \tag{20.1}$$

where $m = 1,\ldots,N$, and α_k^{mn} and β^{mn} are coefficients to be determined. Homogeneity and isotropy of space-time suggest that these coefficients should be dimensionless constants that are independent of $\boldsymbol{r}$ and $\boldsymbol{p}$ and commute with the latter quantities. Assuming this, we can write (20.1) more compactly by introducing the four $N\times N$ matrices α_k and β and the vector

$$\boldsymbol{\alpha} = \alpha_1\hat{i} + \alpha_2\hat{j} + \alpha_3\hat{k}$$

as well as the notation

$$\psi = \begin{pmatrix} \psi_1 \\ \psi_2 \\ \cdots \\ \psi_N \end{pmatrix}$$

Then (20.1) can be written

$$\frac{1}{c}\frac{\partial\psi}{\partial t}+\boldsymbol{\alpha}\boldsymbol{\cdot}\nabla\psi+\frac{imc}{\hbar}\beta\psi=0 \tag{20.2}$$

The corresponding Hermitian conjugate equation is

$$\frac{1}{c}\frac{\partial\psi^{\dagger}}{\partial t}+\nabla\psi^{\dagger}\boldsymbol{\cdot}\boldsymbol{\alpha}^{\dagger}-\frac{imc}{\hbar}\psi^{\dagger}\beta^{\dagger}=0 \tag{20.3}$$

where $\psi^{\dagger}=(\psi_1^*,\psi_2^*,\ldots,\psi_N^*)$.

We now construct an equation of continuity as follows: multiply (20.2) on the left by $\psi^{\dagger}$, multiply (20.3) on the right by ψ, and add the two resulting equations. We then obtain

$$\frac{1}{c}\left(\psi^{\dagger}\frac{\partial\psi}{\partial t}+\frac{\partial\psi^{\dagger}}{\partial t}\psi\right)+\nabla\psi^{\dagger}\boldsymbol{\cdot}\boldsymbol{\alpha}^{\dagger}\psi+\psi^{\dagger}\boldsymbol{\alpha}\boldsymbol{\cdot}\nabla\psi+\frac{imc}{\hbar}(\psi^{\dagger}\beta\psi-\psi^{\dagger}\beta^{\dagger}\psi)=0$$

or

$$\frac{1}{c}\frac{\partial}{\partial t}(\psi^{\dagger}\psi)+\nabla\psi^{\dagger}\boldsymbol{\cdot}\boldsymbol{\alpha}^{\dagger}\psi+\psi^{\dagger}\boldsymbol{\alpha}\boldsymbol{\cdot}\nabla\psi+\frac{imc}{\hbar}\psi^{\dagger}(\beta-\beta^{\dagger})\psi=0 \tag{20.4}$$

This suggests the equation of continuity

$$\nabla\boldsymbol{\cdot}\boldsymbol{j}+\frac{\partial\rho}{\partial t}=0 \tag{20.5}$$

provided that we choose

$$\boldsymbol{\alpha}^{\dagger}=\boldsymbol{\alpha}\quad\text{and}\quad\beta^{\dagger}=\beta \tag{20.6}$$

Then (20.5) is satisfied with

$$\rho=\psi^{\dagger}\psi \tag{20.7}$$

and

$$\boldsymbol{j}=c\psi^{\dagger}\boldsymbol{\alpha}\psi \tag{20.8}$$

The probability density in (20.7) is indeed nonnegative and of the familiar form. However, we shall see that the probability current density (20.8) has some subtle features that are not so easy to understand.

We must still exploit the requirement that each component of ψ should satisfy the Klein-Gordon equation. To do this, we write the Dirac equation

$$\left[\frac{1}{c}\frac{\partial}{\partial t}+\boldsymbol{\alpha}\boldsymbol{\cdot}\nabla+\frac{imc}{\hbar}\beta\right]\psi=0$$

and multiply on the left by the operator

$$\left[\frac{1}{c}\frac{\partial}{\partial t} - \alpha \cdot \nabla - \frac{imc}{\hbar}\beta\right]$$

to obtain

$$\left[\frac{1}{c^2}\frac{\partial^2}{\partial t^2} - \alpha_i \alpha_j \nabla_i \nabla_j - \frac{imc}{\hbar}(\beta\alpha_j + \alpha_j\beta)\nabla_j + \frac{m^2c^2}{\hbar^2}\beta^2\right]\psi = 0 \tag{20.9}$$

In (20.9), we can write

$$\alpha_i \alpha_j \nabla_i \nabla_j = \frac{1}{2}(\alpha_i\alpha_j + \alpha_j\alpha_i)\nabla_i\nabla_j$$

because $\nabla_i \nabla_j = \nabla_j \nabla_i$. For (20.9) to reduce to the Klein-Gordon equation, we require that

$$\alpha_i\alpha_j + \alpha_j\alpha_i = 2\delta_{ij} I \tag{20.10}$$

$$\beta\alpha_j + \alpha_j\beta = 0 \tag{20.11}$$

and

$$\beta^2 = I \tag{20.12}$$

where I is the $N \times N$ identity matrix. Equations (20.10)–(20.12) are the fundamental algebraic relations for the α, β matrices that determine the entire structure of the Dirac theory. Some immediate consequences are as follows: from (20.11), we have

$$\beta\alpha_j = -\alpha_j\beta = -I \cdot \alpha_j\beta \tag{20.13}$$

Thus

$$\det\beta \cdot \det\alpha_j = \det(-I) \cdot \det\alpha_j \cdot \det\beta \tag{20.14}$$

None of the determinants in (20.14) vanish because all the matrices are nonsingular, which follows immediately from (20.10) and (20.12). Hence $\det(-I) = +1$. Because $\det(-I) = (-1)^N$, N must be an even number. Furthermore, from (20.11), we have

$$\alpha_j^{-1}\beta\alpha_j = -\beta \qquad \text{(No summation on index } j\text{)} \tag{20.15}$$

Taking the trace of both sides of (20.15), we have

$$tr\left(\alpha_j^{-1}\beta\alpha_j\right) = tr\left(\alpha_j\alpha_j^{-1}\beta\right) = tr(\beta) = -tr(\beta) \tag{20.16}$$

so $tr(\beta) = 0$ and, similarly, $tr(\alpha_j) = 0$ for each j.

20.2 Hamiltonian form of the Dirac equation

Equation (20.2) is easily written in the form

$$H\psi = i\hbar\dot{\psi} \tag{20.17}$$

provided that we define the free-particle Dirac Hamiltonian as

$$H = -i\hbar c\boldsymbol{\alpha}\cdot\nabla + mc^2\beta \tag{20.18}$$

20.3 Covariant form of the Dirac equation

To proceed further with our understanding of the Dirac equation, it is very convenient to employ new matrices $\gamma^\mu = \gamma^0, \gamma^1, \gamma^2, \gamma^3$ defined as follows:

$$\gamma^0 = \beta \tag{20.19a}$$

$$\gamma^i = \gamma^0\alpha_i = -\alpha_i\gamma^0 \qquad i = 1,2,3 \tag{20.19b}$$

Because $\gamma^0 = \gamma^{0\dagger}$ and $\alpha_i = \alpha_i^\dagger$, we immediately obtain the useful relations

$$\gamma^{i\dagger} = -\gamma^i \tag{20.20a}$$

$$\gamma^0\gamma^{\mu\dagger}\gamma^0 = \gamma^\mu \tag{20.20b}$$

Also, multiplying (20.2) on the left by $i\gamma^0$, we have

$$i\gamma^0\frac{\partial\psi}{\partial x^0} + i\boldsymbol{\gamma}\cdot\nabla\psi - \frac{mc}{\hbar}\psi = 0 \tag{20.21}$$

or

$$i\gamma^\mu\frac{\partial\psi}{\partial x^\mu} - \frac{mc}{\hbar}\psi = 0 \tag{20.22}$$

In natural units, and defining the convenient shorthand $\not{\partial} \equiv \gamma^\mu\partial_\mu$, this is

$$\left(i\gamma^\mu\partial_\mu - m\right)\psi = \left(i\not{\partial} - m\right)\psi = 0 \tag{20.23}$$

which is the free-particle Dirac equation in covariant form. Recalling that $\gamma^{0\dagger} = \gamma^0$ and $\gamma^{i\dagger} = -\gamma^i$, we take the Hermitian conjugate of (20.21); that is,

$$-i\frac{\partial \psi^\dagger}{\partial x^0}\gamma^0 + i\nabla\psi^\dagger\boldsymbol{\cdot}\boldsymbol{\gamma} - \frac{mc}{\hbar}\psi^\dagger = 0 \tag{20.24}$$

We multiply this equation on the right by γ^0, make use of $\gamma^i\gamma^0 = -\gamma^0\gamma^i$, and define the Dirac conjugate wave function as $\bar{\psi} = \psi^\dagger\gamma^0$ to obtain the Dirac conjugate equation

$$i\frac{\partial \bar{\psi}}{\partial x^\mu}\gamma^\mu + \frac{mc}{\hbar}\bar{\psi} = 0 \tag{20.25}$$

or in natural units

$$i\partial_\mu\bar{\psi}\gamma^\mu + m\bar{\psi} = \left(i\not{\partial} + m\right)\bar{\psi} = 0 \tag{20.26}$$

It is also convenient to rewrite the probability density and probability current density in terms of the γ matrices; that is,

$$\boldsymbol{j} = c\psi^\dagger\boldsymbol{\alpha}\psi = c\psi^\dagger\gamma^0\boldsymbol{\gamma}\psi = c\bar{\psi}\boldsymbol{\gamma}\psi$$

and

$$j^0 = c\rho = c\psi^\dagger\psi = c\bar{\psi}\gamma^0\psi$$

which combine to yield

$$j^\mu = c\bar{\psi}\gamma^\mu\psi \tag{20.27}$$

The equation of continuity (20.5) is written in covariant form as

$$\partial^\mu j_\mu = 0 \tag{20.28}$$

The fundamental relations [(20.10)–(20.12)] are also expressed conveniently in terms of the gamma matrices. Employing (20.19a) and (20.19b), we obtain

$$\gamma^\mu\gamma^\nu + \gamma^\nu\gamma^\mu = 2g^{\mu\nu}I \tag{20.29}$$

where, as usual, μ and ν run from 0 through 3.

20.4 A short mathematical digression on gamma matrices

Before we apply the Dirac equation to physical problems, we familiarize ourselves with some mathematical properties of the γ matrices. We can form new matrices by multiplying two or more γ matrices together. Taking into account the fundamental anticommutation relation (20.29) and also including the identity matrix, we find that the total number of distinct matrices is 16, denoted collectively as the Γ^k, $k = 1,\ldots,16$; that is,

$$\begin{array}{cccccc} & & & I & & \\ & \gamma^0 & \gamma^1 & \gamma^2 & \gamma^3 & \\ i\gamma^1\gamma^2 & i\gamma^2\gamma^3 & i\gamma^3\gamma^1 & i\gamma^1\gamma^0 & i\gamma^2\gamma^0 & i\gamma^3\gamma^0 \\ & \gamma^0\gamma^5 & \gamma^1\gamma^5 & \gamma^2\gamma^5 & \gamma^3\gamma^5 & \\ & & & \gamma^5 = i\gamma^0\gamma^1\gamma^2\gamma^3 & & \end{array} \tag{20.30}$$

Note that in the fourth line of (20.30) we can write $\gamma^0\gamma^5 = i\gamma^1\gamma^2\gamma^3$, $\gamma^1\gamma^5 = -i\gamma^0\gamma^2\gamma^3$, and so forth.

We now list without proof several useful mathematical relations for the Γ^k that can be derived by straightforward algebra.

1. $\Gamma^k\Gamma^m = a_{km}\Gamma^n$, where $a_{km} = \pm 1$ or $\pm i$, and given k, n is distinct for each distinct m.
2. $\Gamma^k\Gamma^m = \pm I$ if and only if $k = m$.
3. $\Gamma^k\Gamma^m = \pm\Gamma^m\Gamma^k$.
4. If $\Gamma^k \neq I$, $tr(\Gamma^k) = 0$.
5. The Γ^k are linearly independent. Thus they can be 4×4 matrices, and we shall assume that this is indeed the case. Earlier we proved that N must be an even number, but it cannot be 2 because there are only four linearly independent 2×2 matrices. The choice $N = 4$ is not directly related to four space-time dimensions. Rather, $N = 4$ is connected to the fact that the Dirac equation applies to particles of spin-½ and that there are negative-energy and positive-energy solutions to the Dirac equation. Because the Γ^k span the space of all 4×4 matrices, any arbitrary 4×4 matrix X can be expressed as a linear combination of the Γ^k; that is,

$$X = \sum_{k=1}^{16} x_k \Gamma^k$$

6. Any matrix X that commutes with all four γ^μ is a multiple of the identity.
7. Given two sets of 4×4 matrices $\gamma^\mu, \gamma^{\mu\prime}$, both of which satisfy the fundamental law,

$$\gamma^\mu\gamma^\nu + \gamma^\nu\gamma^\mu = 2g^{\mu\nu}I$$

$$\gamma^{\mu\prime}\gamma^{\nu\prime} + \gamma^{\nu\prime}\gamma^{\mu\prime} = 2g^{\mu\nu}I$$

there exists a nonsingular matrix W such that

$$\gamma^{\mu\prime} = W\gamma^\mu W^{-1}$$

A specific choice of the γ^μ constitutes a representation. Although an infinite number of representations are possible, as a practical matter, only a small number are used. We first consider the standard representation.

20.5 Standard representation: Free-particle plane-wave solutions

Returning to the matrices $\boldsymbol{\alpha}, \beta$, we note that the fundamental relations (20.10)–(20.12) are satisfied by the nonunique choice

$$\boldsymbol{\alpha} = \begin{pmatrix} 0 & \boldsymbol{\sigma} \\ \boldsymbol{\sigma} & 0 \end{pmatrix} \qquad \beta = \gamma^0 = \begin{pmatrix} I & 0 \\ 0 & -I \end{pmatrix} \tag{20.31}$$

This is the standard (Dirac-Pauli) representation, in which

$$\boldsymbol{\gamma} = \gamma^0 \boldsymbol{\alpha} = \begin{pmatrix} 0 & \boldsymbol{\sigma} \\ -\boldsymbol{\sigma} & 0 \end{pmatrix} \tag{20.32}$$

and

$$\gamma^5 = i\gamma^0\gamma^1\gamma^2\gamma^3 = \begin{pmatrix} 0 & I \\ I & 0 \end{pmatrix} \tag{20.33}$$

In (20.31)–(20.33), the σ are the 2×2 Pauli spin matrices, and I is the 2×2 identity matrix. Using the standard representation, we now consider the free-particle Hamiltonian of (20.18); that is,

$$H = -i\hbar c\boldsymbol{\alpha}\bullet\nabla + mc^2\gamma^0 = \begin{pmatrix} mc^2 & -i\hbar c\boldsymbol{\sigma}\bullet\nabla \\ -i\hbar c\boldsymbol{\sigma}\bullet\nabla & -mc^2 \end{pmatrix} \tag{20.34}$$

We seek plane-wave solutions to $H\psi = i\hbar\dot{\psi}$ of the form

$$\psi = u \cdot \exp\left[\frac{i}{\hbar}(\boldsymbol{p}\bullet\boldsymbol{x} - Et)\right] \tag{20.35}$$

where $u = \begin{pmatrix} u_A \\ u_B \end{pmatrix}$ is a four-component spinor, and $u_{A,B}$ are each two-component spinors that might depend on $\boldsymbol{p}$ and E but are independent of $\boldsymbol{x}$ and t. Given (20.34) and (20.35), $H\psi = i\hbar\dot{\psi}$ yields

$$(E - mc^2)u_A = c\boldsymbol{\sigma}\bullet\boldsymbol{p}u_B \tag{20.36}$$

$$(E + mc^2)u_B = c\boldsymbol{\sigma}\bullet\boldsymbol{p}u_A \tag{20.37}$$

We try $u_a = \begin{pmatrix} 1 \\ 0 \end{pmatrix}$. Then (20.37) yields

$$u_B = \frac{c}{E + mc^2}\begin{pmatrix} p_z \\ p_x + ip_y \end{pmatrix} \tag{20.38}$$

If instead we try $u_A = \begin{pmatrix} 0 \\ 1 \end{pmatrix}$, we obtain

$$u_B = \frac{c}{E + mc^2} \begin{pmatrix} p_x - ip_y \\ -p_z \end{pmatrix} \tag{20.39}$$

Each of these two linearly independent solutions corresponds to positive energy ($E > 0$), as can be seen by taking the limit $\boldsymbol{p} \to 0$ in (20.36). In addition, there are two linearly independent negative-energy solutions ($E < 0$):

$$u_B = \begin{pmatrix} 1 \\ 0 \end{pmatrix} \qquad u_A = \frac{c}{E - mc^2} \begin{pmatrix} p_z \\ p_x + ip_y \end{pmatrix} \tag{20.40}$$

and

$$u_B = \begin{pmatrix} 0 \\ 1 \end{pmatrix} \qquad u_A = \frac{c}{E - mc^2} \begin{pmatrix} p_x - ip_y \\ -p_z \end{pmatrix} \tag{20.41}$$

To write all this in a standard form, we define $\varepsilon = |E|$, make use of standard normalization factors, and revert to natural units where $\hbar = c = 1$, to obtain:

$$\psi_{I,II}(x) = \frac{1}{\sqrt{V}} \sqrt{\frac{m}{\varepsilon}} u_{1,2} \exp[i(p \cdot x - \varepsilon t)] \qquad E > 0, \tag{20.42}$$

$$\psi_{III,IV}(x) = \frac{1}{\sqrt{V}} \sqrt{\frac{m}{\varepsilon}} u_{3,4} \exp[i(p \cdot x + \varepsilon t)] \qquad E < 0 \tag{20.43}$$

where

$$u_{1,2} = \sqrt{\frac{\varepsilon + m}{2m}} \begin{pmatrix} \chi_\pm \\ \dfrac{\boldsymbol{\sigma} \bullet \boldsymbol{p}}{\varepsilon + m} \chi_\pm \end{pmatrix} \tag{20.44}$$

and

$$u_{3,4} = \sqrt{\frac{\varepsilon + m}{2m}} \begin{pmatrix} -\dfrac{\boldsymbol{\sigma} \bullet \boldsymbol{p}}{\varepsilon + m} \chi_\pm \\ \chi_\pm \end{pmatrix} \tag{20.45}$$

with $\chi_+ = \begin{pmatrix} 1 \\ 0 \end{pmatrix}$ and $\chi_- = \begin{pmatrix} 0 \\ 1 \end{pmatrix}$. The four 4-spinors $u_{1,2,3,4}$ are mutually orthogonal; that is,

$$u_r^\dagger u_s = \delta_{rs} \frac{\varepsilon}{m} \tag{20.46}$$

An example of (20.46) is

$$\begin{aligned}
u_1^\dagger u_1 &= \frac{\varepsilon+m}{2m}\left(\chi_+^\dagger, \chi_+^\dagger \frac{\boldsymbol{\sigma}\cdot\boldsymbol{p}}{\varepsilon+m}\right)\left(\begin{array}{c}\chi_+ \\ \dfrac{\boldsymbol{\sigma}\cdot\boldsymbol{p}}{\varepsilon+m}\chi_+\end{array}\right) \\
&= \frac{\varepsilon+m}{2m}\chi_+^\dagger\left(1+\frac{\boldsymbol{\sigma}\cdot\boldsymbol{p}}{\varepsilon+m}\frac{\boldsymbol{\sigma}\cdot\boldsymbol{p}}{\varepsilon+m}\right)\chi_+ \\
&= \frac{\varepsilon+m}{2m}\left[1+\frac{p^2}{(\varepsilon+m)^2}\right] \\
&= \frac{\varepsilon+m}{2m}\left[\frac{(\varepsilon+m)^2+\varepsilon^2-m^2}{(\varepsilon+m)^2}\right] = \frac{\varepsilon}{m}
\end{aligned}$$

As in the case of the Klein-Gordon equation, we cannot discard the negative-energy solutions because they are needed for completeness. However, by means of the charge conjugation transformation (to be discussed later), it will be seen that the negative-energy solutions (e.g., for an electron) are transformed into positive-energy solutions (for a positron). In the following section we prove that the Dirac equation applies to particles of spin-½. It thus turns out that ψ_{I} and ψ_{III} correspond to spin-up along the z-axis, whereas ψ_{II} and ψ_{IV} correspond to spin-down.

20.6 Lorentz covariance of the Dirac equation

In classical electrodynamics, consider Maxwell's equations in a given inertial frame with space-time coordinates x and where the electric and magnetic fields are $\boldsymbol{\mathcal{E}}(x)$ and $\boldsymbol{B}(x)$, respectively. Now make a Lorentz transformation to a new inertial frame with space-time coordinates x'. Maxwell's equations take the same form in both inertial frames: they are covariant with respect to Lorentz transformations. However, the electric and magnetic fields themselves are generally not the same in the two frames: $\boldsymbol{\mathcal{E}}'(x') \neq \boldsymbol{\mathcal{E}}(x)$ and $\boldsymbol{B}'(x') \neq \boldsymbol{B}(x)$. Similarly, if Dirac's equation is to be a valid relativistic wave equation, we must require it to be covariant with respect to Lorentz transformations, but we should not expect that the wave functions in the two inertial frames are the same.

To make this point precise, consider a frame F with observer O and space-time coordinates x. Suppose that, according to O, the free-particle Dirac equation is

$$\left(m - i\gamma^\mu \frac{\partial}{\partial x^\mu}\right)\psi(x) = 0 \tag{20.47}$$

with solution $\psi(x)$. In another inertial frame F' with coordinates $x'^\nu = a^\nu{}_\mu x^\mu$, another observer O' describes the same physical situation by means of wave function $\psi'(x')$, which satisfies

$$\left(m - i\gamma^\nu \frac{\partial}{\partial x'^\nu}\right)\psi'(x') = 0 \tag{20.48}$$

Lorentz covariance of the Dirac equation means that we require the gamma matrices to be the same in (20.47) and (20.48).

Now let us assume that there is a nonsingular transformation S that relates $\psi'(x')$ to $\psi(x)$; that is,

$$\psi'(x') = \psi'(ax) = S\psi(x) \tag{20.49}$$

From homogeneity and isotropy of space-time, we assume that S does not depend on the space-time coordinates. We apply S on the left to (20.47), giving

$$mS\psi(x_\mu) - iS\gamma^\mu S^{-1}\frac{\partial}{\partial x^\mu}S\psi(x_\mu) = 0$$

and obtain

$$m\psi'(x') - iS\gamma^\mu S^{-1}\frac{\partial}{\partial x^\mu}\psi'(x') = 0 \tag{20.50}$$

From $x'^\nu = a^\nu{}_\mu x^\mu$, we have

$$\frac{\partial}{\partial x^\mu} = \frac{\partial x'^\nu}{\partial x^\mu}\frac{\partial}{\partial x'^\nu} = a^\nu{}_\mu\frac{\partial}{\partial x'^\nu}$$

Therefore, (20.50) becomes

$$m\psi'(x') - iS\gamma^\mu S^{-1}a^\nu{}_\mu\frac{\partial}{\partial x'^\nu}\psi'(x') = 0 \tag{20.51}$$

We require (20.51) and (20.48) to be the same equation. Thus we must have

$$S\gamma^\mu S^{-1}a^\nu{}_\mu = \gamma^\nu \tag{20.52}$$

Multiplying (20.52) on the left by S^{-1} and on the right by S, we obtain the alternative equation

$$S^{-1}\gamma^\nu S = a^\nu{}_\mu\gamma^\mu \tag{20.53}$$

In what follows, we consider three important examples of (20.53): rotations about a given spatial axis, Lorentz boosts, and spatial inversion.

Rotation of the System about the *z*-Axis by Angle θ

Here the matrix is

$$a = \begin{pmatrix} 1 & 0 & 0 & 0 \\ 0 & \cos\theta & +\sin\theta & 0 \\ 0 & -\sin\theta & \cos\theta & 0 \\ 0 & 0 & 0 & 1 \end{pmatrix}$$

We first consider an infinitesimal rotation where

$$a = \begin{pmatrix} 1 & 0 & 0 & 0 \\ 0 & 1 & \theta & 0 \\ 0 & -\theta & 1 & 0 \\ 0 & 0 & 0 & 1 \end{pmatrix} \tag{20.54}$$

Here we ignore terms of order θ^2. Thus S differs from the identity only by terms of order θ, so we try

$$S = I - i\theta T$$
$$S^{-1} = I + i\theta T$$

Employing (20.54) in (20.53) and discarding all terms of order θ^2, we obtain

$$\begin{aligned} \gamma^1 + \theta\gamma^2 &= \gamma^1 - i\theta\left(T\gamma^1 - \gamma^1 T\right) \\ \gamma^2 - \theta\gamma^1 &= \gamma^2 - i\theta\left(T\gamma^2 - \gamma^2 T\right) \\ 0 &= T\gamma^3 - \gamma^3 T \\ 0 &= T\gamma^0 - \gamma^0 T \end{aligned}$$

It is easy to verify that these conditions are satisfied by $T = +(i/2)\gamma^1\gamma^2$, and thus

$$S_{\text{infinitesimal}} = I + \frac{1}{2}\gamma^1\gamma^2\theta \tag{20.55}$$

For a finite rotation about the z-axis, we thus have:

$$S = \exp\left(\frac{1}{2}\gamma^1\gamma^2\theta\right) \tag{20.56}$$

Now

$$\gamma^1\gamma^2 = \begin{pmatrix} 0 & \sigma_1 \\ -\sigma_1 & 0 \end{pmatrix}\begin{pmatrix} 0 & \sigma_2 \\ -\sigma_2 & 0 \end{pmatrix} = -i\begin{pmatrix} \sigma_3 & 0 \\ 0 & \sigma_3 \end{pmatrix} \equiv -i\Sigma_3 \tag{20.57}$$

Therefore, (20.56) can be written

$$S = \exp\left(-\frac{i}{2}\Sigma_3\theta\right) = I\cos\frac{\theta}{2} - i\Sigma_3\sin\frac{\theta}{2} \tag{20.58}$$

More generally, for a rotation of angle θ about an axis defined by the unit vector $\hat{n}$,

$$S = I\cos\frac{\theta}{2} - i\boldsymbol{\Sigma}\cdot\hat{n}\sin\frac{\theta}{2} \tag{20.59}$$

From (20.59), we draw the important conclusion that solutions to the Dirac equation transform under spatial rotations as particles of spin-½. Thus, although electron spin is attached ad hoc to the nonrelativistic Schroedinger theory to give the Pauli-Schroedinger theory, it emerges naturally from the Dirac equation.

Lorentz Boost along the z-Axis with Velocity β

Here

$$a = \begin{pmatrix} \gamma & 0 & 0 & -\beta\gamma \\ 0 & 1 & 0 & 0 \\ 0 & 0 & 1 & 0 \\ -\beta\gamma & 0 & 0 & \gamma \end{pmatrix}$$

It is convenient to define $\cosh u = \gamma$ and $\sinh u = \beta\gamma$. Then

$$a = \begin{pmatrix} \cosh u & 0 & 0 & -\sinh u \\ 0 & 1 & 0 & 0 \\ 0 & 0 & 1 & 0 \\ -\sinh u & 0 & 0 & \cosh u \end{pmatrix}$$

For small u, $\cosh u \approx 1$ and $\sinh u \approx u$. Thus

$$a = \begin{pmatrix} 1 & 0 & 0 & -u \\ 0 & 1 & 0 & 0 \\ 0 & 0 & 1 & 0 \\ -u & 0 & 0 & 1 \end{pmatrix} \tag{20.60}$$

As was the case for an infinitesimal rotation, here S also differs only infinitesimally from the identity, so we try

$$S = I - iuK$$
$$S^{-1} = I + iuK$$

Making use of (20.53) and (20.60), we then obtain

$$\gamma^0 + iu\left[K\gamma^0 - \gamma^0 K\right] = \gamma^0 - u\gamma^3$$
$$\gamma^3 + iu\left[K\gamma^3 - \gamma^3 K\right] = \gamma^3 - u\gamma^0$$

It is easy to verify that these equations are satisfied by

$$K = -\frac{i}{2}\gamma^0\gamma^3 = -\frac{i}{2}\alpha_3 \tag{20.61}$$

For finite u, this yields

$$S == \exp\left(-\frac{u}{2}\alpha_3\right) = I\cosh\frac{u}{2} - \alpha_3 \sinh\frac{u}{2} \tag{20.62}$$

More generally, if we have a Lorentz boost along some spatial axis defined by unit vector $\hat{n}$ with speed β,

$$S = I\cosh\frac{u}{2} - \boldsymbol{\alpha}\cdot\hat{n}\sinh\frac{u}{2} \tag{20.63}$$

Note that in (20.63), because $\boldsymbol{\alpha}$ is Hermitian, we have $S^{\dagger} = S \neq S^{-1}$. Hence S is not unitary. This is due to the fact that under a Lorentz boost, a spatial volume element does not remain invariant (there is Lorentz contraction). Hence $\psi^{\dagger}\psi$ is also not invariant under the Lorentz transformation.

In the preceding section we presented four linearly independent plane-wave solutions to the Dirac equation for arbitrary 3-momentum $\boldsymbol{p}$ [recall equations (20.42)–(20.45)]. We could have found such solutions for a particle at rest ($\boldsymbol{p}$ = 0) and then applied the transformation S of (20.63) to obtain the results (20.42)–(20.45).

Spatial Inversion

Here

$$a = \begin{pmatrix} 1 & 0 & 0 & 0 \\ 0 & -1 & 0 & 0 \\ 0 & 0 & -1 & 0 \\ 0 & 0 & 0 & -1 \end{pmatrix} \tag{20.64}$$

Thus (20.53) yields $\gamma^{i} = -S^{-1}\gamma^{i}S$ and $\gamma^{0} = S^{-1}\gamma^{0}S$, which imply

$$S = \gamma^{0} \tag{20.65}$$

20.7 Bilinear covariants

Additional useful conclusions can be drawn from (20.53). Starting with $\psi'(ax) = S\psi(x)$, we have

$$\begin{aligned} \overline{\psi'(x')}\psi'(x') &= \psi'^{\dagger}(x')\gamma^{0}\psi'(x') \\ &= \psi^{\dagger}(x)S^{\dagger}\gamma^{0}S\psi(x) \end{aligned}$$

In each of the three cases discussed in the preceding section (i.e., rotations, Lorentz boosts, and spatial inversion), $S^{\dagger}\gamma^{0}S = \gamma^{0}$, as is easily verified. Therefore,

$$\overline{\psi'(x')}\psi'(x') = \overline{\psi(x)}\psi(x) \tag{20.66}$$

In other words, $\bar{\psi}\psi$ is an invariant (or *scalar*) under Lorentz transformations. Next,

$$\begin{aligned}\overline{\psi'}\gamma^\mu\psi' &= \psi'^\dagger\gamma^0\gamma^\mu\psi' \\ &= \psi^\dagger S^\dagger\gamma^0\gamma^\mu S\psi \\ &= \psi^\dagger\gamma^0 S^{-1}\gamma^\mu S\psi \\ &= a^\mu{}_\sigma\bar{\psi}\gamma^\sigma\psi\end{aligned} \tag{20.67}$$

where we have used (20.53). Equation (20.67) reveals that $\bar{\psi}\gamma^\sigma\psi$ is a 4-vector. Because $\bar{\psi}\gamma^\sigma\psi$ changes sign under spatial inversion, it is more appropriate to call it a polar 4-vector. Next, defining

$$\sigma^{\mu\nu} = \frac{i}{2}(\gamma^\mu\gamma^\nu - \gamma^\nu\gamma^\mu) \tag{20.68}$$

we have

$$\begin{aligned}\overline{\psi'}\sigma^{\mu\nu}\psi' &= \psi^\dagger S^\dagger\gamma^0\sigma^{\mu\nu}S\psi \\ &= \frac{i}{2}\psi^\dagger\gamma^0\left[S^{-1}\gamma^\mu SS^{-1}\gamma^\nu S - S^{-1}\gamma^\nu SS^{-1}\gamma^\mu S\right]\psi \\ &= a^\mu{}_\alpha a^\nu{}_\beta\bar{\psi}\sigma^{\alpha\beta}\psi\end{aligned} \tag{20.69}$$

This means that $\bar{\psi}\sigma^{\alpha\beta}\psi$ is a second-rank (antisymmetric) tensor. Next, consider

$$\overline{\psi'}\gamma^5\gamma^\mu\psi' = \bar{\psi}S^{-1}\gamma^5 SS^{-1}\gamma^\mu S\psi \tag{20.70}$$

If we have a proper Lorentz transformation (rotation or boost), S is a sum of products of two gamma matrices, in which case $S\gamma^5 = \gamma^5 S$. However, for a spatial inversion, $S = \gamma^0$, so $S\gamma^5 = -\gamma^5 S$. Thus, under proper Lorentz transformations, $\bar{\psi}\gamma^5\gamma^\mu\psi$ is a 4-vector, but under spatial inversion, it remains invariant. In other words, $\bar{\psi}\gamma^5\gamma^\mu\psi$ is an axial 4-vector. Finally, $\bar{\psi}\gamma^5\psi$ is invariant under proper Lorentz transformations, but it changes sign under spatial inversion: it is a pseudoscalar. To summarize,

$\bar{\psi}\psi$	Scalar
$\bar{\psi}\gamma^\nu\psi$	Polar 4-vector
$\bar{\psi}\sigma^{\mu\nu}\psi$	Second-rank (antisymmetric) tensor
$\bar{\psi}\gamma^5\gamma^\mu\psi$	Axial 4-vector
$\bar{\psi}\gamma^5\psi$	Pseudoscalar

These properties play an important role in later discussions of electromagnetic and weak interactions. For the present, we mention just one very simple consequence. For a free particle at rest (where $E = m$ and $\mathbf{p} = 0$), the following relations are obviously true:

$$\bar{u}_r u_s = \delta_{rs}\eta_r \tag{20.71}$$

where

$$\eta_r = +1 \qquad \text{for } r = 1,2$$
$$\eta_r = -1 \qquad \text{for } r = 3,4$$

Because $\bar{u}u$ is an invariant, (20.71) remains true for a free particle with arbitrary momentum.

20.8 Properties and physical significance of operators in Dirac's theory

In the nonrelativistic Pauli-Schroedinger theory, the spin operator $\boldsymbol{S} = \boldsymbol{\sigma}/2$ and the orbital angular momentum operator $\boldsymbol{L} = \boldsymbol{r} \times \boldsymbol{p}$ independently commute with the free-particle Hamiltonian $H = \boldsymbol{p}^2 / 2m$. However, neither $\boldsymbol{S}$ nor $\boldsymbol{L}$ commutes with the Dirac free-particle Hamiltonian. For the spin operator, we have

$$\begin{aligned}\frac{1}{2}[\Sigma_i, H] &= \frac{1}{2}\left[\Sigma_i, (\alpha_j p_j + m\gamma_4)\right] \\ &= \frac{1}{2}\left[\Sigma_i, \alpha_j\right] p_j\end{aligned}$$

Now $\alpha_j = \gamma^5 \Sigma_j = \Sigma_j \gamma^5$; hence

$$\begin{aligned}\frac{1}{2}[\Sigma_i, H] &= \frac{1}{2}\left[\Sigma_i, \Sigma_j\right]\gamma^5 p_j \\ &= i\varepsilon_{ijk}\Sigma_k \gamma^5 p_j \\ &= i\varepsilon_{ijk}\alpha_k p_j\end{aligned}$$

Therefore,

$$[\boldsymbol{S}, H] = -i\boldsymbol{\alpha} \times \boldsymbol{p} \tag{20.72}$$

Also,

$$\begin{aligned}[L_i, H] &= \left[L_i, p_j\right]\alpha_j \\ &= i\varepsilon_{ijk} p_k \alpha_j\end{aligned}$$

Hence

$$[\boldsymbol{L}, H] = i\boldsymbol{\alpha} \times \boldsymbol{p} \tag{20.73}$$

Thus, although neither $\boldsymbol{S}$ nor $\boldsymbol{L}$ commutes with the free-particle Dirac Hamiltonian, $\boldsymbol{J} = \boldsymbol{L} + \boldsymbol{S}$ does commute with H; that is,

$$[\boldsymbol{J}, H] = 0 \tag{20.74}$$

Next, we consider various properties of the position, momentum, and velocity operators. Here it is convenient to employ the Heisenberg picture, where an operator A with no explicit time dependence satisfies the Heisenberg equation

$$\frac{dA}{dt} = -\frac{i}{\hbar}[A, H] \tag{20.75}$$

To keep the discussion general, we do not assume an explicit form for H at present but merely require that

$$H^2 = c^2 \mathbf{p}^2 + m^2 c^4 \tag{20.76}$$

Thus what follows applies to any relativistic wave equation and not just to the Dirac equation. Setting $\hbar = c = 1$ and employing one spatial dimension for simplicity, we have

$$\left[x, H^2\right] = \left[x, p^2\right] = 2ip \tag{20.77}$$

and

$$i\frac{dx}{dt} = i\dot{x} = \left[x, H\right] = xH - Hx \tag{20.78}$$

Thus

$$i\dot{x}H = xH^2 - HxH$$

and

$$iH\dot{x} = -H^2 x + HxH$$

Combining these last two equations and using (20.77), we obtain

$$i\dot{x}H + iH\dot{x} = 2ip \tag{20.79}$$

However,

$$\ddot{x} = iH\dot{x} - i\dot{x}H \tag{20.80}$$

Hence

$$\ddot{x} = 2iH\dot{x} - 2ip \tag{20.81}$$

and

$$\ddot{x} = -2i\dot{x}H + 2ip \tag{20.82}$$

Differentiating both sides of (20.82) and recalling that for a free particle H and p are constant operators, we have

$$\dddot{x} = -2i\,\ddot{x}H \tag{20.83}$$

This equation may be integrated immediately to yield

$$\ddot{x}(t) = \ddot{x}(0)e^{-2iHt} \tag{20.84}$$

It is important to maintain the order of the operators on the right-hand side of (20.84) because $\ddot{x}$ and H do not commute. Had we used (20.81) instead of (20.82), we would have obtained

$$\ddot{x}(t) = e^{2iHt}\ddot{x}(0) \tag{20.85}$$

Substitution of (20.84) into (20.82) yields

$$-2i\dot{x}H = \ddot{x}(0)e^{-2iHt} - 2ip$$

or

$$\dot{x}(t) = pH^{-1} + \frac{i}{2}\ddot{x}(0)e^{-2iHt}H^{-1} \tag{20.86}$$

As (20.86) reveals, the velocity operator $\dot{x}(t)$ has two parts. The first is the constant operator pH^{-1} and is familiar intuitively from classical relativistic mechanics, where $v = p/E$ ($= pc^2/E$ in ordinary units). However, there is a second part that oscillates extremely rapidly and is called *zitterbewegung* ("jittery motion" in German). For a particle of mass m, the zitterbewegung has an angular frequency of at least $2m = 2mc^2/\hbar$, which corresponds to a frequency of order 10^{21} Hz for an electron. The physical origin of zitterbewegung is as follows: to measure the velocity of a particle, we must determine its position at two different times; then we know its average velocity in the interval between. However, because of the uncertainty principle, a rather precise determination of the position renders the momentum, and thus the energy, correspondingly uncertain. If sufficiently large values of the energy are attained, the particle is relativistic, in which case we can expect all values of the velocity between $-c$ and $+c$ superimposed on an average velocity of pc^2/E. Moreover, if the particle of interest is sufficiently relativistic, various multiparticle processes can occur, and the very validity of the single-particle relativistic wave equation is lost. Thus we can say that the zitterbewegung is a manifestation of this breakdown.

Result (20.86) is valid for any relativistic wave equation. Now, however, we return to the Dirac equation. Here, because the free-particle Hamiltonian is $H = \boldsymbol{\alpha}\cdot\boldsymbol{p} + m\gamma_4$, we have

$$\dot{x}_i = -i\left[x_i, p_j\right]\alpha_j = \alpha_i \tag{20.87}$$

Because the eigenvalues of the matrix α_i are ± 1, the eigenvalues of the velocity operator are also ± 1 ($\pm c$ if c is exhibited explicitly). Given (20.86), this result should not surprise us. We also note that because α_i and α_j do not commute for $i \neq j$, different components of the velocity do not commute. Finally, because α_i connects positive- and negative-energy wave functions, an eigenstate of velocity cannot also be an eigenstate of energy and momentum.

Problems for Chapter 20

20.1. Suppose that there are two spatial dimensions instead of three. What form would the Dirac matrices and Dirac equation take in this case?

20.2. At some instant of time (say, $t = 0$), the normalized Dirac wave function for a free electron is known to be

$$\psi(\boldsymbol{x},0) = \frac{1}{\sqrt{V}}\begin{pmatrix} a \\ b \\ c \\ d \end{pmatrix}\exp\left(\frac{ip_z z}{\hbar}\right)$$

where a, b, c, and d are independent of the space-time coordinates and satisfy

$$|a|^2 + |b|^2 + |c|^2 + |d|^2 = 1$$

(a) Find the probabilities for observing the electron with

- $E > 0$, spin up
- $E > 0$, spin down
- $E < 0$, spin up
- $E < 0$, spin down

(b) Construct the normalized Dirac wave functions for $E > 0$ plane waves that are eigenstates of the helicity operator $h = \boldsymbol{\Sigma}\cdot\boldsymbol{p}/|\boldsymbol{p}| = \boldsymbol{\Sigma}\cdot\hat{\boldsymbol{p}}$. Evaluate the expectation values of $\boldsymbol{\Sigma}\cdot\hat{\boldsymbol{p}}$ and $\gamma^0\boldsymbol{\Sigma}\cdot\hat{\boldsymbol{p}} = -\gamma^5\boldsymbol{\gamma}\cdot\hat{\boldsymbol{p}}$.

(c) Construct the normalized Dirac wave function for an $E > 0$ transversely polarized plane wave whose propagation and spin ($\langle\boldsymbol{\Sigma}\rangle$) directions are along the positive z- and positive x-axes, respectively. Evaluate the expectation values of Σ_1 and $\gamma^0\Sigma_1$.

20.3. Let $\tilde{\gamma}^\mu$ be the transpose of γ^μ. Then

$$\tilde{\gamma}^\mu\tilde{\gamma}^\nu + \tilde{\gamma}^\nu\tilde{\gamma}^\mu = 2g^{\mu\nu}I$$

Thus there exists a nonsingular matrix S such that

$$\tilde{\gamma}^\mu = S\gamma^\mu S^{-1}$$

Find S if the γ^μ are in the standard representation.

20.4. In the Pauli-Dirac (standard) representation,

$$\psi = \begin{pmatrix} \psi_A \\ \psi_B \end{pmatrix}$$

In the Weyl representation,

$$\psi' = \frac{1}{\sqrt{2}}\begin{pmatrix} \psi_A + \psi_B \\ \psi_A - \psi_B \end{pmatrix}.$$

(a) Find the gamma matrices $\gamma'^{0,1,2,3,5}$ in the Weyl representation, and find the matrix S that yields

$$\gamma'^{\mu} = S\gamma^{\mu}S^{-1}$$

(b) Consider the free-particle Dirac equation, which is really four coupled linear equations. Write these in the Weyl representation, and show that in the limit of zero mass these become two decoupled sets of two linear equations. Discuss the role of particle helicity for these two sets of equations.

20.5. In units where $\hbar = c = 1$, the Dirac free-particle Hamiltonian is

$$H = \boldsymbol{\alpha}\cdot\mathbf{p} + \beta m \tag{1}$$

and in the standard representation

$$\boldsymbol{\alpha} = \begin{pmatrix} 0 & \boldsymbol{\sigma} \\ \boldsymbol{\sigma} & 0 \end{pmatrix} \tag{2}$$

and

$$\beta = \begin{pmatrix} I & 0 \\ 0 & -I \end{pmatrix} \tag{3}$$

The $\boldsymbol{\alpha}$ matrices are sometimes called *odd* because they connect the upper two components of a four-component Dirac wave function to the lower two components of another such function. Other odd matrices are γ^5 and $\boldsymbol{\gamma}$ matrices in the standard representation. The matrices $\beta = \gamma^0$, $\boldsymbol{\Sigma}$, and the 4×4 identity matrix are called *even* because in the standard representation they only connect the upper two components to one another and the lower two components to one another. Sometimes it is useful to employ the Foldy-Wouthuysen (F-W) transformation (Foldy and Wouthhuysen 1950) to obtain a representation in which the free-particle Dirac Hamiltonian is expressed entirely in terms of the even matrix β. This problem is concerned with the F-W transformation.

(a) Define the operator

$$U = \frac{\varepsilon + m + \boldsymbol{\alpha}\cdot\mathbf{p}\beta}{\sqrt{2\varepsilon(\varepsilon + m)}} \tag{4}$$

where $\varepsilon = +\sqrt{p^2 + m^2}$, and $\boldsymbol{\alpha}$ and β are given by (2) and (3), respectively. Show that U is unitary and that

$$U^{\dagger}HU = H' = \beta\varepsilon \tag{5}$$

H' is the Hamiltonian in the F-W representation.

(b) For discussions of a free particle of well-defined momentum, it is often convenient to work in momentum space rather than coordinate space. Here the coordinate operator is defined as

$$x_i = i\frac{\partial}{\partial p_i} \tag{6}$$

Using the Heisenberg equation and the F-W representation, show that the velocity operator is

$$\dot{x}_i = \beta\frac{p_i}{\varepsilon} \tag{7}$$

Note that no zitterbewegung appears here, which appears to be a great advantage. However, a price must be paid. Consider the operator X_i in the original representation, which becomes x_i in the F-W representation. X_i turns out to be very complicated; that is,

$$\begin{aligned} X_i &= Ux_iU^\dagger \\ &= x_i + \frac{i}{2\varepsilon}\beta\alpha_i - i\frac{\beta\boldsymbol{\alpha}\cdot\boldsymbol{p}p_i}{2\varepsilon^2(\varepsilon+m)} - \frac{(\boldsymbol{\Sigma}\times\boldsymbol{p})_i}{2\varepsilon(\varepsilon+m)} \end{aligned} \tag{8}$$

The main use of the F-W transformation is its application to problems involving interaction of an electron with an external electromagnetic field near the nonrelativistic limit. Here the F-W transformation can be applied iteratively to eliminate odd operator terms in a Hamiltonian to any desired level of precision. [For example, the F-W transformation is a systematic way of carrying out the two-component reduction of the Dirac equation for a hydrogenic atom to second order and thus to obtain the result given in Section 21.5, equation (21.29).]

21 Interaction of a Relativistic Spin-½ Particle with an External Electromagnetic Field

21.1 The Dirac equation

We now consider the Dirac equation for a spin-½ particle with electric charge q when external electromagnetic potentials are present. As in the case of the Klein-Gordon equation, we make the replacements

$$E \rightarrow E - q\Phi = i\hbar c \frac{\partial}{\partial x^0} - q\Phi$$
$$\boldsymbol{p} \rightarrow \boldsymbol{p} - \frac{q}{c}\boldsymbol{A} = -i\hbar \nabla - \frac{q}{c}\boldsymbol{A}$$

which are equivalent to the replacements

$$\partial_\mu \rightarrow \partial_\mu + \frac{iq}{\hbar c} A_\mu \tag{21.1a}$$

or, in natural units,

$$\partial_\mu \rightarrow \partial_\mu + iqA_\mu \tag{21.1b}$$

Sometimes the right-hand side of (21.1a) or (21.1b) is called the *covariant derivative* D_μ. The Dirac equation now reads

$$\left[m - i\gamma^\mu \left(\partial_\mu + iqA_\mu\right)\right]\psi = \left[m - i\gamma^\mu D_\mu\right]\psi = 0 \tag{21.2}$$

21.2 The second-order equation

It is useful to derive a second-order equation that is similar to the Klein-Gordon equation for a spin-0 particle of charge q. Operating on (21.2) on the left by $\left[m + i\gamma^\nu\left(\partial_\nu + iqA_\nu\right)\right]$, we obtain

$$\left[m + i\gamma^\nu\left(\partial_\nu + iqA_\nu\right)\right]\left[m - i\gamma^\mu\left(\partial_\mu + iqA_\mu\right)\right]\psi = 0$$

which yields

$$\left[m^2 + \gamma^\nu \gamma^\mu (\partial_\nu + iqA_\nu)(\partial_\mu + iqA_\mu)\right]\psi = 0 \tag{21.3}$$

Now

$$\begin{aligned} \gamma^\nu \gamma^\mu &= \frac{1}{2}(\gamma^\nu \gamma^\mu - \gamma^\mu \gamma^\nu) + \frac{1}{2}(\gamma^\nu \gamma^\mu + \gamma^\mu \gamma^\nu) \\ &= -i\sigma^{\nu\mu} + g^{\mu\nu} I \end{aligned} \tag{21.4}$$

Substitution of (21.4) into (21.3) gives

$$\left[m^2 + (\partial^\nu + iqA^\nu)(\partial_\nu + iqA_\nu) - i\sigma^{\nu\mu}(\partial_\nu + iqA_\nu)(\partial_\mu + iqA_\mu)\right]\psi = 0 \tag{21.5}$$

Because $\sigma_{\nu\mu}$ is antisymmetric with respect to interchange of ν and μ, the last term on the left-hand side of (21.5) can be written

$$\begin{aligned} &-\frac{i}{2}\sigma^{\nu\mu}\left[(\partial_\nu + iqA_\nu)(\partial_\mu + iqA_\mu) - (\partial_\mu + iqA_\mu)(\partial_\nu + iqA_\nu)\right]\psi \\ &= \frac{q}{2}\sigma^{\nu\mu}\left(A_\nu \partial_\mu + \partial_\nu A_\mu - A_\mu \partial_\nu - \partial_\mu A_\nu\right)\psi \\ &= \frac{q}{2}\sigma^{\nu\mu}\left[(\partial_\nu A_\mu) - (\partial_\mu A_\nu)\right]\psi \\ &= \frac{q}{2}\sigma^{\nu\mu} F_{\nu\mu}\psi \end{aligned}$$

Therefore, (21.5) becomes

$$\left[m^2 + (\partial^\nu + iqA^\nu)(\partial_\nu + iqA_\nu) + \frac{q}{2}\sigma^{\nu\mu} F_{\nu\mu}\right]\psi = 0 \tag{21.6}$$

or, as is easily verified,

$$\left[m^2 + (\partial^\nu + iqA^\nu)(\partial_\nu + iqA_\nu) - q\boldsymbol{\Sigma}\cdot\boldsymbol{B} + iq\boldsymbol{\alpha}\cdot\boldsymbol{\mathcal{E}}\right]\psi = 0 \tag{21.7}$$

The first two terms on the left-hand side of (21.6) or (21.7) are the same as those that appear in the Klein-Gordon equation for a spin-0 particle of charge q, but the remaining term in (21.6) is unique to a particle of spin-½.

21.3 First-order two-component reduction of Dirac's equation

Now we return to the Dirac equation (21.2) for a particle with definite energy E. Choosing the standard representation and writing $\psi = \begin{pmatrix} \psi_A \\ \psi_B \end{pmatrix}$, where ψ_A and ψ_B are both two-component spinors, we have

$$(E-m-q\Phi)\psi_A = \boldsymbol{\sigma}\cdot(\boldsymbol{p}-q\boldsymbol{A})\psi_B$$
$$(E+m-q\Phi)\psi_B = \boldsymbol{\sigma}\cdot(\boldsymbol{p}-q\boldsymbol{A})\psi_A \tag{21.8}$$

Defining $W = E-m$ and $\boldsymbol{\pi} = \boldsymbol{p}-q\boldsymbol{A} = -i\nabla - q\boldsymbol{A}$ and substituting these quantities into (21.8), we obtain

$$(W-q\Phi)\psi_A = \boldsymbol{\sigma}\cdot\boldsymbol{\pi}\psi_B \tag{21.9}$$

and

$$(W+2m-q\Phi)\psi_B = \boldsymbol{\sigma}\cdot\boldsymbol{\pi}\psi_A \tag{21.10}$$

We write (21.10) as

$$\psi_B = (W+2m-q\Phi)^{-1}\boldsymbol{\sigma}\cdot\boldsymbol{\pi}\psi_A$$

and substitute the latter equation into (21.9) to obtain

$$(W-q\Phi)\psi_A = \boldsymbol{\sigma}\cdot\boldsymbol{\pi}(W+2m-q\Phi)^{-1}\boldsymbol{\sigma}\cdot\boldsymbol{\pi}\psi_A \tag{21.11}$$

In the nonrelativistic limit, $W-q\Phi \ll m$; hence

$$(W-q\Phi+2m)^{-1} = \frac{1}{2m} - \frac{W-q\Phi}{(2m)^2} + \cdots \tag{21.12}$$

In the lowest (first-order) approximation, we retain only the first term on the right-hand side of (21.12), in which case (21.11) becomes

$$(W-q\Phi)\psi_A = \frac{1}{2m}\boldsymbol{\sigma}\cdot\boldsymbol{\pi}\boldsymbol{\sigma}\cdot\boldsymbol{\pi}\psi_A \tag{21.13}$$

Now

$$\begin{aligned}\boldsymbol{\sigma}\cdot\boldsymbol{\pi}\boldsymbol{\sigma}\cdot\boldsymbol{\pi}\psi_A &= (\boldsymbol{\pi}\cdot\boldsymbol{\pi} + i\boldsymbol{\sigma}\cdot\boldsymbol{\pi}\times\boldsymbol{\pi})\psi_A \\ &= (\boldsymbol{\pi}\cdot\boldsymbol{\pi} + q\boldsymbol{\sigma}\cdot\nabla\times\boldsymbol{A})\psi_A \\ &= (\boldsymbol{\pi}\cdot\boldsymbol{\pi} + q\boldsymbol{\sigma}\cdot\boldsymbol{B})\psi_A\end{aligned}$$

Thus, for an electron where $q = -e$, (21.13) becomes

$$\frac{1}{2m_e}(\boldsymbol{p}+e\boldsymbol{A})^2\psi_A - e\Phi\psi_A + g_s\mu_B\boldsymbol{S}\cdot\boldsymbol{B}\psi_A = W\psi_A \tag{21.14}$$

where $g_s = 2$. Note that the latter quantity arises naturally from the Dirac theory. As mentioned in earlier chapters, the actual g_s value, determined by experiment, is

$$g_s = 2(1.0011596) \tag{21.15}$$

The departure of the factor in parentheses from unity is mainly caused by quantum electrodynamic (radiative) effects.

21.4 Pauli moment

As is well known, certain spin-½ particles have very anomalous spin magnetic moments; that is,

$$\mu = \kappa \frac{e\hbar}{2Mc}$$

where M is the particle mass, and κ deviates far from unity. For example,

$$\mu_{\text{proton}} = 2.79 \frac{e\hbar}{2m_p c} \qquad \mu_{\text{neutron}} = -1.91 \frac{e\hbar}{2m_p c}$$

(the latter even though the neutron charge is zero). To describe this, we seek a covariant and gauge invariant if admittedly phenomenologic modification of Dirac's equation (21.2), which reduces in the nonrelativistic limit to

$$\frac{1}{2M}(\boldsymbol{p}-q\boldsymbol{A})^2 \psi_A - \kappa \frac{e\hbar}{2Mc} \boldsymbol{\sigma} \cdot \boldsymbol{B} \psi_A + q\Phi \psi_A = W \psi_A \tag{21.16}$$

where q is the particle charge ($q_{\text{neutron}} = 0$). To find the appropriate modification to (21.2), we can argue intuitively as follows: consider the quantity $\gamma_\mu A_\mu \psi$ appearing in (21.2). Because $\bar{\psi}\gamma^\mu\psi$ and A_μ are both polar 4-vectors, $\bar{\psi}\gamma^\mu\psi A_\mu$ is a scalar. We can form another scalar involving the spin-½ particle and the electromagnetic field, namely, $\bar{\psi}\sigma^{\mu\nu}\psi F_{\mu\nu}$, which suggests that we try the following modification to (21.2):

$$\left[M - i\gamma^\mu \left(\partial_\mu + iqA_\mu\right) + k\sigma^{\mu\nu} F_{\mu\nu}\right]\psi = 0 \tag{21.17}$$

where k is an appropriate constant. Just as in our previous discussion of equations (21.6) and (21.7), (21.17) can be rewritten as

$$\left[M - i\gamma^\mu \left(\partial_\mu + iqA_\mu\right) - 2k\boldsymbol{\Sigma} \cdot \boldsymbol{B} + 2ik\boldsymbol{\alpha} \cdot \boldsymbol{\mathcal{E}}\right]\psi = 0 \tag{21.18}$$

Once again we employ the standard representation to write (21.18) as two coupled equations in ψ_A and ψ_B; that is,

$$\begin{aligned} (W - q\Phi - 2k\boldsymbol{\sigma} \cdot \boldsymbol{B})\psi_A &= (\boldsymbol{\sigma} \cdot \boldsymbol{\pi} - 2ik\boldsymbol{\sigma} \cdot \boldsymbol{\mathcal{E}})\psi_B \\ (W + 2M - q\Phi + 2k\boldsymbol{\sigma} \cdot \boldsymbol{B})\psi_B &= (\boldsymbol{\sigma} \cdot \boldsymbol{\pi} + 2ik\boldsymbol{\sigma} \cdot \boldsymbol{\mathcal{E}})\psi_A \end{aligned}$$

As before, we ignore $W - q\Phi + 2k\boldsymbol{\sigma} \cdot \boldsymbol{B}$ compared with $2M$ in the second of these equations

$$\psi_B \cong \frac{1}{2M}(\boldsymbol{\sigma}\cdot\boldsymbol{\pi}+2ik\,\boldsymbol{\sigma}\cdot\boldsymbol{\mathcal{E}})\,\psi_A$$

and substitute this into the first equation to obtain

$$(W - q\Phi - 2k\boldsymbol{\sigma}\cdot\boldsymbol{B})\,\psi_A = \frac{1}{2M}(\boldsymbol{\sigma}\cdot\boldsymbol{\pi} - 2ik\boldsymbol{\sigma}\cdot\boldsymbol{\mathcal{E}})(\boldsymbol{\sigma}\cdot\boldsymbol{\pi} + 2ik\boldsymbol{\sigma}\cdot\boldsymbol{\mathcal{E}})\,\psi_A \tag{21.19}$$

We now consider two special cases of (21.19), namely,

1. $\Phi = 0,\ \boldsymbol{\mathcal{E}} = 0$. Here (21.19) becomes

$$\begin{aligned} W\psi_A &= \frac{1}{2M}\boldsymbol{\sigma}\cdot\boldsymbol{\pi}\boldsymbol{\sigma}\cdot\boldsymbol{\pi}\,\psi_A + 2k\boldsymbol{\sigma}\cdot\boldsymbol{B}\psi_A \\ &= \frac{\boldsymbol{\pi}^2}{2M}\psi_A + \left(2k - \frac{q}{2M}\right)\boldsymbol{\sigma}\cdot\boldsymbol{B}\psi_A \end{aligned}$$

From this we see that the particle has a spin magnetic moment

$$\mu = \frac{q}{2M} - 2k$$

in natural units. By appropriate choice of k, we can have any magnetic moment we like. If this is so, why should we insist on the value $g_s = 2$ for the electron that emerges from the unmodified Dirac equation, as in (21.14)? There is a compelling reason: the unmodified Dirac equation description is renormalizable, whereas the Pauli moment description with $k \neq 0$ is not renormalizable. To arrive at this conclusion, however, we need more than single-particle relativistic quantum mechanics. Quantum field theory is required.

2. $\boldsymbol{B} = 0,\ \boldsymbol{\mathcal{E}} \neq 0$ for the neutron. Now (21.19) becomes

$$W\psi_A = \frac{1}{2m_n}\boldsymbol{\sigma}\cdot(\boldsymbol{p} + i\mu_n\boldsymbol{\mathcal{E}})\boldsymbol{\sigma}\cdot(\boldsymbol{p} - i\mu_n\boldsymbol{\mathcal{E}})\,\psi_A \tag{21.20}$$

It can be shown that for a slow neutron moving in the Coulomb field of an electron, (21.20) leads to the following effective Schroedinger equation (with $\hbar = c = 1$):

$$-\frac{1}{2m_n}\nabla^2\psi_A + \frac{e\mu_n}{2m_n}\delta^3(\boldsymbol{r})\,\psi_A = W\psi_A. \tag{21.21}$$

Experiments show that in the scattering of slow neutrons, the *Foldy potential*

$$V = \frac{e\mu_n}{2m_n}\delta^3(\boldsymbol{r})$$

actually exists.

21.5 Two-component reduction of Dirac's equation in the second approximation

We now return to (21.11) with $q = -e$, appropriate for an electron; that is,

$$(W + e\Phi)\,\psi_A = \boldsymbol{\sigma}\cdot\boldsymbol{\pi}(W + 2m + e\Phi)^{-1}\boldsymbol{\sigma}\cdot\boldsymbol{\pi}\psi_A \tag{21.22}$$

In an earlier discussion of the expression

$$(W + e\Phi + 2m)^{-1} = \frac{1}{2m} - \frac{W + e\Phi}{(2m)^2} + \cdots \tag{21.23}$$

we employed only the first term on the right-hand side to arrive at the Pauli-Schroedinger equation (21.14) with $g_s = 2$. We now improve the approximation by including in addition the second term on the right-hand side of (21.23). We are interested in particular in the motion of the electron in the hydrogen atom. Accordingly, we set $\boldsymbol{A} = 0$ and choose $\Phi = Ze/(4\pi r)$. Then (21.22) yields

$$(W + e\Phi)\,\psi_A = \boldsymbol{\sigma}\cdot\boldsymbol{p}\frac{1}{2m_e}\left(1 - \frac{W + e\Phi}{2m_e}\right)\boldsymbol{\sigma}\cdot\boldsymbol{p}\psi_A \tag{21.24}$$

In what follows it is necessary to be careful about normalization. Here our starting point is the requirement that

$$\int \psi^\dagger \psi \, d^3\boldsymbol{r} = \int (\psi_A^\dagger \psi_A + \psi_B^\dagger \psi_B)\, d^3\boldsymbol{r} = 1 \tag{21.25}$$

Because $\psi_B \approx (\boldsymbol{\sigma}\cdot\boldsymbol{p}/2m_e)\,\psi_A$, we have $\psi_B^\dagger\psi_B \approx \psi_A^\dagger\,(\boldsymbol{p}^2/4m_e^2)\,\psi_A$,; hence, to sufficient precision, (21.25) can be written

$$\int \psi_A^\dagger \left(1 + \frac{\boldsymbol{p}^2}{4m_e^2}\right)\psi_A \, d^3\boldsymbol{r} = 1$$

Defining $u_A = \Omega\psi_A = \left[1 + (\boldsymbol{p}^2/8m_e^2)\right]\psi_A$, we then have

$$\int u_A^\dagger u_A \, d^3\boldsymbol{r} = 1 \tag{21.26}$$

Now (21.24) may be rewritten as

$$H_{\text{eff}}\,\psi_A = (W + e\Phi)\,\psi_A$$

Hence

$$H_{\text{eff}}\,\Omega^{-1}u_A = (W + e\Phi)\,\Omega^{-1}u_A$$

and thus

$$\Omega^{-1}H_{\text{eff}}\Omega^{-1}u_A = \Omega^{-1}(W+e\Phi)\Omega^{-1}u_A$$

which is

$$\left(1-\frac{\boldsymbol{p}^2}{8m_e^2}\right)\boldsymbol{\sigma}\cdot\boldsymbol{p}\frac{1}{2m_e}\left(1-\frac{W+e\Phi}{2m_e}\right)\boldsymbol{\sigma}\cdot\boldsymbol{p}\left(1-\frac{\boldsymbol{p}^2}{8m_e^2}\right)u_A = \left(1-\frac{\boldsymbol{p}^2}{8m_e^2}\right)(W+e\Phi)\left(1-\frac{\boldsymbol{p}^2}{8m_e^2}\right)u_A \quad (21.27)$$

We now expand the left hand- and right-hand sides of (21.27) and retain only terms up to and including those of order p^4 (which are of relative order α^2 compared with the leading terms). We discard terms of order p^4W, $p^4\Phi$, p^6, and so on. After straightforward algebra, this yields

$$\frac{\boldsymbol{p}^2}{2m_e}u_A - e\Phi u_A - \frac{\boldsymbol{p}^4}{8m_e^3}u_A - \frac{e}{8m_e^2}\left(2\boldsymbol{\sigma}\cdot\boldsymbol{p}\Phi\boldsymbol{\sigma}\cdot\boldsymbol{p}u_A - \boldsymbol{p}^2\Phi u_A - \Phi\boldsymbol{p}^2u_A\right) = Wu_A \quad (21.28)$$

Now

$$\begin{aligned}\boldsymbol{\sigma}\cdot\boldsymbol{p}\Phi\boldsymbol{\sigma}\cdot\boldsymbol{p}u_A &= \boldsymbol{\sigma}\cdot(\boldsymbol{p}\Phi)\boldsymbol{\sigma}\cdot\boldsymbol{p}u_A + \Phi\boldsymbol{\sigma}\cdot\boldsymbol{p}\boldsymbol{\sigma}\cdot\boldsymbol{p}u_A\\ &= (\boldsymbol{p}\Phi)\cdot\boldsymbol{p}u_A + \Phi\boldsymbol{p}^2u_A + i\boldsymbol{\sigma}\cdot\left[(\boldsymbol{p}\Phi)\times\boldsymbol{p}\right]u_A\end{aligned}$$

Thus, finally, (21.28) becomes

$$Wu_A = \frac{\boldsymbol{p}^2}{2m_e}u_A - e\Phi u_A - \frac{\boldsymbol{p}^4}{8m_e^3}u_A - \frac{e}{4m_e^2}\boldsymbol{\sigma}\cdot\left[(\nabla\Phi)\times\boldsymbol{p}\right]u_A - \frac{e}{8m_e^2}(\nabla^2\Phi)u_A \quad (21.29)$$

Reading (21.29) from left to right, we identify the various terms:

1. W is the total energy minus the rest energy.
2. $\boldsymbol{p}^2/2m_e$ is the ordinary nonrelativistic kinetic energy.
3. $-e\Phi$ is the ordinary Coulomb potential energy.
4. $-\boldsymbol{p}^4/8m_e^3$ is the relativistic correction to the kinetic energy.
5. $-(e/4m_e^2)\boldsymbol{\sigma}\cdot\left[(\nabla\Phi)\times\boldsymbol{p}\right] = (Ze^2/8\pi m_e^2)(1/r^3)\boldsymbol{S}\cdot(\boldsymbol{r}\times\boldsymbol{p})$ is the spin-orbit interaction term, which has a nonzero expectation value only for states with $\ell \neq 0$.
6. $-(e/8m_e^2)(\nabla^2\Phi)$ is the Darwin term, which has a nonzero expectation value only for states with $\ell = 0$.

In Section 10.2 we gave a heuristic derivation of terms 5 and 6. Here we see that these terms and the Darwin term (term 6) arise naturally from the Dirac equation. The expectation value of the Darwin term for s states is calculated in the usual manner by first-order perturbation theory using Pauli-Schroedinger wave functions; that is,

$$\left\langle u_{ns}\left|-\frac{e}{8m_e^2}\nabla^2\Phi\right|u_{ns}\right\rangle = \frac{Ze^2\pi}{2m_e^2}\left\langle u_{ns}\left|\delta^3(\boldsymbol{r})\right|u_{ns}\right\rangle = \frac{Z^4\alpha^4}{2n^3} \quad (21.30)$$

in natural units. The fine-structure energy shift due to terms 4, 5, and 6 combined is

$$\Delta E = -\frac{Z^4\alpha^4}{2n^3}\left(\frac{1}{j+1/2} - \frac{3}{4n}\right) \tag{21.31}$$

21.6 Symmetries for the Dirac Hamiltonian with a central potential

We now embark on a discussion that leads to the exact solution to the Dirac equation for the Coulomb potential. We start with the Dirac Hamiltonian for a central potential in natural units

$$H = \boldsymbol{\alpha}\cdot\boldsymbol{p} + m\gamma_4 + V(r) \tag{21.32}$$

and we first concern ourselves with the various operators of physical significance that commute with H.

First,

$$\boldsymbol{J} = \boldsymbol{L} + \frac{1}{2}\boldsymbol{\Sigma}$$

We have already shown that $\boldsymbol{J}$ commutes with the first two terms on the right-hand side of (21.32). We also know that $\boldsymbol{L}$ and $\boldsymbol{\Sigma}$ separately commute with V. Thus $[\boldsymbol{J}, H] = 0$.

Second,

$$K = \gamma^0\left(\Sigma\cdot\boldsymbol{J} - \frac{1}{2}\right)$$

To prove that $[K, H] = 0$, we start with

$$\frac{1}{2}[\gamma^0, H] = \frac{1}{2}[\gamma^0, \boldsymbol{\alpha}\cdot\boldsymbol{p}] = \gamma^0\boldsymbol{\alpha}\cdot\boldsymbol{p} \tag{21.33}$$

Now

$$\begin{aligned}[\gamma^0\boldsymbol{\Sigma}\cdot\boldsymbol{J}, \boldsymbol{\alpha}\cdot\boldsymbol{p}] &= \gamma^0\boldsymbol{\Sigma}\cdot\boldsymbol{J}\boldsymbol{\alpha}\cdot\boldsymbol{p} - \boldsymbol{\alpha}\cdot\boldsymbol{p}\gamma^0\boldsymbol{\Sigma}\cdot\boldsymbol{J} \\ &= \gamma^0\left(\boldsymbol{\Sigma}\boldsymbol{\alpha}\cdot\boldsymbol{p} + \boldsymbol{\alpha}\cdot\boldsymbol{p}\boldsymbol{\Sigma}\right)\cdot\boldsymbol{J}\end{aligned} \tag{21.34}$$

However,

$$\begin{aligned}\boldsymbol{\Sigma}\boldsymbol{\alpha}\cdot\boldsymbol{p} + \boldsymbol{\alpha}\cdot\boldsymbol{p}\boldsymbol{\Sigma} &= \boldsymbol{\Sigma}\boldsymbol{\alpha}\cdot\boldsymbol{p} - \boldsymbol{\alpha}\cdot\boldsymbol{p}\boldsymbol{\Sigma} + 2\boldsymbol{\alpha}\cdot\boldsymbol{p}\boldsymbol{\Sigma} \\ &= -2i\boldsymbol{\alpha}\times\boldsymbol{p} + 2\boldsymbol{\alpha}\cdot\boldsymbol{p}\boldsymbol{\Sigma}\end{aligned}$$

Also,

$$\boldsymbol{\alpha} = \gamma^5\boldsymbol{\Sigma} = \boldsymbol{\Sigma}\gamma^5$$

Hence (21.34) can be written

$$[\gamma^0 \boldsymbol{\Sigma}\cdot\boldsymbol{J}, H] = -2i\gamma^0\gamma^5(\boldsymbol{\Sigma}\times\boldsymbol{p}\cdot\boldsymbol{J}) + 2\gamma^0\gamma^5\boldsymbol{\Sigma}\cdot\boldsymbol{p}\boldsymbol{\Sigma}\cdot\boldsymbol{J} \tag{21.35}$$

Also,

$$\boldsymbol{\Sigma}\cdot\boldsymbol{p}\boldsymbol{\Sigma}\cdot\boldsymbol{J} = \boldsymbol{p}\cdot\boldsymbol{J} + i\boldsymbol{\Sigma}\times\boldsymbol{p}\cdot\boldsymbol{J}$$

Therefore,

$$[\gamma^0 \boldsymbol{\Sigma}\cdot\boldsymbol{J}, H] = +2\gamma^0\gamma^5\boldsymbol{p}\cdot\boldsymbol{J} = 2\gamma^0\gamma^5\boldsymbol{p}\cdot\left(\boldsymbol{L}+\frac{\boldsymbol{\Sigma}}{2}\right)$$

However, $\boldsymbol{p}\cdot\boldsymbol{L} = 0$. Thus

$$[\gamma^0 \boldsymbol{\Sigma}\cdot\boldsymbol{J}, H] = \gamma^0\gamma^5\boldsymbol{p}\cdot\boldsymbol{\Sigma} = \gamma^0\boldsymbol{\alpha}\cdot\boldsymbol{p} \tag{21.36}$$

Comparing (21.36) with (21.33), we see that $[K,H]=0$. In addition, $[K,\boldsymbol{J}]=0$, which can be seen as follows:

$$K = \gamma^0\left(\boldsymbol{\Sigma}\cdot\boldsymbol{L} + \frac{1}{2}\boldsymbol{\Sigma}\cdot\boldsymbol{\Sigma} - \frac{1}{2}\right) = \gamma^0(\boldsymbol{\Sigma}\cdot\boldsymbol{L}+1)$$

Thus

$$\begin{aligned}[\boldsymbol{J},K] &= [\boldsymbol{L},\gamma^0(\boldsymbol{\Sigma}\cdot\boldsymbol{L}+1)] + \frac{1}{2}[\boldsymbol{\Sigma},\gamma^0(\boldsymbol{\Sigma}\cdot\boldsymbol{L}+1)] \\ &= \gamma^0[\boldsymbol{L},\boldsymbol{\Sigma}\cdot\boldsymbol{L}] + \frac{\gamma^0}{2}[\boldsymbol{\Sigma},\boldsymbol{\Sigma}\cdot\boldsymbol{L}]\end{aligned}$$

Now

$$\begin{aligned}[\boldsymbol{L},\boldsymbol{\Sigma}\cdot\boldsymbol{L}]_i &= [L_i,\Sigma_j L_j] = i\Sigma_j L_k \varepsilon_{ijk} = i(\boldsymbol{\Sigma}\times\boldsymbol{L})_i \\ \frac{1}{2}[\boldsymbol{\Sigma},\boldsymbol{\Sigma}\cdot\boldsymbol{L}] &= \frac{1}{2}[\Sigma_i,\Sigma_j L_j] = i\varepsilon_{ijk}\Sigma_k L_j = -i(\boldsymbol{\Sigma}\times\boldsymbol{L})_i\end{aligned}$$

Therefore, $[K,\boldsymbol{J}]=0$. It follows that we can construct simultaneous eigenstates of H, $\boldsymbol{J}^2$, J_z, and K. We already know that the possible values of j are ½, 3/2, 5/2, and so on. What are the possible eigenvalues of K (which are conventionally called $-\kappa$)? We have

$$\begin{aligned}K^2 &= \gamma^0(\boldsymbol{\Sigma}\cdot\boldsymbol{L}+1)\gamma^0(\boldsymbol{\Sigma}\cdot\boldsymbol{L}+1) \\ &= (\boldsymbol{\Sigma}\cdot\boldsymbol{L}+1)^2 = \boldsymbol{\Sigma}\cdot\boldsymbol{L}\boldsymbol{\Sigma}\cdot\boldsymbol{L} + 2\boldsymbol{\Sigma}\cdot\boldsymbol{L}+1 \\ &= \boldsymbol{L}^2 + i\boldsymbol{\Sigma}\cdot\boldsymbol{L}\times\boldsymbol{L} + 2\boldsymbol{\Sigma}\cdot\boldsymbol{L}+1 \\ &= \boldsymbol{L}^2 - \boldsymbol{\Sigma}\cdot\boldsymbol{L} + 2\boldsymbol{\Sigma}\cdot\boldsymbol{L}+1 \\ &= \boldsymbol{L}^2 + \boldsymbol{\Sigma}\cdot\boldsymbol{L}+1\end{aligned} \tag{21.37}$$

Also,

$$\boldsymbol{J}^2 = \left(\boldsymbol{L}+\frac{\boldsymbol{\Sigma}}{2}\right)\cdot\left(\boldsymbol{L}+\frac{\boldsymbol{\Sigma}}{2}\right) = \boldsymbol{L}^2 + \boldsymbol{\Sigma}\cdot\boldsymbol{L} + \frac{3}{4} \tag{21.38}$$

Thus

$$J^2 = K^2 - \frac{1}{4}$$

and therefore

$$\kappa^2 = j(j+1) + \frac{1}{4}$$

Hence

$$\kappa = \pm\left(j + \frac{1}{2}\right) \tag{21.39}$$

Now, in the standard representation, let an eigenfunction of H, J^2, J_z, and K be $\psi = \begin{pmatrix} \psi_A \\ \psi_B \end{pmatrix}$. Then

$$K\psi = \begin{pmatrix} \boldsymbol{\sigma}\cdot\boldsymbol{L}+1 & 0 \\ 0 & -(\boldsymbol{\sigma}\cdot\boldsymbol{L}+1) \end{pmatrix}\begin{pmatrix} \psi_A \\ \psi_B \end{pmatrix} = -\kappa\begin{pmatrix} \psi_A \\ \psi_B \end{pmatrix}$$

or

$$\boldsymbol{\sigma}\cdot\boldsymbol{L}\psi_A = -(\kappa+1)\psi_A \tag{21.40}$$

and

$$\boldsymbol{\sigma}\cdot\boldsymbol{L}\psi_B = (\kappa-1)\psi_B \tag{21.41}$$

These imply that

$$\begin{aligned} J^2\psi_A &= \left(L^2 + \boldsymbol{\sigma}\cdot\boldsymbol{L} + \frac{3}{4}\right)\psi_A = \left(L^2 - \kappa - \frac{1}{4}\right)\psi_A \\ &= j(j+1)\psi_A \end{aligned}$$

and

$$\begin{aligned} J^2\psi_B &= \left(L^2 + \boldsymbol{\sigma}\cdot\boldsymbol{L} + \frac{3}{4}\right)\psi_B = \left(L^2 + \kappa - \frac{1}{4}\right)\psi_B \\ &= j(j+1)\psi_B \end{aligned}$$

Hence

$$L^2\psi_A = \left[\left(j + \frac{1}{2}\right)^2 + \kappa\right]\psi_A \equiv \ell_A(\ell_A+1)\psi_A \tag{21.42}$$

and

Table 21.1 Eigenvalues of the operators J^2, J_z, K, and L^2 for states ψ_A, ψ_B

Operator	Eigenvalue	
	ψ_A	ψ_B
J^2	$j(j+1)$	$j(j+1)$
J_z	m	m
$\sigma\cdot L+1$	$-\kappa$	κ
L^2	$\ell_A(\ell_A+1)$	$\ell_B(\ell_B+1)$
$\kappa=j+1/2$:	$\ell_A=j+1/2$	$\ell_B=j-1/2$
$\kappa=-(j+1/2)$:	$\ell_A=j-1/2$	$\ell_B=j+1/2$

$$L^2\psi_B=\left[\left(j+\frac{1}{2}\right)^2-\kappa\right]\psi_B\equiv\ell_B(\ell_B+1)\psi_B \tag{21.43}$$

where ℓ_A and ℓ_B are defined by (21.42) and (21.43), respectively. When $\kappa=j+1/2$, the last two equations yield $\ell_A=j+1/2$ and $\ell_B=j-1/2$. However, when $\kappa=-(j+1/2)$, we have $\ell_A=j-1/2$ and $\ell_B=j+1/2$. We summarize what has been learned so far in Table 21.1.

Neither ψ_A nor ψ_B is an eigenfunction of L_z or Σ_z because neither of these operators commutes with H. Instead, we may express each of the functions ψ_A and ψ_B as a radial function times an angular-momentum function in the form of a two-component spinor that has well-defined j, m, and κ but is a superposition of eigenstates of L_z and S_z; that is,

$$\psi_A=ig(r)\mathsf{Y}^m_{j\ell_A} \tag{21.44}$$

and

$$\psi_B=f(r)\mathsf{Y}^m_{j\ell_B} \tag{21.45}$$

where

$$\mathsf{Y}^m_{j,j-1/2}=\sqrt{\frac{j+m}{2j}}\begin{pmatrix}1\\0\end{pmatrix}Y^{m-1/2}_{j-1/2}(\theta,\phi)+\sqrt{\frac{j-m}{2j}}\begin{pmatrix}0\\1\end{pmatrix}Y^{m+1/2}_{j-1/2}(\theta,\phi) \tag{21.46}$$

and

$$\mathsf{Y}^m_{j,j+1/2}=\sqrt{\frac{j-m+1}{2j+2}}\begin{pmatrix}1\\0\end{pmatrix}Y^{m-1/2}_{j+1/2}(\theta,\phi)-\sqrt{\frac{j+m+1}{2j+2}}\begin{pmatrix}0\\1\end{pmatrix}Y^{m+1/2}_{j+1/2}(\theta,\phi) \tag{21.47}$$

Here the square-root factors are vector coupling coefficients with the same conventional choice of phases as in Chapter 7 [see, e.g., equations (7.110) and (7.111)].

21.7 Coupled radial equations

The radial functions $ig(r)$ and $f(r)$ are found by solving the Dirac equation for a specific potential $V(r)$. To see how this is done, we start with the coupled Dirac equations

$$\begin{aligned} \boldsymbol{\sigma}\bullet\boldsymbol{p}\psi_B &= [E-m-V(r)]\,\psi_A \\ \boldsymbol{\sigma}\bullet\boldsymbol{p}\psi_A &= [E+m-V(r)]\,\psi_B \end{aligned} \tag{21.48}$$

It is convenient in what follows to employ the identity

$$\boldsymbol{\sigma}\bullet\boldsymbol{p} = \frac{\boldsymbol{\sigma}\bullet\boldsymbol{r}\boldsymbol{\sigma}\bullet\boldsymbol{r}}{r^2}\boldsymbol{\sigma}\bullet\boldsymbol{p}$$

which holds because $\boldsymbol{\sigma}\bullet\boldsymbol{r}\boldsymbol{\sigma}\bullet\boldsymbol{r} = r^2 + i\boldsymbol{\sigma}\bullet\boldsymbol{r}\times\boldsymbol{r} = r^2$. Thus we have

$$\begin{aligned} \boldsymbol{\sigma}\bullet\boldsymbol{p} &= \frac{1}{r^2}\boldsymbol{\sigma}\bullet\boldsymbol{r}\left(\boldsymbol{\sigma}\bullet\boldsymbol{r}\boldsymbol{\sigma}\bullet\boldsymbol{p}\right) \\ &= \frac{1}{r^2}\boldsymbol{\sigma}\bullet\boldsymbol{r}\left(\boldsymbol{r}\bullet\boldsymbol{p} + i\boldsymbol{\sigma}\bullet\boldsymbol{r}\times\boldsymbol{p}\right) \\ &= \frac{1}{r^2}\boldsymbol{\sigma}\bullet\boldsymbol{r}\left(-ir\frac{\partial}{\partial r} + i\boldsymbol{\sigma}\bullet\boldsymbol{L}\right) \end{aligned}$$

Therefore, we obtain

$$\begin{aligned} \boldsymbol{\sigma}\bullet\boldsymbol{p}\psi_A &= \frac{\boldsymbol{\sigma}\bullet\boldsymbol{r}}{r^2}\left(-ir\frac{\partial}{\partial r} + i\boldsymbol{\sigma}\bullet\boldsymbol{L}\right)\psi_A \\ &= \frac{\boldsymbol{\sigma}\bullet\boldsymbol{r}}{r}\mathsf{Y}^m_{j\ell_A}\left(-i\frac{\partial}{\partial r} - i\frac{\kappa+1}{r}\right)ig(r) \end{aligned} \tag{21.49}$$

and

$$\begin{aligned} \boldsymbol{\sigma}\bullet\boldsymbol{p}\psi_B &= \frac{\boldsymbol{\sigma}\bullet\boldsymbol{r}}{r^2}\left(-ir\frac{\partial}{\partial r} + i\boldsymbol{\sigma}\bullet\boldsymbol{L}\right)\psi_B \\ &= \frac{\boldsymbol{\sigma}\bullet\boldsymbol{r}}{r}\mathsf{Y}^m_{j\ell_B}\left(-i\frac{\partial}{\partial r} + i\frac{\kappa-1}{r}\right)f(r) \end{aligned} \tag{21.50}$$

Furthermore, $\Lambda \equiv \boldsymbol{\sigma}\bullet\hat{\boldsymbol{r}}$, the scalar product of an axial 3-vector and a polar 3-vector, is a pseudoscalar operator. When applied to $\mathsf{Y}^m_{j\ell}$, it gives another angular eigenfunction of the same j and m but of opposite parity. In fact, it is easy to verify that with our choice of phases in (21.46) and (21.47),

$$\begin{aligned} \Lambda\mathsf{Y}^m_{j\ell_A} &= \mathsf{Y}^m_{j\ell_B} \\ \Lambda\mathsf{Y}^m_{j\ell_B} &= \mathsf{Y}^m_{j\ell_A} \end{aligned}$$

For example, we apply the operator Λ to $Y^{1/2}_{1/2,0}$:

$$\boldsymbol{\sigma}\boldsymbol{\cdot}\hat{\boldsymbol{r}}Y^{1/2}_{1/2,0} = \left[\sigma_z \cos\theta + \frac{\sigma_-}{2}\sin\theta e^{i\phi} + \frac{\sigma_+}{2}\sin\theta e^{-i\phi}\right]\begin{pmatrix}1\\0\end{pmatrix}Y^0_0$$

$$= \frac{1}{\sqrt{4\pi}}\begin{pmatrix}\cos\theta\\ \sin\theta e^{i\phi}\end{pmatrix} = \begin{pmatrix}3^{-1/2}Y^0_1\\ -(2/3)^{1/2}Y^1_1\end{pmatrix} = Y^{1/2}_{1/2,1}$$

Thus (21.49) and (21.50) become

$$\boldsymbol{\sigma}\boldsymbol{\cdot}\boldsymbol{p}\psi_A = Y^m_{j\ell_B}\left(\frac{\partial}{\partial r} + \frac{\kappa+1}{r}\right)g(r) \tag{21.51}$$

and

$$\boldsymbol{\sigma}\boldsymbol{\cdot}\boldsymbol{p}\psi_B = iY^m_{j\ell_A}\left(-\frac{\partial}{\partial r} + \frac{\kappa-1}{r}\right)f(r) \tag{21.52}$$

respectively. Substitution of these expressions into (21.48) yields the coupled radial equations

$$\left(-\frac{\partial}{\partial r} + \frac{\kappa-1}{r}\right)f(r) = [E - m - V(r)]g(r) \tag{21.53}$$

and

$$\left(\frac{\partial}{\partial r} + \frac{\kappa+1}{r}\right)g(r) = [E + m - V(r)]f(r) \tag{21.54}$$

Frequently it is convenient to employ the substitutions $F(r) = rf(r)$ and $G(r) = rg(r)$ to obtain the alternative radial equations

$$\frac{\partial F}{\partial r} - \frac{\kappa F}{r} = [m - E + V(r)]G \tag{21.55}$$

and

$$\frac{\partial G}{\partial r} + \frac{\kappa G}{r} = [m + E - V(r)]F \tag{21.56}$$

21.8 Dirac radial functions for the Coulomb potential

For the attractive Coulomb potential $V(r) = -Ze^2/4\pi r = -Z\alpha/r$, (21.55) and (21.56) become

$$\frac{\partial F}{\partial r} - \frac{\kappa F}{r} = \left(m - E - \frac{Z\alpha}{r}\right)G \tag{21.57}$$

$$\frac{\partial G}{\partial r} + \frac{\kappa G}{r} = \left(m + E - \frac{Z\alpha}{r}\right)F \tag{21.58}$$

For very large r, $\partial F/\partial r \to (m-E)G$ and $\partial G/\partial r \to (m+E)F$. Thus we have

$$\frac{\partial^2 F}{\partial r^2} = (m^2 - E^2)F \qquad \frac{\partial^2 G}{\partial r^2} = (m^2 - E^2)G$$

which yield the solutions

$$\begin{aligned} \lim_{r\to\infty} F &= a_1 e^{-\lambda r} \\ \lim_{r\to\infty} G &= a_2 e^{-\lambda r} \end{aligned} \tag{21.59}$$

where $\lambda = (m^2 - E^2)^{1/2}$,

$$a_1 = -a\sqrt{1-\frac{E}{m}} \qquad a_2 = +a\sqrt{1+\frac{E}{m}}$$

and a is a constant. Making the substitution $\varepsilon = E/m$, we try to form solutions F and G that are valid for all r by writing

$$\begin{aligned} F &= \sqrt{1-\varepsilon}\,(\phi_1 - \phi_2)e^{-\lambda r} \\ G &= \sqrt{1+\varepsilon}\,(\phi_1 + \phi_2)e^{-\lambda r} \end{aligned} \tag{21.60}$$

where for large r, $\phi_2 \gg \phi_1$. In terms of the new variable $\rho = 2\lambda r$, (21.57) and (21.58) yield

$$\frac{\partial \phi_1}{\partial \rho} = \left(1 - \frac{\alpha\varepsilon}{\sqrt{1-\varepsilon^2}}\frac{Z}{\rho}\right)\phi_1 + \left(-\frac{\kappa}{\rho} - \frac{\alpha}{\sqrt{1-\varepsilon^2}}\frac{Z}{\rho}\right)\phi_2 \tag{21.61}$$

$$\frac{\partial \phi_2}{\partial \rho} = \left(\frac{\alpha\varepsilon}{\sqrt{1-\varepsilon^2}}\frac{Z}{\rho}\right)\phi_2 + \left(-\frac{\kappa}{\rho} + \frac{\alpha}{\sqrt{1-\varepsilon^2}}\frac{Z}{\rho}\right)\phi_1. \tag{21.62}$$

We try solutions of the form

$$\phi_1 = \rho^\gamma \sum_{\nu=0}^{\infty} a_\nu \rho^\nu \qquad \phi_2 = \rho^\gamma \sum_{\nu=0}^{\infty} b_\nu \rho^\nu \tag{21.63}$$

In general, it is impossible to have f and g finite everywhere. However, we can retain the usual requirement for bound states that

$$\int_0^\infty \left(|f|^2 + |g|^2\right) r^2 \, dr$$

is finite. Substituting (21.63) into (21.61) and (21.62) and equating terms with equal powers of ρ, we obtain the following relations:

$$(\gamma + \nu + u)a_\nu = a_{\nu-1} - (\kappa + w)b_\nu \tag{21.64}$$

$$(\gamma + \nu - u)b_\nu = (w - \kappa)a_\nu \tag{21.65}$$

where $w = Z\alpha/\sqrt{1-\varepsilon^2}$ and $u = w\varepsilon$, so $w^2 - u^2 = Z^2\alpha^2$. In particular, because a_{-1} does not exist, (21.64) and (21.65) reduce to the following relations for $\nu = 0$:

$$\begin{aligned}(\gamma + u)a_0 + (\kappa + w)b_0 &= 0\\(w - \kappa)a_0 - (\gamma - u)b_0 &= 0\end{aligned} \tag{21.66}$$

This pair of homogeneous linear equations has a solution only if the determinant of the coefficients vanishes and hence that

$$\gamma^2 = \kappa^2 - Z^2\alpha^2$$

and thus

$$\gamma = \pm\sqrt{\kappa^2 - Z^2\alpha^2}$$

However, the requirement that $\int_0^\infty \left(|f|^2 + |g|^2\right) r^2\, dr$ is finite means that $\gamma > -1/2$, so we choose the positive root

$$\gamma = \sqrt{\kappa^2 - Z^2\alpha^2} \tag{21.67}$$

Returning now to the general equations (21.64) and (21.65), we eliminate b_ν to obtain the relation

$$\frac{a_\nu}{a_{\nu-1}} = \frac{\gamma - u + \nu}{(\gamma + \nu)^2 - \gamma^2} \tag{21.68}$$

If the first sum in (21.63) contains an infinite number of terms, then ν can be arbitrarily large. In this case, (21.68) yields

$$\lim_{\nu\to\infty} \frac{a_\nu}{a_{\nu-1}} \to \frac{1}{\nu}$$

which implies that $\phi_1 \to \text{const}\cdot e^\rho$ and thus that G and F grow like $e^{\rho/2}$ for large ρ. Because this is incompatible with the requirement that $\int_0^\infty \left(|f|^2 + |g|^2\right) r^2\, dr$ be finite, we require that the sums in (21.63) each terminate after a finite number of terms. Then, in the numerator of the right-hand side of (21.68) there exists a nonnegative integer $\nu = n'$ such that

$$n' = u - \gamma = \frac{Z\alpha\varepsilon}{\sqrt{1-\varepsilon^2}} - \gamma \tag{21.69}$$

Furthermore, if $n' = 0$, the second of equations (21.66) implies that $a_0 = 0$, in which case we must have $b_0 \neq 0$ so that F and G do not vanish identically. In this case, the first of the equations

(21.66) requires that $\kappa < 0$; that is: $\kappa = -(\ell_A + 1)$. The quantity n' plays the same role here that the radial quantum number $n_r = n - \ell - 1$ played in the Schroedinger theory. Now we introduce the principal quantum number

$$n = n' + k \tag{21.70}$$

where $k = |\kappa| = j + 1/2 = 1, 2, \ldots, n$ and $n = 1, 2, 3, \ldots$. Then, from (20.69), we obtain

$$\varepsilon = \frac{E}{m} = \frac{1}{\sqrt{1 + \left(\frac{Z\alpha}{n - k + \sqrt{k^2 - Z^2\alpha^2}} \right)^2}} \tag{21.71}$$

Thus the electron binding energy $W = E - m$ is

$$W = \frac{m}{\sqrt{1 + \left(\frac{Z\alpha}{n - k + \sqrt{k^2 - Z^2\alpha^2}} \right)^2}} - m \tag{21.72}$$

This formula gives a correct description of the energy levels of a hydrogenic atom of any Z such that $Z\alpha < 1$, including fine structure but not including hyperfine structure, radiative corrections (e.g., Lamb shift, anomalous part of the electron spin magnetic moment), or finite nuclear size. Expanding (21.72) for small Z, we obtain

$$\frac{W}{m} = -\frac{1}{2}\frac{(Z\alpha)^2}{n^2} - \frac{(Z\alpha)^4}{2n^3}\left(\frac{1}{j + 1/2} - \frac{3}{4n} \right) - \cdots \tag{21.73}$$

The reader will recognize the first term on the right-hand side from the Balmer formula and the second (fine-structure) term that we earlier obtained from the two-component reduction in second order [see equation (21.31)]. The various energy levels are classified in Table 21.2.

For $n = 1$ (the ground state), (21.71) yields the energy

$$E = \frac{m}{\sqrt{1 + \frac{Z^2\alpha^2}{1 - Z^2\alpha^2}}} = m\sqrt{1 - Z^2\alpha^2} \tag{21.74}$$

Although the binding energy $W = E - m$ is small for $Z\alpha \ll 1$, it increases in magnitude as Z increases. For example, for heavy atoms such as Pb ($Z = 82$), the $1s$ electrons (K shell) are so close to the nucleus that screening effects due to the other electrons are very small, and the K-shell electrons can be treated as hydrogenic to a good approximation. Because $Z\alpha = 82/137 \approx 0.6$, we must employ (21.74) rather than the Balmer formula to calculate the energy. If atomic nuclei did exist for $Z \approx 137$, we would have hydrogenic orbitals with $W \to -m$. The solutions to the Dirac equation take on a different character when the limit $Z\alpha = 1$ is passed. For example, when $k = 1$, $\gamma = \sqrt{1 - Z^2\alpha^2}$ becomes imaginary.

We now turn to the Dirac eigenfunctions, the general formula for which is quite complicated [see, e.g., Bethe and Salpeter (1957), pp. 69–70]. The ground-state function is

Table 21.2 Quantum numbers of the lowest states of atomic hydrogen in the Dirac theory

n	j	n'	κ	ℓ_A	ℓ_B	Spectros. notation	Parity π_A
1	1/2	0	−1	0	1	$1^2 s_{1/2}$	+
2	1/2	1	−1	0	1	$2^2 s_{1/2}$	+
2	1/2	1	+1	1	0	$2^2 p_{1/2}$	−
2	3/2	0	−2	1	2	$2^2 p_{3/2}$	−
3	1/2	2	−1	0	1	$3^2 s_{1/2}$	+

$$\psi = \frac{N}{\pi^{1/2}} (Z\alpha)^{3/2} (Z\alpha r)^{\sqrt{1-Z^2\alpha^2}-1} e^{-Z\alpha r} \begin{pmatrix} \chi \\ \dfrac{i\left(1-\sqrt{1-Z^2\alpha^2}\right)}{Z\alpha} \dfrac{\boldsymbol{\sigma}\cdot\mathbf{r}}{r} \chi \end{pmatrix} \tag{21.75}$$

where χ is a two-component spinor with $\chi^\dagger \chi = 1$, and

$$N = 2^{\sqrt{1-Z^2\alpha^2}-1} \sqrt{\frac{1+\sqrt{1-Z^2\alpha^2}}{\Gamma\left(1+2\sqrt{1-Z^2\alpha^2}\right)}} \tag{21.76}$$

The exponential factor in (21.75) is the same as in the Schroedinger theory. However, there is also the factor

$$(Z\alpha r)^{\sqrt{1-Z^2\alpha^2}-1} \tag{21.77}$$

which diverges as $r \to 0$. This factor reaches the value $e = 2.718...$ when r is

$$r = \frac{1}{Z\alpha} \exp\left(-\frac{1}{1-\sqrt{1-Z^2\alpha^2}}\right)$$

For $Z\alpha \ll 1$, this is

$$r \approx \frac{1}{Z\alpha} \exp\left(-\frac{2}{Z^2\alpha^2}\right)$$

The effect is thus totally negligible because of the finite nuclear size for $Z\alpha \ll 1$. However, for large $Z\alpha$, the effect is significant. For example, the radius of the $^{208}_{82}\mathrm{Pb}$ nucleus is approximately 7.1×10^{-13} cm. At this radius, the factor in (21.77) is approximately 2.45. The peculiar singularity at $r = 0$ is a feature of all $j = ½$ hydrogenic bound states.

Next, we consider the "small" component ψ_B in (21.75). As we have already shown, it corresponds to $\ell_B = 1$ (a p angular wave function). According to (21.75), it has the following magnitude relative to ψ_A:

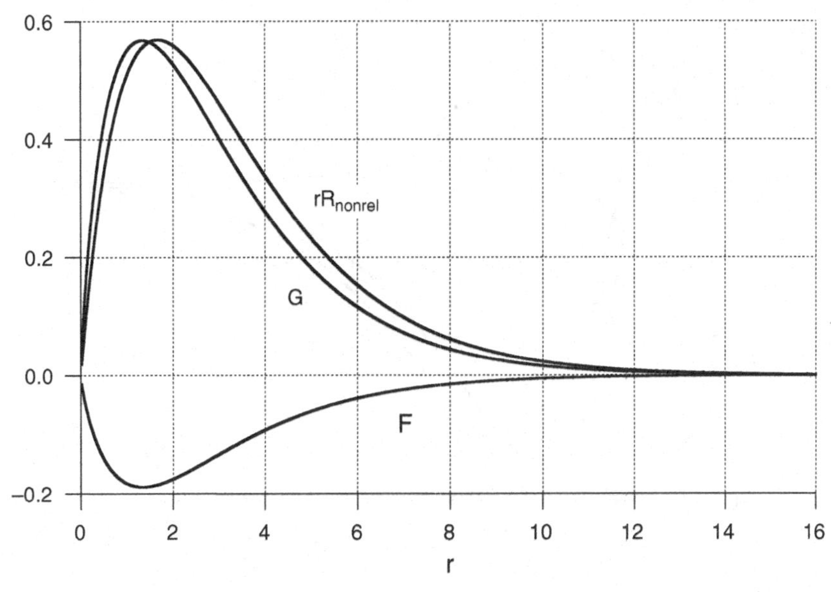

Figure 21·1 Hydrogenic radial functions G, F, and rR_{nonrel} for the $1^2s_{1/2}$ state, $Z=82$. The abscissa is r in units of $\hbar/m_ec$.

$$\frac{1-\sqrt{1-Z^2\alpha^2}}{Z\alpha}\approx\frac{Z\alpha}{2}$$

for $Z\alpha \ll 1$. For hydrogen, this contribution is indeed very small (of order $\alpha/2 \approx 0.0036$). However, for heavy atoms such as mercury, thallium, and lead, where $Z\alpha \approx 0.6$, the "small" component is comparable in magnitude to the "large" component. The consequences are significant for various observable effects, such as hyperfine structure, x-ray spectra, atomic screening in the Coulomb corrections to nuclear beta decay, and parity nonconservation due to the neutral weak interaction.

In Figures 21.1 through 21.4 we plot the functions G and F for low-lying hydrogenic states with $Z = 82$. We see from these figures that except for the behavior at extremely small distances, G is quite similar to the nonrelativistic radial function rR.

21.9 Perturbation calculations with Dirac bound-state wave functions

How do we calculate the Zeeman effect; the Stark effect; hyperfine splittings; emission and absorption of photons; and other physical effects using the Dirac theory? The answer is that we use perturbation theory in very much the same way as with Schroedinger or Pauli-Schroedinger wave mechanics. In fact, in some respects, the Dirac theory is simpler. We now illustrate with two important examples.

21.9.1 Hyperfine structure

Here the most important case is magnetic dipole ($M1$) hyperfine structure, where the magnetic dipole moment of a nucleus with nonzero spin interacts with the spin and orbital magnetic

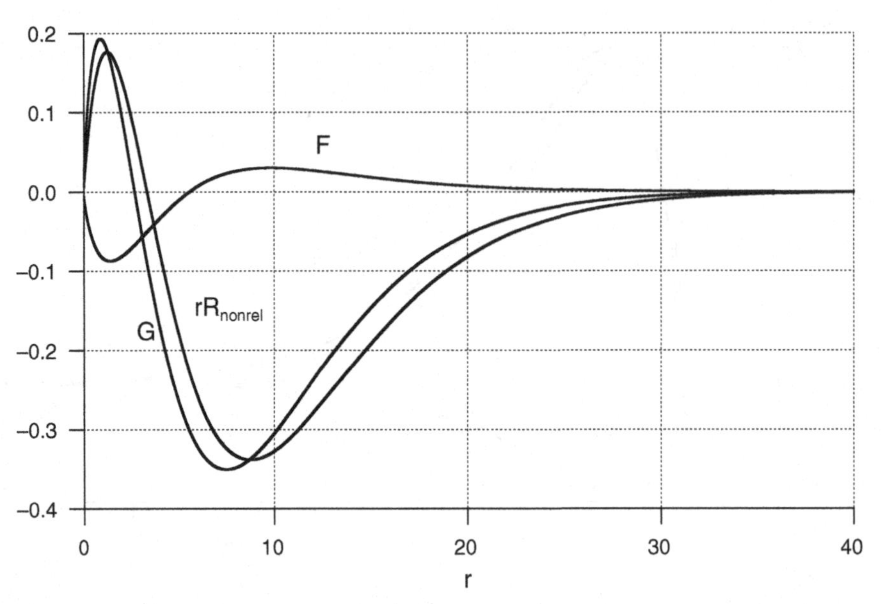

Figure 21.2 Same as Figure 21.1 but for the $2^2 s_{1/2}$ state.

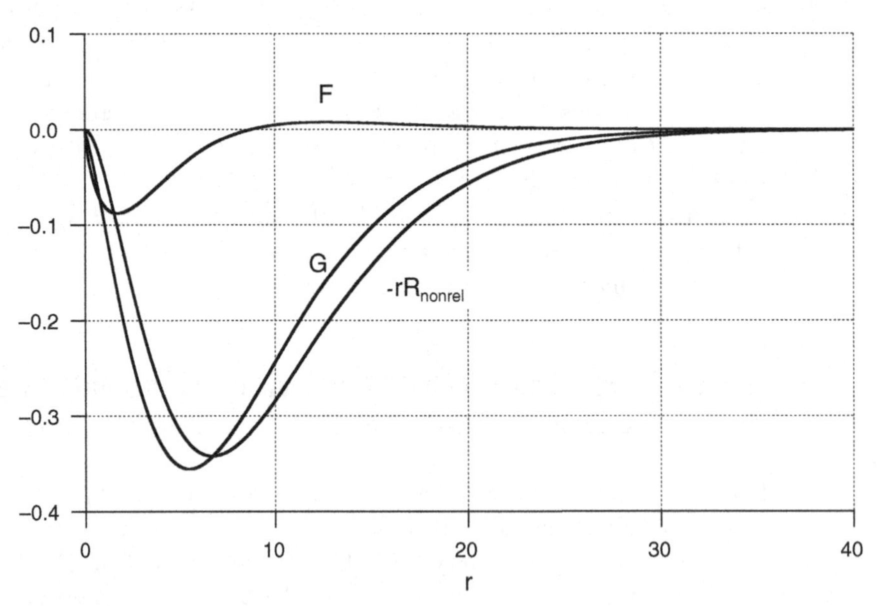

Figure 21.3 Same as Figure 21.1 but for the $2^2 p_{1/2}$ state.

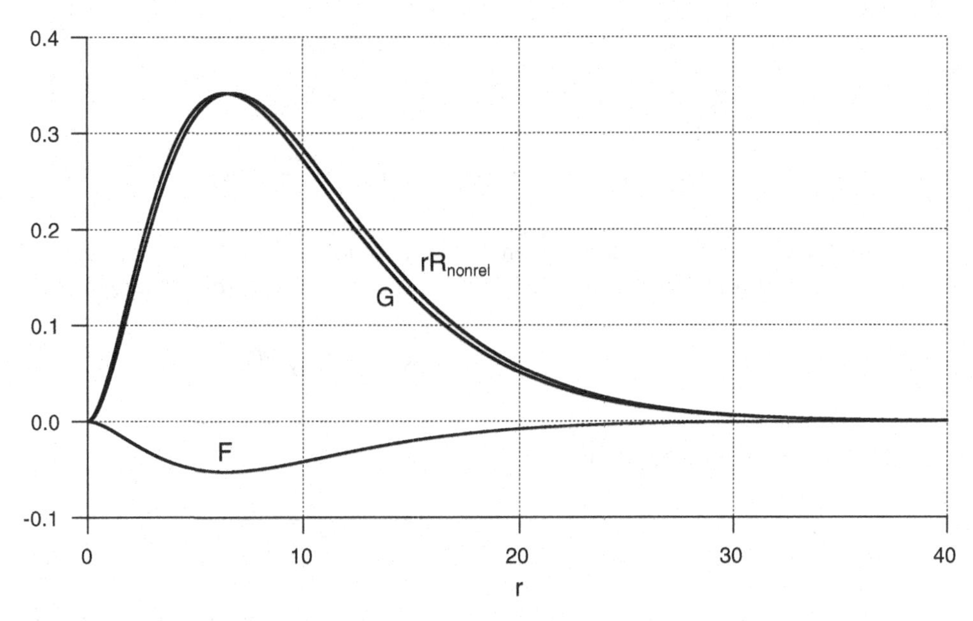

Figure 21.4 Same as Figure 21.1 but for the $2^2p_{3/2}$ state.

dipole moments of an unpaired atomic electron. In nonrelativistic quantum mechanics, the $M1$ hyperfine Hamiltonian for hydrogen was given in Chapter 10 as

$$H_{\text{hfs}} = \left[-\frac{8\pi}{3}\boldsymbol{\mu}_p\cdot\boldsymbol{\mu}_s\delta^3(\boldsymbol{r})\right]_{\ell=0} + \left[\left(\frac{\boldsymbol{\mu}_p\cdot\boldsymbol{\mu}_s}{r^3} - \frac{3\boldsymbol{\mu}_p\cdot\boldsymbol{r}\boldsymbol{\mu}_s\cdot\boldsymbol{r}}{r^5}\right) - \frac{\boldsymbol{\mu}_p\cdot\boldsymbol{\mu}_\ell}{r^3}\right]_{\ell>0} \tag{21.78}$$

Here $\boldsymbol{\mu}_p = g_p\mu_N\boldsymbol{I}$ is the proton nuclear magnetic moment, where $\mu_N = e\hbar/2m_pc$, $g_p = 5.58$, and $\boldsymbol{I}$ is the proton spin (with $I = 1/2$). Also, $\boldsymbol{\mu}_s$ and $\boldsymbol{\mu}_\ell$ are the electron spin and orbital magnetic moments, respectively. The separation of H_{hfs} into three distinct contributions, as shown in (21.78), reflects the fact that in the nonrelativistic theory, electron spin does not arise naturally but is grafted onto the theory in an ad hoc manner.

In the Dirac theory, the Hamiltonian in natural units for a relativistic electron in a Coulomb field and a magnetic field is

$$H = \boldsymbol{\alpha}\cdot\boldsymbol{p} + e\boldsymbol{\alpha}\cdot\boldsymbol{A} + m\gamma^0 - \frac{Z\alpha}{r} \tag{21.79}$$

We treat the second term on the right-hand side of (20.79) as a perturbation and write

$$H = H_0 + H'$$

with

$$H_0 = \boldsymbol{\alpha}\cdot\boldsymbol{p} + m\gamma^0 - \frac{Z\alpha}{r}$$

and

$$H' = e\boldsymbol{\alpha}\cdot\boldsymbol{A} \tag{21.80}$$

For the case of $M1$ hyperfine structure,

$$H' = H_{\text{hfs}} = e\frac{\boldsymbol{\alpha}\cdot\boldsymbol{\mu}_N \times \boldsymbol{r}}{4\pi r^3} = e\boldsymbol{\mu}_N\cdot\frac{\hat{\boldsymbol{r}}\times\boldsymbol{\alpha}}{4\pi r^2}. \tag{21.81}$$

It can be shown (see Problem 21.5) that to lowest order in α, (21.81) yields the following result in natural units for the $M1$ hyperfine splitting in the ground state of hydrogen:

$$\delta = \frac{2}{3}g_p g_s \alpha^4 \frac{m_e}{m_p} \tag{21.82}$$

In atomic units, this is

$$\delta = \frac{2}{3}g_p g_s \alpha^2 \frac{m_e}{m_p}$$

which is the same result obtained by the nonrelativistic calculation in equation (10.32).

21.9.2 Emission and absorption of photons

The emission or absorption of photons by a nonrelativistic atomic electron was discussed at length in Chapter 16. There we saw that the interaction Hamiltonian takes the form

$$H_{\text{int}} = \frac{e}{m_e c}\boldsymbol{A}\cdot\boldsymbol{p} + \frac{e^2}{2m_e c^2}A^2 + g_s\frac{\mu_B}{\hbar}\boldsymbol{S}\cdot\boldsymbol{B} \tag{21.83}$$

where $\boldsymbol{A}$ is the radiation field vector potential in Coulomb gauge, and $\boldsymbol{B} = \nabla\times\boldsymbol{A}$. The term in A^2 makes no contribution for single-photon emission and absorption processes but does play a role when two photons are involved, as in Rayleigh scattering. The third term, involving the coupling of the electron spin to $\boldsymbol{B}$, is added purely phenomenologically because the electron spin itself has no natural explanation in nonrelativistic quantum mechanics. In the Dirac theory, (21.83) is replaced by the perturbation Hamiltonian

$$H' = e\boldsymbol{\alpha}\cdot\boldsymbol{A} \tag{21.84}$$

in natural units. Here it is convenient to describe the radiation field $\boldsymbol{A}$ in vector spherical waves rather than the plane waves employed in Chapter 16. The properties of vector spherical waves are summarized in Appendix C. One finds that $\boldsymbol{A}$ consists of two distinct types of terms: magnetic multipole and electric multipole. Magnetic multipole terms take the form

$$\boldsymbol{A}_\ell^{(m)} = \text{const}\cdot j_\ell(kr)\boldsymbol{Y}_{\ell\ell}^M \tag{21.85}$$

where $\boldsymbol{Y}_{j\ell}^M(\theta,\phi) = \sum_{m'=-1}^{1}\langle \ell,m,1,m'|JM\rangle \hat{\varepsilon}_{m'} Y_\ell^m(\theta,\phi)$ is a vector spherical harmonic with

$$\hat{\varepsilon}_1 = -\frac{\hat{x}+i\hat{y}}{\sqrt{2}}$$
$$\hat{\varepsilon}_0 = \hat{z}$$
$$\hat{\varepsilon}_{-1} = \frac{\hat{x}-i\hat{y}}{\sqrt{2}}$$

and j_ℓ is a spherical Bessel function. Electric multipole terms take the form

$$A_\ell^{(e)} = \text{const}\left\{-\left[\frac{\ell}{2\ell+1}\right]^{1/2} j_{\ell+1}\boldsymbol{Y}_{\ell,\ell+1}^M + \left[\frac{\ell+1}{2\ell+1}\right]^{1/2} j_{\ell-1}\boldsymbol{Y}_{\ell,\ell-1}^M\right\} \tag{21.86}$$

When $\ell = 1$, (21.86) becomes

$$A_1^{(e)} = \text{const}\left\{-\left[\frac{1}{3}\right]^{1/2} j_2\boldsymbol{Y}_{1,2}^M + \left[\frac{2}{3}\right]^{1/2} j_0\boldsymbol{Y}_{1,0}^M\right\} \tag{21.87}$$

In the long-wavelength limit, only the second term on the right-hand side of (21.87) is important because $j_0(kr) \to 1$ and $j_2(kr) \to 0$ for $kr \ll 1$. Also $\boldsymbol{Y}_{1,0}^M = \left(1/\sqrt{4\pi}\right)\hat{\varepsilon}_M$; hence in the long-wavelength limit

$$A_1^{(e)} \to \text{const}\cdot\hat{\varepsilon}$$

Thus we have

$$H' = \text{const}\ \boldsymbol{\alpha}\cdot\hat{\varepsilon} \tag{21.88}$$

This is reduced to a simpler form by noting that

$$[r_i, H_0] = [r_i, \alpha_j p_j] = i\alpha_i$$

Hence (20.88) can be written

$$H' = \text{const}[\boldsymbol{r}\cdot\hat{\varepsilon}, H_0]$$

which is familiar from Chapter 16.

Problems for Chapter 21

21.1. Show that the rate for an electron in the ground state of the hydrogen atom to radiate and fall into empty negative-energy states in the energy interval $-m_ec^2$ to $-2m_ec^2$ is approximately

$$\frac{\alpha^6 m_e c^2}{\pi\hbar} \approx 10^8 \text{ s}^{-1}$$

21.2. Consider an electron in a uniform and constant magnetic field B along the z-axis. Assuming that $g_s = 2$, show that the energy eigenstates are given by the formula

$$E^2 = m_e^2 c^4 + c^2 p_z^2 + 2ne\hbar cB$$

where n is an integer. Compare with the analogous nonrelativistic formula.

21.3. In Section 21.4 we discussed the way in which a spin-½ particle with an anomalous spin magnetic moment is described by a Diraclike equation with a Pauli moment term. It can be shown that the interaction of the Pauli moment with an external magnetic field $\boldsymbol{B}$ is described by the following interaction Hamiltonian:

$$H' = -\mu\gamma^0 \boldsymbol{\Sigma}\boldsymbol{\cdot}\boldsymbol{B}$$

where the magnetic moment operator is $\boldsymbol{\mu} = \mu\boldsymbol{\Sigma}$. Consider a relativistic neutron moving in a homogeneous magnetic field $\boldsymbol{B}$. In zeroth order, we assume that the neutron is a free particle with constant 3-momentum $\boldsymbol{p} = m\boldsymbol{v}\gamma$ and energy $E = m\gamma$ (in units where $\hbar = c = 1$). Now we want to calculate the additional energy ΔE associated with inclusion of H' as a first-order perturbation. Show that

$$\Delta E = -\chi^\dagger \left(\boldsymbol{\mu} - \frac{\gamma}{1+\gamma} \boldsymbol{\mu}\boldsymbol{\cdot}\boldsymbol{v}\boldsymbol{v} \right) \boldsymbol{\cdot}\boldsymbol{B}\chi \tag{1}$$

where the zero-order free-particle Dirac wave function is

$$\psi = \sqrt{\frac{E+m}{2E}} \begin{pmatrix} \chi \\ \dfrac{\boldsymbol{\sigma}\boldsymbol{\cdot}\boldsymbol{p}}{E+m}\chi \end{pmatrix} \exp\left[i(\boldsymbol{p}\boldsymbol{\cdot}\boldsymbol{r} - Et)\right]$$

Equation (1) can be interpreted as the expectation value (for the large-component spinor χ) of the interaction

$$-\left(\boldsymbol{\mu} - \frac{\gamma}{1+\gamma} \boldsymbol{\mu}\boldsymbol{\cdot}\boldsymbol{v}\boldsymbol{v} \right) \boldsymbol{\cdot}\boldsymbol{B} \tag{2}$$

If $\boldsymbol{\mu}$ is the magnetic moment of the neutron in its rest frame and $\boldsymbol{B}$ is the magnetic field in the laboratory frame, how do we interpret the quantity in parentheses in (2)?

(b) In Section 21.4 we discussed the Pauli moment and obtained equation (21.20). We also mentioned that (21.21) can be derived from (21.20). Fill in the steps in this derivation, in the course of which you will have to make some approximations. Please justify the latter: they are in fact legitimate.

21.4. (a) Discuss in detail how the number of nodes (number of zeros for $0 < r < \infty$) of the radial functions $G(r)$ and $F(r)$ of the hydrogen atom Dirac solutions are related to the quantum numbers n, j, and ℓ.

(b) Which hydrogenic bound states satisfy the condition $F(r) = \text{const}\boldsymbol{\cdot}G(r)$, and why?

21.5. Calculate the magnetic dipole hyperfine splitting for the $1^2s_{1/2}$ state of hydrogen according to the Dirac theory, assuming that the hyperfine perturbation Hamiltonian is as given in (21.81). Show that the splitting between the $F = 1$ and $F = 0$ levels in zero external magnetic field is

$$E(F=1)-E(F=0)=-\frac{2\alpha}{3m_p}g_s g_p\int_0^\infty gf\,dr \quad (\hbar = c = 1)$$

Evaluate the integral using the radial wave functions ig and f for the $1^2s_{1/2}$ state, and keep terms only to lowest nonvanishing order in α. Compare your result with that for the nonrelativistic calculation [see, e.g., equation (10.32)].

21.6. In this problem we consider some consequences of the possibility that the electron might have a small intrinsic electric dipole moment (EDM). The present experimental upper limit on the electron EDM is 8.7×10^{-29} e cm, where e = 4.8×10^{-10} esu is the unit of electronic charge in cgs. (Baron et al, 2014). The existence of a measurable (although extremely small) EDM is suggested by current speculative theories (e.g., various types of supersymmetric theories) that attempt to go beyond the standard model. The EDM, like the electron's spin magnetic moment, would be proportional to the electron spin itself. An electron EDM d can be incorporated formally into Dirac's equation in a gauge-invariant, proper-Lorentz invariant way as follows:

$$i\gamma^\mu\left(\partial_\mu + ieA_\mu\right)\psi - i\frac{d}{2}\sigma^{\mu\nu}\gamma^5 F_{\mu\nu}\psi - m_e\psi = 0 \tag{1}$$

where we employ units with $\hbar = c = 1$. The term in d is analogous to the Pauli moment term discussed in Section 21.4 (21.17) except that in equation (1), $\sigma^{\mu\nu}$ is replaced by $i\sigma^{\mu\nu}\gamma^5$.

(a) Show that (1) yields the following effective Dirac Hamiltonian for the electron:

$$H = H_0 + H'$$

where

$$H_0 = \boldsymbol{\alpha}\cdot(\boldsymbol{p} + e\boldsymbol{A}) + m\gamma^0 - e\Phi \tag{2}$$

and

$$H' = -d\gamma^0\left(\boldsymbol{\Sigma}\cdot\boldsymbol{\mathcal{E}} + i\boldsymbol{\alpha}\cdot\boldsymbol{B}\right) \tag{3}$$

(b) Show that H' is odd under space inversion symmetry (parity = P) and time reversal (T). Thus an EDM cannot exist unless parity and time reversal invariance are both violated.

(c) Consider an atomic electron, as in hydrogen. Suppose that $\boldsymbol{B} = 0$ and that the electrostatic potential Φ consists of two parts:

$$\Phi = \Phi_i + \Phi_e$$

where Φ_i is the internal (atomic) contribution, whereas Φ_e is an external potential corresponding to a uniform external electric field $\boldsymbol{\mathcal{E}}_e = -\nabla\Phi_e$. Writing $\boldsymbol{\mathcal{E}} = \boldsymbol{\mathcal{E}}_i + \boldsymbol{\mathcal{E}}_e$, we separate H' into two parts

$$H' = H_1' + H_2'$$

where

$$H_1' = -d\boldsymbol{\Sigma}\boldsymbol{\cdot}\boldsymbol{\mathcal{E}} \text{ and } H_2' = -d\left(\gamma^0 - 1\right)\boldsymbol{\Sigma}\boldsymbol{\cdot}\boldsymbol{\mathcal{E}}$$

Only H_1' contributes in the nonrelativistic limit. Let $|\psi_0\rangle$ be an eigenstate of H_0. Show that if H' is treated as a perturbation on H_0, the first-order energy shift arising from H_1' vanishes. This result is called *Schiff's theorem.*

22 The Dirac Field

22.1 Dirac negative-energy sea

The single-particle Dirac theory is a major improvement over the Pauli-Schroedinger theory, but as we have mentioned previously, it and all other single-particle relativistic theories contain serious difficulties and contradictions. Here are some examples:

1. The one-dimensional step problem was discussed in Chapter 6 as an elementary example of reflection and transmission of de Broglie waves in nonrelativistic quantum mechanics. Let $V = 0$ for $x < 0$ and $V = +V_0$ for $x \geq 0$. In the nonrelativistic case, if a plane wave is incident from the left with energy $E > V_0$, part of the incident wave is reflected and part is transmitted at the step, but if $V_0 > E$, the incident wave is completely reflected, and there is no transmitted wave. The same problem may be studied with relativistic particles, for example, with the Klein-Gordon equation or the Dirac equation. In either of these cases, one finds that when $V_0 > E$ and V_0 is sufficiently large, there is once again an oscillatory transmitted wave (which has a *negative* probability density.) This phenomenon is known as the *Klein paradox*. It cannot be resolved within the single-particle framework.
2. Consider the problem of the lowest bound-state solution to the Dirac equation for an attractive spherically symmetric square-well potential. For fixed well radius, we can examine the binding energy as a function of well depth: $V = -V_0$. If $V_0 < 2mc^2$, the wave function outside the well goes to zero exponentially as r increases, just as we expect. However, if $V_0 > 2mc^2$, the wave function outside the well becomes oscillatory. This is closely related to the Klein paradox, and it cannot be explained sensibly within the single-particle theory. Similarly, the behavior of bound-state solutions to Dirac's equation for the Coulomb potential becomes pathologic if $Z\alpha > 1$.
3. We have seen that there exist negative-energy solutions to the Dirac equation and that these solutions cannot be disregarded. If so, why can't a positive-energy electron in the ground state of a hydrogen atom emit a photon and make a transition to one of these negative-energy states? Using techniques we have already developed, it is not difficult to calculate the probability per unit time W for such a transition. For example, if the final electron has energy in the range $-m_ec^2$ to $-2m_ec^2$, one finds that $W \approx 10^8$ s^{-1}. According to this picture, all ordinary atoms ought to be very unstable and ought to disintegrate with the emission of photons with energies of approximately 1 MeV or more.

How do we resolve these fundamental difficulties, all of which are obviously related to negative-energy states? In 1930, Dirac proposed that positive-energy electrons don't decay to negative-energy states because all the latter states are filled, and the exclusion principle prohibits the coexistence of two or more electrons in the same state. At first, this idea seems preposterous because it requires a negative-energy electron sea with an infinite negative energy density and infinite negative charge density everywhere. Perhaps one might overcome these difficulties, at least superficially, by arguing that a uniform energy density should not have any observable effects on a local scale, and neither should a uniform charge density. However, another serious objection is that relativistic wave equations describing bosons of spin 0, 1, … also exhibit negative-energy solutions. Obviously, Dirac's idea doesn't work for such particles because they are not subject to the exclusion principle, so how do we account for them?

Ignoring these worrisome questions, we consider what happens if a negative-energy electron in the Dirac sea with momentum $-\boldsymbol{p}$ and spin $-\boldsymbol{s}$ absorbs a photon or photons and acquires positive energy. Then we would not only have an electron with positive energy but also a vacancy or *hole* in the negative-energy sea – the absence of a negative-energy electron. No observation could distinguish this hole from the presence of a positively charged electron (a positron) with positive energy, momentum $+\boldsymbol{p}$, and spin $+\boldsymbol{s}$. An observer would describe this phenomenon as the transformation of a photon or photons into a positive-energy electron-positron pair. The reverse process also could occur: the positive-energy electron could emit one or more photons and drop down into the negative-energy sea, thus filling the hole. An observer could equally well describe this as the annihilation of a positive-energy electron-positron pair by photon emission.

Once we start thinking this way, however, it seems only natural to reinterpret the negative-energy electron solutions as positive-energy positron solutions, thereby getting rid of the objectionable concept of the negative-energy sea altogether. Indeed, this can be done, as we see in the next section. However, a price must be paid – we must abandon the single-particle theory and instead embrace the idea of a many-particle relativistic quantum theory, where particles can be created and destroyed.

22.2 Charge-conjugation symmetry of the Dirac equation

We now see how the negative-energy electron solutions to the Dirac equation can be reinterpreted as positive-energy positron solutions. We start with the Dirac equation for an electron in an external electromagnetic field; that is,

$$m\psi - i\gamma^\mu \partial_\mu \psi - e\gamma^\mu A_\mu \psi = 0 \tag{22.1}$$

Now, recognizing that the A_μ are real quantities, we take the complex conjugate of (22.1)

$$m\psi^* + i\gamma^{\mu^*} \partial_\mu \psi^* - e\gamma^{\mu^*} A_\mu \psi^* = 0 \tag{22.2}$$

We show momentarily that one can construct a matrix C such that

$$(C\gamma^0)\gamma^{\mu^*}(C\gamma^0)^{-1} = -\gamma^\mu \tag{22.3}$$

Assuming this, we apply $C\gamma^0$ on the left to both sides of (22.2) to obtain

$$m(C\gamma^0\psi^*) + i(C\gamma^0)\gamma^{\mu^*}(C\gamma^0)^{-1}\partial_\mu(C\gamma^0\psi^*) - e(C\gamma^0)\gamma^{\mu^*}(C\gamma^0)^{-1}A_\mu(C\gamma^0\psi^*) = 0$$

Use of (22.3) simplifies this last equation to

$$\left(m - i\gamma^{\mu}\partial_{\mu} + e\gamma^{\mu}A_{\mu}\right)C\gamma^{0}\psi^{*} = 0 \tag{22.4}$$

Comparison of (22.4) with (22.2) shows that both are Dirac equations for particles of the same mass but opposite charge. In other words, $\psi_C = C\gamma^0\psi^*$ is the solution charge conjugate to ψ. It remains to find an explicit form for C (which is representation dependent). In the standard representation, where

$$\begin{aligned}
\gamma^0 &= \gamma^{0^*} = \tilde{\gamma}^0 \\
\gamma^1 &= \gamma^{1^*} = -\tilde{\gamma}^1 \\
\gamma^2 &= -\gamma^{2^*} = \tilde{\gamma}^2 \\
\gamma^3 &= \gamma^{3^*} = -\tilde{\gamma}^3
\end{aligned}$$

(22.3) may be rewritten as

$$C\tilde{\gamma}^{\mu}C^{-1} = -\gamma^{\mu} \tag{22.5}$$

which has the solution $C = i\gamma^2\gamma^0 = -C^{-1} = -C^{\dagger}$. Therefore, up to an arbitrary phase factor, the solution charge conjugate to ψ is

$$\psi_C = i\gamma^2\psi^* \tag{22.6}$$

To introduce the charge-conjugation transformation, we have just examined the Dirac equation for an electron in an external electromagnetic field. However, now that we have the result (22.6), we can study its consequences for the free-particle solutions listed as follows:

$$\psi_1(\boldsymbol{p}) = \frac{1}{\sqrt{V}}\sqrt{\frac{\varepsilon+m}{2\varepsilon}}\begin{pmatrix} \begin{pmatrix}1\\0\end{pmatrix} \\ \frac{\boldsymbol{\sigma}\cdot\boldsymbol{p}}{\varepsilon+m}\begin{pmatrix}1\\0\end{pmatrix}\end{pmatrix}\exp(i\boldsymbol{p}\cdot\boldsymbol{x} - i\varepsilon t) = \frac{1}{\sqrt{V}}u_1\exp(i\boldsymbol{p}\cdot\boldsymbol{x} - i\varepsilon t) \tag{22.7}$$

$$\psi_2(\boldsymbol{p}) = \frac{1}{\sqrt{V}}\sqrt{\frac{\varepsilon+m}{2\varepsilon}}\begin{pmatrix} \begin{pmatrix}0\\1\end{pmatrix} \\ \frac{\boldsymbol{\sigma}\cdot\boldsymbol{p}}{\varepsilon+m}\begin{pmatrix}0\\1\end{pmatrix}\end{pmatrix}\exp(i\boldsymbol{p}\cdot\boldsymbol{x} - i\varepsilon t) = \frac{1}{\sqrt{V}}u_2\exp(i\boldsymbol{p}\cdot\boldsymbol{x} - i\varepsilon t) \tag{22.8}$$

$$\psi_3(\boldsymbol{p}) = \frac{1}{\sqrt{V}}\sqrt{\frac{\varepsilon+m}{2\varepsilon}}\begin{pmatrix} -\frac{\boldsymbol{\sigma}\cdot\boldsymbol{p}}{\varepsilon+m}\begin{pmatrix}1\\0\end{pmatrix} \\ \begin{pmatrix}1\\0\end{pmatrix}\end{pmatrix}\exp(i\boldsymbol{p}\cdot\boldsymbol{x} + i\varepsilon t) = \frac{1}{\sqrt{V}}u_3\exp(i\boldsymbol{p}\cdot\boldsymbol{x} + i\varepsilon t) \tag{22.9}$$

$$\psi_4(\boldsymbol{p}) = \frac{1}{\sqrt{V}}\sqrt{\frac{\varepsilon+m}{2\varepsilon}}\begin{pmatrix} -\dfrac{\boldsymbol{\sigma}\cdot\boldsymbol{p}}{\varepsilon+m}\begin{pmatrix}0\\1\end{pmatrix} \\ \begin{pmatrix}0\\1\end{pmatrix}\end{pmatrix}\exp(i\boldsymbol{p}\cdot\boldsymbol{x}+i\varepsilon t) = \frac{1}{\sqrt{V}}u_4\exp(i\boldsymbol{p}\cdot\boldsymbol{x}+i\varepsilon t) \tag{22.10}$$

For example, we employ (22.6) to find the charge conjugate of $\psi_1(\boldsymbol{p})$; that is,

$$\psi_{1C} = i\gamma^2\psi_1^* = \frac{1}{\sqrt{V}}\sqrt{\frac{\varepsilon+m}{2\varepsilon}}\begin{pmatrix}0&0&0&1\\0&0&-1&0\\0&-1&0&0\\1&0&0&0\end{pmatrix}\begin{pmatrix}1\\0\\p_z/(\varepsilon+m)\\(p_x-ip_y)/(\varepsilon+m)\end{pmatrix}\exp(-i\boldsymbol{p}\cdot\boldsymbol{x}+i\varepsilon t)$$

$$= \frac{1}{\sqrt{V}}\sqrt{\frac{\varepsilon+m}{2\varepsilon}}\begin{pmatrix}\dfrac{-\boldsymbol{\sigma}\cdot(-\boldsymbol{p})}{\varepsilon+m}\begin{pmatrix}0\\1\end{pmatrix}\\ \begin{pmatrix}0\\1\end{pmatrix}\end{pmatrix}\exp(-i\boldsymbol{p}\cdot\boldsymbol{x}+i\varepsilon t)$$

Therefore,

$$\psi_{1C}(\boldsymbol{p}) = V^{-1/2}u_4(-\boldsymbol{p})e^{i(-\boldsymbol{p})\cdot\boldsymbol{x}}e^{i\varepsilon t} = \psi_4(-\boldsymbol{p}) \tag{22.11}$$

Similarly,

$$\psi_{2C}(\boldsymbol{p}) = -\psi_3(-\boldsymbol{p}) \tag{22.12}$$

$$\psi_{3C}(\boldsymbol{p}) = -\psi_2(-\boldsymbol{p}) \tag{22.13}$$

$$\psi_{4C}(\boldsymbol{p}) = \psi_1(-\boldsymbol{p}) \tag{22.14}$$

In (22.11), $\psi_4(-\boldsymbol{p})$ describes an electron with energy $-\varepsilon$, 3-momentum $-\boldsymbol{p}$, and spin down along z. However, from (22.11), it also can be regarded as describing a positron with energy $+\varepsilon$, 3-momentum $+\boldsymbol{p}$, and spin up along z. All the foregoing is summarized in Table 22.1.

From now on we mention negative-energy electron solutions to the Dirac equation only incidentally because these have been reinterpreted as positive-energy solutions for positrons, as described by the spinors $v_{1,2}$ in Table 22.1. Before we proceed further, we summarize various properties of the u and v spinors that are needed in our future discussions and that are easily derived from previous work. Note the convenient shorthand $\not{a} \equiv \gamma^\mu a_\mu$, which can be employed for any 4-vector a.

$$(\not{p} - m)u_s(p) = 0 \tag{22.15}$$

$$(\not{p} + m)v_s(p) = 0 \tag{22.16}$$

$$\bar{u}_s(p)(\not{p} - m) = 0 \tag{22.17}$$

Table 22.1 Properties of standard positive energy e^-, e^+ spinors

Spinor	Particle	Energy	Momentum	z Spin
$u_1(\boldsymbol{p})$	e^-	$+\varepsilon$	$\boldsymbol{p}$	$+$
$u_2(\boldsymbol{p})$	e^-	$+\varepsilon$	$\boldsymbol{p}$	$-$
$v_1(\boldsymbol{p}) = u_4(-\boldsymbol{p})$	e^+	$+\varepsilon$	$\boldsymbol{p}$	$+$
$v_2(\boldsymbol{p}) = -u_3(-\boldsymbol{p})$	e^+	$+\varepsilon$	$\boldsymbol{p}$	$-$

$$\bar{v}_s(p)\left(\not{p} + m\right) = 0 \tag{22.18}$$

$$u_s^\dagger(p)u_{s'}(p) = \delta_{ss'}\frac{\varepsilon}{m} \tag{22.19}$$

$$v_s^\dagger(p)v_{s'}(p) = \delta_{ss'}\frac{\varepsilon}{m} \tag{22.20}$$

$$\bar{u}_{s'}(p)u_s(p) = \delta_{ss'} \tag{22.21}$$

$$\bar{v}_{s'}(p)v_s(p) = -\delta_{ss'} \tag{22.22}$$

$$\bar{u}_{s'}(p)v_s(p) = 0 \tag{22.23}$$

One more important remark may be made at this stage concerning charge-conjugation symmetry and in particular any electrically neutral spin-½ particle (e.g., the electron neutrino ν_e). Here, even in the presence of an external electromagnetic field, both ψ and ψ_C obviously satisfy the same Dirac equation. This leads to the possibility that the particle of interest and its charge conjugate are identical. Because this possibility was first discussed for neutrinos by E. Majorana in 1937, one speaks of possible self-charge-conjugate neutrinos as *Majorana neutrinos.* So far no experiment has ruled out the possibility that neutrinos of each flavor are of this form.

22.3 A digression on time-reversal symmetry

The Dirac equation exhibits an additional symmetry under time reversal. Before we discuss it, let us consider some general properties that the operator $\boldsymbol{T}$ corresponding to a time-reversal transformation must possess. Consider some physical process, which might be a collision between particles, for example, $e^+ + e^- \to \mu^+ + \mu^-$, or a spontaneous decay, for example, $n \to p + e^- + \bar{\nu}_e$. Imagine that we could make a motion picture of the process that begins before the process takes place and ends after it is completed. Now imagine that we run the motion picture backwards. In the reversed motion picture we see initial and final states interchanged and the spins and linear momenta of all the particles reversed. This is what we mean by a time-reversal transformation. Now consider the interaction Hamiltonian that describes the initial process and others related to it. If the probability of each initial process is the same as that of its time-reversed counterpart, we say that the interaction Hamiltonian is *time-reversal invariant*.

It should be clear from these remarks that $\boldsymbol{T}$ must satisfy the following requirements:

$$\boldsymbol{T}\boldsymbol{x}\boldsymbol{T}^{-1} = \boldsymbol{x} \tag{22.24}$$

$$\boldsymbol{T}\boldsymbol{p}\boldsymbol{T}^{-1} = -\boldsymbol{p} \tag{22.25}$$

$$\boldsymbol{T}\boldsymbol{J}\boldsymbol{T}^{-1} = -\boldsymbol{J} \tag{22.26}$$

where $\boldsymbol{x}$, $\boldsymbol{p}$, and $\boldsymbol{J}$ are the usual operators for position, linear momentum, and angular momentum, respectively. We also want the commutation relations satisfied by these operators as well as the time-dependent Schroedinger equation to be preserved under the time-reversal transformation. However, given (22.24)–(22.26), this means that $\boldsymbol{T}$ must be an antiunitary operator. To see this, consider, for example, the angular-momentum commutation relation

$$J_xJ_y - J_yJ_x = iJ_z$$

We apply the time-reversal transformation to both sides of this relation to get

$$\left(\boldsymbol{T}J_x\boldsymbol{T}^{-1}\right)\left(\boldsymbol{T}J_y\boldsymbol{T}^{-1}\right) - \left(\boldsymbol{T}J_y\boldsymbol{T}^{-1}\right)\left(\boldsymbol{T}J_x\boldsymbol{T}^{-1}\right) = \left(\boldsymbol{T}i\boldsymbol{T}^{-1}\right)\left(\boldsymbol{T}J_z\boldsymbol{T}^{-1}\right)$$

Taking (22.26) into account, we see that this is

$$J_xJ_y - J_yJ_x = -\boldsymbol{T}i\boldsymbol{T}^{-1}J_z \tag{22.27}$$

which requires: $\boldsymbol{T}i\boldsymbol{T}^{-1} = -i$. Thus $\boldsymbol{T}$ must be antiunitary, which means that $\boldsymbol{T} = TK$, where T is a unitary operator, and K is the complex conjugation operator, which acts on every complex number to its right.

Bearing this in mind, we return to the Dirac equation for a particle of mass m and charge q in an external electromagnetic field; that is,

$$m\psi(t') - i\gamma^0\partial_{t'}\psi(t') + q\Phi(t')\gamma^0\psi(t') - i\boldsymbol{\gamma}\boldsymbol{\cdot}\nabla\psi(t') - q\boldsymbol{A}(t')\boldsymbol{\cdot}\boldsymbol{\gamma}\psi(t') = 0 \tag{22.28}$$

Here we have chosen units where $\hbar = c = 1$, not written the dependence of ψ on the spatial variable $\boldsymbol{r}$ for brevity, and labeled the time variable by t'. With the substitution $t = -t'$, (22.28) becomes

$$m\psi(-t) + i\gamma^0\partial_t\psi(-t) + q\Phi(-t)\gamma^0\psi(-t) - i\boldsymbol{\gamma}\boldsymbol{\cdot}\nabla\psi(-t) - q\boldsymbol{A}(-t)\boldsymbol{\cdot}\boldsymbol{\gamma}\psi(-t) = 0 \tag{22.29}$$

Now imagine that at time $t = 0$ we perform a time reversal on the charges and currents that generate Φ and $\boldsymbol{A}$. Because the charges remain invariant but the currents reverse, we have $\Phi(-t) = \Phi(t)$ and $\boldsymbol{A}(t) = -\boldsymbol{A}(-t)$. Thus (22.29) becomes

$$m\psi(-t) + \gamma^0\left[i\partial_t + q\Phi(t)\right]\psi(-t) - \left[i\boldsymbol{\gamma}\boldsymbol{\cdot}\nabla - q\boldsymbol{\gamma}\boldsymbol{\cdot}\boldsymbol{A}(t)\right]\psi(-t) = 0 \tag{22.30}$$

We now apply $\boldsymbol{T} = TK$ from the left to both sides of (22.30). Noting that Φ and $\boldsymbol{A}$ are real functions, we obtain

$$mT\psi^*(-t)-iT\gamma^{0*}T^{-1}[\partial_t+iq\Phi(t)]T\psi^*(-t)+iT\boldsymbol{\gamma}^*T^{-1}\cdot[i\nabla-iq\boldsymbol{A}(t)]T\psi^*(-t)=0 \quad (22.31)$$

This is a Dirac equation for the wave function $\psi'(t)=T\psi^*(-t)$ provided that the following conditions are satisfied:

$$\begin{aligned} T\gamma^{0*}T^{-1} &= \gamma^0 \\ T\boldsymbol{\gamma}^*T^{-1} &= -\boldsymbol{\gamma} \end{aligned} \quad (22.32)$$

In the standard representation, we have

$$\gamma^{0*}=\gamma^0 \quad \gamma^{1*}=\gamma^1 \quad \gamma^{2*}=-\gamma^2 \quad \gamma^{3*}=\gamma^3 \quad (22.33)$$

Thus, in the standard representation, a solution to equations (22.32) is

$$T=\eta_T\gamma^1\gamma^3 \quad (22.34)$$

where η_T is an arbitrary phase factor. It is easy to show from (22.34) that the time-reversal transformation reverses the spin and 3-momentum of a free-particle spinor but does not change its energy.

22.4 Construction of the Dirac field operator

We have emphasized repeatedly that a plausible relativistic quantum theory must account for the fact that particles can be created and destroyed. As we stated in Section 11.9, the method of second quantization is well suited to calculating processes involving the creation and destruction of fermions. Therefore, if we wish to construct a sensible relativistic theory of electrons and positrons, it is only natural that we should join second quantization and the Dirac theory together. That is our goal in this section.

We start with a general fermion field operator, defined as in Section 11.9 by

$$\Phi(\boldsymbol{x},t)=\sum_\alpha \phi_\alpha(\boldsymbol{x},t)b_\alpha \quad (22.35)$$

where $\phi_\alpha(\boldsymbol{x},t)$ is a single-particle wave function with quantum numbers collectively described by the index α, and b_α is the corresponding destruction operator. In the case of the Dirac theory of relativistic electrons, we initially choose ϕ to be the free-particle plane-wave solutions $\psi_r(\boldsymbol{x},\boldsymbol{p},t)$, with r = 1, 2, 3, 4 [recall equations (22.7)–(22.10)]. Then, using the symbol $\Psi(\boldsymbol{x},t)$ to denote the Dirac field operator, we have

$$\Psi(\boldsymbol{x},t)=\sum_{\boldsymbol{p}}\sum_{r=1}^{4}\frac{1}{\sqrt{V}}\sqrt{\frac{m}{\varepsilon}}u_r(\boldsymbol{p})e^{i\boldsymbol{p}\cdot\boldsymbol{x}}e^{-i\eta_r\varepsilon t}b_r(\boldsymbol{p}) \quad (22.36)$$

where the spinors $u_r(\boldsymbol{p})$ are defined in (22.7)–(22.10), and $\eta_{1,2}=+1$ and $\eta_{3,4}=-1$. In view of our discussion of charge-conjugation symmetry in Section 22.2, we know that the single-particle negative-energy electron solutions $\psi_{3,4}$ are equivalent to positron solutions with positive energy. In fact, we have the replacements

$$
\begin{aligned}
u_4(-\boldsymbol{p}) &\to v_1(\boldsymbol{p}): \quad \text{positron with momentum } \boldsymbol{p}\text{, spin up} \\
-u_3(-\boldsymbol{p}) &\to v_2(\boldsymbol{p}): \quad \text{positron with momentum } \boldsymbol{p}\text{, spin down}
\end{aligned}
$$

We can also define new creation and destruction operators

$$
\begin{aligned}
d_1^\dagger(\boldsymbol{p}) &= b_4(-\boldsymbol{p}) \\
d_2^\dagger(\boldsymbol{p}) &= -b_3(-\boldsymbol{p})
\end{aligned}
$$

Here, for example, $b_4(-\boldsymbol{p})$ destroys a negative-energy electron with spin down and momentum $-\boldsymbol{p}$, but this is equivalent to creation of a positron of positive energy with momentum $\boldsymbol{p}$ and spin up [creation operator $d_1^\dagger(\boldsymbol{p})$]. With the changes just mentioned, (22.36) becomes

$$
\Psi(\boldsymbol{x},t) = \sum_{\boldsymbol{p}} \sum_{s=1}^{2} \frac{1}{\sqrt{V}} \sqrt{\frac{m}{\varepsilon}} \left[u_s(\boldsymbol{p}) e^{-ip\cdot x} b_s(\boldsymbol{p}) + v_s(\boldsymbol{p}) e^{ip\cdot x} d_s^\dagger(\boldsymbol{p}) \right] \tag{22.37}
$$

where $p\cdot x \equiv \varepsilon t - \boldsymbol{p}\bullet\boldsymbol{x}$. The Dirac field operator Ψ of (22.37) is quite complicated: it is a function of $\boldsymbol{x}$ and t, it is a 4-component spinor, and it contains destruction operators for electrons and creation operators for positrons.

In addition to Ψ, we also have the *Dirac conjugate field operator*

$$
\bar{\Psi}(\boldsymbol{x},t) = \sum_{\boldsymbol{p}} \sum_{s=1}^{2} \frac{1}{\sqrt{V}} \sqrt{\frac{m}{\varepsilon}} \left[\bar{u}_s(\boldsymbol{p}) e^{ip\cdot x} b_s^\dagger(\boldsymbol{p}) + \bar{v}_s(\boldsymbol{p}) e^{-ip\cdot x} d_s(\boldsymbol{p}) \right] \tag{22.38}
$$

The creation and destruction operators satisfy the following anticommutation relations, the origins of which were explained in Section 11.9:

$$
\begin{aligned}
\left\{ b_s(\boldsymbol{p}), b_{s'}^\dagger(\boldsymbol{p}') \right\} &= \delta_{s,s'}\delta_{\boldsymbol{p},\boldsymbol{p}'} \\
\left\{ d_s(\boldsymbol{p}), d_{s'}^\dagger(\boldsymbol{p}') \right\} &= \delta_{s,s'}\delta_{\boldsymbol{p},\boldsymbol{p}'} \\
\left\{ b_s(\boldsymbol{p}), b_{s'}(\boldsymbol{p}') \right\} &= 0 \\
\left\{ d_s(\boldsymbol{p}), d_{s'}(\boldsymbol{p}') \right\} &= 0 \\
\left\{ b_s(\boldsymbol{p}), d_{s'}^\dagger(\boldsymbol{p}') \right\} &= 0 \\
\left\{ b_s(\boldsymbol{p}), d_{s'}(\boldsymbol{p}') \right\} &= 0
\end{aligned} \tag{22.39}
$$

The Dirac field operator, being a linear superposition of solutions to the Dirac equation, is also a solution; that is,

$$
\left(m - i\gamma^\mu \partial_\mu \right)\Psi = 0 \tag{22.40}
$$

whereas

$$
i\partial_\mu \bar{\Psi}\gamma^\mu + m\bar{\Psi} = 0 \tag{22.41}
$$

From equation (11.75) we know that the Hamiltonian field operator can be expressed as

$$H = \int \Psi^\dagger H_0 \Psi \, d^3x \tag{22.42}$$

where $H_0 = -i\boldsymbol{\alpha}\cdot\nabla + m\gamma^0 = \gamma^0(-i\boldsymbol{\gamma}\cdot\nabla + m)$ is the free single-particle Dirac Hamiltonian. Using (22.37) and (22.38), we obtain

$$\begin{aligned} H = \frac{1}{V}\int d^3x \sum_{\substack{p,s \\ p',s'}} \sqrt{\frac{m^2}{\varepsilon\varepsilon'}} \cdot \left[u_{s'}^\dagger(p')e^{ip'\cdot x}b_{s'}^\dagger(p') + v_{s'}^\dagger(p')e^{-ip'\cdot x}d_{s'}(p')\right] \\ \cdot\gamma^0(-i\boldsymbol{\gamma}\cdot\nabla + m)\cdot\left[u_s(p)e^{-ip\cdot x}b_s(p) + v_s(p)e^{ip\cdot x}d_s^\dagger(p)\right] \end{aligned} \tag{22.43}$$

In this expression, there are four categories of terms involving

1. $b^\dagger b$ terms
2. $b^\dagger d^\dagger$ terms
3. db terms
4. $dd^\dagger$ terms

We consider each of the four sets separately, beginning with category 1. Here, when we integrate over $\boldsymbol{x}$, we obtain

$$\frac{1}{V}\int d^3x e^{i(p-p')\cdot x} = \delta_{p',p}$$

Thus the double sum over $\boldsymbol{p}, \boldsymbol{p}'$ is reduced to a single sum and also $e^{i(\varepsilon'-\varepsilon)t} = 1$ because $\varepsilon' = \varepsilon$. Therefore, in category 1, we obtain the following contribution to H:

$$\sum_{p,s,s'}\left(\frac{m}{\varepsilon}\right)u_{s'}^\dagger(p)\gamma^0(\boldsymbol{\gamma}\cdot\boldsymbol{p} + m)u_s(p)b_{s'}^\dagger(p)b_s(p) \tag{22.44}$$

Now

$$\begin{aligned}(\boldsymbol{\gamma}\cdot\boldsymbol{p} + m)u_s(p) &= (m - \not{p} + \gamma^0\varepsilon)u_s(p) \\ &= \gamma^0\varepsilon u_s(p)\end{aligned}$$

because $(\not{p} - m)u_s(p) = 0$ (the Dirac equation). Thus (22.44) becomes

$$\begin{aligned}\sum_{p,s,s'}\left(\frac{m}{\varepsilon}\right)\varepsilon u_{s'}^\dagger(p)u_s(p)b_{s'}^\dagger(p)b_s(p) &= \sum_{p,s,s'}\left(\frac{m}{\varepsilon}\right)\varepsilon\frac{\varepsilon}{m}\delta_{ss'}b_{s'}^\dagger(p)b_s(p) \\ &= \sum_{p,s}\varepsilon b_s^\dagger(p)b_s(p) \\ &= \sum_{p,s}\varepsilon N_s^-(p)\end{aligned} \tag{22.45}$$

where $N_s^-(p) = b_s^\dagger(p)b_s(p)$ is the number operator for electrons with momentum $\boldsymbol{p}$ and spin s. In category 2, integration over $\boldsymbol{x}$ yields $\boldsymbol{p} = -\boldsymbol{p}'$. However, we then obtain the factor

$$u_s^\dagger(-\boldsymbol{p})v_s(\boldsymbol{p}) = 0$$

Hence there is no contribution from terms of category 2; similarly, there is no contribution from terms of category 3. Finally, we consider category 4. Integration over $\boldsymbol{x}$ yields the following contribution:

$$\sum_{\boldsymbol{p},s,s'} \frac{m}{\varepsilon} v_{s'}^\dagger(\boldsymbol{p})\gamma^0(-\boldsymbol{\gamma}\cdot\boldsymbol{p}+m)v_s(\boldsymbol{p})d_{s'}(\boldsymbol{p})d_s^\dagger(\boldsymbol{p}) \tag{22.46}$$

Note the quantity $-\boldsymbol{\gamma}\cdot\boldsymbol{p}$ that comes from $-i\boldsymbol{\gamma}\cdot\nabla e^{-i\boldsymbol{p}\cdot\boldsymbol{x}}$. Now

$$(-\boldsymbol{\gamma}\cdot\boldsymbol{p}+m)v_s(\boldsymbol{p}) = (\not{p}+m-\gamma^0\varepsilon)v_s(\boldsymbol{p}) = -\gamma^0\varepsilon v_s(\boldsymbol{p})$$

Therefore, (22.46) becomes

$$-\sum_{\boldsymbol{p},s} \varepsilon\left[1-d_s^\dagger(\boldsymbol{p})d_s(\boldsymbol{p})\right] = -\sum_{\boldsymbol{p},s} \varepsilon\left[1-N_s^+(\boldsymbol{p})\right] \tag{22.47}$$

Combining (22.45) with (22.47), we finally obtain the total Hamiltonian

$$H = \sum_{\boldsymbol{p},s} \varepsilon\left[N_s^-(\boldsymbol{p})+N_s^+(\boldsymbol{p})-1\right] \tag{22.48}$$

A similar calculation shows that the electric charge operator, defined as

$$Q = -e\int \Psi^\dagger\Psi\, d^3\boldsymbol{x} \tag{22.49}$$

can be written as

$$Q = -e\sum_{\boldsymbol{p},s}\left[N_s^-(\boldsymbol{p})-N_s^+(\boldsymbol{p})+1\right] \tag{22.50}$$

The first two types of terms on the right-hand side of (22.48) are easy enough to understand: because $N_s^\pm(\boldsymbol{p})$ are number operators for electrons and positrons, respectively, those terms represent the energies of real electrons and positrons. However, $-\sum\varepsilon$ is an infinite negative energy. Similarly, the first and second sets of terms on the right-hand side of (22.50) represent the charges of real electrons and positrons, respectively, whereas $\sum(-e)$ is an infinite negative charge. These third terms in (22.48) and (22.50) play a role that is somewhat analogous to the zero-point energy of the radiation field, which we encountered in Section 15.3. Now the chief practical value of the Dirac field and the quantized radiation field is to facilitate calculation of various real physical processes such as electron-electron scattering, electron-photon scattering, pair production, bremsstrahlung, and so forth. In all such calculations, the zero-point energy of the radiation field, the infinite negative energy appearing in (22.48), and the infinite negative charge appearing in (22.50) play no role, and they may thus be ignored. The formal procedure for discarding such terms is called *normal ordering*, and we briefly describe its implementation for electrons and positrons in the next paragraphs.

We first separate the Dirac field (22.37) and the Dirac conjugate field (22.38) into their positive- and negative-frequency parts, defined as follows:

$$\Psi^{(+)} = \sum_{p}\sum_{s=1}^{2}\frac{1}{\sqrt{V}}\sqrt{\frac{m}{\varepsilon}}\left[u_s(p)e^{-ip\cdot x}b_s(p)\right] \rightarrow \sum bu$$

$$\Psi^{(-)} = \sum_{p}\sum_{s=1}^{2}\frac{1}{\sqrt{V}}\sqrt{\frac{m}{\varepsilon}}\left[v_s(p)e^{ip\cdot x}d_s^{\dagger}(p)\right] \rightarrow \sum d^{\dagger}v$$

$$\overline{\Psi}^{(+)} = \sum_{p}\sum_{s=1}^{2}\frac{1}{\sqrt{V}}\sqrt{\frac{m}{\varepsilon}}\left[\overline{v}_s(p)e^{-ip\cdot x}d_s(p)\right] \rightarrow \sum d\overline{v}$$

$$\overline{\Psi}^{(-)} = \sum_{p}\sum_{s=1}^{2}\frac{1}{\sqrt{V}}\sqrt{\frac{m}{\varepsilon}}\left[\overline{u}_s(p)e^{ip\cdot x}b_s^{\dagger}(p)\right] \rightarrow \sum b^{\dagger}\overline{u}$$

Here, to the right of each formal expression, we have rewritten it in simple schematic fashion to bring out the main point as clearly as possible in what follows. Now, exhibiting the row and column indices α, β explicitly, we write the bilinear form $\overline{\Psi}_{\alpha}F_{\alpha\beta}\Psi_{\beta}$, where F is a 4×4 matrix, in terms of positive- and negative-frequency parts; that is,

$$\overline{\Psi}_{\alpha}F_{\alpha\beta}\Psi_{\beta} = \overline{\Psi}_{\alpha}^{(+)}F_{\alpha\beta}\Psi_{\beta}^{(+)} + \overline{\Psi}_{\alpha}^{(-)}F_{\alpha\beta}\Psi_{\beta}^{(+)} + \overline{\Psi}_{\alpha}^{(+)}F_{\alpha\beta}\Psi_{\beta}^{(-)} + \overline{\Psi}_{\alpha}^{(-)}F_{\alpha\beta}\Psi_{\beta}^{(-)} \tag{22.51}$$

where

$$\overline{\Psi}_{\alpha}^{(+)}F_{\alpha\beta}\Psi_{\beta}^{(+)} \rightarrow \sum db\left(\overline{v}_{\alpha}F_{\alpha\beta}u_{\beta}\right) \tag{22.52a}$$

$$\overline{\Psi}_{\alpha}^{(-)}F_{\alpha\beta}\Psi_{\beta}^{(+)} \rightarrow \sum b^{\dagger}b\left(\overline{u}_{\alpha}F_{\alpha\beta}u_{\beta}\right) \tag{22.52b}$$

$$\overline{\Psi}_{\alpha}^{(+)}F_{\alpha\beta}\Psi_{\beta}^{(-)} \rightarrow \sum dd^{\dagger}\left(\overline{v}_{\alpha}F_{\alpha\beta}v_{\beta}\right) \tag{22.52c}$$

$$\overline{\Psi}_{\alpha}^{(-)}F_{\alpha\beta}\Psi_{\beta}^{(-)} \rightarrow \sum b^{\dagger}d^{\dagger}\left(\overline{u}_{\alpha}F_{\alpha\beta}v_{\beta}\right) \tag{22.52d}$$

Because of the 1 in each combination $dd^{\dagger} = 1 - d^{\dagger}d$ of destruction and creation operators in (22.52c), it is the latter term that causes the undesirable negative infinities in the energy and the charge, as we have seen in the preceding paragraphs. These negative infinities are eliminated by making the change

$$\overline{\Psi}_{\alpha}^{(+)}F_{\alpha\beta}\Psi_{\beta}^{(-)} \rightarrow -\Psi_{\beta}^{(-)}\overline{\Psi}_{\alpha}^{(+)}F_{\alpha\beta}$$

so that (22.51) is replaced by the normally ordered form

$$\overline{\Psi}_{\alpha}F_{\alpha\beta}\Psi_{\beta} = \overline{\Psi}_{\alpha}^{(+)}F_{\alpha\beta}\Psi_{\beta}^{(+)} + \overline{\Psi}_{\alpha}^{(-)}F_{\alpha\beta}\Psi_{\beta}^{(+)} - \Psi_{\beta}^{(-)}\overline{\Psi}_{\alpha}^{(+)}F_{\alpha\beta} + \overline{\Psi}_{\alpha}^{(-)}F_{\alpha\beta}\Psi_{\beta}^{(-)} \tag{22.53}$$

22.5 Lagrangian formulation of electromagnetic and Dirac fields and interactions

It is often useful to formulate a field theory such as Maxwellian electrodynamics or the theory of the Dirac field in terms of the principle of least action. Let us introduce a field $\phi(x_\mu)$; initially, this field could be quite general and not necessarily related to electromagnetism or the Dirac theory. The properties of the field are determined by its Lagrangian density; that is,

$$\mathcal{L} = \mathcal{L}(\phi, \partial_\nu \phi) \tag{22.54}$$

For the physical applications of interest to us, it is sufficient to assume that $\mathcal{L}$ depends on the field and its first space and time derivatives but not on higher derivatives. We form the action by integrating the Lagrangian density over space-time; that is,

$$S = \int \mathcal{L}(\phi, \partial_\nu \phi)\, d^4x \tag{22.55}$$

Variation of ϕ at each space-time point in some arbitrary manner results in a change in S; that is,

$$\begin{aligned} \delta S &= \int \delta\mathcal{L}(\phi, \partial_\nu \phi)\, d^4x \\ &= \int \left[\frac{\partial \mathcal{L}}{\partial \phi} \delta\phi + \frac{\partial \mathcal{L}}{\partial(\partial_\nu \phi)} \delta(\partial_\nu \phi) \right] d^4x \end{aligned} \tag{22.56}$$

Because $\delta(\partial_\nu \phi) = \partial_\nu(\delta\phi)$, the second term in the integrand on the right-hand side of (22.56) can be written

$$\frac{\partial \mathcal{L}}{\partial(\partial_\nu \phi)} \partial_\nu(\delta\phi) = \partial_\nu \left[\frac{\partial \mathcal{L}}{\partial(\partial_\nu \phi)} \delta\phi \right] - \partial_\nu \left[\frac{\partial \mathcal{L}}{\partial(\partial_\nu \phi)} \right] \delta\phi \tag{22.57}$$

The integral over all space-time of the first term on the right-hand side of (22.57) can be expressed as a surface integral via a generalized Gauss's theorem. This vanishes if we assume that $\delta\phi \to 0$ on the infinitely large surface. Hence

$$\delta S = \int \left\{ \frac{\partial \mathcal{L}}{\partial \phi} - \partial_\nu \left[\frac{\partial \mathcal{L}}{\partial(\partial_\nu \phi)} \right] \right\} \delta\phi d^4x$$

The Lagrangian density is constrained by the condition that S has an extreme value, that is, that $\delta S = 0$. Because $\delta\phi$ is arbitrary, this implies that

$$\frac{\partial \mathcal{L}}{\partial \phi} - \partial_\nu \left[\frac{\partial \mathcal{L}}{\partial(\partial_\nu \phi)} \right] = 0 \tag{22.58a}$$

which is called the *Euler-Lagrange equation*. For a multicomponent field ϕ_λ, where $\lambda = 1, 2, \ldots$, there is one Euler-Lagrange equation for each component; that is,

$$\frac{\partial \mathcal{L}}{\partial \phi_\lambda} - \partial_\nu \left[\frac{\partial \mathcal{L}}{\partial(\partial_\nu \phi_\lambda)} \right] = 0 \tag{22.58b}$$

We now apply these general ideas to the case of the electromagnetic field, which has four components A_λ, $\lambda = 0, \ldots, 3$. Thus (22.58b) becomes

$$\frac{\partial \mathcal{L}}{\partial A_\lambda} - \partial_\nu \left[\frac{\partial \mathcal{L}}{\partial(\partial_\nu A_\lambda)} \right] = 0 \tag{22.59}$$

We seek a Lagrangian density $\mathcal{L}$ such that equations (22.59) are the same as the field equations, here written in natural units

$$\partial^\mu F_{\mu\nu} = j_{\mathrm{EM},\nu} \tag{22.60}$$

A straightforward calculation starting from (22.59) shows that this is achieved with

$$\mathcal{L} = \mathcal{L}_{\mathrm{EM}} + \mathcal{L}_{\mathrm{int}} \tag{22.61}$$

where

$$\mathcal{L}_{\mathrm{EM}} = -\frac{1}{4} F^{\mu\nu} F_{\mu\nu} \tag{22.62}$$

and

$$\mathcal{L}_{\mathrm{int}} = -j^\mu_{\mathrm{EM}} A_\mu \tag{22.63}$$

Frequently, $\mathcal{L}_{\mathrm{int}}$ of (22.63) is called the *minimal interaction Lagrangian density*. In classical mechanics one defines the Hamiltonian from the Lagrangian as follows:

$$H = \sum_i \dot{q}_i \frac{\partial L}{\partial \dot{q}_i} - L$$

In an analogous way, we define the Hamiltonian density $\mathcal{H}$ in terms of the Lagrangian density $\mathcal{L}$. For $\mathcal{L}$ of (22.61), we have

$$\mathcal{H} = \frac{\partial \mathcal{L}}{\partial \left(\frac{\partial \mathbf{A}}{\partial t} \right)} \bullet \frac{\partial \mathbf{A}}{\partial t} - \mathcal{L} \tag{22.64}$$

which yields

$$\mathcal{H} = \mathcal{H}_{\mathrm{EM}} + \mathcal{H}_{\mathrm{int}} \tag{22.65}$$

where

$$\mathcal{H}_{\text{EM}} = \frac{1}{2}\left(\boldsymbol{\mathcal{E}}^2 + \boldsymbol{B}^2\right) \tag{22.66a}$$

and

$$\mathcal{H}_{\text{int}} = j^{\mu}_{\text{EM}} A_{\mu} \tag{22.66b}$$

Now we turn our attention to the Dirac field. We have seen that the electron-positron field operator in the Heisenberg picture is

$$\Psi(\boldsymbol{x},t) = \frac{1}{\sqrt{V}} \sum_{\boldsymbol{p}} \sum_{s=1}^{2} \sqrt{\frac{m}{\varepsilon}} \left[u_s(\boldsymbol{p}) e^{-ip\cdot x} b_s(\boldsymbol{p}) + v_s(\boldsymbol{p}) e^{ip\cdot x} d_s^{\dagger}(\boldsymbol{p}) \right] \tag{22.67}$$

It is natural to ask if a Lagrangian density $\mathcal{L}_D$ can be formulated for Ψ and $\bar{\Psi}$ and, if so, whether the Dirac equation

$$\left(m - i\gamma^{\mu}\partial_{\mu}\right)\Psi = 0 \tag{22.68a}$$

and the Dirac-conjugate equation

$$i\partial_{\mu}\bar{\Psi}\gamma^{\mu} + m\bar{\Psi} = 0 \tag{22.68b}$$

can be considered as Euler-Lagrange equations for $\mathcal{L}_D$ with $\Psi, \bar{\Psi}$ regarded as independent components of the field. Indeed, this is the case. An appropriate expression for $\mathcal{L}_D$ is

$$\mathcal{L}_D = \frac{1}{2}\bar{\Psi}\left(i\gamma^{\mu}\partial_{\mu} - m\right)\Psi - \frac{1}{2}\left(i\partial_{\mu}\bar{\Psi}\gamma^{\mu} + m\bar{\Psi}\right)\Psi \tag{22.69}$$

It is easy to verify that with this Lagrangian density, the Euler-Lagrange equations

$$\frac{\partial \mathcal{L}_D}{\partial \bar{\Psi}} - \partial_{\mu} \frac{\partial \mathcal{L}_D}{\partial\left(\partial_{\mu}\bar{\Psi}\right)} = 0 \tag{22.70a}$$

$$\frac{\partial \mathcal{L}_D}{\partial \Psi} - \partial_{\mu} \frac{\partial \mathcal{L}_D}{\partial\left(\partial_{\mu}\Psi\right)} = 0 \tag{22.70b}$$

are just (22.68a) and (22.68b), respectively.

22.6 *U*(1) gauge invariance and the Dirac field

A phase transformation

$$\Psi \to \Psi' = e^{i\alpha}\Psi \qquad \bar{\Psi} \to \bar{\Psi}' = e^{-i\alpha}\bar{\Psi} \tag{22.71}$$

where α is real, is called a $U(1)$ gauge transformation [$U(1)$ because $e^{i\alpha}$ is a 1×1 unitary matrix]. If α is a constant, the transformation is said to be *global*; if α varies from one space-time point to another, one has a *local* gauge transformation. For many purposes it is sufficient to consider the case where α is infinitesimal. Here we can make the replacements

$$e^{\pm i\alpha} \rightarrow 1 \pm i\alpha$$

and thus

$$\delta\Psi = \Psi' - \Psi = i\alpha\Psi \qquad \bar{\Psi} = \bar{\Psi}' - \bar{\Psi} = -i\alpha\bar{\Psi} \tag{22.72}$$

The variations of (22.72) produce a change in the Dirac Lagrangian density (22.69) that is found to be

$$\delta\mathcal{L}_D = -\bar{\Psi}\gamma^{\mu}\Psi\partial_{\mu}\alpha \tag{22.73}$$

If α is a constant, $\delta\mathcal{L}_D = 0$. In other words, $\mathcal{L}_D$ is invariant under global $U(1)$ gauge transformations. If $\alpha = \alpha(x_{\mu})$ is a function of the space-time coordinates, $\delta\mathcal{L}_D \neq 0$. Hence $\mathcal{L}_D$ is not invariant under local $U(1)$ gauge transformations. However, that invariance is restored if we add the interaction Lagrangian (22.63) to $\mathcal{L}_D$; that is,

$$\mathcal{L}'_D = \mathcal{L}_D - q\bar{\Psi}\gamma^{\mu}\Psi\cdot A_{\mu} \tag{22.74}$$

provided that whenever we make the transformation (22.72), we also change the vector potential A_{μ} as follows:

$$A_{\mu} \rightarrow A'_{\mu} = A_{\mu} - \frac{1}{q}\partial_{\mu}\alpha \tag{22.75}$$

The latter is obviously an ordinary gauge transformation of the electromagnetic 4-potential. A convenient way to write $\mathcal{L}'_D$ is to replace the partial derivatives in (22.69) by covariant derivatives; that is,

$$\mathcal{L}_D = \frac{1}{2}\bar{\Psi}\left(i\gamma^{\mu}D_{\mu} - m\right)\Psi - \frac{1}{2}\left(iD_{\mu}\bar{\Psi}\gamma^{\mu} + m\bar{\Psi}\right)\Psi \tag{22.76}$$

where here and henceforth we drop the prime, and

$$\begin{aligned} D_{\mu}\Psi &\equiv \left(\partial_{\mu} + iqA_{\mu}\right)\Psi \\ D_{\mu}\bar{\Psi} &\equiv \left(\partial_{\mu} - iqA_{\mu}\right)\bar{\Psi} \end{aligned} \tag{22.77}$$

It is easy to show that each covariant derivative transforms in the same way as its corresponding field in the transformation (22.72). That is,

$$\delta(D_\mu \Psi) = i\alpha(D_\mu \Psi)$$
$$\delta(D_\mu \bar{\Psi}) = -i\alpha(D_\mu \bar{\Psi}) \tag{22.78}$$

The foregoing analysis reveals the close connection between the minimal electromagnetic interaction and local $U(1)$ gauge invariance of the Dirac Lagrangian density. When the same general principle is extended to the $SU(2)$ group, it has profound consequences in the standard model of electroweak interactions. This is discussed in Chapter 24.

Problems for Chapter 22

22.1. In the standard (Pauli-Dirac) representation, calculate the effect of the time-reversal transformation $\boldsymbol{T} = TK$ on a free-particle Dirac wave function with spin up along the z-axis and positive energy. Show that this time-reversal transformation has the effect of reversing the spin and the 3-momentum but not the energy.

22.2. Carry out the calculation that leads from (22.49) to (22.50).

22.3. Show that the Euler-Lagrange equations (22.59) for the electromagnetic Lagrangian density of (22.61) with (22.62) and (22.63) are the same as the electromagnetic field equations (22.60).

22.4. Using (22.64), derive (22.65) with (22.66a) and (22.66b) from (22.61) with (22.62) and (22.63).

22.5. Given the Dirac Lagrangian density (22.69), show that the Euler Lagrange equations (22.70a) and (22.70b) are the Dirac equations (22.68a) and (22.68b), respectively.

22.6. Starting with the Dirac Lagrangian (22.69) and the infinitesimal gauge transformations (22.72), derive (22.73).

22.7. Show that under the gauge transformations (22.72), the covariant derivatives transform as is indicated in (22.78).

23 Interaction between Relativistic Electrons, Positrons, and Photons

23.1 Interaction density: The *S*-matrix expansion

A state of the combined system of electromagnetic and electron-positron fields is specified by the numbers of photons in various modes $\boldsymbol{k}, \alpha$ and the numbers of electrons and positrons in each momentum-spin state $\boldsymbol{p}, s$. These occupation numbers can change with time only through the interaction between the electrons and positrons on the one hand and photons on the other. The general Hamiltonian, including interaction, takes the following form:

$$H_{\text{total}} = H_{\text{particle}} + H_{\text{radiation}} + H_{\text{int}} \tag{23.1}$$

Here $H_{\text{int}} = \int \mathcal{H}_{\text{int}}\, d^3\boldsymbol{x}$, and the interaction Hamiltonian density $\mathcal{H}_{\text{int}}$ for an electron is given in natural units by

$$\mathcal{H}_{int} = -\mathcal{L}_{\text{int}} = j^{\mu}_{\text{EM}} A_{\mu} = -e\overline{\Psi}\gamma^{\mu}\Psi \cdot A_{\mu} \tag{23.2}$$

Given this form, there exists a standard procedure for calculating physical processes that is called *covariant perturbation theory*. We shall try to explain it in simple terms. Several essential elements have already been worked out in previous chapters, for example, in Section 18.6, but we need to restate the problem in appropriate language.

In the Schroedinger picture, imagine some initial state $|\phi_i^S(t=-\infty)\rangle$ where the superscript S means *Schroedinger* and the subscript i specifies the initial occupation numbers of photons, electrons, and positrons. We imagine that the interaction turns on adiabatically at some remote but finite time $t<0$ and then gradually turns off again at some large but finite time $t>0$. We are interested in the probability amplitude for a transition to a final state $|\phi_f^S(t=+\infty)\rangle$ in which at least one of the occupation numbers has changed. We assume that there exists a time-development operator $U(t,t_0)$ such that

$$|\phi_i^S(t)\rangle = U(t,t_0)|\phi_i^S(t_0)\rangle$$

In the limits $t_0 \to -\infty$, $t \to +\infty$, $U(t,t_0)$ is called the *S-operator*, and its matrix elements constitute the S-matrix. Our goal is to calculate the matrix element

$$A_{fi} = \langle \phi_f^S(t=+\infty)|\phi_i^S(t=+\infty)\rangle = \langle \phi_f^S(+\infty)|S|\phi_i^S(-\infty)\rangle \tag{23.3}$$

In (23.1), let us regard $H_{\text{particle}} + H_{\text{radiation}}$ as a zeroth-order Hamiltonian

$$H_0 = H_{\text{particle}} + H_{\text{radiation}} \tag{23.4}$$

and let us make a unitary transformation from the Schroedinger picture to a new *interaction* picture defined by

$$|\phi\rangle = e^{iH_0 t}\,|\phi^S\rangle \tag{23.5}$$

We also define H by

$$H = e^{iH_0 t}\left(H_0 + H_{\text{int}}\right)e^{-iH_0 t} \tag{23.6}$$

Differentiating both sides of (23.5) with respect to time, we have

$$\begin{aligned} i\partial_t|\phi\rangle &= i\left[iH_0 e^{iH_0 t}\,|\phi^S\rangle + e^{iH_0 t}\partial_t\,|\phi^S\rangle\right] \\ &= -H_0\,|\phi\rangle + e^{iH_0 t}\left(H_0 + H_{\text{int}}\right)e^{-iH_0 t}e^{iH_0 t}\,|\phi^S\rangle \\ &= H_I\,|\phi\rangle \end{aligned} \tag{23.7}$$

where

$$H_I \equiv e^{iH_0 t}H_{\text{int}}e^{-iH_0 t} \tag{23.8}$$

Equation (23.7) determines the time dependence of $|\phi\rangle$ in the interaction picture. If H_I were zero, the interaction picture would reduce to the Heisenberg picture because in that case $|\phi\rangle$ would be constant in time.

Consider the time-development operator in the interaction picture. We have

$$|\phi(t)\rangle = U(t,t_0)|\phi(t_0)\rangle$$

Differentiation of this expression yields

$$i\partial_t\,|\phi(t)\rangle = i\partial_t U(t,t_0)|\phi(t_0)\rangle = H_I U(t,t_0)|\phi(t_0)\rangle$$

Hence

$$i\partial_t U(t,t_0) = H_I U(t,t_0)$$

We integrate this equation subject to the initial condition $U(t_0,t_0) = I$; that is,

$$\begin{aligned} U(t,t_0) &= I - i\int_{t_0}^{t} H_I\left(t_1\right)U(t_1,t_0)\,dt_1 \\ &= I - i\int_{t_0}^{t} H_I\left(t_1\right)\left[I - i\int_{t_0}^{t} H_I\left(t_2\right)U(t_2,t_0)\,dt_2\right]dt_1 \\ &\;\vdots \\ &= I - i\int_{t_0}^{t} H_I\left(t_1\right)dt_1 + (-i)^2\int_{t_0}^{t} H_I\left(t_1\right)dt_1\int_{t_0}^{t} H_I\left(t_2\right)dt_2 \\ &\quad + \cdots + (-i)^n\int_{t_0}^{t} dt_1\int_{t_0}^{t_1} dt_2\,\cdots\int_{t_0}^{t_{n-1}} dt_n H_I(t_1)\cdots H_I(t_n) + \cdots \end{aligned}$$

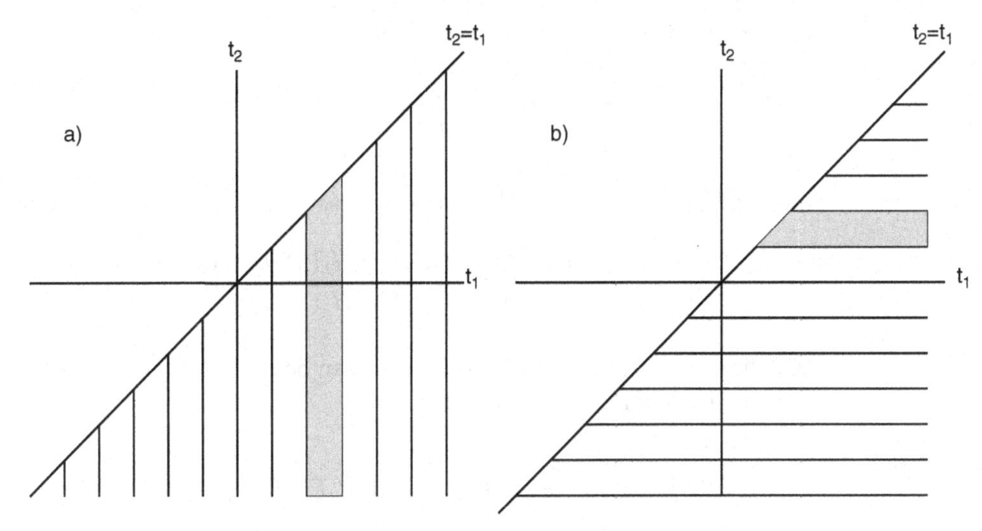

Figure 23.1 Diagrams *a* and *b* show two equivalent ways to evaluate integral I_2 of equation (23.10).

Taking the limits $t \to \infty$, $t_0 \to -\infty$, we have

$$S = I - i\int_{-\infty}^{\infty} H_I(t_1)\,dt_1 + (-i)^2 \int_{-\infty}^{\infty} H_I(t_1)\,dt_1 \int_{-\infty}^{t_1} H_I(t_2)\,dt_2 + \cdots \tag{23.9}$$

which is called the *S-operator expansion* and is another form of the Born series. Now consider the double integral

$$I_2 = \int_{-\infty}^{\infty} H_I(t_1)\,dt_1 \int_{-\infty}^{t_1} H_I(t_2)\,dt_2 \tag{23.10}$$

The region of integration is illustrated in Figure 23.1. In Figure 23.1a, we first integrate on a strip from $t_2 = -\infty$ to $t_2 = t_1$; then we add up the contributions from all the strips. However, the same result is achieved if (as in Figure 23.1b) we integrate along a given strip from $t_1 = t_2$ to $t_1 = +\infty$ and then add up all these strips.

Referring to Figure (23.1b), we write

$$I_2 = \int_{-\infty}^{\infty} dt_2 \int_{t_2}^{\infty} dt_1 H_I(t_1) H_I(t_2)$$

Because in this double integral both t_1 and t_2 are dummy variables of integration, their labels can be interchanged, that is, $t_1 \leftrightarrow t_2$, so I_2 becomes

$$I_2 = \int_{-\infty}^{\infty} dt_1 \int_{t_1}^{\infty} dt_2 H_I(t_2) H_I(t_1) \tag{23.11}$$

Combining (23.10) and (23.11), we obtain

$$I_2 = \frac{1}{2!}\int_{-\infty}^{\infty} dt_1 \int_{-\infty}^{\infty} dt_2 T\{H_I(t_1), H_I(t_2)\}$$

where $T\{H_I(t_1), H_I(t_2)\}$ is called the *time-ordered product* and is defined by

$$\begin{aligned} T\{H_I(t_1), H_I(t_2)\} &= H_I(t_1)H_I(t_2) \qquad t_1 > t_2 \\ T\{H_I(t_1), H_I(t_2)\} &= H_I(t_2)H_I(t_1) \qquad t_2 > t_1 \end{aligned} \tag{23.12}$$

The definition of the time-ordered product can be extended to cover n factors of H_I. Thus we finally obtain S in the useful form

$$S = I + \cdots + \frac{(-i)^n}{n!}\int_{-\infty}^{\infty} dt_1 \int_{-\infty}^{\infty} dt_2 \cdots \int_{-\infty}^{\infty} dt_n T\{H_I(t_1), H_I(t_2), \ldots, H_I(t_n)\} + \cdots \tag{23.13}$$

23.2 Zeroth- and first-order amplitudes

We now consider the S-matrix element

$$\begin{aligned} A_{fi} &= \langle \phi_f(+\infty)|S|\phi_i(-\infty)\rangle \\ &= \langle \phi_f(+\infty)|I|\phi_i(-\infty)\rangle \\ &\quad - i\left\langle \phi_f(+\infty)\middle|\int_{-\infty}^{\infty} H_I(t)\, dt\middle|\phi_i(-\infty)\right\rangle \\ &\quad + \frac{(-i)^2}{2!}\left\langle \phi_f(+\infty)\middle|\int_{-\infty}^{\infty} dt_1 \int_{-\infty}^{\infty} dt_2 T\{H_I(t_1), H_I(t_2)\}\middle|\phi_i(-\infty)\right\rangle \\ &\quad + \cdots + \frac{(-i)^n}{n!}\left\langle \phi_f(+\infty)\middle|\int_{-\infty}^{\infty} dt_1 \cdots \int_{-\infty}^{\infty} dt_n T\{H_I(t_1), H_I(t_2), \ldots, H_I(t_n)\}\middle|\phi_i(-\infty)\right\rangle + \cdots \end{aligned} \tag{23.14}$$

The zeroth-order term in (23.14) is

$$A_{fi}^{(0)} = \langle \phi_f(\infty)|I|\phi_i(-\infty)\rangle = \delta_{fi} \tag{23.15}$$

It is of no great interest because it corresponds to no change in occupation numbers: nothing happens. We thus go on to the first-order term

$$\begin{aligned} A_{fi}^{(1)} &= -i\left\langle \phi_f(+\infty)\middle|\int_{-\infty}^{\infty} H_I(t)\, dt\middle|\phi_i(-\infty)\right\rangle \\ &= -i\left\langle \phi_f(+\infty)\middle|\int_{-\infty}^{\infty} dt \int \mathcal{H}_I(\mathbf{x},t)\, d^3\mathbf{x}\middle|\phi_i(-\infty)\right\rangle \\ &= -i\left\langle \phi_f(+\infty)\middle|\int \mathcal{H}_I\, d^4x\middle|\phi_i(-\infty)\right\rangle \end{aligned} \tag{23.16}$$

where $\mathcal{H}_I = -e\bar{\Psi}\gamma^\mu\Psi \cdot A_\mu$, and the free field operators $\bar{\Psi}, \Psi$, and A_μ are in the Heisenberg picture; that is,

$$A_\mu = \frac{1}{\sqrt{V}}\sum_{k,\alpha}\sqrt{\frac{2\pi}{\omega}}\,\hat{\varepsilon}_\mu^{k\alpha}\left(a_{k\alpha}e^{-ik\cdot x} + a_{k\alpha}^\dagger e^{ik\cdot x}\right) \tag{23.17}$$

$$\Psi(\boldsymbol{x},t) = \frac{1}{\sqrt{V}}\sum_{\boldsymbol{p}}\sum_{s=1}^{2}\sqrt{\frac{m}{E}}\left[u_s(\boldsymbol{p})e^{-ip\cdot x}b_s(\boldsymbol{p}) + v_s(\boldsymbol{p})e^{ip\cdot x}d_s^\dagger(\boldsymbol{p})\right] \tag{23.18}$$

$$\bar{\Psi}(\boldsymbol{x},t) = \frac{1}{\sqrt{V}}\sum_{\boldsymbol{p}}\sum_{s=1}^{2}\sqrt{\frac{m}{E}}\left[\bar{u}_s(\boldsymbol{p})e^{ip\cdot x}b_s^\dagger(\boldsymbol{p}) + \bar{v}_s(\boldsymbol{p})e^{-ip\cdot x}d_s(\boldsymbol{p})\right] \tag{23.19}$$

and where, as usual, $k\cdot x = \omega t - \boldsymbol{k}\cdot\boldsymbol{x}$ and $p\cdot x = Et - \boldsymbol{p}\cdot\boldsymbol{x}$. Separating the field operators into their positive- and negative-frequency parts and using normal ordering, we have

$$\begin{aligned} A_{fi}^{(1)} = ie\left\langle\phi_f(\infty)\right|\int d^4x \Big[&\bar{\Psi}_\alpha^{(+)}\gamma_{\alpha\beta}^\mu\Psi_\beta^{(+)}\left(A_\mu^{(-)} + A_\mu^{(+)}\right) + \bar{\Psi}_\alpha^{(-)}\gamma_{\alpha\beta}^\mu\Psi_\beta^{(+)}\left(A_\mu^{(-)} + A_\mu^{(+)}\right) \\ &- \Psi_\beta^{(-)}\gamma_{\alpha\beta}^\mu\bar{\Psi}_\alpha^{(+)}\left(A_\mu^{(-)} + A_\mu^{(+)}\right) + \bar{\Psi}_\alpha^{(-)}\gamma_{\alpha\beta}^\mu\Psi_\beta^{(-)}\left(A_\mu^{(-)} + A_\mu^{(+)}\right)\Big]\left|\phi_i(-\infty)\right\rangle \end{aligned} \tag{23.20}$$

Each of the eight terms in (23.20) corresponds to a different type of first-order process, and each term is represented by a Feynman diagram in Figure 23.2. Such diagrams, named for Richard Feynman, who first employed them, are exceedingly useful and convenient for constructing quantum electrodynamic amplitudes not only in first order but, more important, in second and higher orders. In what follows, we seek to explain the rules for constructing amplitudes directly from diagrams and also to show how one calculates transition probabilities and cross-sections from the amplitudes.

In each diagram, time increases upward, an electron line is solid with an arrow pointing forward in time, a positron line is also solid with an arrow pointing backward in time, and a photon is represented by a dashed line. In each and every Feynman diagram of any order in quantum electrodynamics, two fermion lines and a photon line meet at a vertex. For example, Figure 23.2a represents photon emission by a free electron. Here an electron with 4-momentum p, spin s is destroyed at the vertex, whereas an electron with 4-momentum p', spin s', and a photon (k', α') are created at the same vertex. Figure 23.2b represents the absorption of photon (k,α), the destruction of electron (p, s), and the creation of electron (p', s'). In Figure 23.2c, a positron (p, s) is destroyed, whereas a positron (p', s') and a photon (k,α) are created at the vertex (even though positron lines have arrows pointing backward in time), and so on.

Before proceeding further, it is important to note that because of energy-momentum conservation, none of the eight first-order processes can occur! For example, consider Figure 23.2a: emission of a photon by a free electron. Energy-momentum conservation requires that $p' + k' = p$. In the rest frame of the initial electron, momentum conservation requires that $\boldsymbol{p}' = -\boldsymbol{k}'$ and thus that the final electron energy is $E' = \sqrt{m_e^2 + \omega^2}$, where $\omega = |\boldsymbol{k}'|$ is the final photon energy. Energy conservation then requires that $m_e = \sqrt{m_e^2 + \omega^2} + \omega$, but this relation cannot be satisfied for any $\omega \neq 0$. Analogous prohibitions hold for the other first-order

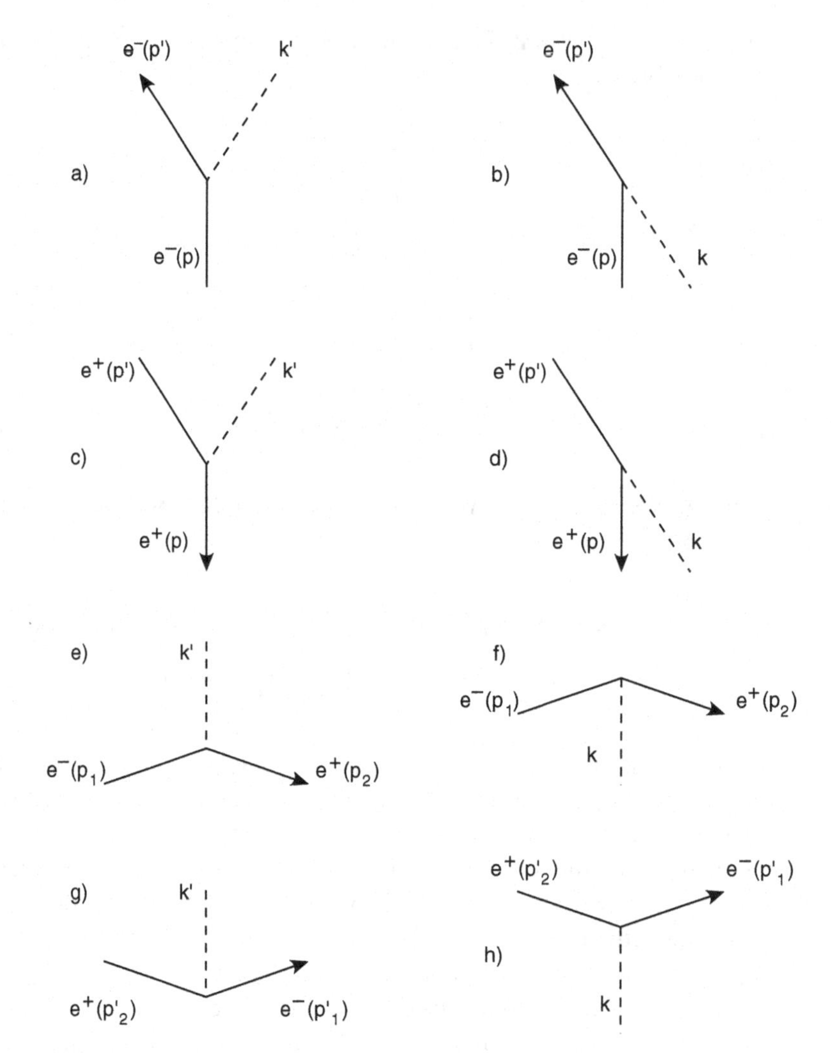

Figure 23.2 Feynman diagrams for first-order processes in quantum electrodynamics.

processes. Nevertheless, we can learn valuable lessons from the first-order processes. Consider Figure 23.2a, where the initial and final states are

$$|\phi_i(-\infty)\rangle = |(ps)_-\rangle \qquad |\phi_f(\infty)\rangle = |(p's')_-,(k'\alpha')_\gamma\rangle$$

From (23.20), the first-order amplitude is

$$\begin{aligned} A^{(1)}_{fi}(a) &= ie\langle (p',s')_-,(k,\alpha')_\gamma | \int d^4x \overline{\Psi}^{(-)}\gamma^\mu \Psi^{(+)} A^{(-)}_\mu | (p,s)_-\rangle \\ &= \frac{ie}{V^{3/2}} \int d^4x \sum_{\substack{p_1,s_1 \\ p_2,s_2 \\ k_1,\alpha_1}} \sqrt{\frac{m^2}{E_1E_2}} \sqrt{\frac{1}{2\omega}} \left(\overline{u}_{p_2s_2}\gamma^\mu \varepsilon_{\mu,\alpha_1} u_{p_1s_1}\right) e^{-i(p_1-p_2-k_1)\cdot x} \\ & \langle 0|a_{k'\alpha'}a^\dagger_{k_1\alpha_1}|0\rangle \langle 0|b_{p's'}b^\dagger_{p_2s_2}b_{p_1s_1}b^\dagger_{p,s}|0\rangle \end{aligned} \tag{23.21}$$

Of all the terms in the six-fold sum on the right-hand side of (23.21), only that single term contributes, for which

$$p_2 = p' \quad p_1 = p \quad k_1 = k' \quad s_2 = s' \quad s_1 = s \quad \alpha_1 = \alpha'$$

And, consequently,

$$\langle 0|a_{k'\alpha'}a^{\dagger}_{k_1\alpha_1}|0\rangle \to \langle 0|a_{k'\alpha'}a^{\dagger}_{k'\alpha'}|0\rangle = 1$$
$$\langle 0|b_{p's'}b^{\dagger}_{p_2s_2}b_{p_1s_1}b^{\dagger}_{ps}|0\rangle \to \langle 0|b_{p's'}b^{\dagger}_{p's'}b_{ps}b^{\dagger}_{ps}|0\rangle = 1$$

In the limit $V \to \infty$, the four-dimensional integral in (23.21) is

$$\int \exp\left[i(p'+k'-p)\cdot x\right] d^4x = (2\pi)^4 \delta^4(p'+k'-p) \tag{23.22}$$

We thus obtain the result

$$A^{(1)}_{fi}(a) = \sqrt{\frac{m}{E'V}}\sqrt{\frac{m}{EV}}\sqrt{\frac{1}{2\omega V}}(2\pi)^4 \delta^4(p'+k'-p) M_a \tag{23.23}$$

where M_a, the Feynman amplitude, is

$$M_a = ie\bar{u}_{p's'} \not{\varepsilon}_{\alpha'} u_{ps} \tag{23.24}$$

The case in Figure 23.2b, photon absorption by a free electron, is described by

$$A^{(1)}_{fi}(b) = \sqrt{\frac{m}{E'V}}\sqrt{\frac{m}{EV}}\sqrt{\frac{1}{2\omega V}}(2\pi)^4 \delta^4(p'-k-p) M_b \tag{23.25}$$

with

$$M_b = M_a = ie\bar{u}_{p's'} \not{\varepsilon}_{\alpha} u_{ps} \tag{23.26}$$

For the case in Figure 23.2c, emission of a photon by a free positron, the initial and final states are

$$|\phi_i(-\infty)\rangle = |(ps)_+\rangle \qquad |\phi_f(\infty)\rangle = |(p's')_+, (k'\alpha')_\gamma\rangle$$

The calculation of $A_{fi}(c)$ is analogous to that of $A_{fi}(a)$, but this time the matrix element of fermion creation and destruction operators is

$$-\langle 0|d_{p's'}d^{\dagger}_{p's'}d_{ps}d^{\dagger}_{ps}|0\rangle = -1$$

where the minus sign originates from the third term on the right-hand side of (23.20). Thus we obtain

$$A^{(1)}_{fi}(c) = \sqrt{\frac{m}{E'V}}\sqrt{\frac{m}{EV}}\sqrt{\frac{1}{2\omega V}}(2\pi)^4 \delta^4(p'+k'-p) M_c \tag{23.27}$$

where

$$M_c = -ie\bar{v}_{ps}\not{\varepsilon}_{\alpha'}v_{p's'} \tag{23.28}$$

Note that in addition to the overall sign change, the ordering of the spinors is reversed compared with the case in Figure 23.2a: $\bar{v}_{ps}$ corresponds to the incoming positron that is destroyed at the vertex, whereas $v_{p's'}$ corresponds to the outgoing positron created at the vertex. The arrow pointing backward in time on the diagram reminds us of this reversal. In the case in Figure 23.2d, photon absorption by a free positron, the sign of M_d is also negative. The signs of $M_{e,f,g,h}$ cannot be determined as easily as for the cases $M_{a,b}$ and $M_{c,d}$, but it can be shown that in cases e, f, g, and h we can consistently choose the signs to be positive.

In (23.23), (23.25), and (23.27) we have deliberately written $A_{fi}^{(1)}$ in a standard form. In addition to M, the right-hand sides of (23.23), (23.25), and (23.27) each contain a factor $\sqrt{m/EV}$ for each external fermion line, a factor $\sqrt{1/2\omega V}$ for each external photon line, and the 4-momentum conservation factor $(2\pi)^4\delta^4\left(P_i - P_f\right)$, where $P_{i,f}$ are the total 4-momenta of the initial (final) state.

The Feynman amplitude M always contains a factor $ie\gamma^\mu$ for each $e^-e^-\gamma$ or $e^-e^+\gamma$ vertex, a factor $-ie\gamma^\mu$ for each $e^+e^+\gamma$ vertex, a factor ε_μ for each external (incoming or outgoing) photon line, a column spinor u for each incoming electron line, a row spinor $\bar{u}'$ for each outgoing electron line, a row spinor $\bar{v}$ for each incoming positron line, and a column spinor v' for each outgoing positron line. The association of these quantities with vertices and external $e^\mp$ and photon lines is generally valid and forms part of a set of Feynman rules for constructing amplitudes directly from the corresponding diagrams for higher-order processes.

23.3 Photon propagator

Because each first-order amplitude in QED is zero because of energy-momentum conservation, we must go to second or higher order in the S-matrix expansion to describe any real physical process. The order of a diagram is the same as the number of its vertices; hence in second order there are two vertices, which must be connected by at least one internal photon or fermion line. Derivations of the Feynman rules for internal photon and fermion lines, starting from basic field-theoretic considerations, are quite lengthy and complicated. We now give those derivations in this and the following section. First, we calculate the amplitude for electron-electron scattering in lowest nonvanishing order, and in so doing, we derive the rule for an internal photon line in this particular case. It is given in (23.51) and (23.52). We trust that after the derivation is digested, the reader will accept the *general* validity of the rule.

To start, we recall that in Coulomb gauge and in natural units, the total Hamiltonian for charged particles, radiation, and their interaction is

$$H = H_{\text{particle}} + \frac{1}{2}\int\left(\mathcal{E}_\perp^2 + \boldsymbol{B}^2\right)d^3\boldsymbol{x} + \frac{1}{8\pi}\iint\frac{\rho(\boldsymbol{x})\rho(\boldsymbol{x}')}{|\boldsymbol{x}-\boldsymbol{x}'|}\,d^3x\,d^3x' - \int \boldsymbol{j}\cdot\boldsymbol{A}\,d^3\boldsymbol{x} \tag{23.29}$$

Here the radiation field is expressed solely in terms of the transverse vector potential A, and the instantaneous Coulomb interaction is expressed entirely in terms of the charge densities; that is,

$$V = \frac{1}{8\pi}\iint \frac{\rho(x)\rho(x')}{|x-x'|}\, d^3x d^3x' \tag{23.30}$$

In lowest order, the interaction between two electrons is proportional to the product of their charges. Therefore, in the amplitude for electron-electron scattering, the instantaneous Coulomb interaction contributes in first order, whereas the last term on the right-hand side of (23.29) contributes in second order. These two contributions turn out to be of comparable significance. We begin with the instantaneous Coulomb interaction. Let the initial and final two-electron states be

$$\begin{aligned} |\phi_i\rangle &= |p_1s_1, p_2s_2\rangle = b_1^\dagger b_2^\dagger |0\rangle \\ |\phi_f\rangle &= |p_1's_1', p_2's_2'\rangle = b_1'^\dagger b_2'^\dagger |0\rangle \end{aligned} \tag{23.31}$$

The first-order amplitude is

$$A_{fi}^{(1)} = -i\left\langle 0 \middle| b_2' b_1' \int_{-\infty}^{\infty} dt \left[\frac{1}{8\pi}\iint \frac{\rho(x,t)\rho(x',t)}{|x-x'|}\, d^3x d^3x'\right] b_1^\dagger b_2^\dagger \middle| 0 \right\rangle \tag{23.32}$$

where $\rho(x,t) = -e\bar{\Psi}(x,t)\gamma^0\Psi(x,t)$ and $\rho(x',t) = -e\bar{\Psi}(x',t)\gamma^0\Psi(x',t)$. The Dirac fields are

$$\bar{\Psi}(x,t) = \frac{1}{V^{1/2}}\sum_{p,s}\sqrt{\frac{m}{E}}\left(\bar{u}_{ps} b_{ps}^\dagger e^{ip\cdot x} + \bar{v}_{ps} d_{ps} e^{-ip\cdot x}\right)$$

$$\Psi(x,t) = \frac{1}{V^{1/2}}\sum_{p',s'}\sqrt{\frac{m}{E'}}\left(u_{p's'} b_{p's'} e^{-ip'\cdot x} + v_{p's'} d_{p's'}^\dagger e^{ip'\cdot x}\right)$$

and

$$\bar{\Psi}(x',t) = \frac{1}{V^{1/2}}\sum_{p'',s''}\sqrt{\frac{m}{E''}}\left(\bar{u}_{p''s''} b_{p''s''}^\dagger e^{ip''\cdot x'} + \bar{v}_{p''s''} d_{p''s''} e^{-ip''\cdot x'}\right)$$

$$\Psi(x',t) = \frac{1}{V^{1/2}}\sum_{p''',s'''}\sqrt{\frac{m}{E'''}}\left(u_{p'''s'''} b_{p'''s'''} e^{-ip'''\cdot x'} + v_{p''s''} d_{p'''s'''}^\dagger e^{ip'''\cdot x'}\right)$$

Consider the various terms in the sum

$$\sum_{\substack{p,s\\p',s'\\p'',s''\\p''',s'''}} (\bullet)$$

To save writing for the moment, we suppress all quantities except for the creation and destruction operators; that is,

$$A_{fi}^{(1)} \to \left\langle 1'2' \middle| \sum \left(b_{ps}^\dagger + d_{ps}\right)\left(b_{p's'} + d_{p's'}^\dagger\right)\left(b_{p''s''}^\dagger + d_{p''s''}\right)\left(b_{p'''s'''} + d_{p'''s'''}^\dagger\right) \middle| 12 \right\rangle$$

Of the 16 types of terms that result from multiplying out the four factors in parentheses, only the quantity $A_{fi}^{(1)} \rightarrow \left\langle 1'2' \middle| \sum b_{ps}^{\dagger} b_{p's'} b_{p''s''}^{\dagger} b_{p'''s'''} \middle| 12 \right\rangle$ contributes to e^-e^- scattering. It yields

$$A_{fi}^{(1)} = \frac{-ie^2}{8\pi V^2}\sqrt{\frac{m^4}{E_1E_2E_1'E_2'}}\int_{-\infty}^{\infty} dt \iint \frac{d^3x d^3x'}{|x-x'|} \bullet \langle 0 | (\bar{u}_{2'}\gamma^0 u_2 \cdot \bar{u}_{1'}\gamma^0 u_1)$$
$$b_{2'}b_{1'}b_{2'}^{\dagger}b_2 b_{1'}^{\dagger}b_1 b_1^{\dagger}b_2^{\dagger} e^{i(p_2'-p_2)\bullet x} e^{i(p_1'-p_1)\bullet x'}$$
$$+(\bar{u}_{1'}\gamma^0 u_1 \bullet \bar{u}_{2'}\gamma^0 u_2) b_{2'}b_{1'}b_{1'}^{\dagger}b_1 b_{2'}^{\dagger}b_2 b_1^{\dagger}b_2^{\dagger} e^{i(p_1'-p_1)\bullet x} e^{i(p_2'-p_2)\bullet x'} \tag{23.33}$$
$$+(\bar{u}_{2'}\gamma^0 u_1 \bullet \bar{u}_{1'}\gamma^0 u_2) b_{2'}b_{1'}b_{2'}^{\dagger}b_1 b_{1'}^{\dagger}b_2 b_1^{\dagger}b_2^{\dagger} e^{i(p_2'-p_1)\bullet x} e^{i(p_1'-p_2)\bullet x'}$$
$$+(\bar{u}_{1'}\gamma^0 u_2 \bullet \bar{u}_{2'}\gamma^0 u_1) b_{2'}b_{1'}b_{1'}^{\dagger}b_2 b_{2'}^{\dagger}b_1 b_1^{\dagger}b_2^{\dagger} e^{i(p_1'-p_2)\bullet x} e^{i(p_2'-p_1)\bullet x'} |0\rangle$$

The matrix elements of the creation and destruction operators are

$$\langle 0 | b_{2'}b_{1'}b_{2'}^{\dagger}b_2 b_{1'}^{\dagger}b_1 b_1^{\dagger}b_2^{\dagger} | 0 \rangle = 1$$
$$\langle 0 | b_{2'}b_{1'}b_{1'}^{\dagger}b_1 b_{2'}^{\dagger}b_2 b_1^{\dagger}b_2^{\dagger} | 0 \rangle = 1$$
$$\langle 0 | b_{2'}b_{1'}b_{2'}^{\dagger}b_1 b_{1'}^{\dagger}b_2 b_1^{\dagger}b_2^{\dagger} | 0 \rangle = -1$$
$$\langle 0 | b_{2'}b_{1'}b_{1'}^{\dagger}b_2 b_{2'}^{\dagger}b_1 b_1^{\dagger}b_2^{\dagger} | 0 \rangle = -1$$

Thus (23.33) becomes

$$A_{fi}^{(1)} = \frac{-ie^2}{8\pi V^2}\sqrt{\frac{m^4}{E_1E_2E_1'E_2'}}\int_{-\infty}^{\infty} dt \iint \frac{d^3x d^3x'}{|x-x'|} \bullet$$
$$\{[\bar{u}_{2'}\gamma^0 u_2 \bullet \bar{u}_{1'}\gamma^0 u_1][e^{-i(p_2'-p_2)\bullet x} e^{-i(p_1'-p_1)\bullet x'} + e^{-i(p_1'-p_1)\bullet x} e^{-i(p_2'-p_2)\bullet x'}] \tag{23.34}$$
$$-[\bar{u}_{2'}\gamma^0 u_1 \bullet \bar{u}_{1'}\gamma^0 u_2][e^{-i(p_2'-p_1)\bullet x} e^{-i(p_1'-p_2)\bullet x'} + e^{-i(p_1'-p_2')\bullet x} e^{-i(p_2'-p_1)\bullet x'}]\} e^{i(E_{2'}-E_2+E_{1'}-E_1)t}$$

Evaluating the integral over t and taking advantage of the symmetry of the remaining integrand under the interchange of variables $x \leftrightarrow x'$, we obtain

$$A_{fi}^{(1)} = \frac{-ie^2\delta(E_1+E_2-E_1'-E_2')}{2V^2}\sqrt{\frac{m^4}{E_1E_2E_1'E_2'}} \bullet [(\bar{u}_{2'}\gamma^0 u_2 \cdot \bar{u}_{1'}\gamma^0 u_1)$$
$$\iint \frac{d^3x d^3x'}{|x-x'|}(e^{-i(p_2'-p_2)\bullet x} e^{-i(p_1'-p_1)\bullet x'}) \tag{23.35}$$
$$-(\bar{u}_{2'}\gamma^0 u_1 \bullet \bar{u}_{1'}\gamma^0 u_2)\iint \frac{d^3x d^3x'}{|x-x'|}(e^{-i(p_2'-p_1)\bullet x} e^{-i(p_1'-p_2)\bullet x'})]$$

To complete the integrations, we let $y = x' - x$. Then the first integral on the right-hand side of (23.35) becomes

$$\iint \frac{d^3x d^3x'}{|x-x'|}(e^{-i(p_2'-p_2)\bullet x} e^{-i(p_1'-p_1)\bullet x'}) = \int d^3x e^{3i(p_2-p_2'+p_1-p_1')\cdot x} \int d^3y \frac{1}{y} e^{i(p_1-p_1')\bullet y}$$
$$= \frac{4\pi}{|q|^2}(2\pi)^3 \delta^3(p_2+p_1-p'_2-p'_1) \tag{23.36}$$

where $\boldsymbol{q} = \boldsymbol{p}_1 - \boldsymbol{p}_1'$. Similarly, the second integral in (23.35) is

$$\iint \frac{d^3x d^3x'}{|\boldsymbol{x}-\boldsymbol{x}'|}\left(e^{i(\boldsymbol{p}_2-\boldsymbol{p}_1')\bullet\boldsymbol{x}} e^{i(\boldsymbol{p}_1-\boldsymbol{p}_2')\bullet\boldsymbol{x}'}\right) = \frac{4\pi}{|\boldsymbol{q}'|^2}(2\pi)^3 \delta^3(\boldsymbol{p}_2+\boldsymbol{p}_1-\boldsymbol{p}_2'-\boldsymbol{p}_1') \tag{23.37}$$

where $\boldsymbol{q}' = \boldsymbol{p}_2 - \boldsymbol{p}_1'$. Therefore, (23.35) can be written as

$$\begin{gathered} A_{fi}^{(1)} = -i(2\pi)^4 \delta^4(p_1+p_2-p_1'-p_2')\sqrt{\frac{m^4}{E_1E_2E_1'E_2'}}\, e^2 \\ \left(\bar{u}_{2'}\gamma^0 u_2 \frac{1}{|\boldsymbol{q}|^2}\bar{u}_{1'}\gamma^0 u_1 - \bar{u}_{2'}\gamma^0 u_1 \frac{1}{|\boldsymbol{q}'|^2}\bar{u}_{1'}\gamma^0 u_2\right) \end{gathered} \tag{23.38}$$

The two terms on the right-hand side of (23.38) are direct and exchange terms. Recall that their relative minus sign arises because of the antisymmetry of the two-electron wave function, which is accounted for by the anticommutation relations satisfied by the electron creation and destruction operators.

Now consider the $\boldsymbol{j}\bullet\boldsymbol{A}$ contribution in second order. The amplitude is

$$A_{fi}^{(2)} = -\frac{1}{2}\left\langle 0 \left| b_{2'}b_{1'}\left[\int_{-\infty}^{\infty} dt_1 \int_{-\infty}^{\infty} T\{H_I(t_1), H_I(t_2)\} dt_2\right] b_1^\dagger b_2^\dagger \right| 0\right\rangle \tag{23.39}$$

where

$$H_I = -\int \boldsymbol{j}^{\mathrm{EM}}\bullet\boldsymbol{A} d^3x = e\int \bar{\Psi}\boldsymbol{\gamma}\bullet\boldsymbol{A}\Psi\, d^3x$$

Equation (23.39), written in full, is

$$\begin{aligned} A_{fi}^{(2)} = -\frac{e^2}{2}\langle 0| b_{2'}b_{1'} \int_{-\infty}^{\infty} dt_1 \int_{-\infty}^{\infty} dt_2 \int d^3x_1 \int d^3x_2 \bullet \big[\bar{\Psi}(x_1)\gamma^i\Psi(x_1)\bar{\Psi}(x_2)\gamma^j\Psi(x_2)A_i(x_1)A_j(x_2) \\ +\bar{\Psi}(x_2)\gamma^j\Psi(x_2)\bar{\Psi}(x_1)\gamma^i\Psi(x_1)A_j(x_2)A_i(x_1)\big] b_1^\dagger b_2^\dagger |0\rangle \end{aligned} \tag{23.40}$$

The vector potential factor in (23.40) is

$$\begin{aligned} &\langle 0|A_i(x_1)A_j(x_2)|0\rangle \qquad t_2 < t_1 \\ &\langle 0|A_j(x_2)A_i(x_1)|0\rangle \qquad t_2 > t_1 \end{aligned}$$

Note that there are no photons in the initial or final state. For $t_2 < t_1$,

$$\begin{aligned} \langle 0|A_i(x_1)A_j(x_2)|0\rangle &= \frac{1}{V}\sum_{\substack{k\lambda \\ k'\lambda'}} \frac{1}{2\sqrt{\omega\omega'}}\hat{\varepsilon}_i(\boldsymbol{k},\lambda)\hat{\varepsilon}_j(\boldsymbol{k}',\lambda') \\ &\quad \langle 0|\left[a_{k\lambda}e^{-ik\bullet x_1} + a_{k\lambda}^\dagger e^{ik\bullet x_1}\right]\left[a_{k'\lambda'}e^{-ik'\bullet x_2} + a_{k'\lambda'}^\dagger e^{ik'\bullet x_2}\right]|0\rangle \\ &= \frac{1}{V}\sum_{k\lambda}\frac{1}{2\omega}\hat{\varepsilon}_i(\boldsymbol{k},\lambda)\hat{\varepsilon}_j(\boldsymbol{k},\lambda)e^{-i\omega\tau}e^{i\boldsymbol{k}\bullet(\boldsymbol{x}_1-\boldsymbol{x}_2)} \end{aligned} \tag{23.41}$$

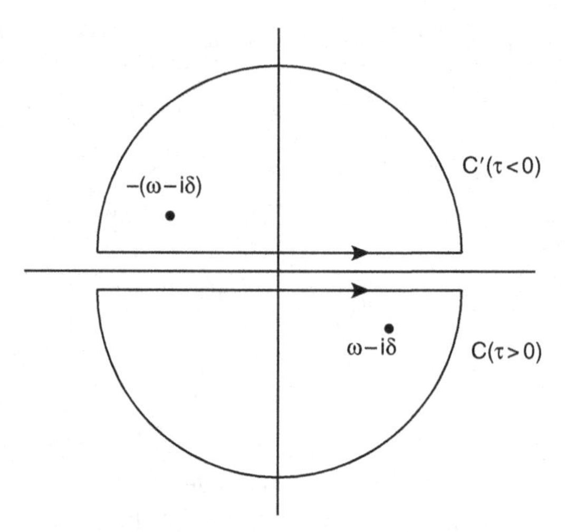

Figure 23.3 Contours for evaluation of integral I defined in (23.43).

where $\tau = t_1 - t_2$. Similarly, for $t_2 > t_1$,

$$\langle 0|A_i(x_1)A_j(x_2)|0\rangle = \frac{1}{V}\sum_{k\lambda}\frac{1}{2\omega}\hat{\varepsilon}_i(\boldsymbol{k},\lambda)\hat{\varepsilon}_j(\boldsymbol{k},\lambda)e^{i\omega\tau}e^{i\boldsymbol{k}\cdot(x_2-x_1)} \tag{23.42}$$

To evaluate the sums in (23.41) and (23.42), we express $\omega^{-1}e^{\pm i\omega\tau}$ as a contour integral. Consider Figure 23.3 and the integral

$$I = \int_{-\infty}^{\infty}\frac{e^{-iz\tau}dz}{(z-\omega+i\delta)(z+\omega-i\delta)} \tag{23.43}$$

where δ is a positive real infinitesimal. Using Figure 23.3, we evaluate I by contour integration, choosing contour C' for $\tau < 0$ and contour C for $\tau > 0$ so that in each case the integrand is exponentially damped on the arc.

From Cauchy's theorem, we have

$$I = -\frac{i\pi}{\omega}e^{i\omega\tau} \qquad \tau < 0$$
$$I = -\frac{i\pi}{\omega}e^{-i\omega\tau} \qquad \tau > 0$$

Now, replacing $(1/V)\sum_k$ by $\left[1/(2\pi)^3\right]\int d^3\boldsymbol{k}$ in (23.41) and (23.42), we obtain

$$\langle 0|A_i(x_1)A_j(x_2)|0\rangle = \frac{i}{16\pi^4}\sum_\lambda\int d^3\boldsymbol{k}\int dz\hat{\varepsilon}_i(\boldsymbol{k}\lambda)\hat{\varepsilon}_j(\boldsymbol{k}\lambda)\frac{e^{i\boldsymbol{k}\cdot(x_1-x_2)}e^{-iz\tau}}{z^2-\omega^2} \qquad \tau > 0$$
$$\langle 0|A_j(x_2)A_i(x_1)|0\rangle = \frac{i}{16\pi^4}\sum_\lambda\int d^3\boldsymbol{k}\int dz\hat{\varepsilon}_i(\boldsymbol{k}\lambda)\hat{\varepsilon}_j(\boldsymbol{k}\lambda)\frac{e^{-i\boldsymbol{k}\cdot(x_1-x_2)}e^{-iz\tau}}{z^2-\omega^2} \qquad \tau < 0$$

Let us define a 4-vector q such that $q = (z, \boldsymbol{k})$. Because $\omega^2 = |\boldsymbol{k}|^2$, we have $q^2 = z^2 - \omega^2$; hence

$$\langle 0|A_i(x_1)A_j(x_2)|0\rangle = \frac{i}{16\pi^4}\sum_\lambda \int d^3\boldsymbol{k}\int dz\hat{\varepsilon}_i(\boldsymbol{k}\lambda)\hat{\varepsilon}_j(\boldsymbol{k}\lambda)\frac{e^{iq\cdot(x_1-x_2)}e^{-iz\tau}}{q^2} \qquad \tau > 0 \tag{23.44}$$

$$\langle 0|A_j(x_2)A_i(x_1)|0\rangle = \frac{i}{16\pi^4}\sum_\lambda \int d^3\boldsymbol{k}\int dz\hat{\varepsilon}_i(\boldsymbol{k}\lambda)\hat{\varepsilon}_j(\boldsymbol{k}\lambda)\frac{e^{-iq\cdot(x_1-x_2)}e^{-iz\tau}}{q^2} \qquad \tau < 0 \tag{23.45}$$

The right-hand sides of (23.44) and (23.45) are identical except for the interchange of $\boldsymbol{x}_1$ and $\boldsymbol{x}_2$. Ultimately, we integrate over $\boldsymbol{x}_1$ and $\boldsymbol{x}_2$ with other factors in the integrand that are symmetric with respect to this interchange. Thus we can make the interchange in (23.45), which renders (23.44) and (23.45) identical. Returning to (23.40), we employ (23.44) and (23.45) to obtain

$$\begin{aligned} A_{fi}^2 = &-\frac{i}{(2\pi)^4}\frac{e^2}{V^2}\sqrt{\frac{m^4}{E_1E_2E_1'E_2'}}\int\frac{d^4q}{q^2}\int d^4x_1\int d^4x_2 \\ &\bullet\sum_\lambda\Big\{\big[\bar{u}_{2'}\boldsymbol{\gamma}\bullet\hat{\boldsymbol{\varepsilon}}u_2\bar{u}_{1'}\boldsymbol{\gamma}\bullet\hat{\boldsymbol{\varepsilon}}u_1\big]e^{i(p_2'-p_2)\bullet x_2}e^{i(p_1'-p_1)\bullet x_1}e^{i(x_2-x_1)\bullet q} \\ &-\big[\bar{u}_{2'}\boldsymbol{\gamma}\bullet\hat{\boldsymbol{\varepsilon}}u_1\bar{u}_{1'}\boldsymbol{\gamma}\bullet\hat{\boldsymbol{\varepsilon}}u_2\big]e^{i(p_1'-p_2)\bullet x_2}e^{i(p_2'-p_1)\bullet x_1}e^{i(x_2-x_1)\bullet q}\Big\} \end{aligned} \tag{23.46}$$

In the first term of (23.46), the integrations give

$$\begin{aligned} \int\frac{d^4q}{q^2}\int d^4x_1\int d^4x_2 e^{i(p_2'-p_2)\bullet x_2}e^{i(p_1'-p_1)\bullet x_1}e^{i(x_2-x_1)\bullet q} &= (2\pi)^8\int\frac{d^4q}{q^2}\delta^4(p_1+q-p_1')\delta^4(p_2+q-p_2') \\ &= (2\pi)^8\frac{1}{(p_1-p_1')^2}\delta^4(p_1+p_2-p_1'-p_2') \end{aligned}$$

with a similar expression for the second term. Note that because of the delta functions, 4-momentum is conserved at each vertex, which is a universal feature of covariant perturbation theory. With the new definitions $q = p_1 - p_1'$ and $q' = p_2 - p_1'$, (23.46) becomes

$$\begin{aligned} A_{fi}^{(2)} = &-i\frac{e^2}{V^2}\sqrt{\frac{m^4}{E_1E_2E_1'E_2'}}(2\pi)^4\delta^4(p_1+p_2-p_1'-p_2') \\ &\bullet\sum_\lambda\left\{\bar{u}_{2'}\boldsymbol{\gamma}\cdot\hat{\boldsymbol{\varepsilon}}u_2\frac{1}{q^2}\bar{u}_{1'}\boldsymbol{\gamma}\cdot\hat{\boldsymbol{\varepsilon}}u_1 - \bar{u}_{2'}\boldsymbol{\gamma}\cdot\hat{\boldsymbol{\varepsilon}}u_1\frac{1}{q'^2}\bar{u}_{1'}\boldsymbol{\gamma}\cdot\hat{\boldsymbol{\varepsilon}}u_2\right\} \end{aligned} \tag{23.47}$$

It remains to evaluate the sum in (23.47). Let $\boldsymbol{q}$ define a z-axis. Because $\hat{\varepsilon}\bullet\boldsymbol{q} = 0$, we can choose one of the two possible unit vectors $\hat{\varepsilon}$ along x and the other along y. Hence

$$\begin{aligned} P \equiv \sum_\lambda \bar{u}_{2'}\boldsymbol{\gamma}\bullet\hat{\boldsymbol{\varepsilon}}u_2\frac{1}{q^2}\bar{u}_{1'}\boldsymbol{\gamma}\bullet\hat{\boldsymbol{\varepsilon}}u_1 &= \frac{1}{q^2}\left(\bar{u}_{2'}\gamma^1u_2\bar{u}_{1'}\gamma^1u_1 + \bar{u}_{2'}\gamma^2u_2\bar{u}_{1'}\gamma^2u_1\right) \\ &= \frac{1}{q^2}\left(\bar{u}_{2'}\gamma u_2\bullet\bar{u}_{1'}\gamma u_1 - \bar{u}_{2'}\gamma^3u_2\bar{u}_{1'}\gamma^3u_1\right) \end{aligned} \tag{23.48}$$

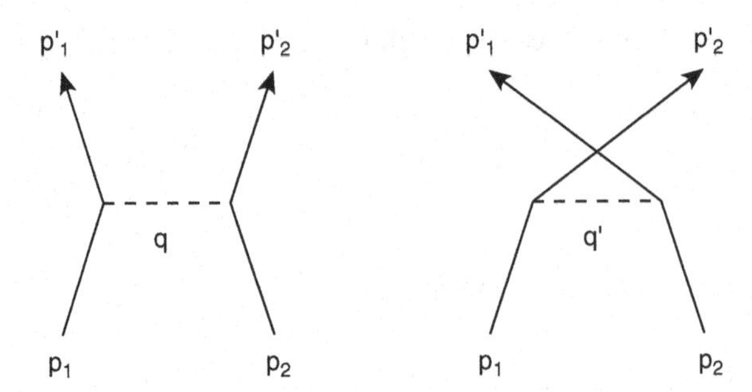

Figure 23.4 Feynman diagrams for electron-electron (Møller) scattering in lowest order.

or, more generally,

$$P = \frac{1}{q^2}\left(\bar{u}_{2'}\boldsymbol{\gamma} u_2 \bullet \bar{u}_{1'}\boldsymbol{\gamma} u_1 - \frac{1}{\boldsymbol{q}^2}\bar{u}_{2'}\boldsymbol{\gamma}\bullet\boldsymbol{q} u_2 \bar{u}_{1'}\boldsymbol{\gamma}\bullet\boldsymbol{q} u_1\right) \tag{23.49}$$

However, $\boldsymbol{\gamma}\bullet\boldsymbol{q} = \gamma^0 q^0 - \not{q}$ and $\bar{u}_{1'}\not{q}u_1 = 0$. Therefore,

$$P = \frac{1}{q^2}\left(\bar{u}_{2'}\boldsymbol{\gamma} u_2 \bullet \bar{u}_{1'}\boldsymbol{\gamma} u_1 - \frac{q_0^2}{\boldsymbol{q}^2}\bar{u}_{2'}\gamma^0 u_2 \bar{u}_{1'}\gamma^0 u_1\right) \tag{23.50}$$

with a similar expression for the exchange term. We now combine the Coulomb interaction and $\boldsymbol{j}\bullet\boldsymbol{A}$ contributions by substituting (23.50) in (23.47), recalling (23.38), and making use of

$$\left(1 - \frac{q_0^2}{q^2}\right) = \frac{q^2 - q_0^2}{q^2} = -\frac{\boldsymbol{q}^2}{q^2}$$

This gives the final result for electron-electron scattering in lowest nonvanishing order; that is,

$$A_{fi} = \sqrt{\frac{m}{VE_1}}\sqrt{\frac{m}{VE_2}}\sqrt{\frac{m}{VE_1'}}\sqrt{\frac{m}{VE_2'}}(2\pi)^4\,\delta^4\left(p_1 + p_2 - p_1' - p_2'\right)M \tag{23.51}$$

where

$$M = e^2\left(\bar{u}_{2'}i\gamma^\mu u_2 \frac{-ig_{\mu\nu}}{q^2}\bar{u}_{1'}i\gamma^\nu u_1 - \bar{u}_{2'}i\gamma^\mu u_1 \frac{-ig_{\mu\nu}}{q'^2}\bar{u}_{1'}i\gamma^\nu u_2\right) \tag{23.52}$$

Result (23.52) is represented by the Feynman diagrams of Figure 23.4.

From (23.52) we extract two general results to add to the list of Feynman rules:

- Each internal photon line is associated with a photon propagator factor

$$-i\frac{g_{\mu\nu}}{q^2}$$

- If two identical fermions are in the final state, there is a relative minus sign between the direct and exchange amplitudes.

The photon propagator can be understood as the Fourier transform in momentum space of the Green function for the inhomogeneous wave equation. Here we recall the discussion in Section 18.6. Consider the inhomogeneous wave equation of electrodynamics in Lorenz gauge

$$\partial^\mu \partial_\mu A_\nu(x) = j_\nu^{\text{EM}}(x)$$

Following the discussion in Section 18.6, we try to express $A_\nu(x)$ in terms of a solution $A_\nu^{(0)}(x)$ to the homogeneous wave equation and an integral involving a Green function; that is,

$$A_\nu(x) = A_\nu^{(0)}(x) - i\int G(x,x') j_\nu^{\text{EM}}(x')\, d^4x'$$

Applying $\partial^\mu \partial_\mu$ to both sides of this equation, where the derivatives are with respect to x and taking into account that $\partial^\mu \partial_\mu A_\nu^{(0)} = 0$, we obtain

$$\partial^\mu \partial_\mu A_\nu(x) = j_\nu^{\text{EM}}(x) = -i\int \left[\partial^\mu \partial_\mu G(x,x')\right] j_\nu^{\text{EM}}(x')\, d^4x'$$

This implies that

$$\partial^\mu \partial_\mu G(x,x') = i\delta^4(x-x')$$

We now Fourier transform both sides of the latter equation. Writing

$$G(x,x') = \frac{1}{(2\pi)^4}\int f(q) e^{iq\cdot(x-x')}\, d^4q, \qquad \delta^4(x-x') = \frac{1}{(2\pi)^4}\int e^{iq\cdot(x-x')}\, d^4q$$

we have

$$\int -q^2 f(q) e^{iq\cdot(x-x')}\, d^4q = i\int e^{iq\cdot(x-x')}\, d^4q$$

which yields $f(q) = -i/q^2$.

23.4 Fermion propagator

We now derive the lowest-order amplitude for electron-positron annihilation to two photons. The result is given in (23.70) and (23.71). In the course of the derivation, we obtain a rule for an internal electron line that is valid not just for this particular case but valid in general. Let the initial and final states be

$$|\phi_i\rangle = |p_+ s_+, p_- s_-\rangle \qquad |\phi_f\rangle = |k_1\alpha_1, k_2\alpha_2\rangle$$

The amplitude is

$$A_{fi} = (-i)^2 \int d^4x_1 \int_{t_1>t_2} d^4x_2 \left\langle \phi_f \left| j_{\text{EM}}^\mu(x_1) A_\mu(x_1) j_{\text{EM}}^\nu(x_2) A_\nu(x_2) \right| \phi_i \right\rangle \tag{23.53}$$

Including in (23.53) only those contributions from the normally ordered current densities that can annihilate an electron and a positron, we obtain

$$\begin{aligned} A_{fi} = &-e^2 \int d^4x_1 \int_{t_1>t_2} d^4x_2 \langle k_1\alpha_1, k_2\alpha_2 | A_\mu(x_1) A_\nu(x_2) | 0 \rangle \gamma^\mu_{\alpha\beta}\gamma^\nu_{\rho\sigma} \\ &\bullet \langle 0 | \bar{\Psi}^{(+)}_\alpha(x_1)\Psi^{(+)}_\beta(x_1)\bar{\Psi}^{(-)}_\rho(x_2)\Psi^{(+)}_\sigma(x_2) - \bar{\Psi}^{(+)}_\alpha(x_1)\Psi^{(+)}_\beta(x_1)\Psi^{(-)}_\sigma(x_2)\bar{\Psi}^{(+)}_\rho(x_2) \\ &+ \bar{\Psi}^{(-)}_\alpha(x_1)\Psi^{(+)}_\beta(x_1)\bar{\Psi}^{(+)}_\rho(x_2)\Psi^{(+)}_\sigma(x_2) - \Psi^{(-)}_\beta(x_1)\bar{\Psi}^{(+)}_\alpha(x_1)\bar{\Psi}^{(+)}_\rho(x_2)\Psi^{(+)}_\sigma(x_2) | p_+s_+, p_-s_- \rangle \end{aligned} \tag{23.54}$$

The photon part of this matrix element is

$$\begin{aligned} a_{\mu\nu} &\equiv \langle k_1\alpha_1, k_2\alpha_2 | A_\mu(x_1) A_\nu(x_2) | 0 \rangle \\ &= \frac{1}{2V}\sqrt{\frac{1}{\omega_1\omega_2}}\left(\hat{\varepsilon}_{1\mu}\hat{\varepsilon}_{2\nu} e^{i(k_1\bullet x_1 + k_2\bullet x_2)} + \hat{\varepsilon}_{1\nu}\hat{\varepsilon}_{2\mu} e^{i(k_1\bullet x_2 + k_2\bullet x_1)}\right) \end{aligned} \tag{23.55}$$

The first and second terms in the parentheses are the direct and exchange terms, respectively. The plus sign between them is a manifestation of the fact that photons are bosons. Concerning the electron-positron part of the matrix element in (23.54), the third and fourth terms give zero by operating on $\langle 0 |$ from the right. The remaining part is

$$\gamma^\mu_{\alpha\beta}\gamma^\nu_{\rho\sigma} \langle 0 | \bar{\Psi}^{(+)}_\alpha(x_1)\Psi^{(+)}_\beta(x_1)\bar{\Psi}^{(-)}_\rho(x_2)\Psi^{(+)}_\sigma(x_2) - \bar{\Psi}^{(+)}_\alpha(x_1)\Psi^{(+)}_\beta(x_1)\Psi^{(-)}_\sigma(x_2)\bar{\Psi}^{(+)}_\rho(x_2) | p_+s_+, p_-s_- \rangle \tag{23.56}$$

The ordering of fermion fields in the first term of (23.56) can be rearranged as follows:

$$\bar{\Psi}^{(+)}_\alpha(x_1)\Psi^{(+)}_\beta(x_1)\bar{\Psi}^{(-)}_\rho(x_2)\Psi^{(+)}_\sigma(x_2) \rightarrow \Psi^{(+)}_\beta(x_1)\bar{\Psi}^{(-)}_\rho(x_2)\bar{\Psi}^{(+)}_\alpha(x_1)\Psi^{(+)}_\sigma(x_2) \tag{23.57}$$

Also, the contribution of the second term in (23.56) to (23.54) is

$$e^2 \int d^4x_1 \int_{t_1>t_2} d^4x_2 a_{\mu\nu}\gamma^\mu_{\alpha\beta}\gamma^\nu_{\rho\sigma} \bullet \langle 0 | \bar{\Psi}^{(+)}_\alpha(x_1)\Psi^{(+)}_\beta(x_1)\Psi^{(-)}_\sigma(x_2)\bar{\Psi}^{(+)}_\rho(x_2) | p_+s_+, p_-s_- \rangle \tag{23.58}$$

Because $\alpha, \beta, \rho, \sigma, \mu$, and ν are all dummy indices and x_1 and x_2 are dummy variables, we can make the exchanges $\rho\sigma \leftrightarrow \alpha\beta$, $\mu \leftrightarrow \nu$, and $x_1 \leftrightarrow x_2$ in (23.58) to obtain

$$\begin{aligned} &e^2 \int d^4x_1 \int_{t_1<t_2} d^4x_2 a_{\mu\nu}\gamma^\nu_{\rho\sigma}\gamma^\mu_{\alpha\beta} \bullet \langle 0 | \bar{\Psi}^{(+)}_\rho(x_2)\Psi^{(+)}_\sigma(x_2)\Psi^{(-)}_\beta(x_1)\bar{\Psi}^{(+)}_\alpha(x_1) | p_+s_+, p_-s_- \rangle \\ &= e^2 \int d^4x_1 \int_{t_1<t_2} d^4x_2 a_{\mu\nu}\gamma^\nu_{\rho\sigma}\gamma^\mu_{\alpha\beta} \bullet \langle 0 | \bar{\Psi}^{(+)}_\rho(x_2)\Psi^{(-)}_\beta(x_1)\bar{\Psi}^{(+)}_\alpha(x_1)\Psi^{(+)}_\sigma(x_2) | p_+s_+, p_-s_- \rangle \end{aligned} \tag{23.59}$$

Therefore, defining $\theta(\tau) = 1$ if $\tau > 0$ and $\theta(\tau) = 0$ if $\tau < 0$, we arrive at

$$A_{fi} = -e^2 \int d^4x_1 \int d^4x_2 a_{\mu\nu}\gamma^\mu_{\alpha\beta}\gamma^\nu_{\rho\sigma} a_{+-} \tag{23.60}$$

where

$$\begin{aligned} a_{+-} = &\langle 0 | \Psi^{(+)}_\beta(x_1)\bar{\Psi}^{(-)}_\rho(x_2)\Psi^{(+)}_\alpha(x_1)\bar{\Psi}^{(+)}_\sigma(x_2)\theta(t_1 - t_2) \\ &- \bar{\Psi}^{(+)}_\rho(x_2)\Psi^{(-)}_\beta(x_1)\Psi^{(+)}_\alpha(x_1)\bar{\Psi}^{(+)}_\sigma(x_2)\theta(t_2 - t_1) | p_+s_+, p_-s_- \rangle \end{aligned} \tag{23.61}$$

Using the completeness relation, where the $|n\rangle$ form a complete set of states of the e^+e^- field, we write (23.61) as follows:

$$\begin{aligned} a_{+-} &= \sum_n \langle 0|\Psi_\beta^{(+)}(x_1)\bar{\Psi}_\rho^{(-)}(x_2)|n\rangle\langle n|\Psi_\alpha^{(+)}(x_1)\bar{\Psi}_\sigma^{(+)}(x_2)|\phi_i\rangle\theta(t_1-t_2) \\ &-\sum_n \langle 0|\bar{\Psi}_\rho^{(+)}(x_2)\Psi_\beta^{(-)}(x_1)|n\rangle\langle n|\Psi_\alpha^{(+)}(x_1)\bar{\Psi}_\sigma^{(+)}(x_2)|\phi_i\rangle\theta(t_2-t_1) \end{aligned} \tag{23.62}$$

The only nonzero matrix elements in this expression are those for which $|n\rangle = 0$ (the vacuum state). Thus (23.62) is

$$\begin{aligned} a_{+-} &= \langle 0|\Psi_\beta^{(+)}(x_1)\bar{\Psi}_\rho^{(-)}(x_2)|0\rangle\langle 0|\Psi_\alpha^{(+)}(x_1)\bar{\Psi}_\sigma^{(+)}(x_2)|\phi_i\rangle\theta(t_1-t_2) \\ &-\langle 0|\bar{\Psi}_\rho^{(+)}(x_2)\Psi_\beta^{(-)}(x_1)|0\rangle\langle 0|\Psi_\alpha^{(+)}(x_1)\bar{\Psi}_\sigma^{(+)}(x_2)|\phi_i\rangle\theta(t_2-t_1) \\ &= \sqrt{\frac{m}{E_+V}}\sqrt{\frac{m}{E_-V}}e^{-i(p_+\bullet x_1+p_-\bullet x_2)}\bar{v}_\alpha(p_+s_+)u_\sigma(p_-s_-) \\ &\bullet\langle 0|\Psi_\beta^{(+)}(x_1)\bar{\Psi}_\rho^{(-)}(x_2)\theta(t_1-t_2)-\bar{\Psi}_\rho^{(+)}(x_2)\Psi_\beta^{(-)}(x_1)\theta(t_2-t_1)|0\rangle \end{aligned} \tag{23.63}$$

The factor on the last line of (23.63) is called the *vacuum expectation value* of the time-ordered product $\left\langle 0\left|T\{\Psi_\beta(x_1),\bar{\Psi}_\rho(x_2)\}\right|0\right\rangle$. It can be expressed as follows, where we make the replacement $V^{-1/2}\sum_p \to (2\pi)^{-3}\int d^3\boldsymbol{p}$:

$$\begin{aligned} \left\langle 0\left|T\{\Psi_\beta(x_1),\bar{\Psi}_\rho(x_2)\}\right|0\right\rangle &= \frac{1}{(2\pi)^3}\int d^3\boldsymbol{p}\left(\frac{m}{E}\right) \\ &\bullet\sum_s\left[u_\beta(ps)\bar{u}_\rho(ps)e^{-ip(x_1-x_2)}\theta(t_1-t_2)-v_\beta(ps)\bar{v}_\rho(ps)e^{ip(x_1-x_2)}\theta(t_2-t_1)\right] \end{aligned} \tag{23.64}$$

Later in this chapter [see (23.94) and (23.96)], we show that

$$\sum_s u_\beta(ps)\bar{u}_\rho(ps) = \frac{(\not{p}+m)_{\beta\rho}}{2m}$$

$$\sum_s v_\beta(ps)\bar{v}_\rho(ps) = \frac{(\not{p}-m)_{\beta\rho}}{2m}$$

Making use of these expressions in (23.64), we arrive at

$$\begin{aligned} \left\langle 0\left|T\{\Psi_\beta(x_1),\bar{\Psi}_\rho(x_2)\}\right|0\right\rangle &= \frac{1}{(2\pi)^3}\int d^3\boldsymbol{p}\frac{1}{2E}e^{i\boldsymbol{p}\bullet(\boldsymbol{x}_1-\boldsymbol{x}_2)}e^{-iE(t_1-t_2)} \\ &\theta(t_1-t_2)(m+\gamma^0E-\boldsymbol{\gamma}\bullet\boldsymbol{p})_{\beta\rho} \\ &+\frac{1}{(2\pi)^3}\int d^3\boldsymbol{p}\frac{1}{2E}e^{-i\boldsymbol{p}\bullet(\boldsymbol{x}_1-\boldsymbol{x}_2)}e^{+iE(t_1-t_2)}\theta(t_2-t_1)(m-\gamma^0E+\boldsymbol{\gamma}\bullet\boldsymbol{p})_{\beta\rho} \end{aligned} \tag{23.65}$$

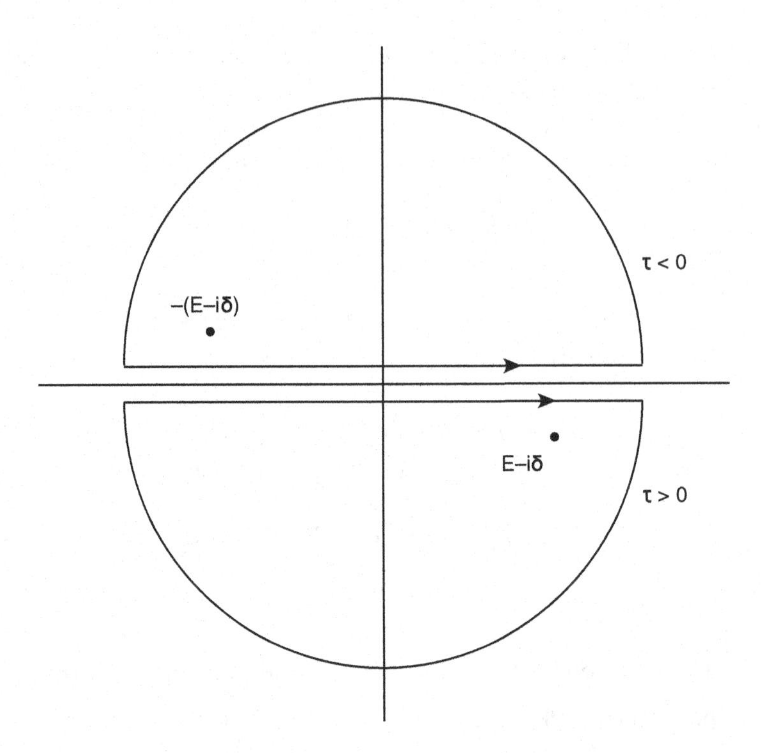

Figure 23.5 Contours for I.

In the last line on the right-hand side, we replace $\boldsymbol{p}$ by $-\boldsymbol{p}$, which does not change the sign of the integral. As a result, (23.65) becomes

$$\langle 0|T\{\Psi_\beta(x_1),\bar{\Psi}_\rho(x_2)\}|0\rangle = \frac{1}{(2\pi)^3}\int d^3\boldsymbol{p}e^{i\boldsymbol{p}\cdot(\boldsymbol{x}_1-\boldsymbol{x}_2)}$$
$$\left\{\frac{1}{2E}e^{-iE(t_1-t_2)}\theta(t_1-t_2)(m+\gamma^0E-\boldsymbol{\gamma}\cdot\boldsymbol{p})_{\beta\rho}\right. \tag{23.66}$$
$$\left.-\frac{1}{2(-E)}e^{-i(-E)(t_1-t_2)}\theta(t_2-t_1)[m-\gamma^0(-E)-\boldsymbol{\gamma}\cdot\boldsymbol{p}]_{\beta\rho}\right\}$$

The factor in curly brackets in the integrand of (23.66) is conveniently expressed by a contour integral. Temporarily, let the 4-vector p be $p=\omega,\boldsymbol{p}$, where ω is a variable quantity, and consider the integral

$$I=\int_{-\infty}^{\infty}\frac{(\not{p}+m)e^{-i\omega t}d\omega}{E^2-\omega^2-i\varepsilon}$$

where $E=+\sqrt{\boldsymbol{p}^2+m^2}$ is a fixed quantity, and ε is a real positive infinitesimal. We write $E^2-\omega^2-i\varepsilon=(E-i\delta-\omega)(E-i\delta+\omega)$, where δ is another real positive infinitesimal, and we use the contours shown in Figure 23.5.

The result for I is

$$\tau > 0: \quad I = \frac{i\pi}{E}\left(E\gamma^0 - \boldsymbol{\gamma}\cdot\mathbf{p} + m\right)e^{-iE\tau}$$
$$\tau < 0: \quad I = \frac{-i\pi}{-E}\left(-E\gamma^0 - \boldsymbol{\gamma}\cdot\mathbf{p} + m\right)e^{-i(-E)\tau}$$

Therefore, with $\tau = t_1 - t_2$, the factor in curly brackets in (23.66) is $I/2\pi i$; hence

$$\left\langle 0\left|T\left\{\Psi_\beta(x_1),\bar{\Psi}_\rho(x_2)\right\}\right|0\right\rangle = \frac{1}{(2\pi)^4}\int e^{-ip\cdot(x_1-x_2)}\frac{i\left(\not{p}+m\right)_{\beta\rho}}{\left(p^2+m^2+i\varepsilon\right)}d^4p \tag{23.67}$$

Now employing (23.67), (23.63), (23.60), and (23.55), we have

$$A_{fi} = -\frac{e^2}{(2\pi)^4}\sqrt{\frac{m}{E_+V}}\sqrt{\frac{m}{E_-V}}\sqrt{\frac{1}{2\omega_1V}}\sqrt{\frac{1}{2\omega_2V}}\int d^4p\int d^4x_1\int d^4x_2$$
$$\bar{v}\left(p_+s_+\right)\left[\not{\varepsilon}_1\frac{i\left(\not{p}+m\right)}{p^2-m^2}e^{-ix_1(p+p_+-k_1)}e^{ix_2(p-p_-+k_2)}\not{\varepsilon}_2 + \not{\varepsilon}_2\frac{i\left(\not{p}+m\right)}{p^2-m^2}e^{-ix_1(p+p_+-k_2)}e^{ix_2(p-p_-+k_1)}\not{\varepsilon}_1\right]u\left(p_-s_-\right) \tag{23.68}$$

The integrals over x_1 and x_2 yield

$$A_{fi} = -e^2(2\pi)^4\sqrt{\frac{m}{E_+V}}\sqrt{\frac{m}{E_-V}}\sqrt{\frac{1}{2\omega_1V}}\sqrt{\frac{1}{2\omega_2V}}\int d^4p$$
$$\left\{\bar{v}\left(p_+s_+\right)\not{\varepsilon}_1\frac{i\left(\not{p}+m\right)}{p^2-m^2}\not{\varepsilon}_2u\left(p_-s_-\right)\delta^4\left(p+p_+-k_1\right)\delta^4\left(p-p_-+k_2\right)\right. \tag{23.69}$$
$$\left.+\bar{v}\left(p_+s_+\right)\not{\varepsilon}_2\frac{i\left(\not{p}+m\right)}{p^2-m^2}\not{\varepsilon}_1u\left(p_-s_-\right)\delta^4\left(p+p_+-k_2\right)\delta^4\left(p-p_-+k_1\right)\right\}$$

The delta functions in (23.69) imply conservation of 4-momentum at each vertex. Finally, integration over p yields the lowest-order amplitude for e^+e^- annihilation to two photons

$$A_{fi} = (2\pi)^4\delta^4\left(p_+ + p_- - k_1 - k_2\right)\sqrt{\frac{m}{E_+V}}\sqrt{\frac{m}{E_-V}}\sqrt{\frac{1}{2\omega_1V}}\sqrt{\frac{1}{2\omega_2V}}M \tag{23.70}$$

with

$$M = e^2\left[\bar{v}\left(p_+s_+\right)i\not{\varepsilon}_1\frac{i\left(\not{p}+m\right)}{p^2-m^2}i\not{\varepsilon}_2u\left(p_-s_-\right)+\bar{v}\left(p_+s_+\right)i\not{\varepsilon}_2\frac{i\left(\not{p}+m\right)}{p'^2-m^2}i\not{\varepsilon}_1u\left(p_-s_-\right)\right] \tag{23.71}$$

and where $p = p_- - k_2 = -p_+ + k_1$ and $p' = p_- - k_1 = -p_+ + k_2$. M is represented by the Feynman diagrams in Figure 23.6.

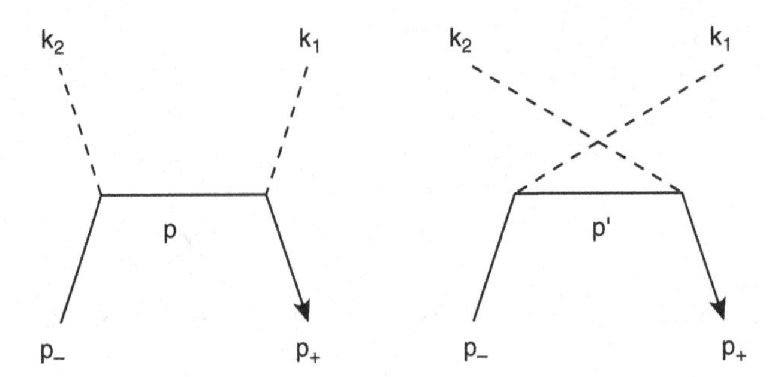

Figure 23.6 Feynman diagrams for e^+e^- annihilation in lowest order.

The foregoing calculation yields two new Feynman rules:

- For each internal spin-½ fermion line, there is a factor

$$i\frac{\not{p}+m}{p^2-m^2}$$

- Note that for an internal fermion line there is no distinction between fermion and antifermion.
- If two identical photons appear in the final state, the direct and exchange amplitudes have the same sign.

23.5 Summary of Feynman rules obtained so far for QED

As we have already suggested, the nth-order S-matrix amplitude A_{fi} for given initial and final states has the following general form:

$$A_{fi}^{(n)} = (2\pi)^4\delta(P_f - P_i)\prod_{\substack{\text{initial}\\\text{fermions}}}\sqrt{\frac{m}{E_iV}}\prod_{\substack{\text{final}\\\text{fermions}}}\sqrt{\frac{m}{E_fV}}\prod_{\substack{\text{initial}\\\text{photons}}}\sqrt{\frac{1}{2\omega_iV}}\prod_{\substack{\text{final}\\\text{photons}}}\sqrt{\frac{1}{2\omega_fV}}M \tag{23.72}$$

The Feynman amplitude M in (23.72) is the sum of Feynman amplitudes (with due regard for relative signs) from all topologically distinct diagrams of nth order. The individual M amplitudes are constructed according to the following rules:

1. For each incoming e^- line, there is a column spinor u_{ps}.
2. For each outgoing e^- line, there is a row spinor $\bar{u}_{p's'}$.
3. For each incoming e^+ line, there is a row spinor $\bar{v}_{ps}$.
4. For each outgoing e^+ line, there is a column spinor $v_{p's'}$.
5. For each incoming or outgoing photon, there is a factor ε_μ.
6. For each $e^-e^-\gamma$ vertex or $e^-e^+\gamma$ vertex, there is a factor $ie\gamma^\mu$.
7. For each $e^+e^+\gamma$ vertex, there is a factor $-ie\gamma^\mu$.
8. For each internal photon line, there is a factor $-ig_{\mu\nu}/q^2$, where q is the 4-momentum of the internal line.

9. For each internal electron line, there is a factor

$$i\frac{(\not{q}+m)}{q^2-m^2}$$

10. Four-momentum is conserved at each vertex. (We postpone to the following section a discussion of what happens in the case of internal loops.)
11. If two identical fermions appear in the final state, the direct and exchange amplitudes have opposite signs.
12. If two (identical) photons appear in the final state, the direct and exchange amplitudes have the same sign.

23.6 Survey of various QED processes in second order

In addition to electron-electron scattering and e^+e^- annihilation, which we discussed in Sections 23.3 and 23.4, respectively, the following second-order processes are of interest:

- Electron-positron (*Bhabha*) scattering (see Figure 23.7). Because an electron and a positron are distinguishable, we do not have an exchange diagram here. Instead, as shown in Figure 23.7b, there is a so-called annihilation diagram. The Feynman amplitude for Bhabha scattering is

$$M = e^2\left[\bar{u}_{1'}(i\gamma^\mu)u_1\frac{-ig_{\mu\nu}}{q^2}\bar{v}_2(-i\gamma^\nu)v_{2'}+\bar{v}_2(i\gamma^\mu)u_1\frac{-ig_{\mu\nu}}{q'^2}\bar{u}_{1'}(i\gamma^\nu)v_{2'}\right] \tag{23.73}$$

- $e^+e^- \to \mu^+\mu^-$ or $\tau^+\tau^-$ by single-photon exchange (see Figure 23.8). The muons μ^-,μ^+ and the tau leptons τ^-,τ^+ are close relatives of e^-,e^+ and interact with photons in the same way, except that in all formulas we must make the appropriate replacements of electron mass m_e by muon mass $(m_\mu = 106\ \text{MeV}/c^2)$ or tau mass $(m_\tau = 1{,}777\ \text{MeV}/c^2)$. The sole lowest-order (annihilation) diagram is Figure 23.8. We calculate the cross section in Section 23.9.
- Compton scattering (see Figure 23.9). Here a photon ($k\varepsilon$) collides with an electron (ps), resulting in a final state consisting of a photon (k', ε') and an electron ($p's'$). The Feynman amplitude is

$$M = e^2\left[\bar{u}_{p's'}(i\not{\varepsilon}')\frac{i(\not{q}+m)}{q^2-m^2}(i\not{\varepsilon})u_{ps}+\bar{u}_{p's'}(i\not{\varepsilon})\frac{i(\not{q}'+m)}{q'^2-m^2}(i\not{\varepsilon}')u_{ps}\right] \tag{23.74}$$

- Pair production by two photons (see Figure 23.10). Photons (k_1,α_1) and (k_2,α_2) convert to an electron-positron pair. The Feynman amplitude is

$$M = e^2\left[\bar{u}(p_-s_-)i\not{\varepsilon}_2\frac{i(\not{p}+m)}{p^2-m^2}i\not{\varepsilon}_1 v(p_+s_+)+\bar{u}(p_-s_-)i\not{\varepsilon}_1\frac{i(\not{p}'+m)}{p'^2-m^2}i\not{\varepsilon}_2 v(p_+s_+)\right] \tag{23.75}$$

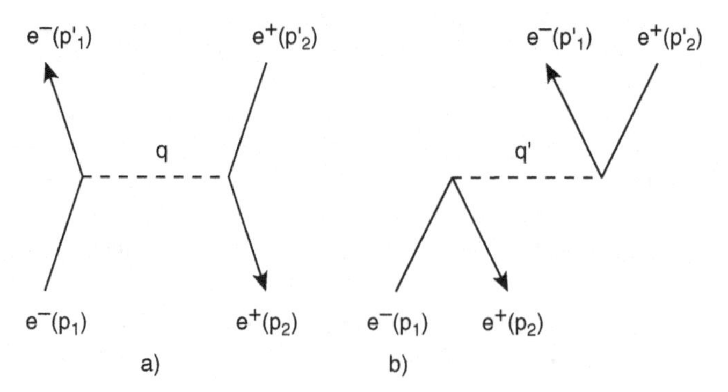

Figure 23.7 Electron-positron scattering in lowest order: (a) direct diagram; (b) annihilation diagram.

- Coulomb (Mott) scattering of an electron by a nucleus (see Figure 23.11). The lowest order for Mott scattering is actually second order (Figure 23.11a). However, if we ignore nuclear recoil, the nucleus generates a fixed Coulomb potential, and the amplitude reduces to first order. This is just Rutherford scattering, except for corrections due to the electron's relativistic motion, which were first analyzed by N. Mott in 1929. We calculate the Mott scattering cross section in Section 23.8.
- Bremsstrahlung (see Figures 23.12 and 23.13). Here an electron scatters from a nucleus and emits a photon. In reality, the lowest nonvanishing order is third, as seen in Figure 23.12. However, as in Mott scattering, it is often sufficient to ignore the nuclear recoil, in which case the nucleus merely acts as the source of a static Coulomb potential. Thus, in effect, we have the second-order diagrams of Figure 23.13.
- Pair production in a Coulomb field (see Figure 23.14). Here an energetic incoming photon interacts with a nucleus to generate an electron-positron pair. The diagrams are analogous to those of bremsstrahlung in Figure 23.13. For photons of very high energy, this is the most important mechanism for absorption of a beam of radiation in its passage through matter.
- Second-order self-energy diagrams (see Figure 23.15). Figure 23.15a and b represents the second-order self energy shifts due to mass and charge, respectively. For the internal loops shown here, conservation of 4-momentum at each vertex is not sufficient to fix the 4-momenta q and k in Figure 23.15a or the 4-momenta q' and p' in Figure 23.15b. Instead, it is necessary in each case to integrate over all 4-momenta, satisfying the conservation law at each vertex. In particular, it can be shown that the Feynman amplitude corresponding to Figure 23.15a is

$$M = -\frac{e^2}{(2\pi)^4}\int d^4k\left[\bar{u}_{ps}\gamma^\alpha \frac{-ig_{\alpha\beta}}{k^2}\frac{i(\not{p}-\not{k}+m)}{(p-k)^2-m^2}\gamma^\beta u_{ps}\right]$$

- which diverges for large k. There is also a divergent integral associated with the case in Figure 23.15b. These infinities are dealt with by renormalization. We have given an elementary discussion of that topic in Section 17.1, but a more detailed treatment is beyond the scope of this book.

Finally, the *disconnected* second-order diagram of Fig. 23.16 contributes nothing to observable transition probabilities, cross sections, or static energies.

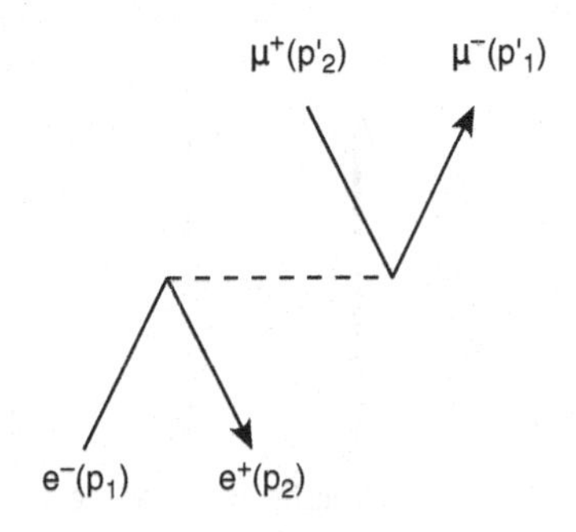

Figure 23.8 Annihilation diagram for $e^+e^- \to \mu^+\mu^-$.

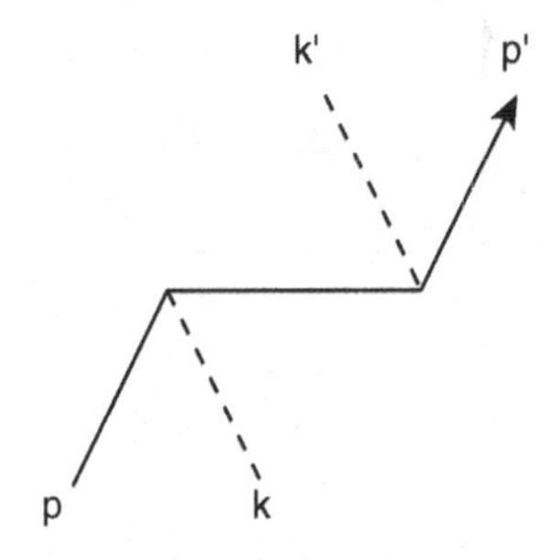

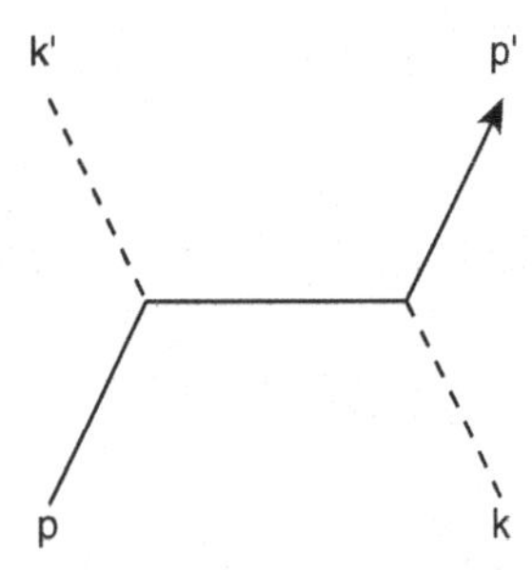

Figure 23.9 Lowest-order diagrams for Compton scattering.

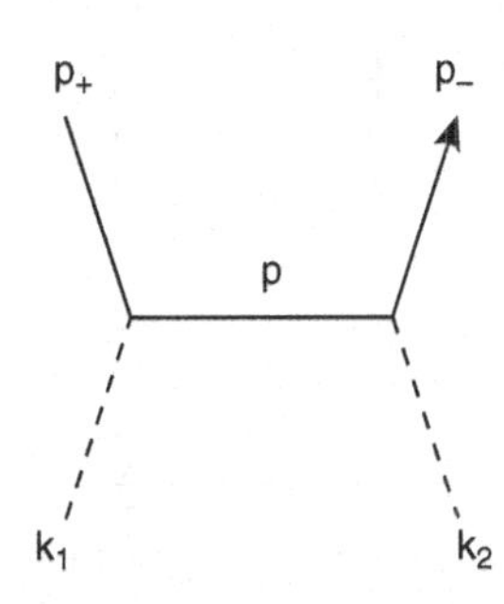

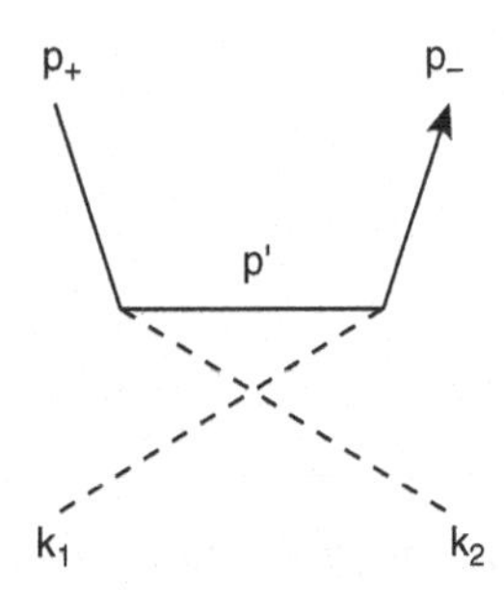

Figure 23.10 Lowest-order diagrams for pair production by two photons.

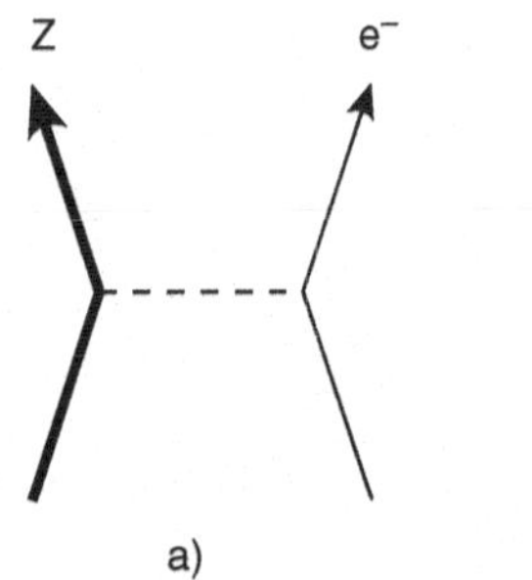

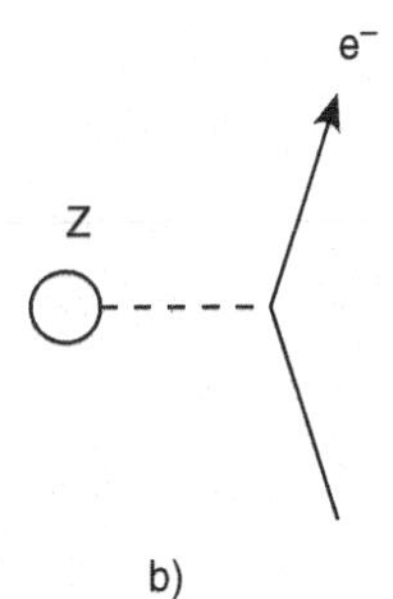

Figure 23.11 (a) Mott scattering of an electron by a nucleus (Z), second order. (b) Mott scattering with no nuclear recoil (first order). In this approximation, the nucleus generates a static Coulomb field.

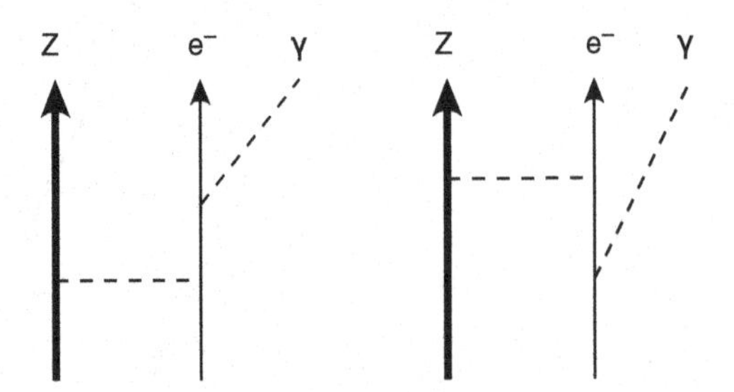

Figure 23.12 Lowest-order diagrams for bremsstrahlung. The heavy solid line represents a nucleus with atomic number Z.

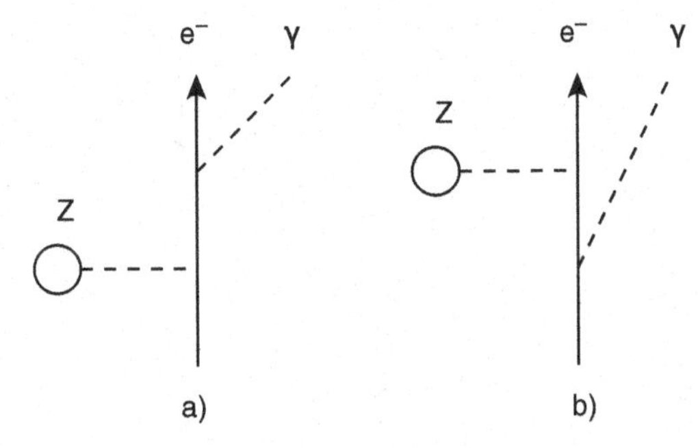

Figure 23.13 Second-order diagrams for bremsstrahlung where nuclear recoil is neglected.

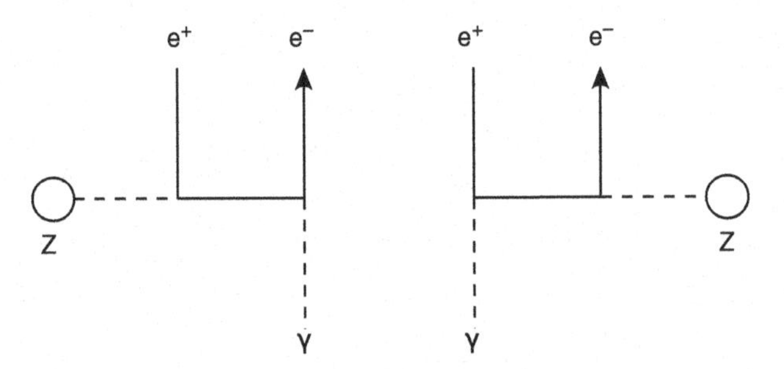

Figure 23.14 Second-order diagrams for pair production in a Coulomb field, where nuclear recoil is neglected.

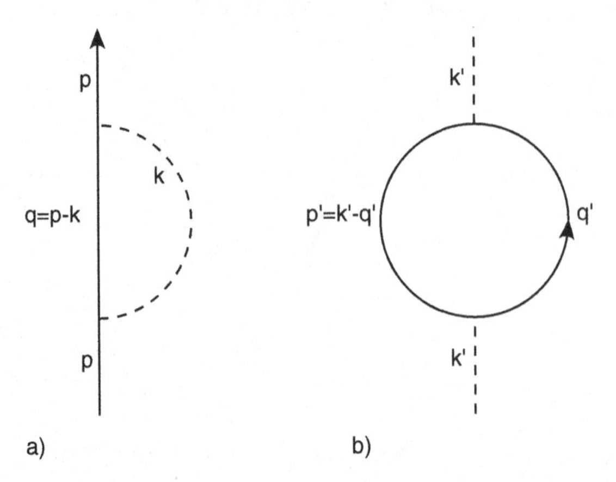

Figure 23.15 Second-order self energy diagrams.

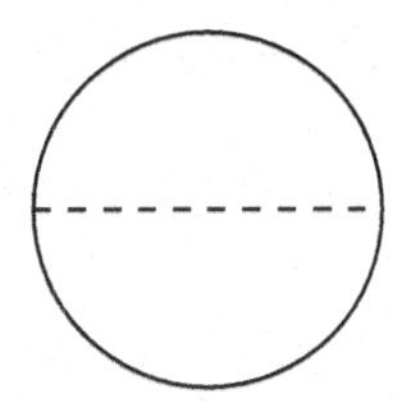

Figure 23.16 Disconnected second-order diagram.

23.7 Transition probabilities and cross sections

The transition probability corresponding to the amplitude of (23.72) is

$$P_{i\to f} = \left|A_{fi}\right|^2 = \left[(2\pi)^4\delta^4(P_f - P_i)\right]^2 \prod_{i,f}\frac{m}{EV}\prod_j\frac{1}{2\omega_j V}\left|M_{fi}\right|^2 \tag{23.76}$$

Now

$$\begin{aligned}\left[(2\pi)^4\delta^4(P_f - P_i)\right]^2 &= (2\pi)^4\delta^4(P_f - P_i)(2\pi)^4\delta^4(0)\\ &= (2\pi)^4\delta^4(P_f - P_i)VT\end{aligned}$$

where $T = t - t_0$. Thus the transition probability per unit time is

$$d\Gamma_0 = V(2\pi)^4\delta^4(P_f - P_i)\prod_{i,f}\frac{m}{EV}\prod_j\frac{1}{2\omega_j V}\left|M_{fi}\right|^2 \tag{23.77}$$

As in past discussions, we are usually interested in the transition probability per unit time to a group of final states, so we must multiply (23.77) by a phase-space factor

$$V\frac{d^3\mathbf{p}_f}{(2\pi)^3}$$

for each final fermion and each final photon. Thus the transition probability per unit time becomes

$$d\Gamma = V(2\pi)^4\delta^4(P_f - P_i)\left(\prod_{i,f}\frac{m}{EV}\prod_j\frac{1}{2\omega_j V}\right)\left[\prod_f V\frac{d^3\mathbf{p}_f}{(2\pi)^3}\right]\left|M_{fi}\right|^2 \tag{23.78}$$

Next, suppose that we are interested in the cross section for some scattering process, for example, $A + B \to C + D$. For definiteness, we assume here that all the initial and final particles are fermions. The modifications if some are photons are obvious. The differential cross section is

$$d\sigma = \frac{d\Gamma}{j} = \frac{V}{v}d\Gamma$$

where $j = v/V$ is the probability current density, with v the relative velocity of A and B. Thus

$$d\sigma = \frac{V^2(2\pi)^4\delta^4(p_C+p_D-p_A-p_B)}{v_{AB}}\frac{m_A}{E_AV}\frac{m_B}{E_BV}\frac{m_C}{E_CV}\frac{m_D}{E_DV}\frac{Vd^3\mathbf{p}_C}{(2\pi)^3}\frac{Vd^3\mathbf{p}_D}{(2\pi)^3}|M_{fi}|^2$$
$$= \frac{\delta^4(p_C+p_D-p_A-p_B)}{(2\pi)^2}\frac{1}{v_{AB}E_AE_B}\frac{m_Am_Bm_Cm_D}{E_CE_D}d^3\mathbf{p}_Cd^3\mathbf{p}_D|M_{fi}|^2 \tag{23.79}$$

The differential cross section $d\sigma$ is an area in the plane perpendicular to the line between A and B; hence it is invariant under Lorentz boosts along this line. It is convenient to go to the rest frame of B, where $v_AE_A = |\mathbf{p}_A|$ and $E_B = m_B$. In any other Lorentz frame obtained by a Lorentz boost along the line between A and B,

$$m_B|\mathbf{p}_A| \to \sqrt{(p_A\bullet p_B)^2 - m_A^2m_B^2}$$

where p_A and p_B are the 4-momenta of A,B. Thus (23.79) can be written as

$$d\sigma = \frac{\delta^4(p_C+p_D-p_A-p_B)}{(2\pi)^2\sqrt{(p_A\bullet p_B)^2 - m_A^2m_B^2}}\frac{m_Am_Bm_Cm_D}{E_CE_D}d^3\mathbf{p}_Cd^3\mathbf{p}_D|M_{fi}|^2 \tag{23.80}$$

Formulas (23.78) and (23.80) are very useful in practical calculations of physical processes, and we refer to them repeatedly in what follows.

23.8 Coulomb scattering of a relativistic electron: Traces of products of gamma matrices

We mentioned that scattering by a fixed Coulomb potential can be treated as a first-order process. The amplitude is

$$A_{fi}^{(1)} = ie\left\langle\phi_f\left|\int d^4x\bar{\Psi}(x)\gamma^\mu\Psi(x)A_\mu(x)\right|\phi_i\right\rangle \tag{23.81}$$

Let the initial and final states be $|\phi_i\rangle = |e^-(p_1,s_1)\rangle$ and $|\phi_f\rangle = |e^-(p_2,s_2)\rangle$, respectively. The vector potential in the present case is

$$A_\mu = \frac{Ze}{4\pi r},0,0,0$$

Thus

$$\bar{\Psi}\gamma^\mu\Psi A_\mu = \frac{Ze}{4\pi|\mathbf{x}|}\Psi^\dagger\Psi$$
$$= \frac{Ze}{4\pi|\mathbf{x}|}\frac{1}{V}\sum_{\substack{p,s\\p',s'}}\sqrt{\frac{m^2}{EE'}}\left(u_{ps}^\dagger b_{ps}^\dagger e^{ip\bullet x} + v_{ps}^\dagger d_{ps}e^{-ip\bullet x}\right)\left(u_{p's'}b_{p's'}e^{-ip'\bullet x} + v_{p's'}d_{p's'}^\dagger e^{ip'\bullet x}\right) \tag{23.82}$$

Inserting (23.82) in (23.81), we see that only a single term contributes; that is,

$$A_{fi} = i\frac{Z\alpha}{V}\sqrt{\frac{m^2}{E_1E_2}}\left(\bar{u}_{p_2s_2}\gamma^0u_{p_1s_1}\right)\int\frac{e^{i(p_2-p_1)\cdot x}}{|\boldsymbol{x}|}d^4x \tag{23.83}$$

Because $e^{i(p_2-p_1)\cdot x} = e^{i(\boldsymbol{p}_1-\boldsymbol{p}_2)\cdot\boldsymbol{x}}e^{-i(E_1-E_2)t}$, the integral on the right-hand side of (23.83) is

$$\begin{aligned}\int\frac{e^{i(p_2-p_1)\cdot x}}{|\boldsymbol{x}|}d^4x &= \int e^{-i(E_1-E_2)t}\,dt\int\frac{e^{i(\boldsymbol{p}_1-\boldsymbol{p}_2)\cdot\boldsymbol{x}}}{|\boldsymbol{x}|}d^3\boldsymbol{x}\\ &= 2\pi\delta(E_1-E_2)\int\frac{e^{i(\boldsymbol{p}_1-\boldsymbol{p}_2)\cdot\boldsymbol{x}}}{|\boldsymbol{x}|}d^3\boldsymbol{x}\end{aligned}$$

The integral over $\boldsymbol{x}$ has been discussed previously [see Section 18.9, equation (18.95)]. Let $\boldsymbol{q} = \boldsymbol{p}_1 - \boldsymbol{p}_2$. Because $E_1 = E_2$, we have $|\boldsymbol{p}_1| = |\boldsymbol{p}_2|$; hence $|\boldsymbol{q}| = 2|\boldsymbol{p}_1|\sin(\theta/2)$, where θ is the scattering angle. From (18.95), we have

$$\int\frac{e^{i(\boldsymbol{p}_1-\boldsymbol{p}_2)\cdot\boldsymbol{x}}}{|\boldsymbol{x}|}d^3\boldsymbol{x} = \frac{4\pi}{\boldsymbol{q}^2}$$

Therefore,

$$|A_{fi}|^2 = \frac{Z^2\alpha^2}{V^2}\frac{m^2}{E_1E_2}(2\pi)^2\left[\delta(E_2-E_1)\right]^2\frac{16\pi^2}{\boldsymbol{q}^4}\left|\bar{u}_{p_2s_2}\gamma^0u_{p_1s_1}\right|^2 \tag{23.84}$$

Now $\left[\delta(E_2-E_1)\right]^2 = \delta(E_2-E_1)\delta(0) = \delta(E_2-E_1)T/(2\pi)$. Thus the transition probability per unit time to a group of final states is

$$d\Gamma = \frac{2^5\pi^3Z^2\alpha^2}{V^2\boldsymbol{q}^4}\frac{m^2}{E_1E_2}\delta(E_1-E_2)\frac{V\boldsymbol{p}_2^2d|\boldsymbol{p}_2|d\Omega_2}{(2\pi)^3}|\bar{u}_2\gamma^0u_1|^2$$

and the differential cross section is

$$\begin{aligned}d\sigma &= \frac{d\Gamma}{\text{inc.flux}} = \frac{d\Gamma}{v_1/V} = \frac{VE_1}{|\boldsymbol{p}_1|}d\Gamma\\ &= \frac{4Z^2\alpha^2m^2}{\boldsymbol{q}^4}\delta(E_1-E_2)\frac{\boldsymbol{p}_2^2d|\boldsymbol{p}_2|d\Omega_2}{|\boldsymbol{p}_1|E_2}|\bar{u}_2\gamma^0u_1|^2\end{aligned} \tag{23.85}$$

Next, we evaluate $|M_0|^2 \equiv |\bar{u}_2\gamma^0u_1|^2$. It is useful to consider the more general form

$$M_0 = \bar{u}_2Fu_1$$

where F is an arbitrary 4×4 matrix. Then

$$\begin{aligned}|M_0|^2 = M_0M_0{}^* &= \bar{u}_2Fu_1u_1^\dagger F^\dagger\left(\bar{u}_2\right)^\dagger\\ &= \bar{u}_2Fu_1u_1^\dagger F^\dagger\left(u_2^\dagger\gamma^0\right)^\dagger\\ &= \bar{u}_2Fu_1\bar{u}_1\left(\gamma^0F^\dagger\gamma^0\right)u_2\end{aligned} \tag{23.86}$$

Here we replace $\gamma^0 F^\dagger \gamma^0$ by F in the last line because F can always be expressed as a sum of products of γ matrices, and $\gamma^0\gamma^{\mu\dagger}\gamma^0 = \gamma^\mu$. Equation (23.86) can be written

$$\begin{aligned} |M_0|^2 &= \bar{u}_{2\alpha} F_{\alpha\beta} (u_1\bar{u}_1)_{\beta\lambda} F_{\lambda\sigma} u_{2\sigma} \\ &= (u_2\bar{u}_2)_{\sigma\alpha} F_{\alpha\beta} (u_1\bar{u}_1)_{\beta\lambda} F_{\lambda\sigma} \\ &= tr\left[(u_2\bar{u}_2)(F)(u_1\bar{u}_1)(F)\right] \end{aligned} \tag{23.87}$$

where in the last line each of the factors in parentheses is a 4×4 matrix. First consider the matrices $u_1\bar{u}_1$ and $u_2\bar{u}_2$. We start with an electron at rest and with spin up along z. Then

$$u\bar{u} = \begin{pmatrix} 1 \\ 0 \\ 0 \\ 0 \end{pmatrix} (1 \;\; 0 \;\; 0 \;\; 0) = \begin{pmatrix} 1 & 0 & 0 & 0 \\ 0 & 0 & 0 & 0 \\ 0 & 0 & 0 & 0 \\ 0 & 0 & 0 & 0 \end{pmatrix}$$

This can be written as

$$u\bar{u} = \frac{1}{2}(I + \gamma^0)\frac{1}{2}(I + \gamma^0\Sigma_3) \tag{23.88}$$

because

$$\frac{1}{2}(I + \gamma^0) = \begin{pmatrix} 1 & 0 & 0 & 0 \\ 0 & 1 & 0 & 0 \\ 0 & 0 & 0 & 0 \\ 0 & 0 & 0 & 0 \end{pmatrix} \qquad \frac{1}{2}(I + \gamma^0\Sigma_3) = \begin{pmatrix} 1 & 0 & 0 & 0 \\ 0 & 0 & 0 & 0 \\ 0 & 0 & 0 & 0 \\ 0 & 0 & 0 & 1 \end{pmatrix}$$

Now $\Sigma_3 = \hat{\boldsymbol{s}}\boldsymbol{\cdot}\boldsymbol{\Sigma}$, where $\hat{\boldsymbol{s}}$ is a unit vector in the direction of the spin expectation value (in the present case, along $+z$). Thus, for an arbitrary spin direction, we can write (23.88) as

$$u\bar{u} = \frac{1}{2m}(mI + m\gamma^0)\frac{1}{2}(I + \gamma^0\hat{\boldsymbol{s}}\boldsymbol{\cdot}\boldsymbol{\Sigma}) \tag{23.89}$$

Now we make a Lorentz transformation to a frame in which the electron has nonzero linear momentum $\boldsymbol{p}$. In this case, $m\gamma^0 \to \not{p}$, but what about $\gamma^0\hat{\boldsymbol{s}}\boldsymbol{\cdot}\boldsymbol{\Sigma}$? Here we define a 4-vector s^μ that reduces to $0, \hat{s}$ in the rest frame. Hence $s\boldsymbol{\cdot}s = -1$ and $s\boldsymbol{\cdot}p = 0$, and because the scalar product of two 4-vectors is an invariant, the last two relations hold in any Lorentz frame. For the special case of a Lorentz boost along the z-axis that yields a particle momentum $\boldsymbol{p} = 0, 0, p_z = m\beta\gamma$, we have

$$a_{\mu\nu} = \begin{pmatrix} \gamma & 0 & 0 & \beta\gamma \\ 0 & 1 & 0 & 0 \\ 0 & 0 & 1 & 0 \\ \beta\gamma & 0 & 0 & \gamma \end{pmatrix} = \begin{pmatrix} E/m & 0 & 0 & |\boldsymbol{p}|/m \\ 0 & 1 & 0 & 0 \\ 0 & 0 & 1 & 0 \\ |\boldsymbol{p}|/m & 0 & 0 & E/m \end{pmatrix}$$

Thus

$$s = \begin{pmatrix} E/m & 0 & 0 & |\boldsymbol{p}|/m \\ 0 & 1 & 0 & 0 \\ 0 & 0 & 1 & 0 \\ |\boldsymbol{p}|/m & 0 & 0 & E/m \end{pmatrix} \begin{pmatrix} 0 \\ \hat{s}_x \\ \hat{s}_y \\ \hat{s}_z \end{pmatrix} = \begin{pmatrix} |\boldsymbol{p}|\hat{s}_z/m \\ \hat{s}_x \\ \hat{s}_y \\ \dfrac{\boldsymbol{E}}{m}\hat{s}_z \end{pmatrix}$$

In the general case where $\boldsymbol{p}$ is in an arbitrary direction, this yields

$$\boldsymbol{s} = \hat{\boldsymbol{s}} + \frac{\hat{\boldsymbol{s}}\cdot\boldsymbol{p}}{m(E+m)}\boldsymbol{p} \tag{23.90}$$

$$s^0 = \frac{\hat{\boldsymbol{s}}\cdot\boldsymbol{p}}{m} \tag{23.91}$$

Now we return to $\gamma^0\hat{\boldsymbol{s}}\cdot\boldsymbol{\Sigma}$. This quantity can be written as

$$\begin{aligned} \gamma^0\hat{\boldsymbol{s}}\cdot(\gamma^5\alpha) &= \gamma^0\gamma^5\hat{\boldsymbol{s}}\cdot\boldsymbol{\alpha} \\ &= \gamma^0\gamma^5\gamma^0\hat{\boldsymbol{s}}\cdot\boldsymbol{\gamma} \\ &= -\gamma^5\hat{\boldsymbol{s}}\cdot\boldsymbol{\gamma} \end{aligned}$$

In the frame where the particle is moving, this last quantity becomes $\gamma_5 \not{s}$. Consequently,

$$\Lambda^-_{ps} \equiv u_{ps}\bar{u}_{ps} = \frac{(m+\not{p})}{2m}\frac{(1+\gamma^5\not{s})}{2} \tag{23.92}$$

When we reverse the spin, all components of s change sign. Hence

$$\Lambda^-_{p-s} = \frac{(m+\not{p})}{2m}\frac{(1-\gamma^5\not{s})}{2} \tag{23.93}$$

If we sum over both signs of spin, we obtain

$$\Lambda^-_p = \frac{(m+\not{p})}{2m} \tag{23.94}$$

Similarly, for positron spinors, one can show that

$$\Lambda^+_{ps} \equiv v_{ps}\bar{v}_{ps} = \frac{(\not{p}-m)}{2m}\frac{(1+\gamma^5\not{s})}{2} \tag{23.95}$$

and

$$\Lambda^+_p = \left(\Lambda^+_{ps} + \Lambda^+_{p-s}\right) = \frac{(\not{p}-m)}{2m} \tag{23.96}$$

We now return to (23.87)

$$\begin{aligned} |M_0|^2 &= tr\left[(u_2\bar{u}_2)(F)(u_1\bar{u}_1)(\gamma^0 F^\dagger \gamma^0)\right] \\ &= tr\left[\Lambda_2^- F \Lambda_1^- (\gamma^0 F^\dagger \gamma^0)\right] \end{aligned} \tag{23.97}$$

which is the trace of a sum of products of gamma matrices. At this point we digress to prove some general theorems about such traces that are useful in calculations of many transition probabilities and cross sections.

1. The trace of a product of an odd number of gamma matrices is zero.

Proof: Because $(\gamma^5)^2 = I$, we can always write

$$\begin{aligned} tr(\gamma_\alpha \gamma_\beta \ldots \gamma_\sigma) &= tr\left((\gamma^5)^2 \gamma_\alpha \gamma_\beta \ldots \gamma_\sigma\right) \\ &= tr(\gamma^5 \gamma_\alpha \gamma_\beta \ldots \gamma_\sigma \gamma^5) \end{aligned}$$

where the last step follows because the trace of a product of matrices is unchanged when these matrices are cyclically permuted. Now move the γ^5 on the far right to the far left by permuting it with each gamma matrix in turn. For each such permutation, we obtain a factor of (–1); hence, for an odd number of gamma matrices, the trace is equal to its negative, and therefore it vanishes.

2. $$tr(\gamma^\mu \gamma^\nu) = \frac{1}{2} tr(\gamma^\mu \gamma^\nu + \gamma^\nu \gamma^\mu) = tr(g^{\mu\nu} I) = 4g^{\mu\nu} \tag{23.98}$$

3. A corollary of (23.98) is

$$tr(\not{a}\not{b}) = 4a \cdot b \tag{23.99}$$

4. $$tr(\gamma^\alpha \gamma^\beta \gamma^\rho \gamma^\sigma) = 4g^{\alpha\beta} g^{\rho\sigma} - 4g^{\alpha\rho} g^{\beta\sigma} + 4g^{\alpha\sigma} g^{\beta\rho} \tag{23.100}$$

Proof:

$$\begin{aligned} \gamma^\alpha \gamma^\beta \gamma^\rho \gamma^\sigma &= (2g^{\alpha\beta} - \gamma^\beta \gamma^\alpha)\gamma^\rho \gamma^\sigma = 2g^{\alpha\beta} \gamma^\rho \gamma^\sigma - \gamma^\beta \gamma^\alpha \gamma^\rho \gamma^\sigma \\ -\gamma^\beta \gamma^\alpha \gamma^\rho \gamma^\sigma &= -2g^{\alpha\rho} \gamma^\beta \gamma^\sigma + \gamma^\beta \gamma^\rho \gamma^\alpha \gamma^\sigma \\ \gamma^\beta \gamma^\rho \gamma^\alpha \gamma^\sigma &= 2g^{\alpha\sigma} \gamma^\beta \gamma^\rho - \gamma^\beta \gamma^\rho \gamma^\sigma \gamma^\alpha \end{aligned}$$

Therefore,

$$\gamma^\alpha \gamma^\beta \gamma^\rho \gamma^\sigma + \gamma^\beta \gamma^\rho \gamma^\sigma \gamma^\alpha = 2g^{\alpha\beta} \gamma^\rho \gamma^\sigma - 2g^{\alpha\rho} \gamma^\beta \gamma^\sigma + 2g^{\alpha\sigma} \gamma^\beta \gamma^\rho \tag{23.101}$$

We now take the trace of both sides of (23.101) and use (23.98) to obtain (23.100).

5. $$tr(\gamma^\alpha \gamma^\beta \gamma^\rho \gamma^\sigma \gamma^5) = -4i\varepsilon^{\alpha\beta\rho\sigma} \tag{23.102}$$

where $\varepsilon^{\alpha\beta\rho\sigma}$ is the completely antisymmetric unit 4-tensor.

Proof: If $\alpha\beta\rho\sigma = 0123$ in (23.102), we have $tr[-i(\gamma^5)^2] = -itr(I) = -4i$. Any even permutation of the numbers 0, 1, 2, 3 obviously yields the same result, and any odd permutation changes the sign. If any two of the indices $\alpha\beta\rho\sigma$ take the same value, this results in a trace of the general form $tr(\gamma^\mu\gamma^\nu\gamma^5)$ that vanishes. If all four of the indices are the same, we have $tr(\gamma^5)$, which also vanishes. If three of the indices are the same, we obtain a trace of an odd number of γ matrices, which also vanishes.

We now return to our calculation of the differential cross section for Coulomb scattering of a relativistic electron, which is given by (23.85); that is,

$$d\sigma = \frac{4Z^2\alpha^2 m^2}{q^4}\delta(E_1 - E_2)\frac{p_2^2 d|p_2|d\Omega_2}{|p_1|E_2}|\bar{u}_2\gamma^0 u_1|^2 \tag{23.85a}$$

If we average over initial spin polarizations (equivalent to the assumption of an unpolarized incoming electron beam) and sum over final polarizations (equivalent to the assumption that the detector is equally sensitive to both final polarization states), then

$$\begin{aligned}|M_0|^2 &= |\bar{u}_2\gamma^0 u_1|^2\\ \rightarrow \overline{|M_0|^2} &= \frac{1}{2}tr\left(\frac{\not{p}_2 + m}{2m}\gamma^0\frac{\not{p}_1 + m}{2m}\gamma^0\right)\end{aligned} \tag{23.103}$$

Because the trace of a product of an odd number of gamma matrices is zero, this expression reduces to

$$\overline{|M_0|^2} = \frac{1}{8m^2}\left\{tr\left[(m\gamma^0)^2\right] + tr\left(\not{p}_2\gamma^0\not{p}_1\gamma^0\right)\right\} \tag{23.104}$$

Now $tr\left[(m\gamma^0)^2\right] = m^2 tr(I) = 4m^2$, and from (23.100),

$$\begin{aligned}tr\left(\not{p}_2\gamma^0\not{p}_1\gamma^0\right) &= 4p_2^0 p_1^0 - 4p_2\cdot p_1 + 4p_2^0 p_1^0\\ &= 8E_1E_2 - 4p_1\cdot p_2\\ &= 4(\mathbf{p}_1\cdot\mathbf{p}_2 + E_1E_2)\end{aligned}$$

Hence

$$\overline{|M_0|^2} = \frac{\mathbf{p}_1\cdot\mathbf{p}_2 + E_1E_2 + m^2}{2m^2} \tag{23.105}$$

We insert (23.105) into (23.85), write $\mathbf{p}_2^2 d|\mathbf{p}_2| = E_2|\mathbf{p}_2|d|E_2|$, and integrate over the delta function to obtain

$$d\sigma = \frac{2Z^2\alpha^2}{q^4}(\mathbf{p}_1\cdot\mathbf{p}_2 + E_1E_2 + m^2)d\Omega \tag{23.106}$$

Because $E_1 = E_2$, we have $|\mathbf{p}_1| = |\mathbf{p}_2|$; thus

$$p_1 \cdot p_2 + E^2 + m^2 = |\boldsymbol{p}|^2 \cos\theta + m^2 + E^2$$
$$= |\boldsymbol{p}|^2 \left(1 - 2\sin^2\frac{\theta}{2}\right) + m^2 + E^2$$
$$= 2E^2\left(1 - v^2 \sin^2\frac{\theta}{2}\right)$$

where θ is the scattering angle, and $v = |\boldsymbol{p}|/E$. Thus (23.106) becomes

$$d\sigma = \frac{4Z^2\alpha^2}{\boldsymbol{q}^4} E^2\left(1 - v^2\sin^2\frac{\theta}{2}\right) d\Omega$$

or, finally,

$$d\sigma = \frac{Z^2\alpha^2}{4\boldsymbol{p}^4 \sin^4\frac{\theta}{2}} E^2\left(1 - v^2\sin^2\frac{\theta}{2}\right) d\Omega \tag{23.107}$$

This formula, which we derived using the first-order approximation in the S-matrix expansion (first-order Born approximation), is called the *Mott scattering formula* for a relativistic electron in a Coulomb potential. It is a modification of the nonrelativistic Rutherford formula that we obtained earlier by two methods:

1. Exact solution of the nonrelativistic Coulomb scattering problem [see Section 18.5, equation (18.66)].
2. First Born approximation solution to the nonrelativistic Coulomb scattering problem [see Section 18.9, equation (18.97)]. That result is

$$d\sigma = \frac{Z^2\alpha^2}{4\boldsymbol{p}^4 \sin^4\frac{\theta}{2}} m^2 d\Omega \tag{23.108}$$

It is easy to see that (23.108) and (23.107) become identical in the nonrelativistic limit, where $v \to 0$ and $E \to m$.

One can solve the Dirac equation exactly to find the wave function of an unbound relativistic electron or positron in the Coulomb potential. [See, e.g., Bethe (1957, pp. 71–76)]. The formula for $d\sigma$ is complicated but can be simplified by expanding in powers of $Z\alpha$. For $Z\alpha \ll 1$, we ignore all terms in this expansion beyond the first two and thus obtain

$$\frac{d\sigma}{d\Omega} = \frac{Z^2\alpha^2 E^2}{4\boldsymbol{p}^4 \sin^4\frac{\theta}{2}}\left[\left(1 - v^2\sin^2\frac{\theta}{2}\right) \pm \pi Z\alpha \sin\frac{\theta}{2}\left(1 - \sin\frac{\theta}{2}\right)\right] \tag{23.109}$$

for scattering from a positively charged nucleus, where the $\pm$ sign applies for an electron (positron). The leading term is identical to the right-hand side of (23.107). Because of the $\pm$ sign in the second term, the electron and positron cross sections are not identical.

23.9 Calculation of the cross section for $e^+e^- \to \mu^+\mu^-$

We give another illustration of calculation methods by working out the cross section for the reaction $e^+e^- \to \mu^+\mu^-$, which is frequently used in high-energy storage-ring experiments. The lowest-order Feynman diagram is given in Figure 23.8. In storage rings, the e^+ and e^- usually (but not always) have the same energies $E \gg m_e$ and equal and opposite linear momenta; that is,

$$\mathbf{p}_1 = \mathbf{p}(e^-)$$
$$\mathbf{p}_2 = \mathbf{p}(e^+) = -\mathbf{p}_1$$

The amplitude is

$$A_{fi} = \sqrt{\frac{m_e^2 m_\mu^2}{V^4 E_1 E_2 E_1' E_2'}} (2\pi)^4 \delta^4 (p_1 + p_2 - p_1' - p_2') M \tag{23.110}$$

where, except for an unimportant phase factor,

$$M = e^2 \left(\bar{v}_2 \gamma^\lambda u_1 \frac{g_{\lambda\sigma}}{q^2} \bar{u}_{1'} \gamma^\sigma v_{2'} \right) \tag{23.111}$$

To calculate the cross section, we employ formula (23.80):

$$d\sigma = \frac{\delta^4 (p_1 + p_2 - p_1' - p_2')}{(2\pi)^2 \sqrt{(p_1 \cdot p_2)^2 - m_e^4}} \frac{m_e^2 m_\mu^2}{E_1' E_2'} |M|^2 \, d^3 p_1' d^3 p_2' \tag{23.80a}$$

In many practical experiments, the incoming electron and positron beams are unpolarized, and the detectors do not discriminate between different muon spin states. To describe such experiments, we average over initial spins and sum over final spins. Thus

$$|M|^2 \to \overline{|M|^2} = \frac{e^4}{q^4} g_{\lambda\sigma} g_{\nu\rho} tr\left[\frac{(m_e - \not{p}_2)}{4m_e} \gamma^\lambda \frac{(m_e + \not{p}_1)}{4m_e} \gamma^\nu \right] \cdot tr\left[\frac{(m_\mu + \not{p}_{1'})}{2m_\mu} \gamma^\sigma \frac{(m_\mu - \not{p}_{2'})}{2m_\mu} \gamma^\rho \right] \tag{23.112}$$

where, as usual, summation over repeated indices is implied. Now consider the traces

$$T^{\lambda\nu}(e) = tr\left[\frac{(m_e - \not{p}_2)}{4m_e} \gamma^\lambda \frac{(m_e + \not{p}_1)}{4m_e} \gamma^\nu \right] \tag{23.113}$$

and

$$T^{\sigma\rho}(\mu) = tr\left[\frac{(m_\mu + \not{p}_{1'})}{2m_\mu} \gamma^\sigma \frac{(m_\mu - \not{p}_{2'})}{2m_\mu} \gamma^\rho \right] \tag{23.114}$$

Assuming that the electron and positron kinetic energies are very large compared with their rest energies, we ignore both m_e in the numerator of (23.113). Then, employing (23.100), we obtain

$$T^{\lambda\nu}(e) = \frac{\left(p_1 \cdot p_2 g^{\lambda\nu} - p_2^{\lambda} p_1^{\nu} - p_2^{\nu} p_1^{\lambda}\right)}{4m_e^2} \tag{23.115}$$

However, we cannot ignore either m_μ in the numerator of (23.114). Thus $T^{\sigma\rho}(\mu)$ is

$$T^{\mu}_{\lambda\sigma} = \frac{\left(m_\mu^2 g^{\sigma\rho} - p_1'^{\sigma} p_2'^{\rho} - p_1'^{\rho} p'^{\nu}_2 + p_1' \cdot p_2' g^{\sigma\rho}\right)}{m_\mu^2} \tag{23.116}$$

Multiplying (23.115) by (23.116) and carrying out the algebra, we find

$$\overline{|M|^2} = \frac{8\pi^2\alpha^2}{q^4}\frac{1}{m_e^2 m_\mu^2}\left(p_1 \cdot p_{2'} p_2 \cdot p_{1'} + p_1 \cdot p_{1'} p_2 \cdot p_{2'} + m_\mu^2 p_1 \cdot p_2\right) \tag{23.117}$$

We now insert (23.117) into (23.80) and integrate over $\boldsymbol{p}_{2'}$. Because in the CM frame $\boldsymbol{p}_1 = \boldsymbol{p}_2$, the integral

$$\int \delta^3(\boldsymbol{p}_1 + \boldsymbol{p}_2 - \boldsymbol{p}_{1'} - \boldsymbol{p}_{2'}) d^3\boldsymbol{p}_{2'} = 1$$

yields $\boldsymbol{p}_{1'} = -\boldsymbol{p}_{2'}$; hence $E_1' = E_2$. Ignoring m_e^4 in the square root in the denominator of (23.80), we obtain

$$d\sigma = \frac{\delta(2E - 2E_1')}{(2\pi)^2 |p_1 \cdot p_2|}\frac{8\pi^2\alpha^2}{q^4 E_1'^2}\left(p_1 \cdot p_{2'} p_2 \cdot p_{1'} + p_1 \cdot p_{1'} p_2 \cdot p_{2'} - m_\mu^2 p_1 \cdot p_2\right)|\boldsymbol{p}_1'| E_1' dE_1' d\Omega_1'$$

Next, we integrate over E_1' to obtain

$$d\sigma = \frac{\alpha^2}{q^4 E}\frac{|\boldsymbol{p}_1|}{|p_1 \cdot p_2|}\left[p_1 \cdot p_{2'} p_2 \cdot p_{1'} + p_1 \cdot p_{1'} p_2 \cdot p_{2'} - m_\mu^2 p_1 \cdot p_2\right] d\Omega_1' \tag{23.118}$$

where now $E_1' = E_2' = E_1 = E_2 = E$. Thus

$$\begin{aligned} p_1 \cdot p_2 &= E^2 + |\boldsymbol{p}_1|^{22} \approx 2E^2 \\ q^2 &= (p_1 + p_2)^2 \simeq 2 p_1 \cdot p_2 = 4E^2 \\ p_1 \cdot p_2' &= p_1' \cdot p_2 = E^2 + \boldsymbol{p}_1 \cdot \boldsymbol{p}_1' \\ p_1 \cdot p_1' &= p_2' \cdot p_2 = E^2 - \boldsymbol{p}_1 \cdot \boldsymbol{p}_1' \end{aligned} \tag{23.119}$$

Let the scattering angle between $\boldsymbol{p}_1'$ and $\boldsymbol{p}_1$ be θ. Inserting the relations (23.119) into (23.118), making use of $|\boldsymbol{p}_1| \approx E$ and $|\boldsymbol{p}_1'| = \beta E$, where β is the muon velocity, and integrating over the solid angle with

$$\int \cos^2\theta d\Omega_1' = \frac{4\pi}{3}$$

we arrive at

$$\sigma = \frac{\pi\alpha^2\beta}{6E^2}(3-\beta^2) \tag{23.120}$$

In the limit where the muons become relativistic, $\beta \to 1$ and

$$\lim_{\beta\to 1}\sigma = \frac{\pi\alpha^2}{3E^2} = \frac{4\pi\alpha^2}{3s} \tag{23.121}$$

where $s = 4E^2$ is the square of the total energy available from e^+ and e^- in the CM frame. Of course, for symmetric colliding beams, the CM frame and the lab frame are identical.

23.10 Further discussion of second-order QED processes

We conclude this chapter with a brief summary of the cross sections for several of the second-order processes. The Feynman amplitude for *Møller* scattering is given in (23.52). In the ultrarelativistic limit, the differential cross section is derived from that amplitude and given by the following formula, expressed in terms of the center-of-mass energy E and scattering angle θ and where we average over initial polarizations and sum over final polarizations:

$$\frac{d\sigma}{d\Omega} = \frac{\alpha^2}{8E^2}\left(\frac{1+\cos^4\frac{\theta}{2}}{\sin^4\frac{\theta}{2}} + \frac{1+\sin^4\frac{\theta}{2}}{\cos^4\frac{\theta}{2}} + \frac{2}{\sin^2\frac{\theta}{2}\cos^2\frac{\theta}{2}}\right) \tag{23.122}$$

The first and second terms on the right-hand side of (23.122) arise from the direct and exchange terms in (23.52), respectively, whereas the third term is due to interference between those terms.

The Feynman amplitude for Bhabha scattering is given in (23.73). The differential cross section in the ultrarelativistic limit is

$$\frac{d\sigma}{d\Omega} = \frac{\alpha^2}{8E^2}\left[\frac{1+\cos^4\frac{\theta}{2}}{\sin^4\frac{\theta}{2}} - 2\frac{\cos^4\frac{\theta}{2}}{\sin^2\frac{\theta}{2}} + \frac{(1+\cos^2\theta)}{2}\right] \tag{23.123}$$

The Feynman amplitude for Compton scattering in lowest nonvanishing order is given in (23.74). Usually we are interested in the scattering of a beam of photons by matter, so we describe Compton scattering in the laboratory frame, where the initial electron is at rest. Here conservation of energy and momentum yield Compton's kinematic formula relating the energies ω, ω' of the initial and final photon to the scattering angle θ of the final photon; that is,

$$\omega' = \frac{\omega}{1+\frac{\omega}{m_e}(1-\cos\theta)} \tag{23.124}$$

The differential cross section is obtained by a straightforward but tedious calculation from (23.74) and is given by the famous Klein-Nishina formula. Averaged over initial spins and summed over final spins, that formula is

$$\frac{d\sigma}{d\Omega} = \frac{\alpha^2}{4m_e^2}\left(\frac{\omega'}{\omega}\right)^2\left[\frac{\omega'}{\omega}+\frac{\omega}{\omega'}+4\left(\hat{\varepsilon}'\boldsymbol{\cdot}\hat{\varepsilon}\right)^2-2\right] \tag{23.125}$$

where $\hat{\varepsilon}$ and $\hat{\varepsilon}'$ are the polarizations of the incident and scattered photons, respectively. In the limit of low-incident photon energy, (23.124) reveals that $\omega'/\omega \to 1$, and thus (23.125) reduces to the Thomson scattering cross section; that is,

$$\left.\frac{d\sigma}{d\Omega}\right|_{\text{Thomson}} = \frac{\alpha^2}{m_e^2}\left(\hat{\varepsilon}\boldsymbol{\cdot}\hat{\varepsilon}'\right)^2 \tag{23.126}$$

Problems for Chapter 23

23.1. Suppose that the Coulomb potential transformed relativistically like a scalar rather than as the zeroth component of a 4-vector. Show that the shape and energy dependence of the differential scattering cross section for Mott scattering would change at high energies, although there would be no modification in the nonrelativistic limit.

23.2. An electron initially moves along the z-axis and has helicity $h = +1$. It undergoes Coulomb scattering into a final state where it moves along z' (which makes an angle θ with respect to z). What is the probability that the electron has positive helicity along z'? (There are two ways to solve this problem: one is rather complicated, and the other is very easy.)

23.3. This problem concerns several important features of charge-conjugation symmetry. Before we pose the problem, we note the following points:

(a) Only a neutral particle or a set of particles with no net charge (not merely electric charge but also baryonic, leptonic, and so on charge) can be in an eigenstate of the charge conjugation operator C because that operator reverses the signs of all charges, electric and otherwise.

(b) If we assume that the electromagnetic interaction is charge-conjugation invariant, then because its Lagrangian density is $\mathcal{L} = e\bar{\Psi}\gamma^\mu\Psi A_\mu$, if $e\bar{\Psi}\gamma^\mu\Psi$ changes sign under charge conjugation, we must also have $CA_\mu C^{-1} = -A_\mu$. Prove that $e\bar{\Psi}\gamma^\mu\Psi$ is indeed odd under charge conjugation.

(c) As we know, A_μ is a linear function of the destruction and creation operators $a_{k\alpha}$ and $a_{k\alpha}^\dagger$; therefore, because $CA_\mu C^{-1} = -A_\mu$, we must have

$$Ca_{k\alpha}C^{-1} = -a_{k\alpha} \qquad Ca_{k\alpha}^\dagger C^{-1} = -a_{k\alpha}^\dagger$$

A photon number eigenstate $|n_{k\alpha}\rangle$ is created by application one or more times of the operator $a_{k\alpha}^\dagger$ to the vacuum state $|0\rangle$. Let us agree to define the vacuum as an eigenstate of C with eigenvalue $c = +1$. Then the charge-conjugation eigenvalue of a state with n photons is $c = (-1)^n$.

(d) Although an electron and a positron are not identical particles, we can generalize the antisymmetrization principle by thinking of them as identical particles in different charge states. This intuitive idea is made more rigorous by a straightforward analysis based on the anticommutation rules for the creation operators $b^\dagger$ and $d^\dagger$. For the purposes of this problem, we shall simply assume that this generalization of the antisymmetrization principle is valid. Then the wave function describing an electron and a positron must be antisymmetric under interchange of spatial coordinates, spins, *and* charges.

Use statements (c) and/or (d) to show that if the electromagnetic interaction is charge-conjugation invariant,

(e) Any nonzero amplitude in quantum electrodynamics, to any order of perturbation theory, that is represented by a Feynman diagram with no external fermion or antifermion lines must have an even number of external photon lines. (This is called *Furry's theorem.*)

23.4. It is intuitively obvious that the first-order amplitude for scattering of a positron by the Coulomb field of a nucleus with charge Ze is equal in magnitude but opposite in sign to the amplitude for scattering of an electron with the same spin and momentum from the same nucleus.

(a) Prove this statement.

(b) From considerations of charge-conjugation invariance, show that the interference terms between the first- and second-order amplitudes are *also* opposite in sign for e^- and e^+ scattering by the same nucleus. Thus e^- and e^+ Coulomb scattering cross sections differ significantly for scattering from a heavy nucleus.

23.5. Employing the Feynman rules we have learned in this chapter, set up the Feynman amplitude for bremsstrahlung in the limit where nuclear recoil is neglected, as illustrated in Figure 23.13a and b.

23.6. Starting from the Feynman amplitude for two-photon annihilation of an electron and positron given by (23.71), one can calculate the total cross section. In the limit where the electron and positron have very low relative velocity β, that cross section, averaged over electron and positron spins, is

$$\sigma = \frac{\pi r_0^2}{\beta}$$

where $r_0 = \alpha^2 a_0$, and a_0 is the Bohr radius. Use this formula for σ to calculate the mean lifetime of the ${}^1S_0$ state of positronium, and check your answer against the observed value

$$\tau\left({}^1S_0\right) = 1.25\times10^{-10}\ \text{s}$$

24 The Quantum Mechanics of Weak Interactions

24.1 The Four interactions: Fundamental fermions and bosons

At the present state of knowledge $\begin{pmatrix} \nu_e \\ e^- \end{pmatrix}$, $\begin{pmatrix} \nu_\mu \\ \mu^- \end{pmatrix}$, and $\begin{pmatrix} \nu_\tau \\ \tau^- \end{pmatrix}$, four distinct physical interactions are known: gravitation, electromagnetism, and the strong and weak interactions. Two classes of objects, presently recognized as fundamental, participate in these interactions: the fundamental fermions and the fundamental bosons. The fermions, in turn, consist of two classes: the leptons and the quarks, both with spin-½, whereas the bosons include the photon (called a *vector boson* because it has spin of unity); the intermediate vector bosons W^+, W^-, and Z; the eight gluons (also vector bosons); and the scalar Higgs boson. In these introductory paragraphs we give a brief review of their most important properties.

24.1.1 The leptons

There are three known generations of leptons

$$\begin{pmatrix} \nu_e \\ e^- \end{pmatrix}, \begin{pmatrix} \nu_\mu \\ \mu^- \end{pmatrix}, \begin{pmatrix} \nu_\tau \\ \tau^- \end{pmatrix} \tag{24.1}$$

and for each lepton listed in (24.1) there is a corresponding antiparticle $\bar{\nu}_e, \bar{\nu}_\mu, \bar{\nu}_\tau$ and e^+, μ^+, τ^+. The neutrinos experience only weak and gravitational interactions and have zero electric charge. Earlier (Chapter 5) we noted that since 1998, convincing experimental evidence has been found for neutrino oscillations and mixing. This implies that each of the neutrino weak interaction eigenstates $|\nu_e\rangle, |\nu_\mu\rangle$, and $|\nu_\tau\rangle$ is a linear combination of neutrino mass eigenstates $|\nu_1\rangle, |\nu_2\rangle$, and $|\nu_3\rangle$, and the latter states are associated with distinct (albeit very small) nonzero masses. Within present experimental accuracy, all neutrinos are created left-handed (spin opposed to momentum), whereas antineutrinos are created right-handed (spin parallel to momentum). This striking fact is a manifestation of parity violation (discussed in more detail later).

The electron, muon, and tau lepton all have electric charge $-e$ and experience electromagnetic as well as weak and gravitational interactions. According to all experimental evidence, the muon and tau lepton are just heavy copies of the electron; that is,

$$m_e = 0.511 \text{ MeV}/c^2 \qquad m_\mu = 106 \text{ MeV}/c^2 \qquad m_\tau = 1{,}777 \text{ MeV}/c^2$$

In Chapter 5 we defined the generational lepton numbers L_e, L_μ, and L_τ, and we noted that all experimental evidence prior to the discovery of neutrino mixing and oscillations was consistent with conservation of each of these numbers in all known interactions. For example, generational lepton number conservation would be violated if the following decay occurred:

$$\mu^- \to e^- \gamma \tag{24.2}$$

In fact, this decay has never been observed, and the branching ratio for it is less than 4×10^{-11}, although it is allowed by conservation of energy, angular momentum, and statistics. This particular decay illustrates a situation in which generational lepton number would be violated, but overall lepton number $L = L_e + L_\mu + L_\tau$ would be conserved. We can also contemplate processes in which overall lepton number conservation is violated. An example is neutrinoless double beta decay $(\beta\beta 0)$; that is,

$$(Z,N) \to (Z+2, N-2) + e^- + e^- \tag{24.3}$$

which has never been observed but is actively searched for in present-day experiments. We know of three lepton generations, but could there be more? Experiments at the large electron-positron collider (LEP) at CERN showed that this is impossible unless the leptons in the additional generation or generations are very massive $(m > 45\ \text{GeV}/c^2)$. No convincing explanation has yet been given for why there are just three lepton generations, and no convincing calculation of the mass of any of the leptons has yet been presented.

24.1.2 The quarks

There are also three generations of quarks

$$\begin{pmatrix} u \\ d \end{pmatrix} \quad \begin{pmatrix} c \\ s \end{pmatrix} \quad \begin{pmatrix} t \\ b \end{pmatrix} \tag{24.4}$$

and corresponding to each there is an antiquark: $\bar{u}, \bar{d}, \bar{c}, \bar{s}, \bar{t}$, and $\bar{b}$. Quarks experience strong as well as electromagnetic, weak, and gravitational interactions. The electric charges of u, c, and t are each $2e/3$, whereas those of d, s, and b are each $-e/3$. We also define the *baryon number* B; that is, $B = +1/3$ for each quark, and $B = -1/3$ for each antiquark. Baryon number is conserved in all known interactions, but this conservation law might break down at some level.

Quarks possess color charge, a degree of freedom of fundamental significance for the strong interaction that does not exist for the leptons. Each quark can exist in three distinct internal states, arbitrarily called the *color* states red (R), green (G), and blue (B), whereas each antiquark can exist in three distinct anticolor states $\bar{R}$, $\bar{G}$, and $\bar{B}$. These are new degrees of freedom in addition to charge, mass, and spin. The color properties of quarks are described by the symmetry group $SU(3)$, which is the group of all 3×3 unitary matrices with determinant equal to plus unity. The physical theory in which this group is employed (the modern theory of strong interactions) is called *quantum chromodynamics* (QCD). It was invented in 1974, has had impressive success, and is generally considered to be the correct theory of strong interactions, at least within the energy range accessible to experiment at present.

The various strongly interacting particles (hadrons) fall into two broad classes: the baryons (p, n, Λ, ...), which have half-integral spin, and the mesons $(\pi^\pm, \pi^0, K^\pm, \ldots)$, which have integral spin. In earlier decades it was thought that at least some of the baryons and mesons are fundamental objects. However, starting in the early 1960s, it became evident that hadrons are composite objects composed of quarks (and/or antiquarks). In particular, a meson consists of a "valence" quark-antiquark pair (together with an "ocean" of virtual quark-antiquark pairs). The valence quark composition of several mesons is as follows:

$$
\begin{aligned}
\pi^+ &= u\bar{d} \\
\pi^- &= \bar{u}d \\
\pi^0 &= \frac{1}{\sqrt{2}}\left(u\bar{u} - d\bar{d}\right) \\
K^+ &= u\bar{s} \\
K^- &= \bar{u}s \\
K^0 &= d\bar{s} \\
\overline{K^0} &= \bar{d}s
\end{aligned}
$$

For all the mesons just listed, the quark and antiquark are in an ${}^1S_0$ state (in an approximate nonrelativistic description). Such mesons have zero net quark-antiquark spin, zero orbital angular momentum for quark-antiquark relative motion, and thus zero total angular momentum. They thus resemble the positronium atom in its ground state. The Dirac theory tells us that the intrinsic parity of a fermion-antifermion pair is negative because $P\psi(x) = \gamma^0\psi(-x)$, and positive-energy antifermion spinors are equivalent to negative-energy fermion spinors. Thus all the mesons just mentioned have odd intrinsic parity ($J^P = 0^-$). The $\phi = s\bar{s}$, $J/\psi = c\bar{c}$, and $\Upsilon = b\bar{b}$ are examples of mesons where the quark-antiquark pair forms an ${}^3S_1$ state.

A baryon consists of three valence quarks and an ocean of virtual quark-antiquark pairs. For example, the valence quark content of several baryons is as follows:

$$
\begin{aligned}
p &= uud \\
n &= udd \\
\Lambda^0 &= uds \\
\Sigma^+ &= uus \\
\Sigma^- &= dds \\
\Sigma^0 &= uds \\
\Xi^0 &= uss \\
\Xi^- &= dss
\end{aligned}
$$

Each of the baryons just listed has spin-½ and belongs to a grouping called the *metastable baryon octet*, which is a multiplet of the approximate symmetry group $SU(3)_F$ (F for flavor). Flavor $SU(3)$ is distinct from the more fundamental $SU(3)$ associated with color.

The baryon number of each baryon is $B = +1$. The corresponding antibaryon, in which each quark is replaced by the corresponding antiquark, has baryon number $B = -1$. Note that the baryon number of any meson is zero and that there is no law of conservation of mesons: they can be created and/or destroyed in any numbers. When quark and antiquark combine to form a meson or when three quarks form a baryon, they do so in such a way that the resulting hadron has no net color: hadron states are color-$SU(3)$ singlets.

Although leptons are observed as free objects, no one has ever observed a free quark, and there are good reasons to believe that quarks must forever remain confined inside hadrons. The force between quarks, while quite weak at short distances (*asymptotic freedom*), increases with

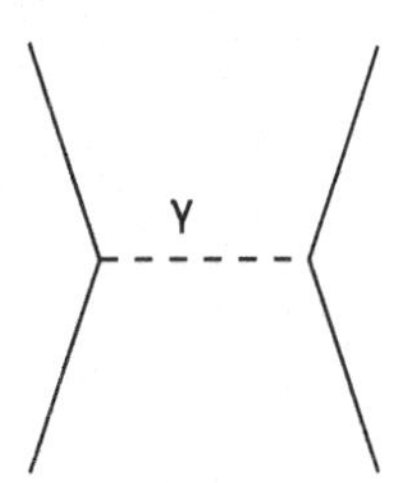

Figure 24.1 Single-photon exchange between two charged particles.

separation, and it appears that an infinite amount of energy would be required to separate two quarks by an infinite distance. When this is attempted, new quark-antiquark pairs are formed from the vacuum, and the net result is the production of new hadrons rather than the liberation of individual quarks.

24.1.3 The photon

As we have discussed at length, the photon is the quantum of the electromagnetic radiation field. It has unit spin (the electromagnetic field is a vector field), and it has zero mass. We have learned that the Coulomb force between two charged particles arises from the exchange of virtual photons. In lowest order of perturbation theory, this is represented by the Feynman diagram of Figure 24.1.

The simpler process of photon emission or absorption by a free electron cannot occur because of energy-momentum conservation, but it is still useful to consider because it is represented by a primitive diagram (Figure 24.2) from which all more complicated diagrams in quantum electrodynamics are constructed.

The Coulomb potential varies with distance as $1/r$. This is a manifestation of the fact that the photon rest mass is zero. Were it greater than zero, the Coulomb potential would take the Yukawa form

$$\frac{e^{-\lambda r}}{4\pi r}$$

where

$$\lambda = \frac{m_\gamma c}{\hbar}$$

Quantum electrodynamics is a highly successful theory. Not only can one calculate any quantum electrodynamic (QED) process (Compton scattering, bremsstrahlung, Møller scattering, etc.) accurately to lowest nonvanishing order, but there is a consistent and successful procedure for calculating higher-order corrections. It is true that, in general, such higher-order corrections involve divergent integrals, but these divergences are removed once and for all by charge and mass renormalization. Although the renormalization process contains questionable mathematical procedures, it achieves remarkable successes: calculations of the Lamb shift and the anomalous magnetic moments of the electron and positron agree with the results of high-precision experiments. There can be little doubt of the essential validity of renormalization in QED.

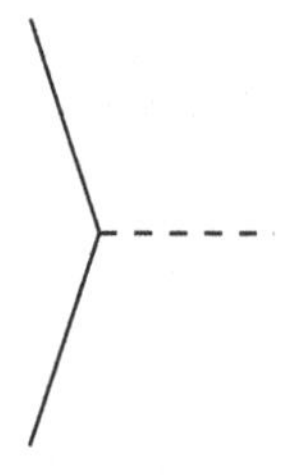

Figure 24.2 The primitive electron-electron-photon vertex of QED. It is the basic building block for all QED Feynman diagrams. The coupling strength of this primitive vertex is $e = \sqrt{4\pi\alpha}$.

24.1.4 Intermediate vector bosons $W^{\pm}$ and Z^0: The Higgs boson

Whereas the photon mediates the electromagnetic interaction, $W^{\pm}$ vector bosons mediate the charged weak interaction, and Z^0 vector bosons mediate the neutral weak interaction. For example, muon decay is represented by the Feynman diagram (Figure 24.3), whereas the neutral weak reaction

$$\nu_\mu + e^- \rightarrow \nu_\mu + e^-$$

is represented by Figure 24.4.

The vector bosons $W^{\pm}$ and Z^0 are analogous to the photon, but there are important differences. First, $W^{\pm}$ carry electric charge, whereas the photon carries no charge, electrical or otherwise. Second, $W^{\pm}$ and Z^0 are massive: $m_W = 80.3$ GeV/c^2 and $m_Z = 91.2$ GeV/c^2. A W or Z decays by semiweak interaction; that is,

$$\begin{aligned} W^- &\rightarrow e^-\bar{\nu}_e \\ &\rightarrow \mu^-\bar{\nu}_\mu \\ &\rightarrow \tau^-\bar{\nu}_\tau \\ &\cdots \end{aligned} \tag{24.5}$$

$$\begin{aligned} Z &\rightarrow \nu\bar{\nu} \\ &\rightarrow e^+e^- \\ &\rightarrow \mu^+\mu^- \\ &\rightarrow \tau^+\tau^- \\ &\cdots \end{aligned} \tag{24.6}$$

Figure 24.5 is a typical diagram for a semiweak decay.

Because $m_W \gg m_e$, there is no prohibition against the first-order process illustrated in Figure 24.5 or the analogous processes listed in (24.5) and (24.6). As in quantum electrodynamics, diagrams such as Figure 24.5 serve as building blocks for constructing higher-order weak interaction diagrams such as Figures 24.3 and 24.4.

The nonzero masses of $W^{\pm}$ and Z^0 mean that the weak interactions, unlike the electromagnetic interaction, have short range. The finite rest mass causes serious problems for creating a renormalizable theory. If one simply starts with a QED-like Lagrangian and adds a mass term, all the attractive features of QED, such as renormalizability and satisfaction of unitarity

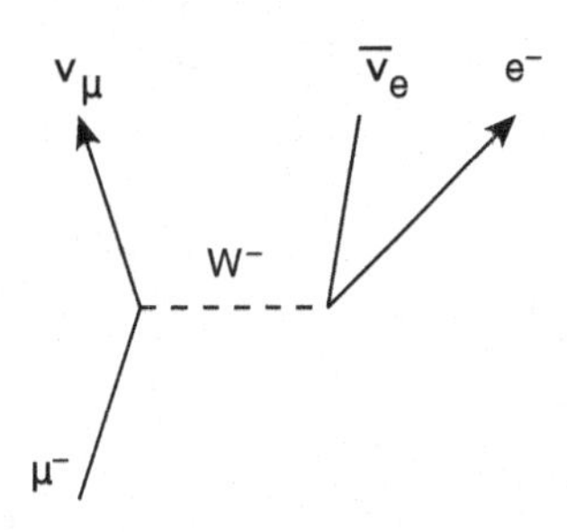

Figure 24.3 Muon decay occurs by virtual W^- exchange.

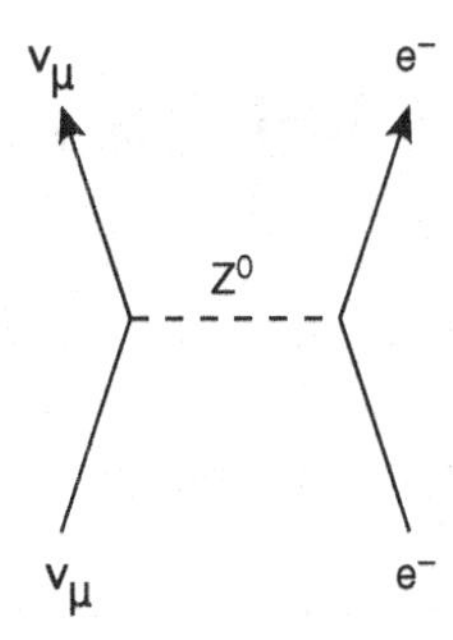

Figure 24.4 $\nu_\mu - e^-$ scattering by Z^0 exchange.

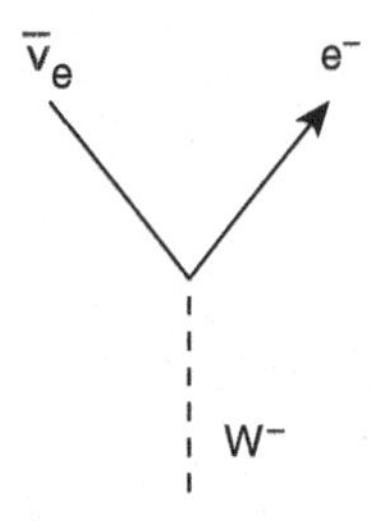

Figure 24.5 First-order diagram for the semiweak decay $W^- \to e^-\bar{\nu}_e$.

in scattering amplitudes, are spoiled. Another way had to be found to impart mass to the W and Z bosons. The solution was found independently by S. Weinberg and A. Salam in 1967–1968 in an unexpected and curious way by means of the Higgs mechanism. Their theory (the *electroweak standard model*) predicts that at least one massive scalar Higgs boson must exist. Experimental evidence for the Higgs boson, with mass $m_H = 126$ GeV/c^2, has been found at the Large Hadron Collider in Geneva in 2012. We discuss the standard model in more detail later.

24.1.5 The Gluons

The eight gluons, like the photon, have zero rest mass, zero electric charge, and unit spin, but unlike the photons or W and Z bosons, they carry color charge as follows:

$$
\begin{array}{lll}
R\bar{G} & R\bar{B} & G\bar{B} \\
\bar{R}G & \bar{R}B & \bar{G}B \\
\multicolumn{3}{l}{\frac{1}{\sqrt{2}}\left(R\bar{R}-G\bar{G}\right)} \\
\multicolumn{3}{l}{\frac{1}{\sqrt{6}}\left(R\bar{R}+G\bar{G}-2B\bar{B}\right)}
\end{array}
$$

Because of color charge, gluons experience nonlinear self-interactions that have no analogue in QED. These lead to unique and peculiar properties of QCD, such as quark confinement in hadrons and the formation of gluon bound states (which are called *glueballs*).

24.2 A Brief history of weak interactions: Early years

We now give a historical sketch of the development of ideas about the weak interaction. We hope that this will enable the reader to place modern theoretical and experimental achievements in an appropriate perspective. Our story begins in 1896 with the discovery of radioactivity by Becquerel. In the first decade of the twentieth century, it was recognized that in beta-radioactivity, a nucleus emits an electron: nuclear beta decay was the first weak process observed. In 1914, James Chadwick demonstrated that the energy spectrum of the electrons in beta decay is continuous. Rutherford and others suggested that this might happen because the emitted electron loses energy stochastically by interaction with atomic electrons. However, in 1927, Ellis and Wooster performed calorimetric measurements of the energy released in beta decay and showed that the energy deposited in the calorimeter is

$$E_{\text{total}} = N\left\langle E\right\rangle$$

where N is the number of decays, and $\langle E\rangle$ is the *average* energy (not the total energy) of the beta spectrum. They thereby eliminated Rutherford's suggestion and created a crisis that left only two possible resolutions: either energy is not conserved in beta decay (which was in fact suggested by Bohr in 1928) or a third particle is also emitted in addition to the beta particle and the recoiling nucleus. This suggestion was made by Pauli in 1930 and again in 1933. In order to be consistent with Ellis and Wooster's results, the third particle would have to be electrically neutral and have extremely feeble interactions with surrounding matter to avoid being absorbed in the calorimeter. By the time Pauli made his suggestion the second time in 1933, the neutron had been discovered, nuclei were correctly understood to be composed of protons and neutrons, and Pauli's particle (now assumed to have half-integral spin and soon to be named the *neutrino* by Fermi) was necessary not only for energy conservation but also for conservation of angular momentum and statistics.

24.3 Fermi's theory of beta decay

Directly stimulated by Pauli's idea, Enrico Fermi created a theory of beta decay in a few days at the end of 1933 (Fermi 1934). Fermi constructed his theory by analogy with quantum

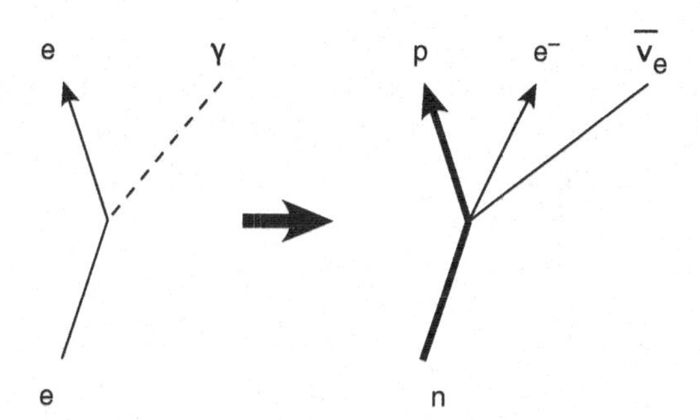

Figure 24.6 Analogy between emission of a photon by a free electron and neutron beta decay in Fermi's theory.

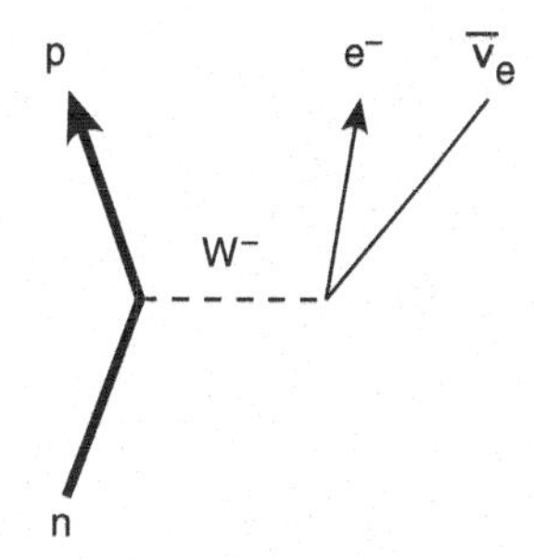

Figure 24.7 Lowest order Feynman diagram for neutron beta decay by W^- exchange.

electrodynamics, which was quite new at the time. In QED, as we know, the Lagrangian density describing the interaction between an electron and the electromagnetic field is

$$\mathcal{L} = -j^{\mu}_{\mathrm{EM}} A_{\mu} = e\bar{\Psi}\gamma^{\mu}\Psi \cdot A_{\mu} \tag{24.7}$$

Because j^{μ} and A_{μ} are both polar 4-vectors, $\mathcal{L}$ is a scalar. As we have noted many times previously, one of the simplest conceivable processes in QED is emission of a photon by a free electron (see Figure 24.2). Although this cannot occur because of energy-momentum conservation, it is still useful conceptually as a starting point in Fermi's theory for consideration of neutron decay (Figure 24.6).

In Fermi's theory, the 4-vector A_{μ} in QED is replaced by the 4-vector $g_{\mu\nu}\bar{\Psi}_e\gamma^{\nu}\Psi_{\nu}$, and $\bar{\Psi}_e\gamma^{\mu}\Psi_e$ in QED is replaced by the 4-vector $\bar{\Psi}_p\gamma^{\mu}\Psi_n$. Thus, as is apparent from Figure 24.6, this theory assumes that four fermions interact at a single space-time point. (Later on we see that this description is really appropriate as the low-energy limit of an intermediate boson description in which the diagram for neutron beta decay is Figure 24.7.)

Fermi's beta-decay Lagrangian density is

$$\mathcal{L}_{\beta} = -\frac{G_F}{\sqrt{2}}\bar{\Psi}_p\gamma^{\mu}\Psi_n \cdot \bar{\Psi}_e\gamma_{\mu}\Psi_{\nu} \tag{24.8}$$

Here the factor $2^{-1/2}$ appears merely for historical reasons; G_F (the Fermi constant) is a coupling constant obtained by fitting to experimental data, and it is found to be

$$G_F = 1.03 \times 10^{-5}\ m_p^{-2} \tag{24.9}$$

in units where $\hbar = c = 1$.

Shortly after Fermi created his theory, positron emission in beta decay was discovered, and several years later, electron capture was observed. These are readily accommodated in Fermi's theory by adding to (24.8) a Hermitian conjugate term

$$\mathcal{L}_\beta = -\frac{G_F}{\sqrt{2}}\left[\bar{\Psi}_p\gamma^\mu\Psi_n \bullet \bar{\Psi}_e\gamma_\mu\Psi_\nu + \bar{\Psi}_n\gamma^\mu\Psi_p \bullet \bar{\Psi}_\nu\gamma_\mu\Psi_e\right] \tag{24.10}$$

In 1936, G. Gamow and E. Teller pointed out that there are other ways to construct a scalar beta-decay Lagrangian density from the Dirac fields for neutron, proton, electron, and neutrino (Gamow and Teller 1936).We only have to recall that under a Lorentz transformation,

- $\bar{\Psi}\Psi$ is a scalar (S).
- $\bar{\Psi}\gamma^\mu\Psi$ is a polar vector (V).
- $\bar{\Psi}\sigma^{\mu\nu}\Psi$ is a second-rank tensor (T).
- $\bar{\Psi}\gamma^\mu\gamma^5\Psi$ is an axial vector (A).
- $\bar{\Psi}\gamma^5\Psi$ is a pseudoscalar (P).

While Fermi's Lagrangian density is the scalar product of two polar vectors (V•V), Gamow and Teller noted that there was no reason a priori to exclude the possibility of such terms as S•S, T•T, A•A, and P•P. Thus the matrix element for beta decay could be written as

$$M = \frac{G_F}{\sqrt{2}}\sum_{\text{nucleons}}\int\sum_{\substack{i=\\ \text{S,V,A,T,P}}} C_i\left(\bar{\psi}_p O_i \psi_n\right)\left(\bar{\psi}_e O_i \psi_\nu\right) d^3\boldsymbol{x} \tag{24.11}$$

where the ψ's are single-particle Dirac wave functions, $O_i = I, \gamma^\mu, \sigma^{\mu\nu}, \gamma^\mu\gamma^5, \gamma^5$, for i = S, V, T, A, P, respectively, and the integration is carried out over the nuclear volume. The coupling constants C_i could only be determined by experiments, but in 1936, Gamow and Teller already recognized that in nuclear beta decay the total energy imparted to the leptons very rarely exceeds a few MeV. Thus the recoiling final nucleus is always nonrelativistic in the rest frame of the initial nucleus. In the standard representation,

$$I = \begin{pmatrix} I & 0 \\ 0 & I \end{pmatrix} \quad \gamma^0 = \begin{pmatrix} I & 0 \\ 0 & -I \end{pmatrix} \quad \boldsymbol{\gamma} = \begin{pmatrix} 0 & \boldsymbol{\sigma} \\ -\boldsymbol{\sigma} & 0 \end{pmatrix} \quad \gamma_5 = \begin{pmatrix} 0 & I \\ I & 0 \end{pmatrix}$$

whereas a Dirac four-component wave function is written

$$\psi = \begin{pmatrix} \chi \\ \phi = \dfrac{\boldsymbol{\sigma}\bullet\boldsymbol{p}}{E+m}\chi \end{pmatrix} \tag{24.12}$$

and χ and ϕ are the *large* and *small* two-component wave functions, respectively. In the nonrelativistic limit, ϕ becomes negligible, and because the matrix γ^5 couples large and small components, we have

$$\bar{\psi}_p\gamma^5\psi_n \to 0 \qquad \text{(NR limit)} \tag{24.13}$$

so the P•P (pseudoscalar) term in (24.11) is negligible. Furthermore, in the nonrelativistic limit,

$$\bar{\psi}_p \gamma^0 \psi_n = \psi_p^\dagger \psi_n \to \chi_p^\dagger \chi_n$$

and

$$\bar{\psi}_p \boldsymbol{\gamma} \psi_n \to 0$$

Hence the nucleon factor in the V•V term in (24.11) reduces to

$$\chi_p^\dagger I \chi_n \tag{24.14}$$

which is the same as the nucleon factor in the S•S term. Also, in the nonrelativistic limit,

$$\bar{\psi}_p \gamma^0 \gamma^5 \psi_n \to 0$$

whereas

$$\bar{\psi}_p \boldsymbol{\gamma} \gamma_5 \psi_n \to \chi_p^\dagger \boldsymbol{\sigma} \chi_n \tag{24.15}$$

which is thus the nucleon factor in the A•A term in (24.11), just as it is for the T•T term. Now, under spatial rotations, I is an invariant, but $\boldsymbol{\sigma}$ is a first-rank tensor. Thus, from the Wigner-Eckart theorem, the angular-momentum selection rules for nuclear spin in allowed beta decay are as follows:

- For S and V: $\Delta J = 0$.
- For A and T: $\Delta J = 0, \pm 1$, but $J_i = 0 \to J_f = 0$ is forbidden.

In 1936, it was already known that there are certain allowed beta decays for which $|\Delta J| = 1$ and others for which $|\Delta J| = 0$. This implied that whereas V and/or S might contribute, there must also be some contribution from A and/or T. However, a complete solution to the problem of the beta-decay constants in (24.11) would have to wait 22 years, and then it came from an unexpected direction.

24.4 Universal Fermi interaction: Discovery of new particles

In the meantime, there were other new discoveries. In 1935, H. Yukawa proposed a meson theory of nuclear forces, also in analogy with QED. Yukawa noted that the nucleon-nucleon force is short ranged and suggested that it is mediated by exchange of a scalar (or pseudoscalar) meson, which would necessarily have a rest mass of approximately 100–200 MeV/c^2. In 1937, Anderson and Neddermeyer discovered the muon in cosmic rays, and at first it was thought to be Yukawa's meson. However, during World War II and in very difficult circumstances in Nazi-occupied Rome, Conversi, Pancini, and Piccioni demonstrated conclusively by simple experiments that the muon could not possibly have strong interactions. This important result became generally known immediately after the conclusion of the war, presenting a mystery that was resolved only when the pion (the real Yukawa particle) was discovered in cosmic rays in 1947 by a British group. Within several years, K mesons and hyperons also were discovered in cosmic rays. Meanwhile, the mean life of the muon and the rate of capture of negative muons by nuclei had been measured. By 1949, it was apparent that muon capture and decay are very similar to beta decay, governed as they are by essentially the same coupling. Thus the idea of a *universal Fermi interaction* gradually began to emerge.

In the early 1950s, the properties of pions, kaons, and hyperons were systematically investigated with new and powerful accelerators. These machines were able to generate far more events in well-controlled conditions than was possible with cosmic-ray observations. Thus, by 1953, the remarkable properties of the neutral K meson system (strangeness oscillations and regeneration) had been discovered and the theory worked out by Gell-Mann, Pais, Nishijima, and others.

24.5 Discovery of parity violation

In 1954, a peculiar problem involving K mesons became apparent. This was the $\tau-\theta$ puzzle. Here, evidently, were two positively charged mesons, θ and τ, each with the same charge, mass, and zero spin. (Do not confuse the name τ for this meson, which is old and has been discarded, for the name of the τ lepton.) It was observed that θ decayed to two pions

$$\theta \to \pi^+\pi^0$$

where the final state has even parity, whereas τ decayed to a three-pion final state with odd parity

$$\tau \to \pi^+\pi^+\pi^-$$

How could two particles exist that were identical in all respects but decayed to states of different parity? If, as almost everyone assumed, the weak interactions conserve parity, the $\tau-\theta$ puzzle seemed inexplicable. However, while studying this problem, T. D. Lee and C. N. Yang realized that there was no experimental evidence to warrant the assumption of parity conservation in weak interactions. They proposed that parity is violated (Lee and Yang 1956) and that τ and θ are really the same particle (as indeed they are; it is now called K^+). Most important, Lee and Yang proposed a variety of practical ways in which the question of parity nonconservation in weak interactions could be investigated.

Within several months, and directly stimulated by Lee and Yang's new ideas, C. S. Wu of Columbia University and her collaborators at the National Bureau of Standards (Wu et al. 1957) observed parity violation in the beta-decay asymmetry of polarized ^{60}Co nuclei. Almost immediately thereafter, R. Garwin, L. Lederman, and G. Weinreich (Garwin et al. 1957) reported observations of parity violation in the chain of decays

$$\begin{aligned} \pi^+ &\to \mu^+\nu_\mu \\ \mu^+ &\to e^+\nu_e\bar{\nu}_\mu \end{aligned}$$

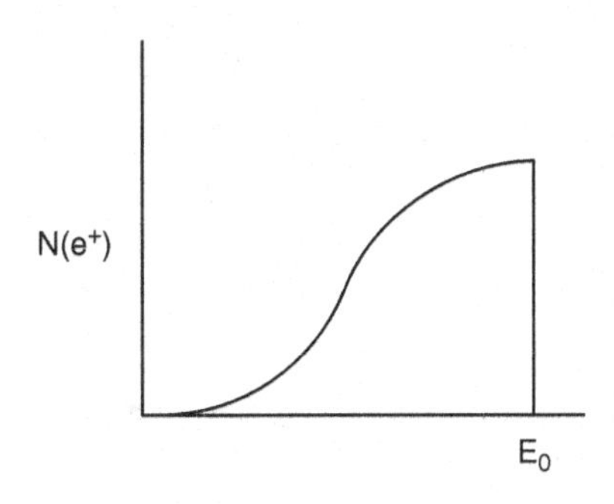

Figure 24.8 Sketch of positron energy spectrum in the decay $\mu^+ \rightarrow e^+\nu_e\bar{\nu}_\mu$

By definition, parity is conserved in a decay or scattering process A if and only if A and the process A' obtained from A by spatial inversion have the same probability. For example, consider the decay

$$\mu^+ \rightarrow e^+\nu_e\bar{\nu}_\mu \tag{24.16}$$

observed in the rest frame of the muon. Because the final state consists of three particles, the positron energy spectrum is continuous, with the shape sketched in Figure 24.8. The maximum positron energy ($E_0 \approx 52$ MeV) occurs when the positron linear momentum $\boldsymbol{p}$ is balanced by linear momenta $-\boldsymbol{p}/2$ for each of the neutrinos. Consider decay of a μ^+ at rest with spin polarized along $+z$ in a case where the positron has energy in the vicinity of E_0. Experiment shows that such positrons are emitted mainly in the $+z$ direction and that they have helicity $h(e^+) = +1$. Conservation of linear momentum requires the two neutrinos to be emitted in the negative z-direction, and conservation of angular momentum requires these neutrinos to have opposite helicities. In fact, experimental observations of pion decay, beta decay, and so on show that one always has $h(\bar{\nu}) = +1$, $h(\nu) = -1$ (Figure 24.9a).

A spatial inversion is equivalent to a mirror reflection in a plane and a rotation by 180° about an axis normal to the plane, and because the laws of nature are invariant under rotations, we may investigate the question of parity invariance by considering the mirror reflection alone. A reflection through the xy-plane containing the muon reverses the linear momenta of the positron and the neutrinos but leaves all the angular momenta along z unaffected; thus this reflection results in $h(e^+) = -1$ and a reversal of the orientation of positron linear momentum with respect to muon spin, which are never observed (Figure 24.9b). Thus parity is violated.

Alternatively, we can perform a charge-conjugation transformation on the process $\mu^+ \rightarrow e^+\nu_e\bar{\nu}_\mu$ of Figure 24.9a. This results in the decay $\mu^- \rightarrow e^-\bar{\nu}_e\nu_\mu$ but does not alter the momenta and spins (and thus the helicities) of Figure 24.9a. We thus have the situation shown in Figure 24.9c. However, experiment shows that in the decay $\mu^- \rightarrow e^-\bar{\nu}_e\nu_\mu$, $h(e^-) = -1$, and the electron is emitted mainly in the $-z$-direction; thus the decay shown in Figure 24.9c is never observed. Hence not only parity (P) but also charge conjugation (C) invariance are violated.

Finally, note that the combination CP of a parity (P) and a charge-conjugation (C) transformation on the decay of Figure 24.9a results in the decay of Figure 23.9d, which *is* observed with the same probability. Thus CP invariance is valid for the decays $\mu^\pm \rightarrow e^\pm\nu_{e(\mu)}\bar{\nu}_{\mu(e)}$. However, as

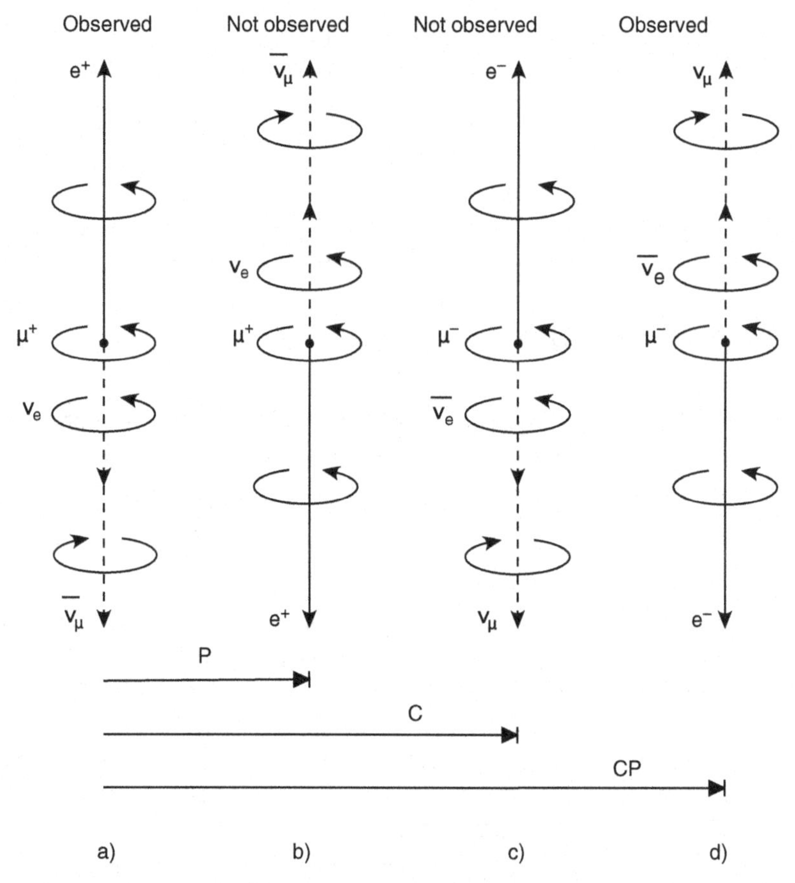

Figure 24.9 Diagrams illustrating linear momenta and spins of the decay products in the decays $\mu^+ \longrightarrow e^+\nu_e\bar{\nu}_\mu, \mu^- \longrightarrow e^-\bar{\nu}_e\nu_\mu$, when the positron or electron energy is near its maximum value. Cases a) and d) are observed, cases b) and c) are never observed.

was discovered in 1964 and will be discussed later, *CP* invariance does break down in other weak decays.

24.6 The *V-A* Law

The discovery of parity violation resulted in vigorous experimental activity that culminated within less than two years in a generalization of Fermi's theory, called the *V-A law*. This was enunciated independently by Sudarshan and Marshak, Feynman and Gell Mann, Sakurai, and Gershtein and Zeldovitch in 1958. Here it was proposed that the Lagrangian density for charged weak interactions (these were the only type known at the time) takes the following form:

$$\mathcal{L}_W = -\frac{G_F}{\sqrt{2}} J_\sigma^\dagger J^\sigma \tag{24.17}$$

where $J^\sigma = J^{\sigma,\text{hadronic}} + J^{\sigma,\text{leptonic}}$ is a charged weak current density containing both vector and axial-vector parts; that is,

$$J^\sigma = V^\sigma - A^\sigma$$

Thus

$$\begin{aligned}\mathcal{L}_W &= -\frac{G_F}{\sqrt{2}}\left(V_\sigma^\dagger - A_\sigma^\dagger\right)\left(V^\sigma - A^\sigma\right) \\ &= -\frac{G_F}{\sqrt{2}}\left(V_\sigma^\dagger V^\sigma + A_\sigma^\dagger A^\sigma\right) + \frac{G_F}{\sqrt{2}}\left(V_\sigma^\dagger A^\sigma + A_\sigma^\dagger V^\sigma\right)\end{aligned} \tag{24.18}$$

The first term in parentheses in the second line of (24.18) is a true scalar, whereas the second term in parentheses is a pseudoscalar. The appearance of both scalar and pseudoscalar terms in the Lagrangian density is a necessary and sufficient condition for parity-violating effects.

Including the τ lepton and its associated neutrino, which were still undiscovered in 1958, the leptonic charged weak current density in the *V-A* law is

$$J^\sigma_{\text{lept}} = \bar{\Psi}_e \gamma^\sigma \left(1-\gamma^5\right)\Psi_{\nu e} + \bar{\Psi}_\mu \gamma^\sigma \left(1-\gamma^5\right)\Psi_{\nu\mu} + \bar{\Psi}_\tau \gamma^\sigma \left(1-\gamma^5\right)\Psi_{\nu\tau} \tag{24.19}$$

As for $J^{\sigma,\text{hadronic}}$, it clearly must contain a term proportional to

$$\bar{D}\gamma^\sigma(1-\gamma^5)U \tag{24.20}$$

where D and U are Dirac fields for the down and up quarks, respectively. This is necessary in order to account for nuclear beta decay and charged pion decay. However, it was known even in the 1950s that there are "strangeness changing" weak decays, for example, $K^+ \rightarrow \pi^0\mu^+\nu_\mu$ and $\Lambda^0 \rightarrow pe^-\bar{\nu}_e$, where it is now known that an up quark transforms into a strange quark or vice versa. (By definition, the *strangeness* of a hadron is the number of $\bar{s}$ quarks minus the number of s quarks contained in the hadron.) Thus the hadronic charged weak current density also must have a component proportional to

$$\bar{S}\gamma^\sigma(1-\gamma^5)U$$

N. Cabibbo (1963) found that many features of hadronic decays could be accounted for by assuming that the hadronic charged weak current density takes the form

$$\begin{aligned}J^{\sigma,\text{had}} &= \cos\theta_C\left[\bar{D}\gamma^\sigma(1-\gamma^5)U\right] + \sin\theta_C\left[\bar{S}\gamma^\sigma(1-\gamma^5)U\right] \\ &= \bar{D}_C\gamma^\sigma(1-\gamma^5)U\end{aligned} \tag{24.21}$$

where $D_C \equiv \cos\theta_C D + \sin\theta_C S$ and the Cabibbo angle $\theta_C \approx 13°$ is an empirically determined parameter. In other words, the quark state participating in the charged weak interaction is neither pure d nor pure s but a linear combination of the two. (The origin of θ_C has still not been explained satisfactorily in terms of more fundamental quantities.) Cabibbo's hypothesis accounted for the relative coupling strength of strangeness-conserving and strangeness-violating baryon semileptonic decays, for the ratio of leptonic decay rates of charged pions and kaons, for the ratio of vector coupling strengths in beta decay and muon decay, and for other important features of the charged weak interaction. Eventually, Cabibbo's hypothesis was generalized to take into account weak transformations of the heavy quarks. When this was done, there emerged three Cabibbo angles instead of one and, in addition, a phase, all contained in the 3×3 Cabibbo-Kobayashi-Maskawa (CKM) matrix. This is discussed in more detail later. The recent experimental results on neutrino oscillations and mixing imply that a somewhat analogous matrix is necessary for the lepton sector.

Although the form (24.21) for the hadronic charged weak current density appears straightforward, to apply it to the calculation of real processes such as nuclear beta decay, hyperon decay, and so on, we must construct matrix elements of this current operator between initial and final hadron states. Here we encounter a difficult problem: hadrons are complex objects in which strong interactions are going on all the time, and this has consequences for the effective weak coupling constants. For example, the amplitude for neutron beta decay in the V-A law turns out to be

$$M = \frac{G_F \cos\theta_C}{\sqrt{2}} \bar{u}_p \gamma^\rho (1-\lambda\gamma^5) u_n \bullet g_{\rho\sigma} \bullet \bar{u}_e \gamma^\sigma (1-\gamma^5) v_{\bar{\nu} e} \tag{24.22}$$

where $\lambda = 1.26$ is determined from experiment. The point here is that while the strength of the vector portion of the hadronic charged weak current is conserved (not renormalized by strong interactions), the axial current is not conserved, and its coefficient is 1.26 rather than unity.

Let us explain the physical meaning of the factor $\gamma^\sigma(1-\gamma^5)$ by considering the leptonic portion of the right-hand side of (24.22). First of all,

$$\left(I-\gamma^5\right)^2 = I - 2\gamma^5 + \left(\gamma^5\right)^2 = 2\left(I-\gamma^5\right)$$

Therefore,

$$\begin{aligned}
\bar{u}\gamma^\sigma\left(I-\gamma^5\right)v &= \frac{1}{2}u^\dagger\gamma^0\gamma^\sigma\left(I-\gamma^5\right)\left(I-\gamma^5\right)v \\
&= \frac{1}{2}u^\dagger\left(I-\gamma^5\right)\gamma^0\gamma^\sigma\left(I-\gamma^5\right)v \\
&= \frac{1}{2}\left[\left(I-\gamma^5\right)u\right]^\dagger\gamma^0\gamma^\sigma\left(I-\gamma^5\right)v \\
&= \frac{1}{2}\overline{\left(I-\gamma^5\right)u}\gamma^\sigma\left(I-\gamma^5\right)v
\end{aligned}$$

Also, because

$$u_\pm = const\begin{pmatrix} \chi_\pm \\ \dfrac{\boldsymbol{\sigma}\bullet\mathbf{p}}{E+m}\chi_\pm \end{pmatrix}$$

and

$$\left(I-\gamma^5\right) = \begin{pmatrix} I & -I \\ -I & I \end{pmatrix}$$

we have

$$(I-\gamma^5)u_\pm = const\begin{pmatrix}\left[1-\dfrac{\sigma\cdot p}{E+m}\right]\chi_\pm \\ -\left[1-\dfrac{\sigma\cdot p}{E+m}\right]\chi_\pm\end{pmatrix} \tag{24.23}$$

where the ± sign refers to spin up (down). The polarization of a sample of electrons from beta decay along an axis parallel to the electron momentum is given by

$$P=\frac{[(I-\gamma^5)u_+]^\dagger[(I-\gamma^5)u_+]-[(I-\gamma^5)u_-]^\dagger[(I-\gamma^5)u_-]}{[(I-\gamma^5)u_+]^\dagger[(I-\gamma^5)u_+]+[(I-\gamma^5)u_-]^\dagger[(I-\gamma^5)u_-]} \tag{24.24}$$

From (24.23), we have

$$\begin{aligned}[(I-\gamma^5)u_\pm]^\dagger[(I-\gamma^5)u_\pm] &= \text{const}\bullet\chi_\pm^\dagger\left(1-\frac{\sigma\cdot p}{E+m}\right)^2\chi_\pm \\ &= \text{const}\bullet\left(1+\frac{p^2}{(E+m)^2}\mp 2\frac{|p|}{E+m}\right)\end{aligned}$$

Thus (24.24) becomes

$$\begin{aligned}P=\frac{-2\dfrac{|p|}{E+m}}{1+\dfrac{p^2}{(E+m)^2}} &= -2\frac{|p|(E+m)}{(E+m)^2+(E+m)(E-m)} \\ &= -\frac{|p|}{E}=-\frac{v}{c}\end{aligned}$$

where in the last step we have exhibited the velocity of light explicitly. A similar analysis performed on the antineutrino spinor $(I-\gamma^5)v$ reveals that the antineutrinos are right handed. Also, in nuclear beta decay the polarization of emitted positrons is $P(e^+)=+v/c$.

24.7 Difficulties with Fermi-type theories

The *V-A* law, supplemented by Cabibbo's hypothesis, provided a very good phenomenological account of the observed charged weak interactions in the 1950s and 1960s within the framework of first-order perturbation theory (= Born approximation). However, it was nothing more than a generalized version of Fermi's original theory, and thus it suffered from the same fundamental difficulties as the latter – difficulties that were already recognized in the 1930s. In order to see the problem in the simplest possible way, we consider the following scattering reaction:

$$\nu_e+e^-\rightarrow\nu_e+e^-$$

which is predicted to exist in the *V-A* law and to be accounted for by the Lagrangian term $-\left(G_F/\sqrt{2}\right)J_\sigma^{e\dagger}J^{\sigma,e}$. Suppose that the energy E of the incoming neutrino is much greater than that of the electron rest mass. Then we can argue on purely dimensional grounds that the cross section must be of order $G_F^2E^2$ as follows: because the scattering amplitude is proportional to G_F, the cross section must be proportional to G_F^2. However, in natural units, G_F^2 has dimensions of (energy)$^{-4}$ = (mass)$^{-4}$ = (length)^4, whereas the cross section itself has dimensions of (length)^2 = (energy)$^{-2}$. Thus the cross section must be expressible as $G_F^2\times(\text{energy})^2$. In the limit where $E \gg m_e$, there is no other energy scale in the problem than the energy E of the neutrino. Thus we expect that the cross section is of order $G_F^2E^2$. In fact, a rather easy explicit calculation starting from the *V-A* law shows that

$$\frac{d\sigma}{d\Omega} = \frac{G_F^2}{\pi^2}E^2 \tag{24.25}$$

On the other hand, we can express a differential cross section as the absolute square of a scattering amplitude $f(\theta)$, and the latter can be expanded in partial waves as

$$f(\theta) = \frac{1}{E}\sum_{J=0}^{\infty}\left(J+\frac{1}{2}\right)M_J P_J(\cos\theta)$$

where M_J is the amplitude of the Jth partial wave. This formula is a generalization of the well-known partial wave expansion for nonrelativistic potential scattering [see equations (18.22a) and (18.22b)]. In a theory of the Fermi type, only the $J = 0$ partial wave enters because we have a contact interaction of zero range. Therefore,

$$\frac{d\sigma}{d\Omega} = |f(\theta)|^2 = \frac{1}{4E^2}|M_0|^2 \tag{24.26}$$

However, unitarity (conservation of probability) requires that $|M_J| \le 1$ for each J. Thus unitarity requires

$$\frac{d\sigma}{d\Omega} \le \frac{1}{4E^2} \tag{24.27}$$

Obviously, (24.25) and (24.27) are in contradiction when

$$E \ge \left(\frac{\pi}{2G_F}\right)^{1/2} \approx 300\text{ GeV} \tag{24.28}$$

Thus our first-order calculation of the cross section fails completely at energies of order $G_F^{-1/2}$. We naturally might suppose at first that this difficulty arises from neglect of higher-order corrections and that if these were included, the cross section might level out and remain within the bounds imposed by unitarity. However, calculation shows that the second-order diagram of Figure 24.10 yields a divergent result.

In quantum electrodynamics one also encounters divergent integrals corresponding to higher-order diagrams. However, the divergences are removed to all orders by mass and charge

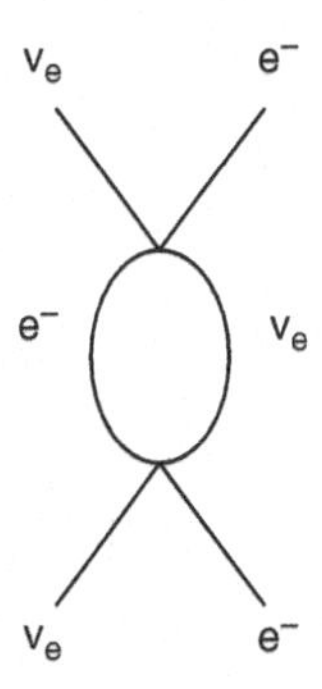

Figure 24.10 Second order Feynman diagram for $e^- \nu_e$ scattering in the V-A theory.

renormalization. In the present case (Figure 24.10), we may attempt to eliminate the divergence by an analogous procedure, but when we go to the next higher order, a new and more severe divergence is encountered that requires new normalization constants. When all diagrams are taken into account, it is found that an infinite set of renormalization constants is required. In short, theories of the Fermi type are not renormalizable.

24.8 Naive intermediate boson theory of charged weak interaction

Is it possible to avoid the difficulties associated with Fermi-type theories by replacing the 4-fermion contact interaction by an intermediate boson theory of charged weak interactions? In view of the success of quantum electrodynamics and the plausibility of Yukawa's meson theory of nuclear forces, it seemed very reasonable and indeed compelling in the late 1950s that the charged weak interaction should proceed by exchange of charged intermediate vector bosons $W^{\pm}$, which would necessarily be quite massive because the weak interaction has a very short range. The idea was even suggested in passing by E. Fermi and discussed at greater length by O. Klein in the 1930s.

Let us try to construct a theory of charged weak interactions in the simplest possible manner by employing such bosons. We choose muon decay $\mu^- \to e^- \bar{\nu}_e \nu_\mu$ as an example. According to the V-A law, the amplitude (apart from a phase factor) is

$$M(V-A) = \frac{G_F}{\sqrt{2}}\left[\bar{u}_{\nu\mu}\gamma^\sigma(1-\gamma^5)u_\mu\right]\left[\bar{u}_e\gamma_\sigma(1-\gamma^5)\mathrm{v}_{\bar{\nu}e}\right] \tag{24.29}$$

In a W-boson theory, the lowest-order diagram is that of Figure 24.11.

The amplitude corresponding to Figure 24.11 is constructed as follows: we need a vertex factor at each lepton-lepton-W-vertex and a propagator factor for the internal W line. Let us assume that each vertex factor is of the form

$$\frac{-ig}{2\sqrt{2}}\gamma^\sigma(1-\gamma^5)$$

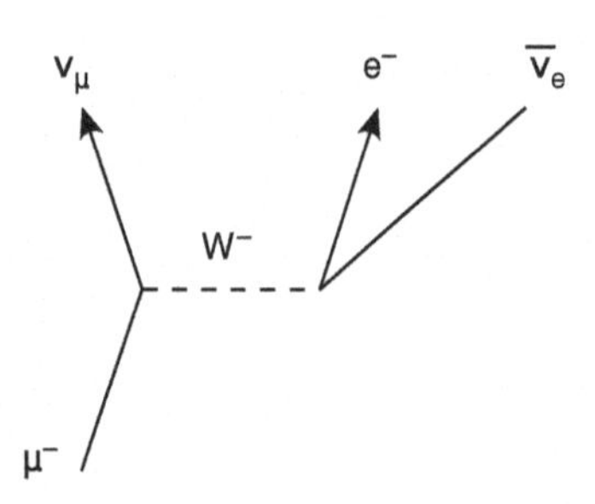

Figure 24.11 Lowest order Feynman diagram for the decay $\mu^- \to e^- \bar{\nu}_e \nu_\mu$ in the naive intermediate vector boson theory.

where g is a *semiweak coupling constant* to be related to G_F in what follows. We also need a propagator factor for the internal W^- line, and we obtain it by an argument resembling that given for the photon propagator at the end of Section 23.3. In electrodynamics, the field equation is

$$\partial^\mu F_{\mu\nu} = \partial^\mu \partial_\mu A_\nu - \partial^\mu \partial_\nu A_\mu = j_\nu^{\mathrm{EM}} \tag{24.30a}$$

Let ϕ_ν be a vector field analogous to A_ν except that the quanta associated with ϕ_ν (the W-bosons) have nonzero mass m_W. Recalling (15.68), we see that the appropriate field equation for ϕ_ν is

$$\partial^\mu \partial_\mu \phi_\nu - \partial^\mu \partial_\nu \phi_\mu + m_W^2 \phi_\nu = J_\nu \tag{24.30b}$$

where J_ν is the charged weak current density that generates the field ϕ. Applying the operator ∂^ν on the left on both sides of (24.30b) and rearranging the order of partial differentiations, we obtain

$$\partial^\mu \partial_\mu \partial^\nu \phi_\nu - \partial^\nu \partial_\nu \partial^\mu \phi_\mu + m_W^2 \partial^\nu \phi_\nu = \partial^\nu J_\nu \tag{24.31}$$

Because the indices μ and ν are both repeated in the first two terms on the left-hand side of (24.31), they are dummy indices. Hence these two terms cancel, yielding

$$\partial^\nu \phi_\nu = \frac{1}{m_W^2} \partial^\nu J_\nu \tag{24.32}$$

If the current were conserved (if $\partial^\nu J_\nu = 0$), then (24.32) would require $\partial^\nu \phi_\nu = 0$ as well. However, in contrast to the case in electrodynamics, $\partial^\nu J_\nu \neq 0$ because the axial portion of the current is not conserved. Thus, using (24.32) in (24.30), we obtain

$$\partial^\mu \partial_\mu \phi_\nu + m_W^2 \phi_\nu = \left(g_{\nu\mu} + \frac{1}{m_W^2} \partial_\nu \partial_\mu \right) J^\mu \tag{24.33}$$

We now employ the Fourier integral transformations

$$\phi_\nu(x) = \int f_\nu(k) e^{-ik\cdot x}\, d^4k \tag{24.34}$$

and

$$J^{\mu}(x)=\int j^{\mu}(k)e^{-ik\cdot x}\,d^4k \tag{24.35}$$

Substitution of (24.34) and (24.35) in (24.33) yields

$$\int(-k^2+m_W^2)f_\nu(k)e^{ik\cdot x}\,d^4k=\int\left(g_{\nu\mu}-\frac{k_\nu k_\mu}{m_W^2}\right)j^{\mu}(k)e^{ik\cdot x}\,d^4k$$

For this equation to be valid generally, the integrands must be equal, so

$$f_\nu(k)=-\left(\frac{g_{\nu\mu}-\dfrac{k_\nu k_\mu}{m_W^2}}{k^2-m_W^2}\right)j^{\mu}(k) \tag{24.36}$$

This yields the W-boson propagator

$$-i\frac{g_{\mu\nu}-\dfrac{k_\mu k_\nu}{m_w^2}}{k^2-m_W^2}$$

and thus the amplitude $M(W)$ is

$$M(W)=i\frac{g^2}{8}\left[\bar{u}_{\nu\mu}\gamma^{\lambda}(1-\gamma_5)u_\mu\right]\frac{g_{\lambda\sigma}-\dfrac{k_\lambda k_\sigma}{m_W^2}}{k^2-m_w^2}\left[\bar{u}_e\gamma^{\sigma}(1-\gamma_5)v_{\bar{\nu}e}\right]\cdot \tag{24.37}$$

where $k=p_\mu-p_{\nu\mu}=p_e+p_{\bar{\nu}e}$ is the 4-momentum transfer. Now it was already known in the 1950s that if the W-boson were to exist, it would have to be quite massive because the weak interaction has a very short range (indeed, subsequently, its mass was found to be 80 GeV/c^2). On the other hand, the momentum transfer in muon decay is small because the muon mass is only 106 MeV/c^2. Therefore, to a very good approximation, in (24.37) we may ignore k^2 in the denominator and $k_\lambda k_\sigma/m_W^2$ in the numerator. In this approximation, (24.37) becomes

$$M(W)=\frac{-ig^2}{8m_W^2}\left[\bar{u}_{\nu\mu}\gamma^{\lambda}(1-\gamma^5)u_\mu\right]\left[\bar{u}_e\gamma_\lambda(1-\gamma^5)v_{\bar{\nu}e}\right] \tag{24.38}$$

which, except for the phase factor, is equivalent to (24.29), provided that we make the identification

$$g=2^{5/4}m_W G_F^{1/2} \tag{24.39}$$

Thus it is clear that instead of employing a Fermi-type theory, we could readily describe all the usual low-energy charged weak interactions by intermediate boson exchange. However, we must still see if the grave difficulties of the Fermi theory associated with violation of unitarity and nonrenormalizability are circumvented. If we were to calculate the amplitude for neutrino-electron scattering in lowest order with W exchange, we would find that in the partial wave expansion of the scattering amplitude there are now many partial waves. The partial cross section associated with each partial wave grows logarithmically with energy; thus unitarity is again violated at sufficiently high energy. (The unitarity limit is violated even more strongly in the cross sections for such processes as

$$\nu+\bar{\nu}\to\left[W^{+}+W^{-}\right]_{\text{longitudinal polarization}}$$

as we see later.) Furthermore, higher-order corrections diverge strongly because of the $k_\lambda k_\sigma / m_W^2$ term in the W propagator numerator. Thus the naive intermediate boson theory of charged weak interactions is also unacceptable.

24.9 The GIM mechanism

Despite these difficulties, it seemed plausible in the 1960s that not only the charged weak interaction should be mediated by $W^{\pm}$ bosons but also that there should exist neutral weak interactions mediated by a neutral massive vector boson. However, there was another major problem: at the time, no neutral weak interactions were observed. Experiments revealed that the decay

$$K_L^0\to\mu^+\mu^- \tag{24.40}$$

proceeds with a branching ratio of less than 10^{-8}, a very perplexing result in view of the fact that the analogous charged decay

$$K^+\to\mu^+\nu_\mu \tag{24.41}$$

is fully allowed with a branching ratio of 63 percent. According to the quark model, these allowed and forbidden decays should be represented by Figure 24.12a and b, respectively. The K^+ is a bound state of $u\bar{s}$; when it decays as in (24.41), the u transforms to an s by W^+ emission (or the $\bar{s}$ transforms to a $\bar{u}$), and the remaining quark-antiquark pair annihilates. The nearly negligible decay rate for (24.40) is accounted for by small radiative corrections. This implies that, unlike the transition $u\to s$, the transition $d\to s$ is forbidden.

The nearly vanishing probability for decay (24.40) seemed even more puzzling when it was recognized that even if for some reason the lowest-order amplitude described by Figure 24.12b should turn out to be zero, there still should be a second-order charged weak contribution, as shown in Figure 24.13a, where the Cabibbo factors are displayed at their respective vertices.

The key to the solution of this puzzle is the charmed quark c (first proposed in 1964 and observed for the first time in 1974). In 1970, Glashow, Iliopoulos, and Maiani (GIM) proposed that although, according to Cabibbo's hypothesis, the u quark is coupled with the linear combination

$$d_C = d\cos\theta_C + s\ \sin\theta_C$$

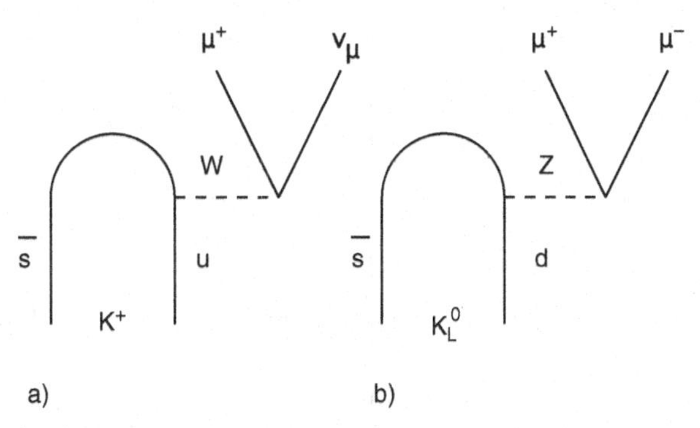

Figure 24.12 Pre GIM concepts of lowest-order Feynman diagrams for: a) $K^+ \rightarrow \mu^+ \nu_\mu$ b) $K_L^0 \rightarrow \mu^+ \mu^-$.

in the charged weak interaction, the c quark is coupled with the orthogonal linear combination

$$s_C = -d \sin\theta_C + s \cos\theta_C$$

Thus, in the GIM framework (Glashow et al. 1970), the Hermitian conjugate charged hadronic weak current is written as

$$J^{\lambda\dagger} = \begin{pmatrix} \bar{U} & \bar{C} \end{pmatrix} \begin{pmatrix} \cos\theta_C & \sin\theta_C \\ -\sin\theta_C & \cos\theta_C \end{pmatrix} \gamma^\lambda (1-\gamma_5) \begin{pmatrix} D \\ S \end{pmatrix} \tag{24.42}$$

With this formulation we can readily understand why there is no second-order charged weak contribution to the rate of decay (24.40). The reason is that along with Figure 24.13a, there is an additional second-order diagram (Figure 24.13b), and the amplitudes corresponding to these two diagrams cancel in the limit where c and u quarks have the same mass. To explain the vanishing first-order neutral weak amplitude associated with Figure 24.12b, GIM proposed that the neutral weak hadronic current density takes the following form (where we suppress space-time indices):

$$\bar{U}U + \bar{C}C + \bar{D}D + \bar{S}S = \bar{U}U + \bar{C}C + \bar{D}_C D_C + \bar{S}_C S_C$$

The significance of this expression is that it contains no cross-terms $\bar{D}S$ or $\bar{S}D$. Thus one expects no neutral weak interactions in which a d quark transforms to an s quark or vice versa and also no transitions in which $u \rightarrow c$ or vice versa.

24.10 *CP* Violation and the CKM matrix

We now come to the phenomenon of CP violation, which was discovered experimentally in 1964 and remains a very important only partially solved problem today. We start by describing the basic properties of neutral K mesons, which are of special interest because of the peculiar particle mixture properties of the $K^0 - \overline{K^0}$ system. Mesons K^0 and $\overline{K^0}$ are charge conjugates of one another and possess definite strangeness +1 and −1, respectively. (The valence quark

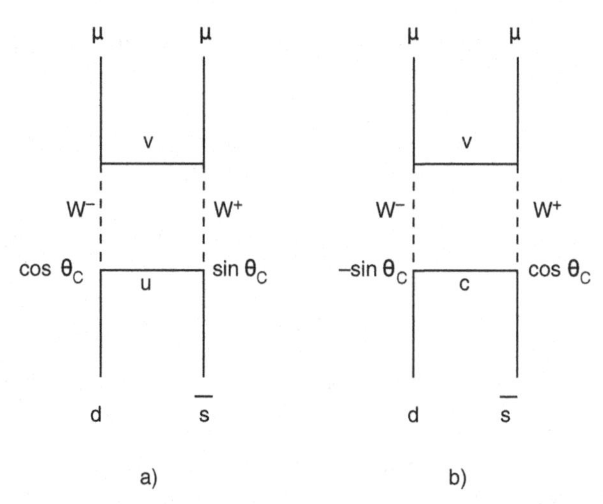

Figure 24.13 Second-order charged weak interaction Feynman diagrams for the decay $K_L^0 \to \mu^+\mu^-$ according to the GIM model.

compositions of K^0 and $\overline{K^0}$ are $d\bar{s}$ and $\bar{d}s$, respectively). It is appropriate to think in terms of the states $|K^0\rangle$ and $|\overline{K^0}\rangle$, which describe these particles at rest, when discussing the strong and electromagnetic interactions, which conserve strangeness. However, neither K^0 nor $\overline{K^0}$ possesses definite mass or a definite decay lifetime because the charged weak interactions do not conserve strangeness. It is useful to imagine a fictitious world in which the weak interactions are "turned off," and $|K^0\rangle$ and $|\overline{K^0}\rangle$ are degenerate eigenstates of the strong and electromagnetic interaction Hamiltonian. Then any independent linear combination of $|K^0\rangle$ and $|\overline{K^0}\rangle$ is also an eigenstate of this zero-order Hamiltonian. When the weak interaction is "turned on," the appropriate linear combinations (called $|K_L^0\rangle$ and $|K_S^0\rangle$ for *long* and *short*, respectively) are those states with definite (and unequal) lifetimes and masses. These states do not have definite strangeness.

The short-lived K_S^0 (mean lifetime 0.89×10^{-10} s) decays in only two significant modes: $\pi^+\pi^-$ and $\pi^0\pi^0$. Each of these final states has CP eigenvalue +1. On the other hand, there are many known modes of decay for K_L^0 (mean lifetime 5.17×10^{-8} s), including the fully allowed decay to $\pi^+\pi^-\pi^0$, a final state that is predominantly an eigenstate of CP with eigenvalue −1. In 1957, L. D. Landau and others suggested that although C and P are separately violated in weak interactions, CP is a valid symmetry for all interactions. This was indeed in accord with all experimental results at the time. If it is assumed to be true, then in light of the foregoing remarks, $|K_L^0\rangle$ and $|K_S^0\rangle$ would be eigenstates of CP with eigenvalues −1 and +1, respectively. Now

$$CP|K^0\rangle = -|\overline{K^0}\rangle \qquad CP|\overline{K^0}\rangle = -|K^0\rangle$$

where the negative signs appear because K^0 and $\overline{K^0}$ each have negative intrinsic parity. Hence, with the assumption of CP conservation, we would have

$$|K_S^0\rangle = \frac{1}{\sqrt{2}}\left(|K^0\rangle - |\overline{K^0}\rangle\right) \tag{24.43}$$

$$|K_L^0\rangle = \frac{1}{\sqrt{2}}\left(|K^0\rangle + |\overline{K^0}\rangle\right) \tag{24.44}$$

According to this, the decays $K_L^0 \to \pi^+\pi^-$ and $K_L^0 \to \pi^0\pi^0$ would be strictly forbidden. However, in 1964, J. Cronin, V. Fitch, and coworkers (Christenson et al. 1964) discovered that these decays actually occur with small but finite probability. Thus expressions (24.43) and (24.44) are incorrect, and it turns out that they must be modified as follows:

$$|K_S^0\rangle = \frac{1}{\sqrt{2(1+|\varepsilon|^2)}}\left[(1+\varepsilon)|K^0\rangle - (1-\varepsilon)|\overline{K^0}\rangle\right] \tag{24.45}$$

$$|K_L^0\rangle = \frac{1}{\sqrt{2(1+|\varepsilon|^2)}}\left[(1+\varepsilon)|K^0\rangle + (1-\varepsilon)|\overline{K^0}\rangle\right] \tag{24.46}$$

where ε is a small complex parameter determined from many precise experiments to be given by $\text{Re}(\varepsilon) = 1.62\times10^{-3}$, $\text{Arg}(\varepsilon) = 43.7°$.

CP violation also has been observed in recent years in the decays of $B^0\overline{B^0}$ meson pairs, where $B^0 = b\bar{d}$ and $\overline{B^0} = \bar{b}d$. The results of all $K^0\overline{K^0}$ and $B^0\overline{B^0}$ *CP* violation experiments so far are consistent with a phenomenological description given by Kobayashi and Maskawa almost four years before discovery of the third quark generation (Kobayashi and Maskawa 1973). This is expressed in terms of the unitary 3×3 Cabibbo-Kobayashi-Maskawa (CKM) mixing matrix. In this description, one writes the Hermitian conjugate charged weak current of quarks as

$$J^{\lambda\dagger} = \overline{P}_L\gamma^\lambda U N_L \tag{24.47}$$

where P_L and N_L are separate column vectors of left-handed quark fields with electric charges $+2/3e$, $-1/3e$, respectively; that is,

$$P_L = \begin{pmatrix} u \\ c \\ t \end{pmatrix}_L \qquad N_L = \begin{pmatrix} d \\ s \\ b \end{pmatrix}_L \tag{24.48}$$

Here *left-handed* means that each three-component quark column vector is preceded by the projection operator $(1-\gamma_5)/2$, and

$$U = \begin{pmatrix} U_{ud} & U_{us} & U_{ub} \\ U_{cd} & U_{cs} & U_{cb} \\ U_{td} & U_{ts} & U_{tb} \end{pmatrix} \tag{24.49}$$

is the CKM mixing matrix, a generalization of the 2×2 GIM (Cabibbo) mixing matrix. Most generally, a complex 3×3 matrix contains $3\times3=9$ complex numbers or 18 real parameters. The unitary condition $U^\dagger U = I$ imposes nine constraints and one overall phase is arbitrary, so the number of independent real parameters in U would at first appear to be 8. However, the relative phases of u, c, t and d, s, b are completely arbitrary. Thus four degrees of freedom remain in U, and most generally it cannot be a real orthogonal 3×3 matrix, which is characterized by only three independent real angles. Instead, we need three Cabibbo angles θ_{12}, θ_{23}, and θ_{13} and an additional real parameter δ, which is interpreted as a CP-violating phase. In standard notation, U is written as

$$U = \begin{pmatrix} c_{12}c_{13} & s_{12}c_{13} & s_{13}e^{-i\delta} \\ -s_{12}c_{23} - c_{12}s_{23}s_{13}e^{i\delta} & c_{12}c_{23} - s_{12}s_{23}s_{13}e^{i\delta} & s_{23}c_{13} \\ s_{12}s_{23} - c_{12}c_{23}s_{13}e^{i\delta} & -c_{12}s_{23} - s_{12}c_{23}s_{13}e^{i\delta} & c_{23}c_{13} \end{pmatrix} \tag{24.50}$$

where $c_{ij} = \cos\theta_{ij}$ and $s_{ij} = \sin\theta_{ij}$, and $i, j = 1, 2, 3$, are generation labels. It can be shown that all CP-violating amplitudes in neutral K and B meson decays are proportional to

$$J = s_{12}s_{13}s_{23}c_{12}c_{13}^2c_{23}\sin\delta \tag{24.51}$$

It can also be shown that CP-violating effects vanish in the limit where any two quarks with the same electric charge (e.g., u and c or d and s) have the same mass; this is due to cancellations in the sum over diagrams containing all quark generations. Various observations of CP violation in K- and B-meson decays yield the value

$$\delta = 1.05 \pm 0.24 \text{ radians} \tag{24.52}$$

Thus δ is a large phase, but $J \approx 3 \times 10^{-5}$ is a very small quantity because of the small values of s_{12}, s_{13}, and s_{23}.

Although all experimental results on CP violation obtained to date are consistent with the formulation we have just described, we cannot say that we understand CP violation, and most physicists familiar with the phenomenon consider it to be one of the major unsolved problems in modern quantum physics. For example, we do not know how to calculate any of the entries in the CKM matrix from first principles.

24.11 Invention of the standard electroweak model

The most important development in the modern history of weak interactions was the invention of the standard electroweak model. This theory was created independently by S. Weinberg (Weinberg 1967) and A. Salam (Salam 1968), was proved to be renormalizable by G. 't Hooft and M. Veltman ('t Hooft and Veltman 1971a, b), and has been verified by numerous precise experiments in the decades that have followed. The theory is based on an intricate combination of subtle ideas, and its complete exposition is beyond the scope of this book. In this section we first give a superficial and heuristic summary of the most significant points, and then we give more details in Sections 24.12–24.16.

The single most important principle assumed in construction of the theory is that its Lagrangian density is invariant under certain local gauge transformations. We have seen in Section 22.6 that in quantum electrodynamics the principle of local $U(1)$ gauge invariance applied to a single Dirac field is equivalent to the familiar minimal coupling of the current density j^{μ}_{EM} and the electromagnetic field A_μ. The gauge quanta that arise here are the massless photons.

In 1954, C. N. Yang and R. Mills (Yang and Mills 1954) proposed an important generalization of this idea to local $SU(2)$ gauge transformations on an isodoublet field. An isodoublet Dirac field Ψ can be expressed as $\Psi = \begin{pmatrix} \Psi_1 \\ \Psi_2 \end{pmatrix}$, where the component fields Ψ_1 and Ψ_2 are associated with quanta having electric charges that differ by unity. For example, Ψ_1 and Ψ_2 could refer to the electron-neutrino and electron fields, respectively, or to the up and down quark fields, respectively. Prior to the introduction of a mechanism for differentiating their masses, one can regard the quanta associated with Ψ_1 and Ψ_2 to be particles of the same mass, identical in all respects except for their electric charge. From this viewpoint, $\Psi = \begin{pmatrix} \Psi_1 \\ \Psi_2 \end{pmatrix}$ closely resembles a two-component spinor in the theory of angular momentum. An $SU(2)$ transformation on such a spinor is equivalent to a rotation that mixes the components of Ψ_1 and Ψ_2 with one another and/or generates opposite-phase shifts in Ψ_1 and Ψ_2. This is to be contrasted with a $U(1)$ gauge transformation, which shifts the phases of both fields Ψ_1 and Ψ_2 by the same amount.

The local gauge invariance principle in the Yang-Mills theory gives rise to massless vector quanta with electric charges +1, −1, and 0. Eventually, it was realized that attractive possibilities are thus suggested for the weak interactions. Might the Yang-Mills quanta be identified as charged and neutral intermediate vector bosons? The difficulty is that the real weak bosons are massive, whereas the Yang-Mills quanta are massless. This defect cannot be repaired simply by adding a mass term to the Yang-Mills Lagrangian density; if we do this, we are back to the naive vector boson theory with all its grave defects. Another way had to be found to impart mass to the Yang-Mills quanta.

At this point, another idea appeared that seemed at first quite unrelated to the problems just discussed but ultimately played a direct role in their solution. This idea was related to the phenomenon of spontaneous symmetry breaking, which is a situation in which the Lagrangian density of a field theory possesses a certain symmetry not shared by the ground or vacuum state of the system. We introduce the phenomenon by means of an example from classical mechanics, first studied by Euler, that involves elastic instability. Consider a thin cylindrical metal rod of length L and radius R. The rod is fixed in bearings attached to clamps at both ends, and the clamps are fixed to a vertical wall, as shown in Figure 24.14. The upper clamp can slide up or down in the wall; the lower clamp cannot move. The clamps are designed with bearings so that the rod can rotate freely in them about the z-axis, much as an automobile crankshaft rotates in its fixed main bearings.

Suppose that we apply a compressional force F directed along the rod axis (z-axis) as in Figure 24.14a. Let the deflections of the rod in the x- and y-directions as a function of z be $X(z)$ and $Y(z)$, respectively. The boundary conditions for the clamped rod are

$$\begin{aligned} &X(0) = X(L) = 0 \\ &X'(0) = X'(L) = 0 \qquad X' = dX/dz \end{aligned}$$

with similar conditions on Y. It can be shown from the theory of elasticity that the conditions for equilibrium are

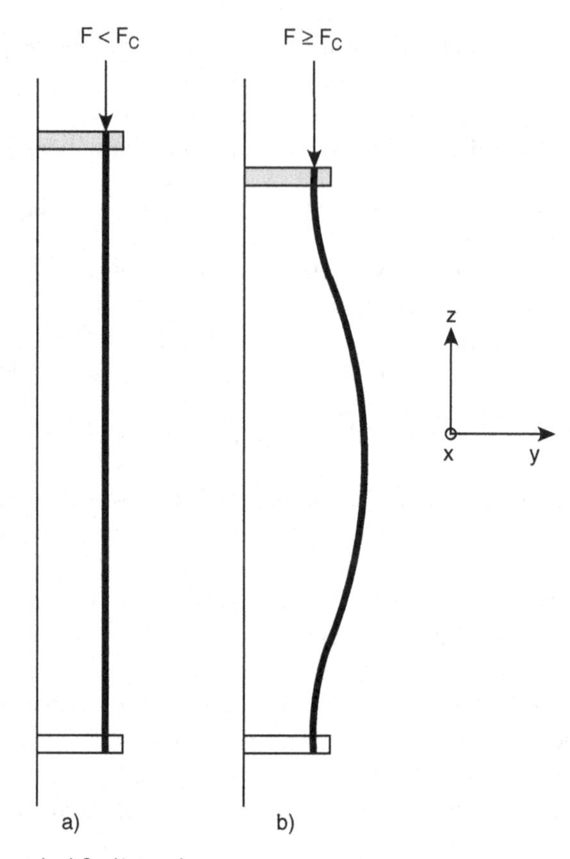

Figure 24.14 The heavy line represents the metal rod, fixed in two clamps.

$$
\begin{aligned}
IE\frac{d^4X}{dz^4}+F\frac{d^2X}{dz^2}=0 \\
IE\frac{d^4Y}{dz^4}+F\frac{d^2Y}{dz^2}=0
\end{aligned}
\tag{24.53}
$$

where E is Young's modulus, and $I=(\pi/4)R^4$. Obviously, if F is very small (Figure 24.14a), the equilibrium condition is that the rod is straight: we have the solutions $X(z) = Y(z) = 0$, and the rod is stable against small perturbations from straightness. Now we increase F until it surpasses a certain critical value F_C. Then the rod becomes unstable against a small transverse perturbation, and it bends to form a new equilibrium shape (Figure 24.14b). Although in general the deflection is large and equations (24.53), which are valid for small deflections, cannot be applied to Figure 24.14b, these equations still can be used to find F_C, the force required for neutral stability. It is not difficult to show that such a solution with $X(z)\neq 0$ is

$$X(z)=\text{const}\bullet\sin^2\frac{\pi z}{L}$$

which corresponds to $F_C = 4\pi^2 IE/L^2$.

We now consider small transverse oscillations of the rod about its equilibrium position. When the rod is straight, it can be shown that the eigenfrequencies ω for small oscillations are given by the formula

$$\cos kL \cosh kL = 1$$

where $k = \omega^{1/2} \left(\mu_0 / EI\right)^{1/4}$, and μ_0 is the mass per unit length. Obviously, when the rod is straight, the frequency spectrum is the same for small oscillations in any plane containing the z-axis. When the rod is bent (Figure 24.14b), we can also have small oscillations. However, the characteristic frequencies are different. In particular, the frequencies for oscillations in the plane of bending are not the same as in the perpendicular direction. Indeed, because the rod can rotate freely about the z-axis in the clamp bearings (like a crankshaft), the lowest frequency of oscillation in the direction perpendicular to the plane of bending is zero.

To summarize, the Lagrangian of this system is axially symmetric. It contains a continuous parameter: the compressional force F. For small values of F, the lowest energy state of the system, which we can call the ground state, also has axial symmetry (the rod is straight). For $F > F_C$, the system, when perturbed infinitesimally, jumps to a new ground state in which the axial symmetry is broken (bent rod). This new state is degenerate: it is but one of an infinite number of possible states because the rod can be bent in any plane containing the z-axis. Small oscillations about the original axially symmetric ground state themselves have axial symmetry, but this is not true of the bent rod.

We now return to spontaneous symmetry breaking in field theory. In place of the rod deflection $X(z)$ or $Y(z)$, we consider a quantum field $\phi(x_\mu)$ with a Lagrangian density possessing a certain symmetry and depending on a continuous parameter (the analogue of the compressional force F_C). For certain values of the parameter, the ground or vacuum state of the field possesses the symmetry of the Lagrangian density. However, if the parameter exceeds a certain critical value, the symmetry is spontaneously broken in the ground state, which becomes degenerate. Small oscillations are analogous to the appearance of particles, the relationship between their energy and momentum being expressed in terms of mass. In general, the mass of the particles will differ for the symmetric ground state, on the one hand, and for the degenerate ground state(s) in which the symmetry is spontaneously broken, on the other hand.

For application to the standard model, one considers a complex isodoublet scalar field (with four real component fields). The Lagrangian density contains a continuous parameter μ^2. For $\mu^2 > 0$, the ground state of the system possesses the full symmetry of the Lagrangian density, and the quanta that appear are four real scalar bosons of identical positive mass μ. However, for $\mu^2 < 0$, the symmetry is broken in the ground state, and we obtain one scalar quantum with positive mass (the Higgs boson) and three massless scalar bosons (the latter are the analogues of the vanishing lowest frequency of small oscillation in the direction perpendicular to the plane of bending of the rod). These massless scalar bosons are called *Goldstone bosons* after the author who first pointed out that for theories of this type, a massless scalar field always appears for each degree of freedom in which the symmetry is spontaneously broken.

Because there is no experimental evidence for such Goldstone bosons, it would appear that we have introduced a new problem with no physical relevance instead of solving our existing problems. However, if we now modify the Lagrangian density of the complex isodoublet scalar field so that it is invariant under *local* $SU(2) \times U(1)$ gauge transformations (and this is done by changing partial derivatives in the Lagrangian density to the appropriate covariant derivatives), the situation is radically altered. In addition to four scalar quanta, we now have four vector gauge quanta [corresponding to three from $SU(2)$ and one from $U(1)$]. Moreover, when μ^2 is chosen to be negative to give spontaneous symmetry breaking, the Goldstone bosons disappear, and the three degrees of freedom to which they correspond appear as additional (longitudinal polarization) degrees of freedom, one for each of three gauge quanta. The appearance of longitudinal polarization in a vector field is equivalent to mass, so we obtain three massive gauge quanta with charges +1, –1, and 0. These are identified as the charged and neutral weak vector bosons. Thus two fundamental difficulties – the appearance of Goldstone bosons and the problem of imparting masses to the weak gauge bosons – very neatly cancel one another. The fourth vector quantum remains massless and is identified as the photon. Of the four original scalar bosons, only the single massive Higgs boson survives. Another outcome of this analysis is that the three massive gauge fields and the Higgs field possess unique and specific nonlinear interactions with one another that have interesting and important physical implications and are an inevitable consequence of this type of theory, which is called *non-Abelian.*

The next step was to introduce leptons and quarks and couple them to the gauge fields. The choices here were very much constrained by the requirement that the new theory reproduce the known and valid results of electrodynamics and the *V-A* law for low-energy charged weak interactions. Thus the *V-A* law, which contains specific predictions concerning *P* and *C* violation, was inserted "by hand" into the new theory. One might argue that this is unsatisfactory because a fundamental theory of weak interactions ought to explain why the striking phenomena of parity violation and charge-conjugation symmetry violation occur. On the other hand, the fermion–gauge field couplings are determined in this manner not only for electrodynamics and the charged weak interactions but also for a whole new class of neutral weak interactions, which were unknown at the time the theory was created. In the years 1973–1981, detailed observations of neutral weak couplings in neutrino-nucleon, neutrino-electron, and electron-nucleon scattering and in low-energy atomic physics began to verify the predictions of the Weinberg-Salam electroweak model in detail. This culminated in the early 1980s in the first direct observations of the $W^{\pm}$ and Z^0 bosons, followed in the early 1990s by extremely precise observations of Z^0 bosons at the large electron-positron collider (LEP) at CERN, with similar observations at the Stanford Linear Accelerator Center (SLAC), together with precise measurements of $W^{\pm}$ properties at Fermilab. Virtually all the results obtained in these experiments are in excellent agreement with predictions of the standard model. In Section 24.17 we summarize many of the Feynman rules that emerge from the Lagrangian density of the standard electroweak model.

There remains the question of the fermion masses. In the standard electroweak theory, this is dealt with by introduction of a gauge-invariant fermion-fermion-Higgs coupling of the Yukawa type, but it remains to be seen whether this portion of the theory is valid. As for the Higgs boson itself, persuasive experimental evidence for its existence, with mass $m_H = 126$ GeV/c^2, was obtained at the CERN Large Hadron Collider in 2012.

24.12 Essential features of the Yang-Mills theory

24.12.1 Review of $U(1)$ gauge invariance in quantum electrodynamics

To set the stage for our discussion of the Yang-Mills theory, we recall that in Section 22.6 we were concerned with a single Dirac field Ψ (and its Dirac-conjugate field $\bar{\Psi}$) governed by the Lagrangian density

$$\mathcal{L} = \frac{1}{2}\bar{\Psi}\left(i\gamma^{\mu}\partial_{\mu} - m\right)\Psi - \frac{1}{2}\left(i\partial_{\mu}\bar{\Psi}\gamma^{\mu} + m\bar{\Psi}\right)\Psi \tag{24.54}$$

We sought the change $\delta\mathcal{L}$ in this Lagrangian density when an infinitesimal gauge transformation was made described by the following equations:

$$\delta\Psi = i\alpha\Psi \qquad \delta\bar{\Psi} = -i\alpha\bar{\Psi} \tag{24.55}$$

We found that

$$\delta\mathcal{L} = -\bar{\Psi}\gamma^{\mu}\Psi\partial_{\mu}\alpha \tag{24.56}$$

If α is a constant, then $\delta\mathcal{L} = 0$. In other words, $\mathcal{L}$ is invariant under global $U(1)$ gauge transformations. If $\alpha = \alpha(x_{\mu})$ is a function of the space-time coordinates, $\delta\mathcal{L} \neq 0$, and $\mathcal{L}$ is not invariant under local $U(1)$ gauge transformations. However, that invariance is restored if we include the interaction Lagrangian (22.63)[1]; that is,

$$\mathcal{L}_{\text{int}} = e\bar{\Psi}\gamma^{\mu}\Psi\cdot A_{\mu} = -j^{\mu}_{\text{EM}}A_{\mu} \tag{24.57}$$

provided that whenever we make the transformation (24.55), we also change the vector potential A_{μ} as follows:

$$A_{\mu} \to A'_{\mu} = A_{\mu} + \frac{1}{e}\partial_{\mu}\alpha \tag{24.58}$$

The combination $\mathcal{L} + \mathcal{L}_{\text{int}}$ is

$$\mathcal{L}_{D} = \frac{1}{2}\bar{\Psi}\left(i\gamma^{\mu}D_{\mu} - m\right)\Psi - \frac{1}{2}\left(iD_{\mu}\bar{\Psi}\gamma^{\mu} + m\bar{\Psi}\right)\Psi \tag{24.59}$$

where the covariant derivatives $D_{\mu}\Psi$ and $D_{\mu}\bar{\Psi}$ are defined as

$$\begin{aligned} D_{\mu}\Psi &\equiv \left(\partial_{\mu} - ieA_{\mu}\right)\Psi \\ D_{\mu}\bar{\Psi} &\equiv \left(\partial_{\mu} + ieA_{\mu}\right)\bar{\Psi} \end{aligned} \tag{24.60}$$

[1] Remember that $e > 0$ and the charge of the electron is $-e$.

The covariant derivative is important because, as is very easy to verify, under the infinitesimal gauge transformation (24.55) it transforms in the same way as the field itself; that is,

$$\delta\left(D_\mu \Psi\right) = i\alpha D_\mu \Psi \tag{24.61}$$

$$\delta\left(D_\mu \bar{\Psi}\right) = -i\alpha D_\mu \bar{\Psi} \tag{24.62}$$

To complete the Lagrangian density of quantum electrodynamics, we add the radiation portion

$$\mathcal{L}_{\text{rad}} = -\frac{1}{4} F^{\mu\nu} F_{\mu\nu} \tag{24.63}$$

which is gauge-invariant.

Sometimes it is convenient to employ the following relations, which are easily derived from the definitions (24.60) and $F_{\mu\nu} \equiv \partial_\mu A_\nu - \partial_\nu A_\mu$:

$$\left(D_\mu D_\nu - D_\nu D_\mu\right)\Psi = -ieF_{\mu\nu}\Psi \tag{24.64}$$

$$\left(D_\mu D_\nu - D_\nu D_\mu\right)\bar{\Psi} = ieF_{\mu\nu}\bar{\Psi} \tag{24.65}$$

24.12.2 *SU*(2) gauge transformations

We now consider a Dirac field with two isospin components:

$$\Psi = \begin{pmatrix} \Psi_1 \\ \Psi_2 \end{pmatrix} \tag{24.66}$$

where Ψ_1 and Ψ_2 are each Dirac fields with four space-time components, but they describe particles with different electric charge. For example, we might have

$$\Psi = \begin{pmatrix} \nu_e \\ e^- \end{pmatrix} \quad \text{or} \quad \Psi = \begin{pmatrix} u \\ d \end{pmatrix}$$

We want to consider infinitesimal $SU(2)$ gauge transformations on this field; that is,

$$\delta\Psi = i\boldsymbol{\varepsilon}\boldsymbol{\cdot}\boldsymbol{t}\Psi \qquad \delta\bar{\Psi} = -i\bar{\Psi}\boldsymbol{\varepsilon}\boldsymbol{\cdot}\boldsymbol{t} \tag{24.67}$$

where $\boldsymbol{\varepsilon} = \varepsilon_1\hat{i} + \varepsilon_2\hat{j} + \varepsilon_3\hat{k}$, $\varepsilon_{1,2,3}$ are three real infinitesimals, and

$$\boldsymbol{t} = \frac{1}{2}\left(\tau_1\hat{i} + \tau_2\hat{j} + \tau_3\hat{k}\right) \tag{24.68}$$

where

$$\tau_1 = \begin{pmatrix} 0 & 1 \\ 1 & 0 \end{pmatrix} \quad \tau_2 = \begin{pmatrix} 0 & -i \\ i & 0 \end{pmatrix} \quad \tau_3 = \begin{pmatrix} 1 & 0 \\ 0 & -1 \end{pmatrix} \tag{24.69}$$

are the Pauli (isospin) matrices. If the quantities ε_i depend on the space-time coordinates, we have a local $SU(2)$ gauge transformation. Once more, we want to investigate the change in the Lagrangian density

$$\mathcal{L} = \frac{1}{2}\bar{\Psi}\left(i\gamma^\mu\partial_\mu - m\right)\Psi - \frac{1}{2}\left(i\partial_\mu\bar{\Psi}\gamma^\mu + m\bar{\Psi}\right)\Psi \tag{24.70}$$

but now we make the transformation (24.67). Following steps similar to those that led to (24.57) and making use of the equation of continuity once again, we find

$$\delta\mathcal{L} = -\left(\bar{\Psi}\gamma^\mu \boldsymbol{t}\Psi\right)\cdot\partial_\mu\boldsymbol{\varepsilon} \tag{24.71}$$

We try to compensate for this by adding an interaction Lagrangian density of the form

$$\mathcal{L}_I = -g\bar{\Psi}\gamma^\mu \boldsymbol{t}\Psi\cdot\boldsymbol{A}_\mu \tag{24.72}$$

where g is a new $SU(2)$ coupling constant, and the $\boldsymbol{A}_\mu$ form a triplet of gauge fields, each with four space-time components.

In quantum electrodynamics, the vector potential transforms according to rule (24.58) in an infinitesimal $U(1)$ gauge transformation. What is the corresponding rule for the $\boldsymbol{A}_\mu$ in the infinitesimal $SU(2)$ gauge transformation? To find it, we start with the covariant derivatives

$$D_\mu\Psi = \left(\partial_\mu + ig\boldsymbol{A}_\mu\cdot\boldsymbol{t}\right)\Psi \tag{24.73}$$

$$D_\mu\bar{\Psi} = \partial_\mu\bar{\Psi} - ig\boldsymbol{A}_\mu\cdot\bar{\Psi}\boldsymbol{t} \tag{24.74}$$

and require that they transform in the same way as the respective fields; that is,

$$\delta\left(D_\mu\Psi\right) = i\boldsymbol{\varepsilon}\cdot\boldsymbol{t}\left(D_\mu\Psi\right) \tag{24.75}$$

$$\delta\left(D_\mu\bar{\Psi}\right) = -i\left(D_\mu\bar{\Psi}\right)\boldsymbol{\varepsilon}\cdot\boldsymbol{t} \tag{24.76}$$

Substituting (24.73) in (24.75), we obtain

$$\delta\left(\partial_\mu\Psi\right) + ig\delta\left(\boldsymbol{A}_\mu\cdot\boldsymbol{t}\Psi\right) = i\boldsymbol{\varepsilon}\cdot\boldsymbol{t}\left(\partial_\mu + ig\boldsymbol{A}_\mu\cdot\boldsymbol{t}\right)\Psi \tag{24.77}$$

However,

$$\delta\left(\partial_\mu\Psi\right) = \partial_\mu\left(\delta\Psi\right) = i\partial_\mu\left(\boldsymbol{\varepsilon}\cdot\boldsymbol{t}\right)\Psi + i\boldsymbol{\varepsilon}\cdot\boldsymbol{t}\partial_\mu\Psi \tag{24.78}$$

Also,

$$\delta\left(A_\mu \boldsymbol{\cdot} t\Psi\right) = \delta\left(A_\mu \boldsymbol{\cdot} t\right)\Psi + iA_\mu \boldsymbol{\cdot} t\varepsilon \boldsymbol{\cdot} t\Psi \tag{24.79}$$

Substituting (24.78) and (24.79) into the left-hand side of (24.77) and comparing both sides of the resulting equation, we arrive at

$$\delta\left(A_\mu \boldsymbol{\cdot} t\right) = -\frac{1}{g}\partial_\mu\left(\varepsilon \boldsymbol{\cdot} t\right) + i\left(\varepsilon \boldsymbol{\cdot} tA_\mu \boldsymbol{\cdot} t - A_\mu \boldsymbol{\cdot} t\varepsilon \boldsymbol{\cdot} t\right) \tag{24.80}$$

Now we make use of the $SU(2)$ commutation relation

$$\left[t_i, t_j\right] = i\varepsilon_{ijk}t_k$$

to simplify the last term on the right-hand side of (24.80). This yields the result

$$\delta A_\mu = -\frac{1}{g}\partial_\mu \varepsilon - \left(\varepsilon \times A_\mu\right) \tag{24.81}$$

On the right-hand side of (24.81), the first term is analogous to that obtained in the $U(1)$ case in (24.58): g replaces $-e$. However, the second term is entirely new; it arises because we are dealing with $SU(2)$ matrices that do not necessarily commute. One refers to this essential new feature by stating that A_μ is a *non-Abelian gauge field.*

The next step is to formulate a gauge invariant "radiation" Lagrangian density analogous to $\mathcal{L}_{\text{rad}} = -\frac{1}{4}F^{\mu\nu}F_{\mu\nu}$ of quantum electrodynamics. It is convenient to start with relations analogous to (24.64) and (24.65); that is,

$$\left(D_\mu D_\nu - D_\nu D_\mu\right)\Psi = igE_{\mu\nu} \boldsymbol{\cdot} t\Psi \tag{24.82}$$

$$\left(D_\mu D_\nu - D_\nu D_\mu\right)\bar{\Psi} = -igE_{\mu\nu} \boldsymbol{\cdot} \bar{\Psi}t \tag{24.83}$$

where $E_{\mu\nu}$ is analogous to $F_{\mu\nu}$. Employing (24.73) in (24.82) and carrying out some simple algebra, we find

$$E_{\mu\nu} = \partial_\mu A_\nu - \partial_\nu A_\mu - g\left(A_\mu \times A_\nu\right) \tag{24.84}$$

The cross-product term on the right-hand side of (24.84) is, once again, a manifestation of the non-Abelian nature of the gauge field A_μ. With (24.84), the gauge invariant "radiation" Lagrangian density $\mathcal{L}_{YM} = -\frac{1}{4}E^{\mu\nu} \boldsymbol{\cdot} E_{\mu\nu}$ contains not only the zeroth-order term similar to that of quantum electrodynamics

$$-\frac{1}{4}\left(\partial^\mu A^\nu - \partial^\nu A^\mu\right) \boldsymbol{\cdot} \left(\partial_\mu A_\nu - \partial_\nu A_\mu\right) \tag{24.85}$$

but also terms of order g, namely,

$$\frac{g}{4}\left[(\partial^\mu \boldsymbol{A}^\nu - \partial^\nu \boldsymbol{A}^\mu)\cdot(\boldsymbol{A}_\mu \times \boldsymbol{A}_\nu) + (\boldsymbol{A}_\mu \times \boldsymbol{A}_\nu)\cdot(\partial^\mu \boldsymbol{A}^\nu - \partial^\nu \boldsymbol{A}^\mu)\right] \tag{24.86}$$

and a term of order g^2, namely,

$$-\frac{g^2}{4}(\boldsymbol{A}^\mu \times \boldsymbol{A}^\nu)\cdot(\boldsymbol{A}_\mu \times \boldsymbol{A}_\nu) \tag{24.87}$$

To summarize, let us compare the quanta of the Yang-Mills field $\boldsymbol{A}_\mu$ to photons, which are the quanta of the electromagnetic field A_μ. There are similarities, but there are also important differences. First, the similarities: the Yang-Mills quanta and the photon all have zero mass, and it can be shown that the Yang-Mills theory, like QED, is renormalizable. As for the differences, the Yang-Mills quanta, described by the combinations

$$-\frac{1}{\sqrt{2}}(A_{1\mu} + iA_{2\mu}) \qquad A_{3\mu} \qquad \frac{1}{\sqrt{2}}(A_{1\mu} - iA_{2\mu})$$

have electric charges $+e$, 0, $-e$, respectively, whereas the photon carries no charge, electrical or otherwise. Also, the terms (24.86) and (24.87) in $\mathcal{L}_{YM}$ have no analogue in electrodynamics. Such terms give rise to self-interaction of the Yang-Mills quanta, with Feynman diagrams having 3-boson vertices from (24.86) and a 4-boson vertex from (24.87).

As we have mentioned, the Yang-Mills quanta appear to be attractive candidates for the charged and neutral weak intermediate vector bosons. However, the latter are massive, whereas the Yang Mills quanta are massless. Thus a way must be found to impart mass to the Yang-Mills quanta without spoiling renormalizability. In the following section we try to explain how this is done.

24.13 Spontaneous symmetry breaking and the Higgs mechanism

The standard model makes use of the simplest possible solution to the problem of intermediate vector boson masses. This solution employs a complex isodoublet scalar field ϕ with self-interaction. However, in order to introduce the ideas as simply and clearly as possible, we start with a complex *isosinglet* scalar field with Lagrangian density

$$\mathcal{L} = \partial^\nu \phi^* \partial_\nu \phi - \mu^2 \phi^* \phi - \lambda(\phi^* \phi)^2 \tag{24.88}$$

and only afterward do we go on to the complex isodoublet case. In (24.88), μ is the mass of a field quantum, and λ is a real positive number that characterizes the strength of the self-interaction term, which is the last term on the right-hand side of (24.88). If λ were equal to zero, the Euler-Lagrange equations for ϕ and ϕ^* would yield the Klein-Gordon equations

$$\partial^\nu \partial_\nu \phi^* + \mu^2 \phi^* = 0$$
$$\partial^\nu \partial_\nu \phi + \mu^2 \phi = 0$$

With $\lambda > 0$, we search for the field configuration that yields the lowest field energy (this is, by definition, the vacuum). This is analogous to finding the configuration of stable equilibrium for the metal rod of Figure 24.14. Of course, the field is an operator, and it is the expectation value of the field that we are speaking of, but for brevity we shall describe the field as if it were an ordinary function. To calculate the field energy, we need the Hamiltonian density, which is

$$\begin{aligned}\mathcal{H} &= \partial_t \phi \frac{\partial \mathcal{L}}{\partial(\partial_t \phi)} + \partial_t \phi^* \frac{\partial \mathcal{L}}{\partial(\partial_t \phi^*)} - \mathcal{L} \\ &= \nabla\phi \cdot \nabla\phi^* + \partial^0 \phi \partial^0 \phi^* + \mu^2 \phi^* \phi + \lambda(\phi^* \phi)^2\end{aligned} \tag{24.89}$$

Because the first two terms in the second line of (24.89) are always positive unless ϕ is a constant (in which case those terms vanish), it is clear that the lowest energy of the field is obtained when ϕ is a constant, independent of the space-time coordinates. In this case, only the last two terms on the right-hand side of (24.89) remain

$$V = \mu^2 y + \lambda y^2 \tag{24.90}$$

where $y = \phi^* \phi \geq 0$. If $\mu^2 > 0$, the minimum of V occurs for $y = 0$, in which case $V = 0$. Now we vary the parameter μ^2, allowing it to become negative. This is analogous to varying the external force F in the case of the cylindrical rod (Figure 24.14). For negative values of μ^2 (analogous to $F > F_C$), we find the minimum of V by differentiating V with respect to y and setting the result equal to zero. Here we obtain

$$y = \frac{-\mu^2}{2\lambda}$$

and thus

$$\phi_{\text{vac}} = \sqrt{\frac{-\mu^2}{2\lambda}} e^{i\theta} \tag{24.91}$$

where θ is an arbitrary real number between 0 and 2π. Each ϕ_{vac} is analogous to a ground state of the rod when $F > F_C$. Thus choice of a particular θ is analogous to choice of a particular vertical plane containing the bent rod. We choose $\theta = 0$, and thus the vacuum state is

$$\phi_0 = \sqrt{\frac{-\mu^2}{2\lambda}} \tag{24.92}$$

By making such a choice, we break the symmetry of the Lagrangian. Of course, in the example of Figure 24.14, an external perturbation must determine that the rod bends in one particular plane rather than another. The same is true for the complex scalar field.

Now we allow the field to vary with the space-time coordinates by a very small amount about the vacuum state. This is analogous to small transverse oscillations of the rod. We write

$$\begin{aligned}\phi &= \phi_0 + \frac{1}{\sqrt{2}}[\phi_1(x) + i\phi_2(x)] \\ &= \frac{1}{\sqrt{2}}[(\phi_1 + v) + i\phi_2]\end{aligned} \tag{24.93}$$

where $\phi_{1,2}$ are two infinitesimal real fields, and $v = \sqrt{2}\phi_0$. Substituting (24.93) into the Lagrangian density (24.88), we obtain

$$\mathcal{L} = \frac{1}{2}\partial^\nu(\phi_1 - i\phi_2)\partial_\nu(\phi_1 + i\phi_2) - \frac{\mu^2}{2}\left[(\phi_1 + v)^2 + \phi_2^2\right] - \frac{\lambda}{4}\left[(\phi_1 + v)^2 + \phi_2^2\right]^2$$

We expand this expression, noting that because $v^2 = -\mu^2/\lambda$, the coefficients of the terms in ϕ_1 and ϕ_2^2 vanish. Discarding a constant term, we obtain

$$\mathcal{L} = \frac{1}{2}\partial^\nu\phi_1\partial_\nu\phi_1 + \frac{1}{2}\partial^\nu\phi_2\partial_\nu\phi_2 + \mu^2\phi_1^2 + (0)\phi_2^2 + (\text{terms in } \phi_1^3, \phi_1^4, \phi_1\phi_2^2, \ldots) \tag{24.94}$$

This Lagrangian density contains two real scalar fields: ϕ_1 and ϕ_2. The mass associated with ϕ_2 is zero. This is analogous to the fact that in the direction perpendicular to the page in Figure 24.14b the lowest frequency of oscillation is zero. The mass associated with ϕ_1 is $(-2\mu^2)^{1/2} > 0$. The field ϕ_1 is analogous to the small oscillations of the bent rod in Figure 24.14b *in* the plane of the paper.

The appearance of zero-mass quanta associated with ϕ_2 is a general phenomenon. As we mentioned, Goldstone showed in 1960 that whenever one has a scalar field theory with a continuous global symmetry [in the present example it is the $U(1)$ global symmetry associated with the factor $e^{i\theta}$] and whenever spontaneous symmetry breaking occurs, there emerge as many zero-mass scalar fields (*Goldstone bosons*) as there are degrees of freedom in which the symmetry is broken.

However, something new and important emerges if we make the Lagrangian density invariant under local as opposed to global gauge transformations. As we know, invariance of the Lagrangian density under a local $U(1)$ gauge transformation is achieved by replacing the partial derivative by an appropriate covariant derivative. Thus, in place of (24.88), we write

$$\mathcal{L} = D^\nu\phi^* D_\nu\phi - \mu^2\phi^*\phi - \lambda(\phi^*\phi)^2 - \frac{1}{4}f^{\alpha\beta}f_{\alpha\beta} \tag{24.95}$$

Here we introduce a $U(1)$ gauge field B_ν analogous to the electromagnetic vector potential by defining the covariant derivatives as

$$\begin{aligned} D_\nu\phi^* &= \left(\partial_\nu - i\frac{g'}{2}B_\nu\right)\phi^* \\ D_\nu\phi &= \left(\partial_\nu + i\frac{g'}{2}B_\nu\right)\phi \end{aligned} \tag{24.96}$$

where $g'/2$ is a coupling constant. We also include a gauge-invariant radiation term, which is the last term on the right-hand side of (24.95).

We know that the field can be expressed as

$$\begin{aligned} \phi &= \phi_0 + \frac{1}{\sqrt{2}}\left[\phi_1(x) + i\phi_2(x)\right] \\ &= \frac{1}{\sqrt{2}}\left[(\phi_1 + v) + i\phi_2\right] \end{aligned}$$

where $\phi_{1,2}$ are infinitesimal real fields that depend on the space-time coordinates and are everywhere very small compared with v. We rewrite ϕ as

$$\phi = \frac{1}{\sqrt{2}}(\phi_1 + v)\left[1 + \frac{i\phi_2}{\phi_1 + v}\right] \approx \frac{1}{\sqrt{2}}(\phi_1 + v)\left[1 + \frac{i\phi_2}{v}\right]$$

and thus we see that ϕ is a real quantity $1/\sqrt{2}(\phi_1 + v)$ multiplied by a phase factor $\exp(i\phi_2/v) \approx 1 + (i\phi_2/v)$ that varies from one space-time point to another. Now we can make an arbitrary local $U(1)$ gauge transformation on ϕ that leaves the Lagrangian density of (24.95) invariant, provided that we make the appropriate accompanying gauge transformation on B_ν. We choose the transformation so that

$$\phi \to \phi' = \phi\left(1 - i\frac{\phi_2}{v}\right) = \frac{1}{\sqrt{2}}(\phi_1 + v)$$

and simultaneously, $B_\nu \to B'_\nu$. Thus the Lagrangian density becomes

$$\mathcal{L} = \frac{1}{2}\left(\partial^\nu - i\frac{g'}{2}B^{\nu\prime}\right)(\phi_1 + v)\left(\partial_\nu + i\frac{g'}{2}B'_\nu\right)(\phi_1 + v) + \mu^2\phi_1^2 + (\cdots) - \frac{1}{4}f'^{\alpha\beta}f'_{\alpha\beta} \tag{24.97}$$

where we have dropped a constant term and where $(\cdots)$ contains terms in ϕ_1^3 and ϕ_1^4. Now

$$\frac{1}{2}\left(\partial^\nu - i\frac{g'}{2}B^{\nu\prime}\right)(\phi_1 + v)\left(\partial_\nu + i\frac{g'}{2}B'_\nu\right)(\phi_1 + v) = \frac{1}{2}\partial^\nu\phi_1\partial_\nu\phi_1 + \frac{g'^2}{8}B'_\nu B'_\nu(v^2 + 2\phi_1 v + \phi_1^2)$$

Thus we have

$$\mathcal{L} = \frac{1}{2}\partial^\nu\phi_1\partial_\nu\phi_1 + \mu^2\phi_1^2 + (\ldots) + \frac{g'^2 v^2}{8}B^\nu B_\nu - \frac{1}{4}f^{\alpha\beta}f_{\alpha\beta} + \frac{g'^2}{8}B^\nu B_\nu(2\phi_1 v + \phi_1^2) \tag{24.98}$$

In this expression, where we have dropped the prime on the gauge field B because it is no longer needed, the right-hand side contains the kinetic energy term, the mass term, and higher-order self-interaction terms for ϕ_1, the gauge field B mass term with

$$m_B = \frac{g'v}{\sqrt{2}} \tag{24.99}$$

and the by now familiar radiation term, and a term that describes higher-order interactions of the gauge field with the ϕ_1 field. This reformulation of the Lagrangian density, in which the massless Goldstone boson field ϕ_2 disappears and the gauge field simultaneously acquires mass, is the *Higgs phenomenon,* and it is the crucial step.

Our discussion is still incomplete because we have not yet included $SU(2)$ gauge transformations. To do this, we replace the complex isosinglet scalar field by a complex isodoublet scalar field

$$\phi = \begin{pmatrix} \phi_1 + i\phi_2 \\ \phi_3 + i\phi_4 \end{pmatrix} \tag{24.100}$$

where $\phi_{1,2,3,4}$ are real scalar fields. In addition, we modify the covariant derivatives to read

$$D_\nu \phi^\dagger = \left(\partial_\nu - i\frac{g}{2} \boldsymbol{A}_\nu \boldsymbol{\cdot} \boldsymbol{\tau} - i\frac{g'}{2} B_\nu \right) \phi^\dagger \tag{24.101}$$

$$D_\nu \phi = \left(\partial_\nu + i\frac{g}{2} \boldsymbol{A}_\nu \boldsymbol{\cdot} \boldsymbol{\tau} + i\frac{g'}{2} B_\nu \right) \phi \tag{24.102}$$

where g is the $SU(2)$ coupling constant, and the $\boldsymbol{A}_\nu$ form a triplet of gauge fields. The Lagrangian density, modified from (24.84), now reads

$$\mathcal{L} = (D_\nu \phi^\dagger)(D_\nu \phi) - \mu^2 \phi^\dagger \phi - \lambda (\phi^\dagger \phi)^2 - \frac{1}{4} f^{\alpha\beta} f_{\alpha\beta} - \frac{1}{4} \boldsymbol{E}^{\alpha\beta} \boldsymbol{\cdot} \boldsymbol{E}_{\alpha\beta} \tag{24.103}$$

We now repeat the steps already carried out with the complex isosinglet scalar field. First, we seek the field configuration that gives the lowest energy (the vacuum state). We assume that this vacuum state is isotropic in space-time; thus we assume that for it the A and B fields are zero. Next, just as before, we assume that the vacuum state of the scalar field must be a constant independent of the space-time coordinates so that derivative terms do not contribute to the Hamiltonian density. And finally, just as before, we vary the parameter μ^2, letting it become negative so that the vacuum state of the field is nonzero. Following steps previously taken, we find that one of the many possible degenerate vacuum states is

$$\phi_0 = \begin{pmatrix} 0 \\ \eta \end{pmatrix} \tag{24.104}$$

where $\eta = \sqrt{-\mu^2/2\lambda}$. The next step is also analogous to what we have done previously: we introduce real infinitesimal scalar fields $\phi_{1,2,3,4}$ that depend on the space-time coordinates and write

$$\phi = \begin{pmatrix} \frac{1}{\sqrt{2}}(\phi_1 + i\phi_2) \\ \eta + \frac{1}{\sqrt{2}}(\phi_3 + i\phi_4) \end{pmatrix}$$

A convenient way to do this is to write

$$\begin{aligned} \phi = \exp(i\boldsymbol{\theta} \boldsymbol{\cdot} \boldsymbol{t}) \begin{pmatrix} 0 \\ \eta + \frac{\sigma}{\sqrt{2}} \end{pmatrix} &\approx (I + i\boldsymbol{\theta} \boldsymbol{\cdot} \boldsymbol{t}) \begin{pmatrix} 0 \\ \eta + \frac{\sigma}{\sqrt{2}} \end{pmatrix} \\ &= \begin{pmatrix} \frac{1}{2}(i\theta_1 - \theta_2)\eta \\ \eta + \frac{\sigma}{\sqrt{2}} - \frac{i}{2}\theta_3 \eta \end{pmatrix} \end{aligned} \tag{24.105}$$

where $\theta_{1,2,3}$ and σ are four infinitesimal real fields, all depending on the space-time coordinates. It is clear from (24.105) that we have obtained ϕ by starting with the real field

$$\begin{pmatrix} 0 \\ \eta + \dfrac{\sigma}{\sqrt{2}} \end{pmatrix} \tag{24.106}$$

and transforming it by means of the local $SU(2)$ gauge transformation defined by $\theta_{1,2,3}$. However, because the Lagrangian density is invariant under local $SU(2)$ and $U(1)$ gauge transformations, we can carry out the inverse transformation on ϕ to form (24.106) as long as we regauge the vector fields A suitably. In this manner we eliminate the Goldstone fields $\theta_{1,2,3}$ and simultaneously impart mass to three of the four vector fields.

To see how this works in detail, we calculate the covariant derivative of ϕ; that is,

$$\begin{aligned} D_\nu \phi &= \left(\partial_\nu + i\frac{g}{2} \boldsymbol{A}_\nu \bullet \boldsymbol{\tau} + i\frac{g'}{2} B_\nu \right)\phi \\ &= \begin{pmatrix} \partial_\nu + i\dfrac{g}{2} A_{3\nu} + i\dfrac{g'}{2} B_\nu & i\dfrac{g}{2}(A_{1\nu} - iA_{2\nu}) \\ i\dfrac{g}{2}(A_{1\nu} + iA_{2\nu}) & \partial_\nu - i\dfrac{g}{2} A_{3\nu} + i\dfrac{g'}{2} B_\nu \end{pmatrix} \begin{pmatrix} 0 \\ \eta + \dfrac{\sigma}{\sqrt{2}} \end{pmatrix} \\ &= \begin{pmatrix} i\dfrac{g}{2}(A_{1\nu} - iA_{2\nu})\left(\eta + \dfrac{\sigma}{\sqrt{2}} \right) \\ \left(\partial_\nu - i\dfrac{g}{2} A_{3\nu} + i\dfrac{g'}{2} B_\nu \right)\left(\eta + \dfrac{\sigma}{\sqrt{2}} \right) \end{pmatrix} \end{aligned} \tag{24.107}$$

Similarly, the covariant derivative of $\phi^\dagger$ is the row isospinor

$$D^\nu \phi^\dagger = \left(-i\frac{g}{2}\left[A_1^\nu + iA_2^\nu \right]\left[\eta + \frac{\sigma}{\sqrt{2}} \right], \left[\partial^\nu + i\frac{g}{2} A_3^\nu - i\frac{g'}{2} B^\nu \right]\left[\eta + \frac{\sigma}{\sqrt{2}} \right] \right) \tag{24.108}$$

Combining (24.107) with (24.108), writing out explicitly the remaining terms of the Lagrangian density (24.103), and dropping an unimportant constant term, we obtain the following expression for $\mathcal{L}$ in terms of the four gauge fields and the field σ (now and henceforth called the *Higgs field*)

$$\begin{aligned} \mathcal{L} &= \frac{1}{2}(\partial^\alpha \sigma)(\partial_\alpha \sigma) + \mu^2 \sigma^2 + \frac{\mu^2}{\eta\sqrt{2}} \sigma^3 + \frac{\mu^2}{8\eta^2} \sigma^4 \\ &+ \frac{g^2}{4}\left[(A_{1\alpha} - iA_{2\alpha})(A_1^\alpha + iA_2^\alpha) + \left(A_{3\alpha} - \frac{g'}{g} B_\alpha \right)\left(A_3^\alpha - \frac{g'}{g} B^\alpha \right) \right]\left(\eta + \frac{\sigma}{\sqrt{2}} \right)^2 \\ &- \frac{1}{4} f_{\alpha\beta} f^{\alpha\beta} - \frac{1}{4} \boldsymbol{E}_{\alpha\beta} \bullet \boldsymbol{E}^{\alpha\beta} \end{aligned} \tag{24.109}$$

At this point it is convenient to define the weak mixing angle θ_W by

$$\frac{g'}{g} = \tan\theta_W \tag{24.110}$$

At the present state of our knowledge, the standard model does not give us a way to calculate θ_W; it must be determined from experiments. Employing (24.110), we write

$$\left(A_{3\alpha} - \frac{g'}{g}B_\alpha\right)\left(A_3^\alpha - \frac{g'}{g}B^\alpha\right) = \frac{1}{\cos^2\theta_W}\left(\cos\theta_W A_{3\alpha} - \sin\theta_W B_\alpha\right)\left(\cos\theta_W A_3^\alpha - \sin\theta_W B^\alpha\right) \tag{24.111}$$

It is also convenient to define the orthogonal linear combinations

$$Z_\alpha = \cos\theta_W A_{3\alpha} - \sin\theta_W B_\alpha \tag{24.112}$$

and

$$A_\alpha = \sin\theta_W A_{3\alpha} + \cos\theta_W B_\alpha \tag{24.113}$$

as well as

$$W_\alpha^{-\dagger} = W_\alpha^+ = \frac{1}{\sqrt{2}}\left(A_{1\alpha} - iA_{2\alpha}\right) \tag{24.114}$$

$$W_\alpha^{+\dagger} = W_\alpha^- = \frac{1}{\sqrt{2}}\left(A_{1\alpha} + iA_{2\alpha}\right) \tag{24.115}$$

With these newly defined quantities, the second line on the right-hand side of (24.109) becomes

$$\frac{g^2}{4}\left(W_\alpha^{+\dagger}W^{+\alpha} + W_\alpha^{-\dagger}W^{-\alpha} + \frac{1}{\cos^2\theta_W}Z_\alpha Z^\alpha\right)\left(\eta^2 + \frac{2\eta\sigma}{\sqrt{2}} + \frac{\sigma^2}{2}\right) \tag{24.116}$$

The terms proportional to the constant η^2 in (24.116) are the mass terms for $W^\pm$ and Z^0. We have

$$m(W^\pm) = \frac{g\eta}{\sqrt{2}} \tag{24.117}$$

and

$$m(Z) = \frac{m_W}{\cos\theta_W} \tag{24.118}$$

The remaining terms in (24.116) involve trilinear and quadrilinear couplings of the gauge fields to the Higgs field. Also, by using (24.112)–(24.115), we can rewrite the last two terms of (24.109) in terms of the electromagnetic field tensor (now labeled $A^{\alpha\beta}$ instead of $F^{\alpha\beta}$), the Z-boson field tensor $Z^{\alpha\beta}$, and the $W^{\pm}$ field tensors $W^{\pm\alpha\beta}$. Also, using the notation $A_{\alpha\beta} \equiv \partial_\alpha A_\beta - \partial_\beta A_\alpha$ and carrying out straightforward algebra, we find that (24.109) is transformed to

$$\begin{aligned}
\mathcal{L} &= \frac{1}{2}(\partial_\alpha\sigma)(\partial^\alpha\sigma) + \mu^2\sigma^2 + \frac{\mu^2}{\eta\sqrt{2}}\sigma^3 + \frac{\mu^2}{8\eta^2}\sigma^4 \\
&+ \frac{m_W^2}{2}\left(W_\alpha^{+\dagger}W^{+\alpha} + W_\alpha^{-\dagger}W^{-\alpha}\right) + \frac{m_Z^2}{2}Z_\alpha Z^\alpha \\
&+ \left(\frac{g^2}{8}\sigma^2 + \frac{g}{2}m_W\sigma\right)\left(W_\alpha^{+\dagger}W^{+\alpha} + W_\alpha^{-\dagger}W^{-\alpha}\right) + \left(\frac{g^2}{8\cos^2\theta_W}\sigma^2 + \frac{g}{2}m_Z\sigma\right)Z_\alpha Z^\alpha \\
&- \frac{1}{4}\left[\left(W_{\alpha\beta}^{+\dagger}W^{+\alpha\beta} + W_{\alpha\beta}^{-\dagger}W^{-\alpha\beta} + Z_{\alpha\beta}Z^{\alpha\beta} + A_{\alpha\beta}A^{\alpha\beta}\right)\right. \\
&+ \frac{g}{4}\left(A^{\alpha\beta}\bullet A_\alpha \times A_\beta + A_\alpha \times A_\beta \bullet A^{\alpha\beta}\right) - \frac{g^2}{4}\left(A^\alpha \times A^\beta \bullet A_\alpha \times A_\beta\right)
\end{aligned} \tag{24.119}$$

On the right-hand side of (24.119), the first line is the Higgs sector: kinetic energy, mass term, and cubic and quartic self-interaction terms. The second line contains the W and Z mass terms. The third line gives the trilinear and quadrilinear Higgs-W and -Z couplings. The fourth line contains the zeroth-order gauge field radiation terms. Note that the electromagnetic field appears in the usual way, i.e., $-\frac{1}{4}A_{\alpha\beta}A^{\alpha\beta}$, but there is no photon mass term anywhere in this Lagrangian density. Finally, the fifth line contains the trilinear and quadrilinear vector boson couplings.

To conclude this section, we go back to the covariant derivative

$$D_\nu\phi = \left(\partial_\nu + i\frac{g}{2}A_\nu \bullet \tau + i\frac{g'}{2}B_\nu\right)\phi$$

and concentrate our attention on the terms $(ig/2)A_{3\nu}\tau_3 + (ig'/2)B_\nu$. Making use of the linear transformation inverse to that given in (24.112) and (24.113) and the definition $\tan\theta_W \equiv g'/g$, we obtain

$$\frac{ig}{2}A_{3\nu}\tau_3 + \frac{ig'}{2}B_\nu = i\left(\frac{I+\tau_3}{2}\right)g\sin\theta_W A_\nu + \frac{ig}{2}(\cos\theta_W\tau_3 - \sin\theta_W\tan\theta_W I)Z_\nu \tag{24.120}$$

The first term on the right-hand side of (24.120) represents the electromagnetic contribution to the covariant derivative. Here $(I+\tau_3)/2$ is the electric charge projection operator: when applied to the upper component of an isodoublet, it yields that component, but when applied to the lower component, it gives zero. Taking into account the fact that for the upper component of the isodoublet scalar field, which has electric charge $+e$, the covariant derivative for electromagnetic interactions is

$$D_\nu = \partial_\nu + ieA_\nu$$

we arrive at a very important result

$$g\sin\theta_W = e \tag{24.121}$$

24.14 The lepton sector

The next step in construction of the standard model is to incorporate the leptons. We start with the first generation; that is,

$$E = \begin{pmatrix} \nu_e \\ e^- \end{pmatrix}$$

As we have mentioned, experiments show that in the charged weak interaction it is the *left-handed* components of these fields that participate; that is,

$$\begin{aligned} \nu_{eL} &= \frac{1}{2}(1-\gamma_5)\nu_e \\ e_L &= \frac{1}{2}(1-\gamma_5)e \end{aligned}$$

These experiments include, but are not limited to, observation of the polarization of electrons and positrons emitted in nuclear beta decay and muon decay, polarization of positive and negative muons in pion decay, asymmetries in the angular distribution of electrons and positrons, and neutrinos emitted in beta decays of polarized nuclei and similar charged weak interaction phenomena. Observations of these phenomena formed the experimental basis for the *V-A* law of 1958.

The ν_{eL} and e_L are transformed into one another by W^+ or W^- emission or absorption. We can say that under $SU(2)$ gauge transformations, ν_e and e^- form a left-handed isodoublet

$$E_L = \begin{pmatrix} \nu_{eL} \\ e_L^- \end{pmatrix}$$

This can be expressed somewhat formally as follows: consider an infinitesimal $SU(2)$ gauge transformation

$$\delta\Psi = i\varepsilon\boldsymbol{\cdot}t\Psi$$

We can also define a *chiral* $SU(2)$ gauge transformation

$$\delta_5\Psi = i\gamma^5\eta\boldsymbol{\cdot}t\Psi$$

where $\eta_{1,2,3}$ are three real infinitesimals. The algebra of the generators is as follows:

$$\begin{aligned} \left[t_i, t_j\right] &= i\varepsilon_{ijk}t_k \\ \left[t_i, \gamma^5 t_j\right] &= i\varepsilon_{ijk}\gamma^5 t_k \\ \left[\gamma^5 t_i, \gamma^5 t_j\right] &= i\varepsilon_{ijk}t_k \end{aligned} \tag{24.122}$$

It is convenient to define left- and right-handed generators by the following formulas:

$$L_i = \frac{1}{2}(1-\gamma^5)t_i$$
$$R_i = \frac{1}{2}(1+\gamma^5)t_i$$

Then it is easy to show that relations (24.122) can be rewritten as

$$\left[L_i, L_j\right] = i\varepsilon_{ijk}L_k$$
$$\left[R_i, R_j\right] = i\varepsilon_{ijk}R_k$$
$$\left[L_i, R_j\right] = 0$$

Thus E_L is a weak isodoublet under $SU(2)$ transformations generated by the L_i. On the other hand, the right-handed components ν_{eR} and e_R do not participate in charged weak interactions, so under $SU(2)_L$ transformations, these fields remain invariant – in other words, they are isosinglets. All this is a convenient formal way of inserting the known facts about parity violation in the charged weak interaction into the theory without explaining these facts in terms of anything more basic.

Next, what about $U(1)$? Let us define a column vector

$$\mathrm{X} = \begin{pmatrix} \nu_{eL} \\ e_L \\ \nu_{eR} \\ e_R \end{pmatrix}$$

and a diagonal 4×4 matrix Y called *weak hypercharge*. Y is introduced so that we can conveniently describe various weak isodoublets and isosinglets that have different electric charges. Let the charge of a given particle be written as Qe. Then we have

$$E_L = \begin{pmatrix} \nu_{eL}(Q=0) \\ e_L(Q=-1) \end{pmatrix} \qquad \nu_{eR}(Q=0) \qquad e_R(Q=-1)$$
$$\begin{pmatrix} u_L(Q=2/3) \\ d_L(Q=-2/3) \end{pmatrix}$$

and so forth.

An infinitesimal $U(1)$ transformation on X can be written

$$\delta\mathrm{X} = i\alpha Y\mathrm{X}$$

Taking into account both $SU(2)$ and $U(1)$ local gauge transformations, we then have a covariant derivative of the form

$$D_\mu \mathrm{X} = \left(\partial_\mu + igA_\mu \cdot L + i\frac{g'}{2}B_\mu Y\right)\mathrm{X} \tag{24.123}$$

Once again, we focus our attention on the A_3 and B terms. Recalling (24.120) and (24.121), we write

Table 24.1 Table of charge, weak isospin, and hypercharge for leptons of first generation

	Q	L_3	Y
ν_{eL}	0	½	−1
e_L	−1	−½	−1
ν_{eR}	0	0	0
e_R	−1	0	−2

$$igA_{3\nu}L_3 + \frac{ig'}{2}YB_\nu = i\left(L_3 + \frac{Y}{2}\right)eA_\nu + ig\left(\cos\theta_W L_3 - \sin\theta_W \tan\theta_W \frac{Y}{2}\right)Z_\nu \tag{24.124}$$

The first term on the right-hand side of (24.124) refers to the electromagnetic interaction; in this term, $L_3 + Y/2 = \hat{Q}$ is the electric charge operator, the eigenvalues of which are the electric charges of the various components of X in units of e. From the formula

$$Y = 2(Q - L_3)$$

we determine the weak hypercharge Y, as indicated in Table 24.1.

We can now write all the covariant derivatives explicitly; that is,

$$D_\mu E_L = \left(\partial_\mu + igA_\mu \cdot L - i\frac{g'}{2}B_\mu\right)E_L \tag{24.125}$$

$$D_\mu \nu_{eR} = \partial_\mu \nu_{eR} \tag{24.126}$$

$$D_\mu e_R = \left(\partial_\mu - ig'B_\mu\right)e_R \tag{24.127}$$

Given these covariant derivatives, we construct the Lagrangian density for the lepton fields. The interaction terms are of interest here. Using (24.124), we obtain

$$\begin{aligned} \mathcal{L}_I^{\text{first gen}} = &-\frac{g}{2\sqrt{2}}\left[\bar{e}\gamma^\mu\left(1-\gamma^5\right)\nu_e\right]W_\mu^- - \frac{g}{2\sqrt{2}}\left[\bar{\nu}_e\gamma^\mu\left(1-\gamma^5\right)e\right]W_\mu^+ \\ &-\frac{g}{4\cos\theta_W}\left[\bar{\nu}_e\gamma^\mu\left(1-\gamma^5\right)\nu_e\right]Z_\mu \\ &+\frac{g}{4\cos\theta_W}\left[\bar{e}\gamma^\mu\left(1-4\sin^2\theta_W-\gamma^5\right)e\right]Z_\mu \\ &+e\left[\bar{e}\gamma^\mu e\right]A_\mu \end{aligned} \tag{24.128}$$

In (24.128), the first line on the right-hand side describes the charged weak coupling of the leptons to $W^\pm$; the second and third lines contain the neutral weak couplings of neutrino and electron, respectively, to Z, and the fourth line is the electromagnetic interaction.

Nature gives us three lepton generations. This mysterious fact is built into the Lagrangian density without further explanation by adding to (24.128) analogous terms for the second and third generations.

Table 24.2 Table of charge, weak isospin, and hypercharge for quarks of first generation

	Q_q	L_{3q}	Y_q
u_L	2/3	½	1/3
d_{CL}	−1/3	−½	1/3
u_R	2/3	0	4/3
d_{CR}	−1/3	0	−2/3

We have already noted that $g \sin\theta_W = e$, and also [in equation (24.39)] that

$$g^2 = 2^{5/2} G_F m_W^2$$

Combining these two formulas, we obtain

$$m_W^2 = \left(\frac{\pi\alpha}{G_F\sqrt{2}}\right)\frac{1}{\sin^2\theta_W} = \frac{(37.5\ \mathrm{GeV}/c^2)^2}{\sin^2\theta_W} \tag{24.129}$$

Diverse experiments establish that $\sin^2\theta_W$ varies slightly with momentum transfer. For example, $\sin^2\theta_W = 0.2312 \pm 0.00015$ at 91 GeV/c^2. Using this value, (24.129) yields $m_W = 79.3$ GeV/c^2. Before comparing theory with the experimentally determined mass, one must correct (24.129) for various radiative effects. When this is done, good agreement is obtained between theory and experiment: $m_W = 80.3$ GeV/c^2. We also recall the prediction $m_Z = m_W / \cos\theta_W$. When this formula is used and radiative corrections are applied, good agreement is obtained with the experimental value $m_Z = 91.2$ GeV/c^2. On the other hand, $\sin^2\theta_W = 0.2397 \pm 0.0012$ at 0.16 GeV/c^2, which is a value more appropriate for relatively low-energy experimental data.

24.15 The quark sector

The coupling of quarks to charged and neutral intermediate vector bosons is handled in a similar way. For simplicity, we consider for the moment just the first quark generation with one Cabibbo angle. Then, assuming that

$$\begin{pmatrix} u_L \\ d_{CL} \end{pmatrix}$$

forms a weak left-handed isodoublet, whereas u_R and d_{CR} are weak isosinglets, we construct Table 24.2 to find the weak hypercharge Y_q in terms of the electric charge Q_q and the weak isospin quantum number L_{3q} according to the formula $Y_q = 2(Q_q - L_{3q})$.

Following steps similar to those in the lepton case, we obtain the following Lagrangian density for interaction of these quark fields with the vector boson fields:

$$
\begin{aligned}
\mathcal{L}_I^{u,dc} = & -\frac{g}{2\sqrt{2}}\left[\bar{u}\gamma^\mu(1-\gamma_5)d_C\right] W_\mu^- - \frac{g}{2\sqrt{2}}\left[\bar{d}_C\gamma^\mu(1-\gamma_5)u\right]W_\mu^+ \\
& -\frac{g}{4\cos\theta_W}\bar{u}\gamma^\mu\left(1-\frac{8}{3}\sin^2\theta_W-\gamma_5\right)uZ_\mu \\
& +\frac{g}{4\cos\theta_W}\bar{d}\gamma^\mu\left(1-\frac{4}{3}\sin^2\theta_W-\gamma_5\right)dZ_\mu \\
& -e\left(\frac{2}{3}\bar{u}\gamma^\mu u-\frac{1}{3}\bar{d}\gamma^\mu d\right)A_\mu
\end{aligned}
\tag{24.130}
$$

We now generalize (24.130) to include three quark generations, taking into account the CKM matrix. Here the first term of the right-hand side of (24.130) becomes

$$
-\frac{g}{2\sqrt{2}}(\bar{u}\ \bar{c}\ \bar{t})\gamma^\mu(1-\gamma_5)\begin{pmatrix} U_{ud} & U_{us} & U_{ub} \\ U_{cd} & U_{cs} & U_{cb} \\ U_{td} & U_{ts} & U_{tb}\end{pmatrix}\begin{pmatrix} d \\ s \\ b\end{pmatrix}W_\mu^+
$$

and the second term on the right-hand side of (24.130) is similarly modified. Of course, the CKM matrix itself is purely empirical; nothing in the standard model tells us how to calculate the various entries in the CKM matrix.

The Lagrangian density has now grown to very large size. It contains the Higgs kinetic energy, Higgs mass, Higgs cubic and quartic interaction terms, vector boson mass terms, Higgs–vector boson interactions, and vector boson radiation terms. It also contains charged and neutral weak interactions, as well as electromagnetic interactions, of the leptons and quarks. We have not even included the lepton or quark kinetic energy terms, nor have we discussed a gauge-invariant and renormalizable prescription for including leptonic and quark masses through coupling to the scalar field.

The reader undoubtedly recognizes by this stage how complicated this construction has become and how much has been inserted by assumption without explanation. First, we assumed a particular form of the Higgs mechanism, based on spontaneous symmetry breaking in a complex isodoublet scalar field. Although the Higgs mechanism could not be simpler, it could be much more complicated, and there could be more than one Higgs boson. Second, we assumed parity violation in the charged weak interaction in the *V-A* form and built it into the theory with the assumption of weak left-handed isodoublets and right-handed isosinglets for leptons and quarks. In a similar vein, the three generations of leptons, as well as of quarks, and the CKM matrix are inserted "by hand" without any explanation of their origin. Thus the standard model is not a fundamental theory. Nevertheless, it constitutes a remarkable advance in our knowledge and understanding of elementary particle physics.

24.16 Summary of Feynman vertex factors in the electroweak standard model

In what follows, we summarize the Feynman vertex factors for the standard model that we subsequently employ in examples of weak interactions. Note from Chapter 23 that an overall factor of $(-i)^n$ appears in the amplitude of nth order in the S-matrix expansion. Because a Feynman diagram of nth order has n vertices, a convenient way to keep track of the factor of $(-i)^n$ is to include a factor of $-i$ that multiplies the relevant coupling constant in each vertex factor. In QED, the coupling constant is $-e$, so each vertex of the following type has the factor $ie\gamma^\mu$:

1.

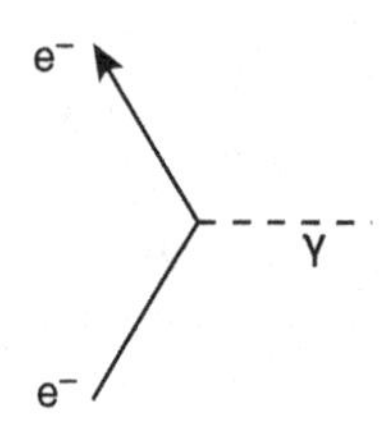

Figure 24.15 $e^-e^-\gamma$ vertex

$$ie\gamma^\mu \tag{24.131}$$

This factor also applies if e^- is replaced by μ^- or τ^-. If in (24.131) e^- is replaced by a u, c, t quark, the vertex factor becomes $-\frac{2}{3}ie\gamma^\mu$. If the replacement is a d, s, b quark, the vertex factor is $\frac{1}{3}ie\gamma^\mu$.

2.

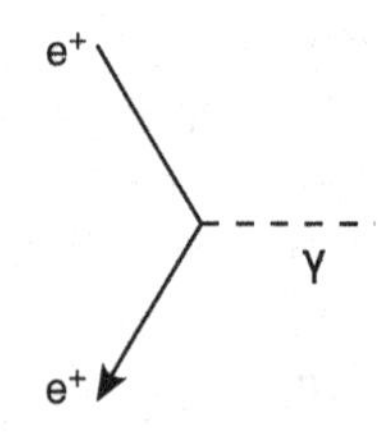

Figure 24.16 $e^+e^+\gamma$ vertex

$$-ie\gamma^\mu \tag{24.132}$$

3.

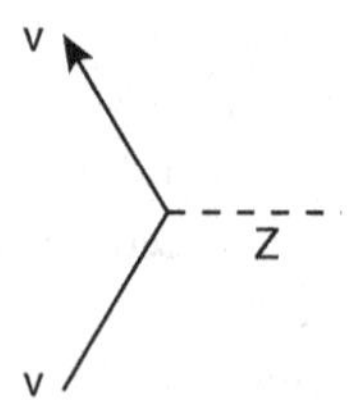

Figure 24.17 $\nu\nu Z$ vertex

$$\frac{ig}{4\cos\theta_W}\gamma^\mu(1-\gamma^5) \tag{24.133}$$

4.

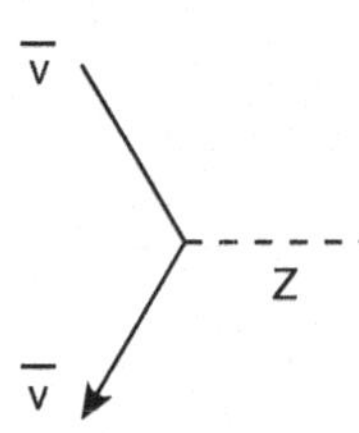

Figure 24.18 $\bar{\nu}\bar{\nu}Z$ vertex

$$\frac{-ig}{4\cos\theta_W}\gamma^\mu(1-\gamma^5) \quad (24.134)$$

5.

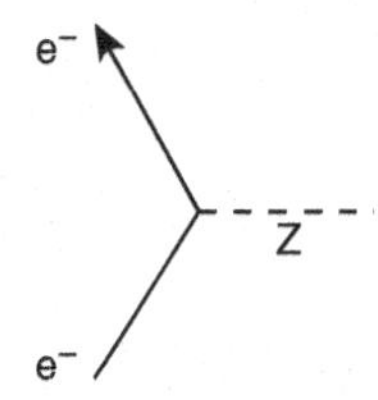

Figure 24.19 e^-e^-Z vertex

$$\frac{ig}{4\cos\theta_W}\gamma^\mu\left[\left(1-4\sin^2\theta_W\right)-\gamma_5\right] \quad (24.135)$$

If e^- is replaced by a u, c, t quark, the vertex factor becomes

$$\frac{-ig}{4\cos\theta_W}\gamma^\mu\left[\left(1-\frac{8}{3}\sin^2\theta_W\right)-\gamma_5\right] \quad (24.136)$$

If in (24.135) the replacement is a d, s, b quark, the vertex factor is

$$\frac{ig}{4\cos\theta_W}\gamma^\mu\left[\left(1-\frac{4}{3}\sin^2\theta_W\right)-\gamma_5\right] \quad (24.137)$$

6.

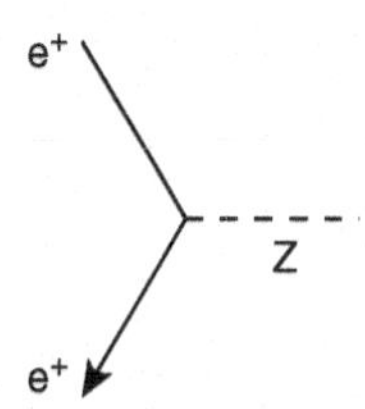

Figure 24.20 e^+e^+Z vertex

$$\frac{-ig}{4\cos\theta_W}\gamma^\mu\left[\left(1-4\sin^2\theta_W\right)-\gamma^5\right] \quad (24.138)$$

7.

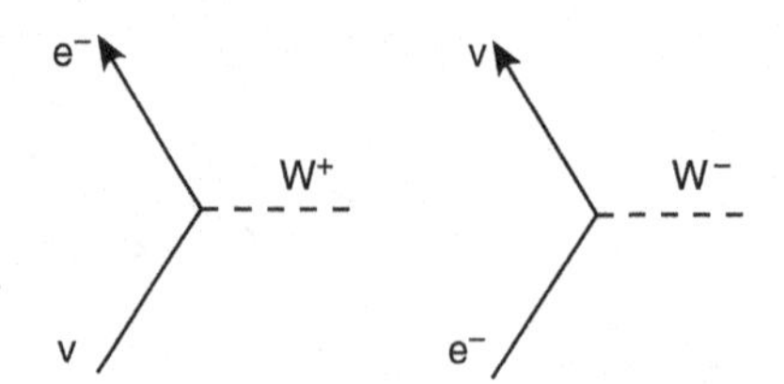

Figure 24.21 $e^-\nu W$ vertices

$$\frac{-ig}{2\sqrt{2}}\gamma^\mu\left(1-\gamma^5\right) \tag{24.139}$$

If in the first diagram ν is replaced by q_i (u,c,t) and e^- is replaced by q_j (d,s,b), the vertex factor becomes

$$\frac{-ig}{2\sqrt{2}\sin\theta_W}U_{ij}\gamma^\mu\left(1-\gamma^5\right) \tag{24.140}$$

where U_{ij} is the CKM matrix.

8.

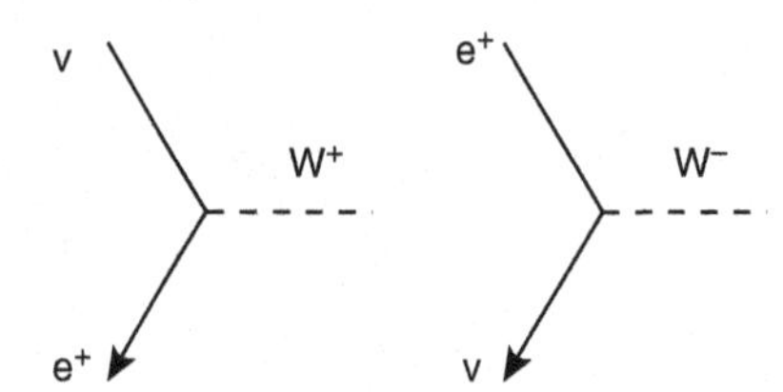

Figure 24.22 $e^+\nu W$ vertices

$$\frac{-ig}{2\sqrt{2}}\gamma^\mu\left(1-\gamma^5\right) \tag{24.141}$$

9.

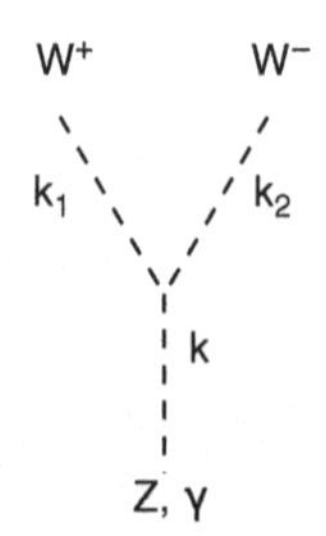

Figure 24.23 W^+W^-Z, $W^+W^-\gamma$ vertices

This is the 3-vector–boson vertex. Its vertex factor is derived from (24.86) and has the following form:

$$V_{\alpha\beta\gamma} = ig\begin{pmatrix}\cos\theta_W \\ \sin\theta_W\end{pmatrix}\left[(k_1 - k_2)_\alpha g_{\beta\gamma} - (2k_1 + k_2)_\gamma g_{\alpha\beta} + (2k_2 + k_1)_\beta g_{\gamma\alpha}\right] \tag{24.142}$$

where $\cos\theta_W$ $(\sin\theta_W)$ is employed for $ZWW(\gamma WW)$, respectively, and k, k_1, k_2 are 4-momenta with $k = k_1 + k_2$. If k is an internal line and $k_{1,2}$ are external lines, the diagram is described by

$$-i\frac{g^{\sigma\alpha} - \dfrac{k^\sigma k^\alpha}{m_Z^2}}{k^2 - m_Z^2} V_{\alpha\beta\gamma}\varepsilon_1^\beta \varepsilon_2^\gamma \quad \text{or} \quad -i\frac{g^{\sigma\alpha}}{k^2} V_{\alpha\beta\gamma}\varepsilon_1^\beta\varepsilon_2^\gamma \tag{24.143}$$

for ZWW (γWW), respectively.

The standard electroweak model also has 4-vector–boson vertices, lepton-lepton-Higgs vertices, quark-quark-Higgs vertices, and so on, but these are not given here. The rules summarized in this section are sufficient for the illustrative calculations of the next section.

24.17 Illustrative calculations of electroweak processes

24.17.1 The decays $W^- \to e^-\bar{\nu}_e$ and $Z \to \nu_\ell\bar{\nu}_\ell$, $Z \to \ell^+\ell^-$, and $Z \to q\bar{q}$

These are the simplest physical processes in the electroweak model because they are just semi-weak interactions with one vertex. We first consider some simple facts concerning the decay $W^- \to e^-\bar{\nu}_e$. The W, like the photon, has spin unity. Suppose that the W at rest is polarized with its spin along the $+z$-direction. Experiment shows that the electron is emitted primarily in the backward direction (along the $-z$-axis); in fact, the angular distribution of electron momentum is proportional to

$$f(\theta) = (1 - \cos\theta)^2 = \left(1 + \cos^2\theta\right) - \left(2\cos\theta\right)$$

where θ is the usual polar angle between the electron momentum and the z-axis. This is represented schematically by Figure 24.24.

By conservation of linear momentum, the electron and antineutrino 3-momenta must be equal and opposite in the W rest frame. Because the W^- has spin unity, conservation of angular momentum demands that both the electron and the antineutrino are polarized along $+z$. Thus the electron spin and its linear momentum are opposed. In other words, the electron has negative helicity (is left-handed), whereas the antineutrino has positive helicity (is right-handed).

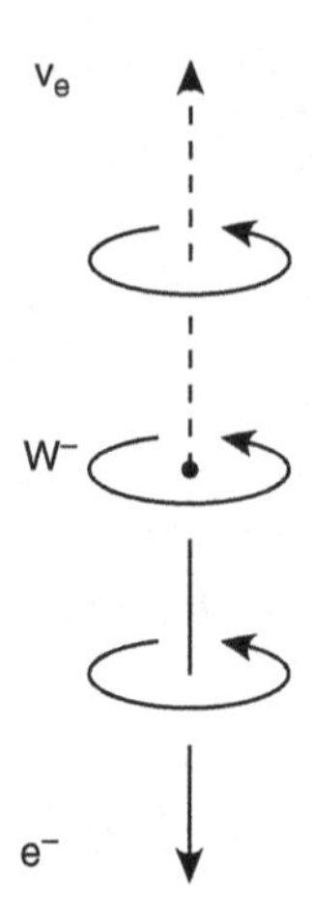

Figure 24.24 Schematic diagram of spins and momenta in the decay $W^- \to e^-\bar{\nu}_e$. The electron is emitted in the backward direction relative to the W spin.

The starting point of our calculation is the invariant amplitude obtained from (24.139); that is,

$$M(W^- \to e^-\bar{\nu}_e) = \frac{-ig}{2\sqrt{2}}\bar{u}_e \not{\varepsilon}(1-\gamma_5)v_{\bar{\nu}} \tag{24.144}$$

To calculate the electron angular distribution, we recall that the differential transition probability per unit time in the W rest frame is

$$d\Gamma = (2\pi)^4 \delta^4\left(p_e + p_\nu - m_W\right)\frac{m_e}{E_e}\frac{m_\nu}{E_\nu}\frac{1}{2m_W}\frac{d^3\mathbf{p}_e}{(2\pi)^3}\frac{d^3\mathbf{p}_\nu}{(2\pi)^3}|M|^2 \tag{24.145}$$

Here

$$|M|^2 = \frac{g^2}{8}\, tr\left[\frac{\not{p}_e + m_e}{2m_e}\boldsymbol{\varepsilon}\cdot\boldsymbol{\gamma}(1-\gamma_5)\frac{\not{p}_\nu - m_\nu}{2m_\nu}\boldsymbol{\varepsilon}^*\cdot\boldsymbol{\gamma}(1-\gamma_5)\right] \tag{24.146}$$

Note that in the W rest frame, ε^μ reduces to a unit 3-vector $\boldsymbol{\varepsilon}$. We choose

$$\boldsymbol{\varepsilon} = \frac{\hat{i} + i\hat{j}}{\sqrt{2}}$$

which corresponds to the spin of the W in the $+z$-direction. Because the electron and antineutrino are both ultrarelativistic, we can ignore their masses in the numerators on the right-hand side of (24.146). Thus the trace is

$$\begin{aligned} T &= \frac{1}{4m_e m_\nu} tr\left[\not{p}_e \boldsymbol{\varepsilon} \cdot \boldsymbol{\gamma} (1-\gamma^5) \not{p}_\nu \boldsymbol{\varepsilon}^* \cdot \boldsymbol{\gamma} (1-\gamma^5) \right] \\ &= \frac{1}{2m_e m_\nu} tr\left[\not{p}_e \boldsymbol{\varepsilon} \cdot \boldsymbol{\gamma} \not{p}_\nu \boldsymbol{\varepsilon}^* \cdot \boldsymbol{\gamma} (1-\gamma^5) \right] \\ &= \frac{1}{2m_e m_\nu} \varepsilon_i \varepsilon_j^* tr\left[\not{p}_e \gamma^i \not{p}_\nu \gamma^j (1-\gamma^5) \right] \\ &= \frac{1}{4m_e m_\nu} tr\left[\left(\not{p}_e \gamma^1 \not{p}_\nu \gamma^1 + \not{p}_e \gamma^2 \not{p}_\nu \gamma^2 + i\not{p}_e \gamma^2 \not{p}_\nu \gamma^1 - i\not{p}_e \gamma^1 \not{p}_\nu \gamma^2 \right) (1-\gamma^5) \right] \end{aligned} \tag{24.147}$$

It is convenient to write $T = T_1 - T_2$, where T_2 contains the γ^5 and T_1 does not. Then

$$\begin{aligned} T_1 &\equiv tr\left[\left(\not{p}_e \gamma^1 \not{p}_\nu \gamma^1 + \not{p}_e \gamma^2 \not{p}_\nu \gamma^2 + i\not{p}_e \gamma^2 \not{p}_\nu \gamma^1 - i\not{p}_e \gamma^1 \not{p}_\nu \gamma^2 \right) \right] \\ &= 8\left(p_e^1 p_\nu^1 + p_e^2 p_\nu^2 + p_e \cdot p_\nu \right) \\ &= 8\left(E_e E_\nu - p_e^3 p_\nu^3 \right) \end{aligned} \tag{24.148}$$

Also, in the W rest frame and where we ignore the masses of the emitted leptons, $E_e = E_\nu = E$ and $p_\nu^3 = -p_e^3 = -E\cos\theta$. Therefore,

$$T_1 = 8E^2\left(1+\cos^2\theta\right) \tag{24.149}$$

It remains to calculate the trace

$$T_2 \equiv tr\left[\left(\not{p}_e \gamma^1 \not{p}_\nu \gamma^1 + \not{p}_e \gamma^2 \not{p}_\nu \gamma^2 + i\not{p}_e \gamma^2 \not{p}_\nu \gamma^1 - i\not{p}_e \gamma^1 \not{p}_\nu \gamma^2 \right) \gamma^5 \right] \tag{24.150}$$

We have previously shown that

$$tr\left(\gamma^\alpha \gamma^\beta \gamma^\rho \gamma^\sigma \gamma^5\right) = -4i\varepsilon^{\alpha\beta\rho\sigma}$$

where $\varepsilon^{\alpha\beta\rho\sigma}$ is the completely antisymmetric unit 4-tensor. The latter is equal to +1 when the indices are 0123 or even permutations thereof, equal to –1 for odd permutations of 0123, and equal to zero if any two indices are the same.

Hence

$$\begin{aligned} T_2 &= 4\left[p_{e3} p_{\nu 0} \left(\varepsilon^{3201} - \varepsilon^{3102} \right) + p_{e0} p_{\nu 3} \left(\varepsilon^{0231} - \varepsilon^{0132} \right) \right] \\ &= 8\left(p_{\nu 3} E - p_{e3} E \right) \\ &= 16E^2 \cos\theta \end{aligned} \tag{24.151}$$

Combining (24.151) and (24.149) and inserting the result into (24.153), we obtain for (24.145)

$$d\Gamma = \frac{1}{32\pi^2} \frac{g^2}{m_W} (1-\cos\theta)^2 \delta^4\left(p_e + p_{\bar{\nu}} - m_W\right) d^3\mathbf{p}_e d^3\mathbf{p}_{\bar{\nu}} \tag{24.152}$$

Table 24.3 Values of a, b for various Z^0 decay modes

	a	b
$\nu\bar{\nu}$	1	1
$\ell^+\ell^-$	$1-4\sin^2\theta_W$	1
$u\bar{u}$	$1-\frac{8}{3}\sin^2\theta_W$	1
$d\bar{d}$	$1-\frac{4}{3}\sin^2\theta_W$	1

which gives the expected electron angular distribution. Integration over $\mathbf{p}_{\bar{\nu}}$ with $\int \delta^3(\mathbf{p}_e+\mathbf{p}_{\bar{\nu}})\,d^3\mathbf{p}_{\bar{\nu}} = 1$ yields

$$
\begin{aligned}
d\Gamma &= \frac{1}{32\pi^2}\frac{g^2}{m_W}(1-\cos\theta)^2\,\delta(2E-m_W)\,E^2 dE d\Omega_e \\
&= \frac{G_F}{\sqrt{2}}\frac{m_W^3}{32\pi^2}(1-\cos\theta)^2\,d\Omega_e
\end{aligned}
\tag{24.153}
$$

where we have used $g^2/8m_W^2 = G_F/\sqrt{2}$. Integration over the electron solid angle yields the total decay rate

$$
\Gamma(W^- \to e^-\bar{\nu}_e) = \frac{G_F}{6\pi\sqrt{2}}m_W^3 \tag{24.154}
$$

Because the masses of μ and τ leptons are also very small compared with m_W, these formulas are valid to a good approximation for all three decay branches: $W^- \to \ell^-\bar{\nu}_\ell$.

A very similar calculation yields the decay rates for the transitions $Z \to \nu_\ell\bar{\nu}_\ell$, $Z \to \ell^+\ell^-$, and $Z \to q\bar{q}$. We only have to replace the vertex factor of (24.139) by those given in (24.133)–(24.138). Thus we obtain

$$
\Gamma = \frac{G_F}{24\pi\sqrt{2}}m_Z^3\left(a^2+b^2\right) \tag{24.155}
$$

where a and b are given in Table 24.3.

The total transition rate is given by

$$
\Gamma_{\text{total}} = 3\Gamma_{\nu\bar{\nu}} + 3\Gamma_{\ell^+\ell^-} + 6\Gamma_{u\bar{u}} + 9\Gamma_{d\bar{d}} \tag{24.156}
$$

Here each of the first two terms on the right-hand side has a factor of 3 for three lepton flavors. The third term has a factor of 3 for three quark colors and a factor of 2 for two quark flavors (u, c); the top quark mass is too large for participation here. Finally, the last term does have three flavors (d, s, b) and three colors. The observed natural width of the Z-boson is 2.490 ± 0.007 GeV, in very good agreement with the numerical value of Γ_{total} calculated from (24.156). From this we conclude that there are three and only three flavors of neutrinos with masses much less than $m_Z/2$.

Table 24.4 Branching fractions for decay modes of the charged pion

Decay mode	Branching fraction
$\mu^+\nu_\mu$	99.9877%
$e^+\nu_e$	1.23×10^{-4}
$\pi^0 e^+\nu_e$	1.03×10^{-8}

24.17.2 Decay of the charged pion: $\pi \rightarrow \ell\ \overline{\nu}_\ell$ (often called $\pi_{\ell 2}$ decay)

This is the simplest weak decay involving a hadron. First, we give some basic experimental facts. The mean lifetime of a charged pion is $\tau(\pi^\pm) = 2.603\times10^{-8}$ s, and the various decay modes are shown in Table 24.4.

Why is the branching fraction for π_{e2} so small compared with that for $\pi_{\mu 2}$? This can be understood intuitively from Figure 24.25, which shows a π^+ at rest at the origin, which decays to a charged lepton ℓ^+ that is emitted in the $+z$-direction. Conservation of linear momentum requires that the neutrino be emitted in the opposite direction. Because the helicity of a neutrino emitted in a weak process is always $h(\nu) = -1$, and because the pion has spin zero, conservation of angular momentum requires that the positive lepton also have negative helicity $h(\ell^+) = -1$, and experiment shows that this is the case.

However, the expected helicity of a positive lepton emitted in a charged weak decay is $+v/c$ according to the V-A law. Because the masses of the relevant particles are $m_{\pi^+} = 139.57$ MeV/c^2, $m_\mu = 105.66$ MeV/c^2, and $m_e = 0.511$ MeV/c^2, the muon in $\pi_{\mu 2}$ decay is quite nonrelativistic (it has kinetic energy of approximately 5 MeV) and is thus rather easily forced into the wrong helicity state, but the electron in π_{e2} decay is very relativistic (kinetic energy of approximately 69 MeV) and is thus forced into the wrong helicity state only with extreme reluctance.

In the decay $\pi^- \rightarrow \ell^-\overline{\nu}_\ell$, the V-A law yields the following amplitude:

$$M = \frac{G_F}{\sqrt{2}}\overline{u}_\ell\gamma^\lambda(1-\gamma_5)v_{\overline{\nu}\ell}\left\langle 0\left|\left(J_\lambda^{\text{had}}\right)^\dagger\right|\pi^-\right\rangle \tag{24.157}$$

Even for the comparatively simple pion, the effects of strong interaction are so complicated and uncertain that it is very difficult to calculate the hadronic portion of the matrix element in (24.157) from first principles. The best we can do in an elementary treatment such as this one is to place restrictions on its possible form with the aid of symmetry. Proper Lorentz invariance requires that M be a linear combination of a scalar and a pseudoscalar. Because the leptonic factor in (24.157) is a vector plus an axial vector, the hadronic matrix element must be a vector, an axial vector, or a linear combination of the two, and it must be constructed from the available kinematic quantities. Because the pion is a pseudoscalar particle, the only available kinematic quantity is the 4-momentum transfer $q = p_\ell + p_{\overline{\nu}}$. Thus we write

$$\begin{aligned}\left\langle 0\left|\left(J_\lambda^{\text{had}}\right)^\dagger\right|\pi^-\right\rangle &= f_\pi \cos\theta_{12}\cos\theta_{13}q_\lambda \\ &= f_\pi\cos\theta_C q_\lambda\end{aligned} \tag{24.158}$$

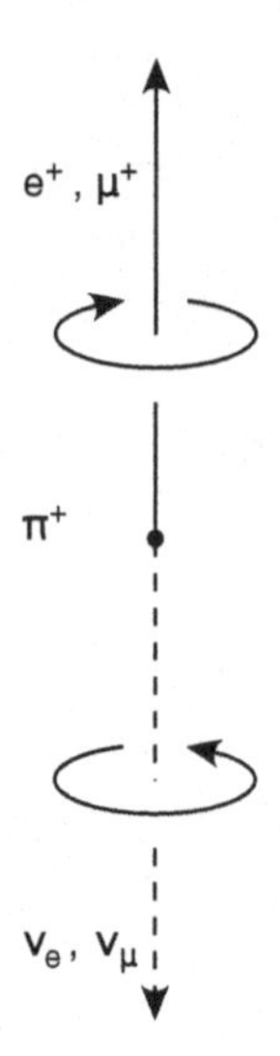

Figure 24.25 Diagram illustrating momenta and spins of decay products in the decay $\pi^+ \longrightarrow \ell^+ \nu_\ell$

where f_π is a numerical constant called the *pion decay constant* that characterizes the effects of strong interactions. We do not have a precise, convincing way to calculate f_π, but for present purposes, we let it be determined by experiment. [Note that because the pion is pseudoscalar and q_λ is a polar 4-vector, it is the *axial* portion of $\left(J_\lambda^{\text{had}}\right)^\dagger$ that is operative here.] Substitution of (24.158) into (24.157) yields

$$M = \frac{G_F}{\sqrt{2}} f_\pi \cos\theta_C \bullet \bar{u}_\ell \left(\not{p}_\ell + \not{p}_{\bar{\nu}}\right)(1-\gamma_5)\mathrm{v}_{\bar{\nu}\ell} \tag{24.159}$$

Now recalling equations (22.16) and (22.17), we have

$$\bar{u}_\ell \not{p}_\ell = m_\ell \bar{u}_\ell$$

and

$$\not{p}_{\bar{\nu}}(1-\gamma_5)v_{\bar{\nu}} = (1+\gamma_5)\not{p}_{\bar{\nu}} v_{\bar{\nu}} = -m_{\bar{\nu}}(1+\gamma_5)v_{\bar{\nu}} \approx 0$$

Hence (24.159) becomes

$$M = \frac{G_F}{\sqrt{2}} f_\pi m_\ell \cos\theta_C \bullet \bar{u}_\ell (1-\gamma_5) v_{\bar{\nu}\ell}$$

Thus

$$|M|^2 = \frac{G_F^2}{2} f_\pi^2 m_\ell^2 \cos^2\theta_C \frac{1}{4m_\ell m_{\bar{\nu}}} T \tag{24.160}$$

where

$$\begin{aligned} T &= tr\left[\left(\not{p}_\ell + m_\ell\right)(1-\gamma^5)\left(\not{p}_{\bar{\nu}}\right)(1+\gamma^5)\right] \\ &= tr\left[\left(\not{p}_\ell\right)(1-\gamma^5)\not{p}_{\bar{\nu}}(1+\gamma^5)\right] \\ &= 2tr\left[\not{p}_\ell \not{p}_{\bar{\nu}}(1+\gamma^5)\right] \\ &= 8p_\ell \bullet p_{\bar{\nu}} \end{aligned}$$

Hence (24.160) becomes

$$|M|^2 = G_F^2 f_\pi^2 \cos^2\theta_C \frac{m_\ell}{m_{\bar{\nu}}} p_\ell \bullet p_{\bar{\nu}} \tag{24.161}$$

The transition probability per unit time in the pion rest frame is

$$d\Gamma = \frac{G_F^2 f_\pi^2 m_\ell^2 \cos^2\theta_C}{2m_\pi} \frac{p_\ell \bullet p_{\bar{\nu}}}{E_\ell E_{\bar{\nu}}} \frac{d^3\boldsymbol{p}_\ell d^3\boldsymbol{p}_{\bar{\nu}}}{(2\pi)^2} \delta(m_\pi - E_\ell - E_{\bar{\nu}})\delta^3(\boldsymbol{p}_\ell + \boldsymbol{p}_{\bar{\nu}}) \tag{24.162}$$

Integrating over $\boldsymbol{p}_\ell$, we obtain

$$\Gamma = \frac{G_F^2 f_\pi^2 m_\ell^2 \cos^2\theta_C}{2\pi m_\pi} \int \frac{(E_\ell E_{\bar{\nu}} + E_{\bar{\nu}}{}^2)E_{\bar{\nu}}}{E_\ell} \delta\left[m_\pi - E_{\bar{\nu}} - \left(E_{\bar{\nu}}{}^2 + m_\ell^2\right)^{1/2}\right] dE_{\bar{\nu}}$$

or, finally,

$$\Gamma = \frac{G_F^2 f_\pi^2 \cos^2\theta_C}{8\pi} m_\ell^2 m_\pi \left(1 - \frac{m_\ell^2}{m_\pi^2}\right)^2 \tag{24.163}$$

Thus we obtain the ratio

$$\frac{\Gamma(\pi^- \to e^-\bar{\nu}_e)}{\Gamma(\pi^- \to \mu^-\bar{\nu}_\mu)} = \left(\frac{m_e}{m_\mu}\right)^2 \left(\frac{m_\pi^2 - m_e^2}{m_\pi^2 - m_\mu^2}\right)^2 = 1.24 \times 10^{-4} \tag{24.164}$$

This result, when corrected for radiative effects, agrees very well with the experimental branching ratio. Finally, comparison of (24.163) with the experimentally determined mean life of $\pi^\pm$ yields $f_\pi = 0.94m_\pi$.

24.17.3 How the neutral and charged weak interactions fit together

Here we discuss as an example the following weak reaction:

$$\nu_e + \bar{\nu}_e \to W^+_{\text{longit. pol.}} + W^-_{\text{longit. pol.}} \tag{24.165}$$

It is interesting because its amplitude according to the naive charged vector boson theory violates unitarity, whereas unitarity is restored to lowest nonvanishing order by adding the effect of neutral weak interactions as prescribed by the standard electroweak model. (In higher orders, we also need additional contributions involving the Higgs boson.) For various practical reasons, an experiment to observe this reaction would be extremely difficult or impossible to perform, but we can still think about it. To begin, we remind ourselves what longitudinal polarization means. We know that if a vector boson has rest mass, it has a rest frame. In that frame its polarization 4-vector ε_μ reduces to a unit 3-vector that we assume lies along the z-axis; that is,

$$\varepsilon^\mu = \begin{pmatrix} 0 \\ 0 \\ 0 \\ 1 \end{pmatrix} \tag{24.166}$$

Also, in the rest frame, the boson 4-momentum k_μ reduces to

$$k^\mu = \begin{pmatrix} m_w \\ 0 \\ 0 \\ 0 \end{pmatrix} \tag{24.167}$$

Now we make a Lorentz boost to a frame where the boson has velocity β, also in the z-direction. The Lorentz transformation matrix for this boost is

$$a = \begin{pmatrix} \gamma & 0 & 0 & \beta\gamma \\ 0 & 1 & 0 & 0 \\ 0 & 0 & 1 & 0 \\ \beta\gamma & 0 & 0 & \gamma \end{pmatrix}$$

Applying a to the 4-vectors in (24.166) and (24.167), we obtain the 4-polarization and the 4-momentum of the moving boson

$$\varepsilon = \begin{pmatrix} \beta\gamma \\ 0 \\ 0 \\ \gamma \end{pmatrix} \qquad k = m_W \begin{pmatrix} \gamma \\ 0 \\ 0 \\ \beta\gamma \end{pmatrix}$$

In the limit where the boson is ultrarelativistic, $\beta \to 1$ and

$$\varepsilon^\mu = \frac{1}{m_W} k_W^\mu \tag{24.168}$$

Now consider reaction (24.165) in lowest nonvanishing order in the charged weak interaction. The Feynman diagram is Figure 24.26.

In this figure, q is the 4-momentum of the internal electron line. Recalling that 4-momentum is conserved at each vertex, we have

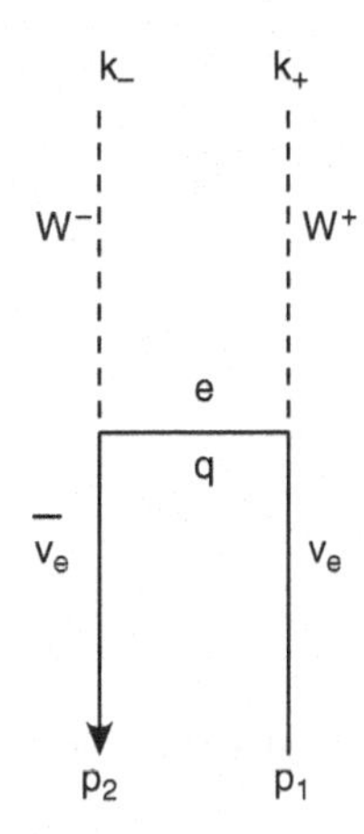

Figure 24.26 Lowest order charged weak interaction Feynman diagram for the reaction $\nu_e + \bar{\nu}_e \rightarrow W^+_{\text{long pol}} + W^-_{\text{long pol}}$

$$p_1 = q + k_+ \qquad p_2 = -q + k_- \tag{24.169}$$

The amplitude corresponding to Figure 24.26 is easily written down from (24.139) and (24.141); that is,

$$M_a = \frac{-ig}{2\sqrt{2}} \bar{v}\not{\varepsilon}_-\left(1-\gamma^5\right)\left[\frac{i\left(\not{q}+m_e\right)}{\left(q^2-m_e^2\right)}\right]\frac{-ig}{2\sqrt{2}}\not{\varepsilon}_+\left(1-\gamma^5\right)u \tag{24.170}$$

Here $\bar{v}$ and u are spinors referring to the antineutrino and neutrino, respectively, $\not{\varepsilon}_-$ and $\not{\varepsilon}_+$ refer to the outgoing W^- and W^+, respectively, and the factor in square brackets is the electron propagator associated with the internal electron line. The right-hand side of (24.170) can be simplified as follows: first, we ignore the electron mass in the denominator because q^2 is extremely large in the ultrarelativistic limit. Then we use (24.168) to obtain

$$M_a = -i\frac{g^2}{8m_W^2}\bar{v}\not{K}_-(1-\gamma^5)\frac{\not{q}+m_e}{q^2}\not{K}_+(1-\gamma^5)u \tag{24.171}$$

Now, using (24.169) and noting from the Dirac equation that

$$\bar{v}\not{p}_2 \approx 0 \qquad \not{p}_1 u \approx 0 \tag{24.172}$$

and also recalling that

$$\left(1-\gamma^5\right)m_e\not{K}_+\left(1-\gamma^5\right) = m_e\not{K}_+\left(1+\gamma^5\right)\left(1-\gamma^5\right) = 0$$

we see that (24.171) becomes

$$\begin{aligned} M_a &= i\frac{g^2}{8m_W^2}\bar{v}\not{q}(1-\gamma^5)\frac{\not{q}}{q^2}\not{q}(1-\gamma_5)u \\ &= \frac{ig^2}{4m_W^2}\bar{v}\not{q}(1-\gamma_5)u \end{aligned} \tag{24.173}$$

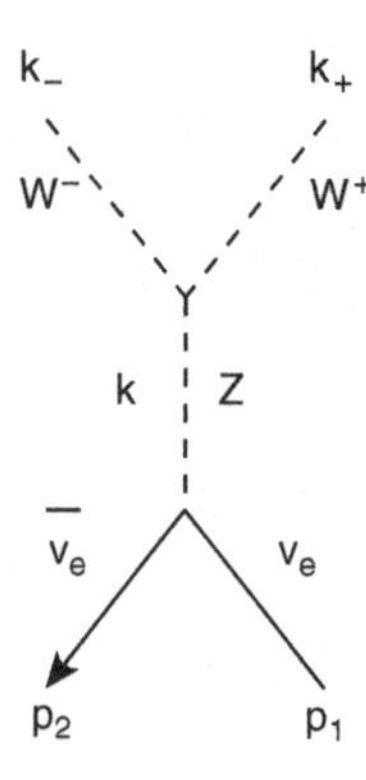

Figure 24.27 Lowest order neutral weak interaction Feynman diagram for the reaction $\nu_e + \bar{\nu}_e \longrightarrow W^+_{\text{long pol}} + W^-_{\text{long pol}}$

Employing (24.173), it is easy to show that the cross section due to this amplitude alone is proportional to E^2 and thus violates unitarity at sufficiently large E. To avoid this difficulty, as already mentioned, we must include another diagram, the amplitude of which cancels the amplitude in (24.173) at large energy. We know that neutral weak interactions exist, and in particular, we have the neutrino-neutrino-Z^0 vertex (24.133) and the 3-boson ZWW vertex (24.142). Thus we consider the diagram in Figure 24.27.

The amplitude corresponding to this diagram takes the following form:

$$M_b = \frac{-ig}{4\cos\theta_W}\bar{v}\gamma_\sigma\left(1-\gamma^5\right)u\left(-i\frac{g^{\sigma\alpha}-\dfrac{k^\sigma k^\alpha}{m_Z^2}}{k^2-m_Z^2}\right)$$
$$\cdot\left\{ig\cos\theta_W\left[\left(k_+-k_-\right)_\alpha g_{\mu\lambda}-(2k_++k_-)_\lambda g_{\alpha\mu}+(2k_-+k_+)_\mu g_{\alpha\lambda}\right]\frac{k_+^\mu k_-^\lambda}{m_W^2}\right\} \tag{24.174}$$

where $k_+ + k_- = k$, and we have assumed (24.168) for k_+ and k_-. After straightforward algebra, (24.174) is simplified in the limit of very large k^2 to the following expression:

$$M_b = -i\frac{g^2}{4m_W^2}\bar{v}\not{q}(1-\gamma^5)u = -M_a \tag{24.175}$$

Thus amplitudes M_a and M_b indeed cancel to lowest nonvanishing order in the high-energy limit, preserving unitarity to lowest nonvanishing order.

24.17.4 Parity nonconservation in atoms

According to the standard electroweak model, an atomic electron interacts with the nucleus and with other electrons of the atom not only by Coulomb interaction (photon exchange) but also by neutral weak interaction (Z^0 exchange). To describe photon exchange in atomic physics, one employs the ordinary atomic Hamiltonian H_0. Because it commutes with the parity operator, its eigenstates (the usual atomic stationary states) have definite parity, even or odd. Now consider the neutral weak interaction between an atomic electron and the nucleus. Because the electronic neutral weak current and the nuclear neutral weak current each have vector and axial vector components, the effective Hamiltonian H' describing their interaction has both scalar and pseudoscalar parts; that is,

$$H' = H_S + H_P$$

where H_S and H_P are expressed schematically as

$$H_S \sim V_e V_N + A_e A_N \qquad H_P \sim A_e V_N + V_e A_N$$

H_S commutes with the parity operator and causes energy shifts so small that they cannot be distinguished in any practical experiment from effects due to H_0 itself because of small uncertainties in the latter that exist even in the case of atomic hydrogen. However, because H_P does not commute with the parity operator, it perturbs an eigenstate $|\psi^0\rangle$ of H_0 by admixing small amounts of states $|\phi_n^0\rangle$ of opposite parity to $|\psi^0\rangle$, and its effects can be and have been observed. According to first-order static perturbation theory,

$$|\psi^0\rangle \to |\psi\rangle = |\psi^0\rangle + \sum_n |\phi_n^0\rangle \frac{\langle \phi_n^0 | H_P | \psi^0 \rangle}{E(\psi^0) - E(\phi_n^0)} \tag{24.176}$$

Of the two terms in $H_P = H_P^{(1)} + H_P^{(2)}$, where $H_P^{(1)} \sim A_e V_N$, $H_P^{(2)} \sim V_e A_N$, $H_P^{(1)}$ is by far the most important, especially for atoms where the atomic number is $Z \gg 1$. There are several reasons for this. First, as we see from (24.135), the electron-electron-Z^0 vertex factor is proportional to $\gamma^\mu(1 - 4\sin^2\theta_W - \gamma^5)$. Thus the vector portion V_e of the electronic neutral weak current is much smaller than the axial portion A_e because of the quantity $(1\text{–}4\sin^2\theta_W) \approx 0.04$. Next, as will be shown momentarily, the vector portion V_N of the nuclear neutral weak current is proportional to

$$Q_W = Z(1 - 4\sin^2\theta_W) - N \tag{24.177}$$

where Z and N are the atomic number and neutron number, respectively. The quantities Z and N in Q_W reflect the fact that because the electronic wave function is spread out over a volume much larger than that of the nucleus, the contributions of the individual protons and neutrons add up coherently to yield V_N. By contrast, the axial nuclear neutral weak current is generated from the spins of the nucleons, which tend to cancel in pairs, leaving at most two unpaired nucleon spins and almost always no more than one such spin. Thus $H_P^{(2)}$ is weaker than $H_P^{(1)}$ by a factor of approximately $Z^{-1}(1 - 4\sin^2\theta_W)$. In what follows, we ignore H_S and $H_P^{(2)}$, concentrate our attention on $H_P^{(1)}$, and derive the form of the latter.

We start by writing the Hamiltonian density $\mathcal{H}_P^{(1)}$ for the interaction by Z exchange between the electron axial current and the up (or down) quark vector current, in the limit of extremely small momentum transfer; that is,

$$\mathcal{H}_P^{(1)} = -\frac{g^2}{16\cos^2\theta_W m_Z^2}\bar{e}\gamma^\mu\gamma^5 e \begin{bmatrix} -\left(1 - \frac{8}{3}\sin^2\theta_W\right)\bar{u}\gamma_\mu u \\ \left(1 - \frac{4}{3}\sin^2\theta_W\right)\bar{d}\gamma_\mu d \end{bmatrix} \tag{24.178}$$

Here e, u, and d are field operators for the electron, up quark, and down quark, respectively, and the various factors in (24.178) are obtained from (24.135), (24.136), (24.137), and the Z propagator in the limit of extremely small momentum transfer. Now

$$\frac{g^2}{16\cos^2\theta_W m_Z^2} = \frac{G_F}{2\sqrt{2}}$$

Also, the valence quark compositions of a proton and a neutron are (uud) and (ddu), respectively. Thus, from the factors multiplying $\bar{u}\gamma_\mu u$ and $\bar{d}\gamma_\mu d$ in (24.178), we obtain

$$-2\left(1-\frac{8}{3}\sin^2\theta_W\right)+\left(1-\frac{4}{3}\sin^2\theta_W\right) = -(1-4\sin^2\theta_W) \qquad \text{for each proton}$$

$$-\left(1-\frac{8}{3}\sin^2\theta_W\right)+2\left(1-\frac{4}{3}\sin^2\theta_W\right) = +1 \qquad \text{for each neutron}$$

Therefore, summing over all nucleons in the nucleus, we obtain the factor Q_W defined in (24.177). Finally, because the protons and neutrons in the nucleus are nonrelativistic, we have

$$\bar{\psi}_{p,n}\gamma^0\psi_{p,n} \cong \chi_{p,n}^\dagger\chi_{p,n} \qquad \bar{\psi}_{p,n}\boldsymbol{\gamma}\psi_{p,n} \to 0$$

where the $\chi_{p,n}$ are large two-component nucleon wave functions. Thus we obtain the following effective single-electron Hamiltonian:

$$H_P^{(1)} = \frac{G_F}{2\sqrt{2}}Q_W\rho(\boldsymbol{r})\gamma^5 \tag{24.181}$$

where $\rho(\boldsymbol{r})$ is the nucleon probability density with $\int \rho\, d^3\boldsymbol{r} = 1$. In the limit of a point nucleus, $\rho(\boldsymbol{r}) \to \delta^3(\boldsymbol{r})$. Also, taking into account that electronic wave functions ψ^0 and ϕ_n^0 in (24.176) can be written as

$$\psi^0 = \begin{pmatrix} \chi \\ \dfrac{\boldsymbol{\sigma}\cdot\boldsymbol{p}}{E+m_e}\chi \end{pmatrix} \qquad \phi_n^0 = \begin{pmatrix} \eta_n \\ \dfrac{\boldsymbol{\sigma}\cdot\boldsymbol{p}}{E+m_e}\eta_n \end{pmatrix}$$

where χ and η_n are the corresponding large components, we see that the matrix element $\left\langle \phi_n^0 \middle| H_P^{(1)} \middle| \psi^0 \right\rangle$ in (24.176) is

$$\left\langle \phi_n^0 \middle| H_P^{(1)} \middle| \psi^0 \right\rangle = \frac{G_F Q_W}{2\sqrt{2}} \int \left(\eta_n^\dagger \quad \left[\frac{\boldsymbol{\sigma}\cdot\boldsymbol{p}}{E+m_e}\eta_n\right]^\dagger \right) \begin{pmatrix} 0 & I \\ I & 0 \end{pmatrix} \begin{pmatrix} \chi \\ \dfrac{\boldsymbol{\sigma}\cdot\boldsymbol{p}}{E+m_e}\chi \end{pmatrix} \delta^3(\boldsymbol{r})d^3\boldsymbol{r}$$

$$= \frac{G_F Q_W}{2\sqrt{2}} \left[\eta_n^\dagger \frac{\boldsymbol{\sigma}\cdot\boldsymbol{p}}{E+m_e}\chi + \left(\frac{\boldsymbol{\sigma}\cdot\boldsymbol{p}}{E+m_e}\eta_n\right)^\dagger \chi \right]_{\boldsymbol{r}=0}$$

If the atomic electron in question is nonrelativistic, we can replace E by m_e. Thus the last formula reduces to

$$\left\langle \phi_n^0 \left| H_P^{(1)} \right| \psi^0 \right\rangle = \frac{-iG_F Q_W}{4m_e\sqrt{2}} \left[\eta_n^\dagger \boldsymbol{\sigma} \cdot \nabla \chi - \left(\boldsymbol{\sigma} \cdot \nabla \eta_n \right)^\dagger \chi \right]_{r=0} \tag{24.182}$$

From this we see that χ and the gradient of η_n (and/or vice versa) must be nonzero at the origin; otherwise, the matrix element vanishes. In other words, single-electron orbitals of the form $s_{1/2}, p_{1/2}$ are required.

Note that for such orbitals, the matrix element in (24.182) scales a bit more steeply than Z^3. One factor of Z arises (approximately) from Q_W; another from the gradient operator; and still another from the delta function. In addition, there are relativistic corrections. Thus, although atomic parity violation effects are extraordinarily small for light elements, they increase rapidly with Z, and they have been measured at the 1 percent level of precision in heavy atoms such as Cs ($Z = 55$, $6^2S_{1/2}$ ground state) and Tl ($Z = 81$, $6^2P_{1/2}$ ground state).

Problems for Chapter 24

24.1. In the very last stages of stellar evolution, the cores of certain stars reach temperatures of the order of 10^{10} K. At such high temperatures the thermal radiation field generates electron-positron pairs. Consequently, the following weak interactions occur:

$$e^+e^- \to \nu_e\bar{\nu}_e \quad \nu_\mu\bar{\nu}_\mu \quad \nu_\tau\bar{\nu}_\tau \tag{1}$$

This has a profound effect on the final stages of stellar evolution because the neutrinos, once generated, depart from the star without further interaction, thus robbing it of vast quantities of energy. In this problem we make a crude but effective order-of-magnitude estimate of the energy converted to $\nu\bar{\nu}$ pairs per unit volume per unit time in stellar material at 10^{10} K by reaction (1) and compare it with the energy density stored in the radiation plus electron-positron field at the same temperature. This yields an estimate of the time scale for evolution due to neutrino losses. Let

$$Q = \sigma v n_+ n_- \left(E_+ + E_- \right) \tag{2}$$

be the energy generated per unit volume per unit time in neutrino-antineutrino pairs. Here σ is the cross section for reaction (1), v is the relative velocity of e^+ and e^-, $n_+ = n_-$ is the number density of positrons (and of electrons), and $E_\pm$ are their energies.

Employing natural units, use a dimensional argument to obtain σ, use Fermi-Dirac statistics for $k_BT \gg m_ec^2$ to obtain $n_\pm$, and replace each energy in the resulting formula (2) by k_BT to obtain the estimate

$$Q \approx G_F^2 \left(k_B T \right)^9 \tag{3}$$

Now compare this result with the thermal energy density in the radiation plus $e^\pm$ field to obtain the time scale t for evolution due to neutrino losses. You should find

$$t \approx 10 \text{ s}$$

For comparison, note that the time scale for evolution of the Sun is $\approx 10^{10}$ years $= 3\times 10^{17}$ s.

24.2. In Section 24.6 we show that according to the *V-A* law, electrons in beta decay have longitudinal polarization $P = -v/c$, whereas antineutrinos have $P = v/c \approx 1$.
(a) Show that in positron beta decay, $P(e^+) = +v/c$ and $P(\nu_e) \approx -1$.
(b) Suppose that instead of the *V-A* law, nature chose an *S-P* law (S = scalar, P = pseudoscalar). What would be the polarizations of e^- and $\bar{\nu}_e$ in electron beta decay?

24.3. In this problem we consider the beta decay of polarized neutrons

$$n \to p e^- \bar{\nu}_e$$

Our starting point is the effective Feynman amplitude

$$M = \frac{G_F \cos\theta_C}{\sqrt{2}} \bar{u}_p \gamma^\mu (1 - \lambda\gamma^5) u_n \bullet \bar{u}_e \gamma_\mu (1 - \gamma^5) v_\nu \qquad (1)$$

It can be shown that if time-reversal invariance is valid, λ must be real. However, no experiment can ever establish that the imaginary part of λ is exactly zero, so we shall assume in this problem that λ might have a small imaginary part. Also, even if λ is real, it departs from unity only because of strong interaction effects in the nucleon; if these effects were absent, we would have $\lambda = 1$.

Let us consider an experiment in which we employ polarized neutrons but do not observe the polarization of any of the final particles (so we can sum over all final polarization states). Show that the transition rate is proportional to

$$1 + \hat{p}_\nu \bullet \mathbf{v}_e \frac{1-|\lambda|^2}{1+3|\lambda|^2} + \mathbf{v}_e \bullet \hat{s}_n \frac{2\,\mathrm{Re}(\lambda) - 2|\lambda|^2}{1+3|\lambda|^2} + \hat{p}_\nu \bullet \hat{s}_n \frac{2\,\mathrm{Re}(\lambda) + 2|\lambda|^2}{1+3|\lambda|^2} - \hat{s}_n \bullet \mathbf{v}_e \times \hat{p}_\nu \frac{2\,\mathrm{Im}(\lambda)}{1+3|\lambda|^2} \qquad (2)$$

Here $\hat{p}_\nu$ and $\hat{s}_n$ are unit vectors in the direction of the antineutrino linear momentum and neutron spin, respectively, and $\mathbf{v}_e$ is the electron velocity in units of the velocity of light.

The second term in (2) describes the antineutrino-electron momentum correlation. In a typical beta decay experiment, it is impractical to observe the neutrino directly, but this correlation can be detected by observing the momentum of the recoil nucleus in coincidence with the electron and making use of conservation of linear momentum. This correlation does not involve the neutron spin and can be observed with an unpolarized sample of initial neutrons. It is also not a parity-violating effect because it involves the scalar product of two polar vectors.

The third term in (2) describes the electron beta decay asymmetry. It requires polarized neutrons, and it is a parity-violating effect because it is proportional to the scalar product of a polar vector and an axial vector.

The fourth term in (2) describes the antineutrino beta decay asymmetry. It also violates parity and is detected by observation of the recoil nucleus in coincidence with the electron.

The fifth term in (2) violates T but not P and is detected by correlating observations of the recoil nucleus in coincidence with the electron with the direction of neutron spin. Measurements of this term yield the best experimental limits on T violation in beta decay.

[*Hint for the solution of this problem:* The recoiling proton is very nonrelativistic in the rest frame of the neutron. Take advantage of this to simplify the matrix element in (1) and the resulting evaluation of traces.]

24.4. Muonium is a hydrogenic atom in which the proton is replaced by a positive muon. There are two ways in which a muonium atom can disintegrate. The muon can decay

$$\mu^+ \to e^+ \nu_e \bar{\nu}_\mu$$

or the positive muon can capture the atomic electron, which we can assume is in the $1^2 s_{1/2}$ state; that is,

$$\mu^+ e^- \to \nu_e \bar{\nu}_\mu$$

Let the symbols 1 and 2 refer to $\bar{\nu}_\mu$ and ν_e respectively. Making use of the formula for the total muon decay rate (in natural units)

$$\Gamma\left(\mu^+ \to e^+ \nu_e \bar{\nu}_\mu\right) = \frac{G_F^2 m_\mu^5}{192\pi^3}$$

and the algebraic formula

$$tr\left[\not{p}_\mu \gamma^\lambda (1-\gamma^5) \not{p}_1 \gamma(1+\gamma_5)\gamma_\sigma \gamma_4\right] tr\left[\not{p}_2 \gamma^\lambda (1-\gamma^5) \not{p}_e \gamma_4 (1+\gamma_5)\gamma_\sigma \gamma_4\right] = 64 p_\mu \ p_2 p_e \ p_1$$

show that the ratio of the rates for the two processes is

$$R = \frac{\Gamma(\mu^+ e^- \to \nu_e \bar{\nu}_\mu)}{\Gamma(\mu^+ \to e^+ \nu_e \bar{\nu}_\mu)} = \frac{48\pi m_e^3 \alpha^3}{m_\mu^3}$$

where α is the fine-structure constant. R is in fact a very small number, so the capture rate is quite negligible.

24.5. In Section 24.12.2 we discussed the Yang-Mills analysis of infinitesimal $SU(2)$ gauge transformations on an isodoublet Dirac field Ψ. Such transformations are characterized by the formulas

$$\delta\Psi = i\boldsymbol{\varepsilon}\cdot\boldsymbol{t}\Psi \quad \text{and} \quad \delta\bar{\Psi} = -i\bar{\Psi}\boldsymbol{\varepsilon}\cdot\boldsymbol{t} \tag{1}$$

The Lagrangian density for the Dirac field and its conjugate is invariant under local $SU(2)$ gauge transformations if we include an interaction with a triplet of gauge fields analogous to the electromagnetic vector potential and collectively denoted by the symbol A_μ. We show with equation (24.81) that the change in A_μ resulting from the infinitesimal gauge transformation described by (1) is

$$\delta A_\mu = -\frac{1}{g}\partial_\mu \boldsymbol{\varepsilon} - \left(\boldsymbol{\varepsilon} \times A_\mu\right) \tag{2}$$

How would we modify (2) if the gauge transformation were finite; that is,

$$\Psi \to \Psi' = \exp\left(i\boldsymbol{\varepsilon}\cdot\boldsymbol{t}\right)\Psi \qquad \bar{\Psi} \to \bar{\Psi}' = \bar{\Psi}\exp\left(-i\boldsymbol{\varepsilon}\cdot\boldsymbol{t}\right)$$

(b) Fill in the steps that lead from (24.73) to (24.84).

24.6. In Section 24.17.3 we considered the reaction

$$\nu_e + \bar{\nu}_e \to W^{+}_{\text{longit pol}} + W^{-}_{\text{longit pol}} \tag{1}$$

and showed that in the ultrarelativistic limit the amplitude associated with Figure 24.26 becomes

$$M_a = -\frac{ig^2}{4m_W^2}\bar{v}\not{q}(1+\gamma_5)u \tag{2}$$

We also remarked in Section 24.17.3 that the amplitude associated with Figure 24.27 is

$$M_b = \frac{-ig}{4\cos\theta_W}\bar{v}\gamma_\sigma(1-\gamma^5)u\left(-i\frac{g^{\sigma\alpha}-\dfrac{k^\sigma k^\alpha}{m_Z^2}}{k^2 m_Z^2}\right)$$
$$\left\{ig\cos\theta_W\left[(k_+-k_-)_\alpha g_{\mu\lambda}-(2k_++k_-)_\lambda g_{\alpha\mu}+(2k_-+k_+)_\mu g_{\alpha\lambda}\right]\frac{k_+^\mu k_-^\nu}{m_W^2}\right\} \tag{3}$$

and we mentioned that in the ultrarelativistic limit, M_b reduces to

$$M_b = -i\frac{g^2}{4m_W^2}\bar{v}\not{q}(1-\gamma^5)u = M_a \tag{4}$$

Because of the cancellation, unitarity is preserved to this order of perturbation theory.
(a) Fill in the steps leading from (3) to (4).
(b) Consider the analogous reaction

$$e^- + e^+ \to W^{-}_{\text{longit pol}} + W^{+}_{\text{longit pol}} \tag{5}$$

In this case there are *three* 2-vertex Feynman diagrams. Draw these diagrams, labeling the various 4-momenta and 4-polarizations carefully. Use the Feynman rules for the standard electroweak model given in Section 24.16 to construct the amplitudes corresponding to the three diagrams, and show that these amplitudes cancel in the ultrarelativistic limit, thereby once again preserving unitarity to this order in perturbation theory.

24.7. In this problem, we ignore *CP* violation, which is a small effect, and write

$$\begin{aligned}|K_S^0\rangle &= \frac{1}{\sqrt{2}}\left(|K^0\rangle - |\bar{K}^0\rangle\right)\\ |K_L^0\rangle &= \frac{1}{\sqrt{2}}\left(|K^0\rangle + |\bar{K}^0\rangle\right)\end{aligned} \tag{1}$$

We also make use of the experimentally determined total decay rates

$$\begin{aligned}\gamma(K_S^0) &= 1.12\times10^{10}\ \text{s}^{-1}\\ \gamma(K_L^0) &= 1.93\times10^{7}\ \text{s}^{-1}\end{aligned}$$

and the experimentally determined mass difference $\Delta m = m_L - m_S = 5.34\times10^9\ \hbar c^{-2}\ \text{s}^{-1}$. Consider an experiment in which a beam of neutral K mesons is formed by some strong interaction collision process. The K beam is assumed to have velocity β in the laboratory frame and to consist of particles in the state $|K^0\rangle$ at initial time $t = 0$.

(a) Find the state vector at time $t > 0$ as measured in the kaon rest frame. Your answer should be expressed in terms of the state vectors $|K_S^0\rangle$ and $|K_L^0\rangle$ and the quantities β, t, $\gamma(K_S^0), \gamma(K_L^0)$, and Δm. Your answer should reveal damped oscillatory behavior, and when t is sufficiently large, the state vector should become proportional to $|K_L^0\rangle$. The phenomenon exhibited here is called *strangeness oscillations*, and it is somewhat analogous to neutrino oscillations.

(b) Let the beam of particles pass through a slab of matter after having traveled in vacuum for a sufficiently long time that the particles are in the state $|K_L^0\rangle$. Noting that the scattering amplitudes for K^0 and $\bar{K}^0$ on matter nuclei are the complex numbers a and b, respectively, with $|b| \gg |a|$, show that the beam emerging from the slab of matter contains a considerable K_S^0 component. This phenomenon is called *regeneration*.

24.8. In the absence of atomic parity nonconservation, the 1s and 2s states of atomic hydrogen both have even (positive) parity, so no electric dipole ($E1$) transition can occur between them. However, when atomic parity nonconservation is taken into account, there is a very small but nonzero $E1$ matrix element connecting these states. Using equations (24.176) and (24.182), calculate this matrix element for atomic deuterium. You may assume that the sum in (24.182) is dominated by a single term, for which the energy denominator is the Lamb shift for deuterium: 1,059 MHz.

25 The Quantum Measurement Problem

25.1 Statement of the problem

The quantum measurement problem can be explained rather simply in terms of a thought experiment designed for observation of the eigenstates of J_z for a spin-1/2 particle. As we know, this system is described by a two-dimensional Hilbert space with two orthonormal state vectors $|\alpha_S\rangle$ and $|\beta_S\rangle$ for spin up and down, respectively, with respect to the z-axis. We assume that the measuring apparatus also obeys quantum mechanics, although it is macroscopic. For reliable measurements, the apparatus must satisfy the following requirements:

- The apparatus is prepared in a "ready" state $|\Phi_0\rangle$ at the beginning of a measurement.
- The system and the apparatus interact for a certain time interval T.
- If at $t = 0$ the system-apparatus state vector is

$$|\Psi(0)\rangle = |\alpha_S\rangle \otimes |\Phi_0\rangle \tag{25.1}$$

then at $t = T$,

$$|\Psi(T)\rangle = U(T,0)(|\alpha_S\rangle \otimes |\Phi_0\rangle) = |\alpha_S\text{'}\rangle|\Phi_+(T)\rangle \tag{25.2}$$

whereas if

$$|\Psi(0)\rangle = |\beta_S\rangle \otimes |\Phi_0\rangle \tag{25.3}$$

then

$$|\Psi(T)\rangle = U(T,0)(|\beta_S\rangle \otimes |\Phi_0\rangle) = |\beta_S\text{'}\rangle|\Phi_-(T)\rangle \tag{25.4}$$

where $U(T, 0)$ is a deterministic unitary time-evolution operator governing the interaction of system and apparatus, and $|\Phi_\pm(T)\rangle$ are apparatus states that are generally macroscopically distinguishable (i.e., they refer to different positions of a pointer on a dial or different arrangements of indicator lights on a counter, etc.).

The requirements we just stated contain the essential features of many experiments that have actually been performed, for example, the Stern-Gerlach experiment and countless variants of it.

So far there is no difficulty, but now suppose that we prepare the system with initial state vector

$$|\psi_S(0)\rangle = a|\alpha_S\rangle + b|\beta_S\rangle \tag{25.5}$$

where a and b are both nonzero, and $|a^2| + |b|^2 = 1$. Then the system-apparatus state vector at $t = 0$ is

$$|\Psi(0)\rangle = \left[a|\alpha_S\rangle + b|\beta_S\rangle\right] \otimes |\Phi_0\rangle \tag{25.6}$$

and because the Schroedinger equation $H|\Psi\rangle = i\hbar|\dot{\Psi}\rangle$ is linear, the system-apparatus state vector at time T must be

$$|\Psi(T)\rangle = U(T,0)|\Psi(0)\rangle = a|\alpha'_S\rangle \otimes |\Phi_+(T)\rangle + b|\beta'_S\rangle \otimes |\Phi_-(T)\rangle \tag{25.7}$$

which is an entangled state of the system and apparatus. This presents a serious problem because superpositions of macroscopically distinguished apparatus states, such as those that appear on the right-hand side of (25.7), are never observed. For example, we never observe a pointer that is simultaneously at two different places on a dial. One always observes

$$|\Phi_+(T)\rangle \quad \text{or} \quad |\Phi_-(T)\rangle \tag{25.8}$$

Furthermore, if both a and b are nonzero, we cannot predict ahead of time which of the states in (25.8) will be obtained on a given measurement; we can only assign a probability for a specific outcome. The contradiction between deterministic unitary evolution described by (25.7), which leads to a superposition of macroscopically distinguishable states and the nonlinear and stochastic state-vector reduction of (25.8), is one important part of the quantum measurement problem.

However, there is more to the problem: nothing in the standard rules of quantum mechanics tells us when it is appropriate to employ deterministic unitary evolution and when we should use state-vector reduction. For example, let us label our spin-½ particle as system 1 (S1) and the apparatus described earlier as apparatus 1 (A1). Then we could consider (S1+A1) as another system (S2) that could be measured by another apparatus (A2). This would shift the boundary between unitary evolution and collapse. Or we could attempt to measure the properties of system (S3) = (S2 + A2) with still another apparatus (A3), which shifts the boundary further. There is no obvious reason why this process cannot be iterated indefinitely, and we are left with the unanswered question: at what point in the chain should we give up unitary evolution and employ state-vector reduction?

Before discussing the problem further, we illustrate the foregoing thought experiment by means of a simple mathematical model **M**. For the system, we use the usual representation; that is,

$$|\alpha_S\rangle = \begin{pmatrix} 1 \\ 0 \end{pmatrix}_S \qquad |\beta_S\rangle = \begin{pmatrix} 0 \\ 1 \end{pmatrix}_S \tag{25.9}$$

As for the apparatus, although it has macroscopically distinguishable final states, we simplify it as much as possible in a schematic description by assuming that

$$|\Phi_0\rangle = \begin{pmatrix} 1 \\ 0 \end{pmatrix}_A \tag{25.10}$$

and

$$|\Phi_+(T)\rangle = \begin{pmatrix} 1 \\ 0 \end{pmatrix}_A \qquad |\Phi_-(T)\rangle = \begin{pmatrix} 0 \\ 1 \end{pmatrix}_A \tag{25.11}$$

It can be seen in this simple model that the apparatus also behaves like a particle with spin-½. The requirements (25.11) are achieved by the following system-apparatus interaction Hamiltonian:

$$H = \frac{1}{4}(\sigma_{xA} - I)(\sigma_{zS} - I)g(t) \tag{25.12}$$

where σ_{xA} and σ_{zS} are Pauli spin matrices referring to the apparatus and system, respectively (each σ_{iA} commutes with every σ_{jS}), whereas $g(t)$ is nonzero only in the interval $0 \le t \le T$ and satisfies

$$\int_0^T g(t)\,dt = \pi\hbar$$

To see that H in (25.12) is the desired Hamiltonian, we write the time-development operator

$$U(T,0) = \exp\left[-\frac{i}{\hbar}\int_0^T H(t)\,dt\right] = \exp\left[-\frac{i\pi}{4}(\sigma_{xA} - I_A)(\sigma_{zS} - I_S)\right] \tag{25.13}$$

and expand the far right-hand side of (25.13) in a power series. Because

$$(\sigma_{xA} - I_A)^2 = \sigma_{xA}^2 - 2\sigma_{xA} + I_A = -2(\sigma_{xA} - I_A)$$

and, similarly,

$$(\sigma_{zS} - I_S)^2 = -2(\sigma_{zS} - I_S)$$

we have

$$\begin{aligned} U(T,0) &= I - \frac{1}{2}(\sigma_{xA} - I)(\sigma_{zS} - I) \\ &= \frac{1}{2}(I + \sigma_{xA} + \sigma_{zS} - \sigma_{xA}\sigma_{zS}) \end{aligned} \tag{25.14}$$

Now let

$$\Psi(0)=\begin{pmatrix}a\\b\end{pmatrix}_S\begin{pmatrix}1\\0\end{pmatrix}_A \tag{25.15}$$

as in (25.6). Then, applying U, we obtain

$$\begin{aligned}\Psi(T)&=\frac{1}{2}\left[\begin{pmatrix}a\\b\end{pmatrix}_S\begin{pmatrix}1\\0\end{pmatrix}_A+\begin{pmatrix}a\\-b\end{pmatrix}_S\begin{pmatrix}1\\0\end{pmatrix}_A+\begin{pmatrix}a\\b\end{pmatrix}_S\begin{pmatrix}0\\1\end{pmatrix}_A-\begin{pmatrix}a\\-b\end{pmatrix}_S\begin{pmatrix}0\\1\end{pmatrix}_A\right]\\&=\begin{pmatrix}a\\0\end{pmatrix}_S\begin{pmatrix}1\\0\end{pmatrix}_A+\begin{pmatrix}0\\b\end{pmatrix}_S\begin{pmatrix}0\\1\end{pmatrix}_A\end{aligned} \tag{25.16}$$

as in (25.7). We will find it helpful to refer to this model **M** in the discussion that follows.

The quantum measurement problem has generated a very extensive (and often controversial) literature, which is surveyed with admirable clarity in Laloe (2012). There are many diverse proposals for resolving the problem. Broadly speaking, these fall into three categories:

1. There is no problem. We have set up false difficulties by giving an oversimplified description of the measurement process.
2. The interpretation of the rules of quantum mechanics must be changed to resolve the quantum measurement problem, but this can be done in ways that are empirically indistinguishable from the standard theory.
3. Deterministic unitary evolution is only an approximation. The time-dependent Schroedinger equation must be modified to include a term that is both nonlinear and stochastic, and/or quantum mechanics must be modified in more fundamental ways. Such a modification could have testable observational consequences, at least in principle.

We now consider each of these categories.

25.2 Is there no problem?

Have we set up false difficulties by giving an oversimplified description of the measurement process? Many persons who believe that there is no problem argue that in any scientific experiment the apparatus is almost always a complex macroscopic piece of equipment with many degrees of freedom, most of these degrees of freedom are coupled uncontrollably to the environment, and such couplings have generally existed for a long time before the start of any measurement at $t = 0$. Consequently, it is unrealistic to describe the apparatus at $t = 0$ in terms of a well-defined ready state $|\Phi_0\rangle$. Even if we permit such a state $|\Phi_0\rangle$, the couplings between the apparatus and the environment cannot be ignored in the time interval $0 \le t \le T$. Many model calculations [see, e.g., Zurek (1982) and Joos (1999)] and several experimental studies [see, e.g., Brune (1996)] have shown that decoherence occurs in this time interval. This means that the relative phases of various apparatus macro states such as $|\Phi_+(t)\rangle, |\Phi_-(t)\rangle$ become completely scrambled by couplings to the environment. Thus, with reference to (25.7), whereas

$$|\Phi_+(0)\rangle = |\Phi_-(0)\rangle = |\Phi_0\rangle$$

so that

$$\langle \Phi_+(0)|\Phi_-(0)\rangle = \langle \Phi_0|\Phi_0\rangle = 1$$

nevertheless,

$$\langle \Phi_+(t)|\Phi_-(t)\rangle \approx e^{-\Lambda t} \tag{25.17}$$

where Λ^{-1} is a decoherence time. Because in virtually all real experimental situations $\Lambda^{-1} \ll T$, any definite phase relation between the two state vectors on the right-hand side of (25.7) is obliterated long before $t = T$, and in general, there are no observable interference effects between these two states.

It has been claimed [see, e.g., Anderson (2001) and Tegmark and Wheeler (2001)] that when these complications are taken into account, the quantum measurement problem disappears. However, this view is not justified. Concerning the apparatus initial state, no calculation has ever been presented that shows how collapse follows from any detailed assumptions about the initial state. On the contrary, a generalized discussion of the initial (and final) states of the apparatus, given by Bassi and Ghirardi (2000) and expanded by Grubl (2002) shows that the basic notions presented in Section 25.1 are still valid (even if they are oversimplified) so far as they concern the apparatus initial state.

The development of decoherence during the time interval $0 \leq t \leq T$ of the measurement process can be represented schematically in model **M** or in other models by the disappearance of off-diagonal matrix elements in the apparatus density matrix. (The latter is obtained from the full density matrix describing system-apparatus entanglement by tracing over system states.) For example, if in (25.16) we have $a = b = 1/\sqrt{2}$, then if there were no decoherence, the apparatus density matrix would be

$$M_A(T) = \frac{1}{2}\begin{pmatrix} 1 & 1 \\ 1 & 1 \end{pmatrix}$$

but decoherence causes the off-diagonal elements of this matrix to vanish, with the result

$$M_A(T) = \frac{1}{2}\begin{pmatrix} 1 & 0 \\ 0 & 1 \end{pmatrix} \tag{25.18}$$

Now it is certainly true that the latter matrix represents an incoherent statistical mixture of the apparatus states $\begin{pmatrix} 1 \\ 0 \end{pmatrix}_A$ and $\begin{pmatrix} 0 \\ 1 \end{pmatrix}_A$ with equal probabilities. However, this does not support the contention that decoherence solves the quantum measurement problem because $M_A(T)$ in (25.18) also represents an incoherent statistical mixture with equal probabilities of the states

$$\frac{1}{\sqrt{2}}\begin{pmatrix}1\\1\end{pmatrix}_A \quad \text{and} \quad \frac{1}{\sqrt{2}}\begin{pmatrix}1\\-1\end{pmatrix}_A$$

and each of the latter states is a superposition of macroscopically distinguishable states. Indeed, infinitely many different combinations of apparatus states correspond to the same density matrix in (25.18). Thus, despite decoherence, we must still introduce a separate probabilistic collapse postulate to explain the fact that in any given measurement we obtain only one or the other of the two states. These points have been emphasized very clearly in Adler (2003, 2004), Schlosshauer (2004), and Bassi (2000, 2003). See also Fine (1969, 1970), Brown (1986), Albert (1992), Bub (1997), and Barrett (1999). Despite significant belief to the contrary, decoherence does not solve the quantum measurement problem.

Next, we consider the following question.

25.3 Can interpretation of the rules of quantum mechanics be changed to resolve the quantum measurement problem, and can this be done so that the modified theory is empirically indistinguishable from the standard theory?

All proposals in this category have a common feature: in the absence of possible future experimental results that would falsify quantum mechanics itself, these proposals are themselves not falsifiable. Here we briefly summarize just a few of the many ideas that have been suggested.

The Ax-Kochen Proposal (Ax and Kochen 1999)

In conventional quantum mechanics, a ray in a Hilbert space is identified with the state of a system; hence any vector belonging to this ray, regardless of its phase, corresponds to the same state. Ax and Kochen proposed that instead, the ray should be identified with the quantum ensemble, and the phase of the vector should be identified with an individual member of the ensemble. In this scheme, if the a priori distribution of phases is assumed to be uniform, then an alternative quantum theory can be set up in such a way that the probabilities of experimental outcomes follow the rules of conventional quantum mechanics.

Bohmian Mechanics

Although the results of photon polarization correlation experiments rule out local hidden-variable theories (with minor caveats), one example of a hidden-variable theory that evades this restriction and has been worked out in some detail is Bohmian mechanics (Bohm 1952). This scheme is designed to resolve the quantum measurement problem for macroscopically distinguishable states. Here the new variables are the "hidden" positions $\boldsymbol{x}_i$ of the particles, which are injected into the theory in addition to the usual generalized coordinates $\boldsymbol{q}_i$ that appear in the wave function. In fact, whereas in conventional quantum mechanics we may say that spatial coordinates and linear momenta play an equal and balanced role, in Bohmian mechanics the spatial coordinates have a special privileged role. The rules of this proposal are as follows:

1. The state of a physical system S at an initial time t_0 is given by the wave function

$$\psi(q_1, q_2, \ldots, q_n; t_0) \tag{25.19}$$

together with the positions $\boldsymbol{x}_1(t_0), \boldsymbol{x}_2(t_0), \ldots, \boldsymbol{x}_n(t_0)$ of all the particles of S.
2. The evolution of the wave function is governed by the Schroedinger equation; that is,

$$i\hbar \frac{\partial \psi(q_1, q_2, \ldots, q_n; t_0)}{\partial t} = H\psi(q_1, q_2, \ldots, q_n; t_0) \tag{25.20}$$

3. The equations of motion for the $\boldsymbol{x}_i$ are

$$\frac{d\boldsymbol{x}_i}{dt} = \frac{\hbar}{m_i} \mathrm{Im} \left. \frac{\psi^*(q_1, q_2, \ldots, q_n; t) \nabla_i \psi(q_1, q_2, \ldots, q_n; t)}{|\psi(q_1, q_2, \ldots, q_n; t)|^2} \right|_{q_i = \boldsymbol{x}_i} \tag{25.21}$$

In this scheme, the Schroedinger equation (25.20) is first solved subject to the initial conditions. Then the solution is inserted into (25.21) to solve for the hidden variables $\boldsymbol{x}_i$. Because this procedure yields definite values of $\boldsymbol{x}_i$ as a function of t, each of the particles always has a definite position; thus macroscopic objects have a definite location in space, and the quantum measurement problem is avoided.

In Bohm's theory, consider an ensemble of physical systems, each described by the wave function

$$\psi(q_1, q_2, \ldots, q_n; t) \tag{25.22}$$

Let each member of the ensemble contain n particles, the positions of which are

$$\boldsymbol{x}_1, \boldsymbol{x}_2, \ldots, \boldsymbol{x}_n$$

and let these positions vary from one member of the ensemble to the next. Assume that the probability distribution of these positions is given at t_0 by

$$\rho(\boldsymbol{x}_1, \boldsymbol{x}_2, \ldots, \boldsymbol{x}_n; t_0) = |\psi(\boldsymbol{x}_1, \boldsymbol{x}_2, \ldots, \boldsymbol{x}_n; t_0)|^2 \tag{25.23}$$

Then it can be shown that the particles in the ensemble follow trajectories such that at any later time t

$$\rho(\boldsymbol{x}_1, \boldsymbol{x}_2, \ldots, \boldsymbol{x}_n; t) = |\psi(\boldsymbol{x}_1, \boldsymbol{x}_2, \ldots, \boldsymbol{x}_n; t)|^2 \tag{25.24}$$

Thus Bohmian mechanics is equivalent to nonrelativistic quantum mechanics with respect to its predictions concerning particle positions. However, the scheme encounters major problems: despite decades of effort by its adherents [see, e.g., Durr et al. (1992)], it has not been possible to incorporate spin into the theory in a convincing way, and there has been no success in constructing a relativistic version of the theory. In particular, there does not seem to be a consistent way to quantize the electromagnetic field to obtain photons.

Two Dynamical Principles

Here one accepts the statements that quantum-mechanical systems *are* governed by deterministic linear evolution and that the measurement process *is* nonlinear and stochastic. How, then, can we deal with the problem stated in Section 25.1 that the boundary between these two contradictory principles is so shifty and arbitrary?

F. London and E. Bauer (1939) and, subsequently, E. Wigner (1967), believing that both principles must be accepted, thought it necessary to go to the extreme end of the chain of observation and to assume that state-vector reduction does not happen until the consciousness of the observer intervenes. Many difficult questions arise here, and most of them cannot be answered given the present state of our knowledge.

1. How can we understand the phenomenon of consciousness in terms of physical-chemical processes in the brain?
2. Computerized robots can manipulate an experiment and record data. They also have memory and can "learn." Should we consider such robots as intermediate parts of the apparatus, or do they qualify as conscious beings at the end of the chain?
3. Does anyone seriously believe that consciousness can affect physical laws?

The Many-Worlds Hypothesis

The advocates of this idea assume that deterministic unitary evolution is the only correct principle. They propose to avoid the quantum measurement problem by assuming that all parts of the state vector, for example, both terms on the right-hand side of (25.7), are realized as follows: each time an interaction occurs that leads to a superposition of macroscopically distinguishable states, the universe splits into replicas of itself, each replica corresponding to one of the terms in the superposition [see, e.g., De Witt and Graham (1973)]. It seems that there are major problems in this proposal. For example, because any given result of a measurement cannot be predicted in advance, the split itself must have a stochastic element. Why should this be any more acceptable than the collapse postulate? One must assume that there is no possibility of communication between one universe and another, and it is difficult to understand how this assumption is any more acceptable than the collapse postulate. Another major problem follows from the remark by J. S. Bell: "Are we not obliged to admit that measurement-like processes are going on more or less all the time, more or less everywhere?" If we take Bell's remark seriously, as it appears that we must, then the process of replication would have to be continuous and also would necessarily include replication of the conscious mind of the observer, so we have not only a many-worlds hypothesis but also a many-minds hypothesis. One wonders how the many-worlds hypothesis could ever be falsifiable and, if it is not falsifiable, whether it can be considered science.

Finally, we consider the following questions discussed in Section 25.4.

25.4 Is deterministic unitary evolution only an approximation? Should the time-dependent schroedinger equation be modified? Might such modifications have testable observational consequences?

The successes of quantum mechanics may generate strong psychological pressure on most contemporary physicists to believe that the theory is complete and correct or at most requires some relatively minor adjustments to its interpretation. However, this is not the first time in history that a major scientific theory was thought to be perfect. Newtonian mechanics was so considered for about 200 years before it was extended by special and general relativity and then supplanted in the micro world by quantum mechanics. Now we realize that Newtonian mechanics, while very useful, is valid only in a restricted domain where velocities are much less than the velocity of light, gravitational fields are weak, and actions are much larger than $\hbar$. Will future physicists recognize that present-day quantum mechanics is also an approximate theory, valid merely in a restricted domain? In fact, there are a number of deep unresolved problems in quantum mechanics and quantum field theory that give us reason to think that it might be just an approximation, albeit an excellent one for all practical purposes so far. In addition to the quantum measurement problem and the well-known problem of incorporating gravitation into quantum theory, here are just two additional difficulties:

- Classical mechanics is understood to be a limiting form of quantum mechanics, but we start with Poisson brackets of various classical variables and then translate them into commutation rules for the corresponding quantum variables. It seems unsatisfactory to derive the rules of the more fundamental theory from the rules governing its less fundamental limiting form. Perhaps the rules for quantization (and an explanation for the existence of $\hbar$) might emerge naturally in a more fundamental theory.
- Quantum field theory contains infinities that arise from the local nature of commutation and anticommutation relations connecting components of quantum fields. In the calculation of higher-order corrections to certain physical phenomena by means of perturbation theory, infinities are removed by the technique of renormalization. An important example involving renormalization is calculation of the Lamb shift in the spectrum of atomic hydrogen, which yields a result in excellent agreement with experiment. However, renormalization is based on a questionable mathematical procedure in which one subtracts one infinite quantity from another to obtain a well-defined finite remainder. One might hope that in a more fundamental nonlocal theory, all the infinities and such mathematical procedures could be avoided.

Any far-reaching attempt to construct a more fundamental theory resolving all the problems of quantum mechanics faces great difficulties, especially because there are at present no obvious experimental guideposts. Nevertheless, some authors have given considerable thought to this task: they include S. L. Adler (2004) and G. 't Hooft (1988, 1997, 1999, 2002, 2003). A less ambitious goal is to modify nonrelativistic quantum mechanics to remove the contradiction between deterministic unitary evolution and nonlinear stochastic collapse. Because collapse is encountered in almost every experimental measurement, many authors in the last several decades have concentrated their efforts on modifying unitary evolution.The list includes Pearle (1976, 1984, 1989); Ghirardi, Rimini, and Weber (1986); Ghirardi, Pearle, and Rimini (1990);

Gisin (1984, 1989); Diosi (1988a, b, 1989), and Percival (1994); and many of these efforts have been described in the detailed review by Bassi and Ghirardi (2003).

Here the following question arises: if we set out to modify the deterministic linear Schroedinger equation, would it be sufficient to replace it with a stochastic linear equation or with a deterministic nonlinear equation, or is it necessary to find an equation that is both stochastic and nonlinear? Gisin (1989) has shown that nonlinearity requires stochasticity, and the reverse is also true [Bassi and Ghirardi (2003)]. Thus it appears necessary to modify the Schroedinger equation by adding stochastic and nonlinear features.

Bearing this in mind, we briefly summarize the main ideas of Ghirardi, Rimini, and Weber (1986), who developed a scheme called *quantum mechanics with spontaneous localization* (QMSL). Here the following goals have been set:

- QMSL must be constructed in such a way as to guarantee a definite position in space to macroscopic objects.
- The modified dynamics must have little impact on microscopic objects so as to avoid conflict with the results of large numbers of precise experiments, but at the same time it must essentially eliminate the superposition of different macroscopic states of macro systems. Hence there must be an "amplification" mechanism when going from the micro to the macro level.

Their theory is thus constructed with the following assumptions:

- Each particle of a system of n distinguishable particles experiences, with mean rate λ_i, a sudden spontaneous localization process, where i refers to the ith particle.
- In the time interval between two successive spontaneous processes, the system evolves according to the usual Schroedinger equation.
- The sudden spontaneous process (called a *hit* in what follows) is a localization described by

$$|\psi\rangle \rightarrow \frac{|\psi_x^i\rangle}{\langle\psi_x^i|\psi_x^i\rangle^{1/2}} \tag{25.25}$$

Here $|\psi_x^i\rangle = L_x^i|\psi\rangle$, where L_x^i is the *localization operator*; that is,

$$L_x^i = \left(\frac{\alpha}{\pi}\right)^{3/4} e^{-(\alpha/2)(q_i-x)^2} \tag{25.26}$$

and the probability density for a hit at x is assumed to be $P_i(x) = \langle\psi_x^i|\psi_x^i\rangle$.

The localization process works as shown by the following example: consider a superposition of two one-dimensional Gaussian functions, one centered at position $-a$ and the other at $+a$; that is,

$$\psi(x) = \frac{1}{N}\left(e^{-(\gamma/2)(x+a)^2} + e^{-(\gamma/2)(x-a)^2}\right) \tag{25.27}$$

where N is a normalization factor. Suppose that $a \gg 1/\alpha^{1/2} \gg 1/\gamma^{1/2}$, which means that the distance between the two Gaussians is much greater than the localization amplitude, which is, in turn, much greater than the width of either Gaussian. Now consider a hit centered at a. The wave function changes as follows:

$$\psi(x) \to \psi_a(x) = \frac{1}{N_a}\left(e^{-2\alpha a^2} e^{-(\gamma/2)(x+a)^2} + e^{-[(\gamma+\alpha)/2](x-a)^2}\right) \tag{25.28}$$

Thus the Gaussian centered at $-a$ is exponentially suppressed relative to the other term, the width of which remains essentially unchanged. The new wave function describes a particle that is well localized around $+a$. Because a hit is more likely to occur where the probability to find a particle is greater, the probability of a hit at $x = 0$ is very small; if it does occur, the effect on the wave function (25.27) is quite negligible.

If a single particle suffers a hit, its wave function ψ changes into the wave function ψ_x. Because we do not know where the hit occurs but only the probability for it to occur around x, the pure state is transformed into the statistical mixture

$$\begin{aligned} |\psi\rangle\langle\psi| &\to \int d^3x P(x) \frac{|\psi_x\rangle\langle\psi_x|}{\langle\psi_x|\psi_x\rangle} \\ &= \int d^3x L_x |\psi\rangle\langle\psi| L_x \equiv T\left[|\psi\rangle\langle\psi|\right] \end{aligned} \tag{25.29}$$

It can be shown that this implies the following equation for the time evolution of the one-particle density operator ρ:

$$\frac{d\rho(t)}{dt} = \frac{1}{i\hbar}[H, \rho(t)] - \lambda\{\rho(t) - T[\rho(t)]\} \tag{25.30}$$

This master equation, which in effect replaces the time-dependent Schroedinger equation and is of central importance for QMSL, differs from the equation of motion for the density matrix in standard quantum mechanics because it includes the term proportional to λ. The nonlinear contribution is, of course, the quantity $\lambda T(\rho)$.

It is found that if λ is chosen so that a typical microsystem (e.g., an electron or an H atom) undergoes a spontaneous localization once every 10^{16} s $\approx 3\times10^8$ years, then no known experimental result in atomic, nuclear, condensed-matter, or elementary particle physics appears to be contradicted, but the quantum measurement problem is avoided. An additional attractive feature of this scheme is that it does not introduce any new hidden variables (as is done in Bohmian mechanics); rather, QMSL relies on conventional notions of state vectors as rays in Hilbert space.

A continuous stochastic version of QMSL, applicable to systems of identical particles and known as *continuous spontaneous localization* (CSL), was developed by Pearle (1989) and by Ghirardi, Pearle, and Rimini (1990). This and related models are discussed in Adler (2004) and Bassi and Ghirardi (2003). Various experimental tests of spontaneous localization have been suggested, but so far none appear to be viable.

In conclusion, although many interesting suggestions have been made for overcoming the quantum measurement problem, it remains unsolved. We can only hope that some future experimental observation may guide us toward a solution.

A Appendix Useful Inequalities for Quantum Mechanics

A.1 The Arithmetic Mean–Geometric Mean Inequality and Some Consequences

Let $a_1, a_2, ..., a_{2^m}$ be 2^m positive real numbers, where m is a positive integer. Then

$$a_1 a_2 = \left(\frac{a_1 + a_2}{2}\right)^2 - \left(\frac{a_1 - a_2}{2}\right)^2 \leq \left(\frac{a_1 + a_2}{2}\right)^2$$

with equality only if $a_1 = a_2$. Similarly,

$$a_1 a_2 a_3 a_4 \leq \left(\frac{a_1 + a_2}{2}\right)^2 \left(\frac{a_3 + a_4}{2}\right)^2 \leq \left(\frac{a_1 + a_2 + a_3 + a_4}{4}\right)^4$$

with equality only if $a_1 = a_2 = a_3 = a_4$. Repeating this same procedure m times, we have

$$a_1 a_2 \ldots a_{2^m} \leq \left(\frac{a_1 + \cdots + a_{2^m}}{2^m}\right)^{2^m} \tag{A.1}$$

Now suppose that n is any integer less than 2^m. Let $b_1 = a_1,\ \ldots,\ b_n = a_n$ and

$$b_{n+1} = \cdots = b_{2^m} = \frac{a_1 + \cdots + a_n}{n} \equiv A \tag{A.2}$$

Then

$$b_1 b_2 \cdots b_{2^m} = a_1 a_2 \cdots\ a_n \left(\frac{a_1 + \cdots + a_n}{n}\right)^{2^m - n} \leq \left(\frac{b_1 + b_2 + \cdots + b_{2^m}}{2^m}\right)^{2^m}$$

$$= \left(\frac{nA + (2^m - n)A}{2^m}\right)^{2^m} = A^{2^m}$$

Thus, for any n,

$$a_1 a_2 \cdots\ a_n \leq A^{2^m} A^{n-2^m} = A^n$$

or

$$(a_1 a_2 \cdots a_n)^{1/n} \le A \tag{A.3}$$

Expression (A.3) states that the geometric mean is always less than or equal to the arithmetic mean, and equality of these means occurs only when all the numbers a are the same. If a_i, b_i, c_i with $i = 1,\ldots,n$ are three sets of positive numbers, then from (A.3) we have

$$\left(\sum_{i=1}^{n} a_i \sum_{i=1}^{n} b_i \sum_{i=1}^{n} c_i\right)^{1/3} \le \frac{1}{3}\sum_{i=1}^{n}(a_i + b_i + c_i) \tag{A.4}$$

This is easily generalized to the following inequality for integrals:

$$\left(\int f\, d\tau \int g\, d\tau \int h\, d\tau\right)^{1/3} \le \frac{1}{3}\int (f + g + h)\, d\tau \tag{A.5}$$

where f, g, and h are nonnegative real functions, and $d\tau = dxdydz$. Also, if a, b, and c are three real numbers, we have

$$(a-b)^2 + (b-c)^2 + (c-a)^2 \ge 0$$

so

$$a^2 + b^2 + c^2 \ge \frac{1}{3}(a^2 + b^2 + c^2 + 2ab + 2bc + 2ca)$$

or

$$a^2 + b^2 + c^2 \ge \frac{1}{3}(a + b + c)^2$$

Hence, employing (A.5), we obtain

$$\int \sqrt{f^2 + g^2 + h^2}\, d\tau \ge \frac{1}{\sqrt{3}}\int (f + g + h)\, dxdydz \ge \sqrt{3}\left(\int f\, d\tau \int g\, d\tau \int h\, d\tau\right)^{1/3} \tag{A.6}$$

A.2 Hölder's Inequality

Consider two real nonnegative numbers u and w. We now show that

$$uw \le \frac{1}{p}u^p + \frac{1}{q}w^q \tag{A.7}$$

where p and q are two positive real numbers satisfying

$$\frac{1}{p}+\frac{1}{q}=1 \tag{A.8}$$

Because (A.7) is obviously satisfied when $u=0$ or $w=0$, we prove (A.7) in what follows for $uw>0$. Also, if (A.7) holds, then

$$ut^{1/p}wt^{1/q}=uwt\le\frac{1}{p}u^p t+\frac{1}{q}w^q t \tag{A.9}$$

for any t, and conversely, (A.9) implies (A.7). Thus, in what follows, we restrict ourselves to values of u and w such that $uw=1$. We then wish to show that

$$\frac{1}{p}u^p+\frac{1}{q}w^q\ge 1$$

Consider the function

$$f(u,w)=\frac{1}{p}u^p+\frac{1}{q}w^q$$

To minimize f subject to the constraint $uw=1$, we write

$$h(u,w)=f(u,w)-\lambda uw$$

where λ is an undetermined multiplier. Then

$$\frac{\partial h}{\partial u}=u^{p-1}-\lambda w$$
$$\frac{\partial h}{\partial w}=w^{q-1}-\lambda u$$

These partial derivatives vanish when $\lambda=u^p=w^q$ and hence when $u^{p+q}=w^{p+q}=1$ and when $u=w=1$. Thus the minimum value of f is $f(1,1)=1$, so inequality (A.7) is demonstrated.

Now let us make the replacements; that is,

$$u\to\frac{u_i}{\left(\sum_{i=1}^{n}u_i^p\right)^{1/p}}\qquad w\to\frac{w_i}{\left(\sum_{i=1}^{n}w_i^q\right)^{1/q}}$$

where the u_i and w_i are nonnegative real numbers with at least one $u_i>0$ and at least one $w_i>0$. Then

$$\frac{u_i}{\left(\sum_{i=1}^{n}u_i^p\right)^{1/p}}\frac{w_i}{\left(\sum_{i=1}^{n}w_i^q\right)^{1/q}}\le\frac{1}{p}\frac{u_i^p}{\left(\sum_{i=1}^{n}u_i^p\right)}+\frac{1}{q}\frac{w_i^q}{\left(\sum_{i=1}^{n}w_i^q\right)}\qquad i=1,\ldots,n$$

Summing these inequalities over i from 1 to n, we obtain

$$\frac{\sum_{i=1}^{n} u_i w_i}{\left(\sum_{i=1}^{n} u_i^p\right)^{1/p}\left(\sum_{i=1}^{n} w_i^q\right)^{1/q}} \leq \frac{1}{p}+\frac{1}{q}=1$$

Hence we obtain Hölder's inequality

$$\sum_{i=1}^{n} u_i w_i \leq \left(\sum_{i=1}^{n} u_i^p\right)^{1/p}\left(\sum_{i=1}^{n} w_i^q\right)^{1/q} \tag{A.10}$$

Let $f(x,y,z)$ and $g(x,y,z)$ be two nonnegative real functions of x, y, z. Then (A.10) is easily generalized to give Hölder's integral inequality; that is,

$$\left(\int f^p \, d\tau\right)^{1/p}\left(\int g^q \, d\tau\right)^{1/q} \geq \int fg \, d\tau \tag{A.11}$$

When $p = q = 2$, (A.11) reduces to the Cauchy-Schwarz integral inequality; that is,

$$\int f^2 \, d\tau \int g^2 \, d\tau \geq \left(\int fg \, d\tau\right)^2 \tag{A.12}$$

A.3 Sobolev's Inequality in Three Dimensions

A differentiable function $F(x,y,z)$ can be written as

$$F(x,y,z) = \int_{-\infty}^{x} \frac{\partial F(r,y,z)}{\partial r} \, dr$$

and we have the obvious inequality

$$|F(x,y,z)| \leq \int_{-\infty}^{\infty} |\partial_r F(r,y,z)| \, dr \equiv g_1(y,z)$$

Repeating this in y and in z, we obtain

$$|F(x,y,z)|^3 \leq g_1(y,z) g_2(x,z) g_3(x,y) \tag{A.13}$$

where $g_2(x,z) = \int_{-\infty}^{\infty} |\partial_s F(x,s,z)| \, ds$ and $g_3(x,y) = \int_{-\infty}^{\infty} |\partial_t F(x,y,t)| \, dt$. From (A.13) we obtain

$$\int |F|^{3/2} \, d\tau \leq \int \sqrt{g_1(y,z) g_2(x,z) g_3(x,y)} \, d\tau$$

However,

$$\int \sqrt{g_1(y,z)g_2(x,z)g_3(x,y)}\, d\tau = \int \sqrt{g_1(y,z)}\left[\int \sqrt{g_2(x,z)g_3(x,y)}\, dx\right] dydz$$
$$\leq \int \sqrt{g_1(y,z)}\left[\sqrt{\int g_2(x,z)\, dx \int g_3(x,y)\, dx}\right] dydz$$
$$\equiv Q_1$$

where the last step follows from the Cauchy-Schwarz integral inequality (A.12). Now, by another application of the latter inequality,

$$Q_1 = \int dz \sqrt{\int g_2(x,z)\, dx}\left[\int dy \left(\sqrt{g_1(y,z)}\sqrt{\int g_3(x,y)\, dx}\right)\right]$$
$$\leq \int dz \sqrt{\int g_2(x,z)\, dx}\sqrt{\int g_1(y,z)\, dy}\sqrt{\iint g_3(x,y)\, dxdy}$$
$$\equiv Q_2$$

and by a third application of the Cauchy-Schwarz inequality,

$$Q_2 \leq \sqrt{\iint g_1(y,z)\, dydz \iint g_2(x,z)\, dxdz \iint g_3(x,y)\, dxdy} \equiv Q_3$$

Thus

$$\left[\int |F(x,y,z)|^{3/2}\, d\tau\right]^{2/3} \leq \left(\int |\partial_x F|\, d\tau \cdot \int |\partial_y F|\, d\tau \cdot \int |\partial_z F|\, d\tau\right)^{1/3}$$

Combining this last inequality with (A.6), we obtain

$$\left(\int |F|^{3/2}\, d\tau\right)^{2/3} \leq \int \sqrt{(\partial_x F)^2 + (\partial_y F)^2 + (\partial_z F)^2}\, d\tau \tag{A.14}$$

If in the calculation just completed we had replaced F by F^s, where $s > 0$, we would have obtained the following inequality:

$$\left(\int |F^s|^{3/2}\, d\tau\right)^{2/3} \leq s \int |F^{s-1}| \sqrt{(\partial_x F)^2 + (\partial_y F)^2 + (\partial_z F)^2}\, d\tau \tag{A.15}$$

We now employ Hölder's integral inequality (A.11) on the right-hand side of (A.15); choosing $s = 2p/(3-p)$, we obtain

$$\left(\int |F^{2p/(3-p)}|^{3/2}\, d\tau\right)^{2/3} \leq \frac{2p}{3-p}\left\{\int \left[(\partial_x F)^2 + (\partial_y F)^2 + (\partial_z F)^2\right]^{p/2} d\tau\right\}^{1/p} \left(\int |F|^{[(3p-3)/3-p]q}\, d\tau\right)^{1/q} \tag{A.16}$$

where $q = p/(p-1)$. Let us choose $p = 2$, in which case $q = 2$ and $s = 4$. Then (A.16) becomes

$$\left(\int |F|^6\, d\tau\right)^{2/3} \leq 4\left[\int (\nabla F)^2\, d\tau\right]^{1/2} \left(\int |F|^6\, d\tau\right)^{1/2}$$

or

$$\int (\nabla F)^2 \, d\tau \geq C\left(\int |F|^6 \, d\tau\right)^{1/3} \tag{A.17}$$

where C is a positive constant. Equation (A.17) is Sobolev's inequality in three spatial dimensions, expressed in a form that is useful for applications to quantum mechanics.

A.4 Application of Sobolev and Hölder Inequalities to Hydrogenic Atoms

Now consider the Hamiltonian for a hydrogenic atom with atomic number Z. In atomic units, we have

$$H = -\frac{1}{2}\nabla^2 - \frac{Z}{r}$$

Of course, we know that the ground-state wave function is

$$\psi_{1s} = \frac{Z^{3/2}}{\pi^{1/2}} e^{-Zr}$$

and the ground-state energy is $E_{1s} = -Z^2/2$. However, in what follows we ignore these well-known facts and try to bound the ground-state energy from below, simply by using Sobolev's inequality (A.17). Now the expectation value of the kinetic energy for any given wave function ψ with unit norm is

$$\langle T \rangle = -\frac{1}{2}\int \psi \nabla^2 \psi \, d\tau = \frac{1}{2}\int (\nabla \psi)^2 \, d\tau$$

Thus (A.17) implies that

$$2\langle T \rangle = \int (\nabla \psi)^2 \, d\tau \geq C\left(\int \psi^6 \, d\tau\right)^{1/3} \tag{A.18}$$

or

$$\left[\int (\nabla \psi)^2 \, d\tau\right]^3 \geq C^3 \int \psi^6 \, d\tau \tag{A.19}$$

By varying ψ, we try to find the "sharpest" (i.e., the smallest) value of C, which we shall call C_0. Thus we write

$$K = \frac{A}{B} = \frac{\left[\int (\nabla \psi)^2 \, d\tau\right]^3}{\int \psi^6 \, d\tau}$$

K reaches an extreme value (which can be shown to be a minimum) when

$$\frac{\delta A}{A} = \frac{\delta B}{B}$$

that is, when

$$\frac{\int \nabla^2 \psi \delta \psi \, d\tau}{\int \psi \nabla^2 \psi \, d\tau} = \frac{\int \psi^5 \delta \psi \, d\tau}{\int \psi^6 \, d\tau}$$

This relation is satisfied if

$$\nabla^2 \psi = -3\alpha \psi^5 \tag{A.20}$$

where α is a constant. Assuming spherical symmetry, (A.20) becomes

$$\frac{1}{r^2}\frac{\partial}{\partial r}\left(r^2 \frac{\partial \psi}{\partial r}\right) = -3\alpha \psi^5$$

With the substitution $u = (3\alpha)^{1/2} r$, we obtain the equation

$$\frac{1}{u^2}\frac{\partial}{\partial u}\left(u^2 \frac{\partial \psi}{\partial u}\right) = -\psi^5$$

which has the solution:

$$\psi = \frac{1}{\sqrt{1 + \frac{u^2}{3}}} = \frac{1}{\sqrt{1 + \alpha r^2}} \tag{A.21}$$

Employing (A.21) in (A.19), we obtain

$$K = C_0^3 = (3\alpha)^3 \left[4\pi \int_0^\infty \frac{r^2 \, dr}{(1 + \alpha r^2)^3} \right]^2 = \frac{27\pi^4}{16}$$

Hence

$$C_0 = 3\left(\frac{\pi}{2}\right)^{4/3} = 5.4779$$

and the sharpest Sobolev inequality for the kinetic energy is

$$\langle T \rangle \geq \frac{C_0}{2}\left(\int \rho^3 \, d\tau\right)^{1/3} \tag{A.22}$$

where $\rho = |\psi|^2$.

The expectation value of the potential energy for the hydrogenic atom is

$$\langle V \rangle = -Z\int \frac{\rho}{r}\, d\tau$$

Thus the energy functional is

$$\begin{aligned} E[\rho] &= \langle T \rangle + \langle V \rangle \\ &\geq \frac{C_0}{2}\left(\int \rho^3\, d\tau\right)^{1/3} - Z\int \frac{\rho}{r}\, d\tau = h[\rho] \end{aligned} \tag{A.23}$$

We want to minimize $h[\rho]$ by varying ρ subject to the constraint $\int \rho\, d\tau = 1$. Thus we employ an undetermined multiplier λ and consider

$$h_1[\rho] = \frac{C_0}{2}\left(\int \rho^3\, d\tau\right)^{1/3} - Z\int \frac{\rho}{r}\, d\tau - \lambda \int \rho\, d\tau$$

Making arbitrary small variations of ρ, we see that an extreme value of h_1 is reached when

$$\delta h_1 = \frac{C_0}{2}\frac{\int \rho^2 \delta\rho\, d\tau}{\left(\int \rho^3\right)^{2/3}} - Z\int \frac{\delta\rho}{r}\, d\tau - \lambda \int \delta\rho\, d\tau = 0$$

This implies that

$$\frac{C_0}{2I}\rho^2 - \frac{Z}{r} - \lambda = 0 \tag{A.24}$$

where

$$I \equiv \left(\int \rho^3\right)^{2/3}$$

Because $\rho \geq 0$ and $\lim_{r\to\infty} \rho = 0$ is required for a bound state, (A.24) gives the following result after a routine calculation:

$$\begin{aligned} \rho &= \frac{2^{9/2}}{3^{5/2}\pi^2} Z^{5/2}\left(\frac{1}{r} - \frac{2Z}{3}\right)^{1/2} & r &< \frac{3}{2Z} \\ \rho &= 0 & r &> \frac{3}{2Z} \end{aligned} \tag{A.25}$$

and

$$h = -\frac{2Z^2}{3} = -0.6667Z^2 \tag{A.26}$$

Thus Sobolev's inequality yields the lower bound (A.26) on the ground-state energy, a bound that is not too far from the actual energy $E_{1s} = -(Z^2/2)$.

There is a weaker but more useful inequality than (A.18). We recall Hölder's inequality (A.11) and choose $f = \rho$, $g = \rho^{2/3}$, $p = 3$, and $q = 3/2$ to obtain

$$\left(\int \rho^3 \, d\tau\right)^{1/3} \left[\int \left(\rho^{2/3}\right)^{3/2} d\tau\right]^{2/3} \geq \int \rho^{5/3} \, d\tau$$

The second integral on the left-hand side of this expression is just unity because we assume that ψ is normalized. Hence

$$\int (\nabla \psi)^2 \, d\tau \geq D \int \rho^{5/3} \, d\tau \tag{A.27}$$

where D is a positive constant. Inequality (A.27) plays an important role in analysis of the stability of matter, discussed in Section 14.1. Let us find the sharpest value of D by writing

$$D_0 = \min\{K\}$$

where

$$K = \frac{\int (\nabla \psi)^2}{\int \psi^{10/3}} - \lambda \int \psi^2$$

and, as before, λ is an undetermined multiplier used to account for the normalization constraint. Then

$$\delta K = -2 \frac{\int \delta\psi \nabla^2 \psi}{\int \psi^{10/3}} - \frac{10}{3} \frac{\int (\nabla \psi)^2}{\left(\int \psi^{10/3}\right)^2} \int \delta\psi \cdot \psi^{7/3} - 2\lambda \int \psi \delta\psi \tag{A.28}$$

Setting $\delta K = 0$, we obtain the following nonlinear PDE after a few steps of algebra:

$$\nabla^2 \psi + \frac{5}{3} \frac{A}{B} \psi^{7/3} = \frac{2}{3} A \psi \tag{A.29}$$

where

$$A = \int (\nabla \psi)^2 \qquad B = \int \psi^{10/3}$$

Again assuming spherical symmetry, we define $\phi = r\psi$. Then (A.29) yields

$$\frac{d^2 \phi}{dr^2} + \frac{5}{3} \frac{A}{B} \frac{\phi^{7/3}}{r^{4/3}} = \frac{2}{3} A \phi \tag{A.30}$$

We make the additional substitutions $r = ax,\ \phi = b\chi$, where

$$a = \left(\frac{3}{2A}\right)^{1/2} \qquad b = \frac{2^{1/4}3^{1/2}}{5^{3/4}}\frac{B^{3/4}}{A^{1/2}}$$

to find after a few steps of algebra that (A.30) becomes

$$\frac{d^2\chi}{dx^2} = \chi\left[1 - \left(\frac{\chi}{x}\right)^{4/3}\right] \tag{A.31}$$

with the boundary conditions $\chi(0) = \chi(\infty) = 0$. The same analysis also shows that

$$D_0 = \frac{A}{B} = \left(\frac{8\pi \bullet 3^{1/2}}{5^{3/2}}\int_0^\infty \chi'^2\, dx\right)^{2/3} = \left(\frac{8\pi \bullet 3^{3/2}}{5^{5/2}}\int_0^\infty \chi^{10/3}x^{-4/3}\, dx\right)^{2/3} \tag{A.32}$$

Numerical integration of (A.31) with use of (A.32) yields

$$D_0 = 9.578$$

Thus

$$\langle T\rangle \geq \frac{D_0}{2}\int \rho^{5/3}\, d\tau = 4.789\int \rho^{5/3}\, d\tau \tag{A.33}$$

It is convenient to replace the number 4.789 in (A.33) by the slightly small\r number $D_C = 3(6\pi^2)^{2/3}/10 = 4.5578$ because D_C appears naturally in an analysis that can be generalized to the case of many electrons. Consider a perfect nonrelativistic Fermi gas of electrons at zero temperature. The number density of this gas is

$$n = \frac{q}{2\pi^2\hbar^3}\int_0^{p_F} p^2\, dp = \frac{q}{6\pi^2\hbar^3}p_F^3$$

where q is the spin statistical weight ($q = 2$ for electrons), and p_F is the Fermi momentum. The kinetic energy density is

$$\begin{aligned}\varepsilon = \frac{q}{2\pi^2\hbar^3}\int_0^{p_F}\frac{p^2}{2m}p^2\, dp &= \frac{q}{20\pi^2\hbar^3 m}p_F^5 \\ &= \frac{1}{q^{2/3}}\frac{3}{10}(6\pi^2)^{2/3}\frac{\hbar^2}{m}n^{5/3} \\ &= \frac{4.5578}{q^{2/3}}n^{5/3}\end{aligned}$$

where we employ atomic units in the last line. Thus D_C is the numerical coefficient that would multiply $n^{5/3}$ to yield the kinetic energy density in atomic units if the spin statistical weight were $q = 1$.

We now outline steps similar to those that led to (A.26). For a hydrogenic atom, the energy functional satisfies the following inequality:

$$\boldsymbol{E}[\rho] \geq D_C \int \rho^{5/3}\, d\tau - Z \int \frac{\rho}{r}\, d\tau = h[\rho] \tag{A.34}$$

Once again we introduce an undetermined multiplier λ and write

$$\delta h_1 = 0 = \frac{5}{3} D_C \int \rho^{2/3}\, \delta\rho - Z \int \frac{\delta\rho}{r} - \lambda \int \delta\rho$$

After some routine algebra, this yields the following lower bound on the hydrogenic atom ground-state energy:

$$h = -\frac{3^{1/3} Z^2}{2} = -0.7211 Z^2 \tag{A.35}$$

that is only slightly weaker than the lower bound (A.26).

B Appendix Bell's Inequality

Here we derive Bell's inequality as applied to the calcium two-photon polarization correlation experiment described in Section 3.3. For simplicity, we limit ourselves to the case of ideal polarizers, which transmit 100 percent of the right polarization and 0 percent of the wrong one. Actual laboratory polarizers fall short of this ideal, and significant corrections thus must be made. However, the essential idea is clearly revealed in the present simplified case. We also assume that each detector operates with 100 percent efficiency.

We consider an ensemble of two-photon pairs and discuss the consequences of assuming that the result of each coincidence measurement is determined by the value of a local hidden variable u corresponding to each pair. We associate with the ensemble a distribution function $\rho(u)$ that gives the probability that a particular pair has the value u. This function is normalized to unity; that is,

$$\int_{\Gamma} \rho(u)\, du = 1 \tag{B.1}$$

where Γ is the space of the hidden variable (or variables). Let $A(\boldsymbol{a},u)$ and $B(\boldsymbol{b},u)$ refer to signals in detectors 1 and 2, respectively. Here $\boldsymbol{a}$ and $\boldsymbol{b}$ are unit vectors describing the orientations of polarizers 1 and 2, respectively. Quantities A and B are defined as follows: if photon 1 passes through polarizer 1, it registers a count in detector 1. In this case we have $A(\boldsymbol{a},u) = +1$. If photon 1 is rejected by polarizer 1, we have $A(\boldsymbol{a},u) = -1$. Similar remarks hold for B in the case of photon 2. Note that A is assumed to depend only on $\boldsymbol{a}$ and B only on $\boldsymbol{b}$. In other words, the outcome in one detector does not depend on the conditions that apply in the other polarizer. In this way we restrict ourselves to a *local* hidden variable theory.

Let us define the correlation function $P(\boldsymbol{a},\boldsymbol{b})$ that determines the coincidence rate as a function of $\boldsymbol{a}$ and $\boldsymbol{b}$; that is,

$$P(\boldsymbol{a},\boldsymbol{b}) = \int_{\Gamma} \rho(u) A(\boldsymbol{a},u)\, B(\boldsymbol{b},u)\, du \tag{B.2}$$

We have

$$\begin{aligned}|P(\boldsymbol{a},\boldsymbol{b}) - P(\boldsymbol{a},\boldsymbol{c})| = \Big| &\int_{\Gamma} \rho(u) A(\boldsymbol{a},u) B(\boldsymbol{b},u)\left[1 \pm A(\boldsymbol{d},u)\, B(\boldsymbol{c},u)\right] du \\ &- \int_{\Gamma} \rho(u) A(\boldsymbol{a},u)\, B(\boldsymbol{c},u)\left[1 \pm A(\boldsymbol{d},u)\, B(\boldsymbol{b},u)\right] du\Big|\end{aligned}$$

Therefore,

$$\left|P(\boldsymbol{a},\boldsymbol{b})-P(\boldsymbol{a},\boldsymbol{c})\right| \le \int_{\Gamma} \rho(\mathrm{u})\left|A(\boldsymbol{a},u)\,B(\boldsymbol{b},u)\right| \bullet \left|\left[1 \pm A(\boldsymbol{d},u)\,B(\boldsymbol{c},u)\right]\right| du$$
$$+\int_{\Gamma} \rho(u)\left|A(\boldsymbol{a},u)\,B(\boldsymbol{c},u)\right| \bullet \left|\left[1 \pm A(\boldsymbol{d},u)\,B(\boldsymbol{b},u)\right]\right| du$$

which yields

$$\left|P(\boldsymbol{a},\boldsymbol{b})-P(\boldsymbol{a},\boldsymbol{c})\right| \le \int_{\Gamma} \rho(\mathrm{u})\left[2 \pm A(\boldsymbol{d},u)\,B(\boldsymbol{c},u) \pm A(\boldsymbol{d},u)\,B(\boldsymbol{b},u)\right] du$$

It follows that

$$\left|P(\boldsymbol{a},\boldsymbol{b})-P(\boldsymbol{a},\boldsymbol{c})\right| \le 2 \pm P(\boldsymbol{d},\boldsymbol{c}) \pm P(\boldsymbol{d},\boldsymbol{b})$$

Hence

$$-2 \le P(\boldsymbol{a},\boldsymbol{b})-P(\boldsymbol{a},\boldsymbol{c})+P(\boldsymbol{d},\boldsymbol{c})+P(\boldsymbol{d},\boldsymbol{b}) \le 2 \tag{B.3}$$

and also

$$-2 \le P(\boldsymbol{a},\boldsymbol{b})-P(\boldsymbol{a},\boldsymbol{c})-P(\boldsymbol{d},\boldsymbol{c})-P(\boldsymbol{d},\boldsymbol{b}) \le 2 \tag{B.4}$$

At this point it is convenient to divide the space Γ into the following sectors:

$$\begin{array}{ll} \Gamma = \Gamma_{++} & \text{if } A=+1, B=+1 \\ \Gamma = \Gamma_{+-} & \text{if } A=+1, B=-1 \\ \Gamma = \Gamma_{-+} & \text{if } A=-1, B=+1 \\ \Gamma = \Gamma_{--} & \text{if } A=-1, B=-1 \end{array}$$

Also, we define

$$P_{ij}(\boldsymbol{a},\boldsymbol{b}) = \int_{\Gamma_{ij}(\boldsymbol{a},\boldsymbol{b})} \rho(u)\,du$$

Then it is clear that

$$P(\boldsymbol{a},\boldsymbol{b}) = P_{++}(\boldsymbol{a},\boldsymbol{b})+P_{--}(\boldsymbol{a},\boldsymbol{b})-P_{+-}(\boldsymbol{a},\boldsymbol{b})-P_{-+}(\boldsymbol{a},\boldsymbol{b}) \tag{B.5}$$

and also

$$1 = P_{++}(\boldsymbol{a},\boldsymbol{b})+P_{--}(\boldsymbol{a},\boldsymbol{b})+P_{+-}(\boldsymbol{a},\boldsymbol{b})+P_{-+}(\boldsymbol{a},\boldsymbol{b}) \tag{B.6}$$

It is also convenient to define $P_1(\boldsymbol{a})$ and $P_2(\boldsymbol{b})$ as the probabilities of joint passage of the photons when one of the polarizers (2, 1 respectively) is absent. It is easy to see that

$$P_1(\boldsymbol{a}) = P_{++}(\boldsymbol{a},\boldsymbol{b})+P_{+-}(\boldsymbol{a},\boldsymbol{b}) \tag{B.7}$$

and

$$P_2(\boldsymbol{b}) = P_{++}(\boldsymbol{a},\boldsymbol{b}) + P_{-+}(\boldsymbol{a},\boldsymbol{b}) \tag{B.8}$$

It follows from (B.5)–(B.8) that

$$P(\boldsymbol{a},\boldsymbol{b}) = 4P_{++}(\boldsymbol{a},\boldsymbol{b}) - 2P_1(\boldsymbol{a}) - 2P_2(\boldsymbol{b}) + 1 \tag{B.9}$$

Then, after some algebra, (B.3), (B.4), and (B.9) yield

$$-1 \le P_{++}(\boldsymbol{a},\boldsymbol{b}) - P_{++}(\boldsymbol{a},\boldsymbol{c}) + P_{++}(\boldsymbol{d},\boldsymbol{c}) + P_{++}(\boldsymbol{d},\boldsymbol{b}) - P_1(\boldsymbol{d}) - P_2(\boldsymbol{b}) \le 0 \tag{B.10}$$

Recalling that all transmitted photons are assumed to be detected, we express the quantities in (B.10) as follows:

$$P_{++}(\boldsymbol{a},\boldsymbol{b}) = \frac{R(\boldsymbol{a},\boldsymbol{b})}{R_0}$$

$$P_1(\boldsymbol{a}) = \frac{R_1(\boldsymbol{a})}{R_0}$$

$$P_2(\boldsymbol{b}) = \frac{R_2(\boldsymbol{b})}{R_0}$$

where R_0 is the coincidence rate with both polarizers removed, $R(\boldsymbol{a}, \boldsymbol{b})$ is that with both polarizers in place, and $R_1(\boldsymbol{a})$ and $R_2(\boldsymbol{b})$ are the coincidence rates with polarizer 2 (1) removed, respectively.

Merely because of axial symmetry and quite apart from any detailed theory, $R_1(\boldsymbol{a})$ must be independent of $\boldsymbol{a}$, and $R_2(\boldsymbol{b})$ must be independent of $\boldsymbol{b}$. Also, $R(\boldsymbol{a}, \boldsymbol{b})$ can only depend on the relative orientation of $\boldsymbol{a}$ and $\boldsymbol{b}$: $R(\boldsymbol{a},\boldsymbol{b}) = R(\boldsymbol{a}\cdot\boldsymbol{b})$. Taking all this into account, we see that (B.10) can be written as

$$-1 \le \frac{R(\boldsymbol{a}\cdot\boldsymbol{b})}{R_0} - \frac{R(\boldsymbol{a}\cdot\boldsymbol{c})}{R_0} + \frac{R(\boldsymbol{d}\cdot\boldsymbol{c})}{R_0} + \frac{R(\boldsymbol{d}\cdot\boldsymbol{b})}{R_0} - \frac{R_1}{R_0} - \frac{R_2}{R_0} \le 0 \tag{B.11}$$

The unit vectors $\boldsymbol{a}$, $\boldsymbol{b}$, $\boldsymbol{c}$, and $\boldsymbol{d}$ are all in the same (xy) plane. We define the angles between them as in Figure B.1, and we also define $\phi_4 = \phi_1 + \phi_2 + \phi_3$. Then (B.11) can be rewritten as

$$-1 \le \frac{R(\phi_1)}{R_0} - \frac{R(\phi_4)}{R_0} + \frac{R(\phi_2)}{R_0} + \frac{R(\phi_3)}{R_0} - \frac{R_1}{R_0} - \frac{R_2}{R_0} \le 0 \tag{B.12}$$

According to quantum mechanics,

$$\frac{R(\phi)}{R_0} = \frac{1}{2}\cos^2\phi = \frac{1}{4}(1 + \cos 2\phi) \tag{B.13}$$

and also

$$\frac{R_1}{R_0} = \frac{R_2}{R_0} = \frac{1}{2} \tag{B.14}$$

Thus, defining the function

$$G(\phi_1, \phi_2, \phi_3) = \frac{R(\phi_1)}{R_0} - \frac{R(\phi_4)}{R_0} + \frac{R(\phi_2)}{R_0} + \frac{R(\phi_3)}{R_0} - \frac{R_1}{R_0} - \frac{R_2}{R_0} \tag{B.15}$$

we see that according to quantum mechanics,

$$G_{\text{QM}}(\phi_1, \phi_2, \phi_3) = -1 + \frac{1}{4}(\cos 2\phi_1 + \cos 2\phi_2 + \cos 2\phi_3) - \frac{1}{4}\cos\left[2(\phi_1 + \phi_2 + \phi_3)\right] \tag{B.16}$$

We find the extreme values of this function by differentiating with respect to the independent variables ϕ_1, ϕ_2, and ϕ_3 and setting the results equal to zero. Thus we find that G_{QM} reaches an extreme value when

$$\phi_1 = \phi_2 = \phi_3 = \frac{\pi}{8} \quad \text{or} \quad \frac{3\pi}{8}$$

We therefore make use of inequalities (B.12) at these specific angles. Noting that, by definition, the angle between polarization vectors cannot exceed $\pi/2$, we obtain from (B.12)

$$-1 \leq \frac{3R\left(\frac{\pi}{8}\right)}{R_0} - \frac{R\left(\frac{3\pi}{8}\right)}{R_0} - \frac{R_1}{R_0} - \frac{R_2}{R_0} \leq 0 \tag{B.17}$$

and

$$-1 \leq \frac{3R\left(\frac{3\pi}{8}\right)}{R_0} - \frac{R\left(\frac{\pi}{8}\right)}{R_0} - \frac{R_1}{R_0} - \frac{R_2}{R_0} \leq 0 \tag{B.18}$$

Thus, in particular,

$$-\frac{3R\left(\frac{\pi}{8}\right)}{R_0} + \frac{R\left(\frac{3\pi}{8}\right)}{R_0} + \frac{R_1}{R_0} + \frac{R_2}{R_0} \geq 0 \tag{B.19}$$

and

$$\frac{3R\left(\frac{3\pi}{8}\right)}{R_0} - \frac{R\left(\frac{\pi}{8}\right)}{R_0} - \frac{R_1}{R_0} - \frac{R_2}{R_0} \geq -1 \tag{B.20}$$

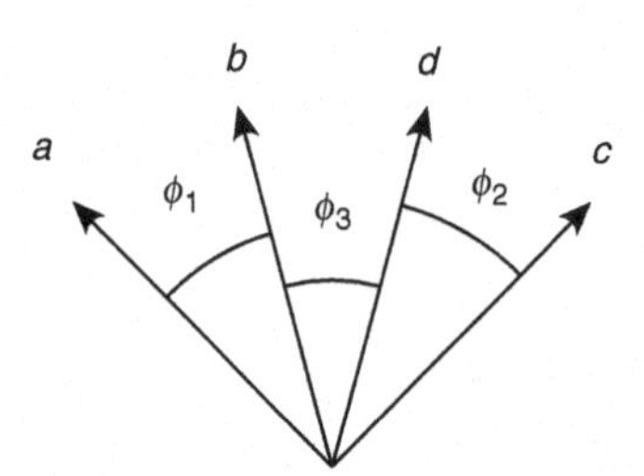

Figure B.1 Angles $\varphi_{1,2,3}$ are defined by the unit vectors a, b, c, and d.

Adding (B.19) and (B.20), we obtain

$$-\frac{4R\left(\frac{\pi}{8}\right)}{R_0}+\frac{4R\left(\frac{3\pi}{8}\right)}{R_0}\geq -1$$

which finally yields Bell's inequality for the calcium two-photon polarization correlation experiment; that is,

$$\frac{R\left(\frac{\pi}{8}\right)}{R_0}-\frac{R\left(\frac{3\pi}{8}\right)}{R_0}-\frac{1}{4}\leq 0 \tag{B.21}$$

According to quantum mechanics,

$$\frac{R\left(\frac{\pi}{8}\right)}{R_0}=\frac{1}{4}\left(1+\cos\frac{\pi}{4}\right)=\frac{1}{4}\left(1+\frac{1}{\sqrt{2}}\right)$$

and

$$\frac{R\left(\frac{3\pi}{8}\right)}{R_0}=\frac{1}{4}\left(1+\cos\frac{3\pi}{4}\right)=\frac{1}{4}\left(1-\frac{1}{\sqrt{2}}\right)$$

Obviously, the quantum-mechanical values violate Bell's inequality (B.21).

C

Appendix
Spin of the Photon: Vector Spherical Waves

To introduce the subjects of this appendix, we consider the rotational symmetry of a general vector field $\boldsymbol{V}(\boldsymbol{r})$. A rotation in three-dimensional space corresponds to a 3×3 orthogonal matrix M; let us apply this matrix to $\boldsymbol{V}(\boldsymbol{r})$. Then at position $\boldsymbol{r}$ the rotated field $\boldsymbol{V}'(\boldsymbol{r})$ is

$$\boldsymbol{V}'(\boldsymbol{r}) = M\boldsymbol{V}\left(M^{-1}\boldsymbol{r}\right) \tag{C.1}$$

Suppose that M corresponds to an infinitesimal rotation by angle ε about the z-axis. Then, to order ε,

$$M = \begin{pmatrix} 1 & -\varepsilon & 0 \\ \varepsilon & 1 & 0 \\ 0 & 0 & 1 \end{pmatrix} \qquad M^{-1} = \begin{pmatrix} 1 & \varepsilon & 0 \\ -\varepsilon & 1 & 0 \\ 0 & 0 & 1 \end{pmatrix}$$

Hence

$$M^{-1}\boldsymbol{r} = (x+\varepsilon y)\hat{i} + (y-\varepsilon x)\hat{j} + z\hat{k}$$

and

$$\boldsymbol{V}'(\boldsymbol{r}) = V_x'\hat{i} + V_y'\hat{j} + V_z'\hat{k}$$

where, from (C.1),

$$\begin{aligned} V_x' &= V_x(x+\varepsilon y, y-\varepsilon x, z) - \varepsilon V_y(x+\varepsilon y, y-\varepsilon x, z) \\ V_y' &= V_y(x+\varepsilon y, y-\varepsilon x, z) + \varepsilon V_x(x+\varepsilon y, y-\varepsilon x, z) \\ V_z' &= V_z(x+\varepsilon y, y-\varepsilon x, z) \end{aligned}$$

Writing Taylor expansions in ε for each of these quantities and discarding all terms of order ε^2 and higher, we obtain

$$\begin{aligned} V_x'(x,y,z) &= V_x(x,y,z) - \varepsilon\left(x\frac{\partial V_x}{\partial y} - y\frac{\partial V_x}{\partial x}\right) - \varepsilon V_y(x,y,z) \\ V_y'(x,y,z) &= V_y(x,y,z) - \varepsilon\left(x\frac{\partial V_y}{\partial y} - y\frac{\partial V_y}{\partial x}\right) + \varepsilon V_x(x,y,z) \\ V_z'(x,y,z) &= V_z(x,y,z) - \varepsilon\left(x\frac{\partial V_z}{\partial y} - y\frac{\partial V_z}{\partial x}\right) \end{aligned} \tag{C.2}$$

Now we define an angular momentum operator J_z in the usual way as the generator of the infinitesimal rotation; that is,

$$V' = \exp(-i\varepsilon J_z)V \approx (1 - i\varepsilon J_z)V \tag{C.3}$$

Comparing (C.3) with (C.2), we see that

$$J_z = L_z + S_z \tag{C.4}$$

where L_z is the usual orbital angular momentum operator

$$L_z = -i\left(x\frac{\partial}{\partial y} - y\frac{\partial}{\partial x}\right) \tag{C.5}$$

and

$$S_z V = i\hat{k} \times V \tag{C.6}$$

By cyclically permuting the coordinates, we generalize (C.4) to read

$$\boldsymbol{J} = \boldsymbol{L} + \boldsymbol{S} \tag{C.7}$$

where

$$\boldsymbol{L} = -i\boldsymbol{r} \times \nabla \tag{C.8}$$

and

$$\boldsymbol{S} = S_x\hat{i} + S_y\hat{j} + S_z\hat{k}$$

with

$$S_x = i\,\hat{i}\times \qquad S_y = i\,\hat{j}\times \qquad S_z = i\,\hat{k}\times \tag{C.9}$$

Equation (C.7) expresses the fact that the angular momentum operator for a vector field is the sum of an orbital contribution $\boldsymbol{L}$ and an additional contribution $\boldsymbol{S}$, which has the attributes of spin. It is easy to verify that each component of $\boldsymbol{L}$ commutes with every component of $\boldsymbol{S}$. It is also easy to show from (C.9) that the eigenvectors of $\boldsymbol{S}_z$ are

$$\chi_1 = -\frac{1}{\sqrt{2}}\left(\hat{i} + i\,\hat{j}\right) \qquad \chi_0 = \hat{k} \qquad \chi_{-1} = +\frac{1}{\sqrt{2}}\left(\hat{i} - i\,\hat{j}\right) \tag{C.10}$$

with eigenvalues 1, 0, and –1, respectively, and that

$$\boldsymbol{S}^2 V = 2V \tag{C.11}$$

Comparing (C.11) with the eigenvalue equation $\boldsymbol{S}^2 V = S(S+1)V$, we see that the spin S of a vector field must be $S = 1$. These general remarks apply to the vector potential $\boldsymbol{A}$ of the

Table C.1 Vector coupling coefficients appearing in equation (C.12)

	$m'=-1$	$m'=0$	$m'=1$
$l=J+1$	$\left[\frac{(J+1+M)(J+2+M)}{2(J+1)(2J+3)}\right]^{1/2}$	$-\left[\frac{(J+1-M)(J+1-M)}{(J+1)(2J+3)}\right]^{1/2}$	$\left[\frac{(J+1-M)(J+2-M)}{2(J+1)(2J+3)}\right]^{1/2}$
$l=J$	$\left[\frac{(J+1+M)(J-M)}{2J(J+1)}\right]^{1/2}$	$\frac{M}{\sqrt{J(J+1)}}$	$-\left[\frac{(J+M)(J+1-M)}{2J(J+1)}\right]^{1/2}$
$l=J-1$	$\left[\frac{(J-1-M)(J-M)}{2J(2J-1)}\right]^{1/2}$	$\left[\frac{(J+M)(J-M)}{J(2J-1)}\right]^{1/2}$	$\left[\frac{(J-1+M)(J+M)}{2J(2J-1)}\right]^{1/2}$

radiation field in Coulomb gauge. They imply that the quantum of the radiation field, the photon, has a spin of unity.

Because $\boldsymbol{L}$ and $\boldsymbol{S}$ commute, we can employ the standard methods for adding angular momenta to build up vector eigenstates of $\boldsymbol{J}^2$ and J_z, denoted by $\boldsymbol{Y}_{J\ell}^M$, from the eigenfunctions of orbital angular momentum (the spherical harmonics Y_ℓ^m) and the spin eigenfunctions of (C.10). The quantities $\boldsymbol{Y}_{J\ell}^M$ are called *vector spherical harmonics* and are given by the formula

$$\boldsymbol{Y}_{J\ell}^M(\theta,\phi) = \sum_{m'=-1}^{1} \langle \ell,m,1,m'|JM\rangle \chi_{m'} Y_\ell^m(\theta,\phi) \qquad \text{(C.12)}$$

where the $\langle \ell,m,1,m'|J,M\rangle$ are vector coupling coefficents with $m+m'=M$ (see Table C.1). Possible values of J are $\ell+1$, ℓ, and $\ell-1$ for $\ell>0$, and possible values of M range from $-J$ to J.

The vector spherical harmonics play a significant role in solutions to the homogeneous vector wave equation in spherical polar coordinates; that is,

$$(\nabla^2+k^2)\boldsymbol{A}=0 \qquad \text{(C.13)}$$

where we assume the time dependence $\exp(-i\omega t)$ and the condition $\nabla\cdot\boldsymbol{A}=0$ in Coulomb gauge. We begin with a simpler equation: the homogeneous scalar wave equation in spherical polar coordinates

$$(\nabla^2+k^2)w=0 \qquad \text{(C.14)}$$

We have seen in Section 18.3 that fundamental solutions of (C.14) take the form

$$w(r,\theta,\phi)=g_\ell(r)Y_\ell^m(\theta,\phi) \qquad \text{(C.15)}$$

where the $g_\ell(r)$ satisfy

$$\frac{\partial^2 g_\ell}{\partial r^2}+\frac{2}{r}\frac{\partial g_\ell}{\partial r}+\left[k^2-\frac{\ell(\ell+1)}{r^2}\right]g_\ell=0 \qquad \text{(C.16)}$$

The latter equation has two independent solutions: the spherical Bessel functions $j_\ell(kr)$ and $n_\ell(kr)$. The spherical Hankel functions

$$h_\ell^{(1)}(kr) = j_\ell(kr) + in_\ell(kr) \tag{C.17}$$

and

$$h_\ell^{(2)}(kr) = j_\ell(kr) - in_\ell(kr) \tag{C.18}$$

describe outgoing (incoming) spherical waves, respectively.

We want to construct divergence-free solutions to the vector wave equation (C.13) starting from solutions (C.15) to the scalar wave equation. First we try

$$\boldsymbol{F}_1 = \nabla w \tag{C.19}$$

$\boldsymbol{F}_1$ is indeed a solution to (C.13): $\nabla^2 F_{1j} = \partial_i\partial_i\partial_j w = \partial_j\partial_i\partial_i w = -k^2\partial_j w = -k^2 F_{1j}$. However, $\nabla\boldsymbol{\cdot}\boldsymbol{F}_1 \neq 0$ and $\nabla\times\boldsymbol{F}_1 = \nabla\times\nabla w = 0$, whereas $\nabla\boldsymbol{\cdot}\boldsymbol{A} = 0$ and $\nabla\times\boldsymbol{A}\neq 0$. To find the explicit form of $\boldsymbol{F}_1$, we employ the following well-known gradient formula that can be derived with the aid of the Wigner-Eckart theorem:

$$\nabla\left[f(r)Y_\ell^m\right] = -\left(\frac{\ell+1}{2\ell+1}\right)^{1/2}\boldsymbol{Y}_{\ell,\ell+1}^m\boldsymbol{\cdot}\left(\frac{d}{dr}-\frac{\ell}{r}\right)f(r) + \left(\frac{\ell}{2\ell+1}\right)^{1/2}\boldsymbol{Y}_{\ell,\ell-1}^m\boldsymbol{\cdot}\left(\frac{d}{dr}+\frac{\ell+1}{r}\right)f(r) \tag{C.20}$$

Here $f(r)$ is any continuous function of r. Using the identities

$$\frac{dg_\ell(kr)}{dr} = \frac{\ell}{r}g_\ell - kg_{\ell+1}$$
$$\frac{dg_\ell(kr)}{dr} = -\frac{\ell+1}{r}g_\ell + kg_{\ell-1}$$

in (C.20), we obtain

$$\boldsymbol{F}_1 = k\left[\left(\frac{\ell+1}{2\ell+1}\right)^{1/2} g_{\ell+1}\boldsymbol{Y}_{\ell,\ell+1}^m + \left(\frac{\ell}{2\ell+1}\right)^{1/2} g_{\ell-1}\boldsymbol{Y}_{\ell,\ell-1}^m\right] \tag{C.21}$$

We next try

$$\boldsymbol{F}_2 = -i\boldsymbol{r}\times\nabla w = \boldsymbol{L}w \tag{C.22}$$

We now show that $\boldsymbol{F}_2$ is a transverse divergence-free solution to (C.13). First, it is obvious that $\boldsymbol{r}\boldsymbol{\cdot}\boldsymbol{F}_2 = 0$ ($\boldsymbol{F}_2$ is transverse) from the definition $\boldsymbol{F}_2 = -i\boldsymbol{r}\times\nabla w = \boldsymbol{L}w$. Next, $\left(\nabla^2 + k^2\right)\boldsymbol{F}_2 = 0$ because $\left(\nabla^2+k^2\right)\boldsymbol{F}_2 = \left(\nabla^2+k^2\right)\boldsymbol{L}w$, and the operator ∇^2 contains radial derivatives that commute with $\boldsymbol{L}$ and also a term proportional to $\boldsymbol{L}^2/r^2$. However, $\boldsymbol{L}^2$ commutes with each component of $\boldsymbol{L}$, and therefore it commutes with $\boldsymbol{L}$. Thus $\nabla^2\boldsymbol{L}w = \boldsymbol{L}\nabla^2 w = -k^2\boldsymbol{L}w$.

Finally, $\nabla \cdot \boldsymbol{F}_2 = 0$ because

$$\nabla \cdot \boldsymbol{F}_2 = -i\partial_i \left(x_j \partial_k w \right) \varepsilon_{ijk} = -i\left[\left(\partial_i x_j \right) \partial_k w + x_j \partial_i \partial_k w \right] \varepsilon_{ijk}$$
$$= -i\left(\delta_{ij} \partial_k w + x_j \partial_i \partial_k w \right) \varepsilon_{ijk}$$

The first term in the last expression is zero because $\varepsilon_{ijk} = 0$ whenever two indices are the same. The second term vanishes because ε_{ijk} is antisymmetric, whereas $\partial_i \partial_k$ is symmetric with respect to exchange of indices i and k.

With the aid of (C.12) and Table C.1, it can be shown that

$$\boldsymbol{F}_2 = -i\boldsymbol{r} \times \nabla w = \sqrt{\ell(\ell+1)}\, g_\ell(r) \boldsymbol{Y}_{\ell\ell}^{M} \tag{C.23}$$

There is one additional type of divergence-free solution to the vector wave equation (C.13); that is,

$$\boldsymbol{F}_3 = \nabla \times \boldsymbol{F}_2 = \nabla \times \boldsymbol{L} w \tag{C.24}$$

By means of the gradient formula and other manipulations similar to those already described, it can be shown that

$$\boldsymbol{F}_3 = ik\sqrt{\ell(\ell+1)} \left[-\left(\frac{\ell}{2\ell+1} \right)^{1/2} g_{\ell+1} \boldsymbol{Y}_{\ell,\ell+1}^{M} + \left(\frac{\ell+1}{2\ell+1} \right)^{1/2} g_{\ell-1} \boldsymbol{Y}_{\ell,\ell-1}^{M} \right] \tag{C.25}$$

The most general divergence-free vector potential is a linear combination of quantities $\boldsymbol{F}_2$ and $\boldsymbol{F}_3$ given by (C.23) and (C.25).

Now consider a vector potential $\boldsymbol{A}$ of the form $\boldsymbol{F}_2$. Given the time dependence $e^{-i\omega t}$, the electric field is $\boldsymbol{\mathcal{E}} = ik\boldsymbol{A}$ and is also transverse and of the form $\boldsymbol{F}_2$. The magnetic field $\boldsymbol{B} = \nabla \times \boldsymbol{A}$ is then of the form $\boldsymbol{F}_3$, which is not transverse. For given ℓ and M, $\boldsymbol{\mathcal{E}}$ and $\boldsymbol{B}$ are said to form a *magnetic multipole* and are denoted by the symbols $\boldsymbol{\mathcal{E}}_{\ell M}^{(m)}$ and $\boldsymbol{B}_{\ell M}^{(m)}$. The nomenclature derives from the behavior of the fields close to the origin (where $kr \ll 1$). There $\boldsymbol{B}$ resembles a quasi-static magnetic $\ell-$**pole** field and greatly dominates over the electric field. Another term used to describe this arrangement of fields is *TE* for "transverse electric." From (C.23) and (C.12) and the fact that the parity of w is that of the spherical harmonic Y_ℓ^m, namely, $(-1)^\ell$, we see that the vector potential and electric field of a magnetic multipole field have parity $(-1)^\ell$, whereas the magnetic field has parity $(-1)^{\ell+1}$.

Suppose instead that the magnetic field is of form $\boldsymbol{F}_2$ (transverse). Then, because $\nabla \times \boldsymbol{B} = -ik\boldsymbol{\mathcal{E}}$, the electric field and the vector potential are of form $\boldsymbol{F}_3$. Such an arrangement for given ℓ and M is called an *electric multipole field*, for which the following symbols are used: $\boldsymbol{\mathcal{E}}_{\ell M}^{(e)}$ and $\boldsymbol{B}_{\ell M}^{(e)}$. Here the magnetic field has parity $(-1)^\ell$, whereas the vector potential and electric field have parity $(-1)^{\ell+1}$. An electric multipole is so named because in the near-field region, $\boldsymbol{\mathcal{E}}_{\ell M}^{(e)}$ is like that of a quasi-static $\ell-$pole electric field, which greatly dominates in magnitude over the magnetic field.

Works Cited

Adler, S. L. Why decoherence has not solved the measurement problem: A response to P. W. Anderson. *Stud. Hist. Philos. Mod. Phys.* 34 (2003): 135.

Adler, S. L. *Quantum Theory as an Emergent Phenomenon*. Cambridge, UK: Cambridge University Press, 2004.

Aharonov, Y., and Bohm, D. Significance of electromagnetic potentials in quantum theory. *Phys. Rev.* 115 (1959): 485.

Aharonov, Y., and Casher, A. Topological quantum effects for neutral particles. *Phys. Rev. Lett.* 53 (1984): 319.

Albert, D. Z. *Quantum Mechanics and Experience*. Cambridge, MA: Harvard University Press, 1992.

Anderson, P. W. Science: A "dappled world" or a "seamless web"? *Stud. Hist. Philos. Mod. Phys.* 32 (2001): 487, 499.

Aspect, A., Dalibard, J., and Roger, G. Experimental test of Bell's inequalities using time-varying analyzers. *Phys. Rev. Lett.* 49 (1982): 1804.

Atkins, P. W., and Friedman, R. *Molecular Quantum Mechanics*, 5th ed. Oxford, UK: Oxford University Press, 2011.

Ax, J., and Kochen, S. Extension of quantum mechanics to individual systems. *arXiv:quant-ph/9905077*, 1999.

Baron, J., et al (ACME Collaboration). Order of magnitude smaller limit on the electric dipole moment of the electron. *Science* 343 (2014):269.

Barrett, J. A. *The Quantum Mechanics of Minds and Worlds*. Oxford, UK: Oxford University Press, 1999.

Bassi, A., and Ghirardi, G. C. A general argument against the universal validity of the superposition principle. *Phys. Lett. A* 275 (2000): 373.

Bassi, A., and Ghirardi, G. C. Dynamical reduction models. *Phys. Rep.* 379 (2003): 257.

Bates, D. R., Ledsham, K., and Stewart, A. L. Wave functions of the hydrogen molecular ion. *Phil. Trans. R. Soc. Lond. A* 246 (1953): 215.

Bell, J. S. On the problem of hidden variables in quantum mechanics. *Rev. Mod. Phys.* 38 (1966): 447.

Bethe, H. A. The electromagnetic shift of energy levels. *Phys. Rev.* 72 (1947): 339.

Bethe, H. A., and Salpeter, E. E. *Quantum Mechanics of One- and Two-Electron Atoms*. New York: Academic Press, 1957.

Bohr, N. On the constitution of atoms and molecules. *Philos. Mag.* 26 (1913): 1–25, 476–502, 857–75.

Born, M., and Oppenheimer, J. R. Zur quantentheorie der molekeln. *Ann. Phys.* 84 (1927): 457.

Born, M., and von Karman, T. On fluctuations in spatial lattices. *Phys. Z.* 13 (1912): 297–309.

Born, M., and von Karman, T. On the theory of specific heat. *Phys. Z.* 14 (1913): 15–19.

Bressi, G., et al. Measurement of the Casimir force between parallel metallic surfaces. *Phys. Rev. Lett.* 88 (2002): 041804.

Brown, H. R. The insolubility proof of the quantum measurement problem. *Found. Phys.* 16 (1986): 857.

Brune, M. E., et al. Observing the progressive decoherence of the "meter" in quantum mechanics. *Phys. Rev. Lett.* 77 (1996): 4887.

Bub, J. *Interpreting the Quantum World.* Cambridge, UK: Cambridge University Press, 1997.

Cabibbo, N. Unitary symmetry and leptonic decays. *Phys. Rev. Lett.* 10 (1963): 531.

Casimir, H. B. G. On the attraction between two perfectly conducting plates. *Proc. Kon. Nederland Akad. Wetensch.* B51 (1948): 793.

Casimir, H. B. G., and Polder, D. The influence of retardation on the London–van der Waals forces. *Phys. Rev.* 73 (1948): 360.

Chambers, R. G. Shift of an electron interference pattern by enclosed magnetic flux. *Phys. Rev. Lett.* 5 (1960): 3.

Chandrasekhar, S. *An Introduction to the Study of Stellar Structure.* Chicago: University of Chicago Press, 1939.

Christenson, J. H., Cronin, J. W., Fitch, V. W., and Turlay, R. Evidence for the 2π decay of the K_2^0 meson. *Phys. Rev. Lett.* 13 (1964): 138.

Cimmino, A., et al. Observation of the topological Aharonov-Casher phase shift by neutron interferometry. *Phys. Rev. Lett.* 63 (1989): 380.

Commins, E. D. Electron spin and its history. *Annu. Rev. Nucl. Part. Sci.* 62 (2012): 133–57

Compton, A. H. A quantum theory of the scattering of x-rays by light elements. *Phys. Rev.* 21 (1923): 483–502.

Condon, E. U., and Shortley, G. H. *The Theory of Atomic Spectra.* Cambridge, UK: Cambridge University Press, 1953.

Davies, W. T., and Grace, M. A. On the identity of β rays with electrons. *Proc. R. Soc. Phys. Lond. A* 64 (1951): 846.

Davisson, C., and Germer, L. H. Diffraction of electrons by a crystal of nickel. *Phys. Rev.* 30 (1927): 705–740.

Davisson, C., and Kunsman, C. H. The scattering of electrons by nickel. *Science* 54 (1921): 522–4.

de Broglie, L. Radiation-waves and quanta. *Comp. Rendus* 177 (1923): 507–10.

de Broglie, L. A tentative theory of light quanta. *Philos. Mag.* 47 (1924): 446–58.

De Witt, B. S., and Graham, N. *The Many-Worlds Interpretation of Quantum Mechanics.* Princeton, NJ: Prineton University Press, 1973.

Debye, P. Zur theorie der spezifischen warmen. *Ann. Phys.* 39 (1912): 789–839.

Dennison, D. M. A note on the specific heat of the hydrogen molecule. *Proc. R. Soc. Lond. A* 115 (1927): 483.

Diosi, L. Quantum stochastic processes as models for state vector reduction. *J. Phys. A* 21 (1988a): 2885.

Diosi, L. Continuous quantum measurement and the Ito formalism. *Phys. Lett. A* 129 (1988b): 419.

Diosi, L. Models for universal reduction of macroscopic quantum fluctuations. *Phys. Rev. A* 40 (1989): 1165.

Dirac, P. A. M. The quantum theory of the electron. *Proc. R. Soc. Lond. A* 117 (1928a): 610.

Dirac, P. A. M. The quantum theory of the electron, part II. *Proc. R. Soc. Lond. A* 118 (1928b): 351.

Dirac, P. A. M. Note on exchange phenomena in the Thomas atom. *Proc. Cambridge Philos. Soc.* 26 (1930): 376.

Durr, D., Goldstein, S., and Zanghi, N. Quantum equilibrium and the origin of absolute uncertainty. *J. Stat. Phys.* 67 (1992): 843.

Dyson, F. J., and Lenard, A. Stability of matter I. *J. Math. Phys.* 8 (1967): 423.

Ehrenberg, W., and Siday, R. The refractive index in electron optics and the principles of dynamics. *Proc. Phys. Soc.* B62 (1949): 8.

Einstein, A. Generation and transformation of light. *Ann. Phys.* 17 (1905): 132–48.

Einstein, A. Elementare betrachtungen uber die thermische molekularbewegung in festen korpern. *Ann. Phys.* 35 (1911): 679–94.

Einstein, A., Podolsky, B., and Rosen, N. Can quantum-mechanical description of physical reality be considered complete? *Phys. Rev.* 47 (1935): 777.

Esposito, S. Majorana solution of the Thomas-Fermi equation. *Am. J. Phys.* 70 (2002): 852.

Fermi, E. Statistical method of investigating electrons in atoms. *Z. Phys.* 48 (1928): 73.

Fermi, E. Versuch einer theorie der β-strahlen. *Z. Phys.* 88 (1934): 161.

Fine, A. On the general quantum theory of measurement. *Proc. Cambridge Philos. Soc.* 65 (1969): 111.

Fine, A. Insolubility of the quantum measurement problem. *Phys. Rev. D* 2 (1970): 2783.

Fock, V. Naherungsmethode zur Losung des quantenmechanischen Mehrkorperproblems. *Z. Phys.* 61 (1930): 126.

Foldy, L. L., and Wouthuysen, S. A. On the Dirac theory of spin-½ particles and its nonrelativistic limit. *Phys. Rev.* 78 (1950): 29.

Franck, J., and Hertz, G. On the collisions between electrons and molecules of mercury vapor and the ionization potential of the same. *Verhand. Deutsch. Phys. Ges.* 16 (1914): 457–67.

Freedman, S. J., and Clauser, J. F. Experimental test of local hidden-variable theories. *Phys. Rev. Lett.* 28 (1972): 938.

Fry, E. S., and Thompson, R. C. Experimental test of local hidden-variable theories. *Phys. Rev. Lett.* 37 (1976):465

Furry, W. H., and Ramsey, N. F. Significance of potentials in quantum theory. *Phys. Rev.* 118 (1960): 623.

Gamow, G., and Teller, E. Selection rules for the β–disintegration. *Phys. Rev.* 49 (1936): 895.

Garwin, R., Lederman, L., and Weinreich, M. Observation of the failure of conservation of parity and charge conservation in meson decays: The magnetic moment of the free muon. *Phys. Rev.* 105 (1957): 1415.

Ghirardi, G. C., Pearle, P., and Rimini, A. Markov processes in Hilbert space and continuous spontaneous localization of systems of identical particles. *Phys. Rev. A* 42 (1990): 78.

Ghirardi, G. C., Rimini, A., and Weber, T. Unified dynamics for microscopic and macroscopic systems. *Phys. Rev. D* 34 (1986): 470.

Gisin, N. Quantum measurements and stochastic processes. *Phys. Rev. Lett.* 52 (1984): 1657.

Gisin, N. Stochastic quantum dynamics and relativity. *Helv. Phys. Acta* 62 (1989): 363.

Glashow, S., Iliopoulos, J., and Maiani, L. Weak interactions with lepto-hadron symmetry. *Phys. Rev. D* 2 (1970): 1285.

Goldberger, M. L., and Watson, K. M. *Scattering Theory*. New York: Wiley, 1964.

Goldhaber, M., and Scharff-Goldhaber, G. Identification of beta rays with atomic electrons. *Phys. Rev.* 73 (1948): 1472.

Grubl, G. The quantum measurement problem enhanced. *arXiv:quant-ph/0202101*, 2002.

Hartree, D. R. The wave mechanics of an atom with a non-Coulomb central field. *Proc. Cambridge Philos. Soc.* 24 (1928): 89, 111.

Heisenberg, W. Uber den anschaulichen Inhalt der quantentheoretischen Kinematik und Mechanik. *Z. Phys.* 43 (1927): 172–98.

Heitler, W., and London, F. Wechselwirkung neutraler Atome und homopolare Bindung nach der Quantenmechanik. *Z. Phys.* 44 (1927): 455.

Holstein, B. R. Second Born approximation and Coulomb scattering. *Am. J. Phys.* 75 (2007): 537.

Jackson, J. D. *Classical Electrodynamics*, 3rd ed. New York: Wiley, 1998.

James, H., and Coolidge, A. The ground state of the hydrogen molecule. *J. Chem. Phys.* 1 (1933): 825.

James, H., and Coolidge, A. A correction and addition to the discussion of the ground state of H_2. *J. Chem. Phys.* 3 (1935): 129.

Jammer, M. *The Conceptual Development of Quantum Mechanics*. New York: McGraw-Hill, 1966.

Johnson, W. R. *Atomic Structure Theory: Lectures on Atomic Physics*. New York: Springer, 2007.

Joos, E. Elements of environmental decoherence. In *Decoherence: Theoretical, experimental, and conceptual problems*, ed. by P. Blanchard et al. New York: Springer, 1999.

Jordan, P., and Pauli, W. Zur Quantenelektrodynamik ladungsfreier Felder. *Z. Phys.* 47 (1928): 151.

Khalfin, L. A. Theory of the decay of a quasi-stable state. *Soviet Phys. JETP* 6 (1958): 1053.

Kinoshita, T. Ground state of the helium atom. *Phys. Rev.* 105 (1957): 1490.

Kobayashi, M., and Maskawa, T. *CP* violation in the renormalizable theory of weak interaction. *Prog. Theor. Phys. Japan* 49 (1973): 652.

Laloe, F. *Do We Really Understand Quantum Mechanics?* Cambridge, UK: Cambridge University Press, 2012.

Lamb, W. E. Anomalous fine structure of hydrogen and singly ionized helium. *Rep. Prog. Phys.* 14 (1951): 19.

Lamb, W. E. Fine structure of the hydrogen atom, part III. *Phys. Rev.* 85 (1952): 259.

Lamb, W. E., and Retherford, R. C. Fine structure of the hydrogen atom by a microwave method. *Phys. Rev.* 72 (1947): 241.

Lamb, W. E., and Retherford, R. C. Fine structure of the hydrogen atom, part I. *Phys. Rev.* 79 (1950): 549.

Lamb, W. E., and Retherford, R. C. Fine structure of the hydrogen atom, part II. *Phys. Rev.* 81 (1951): 222.

Lamoreaux, S. K. Demonstration of the Casimir force in the 0.6 to 6 μm range. *Phys. Rev. Lett.* 78 (1997): 5.

Lamoreaux, S. K. The Casimir force and related effects: The status of the finite temperature correction and limits on new long-range forces. *Ann. Rev. Nuc. Part. Sci.* 62 (2012): 37.

Leaf, B. Momentum operators for curvilinear coordinate systems. *Am. J. Phys.* 47 (1979): 811.

Lee, T. D., and Yang, C. N. Question of parity conservation in weak interaction. *Phys. Rev.* 104 (1956): 254.

Lenard, A., and Dyson, F. Stability of matter I. *J. Math. Phys.* 9 (1968): 698.

Lieb, E. The stability of matter. *Rev. Mod. Phys.* 48 (1976): 553.

Lieb, E. H., and Seiringer, R. *The Stability of Matter in Quantum Mechanics*. Cambridge, UK: Cambridge University Press, 2010.

Lieb, E. H., and Simon, B. Thomas-Fermi theory revisited. *Phys. Rev. Lett.* 31 (1973): 681.

Lieb, E. H., and Thirring, W. In *Studies in Mathematical Physics*, ed. By E. Lieb, B. Simon, and A. Wightman, pp. 301–2. Princeton, NJ: Princeton University Press, 1976.

London, F., and Bauer, E. *La theorie de l'osbervation en mecanique quantique*. Paris: Hermann, 1939.

Lorentz, H. A. Einfluss magnetischer Krafte auf die Emission des Lichtes. *Wied. Ann. Phys.* 63 (1897): 278–84.

Majorana, E. Oriented atoms in a variable magnetic field. *Nuovo Cimento* 9 (1932): 43.

Millikan, R. A direct photoelectric determination of Planck's "*h*." *Phys. Rev.* 7 (1916): 355–88.

Newton, R. G. *Scattering Theory of Waves and Particles*, 2nd ed. New York: McGraw-Hill, 1982.

Pauli, W. Uber den Zusammenhang des Abschlusses der Elektronengruppen im Atom mit der Komplexstruktur der Spektren. *Z. Phys.* 31 (1925): 765–85.

Pauli, W. Zur Quantenmechanik des magnetischen Elektrons. *Z. Phys.* 43 (1927): 601.

Pauli, W. The connection between spin and statistics. *Phys. Rev.* 58 (140): 716.

Pearle, P. Reduction of the state vector by a nonlinear Schroedinger equation. *Phys. Rev. D* 13 (1976): 857.

Pearle, P. Comment on "Quantum measurements and stochastic processes." *Phys. Rev. Lett.* 53 (1984): 1775.

Pearle, P. Combining stochastic dynamical state-vector reduction with spontaneous localization. *Phys. Rev. A* 39 (1989): 2277.

Peckeris, C. Ground state of two-electron atoms. *Phys. Rev.* 112 (1958): 1649.

Percival, I. Primary state diffusion. *Proc. R. Soc. Lond. A* 447 (1994): 189.

Planck, M. Correction of Wien's equation. *Verhand. Deutsch. Phys. Ges.* 2 (1900a): 202–4.

Planck, M. Distribution of energy in the normal spectrum. *Verhand. Deutsch. Phys. Ges.* 2 (1900b): 237–45.

Rabi, I. I. Space quantization in a gyrating magnetic field. *Phys. Rev.* 51 (1937): 652.

Racah, G. Theory of complex spectra, part II. *Phys. Rev.* 62 (1942): 438.

Ramsey, N. F. *Molecular Beams*. Oxford, UK: Oxford University Press, 1956.

Rosen, N. The normal state of the hydrogen molecule. *Phys. Rev.* 38 (1931): 2099.

Rothe, C., Hintschich, S. I., and Monkman, A. P. Violation of the exponential-decay law at long times. *Phys. Rev. Lett.* 96 (2006): 163601.

Rutherford, E. The scattering of alpha and beta particles by matter and the structure of the atom. *Philos. Mag.* 21 (1911): 668–88.

Salam, A. *Weak and Electromagnetic Interactions: Nobel Symposium No.8*. Stockholm: Almquist and Wiksell, 1968, p. 367.

Schlosshauer, M. Decoherence, the measurement problem, and interpretations of quantum mechanics. *Rev. Mod. Phys.* 76 (2004): 1267.

Schwinger, J. Thomas-Fermi model: The leading correction. *Phys. Rev. A* 22 (1980): 1827.
Schwinger, J. Thomas-Fermi model: The second correction. *Phys. Rev. A* 24 (1981): 2353.
Scott, J. M. C. The binding energy of the Thomas-Fermi atom. *Philos. Mag.* 43 (1952): 859.
Slater, J. C. Note on Hartree's method. *Phys. Rev.* 35 (1930): 210.
Sommerfeld, A. Zur Quantentheorie der Spektrallinien. *Ann. Phys.* 51 (1916): 1–94, 125–67.
Stoner, E. C. The distribution of electrons among energy levels. *Philos. Mag.* 48 (1924): 719–26.
't Hooft, G. Renormalization of massless Yang- Mills fields. *Nucl. Phys. B* 33 (1971a): 173.
't Hooft, G. Renormalizable Lagrangians for massive Yang-Mills Fields. *Nucl. Phys. B* 35 (1971b): 167.
't Hooft, G. Equivalence relations between deterministic and quantum mechanical systems. *J. Stat. Phys.* 53 (1988): 323.
't Hooft, G. Quantum mechanical behavior in a deterministic model. *Found. Phys. Lett.* 10 (1997): 105.
't Hooft, G. Quantum gravity as a dissipative deterministic system. *Class. Quant. Grav.* 16 (1999): 3263.
't Hooft, G. Determinism beneath quantum mechanics. *arXiv:quant-ph/0212095*, 2002.
't Hooft, G. Determinism in free bosons. *Int. J. Theol. Phys.* 42 (2003): 355.
Tegmark, M., and Wheeler, J. 100 years of the quantum. *Sci. Am.* 2001: 68.
Teller, E. On the stability of molecules in the Thomas-Fermi theory. *Rev. Mod. Phys.* 34 (1962): 627.
Thomas, L. H. The motion of the spinning electron. *Nature* 117 (1926): 514.
Thomas, L. H. The kinematics of an electron with an axis. *Philos. Mag.* 3 (1927a): 1.
Thomas, L. H. The calculation of atomic fields. *Proc. Cambridge Philos. Soc.* 23 (1927b): 542.
Thompson, G. P. Experiments on the diffraction of cathode rays. *Proc. R. Soc. Lond. A* 117 (1928): 600–9.
Thompson, G. P., and Reid, A. Diffraction of cathode rays by a thin film. *Nature* 119 (1927): 890.
Uhlenbeck, G. E., and Goudsmit, S. Ersetzung der Hypothese vom unmechanischen Zwang durch eine Forderung bezuglich des inneren Verhaltens jedes einzelnen Elektrons. *Die Naturwiss.* 13 (1925): 953–4.
Uhlenbeck, G. E., and Goudsmit, S. Spinning electrons and the structure of spectra. *Nature* 117 (1926): 264–5.
Wang, S. C. The problem of the normal hydrogen molecule in the new quantum mechanics. *Phys. Rev.* 31 (1928): 579.
Weinberg, S. A model of leptons. *Phys. Rev. Lett.* 19 (1967): 1264.
Weisskopf, V. F., and Wigner, E. P. Berechnung der naturlichen Linienbreite auf Grund der Diracschen Lichttheorie. *Z. Phys.* 63 (1930): 54.
Werner, S., et al. Observation of the phase shift of a neutron due to precession in a magnetic field. *Phys. Rev. Lett.* 35 (1975): 1053.
Wigner, E. *Symmetries and Reflections*. Bloomington, IN: Indiana University Press, 1967.
Wigner, E. P. *Group Theory*. New York: Academic Press, 1959.
Wolfenstein, L. Neutrino oscillations in matter. *Phys. Rev. D.* 17 (1978): 2369.

Wu, C. S., et al. Experimental test of parity conservation in beta decay. *Phys. Rev.* 105 (1957): 1413.

Yang, C. N., and Mills, R. Conservation of isotopic spin and isotopic gauge invariance. *Phys. Rev.* 96 (1954): 191.

Zeeman, P. On the influence of magnetism on the nature of the light emitted by a substance. *Philos. Mag.* 43 (1897): 226–39.

Zurek, W. H. Environment-induced superselection rules. *Phys. Rev. D* 26 (1982): 1862.

Bibliography

Atkins, P. W., and Friedman, R. *Molecular Quantum Mechanics*, 5th ed. Oxford, UK: Oxford University Press, 2011.

Bell, J. S. *Speakable and Unspeakable in Quantum Mechanics*. Cambridge, UK: Cambridge University Press, 2004.

Bethe, H. A., and Jackiw, R. *Intermediate Quantum Mechanics*, 2nd ed. New York: Benjamin, 1968.

Bethe, H. A., and Salpeter, E. E. *Quantum Mechanics of One- and Two-Electron Atoms*. New York: Academic Press, 1957.

Bjorken, J. D., and Drell, S. D. *Relativistic Quantum Fields*. New York: McGraw-Hill, 1965.

Bjorken, J. D., and Drell, S. D. *Relativistic Quantum Mechanics*. New York: McGraw-Hill, 1964.

Cohen-Tannoudji, C., Diu, B., and Laloe, F. *Quantum Mechanics*, Vols. I and II. New York: Wiley, 1977.

Condon, E. U., and Shortley, G. H. *The Theory of Atomic Spectra*. Cambridge, UK: Cambridge University Press, 1953.

Dirac, P. A. M. *The Principles of Quantum Mechanics*, 4th ed. Oxford, UK: Oxford University Press, 1958.

Edmonds, A. R. *Angular Momentum in Quantum Mechanics*. Princeton, NJ: Princeton University Press, 1996.

Fetter, A. L., and Walecka, J. D. *Quantum Theory of Many-Particle Systems*. New York: McGraw-Hill, 1971.

Feynman, R. P., Hibbs, A. R., and Styer, D. F. *Quantum Mechanics and Path Integrals*. Mineola, NY: Dover, 2010.

Goldberger, M. L., and Watson, K. M. *Scattering Theory*. New York: Wiley, 1964.

Gottfried, K., and Yan, T.-M. *Quantum Mechanics: Fundamentals*. New York: Springer, 2003.

Heitler, W. *The Quantum Theory of Radiation*, 3rd ed. Oxford, UK: Oxford University Press, 1954.

Johnson, W. R. *Atomic Structure Theory: Lectures on Atomic Physics*. New York: Springer, 2007.

Laloe, F. *Do We Really Understand Quantum Mechanics?* Cambridge, UK: Cambridge University Press, 2012.

Landau, L. D., and Lifshitz, E. M. *Quantum Mechanics, Non-Relativistic Theory*. Oxford, UK: Pergamon, 1977.

Lieb, E. H., and Seiringer, R. *The Stability of Matter in Quantum Mechanics*. Cambridge, UK: Cambridge University Press, 2010.

Mandl, F., and Shaw, G. *Quantum Field Theory*. Hoboken, NJ: Wiley, 2010.

Messiah, A. *Quantum Mechanics*, Vols. I and II. Amsterdam: North Holland, 1961.

Newton, R. G. *Scattering Theory of Waves and Particles*, 2nd ed. New York: McGraw-Hill, 1982.

Sakurai, J. J. *Advanced Quantum Mechanics*. Reading, MA: Addison-Wesley, 1978.

Sakurai, J. J., and Napolitano, J. *Modern Quantum Mechanics*. Reading, MA: Addison-Wesley, 2011.

Schiff, L. I. *Quantum Mechanics*, 3rd ed. New York: McGraw-Hill, 1968.

Slater, J. C. *Quantum Theory of Matter*, 2nd ed. New York: McGraw-Hill, 1968.

Weinberg, S. *Lectures on Quantum Mechanics*. Cambridge, UK: Cambridge University Press, 2013.

Wigner, E. P. *Group Theory*. New York: Academic Press, 1959.

Index

Printed in the United States
by Baker & Taylor Publisher Services